Lecture Notes in Computational Science and Engineering

Volume 141

This series contains monographs of lecture notes type, lecture course material, and high-quality proceedings on topics described by the term "computational science and engineering". This includes theoretical aspects of scientific computing such as mathematical modeling, optimization methods, discretization techniques, multiscale approaches, fast solution algorithms, parallelization, and visualization methods as well as the application of these approaches throughout the disciplines of biology, chemistry, physics, engineering, earth sciences, and economics.

Ildar B. Badriev • Victor Banderov • Sergey A. Lapin
Editors

Mesh Methods for Boundary-Value Problems and Applications

13th International Conference, Kazan, Russia, October 20-25, 2020

Springer

Editors
Ildar B. Badriev
Kazan Federal University
Kazan, Russia

Victor Banderov
Kazan Federal University
Kazan, Russia

Sergey A. Lapin
Department of Mathematics and Statistics
Washington State University
Pullman, Washington, USA

ISSN 1439-7358 ISSN 2197-7100 (electronic)
Lecture Notes in Computational Science and Engineering
ISBN 978-3-030-87808-5 ISBN 978-3-030-87809-2 (eBook)
https://doi.org/10.1007/978-3-030-87809-2

Mathematics Subject Classification: 65-XX, 65Mxx, 65Nxx, 65Yxx, 65Zxx

This Springer imprint is published by the registered company Springer Nature Switzerland AG.
The registered company address is: Gewerbestrasse 11, 6330 Cham, Switzerland

Preface

This volume presents selected papers from the 13th International Conference "Mesh Methods for Boundary-Value Problems and Applications" that was held in Kazan, Russia, during October 20–25, 2020. The conference was attended by scientists from leading scientific centers engaged in mathematical modeling of nonlinear problems, theory of numerical methods for solving differential equations and inequalities, computer modeling, development training systems, and computational experiments. The tradition of holding this conference originates from conferences and schools conducted under the leadership of academician A. A. Samarskiy and has had a significant impact on the development of computational mathematics and its applications in various fields of knowledge, especially the theory of grid methods, in global scientific centers. The conference is biannual (since 1996), and it is one of the well-known international conferences in the area of mesh methods for boundary-value problems.

With over 300 attendees, "Mesh Methods for Boundary-Value Problems and Applications 2020" has been the largest edition of the conference series to date. The program consisted of 7 invited speakers across the week, who are internationally renowned researchers, along with 7 minisymposiums (of around 250 presentations) dedicated to specialized topics in mathematical modeling of nonlinear problems, the theory of numerical methods for solving differential equations and inequalities, computer modeling, development training systems, and computational experiments as well as 190 contributed talks. The goal of this book is to provide a good balance between engineering algorithms and mathematical foundations. The content of these proceedings is organized as follows. The main section criteria are based on the recommendations of anonymous peer reviews from experts of the corresponding fields. The content of these proceedings consists of refereed selected papers highlighting the broad spectrum of topics presented at Mesh Methods 2020.

We would like to give special thanks to our local organizing committee for their efforts in organizing and promoting the event. In particular, we would also like to thank Mr. Ildar Badriev for his organizational efforts leading up to the conference, as well as the administrative staff of the Institute of Computational Mathematics and Information Technologies at Kazan Federal University for their help in coordinating the logistics of the event. We also thank many student helpers for their advice, help, and support given to the delegates during the event itself, all of whom contributed to the smooth running of the event.

Kazan, Russia — Ildar B. Badriev
Kazan, Russia — Victor Banderov
Pullman, Washington, USA — Sergey A. Lapin

Contents

Quantum Algorithms for String Processing

Farid Ablayev, Marat Ablayev, Kamil Khadiev, Nailya Salihova, and Alexander Vasiliev

Abstract In the paper, we investigate two problems on strings. The first one is the String matching problem, and the second one is the String comparing problem. We provide a quantum algorithm for the String matching problem that uses exponentially less quantum memory than existing ones. The algorithm uses the hashing technique for string matching, quantum parallelism, and ideas of Grover's search algorithm. Using the same ideas, we provide two algorithms for the String comparing problem. These algorithms also use exponentially less quantum memory than existing ones. Additionally, the second algorithm works exponentially faster than the existing one.

1 Introduction

Possibilities of quantum speedup for string matching problem have been investigated during the last decades by different authors [19, 21, 22]. Most of these algorithms are based on Grover's algorithm [4, 9] for search through unstructured data.

In the paper we consider a problem of searching any occurrence of a string w of length m in a string s of length n. The best known classical algorithm for this problem is Knuth-Morris-Pratt algorithm [15]. Time complexity of this algorithm is $O(n + m)$. Quantum algorithms for this problem are typically considered in the query model [1, 2, 20]. Here the algorithm has access to an oracle (the unchangeable part of memory that holds input data) and complexity is a number of queries to this oracle. In the early 2000s researchers obtained one of the first

F. Ablayev · M. Ablayev · K. Khadiev · A. Vasiliev (✉)
Kazan Federal University, Kazan, Russian Federation

Kazan E. K. Zavoisky Physical-Technical Institute, Kazan, Russian Federation

N. Salihova
Kazan Federal University, Kazan, Russian Federation

I. B. Badriev et al. (eds.), *Mesh Methods for Boundary-Value Problems and Applications*, Lecture Notes in Computational Science and Engineering 141,
https://doi.org/10.1007/978-3-030-87809-2_1

results on quantum algorithms for the problem [21]. This algorithm has query complexity $O(\sqrt{n}\log\sqrt{\frac{n}{m}}\log m + \sqrt{m}\log^2 m)$. Later in 2017 the algorithm [19] with query complexity $\tilde{O}(\sqrt{\frac{n}{m}}2^{O(\sqrt{\log m})})$ was presented. In 2020, Soni and Rasool [22] suggested an algorithm with $O(n \log n)$ query complexity. Note that, these algorithms have time complexity (or circuit complexity) logarithmic times larger than query complexity.

At the same time, researchers did not pay attention to the size of the quantum memory that we should use for algorithms (a changeable by algorithm part and the unchangeable part that oracle holds). Due to our analysis, all of these algorithms use $O(n+m)$ quantum bits (including the unchangeable part of the quantum memory). Due to the restricted resources of the current and near-future devices, the quantum memory size, even unchangeable, is an important question.

In the paper, we provide a quantum algorithm for solving string matching problem with $O(\sqrt{n}(\log n + \sqrt{\log n + \log m} \cdot (\log\log n + \log\log m)))$ time complexity and $O((\log n)^2 + \log n \cdot \log m)$ qubits of memory (Theorem 1). The algorithm is based on Grover's search algorithm, the idea of hashing (or fingerprinting method [8]) and ideas of Rabin-Karp algorithm [13]. The algorithm is not a query model algorithm but a quantum circuit algorithm that can be used as a part of other more complex algorithms for other problems. Many known algorithms like [10] use similar motivation. Our algorithm assumes that the initial state is prepared. At the same time, this initial state can be prepared approximately as fast as loading data to unchangeable memory for oracle.

Additionally, we use the same ideas for comparing two strings u and v in lexicographical order. The existing algorithm [14] uses modifications [16–18] of Grover's search [9] and compares two strings with query complexity $O(\sqrt{k})$, time complexity $O(\sqrt{k}\log k)$ and uses $O(k)$ qubits of memory, where k is the minimum of lengths of two strings. Here we use the idea with hashing and provide two algorithms. The first one has $O(\sqrt{k}\log k)$ time complexity and uses $O((\log k)^2)$ qubits (Theorem 2). The second one has $O((\log k)^2 \log\log k)$ time complexity and uses $O((\log k)^2)$ qubits (Theorem 3). Both algorithms have an exponential advantage in memory and the second one has an exponential advantage in speed. At the same time, the second algorithm is more complex.

The structure of the paper is the following. Section 2 contains preliminaries. We present an algorithm for string matching in Sect. 3. Section 4 contains algorithms for comparing two strings. The conclusion is presented in Sect. 5.

2 Preliminaries

Let us consider a string $u = (u_1, \ldots, u_\ell)$ for some integer ℓ. Then, $|u| = \ell$ is the length of the string. $u[i, j] = (u_i, \ldots, u_j)$ is a substring of u.

In the paper, we compare strings in the lexicographical order. For two strings u and v, the notation $u < v$ means u precedes v in the lexicographical order.

In the paper, we consider only binary strings. At the same time, all results can be easily modified for a non-binary alphabet.

2.1 *Rolling Hash for Strings Comparing*

2.1.1 Rolling Hash

The rolling hash was presented in [8, 13]. For a string $u = (u_1, \ldots, u_{|u|})$ we define a rolling hash function $h_p(u) = \left(\sum_{i=1}^{|u|} u_i \cdot 2^{i-1}\right) \bmod p$, where p is a prime integer.

2.1.2 Fingerprinting Technique for Comparing Strings

We can use the rolling hash and the fingerprinting method [8] for comparing two strings u and v. Let us randomly choose p from the set of the first r primes, such that $r \leq \frac{\max(|u|,|v|)}{\varepsilon}$ for some $\varepsilon > 0$. According to the Chinese Remainder Theorem and [8], if we have $h_p(u) = h_p(v)$, then $u = v$ with error probability at most ε. If we invoke a comparing procedure δ times, then we should choose a prime number from the first $\frac{\delta \cdot \max(|u|,|v|)}{\varepsilon}$ primes for getting the error probability ε for the whole algorithm. Due to Chebyshev's theorem, the r-th prime number $p_r \approx r \ln r$. If $r = \frac{\delta \cdot \max(|u|,|v|)}{\varepsilon}$, then $p_r = \frac{\delta \cdot \max(|u|,|v|)}{\varepsilon} \cdot (\ln(\delta) + \ln(\max(|u|, |v|)) - \ln(\varepsilon))$ and it can be encoded using $O(\log(\delta) + \log(\max(|u|, |v|)) - \log(\varepsilon))$ bits.

2.1.3 Comparing Strings Using a Rolling Hash

For a string u, we can compute a prefix rolling hash, that is $h_p(u[1, i])$. It can be computed in $O(|u|)$ running time using formula

$$h_p(u[1,i]) = \left(h_p(u[1, i-1]) + (2^{i-1} \bmod p) \cdot u_i\right) \bmod p \text{ and } h_p(u[1:0]) = 0.$$

Assume, that we have computed prefix rolling hashes for two strings u and v. Then, we can compare these strings in the lexicographical order in $O(\log \min(|u|, |v|))$ running time. The algorithm is following. We search the longest common prefix of u and v. Let $lcp(u, v)$ be an integer x such that $u_1 = v_1, \ldots, u_x = v_x$ and $u_{x+1} \neq v_{x+1}$. In the case of u is a prefix of v, then $lcp(u, v) = |u|$. In the case of v is a prefix of u, we have $lcp(u, v) = |v|$. Notice, that for any integer $mid \in \{1, \ldots, \min(|u|, |v|)\}$ the following two statements are true.

- If $mid \leq lcp(u, v)$, then $u[1, mid] = v[1, mid]$, and $h_p(u[1, mid]) = h_p(v[1, mid])$.

- If $mid > lcp(u, v)$, then $u[1, mid] \neq v[1, mid]$, and $h_p(u[1, mid]) \neq h_p(v[1, mid])$ with high probability.

Using binary search we find the index x such that $h_p(u[1, x]) = h_p(v[1, x])$ and $h_p(u[1, x + 1]) \neq h_p(v[1, x + 1])$. In that case $lcp(u, v) = x$. After that, we compare u_t and v_t for $t = lcp(u, v) + 1$. Then, we get the following cases:

- If $u_t < v_t$ or $t = |u| < |v|$, then $u < v$.
- If $u_t > v_t$ or $t = |v| < |u|$, then $u > v$.
- If $|u| = |v| = t - 1$, then $u = v$.

Binary search works in $O(\log(\min(|u|, |v|)))$ running time.

2.2 Problems

String Matching Problem
Given a string (text) $s = (s_1, \dots, s_n)$ of length n and a string w of length m, where $m \leq n$, one needs to determine the index of the string w occurrence in the text s. Formally, the task is to find the index d such that $w = (s_d \dots s_{d+m-1})$.

We use the following notations. Let $T(s) = (s^1, \dots, s^{n-m+1})$, where $s^i = s[i, i + m - 1]$ for $i \in \{1, \dots, n - m + 1\}$. $T(s)$ is a sequence of substrings of length m. Let $N = n - m + 1$.

String Comparing Problem
Given two strings u and v. The problem is comparing these two strings in lexicographical order. Formally, we want to determine one of three options:

- If $u < v$, then the result is -1.
- If $u > v$, then the result is $+1$.
- If $u = v$, then the result is 0.

2.3 Basics of Quantum Computation and Computational Model

The main difference between quantum computation and the classical one is manipulations with quantum bits (qubits). A state of a qubit is a vector from two-dimensional complex Hilbert space. We can represent it using Dirac notation as $|\psi\rangle = a|0\rangle + b|1\rangle$, where $|0\rangle$ and $|1\rangle$ are unit vectors, and a and b are complex numbers such that $|a|^2 + |b|^2 = 1$. We can use two kinds of transformations: *transition* and *measurement*. The transition is multiplying a vector of state to 2×2 unitary matrix. The measurement is obtaining 0-result with probability $|a|^2$ and 1-result with probability $|b|^2$. Similarly, a state of a register of q qubits is a vector from 2^q-dimensional complex Hilbert space, and is traditionally denoted as

$|\psi\rangle = \sum_{i=0}^{2^q-1} a_i|i\rangle$, where $\sum_{i=0}^{2^q-1} |a_i|^2 = 1$. Transformations are defined in an analogous manner.

A quantum circuit is a circuit that uses four types of gates that are 1-qubit Hadamard gate (H-gate), T-gate and S-gate; and 2-qubit $CNOT$-gate. That are

$$T = \begin{pmatrix} 1 & 0 \\ 0 & e^{i\pi/4} \end{pmatrix}, \quad S = \begin{pmatrix} 1 & 0 \\ 0 & j \end{pmatrix}, \quad H = \frac{1}{\sqrt{2}}\begin{pmatrix} 1 & 1 \\ 1 & -1 \end{pmatrix}, \quad CNOT = \begin{pmatrix} 1 & 0 & 0 & 0 \\ 0 & 1 & 0 & 0 \\ 0 & 0 & 0 & 1 \\ 0 & 0 & 1 & 0 \end{pmatrix}.$$

An algorithm's time complexity is the size of a circuit that uses only presented gates and implements the algorithm.

The standard form of the quantum query model is a generalization of the decision tree model of classical computation that is commonly used to lower-bound the amount of time required by a computation. Let $f : D \to \{0, 1\}$, $D \subseteq \{0, 1\}^M$ be an M variable function we wish to compute on an input $x = (x_0, \ldots, x_{M-1}) \in D$. We have an oracle access to the input. That is implemented by storing the input into an unchangeable part of quantum memory $|x\rangle$. The oracle access is realized by a specific unitary transformation usually defined as $|i\rangle|\phi\rangle|\psi\rangle|x\rangle \to |i\rangle|\phi \oplus x_i\rangle|\psi\rangle|x\rangle$ where the $|i\rangle$ register indicates the index of the variable we are querying, $|\phi\rangle$ is the output register, and $|\psi\rangle$ is some auxiliary work-space. An algorithm in the query model consists of alternating applications of arbitrary unitaries (that are independent of the input) and the input-dependent query unitary, and a measurement in the end. The smallest number of queries for an algorithm that outputs $f(x)$ with probability $\geq \frac{2}{3}$ on all x is called the quantum query complexity of the function f.

More information on quantum computation and computational models can be found in [1, 2, 20].

3 Quantum Algorithm for String Matching Problem

Firstly, let us present Grover's search algorithm because we use its ideas as a base for our algorithm.

3.1 Grover's Search Algorithm

Definition 1 (Search Problem) Suppose we have a set of objects named $\{1, 2, \ldots, M\}$, of which some are targets. Suppose O is an oracle that identifies the targets. The goal of a search problem is to find a target $i \in \{1, 2, \ldots, M\}$ by making queries to the oracle O.

Remind that Oracle is implemented by accessing an unchangeable (by the algorithm) part of the quantum memory.

In search problems, one will try to minimize the number of queries to the oracle. In the classical setting, one needs $O(M)$ queries to solve such a problem. Grover, on the other hand, constructed a quantum algorithm that solves the search problem with only $O(\sqrt{M})$ queries [9], provided that there is a unique target.

The algorithm uses additional $\log M$ qubits for indexing element in a state $\frac{1}{\sqrt{M}}\sum_{t=0}^{M-1}|t\rangle$ and one additional qubit $|\xi\rangle$ in a state $\frac{1}{\sqrt{2}}(|0\rangle - |1\rangle)$. On step of the algorithm is applying two operations: Grover's diffusion D and a query to oracle Q. The matrix D can be implemented using $\log M$ gates due to [9].

The matrix Q is a transformation that converts $|t\rangle|\xi \oplus f(t)\rangle = (-1)^{f(t)}|t\rangle|\xi\rangle$, where $f(t)$ is a Boolean function that shows whether t-th object is target.

After $O(\sqrt{M})$ iterations, the algorithm measures the quantum register and obtains the index of the target object with high probability. If there are no target objects, then the algorithm returns any object with equal probability.

When the number of targets is unknown, Brassard et al. designed a modified Grover algorithm that solves the search problem with $O(\sqrt{M})$ queries [4], which is of the same order as the query complexity of the Grover search.

The algorithm repeats Grover's search algorithm for $\log_2(\sqrt{M})$ times. It does 2^j iterations on j-th repetition. Such behavior allows us to obtain one of the target objects with a probability at least $1/2$.

3.2 *Our Algorithm*

Let us choose a prime p from the first $\frac{\delta \cdot m}{\varepsilon}$ prime numbers, where $0 < \varepsilon < 1$ is some constant and $\delta = N$ because we will have N hashes of substrings of the string s. Additionally, we will use a hash function h_p, that is discussed in Sect. 2.1.

Assume that the initial state for our algorithm is the following one

$$|\varphi\rangle = \left|h_p(w)\right\rangle \otimes \bigotimes_{t=1}^{\log_2 N} \frac{1}{\sqrt{N}} \sum_{a=0}^{N-1} |a\rangle \otimes \left|h(s^{a+1})\right\rangle. \quad (1)$$

3.2.1 Unique Target Case

Firstly, assume that there is only one position d such that $s^d = w$. In that case, we use only the following part of the quantum register.

$$|\varphi'\rangle = \left|h_p(w)\right\rangle \otimes \left(\frac{1}{\sqrt{N}} \sum_{i=0}^{N-1} |i\rangle \otimes \left|h(s^{i+1})\right\rangle\right).$$

Let us have a function $f : \{0, \ldots, N-1\} \rightarrow \{0, 1\}$, such that $f(i) = 1$ iff $h_p(w) = h(s^{i+1})$. We will discuss the implementation of the algorithm for f later.

Then, we can add additional qubit $|\xi\rangle$ in a state $\frac{1}{\sqrt{2}}(|0\rangle - |1\rangle)$ and apply $O(\sqrt{N})$ times D and Q matrices to. $|\varphi'\rangle$ state. Here D is the Grover's diffusion and Q is the transformation that works in the following way:

$$Q : |i\rangle \left|h(s^{i+1})\right\rangle |h_p(w)\rangle|\xi\rangle \rightarrow |i\rangle \left|h(s^{i+1})\right\rangle |h_p(w)\rangle|\xi \oplus f(i)\rangle =$$

$$= (-1)^{f(i)}|i\rangle \left|h(s^{i+1})\right\rangle |h_p(w)\rangle|\xi\rangle$$

So, using the idea of Grover's search algorithm, we can do $O(\sqrt{N})$ iterations of D and Q and after that measure the quantum register and obtain the index d such that $h_p(w) = h(s^{d+1})$.

Let us discuss, the implementation of the function f. The computation of $f(i)$ is equivalent to the problem of checking equality of two strings $z = h_p(w)$ and $z' = h(s^{i+1})$ that are stored in quantum memory. Let us define a function $g_i : \{0, \ldots, \lceil \log_2 p \rceil - 1\} \rightarrow \{0, 1\}$, where $g_i(j) = 1$ iff $z_j \neq z'_j$. In other words, we mark the indexes of unequal symbols of two strings. We can say that $f(i) = 0$ iff there is $j \in \{0, \ldots, \log_2 p\}$ such that $g_i(j) = 1$.

Let us solve this problem using Grover's search algorithm. In fact, we have two strings in unchangeable memory, and using additional $O(\log \log p)$ qubits can implement Grover's search algorithm for searching an index j_1 such that $g_i(j_1) = 1$.

If the Grover's search algorithm finds j_1 and $g_i(j_1) = 1$, then $f(i) = 0$. If the result index j_1 is such that $g_i(j_1) = 0$, then $f(i) = 1$.

Note that for a standard version of Grover's search algorithm, function f should be computed with no error. At the same time, our version of the implementation of f can return a result with constant error probability. That is why we should use the modification of Grover's search algorithm [11] for bounded-error oracle. This algorithm uses the generalization of Grover's search algorithm that is called Amplitude Amplification [5].

Lemma 1 *The presented algorithm solves string matching problem for unique target with bounded error, has* $O(\sqrt{n}(\log n + \sqrt{\log n + \log m} \cdot (\log \log n + \log \log m)))$ *time complexity and uses* $O(\log n + \log m)$ *qubits of memory.*

Proof Due to description of the algorithm, it finds the index i such that $f(i) = 1$, i.e. $h_p(w) = h(s^i)$ with constant probability. Let us say that the probability of success is at least 0.5. Due to choice of p and results that discussed in Sect. 2.1, The fact $h_p(w) = h(s^i)$ means $w = s^i$ with probability at least $1 - \varepsilon$. So the total probability of success is $0.5 \cdot (1 - \varepsilon)$. If we want a bigger success probability, then we can repeat the process several times and choose the major result. A similar technique was used, for example, in [3, 12]. Constant times repetitions increase the total time complexity and memory size only in constant times.

Let us discuss the time complexity of the algorithm. Due to [9], the time complexity of Grover's algorithm is $O(\sqrt{M}\log M)$ in a case of searching the target element among M elements and constant time implementation of the oracle. In fact, the time complexity of all Grover's diffusion operators is $O(\sqrt{M}\log M)$ and all Oracle operators is $O(\sqrt{M})$. The modification with bounded-error oracle [11] has only constant time bigger time complexity. In our case $M = N$ and the oracle is complex. The implementation of f function has $O(\sqrt{\log p}\log\log p)$ time complexity because there are $\log_2 p$ objects for search. Hence, the total time complexity is $O(\sqrt{N}\log N + \sqrt{N}\cdot\sqrt{\log p}\log\log p)$. Here $p = \frac{\delta\cdot m}{\varepsilon} = \frac{N\cdot m}{\varepsilon}$ and $O(\sqrt{\log p}\log\log p) = O((\log N + \log m - \log\varepsilon)^{0.5}\cdot(\log(\log N + \log m - \log\varepsilon)) = O((\log N + \log m)^{0.5}\cdot(\log\log N + \log\log m))$. Therefore, the time complexity is

$$O(\sqrt{N}\log N + \sqrt{N\cdot(\log N + \log m)}\cdot\sqrt{\log p}\log\log p) =$$

$$= O(\sqrt{N}\log N + \sqrt{N(\log N + \log m)}\cdot(\log\log N + \log\log m)) =$$

remember that $N = n - m$, therefore

$$= O(\sqrt{n}(\log n + \sqrt{\log n + \log m}\cdot(\log\log n + \log\log m))).$$

Let us discuss the memory complexity. The main part of the algorithm requires $O(\log N + \log p)$ qubits. Additionally, we need $O(\log\log p)$ for Grover's Search that implements the function f. Therefore, the total complexity is

$$O(\log N + \log p + \log\log p) = O(\log N + \log N + \log m + \log\log N + \log\log m) =$$

$$= O(\log N + \log m) = O(\log n + \log m).$$

Finally, we have proved the claim. □

3.2.2 Multi-Target Case

Let us consider the general case when the string w can occur in s several times. As mentioned in Sect. 3.1, we should repeat our algorithm $\log_2 N$ times with a different number of iterations. For several repetitions of the algorithm, we should have an unchangeable part of a quantum memory that holds all hashes $h_p(s^i)$. At the same time, our algorithm destroys the quantum state that holds the required data.

Therefore, we should have $\log_2 N$ copies of our state that allow us to repeat the process several times. That is why we use the initial state in the (1) form.

Let us analyze the complexity of the algorithm.

Theorem 1 *The presented algorithm solves string matching problem with bounded error, has* $O(\sqrt{n}(\log n + \sqrt{\log n + \log m}\cdot(\log\log n + \log\log m)))$ *time complexity and uses* $O((\log n)^2 + \log n\cdot\log m)$ *qubits of memory.*

Proof Due to Lemma 1, the algorithm for unique target solves the problem with bounded error, has $O(\sqrt{n}(\log n + \sqrt{\log n + \log m} \cdot (\log\log n + \log\log m)))$ time complexity and uses $O(\log n + \log m)$ qubits of memory.

Let us discuss time complexity. Due to [4, 9], if we do b iterations of the Grover's search algorithm for M elements, then time complexity is $O(b \log M)$. In our algorithm we do 2^j iterations for j-th step, where $j \in \{0, \dots \lceil \log_2 \sqrt{M} \rceil\}$ and $M = N$. Therefore, similar to the proof of Lemma 1, we can show that time complexity is

$$O\left(\sum_{j=1}^{\log_2 \sqrt{N}} 2^j \left(\log n + \sqrt{\log n + \log m} \cdot (\log\log n + \log\log m)\right)\right) =$$

Due to the properties of the sum of geometric progression and $N = n - m$, we have

$$= O\left(\sqrt{n}\left(\log n + \sqrt{\log n + \log m} \cdot (\log\log n + \log\log m)\right)\right).$$

We have $\log_2 N$ copies of qubits. Each of them is used for a single invocation of the algorithm for a unique target. Therefore, the total memory complexity is $O(\log n(\log n + \log m)) = O((\log n)^2 + \log n \cdot \log m)$. □

4 Quantum Algorithm for String Comparing Problem

Let us discuss the algorithm for String Comparing Problem. There are two algorithms. The first one is based on Grover's search algorithm that was discussed in Sect. 3.1. The second one is faster and based on comparing strings using Binary search algorithm and rolling hash that was discussed in Sect. 2.1, but it requires a more complex initial state.

4.1 The Algorithm Based on Grover's Search Algorithm

Let $k = \min(|u|, |v|)$ for strings u and v. As it was discussed in Sect. 2.1, for comparing two string u and v, it is enough to find the Longest common prefix. We can use an idea similar to [12, 14]. Let us consider a function $g' : \{1, \dots, k\} \to \{0, 1\}$ such that $k = \min(|u|, |v|)$, $g'(i) = 1$ iff $u_i \neq v_i$. If we found the smallest lexicographical element of the sequence $(1 - g'(i), i)$ for $i \in \{1, \dots, k\}$, then it corresponds to the minimal argument i_1 such that $g'(i_1) = 1$. Such idea is used in [12, 16–18] algorithms for searching the first target object.

We can use the Dürr-Høyer algorithm for minimum search [6, 7]. Let us briefly present its idea in Sect. 4.1.1 and then present algorithm itself in Sect. 4.1.2.

4.1.1 Dürr-Høyer Minimum Search Algorithm

The problem is searching for the index of minimal element among $(a_1, \ldots, a_M\}$ for some positive integer M.

The algorithm contains several phases. The 0-th phase is an assumption that minimal element y^0 is any element. On i-th phase, we have an assumption that the minimal element is y^i. Then, we run the Grover's search algorithm for searching for a smaller than y^i element. We consider a function $g^i : \{1, \ldots, M\} \to \{0, 1\}$ such that $g^i(j) = 1$ iff $a_j < y^i$. The algorithm finds any argument j such that $g^i(j) = 1$ and updates the assumption of the minimum by assigning $y^{i+1} \leftarrow a_j$, where $g^i(j) = 1$.

Due to [7], the expected number of phases is $O(\log M)$. At the same time, the expected number of all iterations of all invocations of Grover's search algorithm is $O(\sqrt{M})$. Due to Markov's inequality, if the algorithm stops after 3 times more phases than the expectation, then we get a result with bounded error.

Note that used Grover's search implementation is the algorithm for multi-target case.

4.1.2 The Main Part of the Algorithm

Assume that the initial state for our algorithm is the following one

$$|\varphi\rangle = |\xi\rangle \bigotimes_{i=1}^{3\log_2 k} \bigotimes_{t=1}^{\log_2 k} \frac{1}{\sqrt{k}} \sum_{a=0}^{k-1} |a\rangle \otimes |u_a\rangle \otimes |v_a\rangle, \quad |\xi\rangle = \frac{1}{\sqrt{k}} \sum_{a=0}^{k-1} |a\rangle \otimes |u_a\rangle \otimes |v_a\rangle. \tag{2}$$

As in Sect. 3, we implement Grover's search algorithm on our state. Let us discuss i-th phase of the algorithm. We use

$$\bigotimes_{t=1}^{\log_2 k} \frac{1}{\sqrt{k}} \sum_{a=0}^{k-1} |a\rangle \otimes |u_a\rangle \otimes |v_a\rangle. \tag{3}$$

Let $i = 0$. We invoke Grover's search algorithm on the quantum state and find any j_1 such that $g(j_1) = 1$. Note, that computing g have constant time and memory complexity because it is comparing two qubits for equality. Then, we store $1 - g(j_1)$ to a qubit $|\phi^0\rangle$ and we denote the obtained index as a qubit $|\psi^0\rangle$.

Let us consider the case of $i > 0$. Assume that we have a function $comp$: $\{0, 1\} \times \{1, \ldots, k\} \times \{0, 1\} \times \{1, \ldots, k\} \to \{0, 1\}$ that compares two pairs (q, i) and (q', i') in lexicographical order, i.e. $comp(q, i, q', i') = 1$ iff $q < q'$ or $(q = q')\&(i < i')$. The function can be implemented in constant time and memory complexity because each value is a single qubit. We can say that $g^i(j) = comp(|\phi^i\rangle, |\psi^i\rangle, |1 - g'(j)\rangle |j\rangle)$. Using the state (3) and $|\phi^i\rangle|\psi^i\rangle$ we can implement multi-target Grover's search as in Sect. 3. After measurement we obtain a result index j and value of the function $g(j)$. Then, we store them to the register $|\phi^{i+1}\rangle|\psi^{i+1}\rangle$.

Then, we do $3\log_2 k$ phases using new copies of the state (3). Finally, we obtain the minimal index i_0 of unequal symbols. We can access to i_0-th element of state $|\xi\rangle$, compare u_{i_0} and v_{i_0}, and return the answer according to the discussion in Sect. 2.1. For accessing to i-th element, we can swap it with 0-th element using CNOT gates and then apply Hadamard transformation for collecting whole amplitude in 0-th element. These operations require $O(\log k)$ time complexity.

Theorem 2 *The presented algorithm solves string comparing problem with bounded error. It has $O(\sqrt{k}\log k)$ time complexity and uses $O((\log k)^3)$ qubits of memory.*

Proof The algorithm solves the problem because it implements the idea from [14]. The probability of success is constant because of the properties of the Dürr-Høyer algorithm for minimum search [6, 7].

Let us compute time complexity of the algorithm. Due to the properties of the Dürr-Høyer algorithm, the total number of iterations of all invocations of Grover's search is $O(\sqrt{k})$. Time complexity of computing $g(i)$ and $comp$ are constant. Therefore, the total time complexity is $O(\sqrt{k}\log k)$

Let us consider the memory complexity. We need $O((\log k)^3)$ qubits for state (2). □

4.2 The Algorithm Based on Binary Search

Let us implement the idea with the Binary search algorithm that was discussed in Sect. 2.1.3.

Let $k = \min(|u|, |v|)$ for two comparing strings u and v. Let us choose a prime p from the first $\frac{\delta \cdot k}{\varepsilon}$ prime numbers, where $0 < \varepsilon < 1$ is some constant and $\delta = k$ because we will have k hashes of substrings of the string u and v.

Assume that the initial state for our algorithm is the following.

$$|\phi\rangle \otimes \bigotimes_{t=1}^{\log_2 k} \frac{1}{\sqrt{k}} \sum_{a=0}^{k-1} |a\rangle \otimes |h(u[1, a+1])\rangle \otimes |h(v[1, a+1])\rangle, \quad |\phi\rangle = \sum_{a=0}^{k-1} |a\rangle|u_a\rangle \tag{4}$$

We can find the first a_0 such that $h(u[1, a_0 + 1]) \neq h(v[1, a_0 + 1])$ using Binary search algorithm because of arguments from Sect. 2.1.3. On each phase, we should access to some middle element with an index mid and compare $h(u[1, mid + 1])$ and $h(v[1, mid+1])$. For accessing to i-th element we can swap it with 0-th element using CNOT gates, and then apply Hadamard transformation for collecting whole amplitude in 0-th element. These operations require $O(\log k)$ time complexity.

Next, we should compare two strings of length $O(\log p)$ for equality. We can do it using Grover's search algorithm as it was done in Sect. 3. The time complexity of this algorithm is $O(\sqrt{\log p}\log\log p)$ and memory complexity is $O(\log\log p)$ qubits.

Therefore, after $\log_2 k$ steps of the Binary search algorithm, we obtain the minimal index a_0 such that $h(u[1, a_0+1]) \neq h(v[1, a_0+1])$. If $h(u[1, a_0+1]) = h(v[1, a_0+1])$, then $u = v$. If $h(u[1, a_0+1]) \neq h(v[1, a_0+1])$, then we can access to a_0-th element of $|\phi\rangle$ for accessing to u_{a_0}. If $u_{a_0} = 0$, then $u < v$ and $u > v$ otherwise.

Theorem 3 *The presented algorithm solves string comparing problem with bounded error. It has $O((\log k)^2 \log\log k)$ time complexity and uses $O((\log k)^2)$ qubits of memory.*

Proof The algorithm solves the problem because it implements the idea from Sect. 2.1.3.

Let us compute time complexity of the algorithm. There are $O(\log k)$ phases of Binary search. Each phase requires comparing hashes in $O(\sqrt{\log p}\log\log p)$ and accessing to mid-th element in $O(\log k)$. The final step is accessing to element with $O(\log k)$ time complexity. The final time complexity is

$O(\log k \cdot (\sqrt{\log p}\log\log p + \log k) + \log k) = O((\log k) \cdot (\sqrt{\log k}\log\log k + \log k)) = O((\log k)^2 \log\log k)$.

Let us consider the memory complexity. We need $O(\log k \cdot (\log k + \log p) + \log k) = O((\log k)^2)$ qubits for state (4) and $O(\log\log p) = O(\log\log k)$ states for the implementation of two hashes comparing. So, the total memory complexity is $O((\log k)^2)$. □

5 Conclusion

In the paper, we presented algorithms for two problems—String matching problem and String comparing problem. The algorithm for the String matching problem works as fast as the best-known quantum algorithm up to a log factor. At the same time, our algorithm uses exponentially fewer qubits of memory. We have presented two algorithms for string comparing problem. Both use exponentially fewer qubits comparing to the best-known algorithm for the problem. The first one is based on Grover's search algorithm and uses more qubits than the second one based on Binary search. The second algorithm works exponentially faster than the first one and than the existing algorithm [14]. At the same time, the initial state of the second algorithm is more complex compared to the initial state of the first algorithm.

The initial state of all algorithms is not just stored input in quantum memory. At the same time, preparing this state is not much harder than storing input data in quantum memory as is. Additionally, these algorithms can be used as a part of other algorithms. A similar motivation is presented in different papers, for example, in [10].

Acknowledgments The research is funded by the subsidy allocated to Kazan Federal University for the state assignment in the sphere of scientific activities, project No. 0671-2020-0065 and by the Kazan Federal University Strategic Academic Leadership Program ("PRIORITY-2030"). A part of the reported study (Sect. 4) was funded by RFBR according to the research project No. 20-37-70080.

References

1. Ablayev, F., Ablayev, M., Huang, J.Z., Khadiev, K., Salikhova, N., Wu, D.: On quantum methods for machine learning problems part I: Quantum tools. Big Data Mining and Analytics. **3(1)**, 41–55 (2019)
2. Ambainis, A.: Understanding quantum algorithms via query complexity. In: Proc. Int. Conf. of Math. **4**, 3283–3304 (2018)
3. Ambainis, A., Balodis, K., Iraids, J., Khadiev, K., Kļevickis, V., Prūsis, K., Shen, Y., Smotrovs, J., Vihrovs, J.: Quantum Lower and Upper Bounds for 2D-Grid and Dyck Language. In: 45th International Symposium on Mathematical Foundations of Computer Science (MFCS 2020), Leibniz International Proceedings in Informatics (LIPIcs). **170**, 8:1–8:14 (2020)
4. Boyer, M., Brassard, G., Høyer, P., Tapp, A.: Tight bounds on quantum searching. Fortschritte der Physik. **46(4–5)**, 493–505 (1998)
5. Brassard, G., Høyer, P., Mosca, M., Tapp, A.: Quantum amplitude amplification and estimation. Contemporary Mathematics. **305**, 53–74 (2002)
6. Dürr, C., Høyer, P.: A quantum algorithm for finding the minimum. arXiv:quant-ph/9607014 (1996) Available via arXiv. https://arxiv.org/abs/quant-ph/9607014. Cited 23 Nov 2020
7. Dürr, C., Heiligman, M., Høyer, P., Mhalla, M.: Quantum query complexity of some graph problems. SIAM Journal on Computing. **35(6)**, 1310–1328 (2006)
8. Freivalds, R.: Fast probabilistic algorithms. In: Mathematical Foundations of Computer Science 1979, LNCS. **74**, 57–69 (1979)
9. Grover, L.K.: A fast quantum mechanical algorithm for database search. In: Proceedings of the twenty-eighth annual ACM symposium on Theory of computing. 212–219 (1996)
10. Harrow, A.W., Hassidim, A., Lloyd, S.: Quantum algorithm for linear systems of equations. Physical review letters. **103(15)**, 150502 (2009)
11. Høyer, P., Mosca, M., de Wolf, R.: Quantum search on bounded-error inputs. In: Baeten, J.C.M., Lenstra, J.K., Parrow, J., Woeginger, G.J. (eds) Automata, Languages and Programming, pp. 291–299. Springer, Berlin, Heidelberg (2003)
12. Kapralov, R., Khadiev, K., Mokut, J., Shen, Y., Yagafarov, M.: Fast classical and quantum algorithms for online k-server problem on trees. arXiv:200800270 (2020) Available via arXiv. https://arxiv.org/abs/2008.00270. Cited 23 Nov 2020
13. Karp, R.M., Rabin, M.O.: Efficient randomized pattern-matching algorithms. IBM journal of research and development. **31(2)**, 249–260 (1987)
14. Khadiev, K., Ilikaev, A.: Quantum algorithms for the most frequently string search, intersection of two string sequences and sorting of strings problems. In: International Conference on Theory and Practice of Natural Computing, pp. 234–245 (2019)
15. Knuth, D.E., Morris, J.H. Jr., Pratt, V.R.: Fast pattern matching in strings. SIAM journal on computing. **6(2)**, 323–350 (1977)
16. Kothari, R.: An optimal quantum algorithm for the oracle identification problem. In: 31st International Symposium on Theoretical Aspects of Computer Science, pp. 482–482 (2014)
17. Lin, C.Y.Y., Lin, H.H.: Upper bounds on quantum query complexity inspired by the Elitzur-Vaidman bomb tester. In: 30th Conference on Computational Complexity (CCC 2015), Schloss Dagstuhl-Leibniz-Zentrum fuer Informatik (2015)
18. Lin, C.Y.Y., Lin, H.H.: Upper bounds on quantum query complexity inspired by the Elitzur–Vaidman bomb tester. Theory of Computing. **12(18)**, 1–35 (2016)

19. Montanaro, A.: Quantum pattern matching fast on average. Algorithmica. **77(1)**, 16–39 (2017)
20. Nielsen, M.A., Chuang, I.L.: Quantum computation and quantum information. Cambridge univ. press (2010)
21. Ramesh, H., Vinay, V.: String matching in $o(\sqrt{n} + \sqrt{m})$ quantum time. Journal of Discrete Algorithms. **1(1)**, 103–110 (2003)
22. Soni, K.K., Rasool, A.: Pattern matching: A quantum oriented approach. Procedia Computer Science. **167**, 1991–2002 (2020)

Multicriteria Optimization Techniques in SVM Method for the Classification Problem

Anastasia A. Andrianova

Abstract The paper considers the procedures for solving the multiclass classification problem using a series of support vector machine optimization problems for the binary classification problem in the multicriteria formulation. In the traditional formulation, the objective function takes into account the width of the separating the classes hyperplane and the function of the penalty for classification errors. Multicriteria modifications of this problem allow us to study the influence of each criterion separately on the classification error. For the problem of multiclass classification, the use of the optimization problem of binary classification and its modifications is carried out within the strategies of the "one against many" or "elimination tournament". The research of various procedures for solving a multiclass problem is carried out using the example of the intrusion detection problem.

1 Introduction

The classification problem is one of the most popular big data problems. One of the ways to solve it, both in the case of binary classification and in the case of multiclass classification, is the Support Vector Machine method (SVM) [1–3].

The main idea of the SVM-method consists of constructing the classes separating hyperplane (linear separation case) or the nonlinear surface (for kernel approach). To find it, the optimization problem is used, which is based on maximizing the width of the classes separating "strip". The constraint system of this problem includes inequalities for all training samples, which provides the conditions for belonging to the "correct side" of the separation, taking into account the sample class labels and the possible classification error. In order to take into account classification

A. A. Andrianova (✉)
Department of System Analysis and Information Technologies, Institute of Computational Mathematics and Information Technologies, Kazan Federal University, Kazan, Russia
e-mail: Anastasiya.Andrianova@kpfu.ru

I. B. Badriev et al. (eds.), *Mesh Methods for Boundary-Value Problems and Applications*, Lecture Notes in Computational Science and Engineering 141,
https://doi.org/10.1007/978-3-030-87809-2_2

errors, which is especially important when classes are not separable by the using surface, the objective function is complemented by a penalty term, which depends on the errors for each of the samples of the training dataset. Thus, we obtain an optimization problem of a large dimension. In this paper, other ways of formulating the objective function of SVM optimization problem will be considered. The main attention will be paid to the minimization of the different types of the classification error functions including multicriteria models for binary classification problem [4].

Several optimizing classification models will be formulated. For them, the results of an experimental study are given to obtain conclusions on the possibility of improving the accuracy of a classification and reducing the computational complexity of the optimization problem. The experiments were performed on known datasets [5, 6].

Also, methods of applying the multicriteria approach to solving the problem of multiclass classification [7] are considered by the example of solving the problem of intrusion detection [8].

2 Optimization Problem of SVM in a Multicriteria Formulation

In the general form for an arbitrary dataset, the classical classification method C-SVC (Common-Support Vector Classification) defines a hyperplane as a solution to the following optimization problem:

$$\min \rightarrow 0.5\,\|w\|^2 + C\sum_{i=1}^{K} \varepsilon_i \tag{1}$$

with the following constraints:

$$y_i(w^T \varphi(x_i) + b) \geq 1 - \varepsilon_i, \quad i = 1, \ldots, K, \quad \varepsilon_i \geq 0 \quad i = 1, \ldots, K,$$

where K is the size of the training sample, $\{x_i, y_i\}$ $i \in K$—training sample where $y_i \in \{-1, 1\}$—class labels. The variables of the problem are the parameters of the separating hyperplane w and variables of the error that estimate the i-th example of the training sample is assigned to the wrong class ε_i, $C > 0$ is the penalty constant. Objective function (1) combines maximizing dividing bandwidth and minimizing error.

The following cases, which depend on the value ε_i, are possible. If the class is correctly defined then $\varepsilon_i = 0$. Another case is when the sample lies in the separating strip, here $0 < \varepsilon \leq 1$. If the class is defined incorrectly $\varepsilon_i > 1$, then the error estimation is proportional to the distance from the classified object x_i to the hyperplane. The function $\varphi(x)$ is called the kernel and allows us to consider not only the linear separability of classes.

Let us describe four models in which used combinations of two criteria instead of the classical objective function. In this modification (2), (3) of the method, the main criteria are separating strip width and total error. The second most important criterion is the maximum of the examples errors.

Modification 1 (Modif1):

$$\min \to 0.5\,\|w\|^2 + \sum_{i=1}^{K} \varepsilon_i \tag{2}$$

$$\min \to \max(\varepsilon_i) \tag{3}$$

with the following constraints:

$$y_i(w^T\varphi(x_i) + b) \geq 1 - \varepsilon_i, \quad i = 1, \ldots, K, \quad \varepsilon_i \geq 0 \quad i = 1, \ldots, K,$$

Let us consider the second modification (4)–(5) of the classical method. Let us pass from the problem of maximization the width of the separating strip to minimizing the error function. The second most important criterion is the maximum error.

Modification 2 (Modif2):

$$\min \to \sum_{i=1}^{K} \varepsilon_i \tag{4}$$

$$\min \to \max(\varepsilon_i) \tag{5}$$

with the following constraints:

$$y_i(w^T\varphi(x_i) + b) \geq 1 - \varepsilon_i, \quad i = 1, \ldots, K, \quad \varepsilon_i \geq 0 \quad i = 1, \ldots, K,$$

Let us show another modification which is based on the separation of two types of classification errors. As discussed above, a positive error is possible in two cases:

1. $0 < \varepsilon_i \leq 1$, if the sample lies in the separating strip. This is equivalent to a state of uncertainty, the classifier is more likely to choose the class, closer to the border of which the object is located;
2. $\varepsilon_i > 1$, if the class is defined incorrectly. This is a more serious error compared to the first type of error.

Modification 3 (Modif3):

$$\min \to c \sum_{i=1}^{K} \varepsilon_i + C \sum_{i=1}^{K} \mu_i \tag{6}$$

$$\min \to \max(\mu_i) \tag{7}$$

with the following constraints:

$$y_i(w^T \varphi(x_i) + b) \geq 1 - \varepsilon_i - \mu_i,\ i = 1, \ldots, K,\ 0 \leq \delta_i \leq 1,\ \mu_i \geq 0\ i = 1, \ldots, K,$$

Let us denote the error variable as the sum of two variables $\varepsilon_i = \delta_i + \mu_i$, where $0 \leq \delta_i \leq 1$, $\mu_i \geq 0$. In this case, the following outcomes are possible:

1. $\varepsilon_i = 0$, the object is classified correctly. Then $\delta_i = 0$, $\mu_i = 0$ (the object is classified correctly);
2. $0 < \varepsilon_i \leq 1$, the example falls into the dividing strip. Three cases are possible $0 < \delta_i \leq 1$ and $\mu_i = 0$; another one $0 \leq \delta_i \leq 1$ and $0 < \mu_i \leq 1$ with the constraint $\delta_i + \mu_i \leq 1$; and the last case $\delta_i = 0$ and $0 < \mu_i \leq 1$;
3. $\varepsilon_i > 1$, the object is classified incorrectly for $\delta_i = 0$ and $\mu_i > 1$; $0 < \delta_i$ and $0 < \mu_i$.

A positive value of μ_i is possible in the third case. Such a multicriteria model has a more complex structure, since the number of error variables doubles and the number of constraints increases significantly. By increasing the constant C, we significantly reduce the magnitude of the error function.

In the following modification, the main criterion is the width of the separating strip and total error. The second criterion is the sum of the moduli of the components of the normal vector w to the separating hyperplane.

Modification 4 (Modif4):

$$\min \to 0.5\, \|w\|^2 + \sum_{i=1}^{K} \varepsilon_i \tag{8}$$

$$\min \to \sum_{i=1}^{L} |w_i| \tag{9}$$

with the following constraints:

$$y_i(w^T \varphi(x_i) + b) \geq 1 - \varepsilon_i, \quad i = 1, \ldots, K, \quad \varepsilon_i \geq 0 \quad i = 1, \ldots, K,$$

It should be noted that criteria (3), (5), (7), (9) are non-differentiable, which complicates the methods of their solution.

One of the models of multicriteria optimization was investigated on the datasets a1a, a1a, a2a, a3a, a4a, a5a, a6a, a7a, a8a, a9a for binary classification problem [5], in which it showed an improvement in accuracy in 33% of tests (from 120 tasks) compared to the classical SVM-method. The improvement was just under 1% accuracy.

3 Application of the SVM Method for Multiclass Classification Problem

This section presents a theoretical description of the transition from a binary classification problem to a multiclass one. Suppose there are M classes, then the set of class labels Y has the form: $Y = \{0, 1, 2, 3 \ldots M - 1\}$.

Nowadays, there are two main approaches to solving the problem of multiclass classification in SVM. The first is called "one-step solution" or "all-together". This approach is used for multiclass classifiers such as "decision trees".

In another approach, the solution of a multiclass problem is reduced to the solution of a sequence of problems with two classes. In this case, there can be several strategies for generating such a sequence. "one against many", "one against one", "elimination tournament". Thus, in this approach, the multiclass problem is divided into a set of binary problems that are solved independently using binary classification algorithms. The partitioning process itself is usually called the reduction of a multiclass problem to a sequence of binary ones. Methods for reducing a multiclass classification problem to a sequence of binaries problems are trained faster and give fewer errors, while the one-step solution approach results in fewer support vectors.

Let us consider the "one against many" strategy. K classifiers are trained for M classes, each of which separates "its" class from all other classes. Thus, a classifier is built for each class. During recognition, the unknown vector X is fed independently to all M classifiers. The class to which the vector X belongs is determined by the classifier that gives the highest estimate $f(x) = argmax(< w_k, x > +b_k), k = \{1, \ldots, M\}$. The disadvantages of this approach include the fact that each of the K classifiers trains on its own sample, from which the obtained values may have different scales, so it would be incorrect to compare them. It will also be incorrect to normalize the weight vectors so that the answers are on the same scale since this procedure will change the weight norm, as a result of which the weights will no longer be solutions of the support vector classification problem. This problem is called the problem of the incommensurability of quantities.

Here, we consider the second (paired) approach "one against one". Let us construct $C_M^2 = M(M-1)/2$ binary classifiers $a_{ij}(x)$, where $i, j = \overline{1, M}, i \neq j$, learners to distinguish all possible pairs of classes from each other. We will adjust the classifier $a_{ij}(x)$ for that part of the sample that contains only objects of classes i and j. For the recognition of the vector, each classifier produces an estimate $f_{ij}(x)$, which reflect the belonging to classes i and j. The result is a class with the maximum sum $\sum_{i \neq j} f_{ij}(x)$, where g is a monotonically non-decreasing function, namely, the identity or logistic function. The "elimination tournament" strategy is similar to the "one against one" strategy. This strategy also constructs $C_M^2 = M(M-1)/2$ binary classifiers capable of distinguishing all possible pairs of classes. The difference is that at the stage of choosing whether the input vector belongs to a class, a tournament is held between the two classes: at each step of recognizing the vector, a single classifier is selected, the winning class determines which classifier will be used at the next step.

4 Dataset UNSW-NB15

The UNSW-NB15 dataset [8] is network packets generated by the IXIA PerfectStorm tool at the Australian Cyber Security Center (ACCS) Cyber Range Lab. They are a combination of real normal actions and artificial attacks. The selected dataset includes nine types of attacks: Fuzzers, Analysis, Backdoor, DoS, Exploits, Generic, Reconnaissance, Shellcode, and Worms.

1. Fuzzers is an attack in which a large amount of randomly generated data is fed to the program/network;
2. Dos is a denial of service attack, it consists in the difficulty/refusal of providing access to system resources;
3. Analysis is an attack that is carried out by scanning ports and sending spam;
4. Bakdoors is a bypass of protection in order to gain unauthorized access to a computer;
5. Exploits are the exploitation of system errors and vulnerabilities, leading to unexpected behavior of the network or host;
6. Generic is an attack that uses a hash function to cause a collision;
7. Reconnaissance are attacks that collect information about a computer;
8. Shellcode is the collection of information about the network for further bypassing the protection of the studied system;
9. Worms is an attack during which the attacking code copies itself for further transmission over a computer network.

Thus, when using the dataset in relation to the classification problem, ten classes can be distinguished: nine classes correspond to the above attacks, the 10th class corresponds to normal data transactions, 43 characteristics with a class label: 42 signs of network traffic of five types: integer, string, double, boolean, time, and 43 contain information about the class label (0—normal actions, 1—attacks). All characteristics have been converted to float64 type.

The developers of this dataset have made it easier to split the sample and provide two generated datasets for use. The training set contains 175,341 records, and the test set contains 82,332 records, including various types of attacks and normal

Table 1 Distribution of the examples

Category	Training set	Testing set
Normal	56,000	37,000
Analysis	2000	677
Backdoor	1746	583
Dos	12,264	4089
Exploit	33,393	11,132
Fuzzers	18,184	6062
Generic	40,000	18,871
Reconnaissance	10,491	3496
Shellcode	1133	378
Worm	130	44
Total amount	175,341	82,332

actions. Table 1 shows the distribution of the number of examples of different classes in the training and test samples.

Evaluation of the UNSW-NB15 dataset in existing classification systems has shown that this dataset is close to real traffic data. It is important to say that this dataset can be used to effectively evaluate existing and new classification methods. The UNSW-NB15 dataset was used to conduct experiments on both binary and multiclass classification. For the problem of binary classification, the 43rd component of the characteristic vector was used as class labels, which has the values 0—if the action is normal, and 1—if the action is related to attacks. For a multiclass problem, the 42nd component of the input vector was used as class labels, which can take one of ten values: Normal, Fuzzers, Analysis, Backdoor, DoS, Exploits, Generic, Reconnaissance, Shellcode and Worms.

5 Multiclass Classification Experiments

In this section, we describe the comparative analysis of data obtained as a result of multiclass classification experiments using the UNSW-NB15 dataset. Comparison of algorithm modifications and selection of the "best" one will be made according to the following indicators: time, Accuracy (Ac), Precision (Pr) and Recall (Rec). The last two metrics are defined as the average over a series of binary classification problems.

The main technique for solving a multiclass classification problem during experiments was the "one against many".

Experiment 1 During the experiment, 30 different samples were generated from the original dataset. Each sample consists of 3540 examples, of which: 70% for training, 15% examples for each validation and testing. It is guaranteed that there are no training sample examples in the validation and training samples.

Table 2 Methods performance indicators

	Ac (%)	Time (s)
C-SVM	62.83	612
Modif1	64.1	640
Modif2	69.6	2110
Modif3	67.4	19,322 (near 5 h)
Modif4	64.1	630

Table 3 Comparison of the accuracy of the classical method and modifications

	$\Delta Ac_{(modif\text{-}class)}$ (%)
Modif1	1.27
Modif2	6.77
Modif3	4.57
Modif4	1.27

Using the generated samples, it was possible to build a linear classifier with an Accuracy of 59.32–69.74%. To solve multicriteria problems, the concessions method is used with the concession size of the main criterion—0.1. Table 2 shows the average values of the accuracy of the classical method and its four modifications, as well as the average running time of the algorithm for all series of this experiment.

Table 3 shows the difference between the average accuracy of the algorithm modification and the average accuracy of its classical interpretation within the series of this experiment. We analyzed the results obtained during the experiment and now we can draw the following conclusions. The usage of modifications in the form of multicriteria formulations of optimization problems made it possible to improve the classification accuracy in all cases. The highest accuracy was shown by the Modif3 modification. Modifications Modif1 and Modif4, in which the width of the separating strip is the main criterion, operate with the same average accuracy, while the operating time of these modifications is comparable to the operating time of the classical SVM.

Modif3 modification is much more laborious due to the complication of its structure and doubling of the number of error variables. Nevertheless, it shows relatively high rates of classification accuracy, which is of interest from the point of view of finding more effective methods for its solution. The average difference between the accuracy of the Modif3 modification and the classical method is 4.57%, but more time is spent, the operating time of the Modif3 modification is 31 times longer than the classical method. If the bandwidth is not taken into account (in particular, the modification of Modif 2), then the accuracy of the algorithm does not decrease.

Experiment 2 The experiment contained, as a training sample, a dataset with the same number of examples for each class. In this case, the classification accuracy dropped quite dramatically—to 17–39%.

In order for us to understand in which classes the classifier is mistaken, a Confusion matrix was displayed for each method. Figure 1 shows an example of

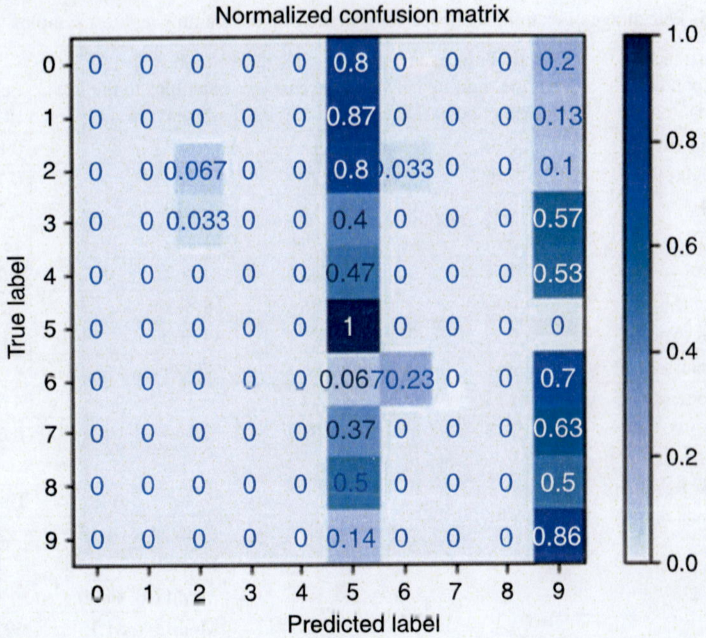

Fig. 1 Confusion matrix

Table 4 Results of experiment 2.3

	Acc (%)	Time (s)	Classes with maximum recall value
C-SVM	58.63	588.31	5.6
Modif1	59.33	510.67	5
Modif2	65.12	1109.52	5
Modif3	62.77	15,707.8	5
Modif4	61.78	523.76	5

a confusion matrix for the library version of the SVM-method. It should be noted that the library version of the SVM-method has the lowest accuracy—17.9%. You can see that the classifier defines the instances of the fifth class as well as possible, but at the same time it is mistaken, mistakenly attributing instances of other classes to the fifth and ninth classes.

After that the experiment was conducted in which the percentage of examples of each class coincides with the ratio in the original data set. Thus, conditions have been artificially created that correspond to the frequency and importance of the classification of individual private classes (Table 4).

Table 5 shows the percentage of examples of each class in the original training and test samples.

Table 5 Percentage of examples of each class in the original training and test samples

Class name (corresponding number)	Number of examples in the training sample, pcs.	Percentage (%)	Number of examples in the test sample, pcs.	Percentage (%)
Analysis (0)	2000	1.1	677	0.8
Backdoor (1)	1746	0.99	583	0.7
DoS (2)	12,264	7	4089	4.2
Exploit (3)	33,393	19.04	11,132	13
Fuzzers (4)	18,184	10.37	6062	7.25
Generic (5)	40,000	22.81	18,871	22
Normal (6)	56,000	32	37,000	44
Reconnaissance (7)	10,491	5.98	3496	4
Shellcode (8)	1133	0.64	378	4
Worm (9)	130	0.07	44	0.05

Table 6 Results of experiment 4

	Acc (%)	Time (s)
C-SVM	61.12	2961.08
Modif1	63.70	4475.71
Modif2	84.09	7843.45
Modif3	81.15	69,544.02
Modif4	63.58	5304.42

After our analysis of the obtained results, the following conclusions were drawn. It was noted that all classifiers define the fifth class well, but at the same time they mistakenly attribute objects of other classes to it. When using the same number of examples of each class in the training sample and in the test and validation samples, the accuracy of the methods is significantly reduced. Also, an experiment that minimized the presence of fifth class also did not improve the classification accuracy. Examples erroneously refer to normal traffic, which poses a threat to computer security; when using the percentage of examples similar to the original data sets, the accuracy of the methods practically does not differ from the results obtained in experiment No. 1, in which the samples were created by randomly choosing the required number of examples. Methods are best at recognizing normal traffic, with most examples of other classes erroneously categorizing as "Generic" attacks. This may be due to the similarity of signs of different types of attacks.

Experiment 3 This experiment consisted of a multi-stage classification procedure, in which at each step a classification of three classes was made, with the fifth and sixth classes present in all tests. Table 6 shows the results obtained during this experiment.

Tables 7 and 8 show the average values of the Precision and Recall characteristics for Modifications 2 and 3, respectively.

According to Table 7, it can be seen that Modification 2 (Modif2) does not predict the ninth class, Modification 3 (Modif3) determines them, albeit with low

Table 7 Average value of metrics for Modification 2 (Modif2)

Class	Precision	Recall	Class	Precision	Recall
0	95.6	90.5	5	78.8	89.9
1	96	93.3	6	75.4	80.3
2	94.1	65.3	7	84.3	51.2
3	67.4	95.6	8	53.8	56.3
4	70.1	94.2	9	0	0

Table 8 Average metrics for Modification 3 (Modif3)

Class	Precision	Recall	Class	Precision	Recall
0	74.1	92.1	5	71.4	93.2
1	91.2	94.1	6	76.5	64.2
2	86.4	73.8	7	64.2	63.7
3	67.4	74.6	8	64.4	70.3
4	70.1	77.8	9	8.2	14.4

Table 9 Comparison of the accuracy of the methods

Classes	Acc. Modif2 (%)	Acc. Modif3 (%)	Acc. C-SVM (%)
0,5,6	91.46	84.67	34.63
1,5,6	92.34	90.85	62.65
2,5,6	82.68	82.37	61.68

accuracy (Table 8). Moreover, in both modifications the zero, first and second classes differ best from the fifth and sixth classes. Table 9 shows the average value of the accuracies for the series of experiments with the best result of Modification 2 (Modif2) and Modification 3 (Modif3). In this experiment, Modifications 2 and 3 (Modif 2 and Modif 3) showed good results, with a low accuracy of the classical SVM-method (about 60%), the accuracy of the modifications reached 95%. The use of such a multi-stage scheme has significantly improved the classification accuracy, the best results were shown by Modif2 and Modif3. Both modifications only take into account the error without considering the width of separating strip. This may be due to the fact that features of different classes have similar meanings and therefore classifiers easily mistakenly classify objects of different classes as the most common classes. With the sequential classification of the three classes, classifiers manage to more accurately determine the predicted class.

6 Conclusion

Thus, on the basis of the of the experiments, it was found that the approaches of multicriteria optimization in the formulation of optimization models of the SVM-method make it possible to increase the accuracy of the classifier even in cases of poor class separation. Despite the formulation of an optimization model based on the binary classification problem, this approach is applicable to the multiclass classification problem. And also on its basis, it is realistic to make multi-stage classification algorithms, the use of which, on the example of the intrusion detection problem, showed a significant increase in the classification accuracy.

Acknowledgment This paper has been supported by the Kazan Federal University Strategic Academic Leadership Program ("PRIORITY-2030").

References

1. Burges, C.J.: A Tutorial on Support Vector Machines for Pattern Recognition. Data Mining and Knowledge Discovery **2**, 121–167 (1998)
2. Cortes, C., Vapnik, V.: Support-vector networks. Mach Learn **20**, 273–297 (1995)
3. Munoz, A.: Machine learning and optimization (2014) https://www.cims.nyu.edu/~munoz/files/ml_optimization.pdf. Cited 01 Oct 2020
4. Andrianova, A.A.: Comparative analysis of optimization models for the binary classification problem by the SVM method. Journal of Physics: Conference Series. **1158(2)** (2019)
5. Chang, C.C., Lin, C.J.: LIBSVMdata: Classification, regression and multi-label (2011). https://www.csie.ntu.edu.tw/~cjlin/libsvmtools/datasets/. Cited 01 Oct 2020
6. Chang, C.C., Lin, C.J.: LIBSVM: A library for support vector machines (2011) https://doi.org/10.1145/1961189.1961199
7. Weston, J., Watkins, C.: Multi-class support vector machines. Technical Report (2002) https://doi.org/10.1109/ICPR.2002.1048282
8. DataSet UNSW-NB15 (2018). https://www.unsw.adfa.edu.au/unsw-canberra-cyber/cybersecurity/ADFA-NB15-Datasets/. Cited 01 Oct 2020

Developing Experimental–Numerical Methods for Constructing True Deformation Diagrams of Elastoplastic Materials

Valentin G. Bazhenov, Elena V. Nagornykh, Sergey L. Osetrov, and Dmitry L. Osetrov

Abstract True deformation diagrams are constructed using an iterative procedure of updating the strain intensity–stress intensity relation proportionally to the relative difference in the values of axial forces as obtained numerically and experimentally for an inhomogeneous stress-strain state, accounting for necking, up to rupture. The procedure requires multiply solving the problem, which is a time-consuming computational task. Two scenarios of analyzing a boundary-value problem are considered. The first scenario involves analyzing the entire direct problem over the whole loading interval; in the second one, the entire loading process is subdivided into several intervals defined by discrete values of an experimentally found generalized displacement–generalized force relation. At each small interval, a deformation diagram is constructed, using a nonlinear extrapolation procedure. At the end of each interval, the difference between the calculated and experimentally determined generalized forces is checked, and the stress intensity value is iteratively updated. The presented numerical studies show that constructing a deformation diagram with accuracy less than 1% according to the first scenario required 5–10 repeated analyses of the direct problem, whereas in the second scenario not more than two direct analyses suffice. Monotone convergence of the considered algorithms is examined using a number of problems.

V. G. Bazhenov (✉) · S. L. Osetrov · D. L. Osetrov
Research Institute of Mechanics of National Research Lobachevsky State University of Nizhny Novgorod, Nizhny Novgorod, Russian Federation
e-mail: bazhenov@mech.unn.ru

E. V. Nagornykh
National Research Lobachevsky State University of Nizhny Novgorod, Nizhny Novgorod, Russian Federation
e-mail: pavlyonkova@mech.unn.ru

I. B. Badriev et al. (eds.), *Mesh Methods for Boundary-Value Problems and Applications*, Lecture Notes in Computational Science and Engineering 141,
https://doi.org/10.1007/978-3-030-87809-2_3

1 Introduction

The current state of the art in numerical analyses of strength of structural parts and elements requires reliable data on material behavior (deformation diagrams, ultimate strain and strength characteristics etc.). Obtaining such data with the currently available instrumental means for high elastoplastic deformations of material by way of direct experimental measurements is hampered by non-uniaxial and non-uniform stress–strain states (SSS) in laboratory specimens due to high deformations, as well as the presence of boundary effects etc. The identification of deformational and strength properties of materials, in this case, is done using analytical [1, 2] or numerical approaches [3–13] which make it possible to determine the characteristics of SSS and representing the deformation diagram as exponential functions, using indirect experimental data (forces and displacements). However, the use of analytical methods often strongly limits specimen geometry and type of loading and involves force and kinematic assumptions for the parameters of SSS [14]. The issue of describing diagrams of elastoplastic deformation up to failure has not been studied well enough by now.

In this connection, the study of material properties under high elastoplastic deformations calls for developing an experimental–numerical approach devoid, to a large degree, of the limitations of experimental–analytical methods. An experimental–numerical approach involves experimenting and full-scale (in the framework of mechanics of solids) computer modeling of deformational processes in a laboratory specimens or structural elements and iteratively elaborating on the deformation diagram.

All the above-said makes topical the studies aimed at developing methods of computer modeling of deformation and failure processes of standard laboratory specimens, as well as effective algorithms of identification of deformational and strength characteristics of elastoplastic materials in the conditions of high strains.

2 The Experimental–Numerical Approach

In a general case, to determine mechanical constants and to construct a deformation diagram of a material, a goal function is formed that describes the differences between full-scale and numerical experiments. Then, an iterative process of determining mechanical constants and material relations is constructed.

It is required to find a set of parameters of the equation of state $b = (b_1, b_2, \ldots, b_n)$ providing the best agreement between the solution of the problem and the numerical results. To this end, such parameters as modulus of dilatation K, shear modulus G, yield strength σ_T, stress intensity for a fixed value of accumulated plastic strains $\sigma_i(\varkappa)$ and others can be used. To find the sought parameters, it is necessary to minimize the function that is the value of mean square deviation between the numerical and experimental results:

$$c(b) = \sum_{j=1}^{N} (p_j^{exp} - p_j^{calc})^2 \tag{1}$$

where p_j^{exp} is the value of the comparison parameters determined experimentally and p_j^{calc} is the value of the comparison parameters obtained numerically. Forces, displacements, stresses, strains, etc. can be used as comparison parameters. The limits for sought parameters b_i in the search area are defined by equations of an initial boundary-value problem, whereas the boundaries of the search area are determined based on experimental facts and physical principles. The analysis of the problem in question depends, to a large degree, on the choice of the optimization algorithm. When choosing among available algorithms and developing new ones, special features of problems being analyzed must be accounted for. In the case considered, the limits are verified by numerically analyzing a nonlinear initial boundary-value problem, which is a huge computational task. That is why time spent to analyze one version, the choice of the initial approximation, and the convergence rate of the numerical process are all very important. It is appropriate to reduce the general problem to a succession of particular problems with one or two comparison parameters. An initial approximation is determined by analytically or numerically analyzing an idealized problem. One of the possible optimization algorithms is the method of successive approximations of the sought parameters that is based on parameter updating according to the relative difference between the experimental and numerical values.

Based on the experimental–numerical approach [15], the present authors have developed methodologies and algorithms of analyzing deformational and strength characteristics of elastoplastic materials subjected to various loading types: tension of cylindrical rods and shells [15, 16], torsion of rods [17], kinetic indentation of specimens with a sphere [18] and dynamic compression of tablet-shaped specimens [19]. In what follows, the efficiency of using experimental–numerical approach [15] is considered, as applied to analyzing solid cylindrical specimens loaded in tension.

3 Constructing Deformation Diagrams of Elastoplastic Rods Loaded in Tension

True material deformation diagrams are constructed by iteratively updating the relation between e_i (strain intensity) and σ_i (stress intensity) over the entire deformation process in the specimen. To this end, at each iteration of the numerical analysis of a tensile specimen problem, the relation between axial forces determined experimentally, F_{exp} and numerically, F_{calc}, $\beta = F_{exp}/F_{calc}$, for the same elongation of the specimen is analyzed. Then a functional relation between maximal strain intensity in the bulk of the specimen e_i^* and the corresponding elongation is established. The diagram is iteratively updated according to the formula $\overline{\sigma}_i(e_i^*) = \beta\sigma_i(e_i^*)$ until the experimentally and numerically determined relations of axial forces agree to a prescribed accuracy. The entire deformation diagram is updated at a time. In this connection, it is necessary to analyze multiply the direct problem and to process the results obtained, which is quite a time-consuming computational chore. The introduced algorithm makes it possible to use any available solver of direct problems without any modifications. The studies showed that, for the iteration procedure to converge, it suffices to assign any convex deformation diagram of a hardening material.

Let us consider the application of the algorithm to the problem of cylindrical bar tension. The experimental specimen was made of class 12H18N10T austenitic steel with the following dimensions: initial radius of the working part $R_0 = 5$ mm, the initial length of the working part $L_0 = 60$ mm. In the finite element model, one end of the rod was assumed rigidly fixed and the other one moved at a constant velocity.

An initial approximation of the true deformation diagram was defined by σ_i e_i in the assumption of incompressibility of the material and uniform deformation of the working part according to the experimentally determined relation between axial force F and elongation ΔL of the rod according to the following formulas:

$$e_i = ln\left(1 + \frac{\Delta L}{L_0}\right), \sigma_i = \frac{F}{A}\left(1 + \frac{\Delta L}{L_0}\right), \qquad (2)$$

where F is the axial force at the end, A is an initial cross-section area of the specimen, ΔL is the displacement of the end in the course of loading.

The deformation diagram was iteratively updated until the experimentally and numerically determined axial forces agreed to the accuracy of 1%. The true deformation diagram (curve 1) obtained in the process of updating is depicted in Fig. 1. The studies showed that one iteration step suffices up to the moment of necking and five iterations are sufficient after necking and up to rupture. The following designations are used in the figure: $q = \sigma_i/\sigma_T$, σ_T is the yield strength of the material.

The deformation diagrams constructed using experimental–analytical methods [1] and [2] practically coincide and are represented in Fig. 1 by curve 2. Figures 2 and 3 show the displacement of axial circumferential and radial stresses ($\sigma_z, \sigma_\theta, \sigma_r$) and strains ($e_z, e_\theta, e_r$) over a minimal cross-section of the specimen after the loss of stability of plastic deformation, as determined using the experimental–numerical method (black curves) and using [1, 2] (gray curves).

It is to be noted that experimental–analytical approaches [1, 2] are based on the assumption that the values of the circumferential and radial strains along the radius of the minimal cross-section of the neck are the same and equal to a constant value. The presented numerical analyze reveal (Fig. 2) that the difference between the axial stresses along the radius of the minimal cross-section of the rod is less

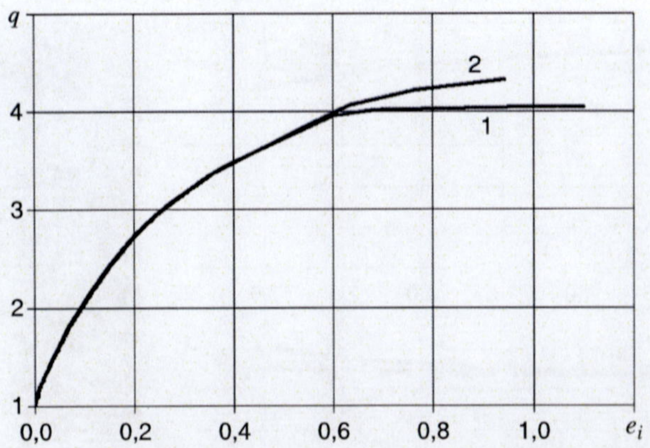

Fig. 1 True deformation diagrams constructed using the experimental–numerical method (curve 1) and the experimental–analytical method (curve 2)

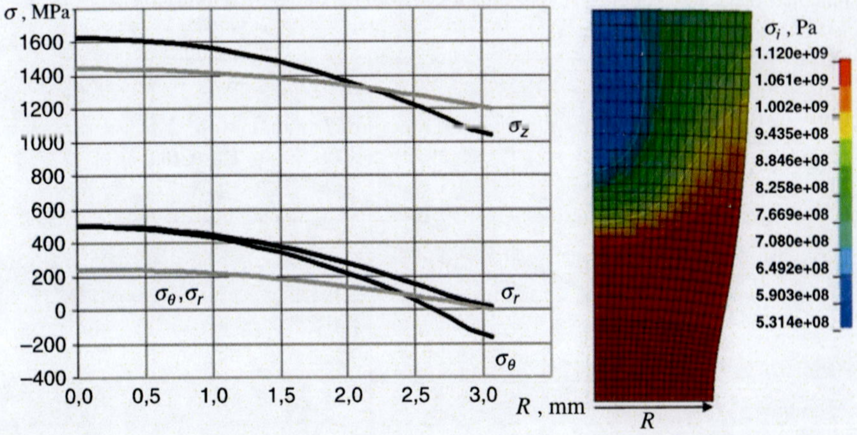

Fig. 2 Relations for axial, circumferential and radial stresses ($\sigma_z, \sigma_\theta, \sigma_r$) obtained using the experimental–numerical method (black curves) and methodologies [1, 2] (grey curves) over the minimal cross-section of the specimen after the loss of stability of plastic deformation

than 11%, whereas the difference between the circumferential and radial stresses is more than 50%. The radial and circumferential strains in the neck are not constant and not equal to each other (Fig. 3), as was assumed in analytical approaches [1, 2]. The circumferential stresses over the free surface of the specimen become negative after necking (Fig. 2). Thus, the use of experimental–analytical methods results in pronounced inaccuracies in constructing deformation diagrams of materials with large deformations at the pre-failure stage.

The use of the above experimental–numerical method is a fairly time-consuming computational task, several times longer than the time for analyzing a direct problem

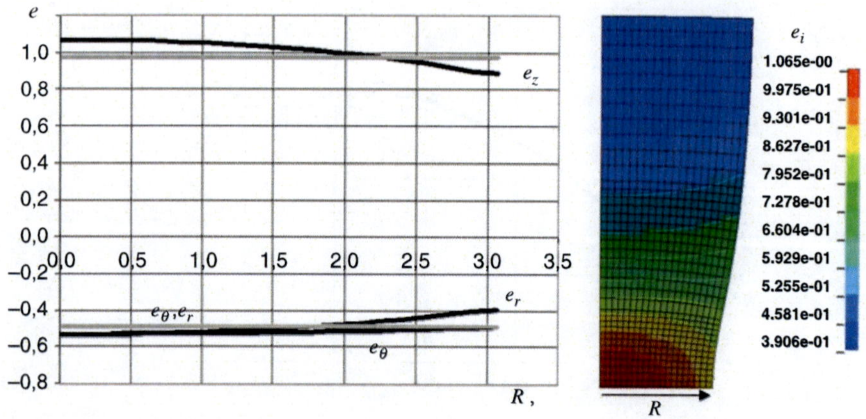

Fig. 3 Relations for axial, circumferential and radial strains (e_z, e_θ, e_r) obtained using the experimental–numerical method (black curves) and methodologies [1, 2] (gray curves) in the minimal cross-section of the specimen after loss of stability of plastic deformation

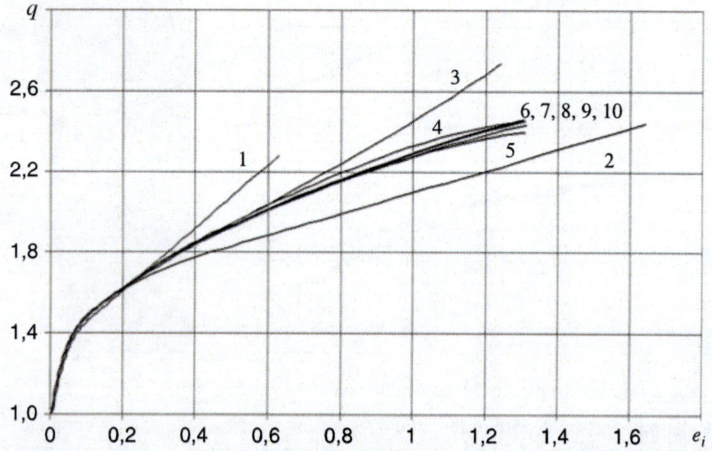

Fig. 4 Iterative construction of the deformation diagram (grey line): 1—initial approximation; 2–10—iterations 1–9, respectively

of a tensile rod once. To assess the effect of accuracy of constructing deformation diagrams on the time consumption of the computational process, the problem of tensile loading of a solid cylindrical rod is considered. The rod is made of the perlite class steel with the following dimensions: initial radius of the working part $R_0 = 5$ mm, the initial length of the working part $L_0 = 60$ mm. A deformation diagram was constructed to the accuracy of axial forces of 0.1% and 1%. Figures 4 and 5 depict the process of constructing the diagram and the related variation of the axial force for the accuracy of 0.1%.

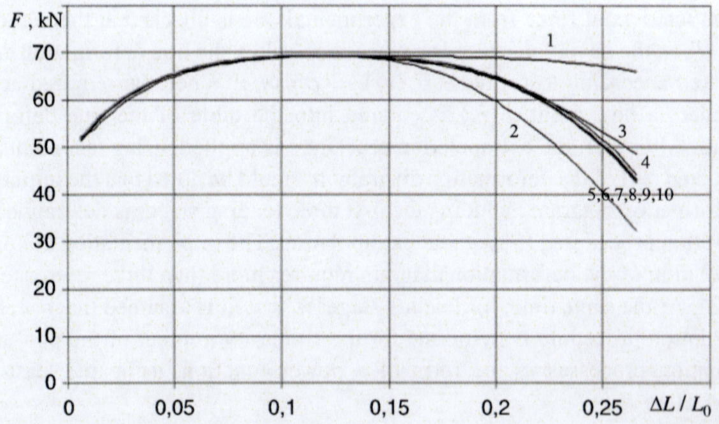

Fig. 5 Variation of the axial forces obtained with the iteration procedure: 1—initial approximation; 2–10—iterations 1–9, respectively

The problem was numerically analyzed five times after achieving the accuracy of 1% on the axial forces, and ten times with the accuracy of 0.1%. It is to be noted that, to construct the before-necking part of the diagram with an accuracy of 1%, only one iteration step is sufficient with the initial approximation defined by formulas (2). Thus, computational expenses on constructing a deformation diagram are doubled to reduce the inaccuracy by ten times. Updating the entire deformation diagram and multiply repeated direct numerical analysis of the problem results in large consumption of computation time. The authors of [20, 21] use piecewise-linear approximation for constructing true deformation diagrams of elastoplastic materials. This involves multiplying repeated analyses of a direct problem at each stage of piecewise-linear loading. The authors do not characterize the accuracy of constructing deformation diagrams but give the number of direct analyses of the order of five iterations. The successive approximation algorithm itself is not presented. It is evident that, in the sense of computational expenses, it has no advantages over the above considered one.

4 Modification of the Algorithm of Constructing Deformation Diagrams

To increase the efficiency of the algorithm, it is appropriate to use a nonlinear extrapolation procedure. The computational process of modeling the loading is subdivided into several stages $n = \overline{1, N}$. The number of states N is equal to the number of points in the tabular representation of the experimental axial elongation–axial force relation, and the constructed true deformation diagram will comprise the same number of points. In the course of computations, the value of the deviation of the computed axial force from the experimental one is checked at the end of each stage. When the deviation exceeds a prescribed value, the true deformation diagram is updated according to formulas $\overline{\sigma}_i(e_i^*) = \beta\sigma_i(e_i^*)$. When the required accuracy is reached, a new point $\sigma_i(\hat{e}_i)$ is entered into the table of the true deformation diagram. After that, the extrapolation procedure is applied, using last points of the constructed part of the deformation diagram. It should be noted that the initial part of the deformation diagram including the first three loading stages is determined using the iteration procedure [15] without extrapolation. The implementation of nonlinear extrapolation of the deformation diagram requires more than three reference points ($m \geqslant 3$). At the same time, for loading stages $n < m$ it is assumed that $n = m$. The extrapolation procedure is first done for the similarity number of the non-uniform deformation processes in the form of a power function, using the least-square method. Here K is

$$K(e_i) = \frac{1}{\sigma_i(e_i)} \frac{d\sigma_i(e_i)}{de_i} \tag{3}$$

Then the deformation diagram is extrapolated with a given exponential relation, $\sigma_i(e_i) = \sigma_i(\hat{e}_i) exp\left(\int_{\hat{e}_i}^{e_i} K de_i\right)$. Figure 6 presents the results of extrapolation of the deformation diagram and parameter K.

It is noted that the deformation diagrams are defined by monotone increasing functions with a decreasing derivative, which makes it possible to determine to a high accuracy the initial approximation of the diagram for the next loading stages.

The effect of the number of extrapolation points m on time consumption of constructing deformation diagrams after reaching the accuracy of 0.1% for the

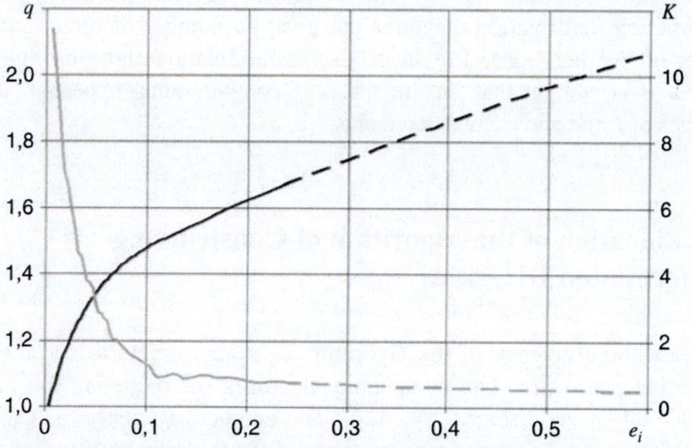

Fig. 6 Extrapolation (dashed lines) of the relation for K (grey curve) and the deformation diagram (black curve)

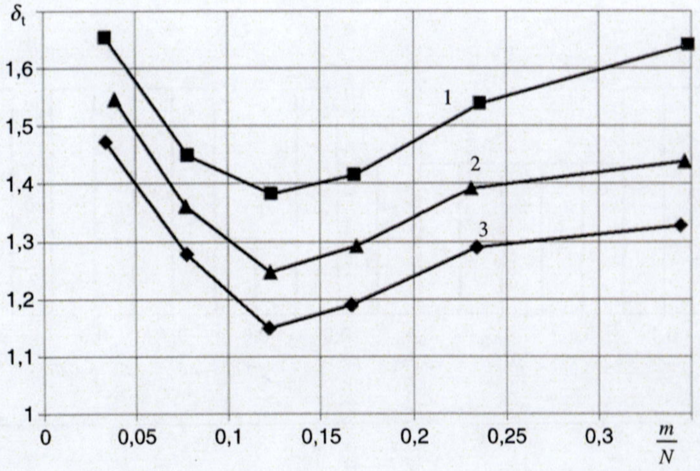

Fig. 7 Variation of relative parameter δ_t characterizing the degree of time spent on constructing a deformation diagram as a function of number of extrapolation points m/N for $N = 90$ (curve 1), $N = 130$ (curve 2) and $N = 180$ (curve 3)

axial force was numerically investigated. The experimental relation for axial forces was approximated by $N = 90$, $N = 130$ and $N = 180$ points. Figure 7 shows the variation of parameter $\delta_t = t/T$ characterizing the degree of time spent on constructing a deformation diagram in relation with time spent on one direct analysis of the problem as a function of the used number of extrapolation points m/N (T is time spent on one direct numerical analysis of the problem, t is time spent on constructing a deformation diagram).

Maximal efficiency of the introduced algorithm is achieved in the case when number of extrapolation points m amounts to about 12% of total number of points approximating the diagram. With the increasing number of stages (points approximating the diagram) the iterative procedure of constructing a deformation diagram is practically reduced to a single direct analysis without using the iteration procedure ($\delta_t \to 1$), which increases the effectiveness of the present algorithm.

Figure 8 presents the results of variation of the number of direct analyses of the problem r_n at the n-th stage as a function of parameter (n/N) for $N = 20$ and $N = 110$. The columns in Fig. 8 characterize the corresponding loading stage.

For numerous of loading stages (more than $N = 100$ points approximating the diagram) any iteration procedure becomes unnecessary in view of high accuracy of extrapolation. If it is necessary (for $r_n = 2$, Fig. 8), only one updating of the true deformation diagram is sufficient at a current loading stage without repeating the analysis of the problem. It is to be noted that the present algorithm substantially (up to 10 times) increases the efficiency of the earlier developed methodologies of constructing diagrams of deformation of elastoplastic materials. This, in its

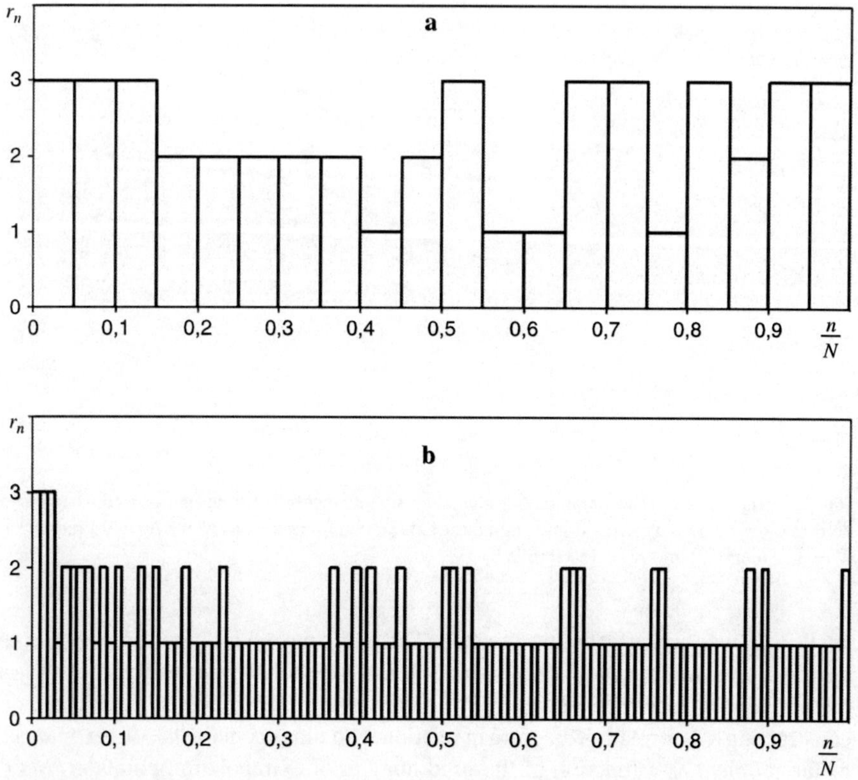

Fig. 8 Distribution of the number of direct numerical calculations r_n at the n-th stage for $N = 20$ (**a**) and $N = 110$ (**b**)

turn, considerably increases the potential of studying deformational and strength characteristics of elastoplastic materials for various types of loading: the tension of cylindrical rods and shells, torsion of rods, kinetic indentation of specimens with a sphere, dynamic compression of tablet-shaped specimens and a number of other problems. In view of monotone convergence of the iteration process of constructing deformation diagrams, inaccuracy of the experimental-numerical methodology is mainly determined by the field of application of the mathematical model of elastoplastic material being used and the type of loading (simple, complex).

Acknowledgment Work was supported by the Russian Foundation for Basic Research (project No. 20-08-00667-a).

References

1. Bridgeman, P. Issledovaniya bolshih plasticheskih deformatsiy i razryva. Izd-vo inostr. lit. Moscow (1955) (in Russian)
2. Davidenkov, N.A., Spiridonova, N.I.: Analiz napryazhennogo sostoyaniya v sheyke rastyanutogo obraztsa. Zavodskaya laboratoriya. **6**, 583–593 (1945) (in Russian)
3. Zhang, Z.L., Odegard, J., Hauge, M.P., Thaulow C. Determining material true stress-strain curve from tensile specimens with rectangular cross-section. Int. J. Solids and Struct. **36**, 3497–3516 (1999)
4. Zhang, Z.L., Odegard, J., Sovik, O.P. Determining true stress-strain curve for isotropic and anisotropic materials with rectangular tensile bars: method and verifications. Comput. Mater. Sci. **69**(1), 77–85 (2001)
5. Zhang, Z.L., Odegard, J., Sovik, O.P., Thaulow C. A study on determining true stress-strain curve for anisotropic materials with rectangular tensile bars. Int. J. Solids and Struct. **38**(26–27), 4489–4505 (2001)
6. Zhang, Z.L., Odegard, J., Hauge, M.P., Thaulow, C.: A notches cross weld tensile testing method for determining true stress-strain curves for weldments. Engineering Fracture Mech. **69**, 353–366 (2002)
7. Cabezas, E.E., Celentano, D.J.: Experimental and numerical analysis of the tensile test using sheet specimens. Finite Elements in Analysis and Design. **40**, 555–575 (2004)
8. Ling, Y.: Uniaxial True Stress–Strain after Necking. AMP Journal of Technology. **5**, 37–48 (1996)
9. Mirone, G.: A new model for the elastoplastic characterization and the stress–strain determination on the necking section of a tensile specimen. International Journal of Solids and Structures. **41**, 3545–3564 (2004)
10. Choung, J.M., and Cho, S.R.: Study on true stress correction from tensile tests. Journal of Mechanical Science and Technology. **22**, 1039–1051 (2008)
11. Joun, M., Eom, J.G., Lee, M.C.: A new method for acquiring true stress–strain curves over a large range of strains using a tensile test and finite element method. Mechanics of Materials. **40**, 586–593 (2009)
12. Lee, J.H., Lim, D., Hyun, H., Lee, H.: A numerical approach to indentation technique to evaluate material properties of film-on-substrate systems. International Journal of Solids and Structures. **49**(7–8), 1033–1043 (2012)
13. Nayebi, A., Abdi, R.E., Bartier, O., Mauvoisin, G.: New procedure to determine steel mechanical parameters from the spherical indentation technique. Mechanics of Materials. **34**, 243–254 (2002)
14. Vasin, R.A., Il'yushin, A.A., and Mossakovskii, P.A.: Issledovaniye opredelyayushchikh sootnosheniy i kriteriyev razrusheniya na sploshnykh i tolstostennykh trubchatykh tsilindricheskikh obraztsakh. Izv. Akad. Nauk. Mekh. Tverd. Tela. **2**, 177–184 (1994) (in Russian)
15. Bazhenov, V.G., Zefirov, S.V., Osetrov, S.L.: Eksperimentalno–raschetnyi metod identifikatsii deformatsyonnyh i prochnostnyh svoistv materialov. Zavodskaya laboratoriya. Diagnostika materialov. **72(9)**, 39–45 (2006) (in Russian)
16. Bazhenov, V.G., Lomunov, V.K., Osetrov, S.L., Pavlenkova, E.V.: Experimental and computational method of studying large elastoplastic deformations of cylindrical shells in tension to rupture and constructing strain diagrams for an inhomogeneous stress-strain state. Journal of Applied Mechanics and Technical Physics. **54(1)**, 100–107 (2013)
17. Bazhenov, V.G., Zefirov, S.V., Kramarev, L.N., Pavlenkova, E.V.: Modelling of the deformation processes and the localization of plastic deformations in the torsion-tension of solids of revolution. Journal of Applied Mathematics and Mechanics. **72(2)**, 226–232 (2008)
18. Bazhenov V.G., Zefirov S.V., Osetrov S.L.: Experimental and computing method for constructing true deformation diagrams at large strains on the basis of tests for hardness. Doklady Physics. **51(3)**. 118–121 (2006)

19. Bazhenov, V.G., Osetrov, D.L.: Method of identification of dry and viscous friction forces and construction of dynamic deformation diagrams of metals in experiments with impact compression. Lobachevskii Journal of Mathematics. **40(3)**, 278–283 (2019)
20. Kamaya Masayuki, Kawakubo Masahiro: A procedure for determining the true stress–strain curve over a large range of strains using digital image correlation and finite element analysis. Mechanics of Materials. **43**, 243–253 (2011)
21. Vladimirov, S.A., Trefilov, S.I.: Issledovaniye protsessa glubokogo deformirovaniya obraztsov s koltsevoy vytachkoy pri ih rastyazhenii. Kosmonavtika i Raketostroenie. **(82)**, 81–85 (2015) (in Russian)

Cubic Spline on a Bakhvalov Mesh in the Presence of a Boundary Layer

Igor Blatov, Elena Kitaeva, and Nikita Zadorin

Abstract The problem of cubic spline interpolation on the Bakhvalov mesh of functions with region of large gradients is considered. Asymptotically accurate two-side error estimates are obtained for a class of functions with an exponential boundary layer. It is proved that the error estimates of traditional spline interpolation are not uniform in a small parameter, and the error itself can increase indefinitely when the small parameter tends to zero at a fixed number of nodes N. A modified cubic spline is proposed for which uniform estimates of the order $O(N^{-4})$ have been experimentally confirmed.

1 Introduction

Cubic splines are widely used for smooth interpolation of functions [1, 2]. When using difference methods to solve singularly perturbed problems are used strongly nonuniform grids. In this case, there is a need to restore function for all values of the independent variable. In the case of a piecewise uniform grid of G. I. Shishkin [3], in [4] error estimates of cubic spline are obtained. It is shown that the convergence of the interpolation process is nonuniform in a small parameter. To achieve uniform accuracy with respect to a small parameter, it is proposed to shift one of the interpolation points.

In this paper, we study the cubic spline interpolation [2] on the mesh of N. S. Bakhvalov [5], which dense in the boundary layer. Error estimates are obtained,

I. Blatov (✉)
Povolzhskiy State University of Telecommunications and Informatics, Samara, Russia

E. Kitaeva
Korolev Samara State University, Samara, Russia

N. Zadorin
Sobolev Institute of Mathematics, Siberian Branch RAS, Novosibirsk, Russia
e-mail: nik-zadorin@yandex.ru

I. B. Badriev et al. (eds.), *Mesh Methods for Boundary-Value Problems and Applications*, Lecture Notes in Computational Science and Engineering 141,
https://doi.org/10.1007/978-3-030-87809-2_4

which, however, are not uniform in a small parameter ε. It is shown that for $\varepsilon \to 0$ the interpolation error has unlimited growth, and the development of special interpolation methods for this class of problems is important. There is offered a modified interpolation spline for which experimentally established uniform in ε convergence.

Introduce the notations. Set the mesh of interval [0, 1] :

$$\Omega = \{x_n : x_n = x_{n-1} + h_n,\ n = 1, 2, \ldots, N,\ x_0 = 0, x_N = 1\}.$$

Denote by $S(\Omega, k, 1)$ the space of polynomial splines of degree k of defect 1 [2] on the mesh Ω. If necessary, we consider the partition Ω extended to the left of the point 0 with the step $h_1 = x_1 - x_0$ and to the right of the point 1 with the step $h_N = x_N - x_{N-1}$. We set $h = 1/N$. By C and C_j we mean positive constants independent of the parameter ε and the number of grid nodes. In this case, the same symbol C_j can denote different constants. Will write $f = O(g)$ if the estimate $|f| \le C|g|$ and $f = O^*(g)$ if $f = O(g)$ and $g = O(f)$. $C[a, b]$, $L_2[a, b]$ – spaces of continuous and quadratically summable on $[a, b]$ functions with the norms $\|\cdot\|_{C[a,b]}$ and $\|\cdot\|_{L_2[a,b]}$ accordingly, $(\cdot, \cdot)$ is the scalar product in $L_2[0, 1]$.

2 Formulation of the Problem and Main Results

Let us a function $u(x)$ be decomposed in the form of the sum of regular and singular components:

$$u(x) = q(x) + \Phi(x),\ x \in [0, 1], \tag{1}$$

where for some C_1

$$|q^{(j)}(x)| \le C_1,\ |\Phi^{(j)}(x)| \le \frac{C_1}{\varepsilon^j} e^{-\alpha x/\varepsilon},\ 0 \le j \le 4, \tag{2}$$

where the functions $q(x)$ and $\Phi(x)$ do not explicitly given, $\alpha > 0, \varepsilon > 0$. Decomposition (1) holds for the solution of a singularly perturbed boundary value problem [3].

We set the grid of the interval [0, 1] based on [5].

Let us

$$\sigma = \min\left\{\frac{1}{2}, \frac{4\varepsilon}{\alpha} \ln\frac{1}{\varepsilon}\right\}$$

if $\varepsilon \le e^{-1}$ and $\sigma = 1/2$ if $\varepsilon > e^{-1}$.

When $\sigma < 1/2$, we define mesh nodes of Ω as

$$x_n = g(n/N),\ n = 0, 1, \ldots, N, \tag{3}$$

where

$$g(t) = \begin{cases} -\frac{4\varepsilon}{\alpha}\ln\left[1-2(1-\varepsilon)t\right], 0 \le t \le \frac{1}{2}, \\ \sigma + (2t-1)(1-\sigma), \ \ 1/2 \le t \le 1. \end{cases}$$

When $\sigma = 1/2$, we define a mesh Ω as uniform with the step $h = 1/N$.

Let us estimate the error of the cubic spline $g_3(x, u) \in S(\Omega, 3, 1)$ on the mesh Ω defined from interpolation conditions

$$g_3(x_n, u) = u(x_n),\ 0 \le n \le N,\ g_3'(0, u) = u'(0), g_3'(1, u) = u'(1).$$

We state the main results in the form of theorems.

Theorem 1 *There are constants C_4, C_5 and $\beta > 0$ that are independent of ε, N such that for $\varepsilon \le C_4 N^{-1}$ the following estimates hold*

$$\| g_3(x,u) - u(x) \|_{C[x_n, x_{n+1}]} \le C_5 \begin{cases} N^{-4},\ \ 0 \le n \le N/2-2, \\ N^{-4}\ln\left(1+\frac{1}{\varepsilon N}\right) + \frac{1}{N^4},\ n = \frac{N}{2}-1, \\ \frac{N^{-5}}{\varepsilon} e^{-\beta(n-N/2)} + 1/N^4,\ \ N/2 \le n. \end{cases} \quad (4)$$

The following theorem shows that the estimates (4) are unimprovable.

Theorem 2 *Let $\Phi(x) = e^{-x/\varepsilon}$. Then there are constants C_4, C_6, $\beta_1 > 0$ independent of ε, N such that for $\varepsilon \le C_4 N^{-1}$ lower bounds will be valid*

$$\| g_3(x,\Phi) - \Phi(x) \|_{C[x_n, x_{n+1}]} \ge C_6 \frac{N^{-5}}{\varepsilon} e^{-\beta_1(n-N/2)},\quad \frac{N}{2} \le n \le N-1. \quad (5)$$

3 Auxiliary Results

Below, without loss of generality, we assume that in (2) $\alpha = 1$, since the general case reduces to this by replacing $\alpha x = y$ with preservation of estimates of the form (2).

Lemma 1 *With $\sigma < 1/2$ sequence h_n for $n \le N/2$ monotonically increases and*

$$h_n = \begin{cases} O^*(\frac{\varepsilon}{N/2-n}),\ 1 \le n \le N/2-1, \\ O^*(\varepsilon \ln(1+\frac{1}{N\varepsilon})),\ n = N/2, \\ O^*(1/N),\ \ N/2+1 \le n \le N. \end{cases} \quad (6)$$

The proof follows from (3) and the definition of $g(t)$.

Let

$$N_{n,1}(x) = \begin{cases} \frac{x-x_n}{x_{n+1}-x_n}, \ x \in [x_n, x_{n+1}], \\ \frac{x_{n+2}-x}{x_{n+2}-x_{n+1}}, \ x \in [x_{n+1}, x_{n+2}], \ -1 \le n \le N-1, \\ 0, \ x \notin (x_n, x_{n+2}) \end{cases}$$

is B-spline of first degree, then

$$\| N_{n,1} \|_{L_2[0,1]} = \frac{1}{\sqrt{3}}(h_{n+1} + h_{n+2})^{1/2},$$

thus, in view of Lemma 1

$$\| N_{n,1} \|_{L_2[0,1]} = \begin{cases} O^*((\varepsilon/(N/2-n))^{1/2}), \ 0 \le n \le N/2-3, \\ O^*((\varepsilon \ln(1+\frac{h}{\varepsilon}))^{1/2}), \ n = N/2-2, \\ O^*(h^{1/2}), \ N/2-1 \le n \le N-1. \end{cases}$$

Let $\tilde{N}_{n,1}(x) = N_{n,1}(x)/ \| N_{n,1} \|_{L_2[0,1]}, 0 \le n \le N-2$. For $n = -1$ and $n = N-1$ we set $\tilde{N}_{-1,1}(x) = \tilde{N}_{0,1}(x+h_1), \tilde{N}_{N-1,1}(x) = \tilde{N}_{N-2,1}(x-h_N)$.
Then, taking into account the last two formulas we get

$$\| \tilde{N}_{n,1} \|_{C[0,1]} = \begin{cases} O^*((\varepsilon/(N/2-n))^{-1/2}), \ 0 \le n \le N/2-3, \\ O^*((\varepsilon \ln(1+\frac{h}{\varepsilon}))^{-1/2}), \ n = N/2-2, \\ O^*(h^{-1/2}), \ N/2-1 \le n \le N-1. \end{cases} \tag{7}$$

Let $e(x) = g_3(x, \Phi) - \Phi(x)$. We study the function $e''(x) = g_3''(x, \Phi) - \Phi''(x)$. According to [6, chapter 5] $g_3''(x, \Phi) = P\Phi''(x)$, where P is the projector on $L_2[0, 1]$ orthogonal to $S(\Omega, 1, 1)$. Denote by $\tilde{gI}(x) \in S(\Omega, 1, 1)$ the linear interpolant $\Phi''(x)$ at the nodes of the mesh, and through $gI(x)$ a function from $S(\Omega, 1, 1)$ equal to $\tilde{gI}(x)$ for $x \in [0, x_{N/2-2}]$ and zero for $x \in [x_{N/2-1}, 1]$. It's obvious that $gI(x) \in S(\Omega, 1, 1)$. Then we have

$$e''(x) = P(\Phi''(x) - gI(x)) + (gI(x) - \Phi''(x)). \tag{8}$$

We represent the function $P(\Phi''(x) - gI(x))$ in the form

$$P(\Phi''(x) - gI(x)) = \sum_{n=-1}^{N-1} \alpha_n \tilde{N}_{n,1}(x).$$

From the conditions of orthogonality of the difference $g_3''(x, \Phi) - \Phi''(x)$ and the space $S(\Omega, 1, 1)$ we obtain a system of linear equations for coefficients

$$\sum_{n=-1}^{N-1} \alpha_n(\tilde{N}_{n,1}, \tilde{N}_{k,1}) = (\Phi'' - gI, \tilde{N}_{k,1}), -1 \le k \le N-1,$$

or in matrix form $\Gamma\alpha = F$, where $\Gamma = \{\gamma_{nk}\} = \{(\tilde{N}_{n,1}, \tilde{N}_{k,1})\}$—Gram matrix of normalized B-splines, $F = (F_{-1}, F_0, \cdots, F_{N-1})^T$, $F_j = (\Phi'' - gI, \tilde{N}_j)$. It's obvious that $0 \le \gamma_{nk} \le 1$.

Lemma 2 *The matrix* Γ *has the form*

$$\Gamma = tridiag\{a_n, c_n, b_n\}, \ -1 \le n \le N-1, a_{-1} = b_{N-1} = 0, \tag{9}$$

$$a_{n+1} = b_n = O^*(1) > 0, \ 0 \le n \le N-2, \ n \ne N/2-3, n \ne N/2-2, \tag{10}$$

$$a_{N/2-1} = b_{N/2-2} = O^*\Big(\Big(\frac{\varepsilon \ln(1+h/\varepsilon)}{h}\Big)^{1/2}\Big),$$

$$a_{N/2-2} = b_{N/2-3} = O^*\Big(\Big(\frac{1}{\ln(1+h/\varepsilon)}\Big)^{1/2}\Big), \tag{11}$$

$$c_n = 1, \ 0 \le n \le N-2, c_{-1} = c_{N-1} = 1/\sqrt{2}. \tag{12}$$

The matrix Γ *has strict diagonal dominance over rows with a prevalence index* $1/\sqrt{2}$.

The proof is obtained by direct calculation of the integrals taking into account (6)–(7).

Denote by $cond_2\Gamma$ the spectral condition number of Γ.

Corollary 1 *The matrix* Γ *has the form*

$$\Gamma = \begin{pmatrix} \Gamma_{11} & \Gamma_{12} \\ \Gamma_{21} & \Gamma_{22} \end{pmatrix},$$

where Γ_{11}, Γ_{22} *are tridiagonal square matrices of order* $(N/2) \times (N/2)$ *and* $(N/2+1) \times (N/2+1)$ *respectively, with strict diagonal line prevalence with a prevalence index of* $1/\sqrt{2}$, $cond_2\Gamma = O(1)$, $cond_2\Gamma_{ii} = O(1), i = 1, 2$; *matrices* Γ_{12} *and* Γ_{21} *are rectangular matrices with the only nonzero element of order* $O^*((\varepsilon \ln(1+h/\varepsilon)/h)^{1/2})$ *in the left lower and upper right corners respectively.*

The matrix Γ_{11} has the form

$$\Gamma_{11} = \begin{pmatrix} \hat{\Gamma}_{11} & \hat{\Gamma}_{12} \\ \hat{\Gamma}_{21} & \hat{\Gamma}_{22} \end{pmatrix},$$

where $\hat{\Gamma}_{11}$, is the tridiagonal square matrix of order $(N/2-1)\times(N/2-1)$ with strict diagonal predominance in rows with predominance of $1/\sqrt{2}$, $\hat{\Gamma}_{22}=1$, $\hat{\Gamma}_{21}=(0\cdots 0\ a_{N/2-2})$, $\hat{\Gamma}_{12}=\hat{\Gamma}_{21}^{T}$—matrices with the only nonzero element of order $O^*((\ln(1+h/\varepsilon))^{-1/2})$.

Lemma 3 *The matrices Γ_{11}, Γ_{22}, $\hat{\Gamma}_{11}$ are invertible, and for elements $\check{\gamma}_{nk}^{ii}$, $i=1,2$ of inverse matrix the estimates $|\check{\gamma}_{nk}^{ii}|\le Ce^{-\beta|n-k|}$ hold and similar estimates hold for elements of $\hat{\Gamma}_{11}$. Here C, β are independent of N, ε.*

Proof Invertibility of the matrices Γ_{11}, Γ_{22} and element estimates follow from strict diagonal prevalence with a prevalence index of $1/\sqrt{2}$ and Demko's theorem [7]. □

Lemma 4 *For the matrix Γ_{11}^{-1}, following representation holds*

$$\Gamma_{11}^{-1}=\begin{pmatrix}\bar{\Gamma}_{11} & \bar{\Gamma}_{12}\\ \bar{\Gamma}_{21} & \bar{\Gamma}_{22}\end{pmatrix},$$

where elements $\bar{\gamma}_{nk}^{ij}$ of matrix $\bar{\Gamma}_{ij}$ for some $\beta>0$, independent of ε, N, satisfy the estimates

$$|\bar{\gamma}_{nk}^{11}|\le Ce^{-\beta|n-k|},\ -1\le n,k\le N/2-3;\ |\bar{\Gamma}_{22}|\le C, \tag{13}$$

$$|\bar{\gamma}_{nk}^{ij}|\le(\ln(1+h/\varepsilon))^{-1/2}Ce^{-\beta|n-k|},\ n=N/2-2,-1\le k\le N/2-3,i=1,j=2;$$

$$k=N/2-2,-1\le n\le N/2-3,\ i=2,j=1. \tag{14}$$

Proof Using the Gauss block method, we find

$$\Gamma_{11}^{-1}=\begin{pmatrix}\hat{\Gamma}_{11}^{-1}+\hat{\Gamma}_{11}^{-1}\hat{\Gamma}_{12}\tilde{\Gamma}^{-1}\hat{\Gamma}_{21}\hat{\Gamma}_{11}^{-1} & -\hat{\Gamma}_{11}^{-1}\hat{\Gamma}_{12}\tilde{\Gamma}^{-1}\\ -\tilde{\Gamma}^{-1}\hat{\Gamma}_{21}\hat{\Gamma}_{11}^{-1} & \tilde{\Gamma}^{-1}\end{pmatrix}, \tag{15}$$

where $\tilde{\Gamma}=\hat{\Gamma}_{22}-\hat{\Gamma}_{21}\hat{\Gamma}_{11}^{-1}\hat{\Gamma}_{12}$. Here, the reversibility of all blocks and uniform in ε, N the boundedness of the norms of all inverse matrices follows from the corollary 1. This implies that $\tilde{\Gamma}^{-1}$ is also uniformly bounded in the norm. From the Demko's theorem [7] we obtain that the elements of the matrix $\hat{\Gamma}_{11}^{-1}$ satisfy estimates of the form (13). With this estimates (13)–(14) follow from (15) and the corollary 1. □

Lemma 5 *The following representation is valid*

$$\Gamma^{-1}=\begin{pmatrix}\tilde{\Gamma}_{11} & \tilde{\Gamma}_{12}\\ \tilde{\Gamma}_{21} & \tilde{\Gamma}_{22}\end{pmatrix},$$

where the elements $\tilde{\gamma}_{nk}^{ij}$ *of matrices* $\tilde{\Gamma}_{ij}$ *for some* $\beta > 0$*, independent of* ε, N*, satisfy estimates*

$$|\tilde{\gamma}_{nk}^{11}| \le Ce^{-\beta|n-k|},\ -1 \le n, k \le N/2-3;\ |\tilde{\gamma}_{nk}^{22}| \le Ce^{-\beta|n-k|},$$

$$N/2-1 \le n, k \le N-1, \tag{16}$$

$$|\tilde{\gamma}_{nk}^{11}| \le (\ln(1+h/\varepsilon))^{-1/2}Ce^{-\beta|n-k|},\ n = N/2-2, -1 \le k \le N/2-3$$

$$\text{or } k = N/2-2, -1 \le n \le N/2-3, \tag{17}$$

$$|\tilde{\gamma}_{nk}^{ij}| \le C(\varepsilon/h)^{1/2}e^{-\beta|n-k|}, \tag{18}$$

where $-1 \le n \le N/2-2, N/2-1 \le k \le N-1$ *for* $i = 1, j = 2$*;* $-1 \le k \le N/2-2, N/2-1 \le n \le N-1$ *for* $i = 2, j = 1$.

Proof Using the Gauss block method similarly (15), we find

$$\Gamma^{-1} = \begin{pmatrix} \Gamma_{11}^{-1} + \Gamma_{11}^{-1}\Gamma_{12}\tilde{\Gamma}^{-1}\Gamma_{21}\Gamma_{11}^{-1} & -\Gamma_{11}^{-1}\Gamma_{12}\tilde{\Gamma}^{-1} \\ -\tilde{\Gamma}^{-1}\Gamma_{21}\Gamma_{11}^{-1} & \tilde{\Gamma}^{-1} \end{pmatrix}, \tag{19}$$

where $\tilde{\Gamma} = \Gamma_{22} - \Gamma_{21}\Gamma_{11}^{-1}\Gamma_{12}$. Here, the reversibility of all blocks and uniform in ε, N the boundedness of all inverse matrices follows from the corollary 1. From [7] it follows that the elements of the matrix Γ_{11}^{-1} satisfy estimates of the form (16), therefore, by the form of the matrices Γ_{12}, Γ_{21}, the elements of the matrix $\tilde{\Gamma}$ satisfy the same estimates. But for matrices having an inverse matrix, bounded in the spectral norm by a constant independent of the order of the matrix and the parameters that determine its elements, in [8] proved that the elements of the inverse matrix $\tilde{\Gamma}^{-1}$ satisfy the same estimates, possibly with other constant $\beta_1 \in (0, 1)$, which is also independent of N, ε. It was also proved that elements of the product of two matrices, satisfying estimates of the form (16), satisfy the same estimates. From here estimates (16) follow.

The estimates (17) follow from (19), Lemma 4, corollary 1 and estimates of the form (16) for the elements of $\tilde{\Gamma}^{-1}$. Let us prove the estimates (18) for $i = 1, j = 2$. Let

$$\tilde{\Gamma}^{-1} = \{\tilde{\gamma}_{nk}, N/2-1 \le n, k \le N-1\},$$

$$\Gamma_{12} = \{\gamma_{nk}, 1 \le n \le N/2-2, N/2-1 \le k \le N-1\},$$

$\Gamma_{11}^{-1} = \{\tilde{\gamma}_{nk}^{11}, 1 \le n, k \le N/2-2\}$. Since the matrix Γ_{12} have only nonzero element $\gamma_{(N/2-2)(N/2-1)}$, then, multiplying matrices, we find elements of matrix $\tilde{\Gamma}_{12}$: $\tilde{\gamma}_{nk}^{12} = \tilde{\gamma}_{n(N/2-2)}^{11}\gamma_{(N/2-2)(N/2-1)}\tilde{\gamma}_{(N/2-2)k}$. Hence, given the estimates (14), (11), (16) for the first, second and third factors, respectively, we get (18). For $i = 2, j = 1$, the estimates are obtained by virtue of symmetries of Γ^{-1}. The lemma is proved. □

Lemma 6 *For any* $\varepsilon \in (0, 1)$, N, *the following estimates hold*

$$F_n = \begin{cases} O(h_{n+1}^{5/2}\varepsilon^{-4}e^{-x_{n+1}/\varepsilon}), & -1 \le n \le N/2-3, \\ O((\varepsilon \ln(1+h/\varepsilon))^{-1/2}\varepsilon^{-1}e^{-x_{N/2-1}/\varepsilon}), & n = N/2-2, \\ O(h^{-1/2}\varepsilon^{-1}e^{-x_n/\varepsilon}), & N/2-1 \le n \le N-1. \end{cases} \tag{20}$$

The proof is obtained by direct calculation of the integrals with taking into account (7) and estimates of the error of linear interpolation.

Lemma 7 *For the coefficients* α_n *in the decomposition of* $P(\Phi''(x) - gI(x))$ *through the basis of* $\tilde{N}_{n,1}(x)$ *the following estimates hold*

$$\alpha_n = \begin{cases} O(h_{n+1}^{5/2}\varepsilon^{-4}e^{-x_{n+1}/\varepsilon}), & -1 \le n \le N/2-3, \\ O((\varepsilon \ln(1+h/\varepsilon))^{-1/2}\varepsilon^{-1}e^{-x_{N/2-1}/\varepsilon}), & n = N/2-2, \\ O(h^{-1/2}\varepsilon^{-1}e^{-x_{N/2-1}/\varepsilon}e^{-\beta(n-N/2)}), & N/2-1 \le n \le N-1. \end{cases} \tag{21}$$

Proof We have $\alpha = \Gamma^{-1}F$. Set $\alpha = (\alpha^{(1)}, \alpha^{(2)}, \alpha^{(1)})$, where $dim(\alpha^{(1)}) = N/2-1$, $dim(\alpha^{(2)}) = 1$. Then according to Lemma 5, for any $n \in [-1, N/2-3]$

$$\alpha_n^{(1)} = \alpha_n = \sum_{k=-1}^{N/2-3} \tilde{\gamma}_{nk}^{11} F_k + \tilde{\gamma}_{n(N/2-2)}^{11} F_{N/2-2} + \sum_{k=N/2-1}^{N-1} \tilde{\gamma}_{nk}^{12} F_k. \tag{22}$$

By virtue of (16)

$$\left|\sum_{k=-1}^{N/2-3} \tilde{\gamma}_{nk}^{11} F_k\right| \le \frac{C}{\varepsilon^4} h_{n+1}^{5/2} e^{-x_{n+1}/\varepsilon} \sum_{k=-1}^{N/2-3} e^{-\beta|n-k|} e^{-(x_{k+1}-x_{n+1})/\varepsilon} \cdot \left(\frac{h_{k+1}}{h_{n+1}}\right)^{5/2}. \tag{23}$$

Further, since $h_k/h_n \le 1$ and $k \le n$, taking into account (6), we have

$$\sum_{k=-1}^{n} e^{-\beta|n-k|} e^{-(x_{k+1}-x_{n+1})/\varepsilon} \left(\frac{h_{k+1}}{h_{n+1}}\right)^{5/2} \le \sum_{k=-1}^{n} e^{\beta(k-n)} e^{\sum_{s=k+1}^{n+1} h_s/\varepsilon} \left(\frac{h_{k+1}}{h_{n+1}}\right)^{5/2}$$

$$\le \sum_{k=-1}^{n} e^{\beta(k-n)} e^{C \ln \frac{N/2-n+1}{N/2-k+1}}$$

$$= \sum_{k=-1}^{n} e^{\beta(k-n)} \left(\frac{N/2-n+1}{N/2-k+1}\right)^C \le \sum_{k=-1}^{n} e^{\beta(k-n)} (n-k+1)^C \le C_1; \tag{24}$$

$$\sum_{k=n+1}^{N/2-3} e^{-\beta|n-k|} e^{-(x_{k+1}-x_{n+1})/\varepsilon} \left(\frac{h_{k+1}}{h_{n+1}}\right)^{5/2}$$

$$\leq \sum_{k=n+1}^{N/2-3} e^{-\beta(n-k)} \left(\frac{N/2-n}{N/2-k}\right)^{5/2} \leq C_2. \tag{25}$$

By virtue of (17), (20) we have

$$|\tilde{\gamma}_{n(N/2-2)}^{11} F_{N/2-2}| \leq \frac{C}{\varepsilon^4} h_{n+1}^{5/2} e^{-x_{n+1}/\varepsilon} \varepsilon^4 e^{x_{n+1}/\varepsilon} \left(\frac{\varepsilon}{h}\right)^{1/2} e^{\beta(n-N/2)} \left(\varepsilon \ln \frac{h}{\varepsilon}\right)^{-1/2}$$

$$\times \frac{1}{\varepsilon} e^{-x_{N/2-1}/\varepsilon} = \frac{C}{\varepsilon^4} h_{n+1}^{5/2} e^{-x_{n+1}/\varepsilon} \varepsilon^3 h_{n+1}^{-5/2} e^{-(x_{n+1}-x_{N/2-1})/\varepsilon} \left(\varepsilon \ln \frac{h}{\varepsilon}\right)^{-1/2}$$

$$\times e^{-\beta|n-N/2|} \leq \frac{C}{\varepsilon^4} h_{n+1}^{5/2} e^{-x_{n+1}/\varepsilon} \left(\frac{\varepsilon}{h \ln(h/\varepsilon)}\right)^{1/2}$$

$$\times (N/2-n)^{5/2} e^{-\beta|n-N/2|} e^{-(x_{n+1}-x_{N/2-1})/\varepsilon} \leq \frac{C_1}{\varepsilon^4} h_{n+1}^{5/2} e^{-x_{n+1}/\varepsilon}. \tag{26}$$

Taking into account (18), (20), we have

$$\left|\sum_{k=N/2-1}^{N-1} \tilde{\gamma}_{nk}^{12} F_k\right| \leq \frac{C}{\varepsilon^4} h_{n+1}^{5/2} e^{-x_{n+1}/\varepsilon} \sum_{k=N/2-1}^{N-1} \varepsilon^4 h_{n+1}^{-5/2} e^{x_{n+1}/\varepsilon} \left(\frac{\varepsilon}{h}\right)^{1/2}$$

$$\times e^{\beta|n-k|} h^{-1/2} \varepsilon^{-1} e^{-x_k/\varepsilon} \leq \frac{C}{\varepsilon^4} h_{n+1}^{5/2} e^{-x_{n+1}/\varepsilon}$$

$$\times \sum_{k=N/2-1}^{N-1} \left(\frac{\varepsilon}{h}\right)^{1/2} (N/2-n)^{5/2} e^{(x_{n+1}-x_k)/\varepsilon} e^{-\beta|n-k|} = \frac{C_1}{\varepsilon^4} h_{n+1}^{5/2} e^{-x_{n+1}/\varepsilon}$$

$$\times \sum_{k=N/2-1}^{N-1} \left(\frac{\varepsilon}{h}\right)^{1/2} (N/2-n)^{5/2} e^{-(\beta/2)|n-N/2|} e^{(x_{n+1}-x_k)/\varepsilon} e^{(-\beta/2)|n-k|}$$

$$\leq \frac{C_2}{\varepsilon^4} h_{n+1}^{5/2} e^{-x_{n+1}/\varepsilon} \sum_{k=N/2-1}^{N-1} e^{-\beta/2|n-k|} \leq \frac{C_3}{\varepsilon^4} h_{n+1}^{5/2} e^{-x_{n+1}/\varepsilon}. \tag{27}$$

Further,

$$\alpha^{(2)} = \sum_{k=-1}^{N/2-3} \tilde{\gamma}^{11}_{(N/2-2)k} F_k + \tilde{\gamma}^{11}_{(N/2-2)(N/2-2)} F_{N/2-2} + \sum_{k=N/2-1}^{N-1} \tilde{\gamma}^{12}_{(N/2-2)k} F_k. \tag{28}$$

Similarly, we have

$$\Big|\sum_{k=-1}^{N/2-3} \tilde{\gamma}^{11}_{(N/2-2)k} F_k\Big| \le C\Big(\varepsilon \ln \frac{h}{\varepsilon}\Big)^{-1/2} \varepsilon^{-1} e^{-x_{N/2-1}/\varepsilon}$$

$$\times \sum_{k=-1}^{N/2-3} \Big(\varepsilon \ln \frac{h}{\varepsilon}\Big)^{1/2} e^{x_{N/2-1}/\varepsilon} \varepsilon \Big(\ln \frac{h}{\varepsilon}\Big)^{-1/2} e^{-\beta|N/2-k|} h_{k+1}^{5/2} \frac{1}{\varepsilon^4} e^{-x_{k+1}/\varepsilon} =$$

$$C\Big(\varepsilon \ln \frac{h}{\varepsilon}\Big)^{-1/2} \varepsilon^{-1} e^{-x_{N/2-1}/\varepsilon} \sum_{k=-1}^{N/2-3} \varepsilon^{-5/2} h_{k+1}^{5/2} e^{-\beta|N/2-k|}$$

$$\times e^{-(x_{k+1}-x_{N/2-1})/\varepsilon} \le C_1\Big(\varepsilon \ln \frac{h}{\varepsilon}\Big)^{-1/2} \varepsilon^{-1} e^{-x_{N/2-1}/\varepsilon} \sum_{k=-1}^{N/2-3} (N/2-1-k)^{-5/2}$$

$$\times e^{-\beta|N/2-k|} (N/2-k+1)^C \le C_2\Big(\varepsilon \ln \frac{h}{\varepsilon}\Big)^{-1/2} \varepsilon^{-1} e^{-x_{N/2-1}/\varepsilon}. \tag{29}$$

Since the norm Γ^{-1} is uniformly bounded, we have $|\gamma^{11}_{(N/2-1)(N/2-1)}| \le C$. Therefore, taking into account (20), we have

$$\Big|\tilde{\gamma}^{11}_{(N/2-2)(N/2-2)} F_{N/2-2}\Big| \le C\Big(\varepsilon \ln \frac{h}{\varepsilon}\Big)^{-1/2} \varepsilon^{-1} e^{-x_{N/2-1}/\varepsilon}, \tag{30}$$

$$\Big|\sum_{k=N/2-1}^{N-1} \tilde{\gamma}^{12}_{(N/2-2)k} F_k\Big| \le C\Big(\varepsilon \ln \frac{h}{\varepsilon}\Big)^{-1/2} \varepsilon^{-1} e^{-x_{N/2-1}/\varepsilon} \times$$

$$\sum_{k=N/2-1}^{N-1} \Big(\varepsilon \ln \frac{h}{\varepsilon}\Big)^{1/2} \varepsilon e^{x_{N/2-1}/\varepsilon} \Big(\varepsilon \ln \frac{h}{\varepsilon}\Big)^{1/2}$$

$$\times e^{x_{N/2-1}/\varepsilon} \Big(\frac{\varepsilon}{h}\Big)^{1/2} e^{-\beta|N/2-k|} h^{-1/2} \varepsilon^{-1} e^{-x_k/\varepsilon} = C\Big(\varepsilon \ln \frac{h}{\varepsilon}\Big)^{-1/2} \varepsilon^{-1} e^{-x_{N/2-1}/\varepsilon}$$

$$\times \sum_{k=N/2-1}^{N-1} e^{-(x_k - x_{N/2-1})/\varepsilon}\left(\frac{\varepsilon}{h}\ln\frac{\varepsilon}{h}\right)^{1/2} e^{-\beta|N/2-k|} \le C_1\left(\varepsilon\ln\frac{h}{\varepsilon}\right)^{-1/2}\varepsilon^{-1}e^{-x_{N/2-1}/\varepsilon}$$

$$\times \sum_{k=N/2-1}^{N-1} e^{-\beta|N/2-k|}(N/2-k-1)^C \le C_2\left(\varepsilon\ln\frac{h}{\varepsilon}\right)^{-1/2}\varepsilon^{-1}e^{-x_{N/2-1}/\varepsilon}. \tag{31}$$

Finally,

$$\alpha^{(k)} = \sum_{k=-1}^{N/2-3}\tilde{\gamma}_{nk}^{21}F_k + \tilde{\gamma}_{n(N/2-2)}^{21}F_{N/2-2} + \sum_{k=N/2-1}^{N-1}\tilde{\gamma}_{nk}^{22}F_k, \tag{32}$$

$$\left|\sum_{k=-1}^{N/2-3}\tilde{\gamma}_{nk}^{21}F_k\right| \le Ch^{-1/2}\varepsilon^{-1}e^{-x_{N/2-1}/\varepsilon}e^{-\beta(n-N/2)}$$

$$\times \sum_{k=-1}^{N/2-3} h^{1/2}\varepsilon e^{x_{N/2-1}/\varepsilon}e^{\beta(n-N/2)}\left(\frac{\varepsilon}{h}\right)^{1/2}h_k^{5/2}\varepsilon^{-4}e^{-x_{k+1}/\varepsilon}; \tag{33}$$

$$\sum_{k=-1}^{N/2-3} h^{1/2}\varepsilon e^{x_{N/2-1}/\varepsilon}e^{\beta(n-N/2)}\left(\frac{\varepsilon}{h}\right)^{1/2}h_k^{5/2}\varepsilon^{-4}e^{-x_{k+1}/\varepsilon}$$

$$= \sum_{k=-1}^{N/2-3} h_k^{5/2}\varepsilon^{-5/2}e^{\beta(k-N/2)}e^{(x_{N/2-1}-x_{k+1})/\varepsilon}$$

$$\le C_1\sum_{k=-1}^{N/2-3}(n/2-k)^{-5/2}e^{\beta(k-N/2)}(N/2-k-1)^C \le C_2, \tag{34}$$

$$|\tilde{\gamma}_{n(N/2-2)}^{21}F_{N/2-2}| \le Ch^{-1/2}\varepsilon^{-1}e^{-x_{N/2-1}/\varepsilon}e^{-\beta(n-N/2)}h^{1/2}\varepsilon\times$$

$$e^{x_{N/2-1}/\varepsilon}e^{\beta(n-N/2)}\left(\frac{\varepsilon}{h}\right)^{1/2}e^{-\beta(n-N/2)}\times$$

$$\left(\varepsilon\ln\frac{h}{\varepsilon}\right)^{-1/2}\varepsilon^{-1}e^{-x_{N/2-1}/\varepsilon} = Ch^{-1/2}\varepsilon^{-1}e^{-x_{N/2-1}/\varepsilon}e^{-\beta(n-N/2)}\left(\ln\left(\frac{h}{\varepsilon}\right)\right)^{-1/2} \le$$

$$C_1h^{-1/2}\varepsilon^{-1}e^{-x_{N/2-1}/\varepsilon}e^{-\beta(n-N/2)}; \tag{35}$$

$$\left|\sum_{k=N/2-1}^{N-1} \tilde{\gamma}_{nk}^{22} F_k\right| \le C_1 h^{-1/2} \varepsilon^{-1} e^{-x_{N/2-1}/\varepsilon} e^{-\beta(n-N/2)} \times$$

$$\sum_{k=N/2-1}^{N-1} h^{1/2} \varepsilon e^{x_{N/2-1}/\varepsilon} e^{\beta(n-N/2)} e^{-\beta|n-k|} h^{-1/2} \varepsilon^{-1} e^{-x_k/\varepsilon}, \tag{36}$$

$$\sum_{k=N/2-1}^{N-1} h^{1/2} \varepsilon e^{x_{N/2-1}/\varepsilon} e^{\beta(n-N/2)} e^{-\beta|n-k|} h^{-1/2} \varepsilon^{-1} e^{-x_k/\varepsilon} =$$

$$\sum_{k=N/2-1}^{N-1} e^{\frac{x_{N/2-1}-x_k}{\varepsilon}} e^{\beta(n-N/2)} e^{-\beta|n-k|}. \tag{37}$$

We represent the last sum as

$$\sum_{k=N/2-1}^{N-1} e^{\frac{x_{N/2-1}-x_k}{\varepsilon}} e^{\beta(n-N/2)} e^{-\beta|n-k|} = \sum_{k=N/2-1}^{n} (\cdots) + \sum_{k=n+1}^{N-1} (\cdots) = \Sigma_1 + \Sigma_2.$$

Then

$$\Sigma_1 = \sum_{k=N/2-1}^{n} e^{\frac{x_{N/2-1}-x_k}{\varepsilon}} e^{\beta(k-N/2)} = \sum_{k=N/2-1}^{n} e^{-(k-N/2)\frac{h}{\varepsilon}+\beta(k-N/2)}$$

$$= \sum_{k=N/2-1}^{n} e^{-(k-N/2)\frac{h}{\varepsilon}} \le C_1 \tag{38}$$

if $h/\varepsilon \ge 2\beta$. Further,

$$\Sigma_2 = \sum_{k=n+1}^{N-1} e^{\frac{x_{N/2-1}-x_k}{\varepsilon}} e^{\beta(2n-k-N/2)} = \sum_{k=n+1}^{N-1} e^{-(k-N/2)\frac{h}{\varepsilon}+k+N/2-2n}.$$

If $h/\varepsilon \ge \beta$, then we get

$$\Sigma_2 \le \sum_{k=n+1}^{N-1} e^{-2\beta(k-n)} \le C_1. \tag{39}$$

The statement of the lemma follows from (22)–(39). □

Lemma 8 *There are constants $C > 0, \beta > 0$ that are independent of ε, N, such that the estimates hold*

$$\| P(\Phi'' - gI)(x) \|_{C[x_n,x_{n+1}]} = \begin{cases} O\left(\frac{C}{\varepsilon^4} h_{n+1}^2 e^{-x_n/\varepsilon}\right), & 0 \le n \le \frac{N}{2} - 2, \\ O\left(\frac{C}{\varepsilon^2 \ln(1+h/\varepsilon)} e^{-x_{N/2-1}/\varepsilon}\right), & n = \frac{N}{2} - 1, \\ O\left(\frac{1}{\varepsilon h} e^{-\frac{x_{N/2-1}}{\varepsilon}} e^{-\beta|n-\frac{N}{2}|}\right), & \frac{N}{2} \le n \le N. \end{cases} \quad (40)$$

Proof Since at each node x_n there is different from zero only one B-spline $N_{n-1,1}$, then the equality holds

$$P(\Phi'' - gI)(x_n) = \alpha_{n-1} \tilde{N}_{n-1,1}(x_n).$$

Hence, from Lemma 7 and estimates (7), the assertion of the lemma follows. □

Lemma 9 *The following estimates hold*

$$\| e''(x) \|_{C[x_n,x_{n+1}]} \le \frac{C}{\varepsilon^4} h_{n+1}^2 e^{-x_n/\varepsilon}, \; 0 \le n \le \frac{N}{2} - 2. \quad (41)$$

Proof By virtue of (8), (40) is enough to evaluate $\| gI(x) - \Phi''(x) \|_{C[x_n,x_{n+1}]}$. But an estimate of this expression of the form (41) follows from the estimate of linear interpolation errors on the segment $[x_n, x_{n+1}]$. □

4 Proof of Theorems

Theorem 1

Proof According to [2] for the interpolating cubic spline $g_3(x, u) \in S(\Omega, 3, 1)$, the estimate holds

$$|g_3(x, u) - u(x)| \le \frac{5}{384} \| u^{(4)} \|_{C[0,1]} \max_n h_n^4. \quad (42)$$

According to (1) $g_3(x, u) = g_3(x, q) + g_3(x, \Phi)$, and by virtue of the conditions (2) and (42) we have

$$\| g_3(x, q) - q(x) \|_{C[0,1]} \le C \max_n h_n^4 \le CN^{-4}. \quad (43)$$

It remains to evaluate $\| g_3(x, \Phi) - \Phi(x) \|_{C[x_n,x_{n+1}]}$ for each grid interval. When $\sigma = 1/2$, the parameter ε is limited by positive constant below, so according to (42) the spline $g_3(x, \Phi)$ has an error of the order of $O(N^{-4})$ uniformly in ε. Therefore, we will assume below that $\sigma < 1/2$.

First, we prove the estimates (4) for $n \le \frac{N}{2} - 2$. Set $e(x) = g_3(x, \Phi) - \Phi(x)$. Since $e(x_n) = e(x_{n+1}) = 0$, then considering $e(x)$ as the solution of the problem $e''(x) = e''(x)$ with zero boundary conditions on interval $[x_n, x_{n+1}]$, we get

$$e(x) = \int_{x_n}^{x_{n+1}} G(x,s)e''(s)ds,$$

where
$G(x,s) = \frac{1}{x_{n+1}-x_n}\begin{cases} (x - x_n)(x_{n+1} - s), & x_n \le x \le s, \\ (s - x_n)(x_{n+1} - x), & s < x \le x_{n+1} \end{cases}$
is Green function. Since $|G(x,s)| \le x_{n+1} - x_n = h_{n+1}$, from (41), (6). (3) we get

$$\| e(x) \|)_{C[x_n,x_{n+1}]} \le h_{n+1} \int_{x_n}^{x_{n+1}} |e''(s)|ds \le h_{n+1}^2 \| e''(s) \|_{C[x_n,x_{n+1}]} \le$$

$$\frac{C}{\varepsilon^4} e^{-x_n/\varepsilon} h_{n+1}^4 \le \frac{C}{(N/2-n)^4}(1 - 2(1-\varepsilon)\frac{n}{N})^4 = \frac{16C}{N^4}\frac{(N/2-n+\varepsilon N)^4}{(N/2-n-1)^4} \le \frac{C_1}{N^4}.$$

Taking into account the estimate (43), we obtain the estimate (4) for $n \le \frac{N}{2} - 2$.

For $n \ge N/2 - 1$ we have

$$\| e(x) \|_{C[x_n,x_{n+1}]} \le Ch_{n+1} \int_{x_n}^{x_{n+1}} |e''(s)|ds \le$$

$$Ch_{n+1}\Big(\int_{x_n}^{x_{n+1}} |\Phi''(s)|ds + \int_{x_n}^{x_{n+1}} |g_3''(s, \Phi)|ds \Big). \tag{44}$$

For $n = N/2 - 1$ we get

$$\int_{x_n}^{x_{n+1}} |\Phi''(s)|ds \le \frac{C}{\varepsilon^2} \int_{x_n}^{x_{n+1}} e^{-\frac{x}{\varepsilon}} ds \le \frac{C}{\varepsilon} e^{-\frac{x_n}{\varepsilon}} \le \frac{C}{\varepsilon N^4}, \; n = N/2 - 1. \tag{45}$$

Considering (40) and $gI(x) = 0$ for $x \ge x_{N/2-1}$, thus $P(\Phi'' - gI)(x) = g_3''(x, \Phi)$, we obtain

$$\int_{x_n}^{x_{n+1}} |g_3''(s, \Phi)|ds \le$$

$$C\varepsilon \ln(1 + h/\varepsilon) \frac{1}{\varepsilon^2 \ln(1 + h/\varepsilon)} e^{-x_{N/2-1}/\varepsilon} \le C\varepsilon \cdot \frac{1}{\varepsilon^2 N^4} = \frac{C}{\varepsilon N^4}, \; n = N/2 - 1. \tag{46}$$

Similarly for $n \geq N/2$ we have

$$\int_{x_n}^{x_{n+1}} |\Phi''(s)|ds \leq \frac{C}{\varepsilon^2}\int_{x_n}^{x_{n+1}} e^{-\frac{x}{\varepsilon}}ds \leq \frac{C}{\varepsilon}e^{-\frac{x_n}{\varepsilon}} = \frac{C}{\varepsilon}e^{-\frac{x_{N/2}}{\varepsilon}} \cdot e^{-\frac{x_n - x_{N/2}}{\varepsilon}}$$

$$\leq \frac{C_1}{\varepsilon}N^{-4}e^{-(n-N/2)\frac{h}{\varepsilon}} \leq \frac{C_1}{\varepsilon N^4}e^{-\beta(n-N/2)}. \tag{47}$$

$$\int_{x_n}^{x_{n+1}} |g_3''(s,\Phi)|ds \leq Ch\frac{1}{\varepsilon h}e^{-x_{N/2-1}/\varepsilon}e^{-\beta|n-\frac{N}{2}|} \leq C\varepsilon \cdot \frac{1}{\varepsilon^2 N^4}e^{-\beta|n-\frac{N}{2}|} =$$

$$\frac{C}{\varepsilon N^4}e^{-\beta|n-\frac{N}{2}|}. \tag{48}$$

From (44)–(48), Lemmas 1 and (43) estimates (4) for $N/2 - 1 \leq n \leq N - 1$ follow. Theorem 1 is proved. □

The proof of Theorem 2 is based on lower bounds for $\| \ e(x) \ \|_{C[x_n,x_{n+1}]}$ for $n = N/2 - 2$ and $n = N/2 - 1$ in the case $\Phi(x) = e^{-x/\varepsilon}$ and is carried out similarly to the proof of Theorem 4 from [4]. This proof is based on lower bounds for the elements of the matrix Γ_{22}. The matrix Γ_{22} in the case of a Bakhvalov mesh corresponds to a segment of a uniform partition and for $\varepsilon \leq CN^{-1}$ is completely similar in properties to the matrix Γ_{22} from [4].

5 Results of Numerical Experiments

We define the function of the form (1):

$$u(x) = \cos\frac{\pi x}{2} + e^{-\frac{x}{\varepsilon}}, \ x \in [0, 1].$$

The tables show the maximum errors of spline interpolation calculated at nodes of the condensed mesh obtained from the original computational mesh by dividing each of its mesh intervals into 10 equal parts. Table 1 shows the errors for the traditional cubic spline $g_3(x, u)$. The errors confirm the estimates of Theorems 1 and 2. The table shows that the error increases with decreasing ε for fixed N.

Due to the non-uniform in ε convergence of the cubic spline $g_3(x, u)$, we construct a modified interpolation spline. We use an approach [4], where cubic spline on the Shishkin grid is considered. We define $\bar{x}_n = (x_n + x_{n+1})/2, n \in [N/2 - 1, N/2]$, $\bar{x}_n = x_n, n \in [0, N/2 - 2] \cup [N/2 + 1, N]$. Let us $gm_3(x, u) \in S(\Omega, 3, 1)$ be cubic spline determined from conditions $gm_3(\bar{x}_n, u) = u(\bar{x}_n), \ n \in [0, N]$, $gm_3'(0, u) = u'(0), \ gm_3'(1, u) = u'(1)$.

The results of Table 2 for the modified spline $gm_3(x, u)$ show the uniform in ε error of order $O(1/N^4)$.

Table 1 The error of cubic interpolation spline $g_3(x, u)$

	N					
ε	16	32	64	128	256	512
10^{-1}	$1.33 \cdot 10^{-4}$	$1.02 \cdot 10^{-5}$	$6.99 \cdot 10^{-7}$	$4.52 \cdot 10^{-8}$	$2.89 \cdot 10^{-9}$	$1.82 \cdot 10^{-10}$
10^{-2}	$1.72 \cdot 10^{-4}$	$1.06 \cdot 10^{-5}$	$6.74 \cdot 10^{-7}$	$7.95 \cdot 10^{-8}$	$8.80 \cdot 10^{-9}$	$8.12 \cdot 10^{-9}$
10^{-3}	$4.82 \cdot 10^{-4}$	$1.37 \cdot 10^{-5}$	$7.04 \cdot 10^{-7}$	$4.38 \cdot 10^{-8}$	$2.71 \cdot 10^{-9}$	$1.64 \cdot 10^{-10}$
10^{-4}	$6.35 \cdot 10^{-3}$	$1.88 \cdot 10^{-4}$	$5.45 \cdot 10^{-6}$	$1.56 \cdot 10^{-7}$	$4.45 \cdot 10^{-9}$	$1.72 \cdot 10^{-10}$
10^{-5}	$7.22 \cdot 10^{-2}$	$2.19 \cdot 10^{-3}$	$6.62 \cdot 10^{-5}$	$1.98 \cdot 10^{-6}$	$5.86 \cdot 10^{-8}$	$1.71 \cdot 10^{-9}$
10^{-6}	$7.73 \cdot 10^{-1}$	$2.38 \cdot 10^{-2}$	$7.28 \cdot 10^{-4}$	$2.22 \cdot 10^{-5}$	$6.76 \cdot 10^{-7}$	$2.05 \cdot 10^{-8}$
10^{-7}	8.06	$2.49 \cdot 10^{-1}$	$7.70 \cdot 10^{-3}$	$2.37 \cdot 10^{-4}$	$7.29 \cdot 10^{-6}$	$2.24 \cdot 10^{-7}$
10^{-8}	83.1	2.58	$7.98 \cdot 10^{-2}$	$2.47 \cdot 10^{-3}$	$7.64 \cdot 10^{-5}$	$2.36 \cdot 10^{-6}$

Table 2 The error of modified cubic spline $gm_3(x, u)$

	N					
ε	16	32	64	128	256	512
10^{-1}	$2.21 \cdot 10^{-4}$	$1.85 \cdot 10^{-5}$	$1.35 \cdot 10^{-6}$	$9.09 \cdot 10^{-8}$	$5.02 \cdot 10^{-9}$	$3.78 \cdot 10^{-10}$
10^{-2}	$1.67 \cdot 10^{-4}$	$1.09 \cdot 10^{-5}$	$1.00 \cdot 10^{-6}$	$1.16 \cdot 10^{-7}$	$1.45 \cdot 10^{-8}$	$1.47 \cdot 10^{-9}$
10^{-3}	$1.89 \cdot 10^{-4}$	$1.10 \cdot 10^{-5}$	$6.49 \cdot 10^{-7}$	$4.06 \cdot 10^{-8}$	$2.59 \cdot 10^{-9}$	$2.16 \cdot 10^{-10}$
10^{-4}	$2.25 \cdot 10^{-4}$	$1.35 \cdot 10^{-5}$	$8.05 \cdot 10^{-7}$	$4.74 \cdot 10^{-8}$	$2.77 \cdot 10^{-9}$	$1.61 \cdot 10^{-10}$
10^{-5}	$2.46 \cdot 10^{-4}$	$1.51 \cdot 10^{-5}$	$9.19 \cdot 10^{-7}$	$5.57 \cdot 10^{-8}$	$3.35 \cdot 10^{-9}$	$2.00 \cdot 10^{-10}$
10^{-6}	$2.58 \cdot 10^{-4}$	$1.60 \cdot 10^{-5}$	$9.84 \cdot 10^{-7}$	$6.05 \cdot 10^{-8}$	$3.71 \cdot 10^{-9}$	$2.26 \cdot 10^{-10}$
10^{-7}	$2.66 \cdot 10^{-4}$	$1.65 \cdot 10^{-5}$	$1.02 \cdot 10^{-6}$	$6.33 \cdot 10^{-8}$	$3.91 \cdot 10^{-9}$	$2.41 \cdot 10^{-10}$
10^{-8}	$2.71 \cdot 10^{-4}$	$1.68 \cdot 10^{-5}$	$1.05 \cdot 10^{-6}$	$6.50 \cdot 10^{-8}$	$4.03 \cdot 10^{-9}$	$2.5 \cdot 10^{-10}$

6 Conclusion

The error of interpolation by a cubic spline on the Bakhvalov mesh in the presence of an exponential boundary layer is estimated. It is proved that for a given number of mesh nodes, the interpolation error can grow unlimitedly with decreasing value of a small parameter. The results of computational experiments are consistent with the obtained error estimates. The cubic spline was modified on a Bakhvalov mesh and numerically shown that the resulting spline has an error of the order of $O(1/N^4)$ uniformly in ε and N.

Acknowledgments The work of Blatov I.A. and Kitaeva E.V. was funded by RFBR, project number 20-01-00650 and the work of Zadorin N.A. was funded by RFBR, project number 19-31-60009.

References

1. Ahlberg, J.H., Nilson, E.N., Walsh, J.L.: The Theory of Splines and their Applications. Academic Press, New York (1967)
2. Zav'yalov, Yu.S., Kvasov, B.I., Miroshnichenko, V.L.: Methods of Spline Functions. Nauka, Moscow (1981) (in Russian)
3. Shishkin, G.I.: Grid Approximations of Singular Perturbation Elliptic and Parabolic Equations. UB RAS, Yekaterinburg (1992) (in Russian)
4. Blatov, I.A., Zadorin, A.I., Kitaeva, E.V.: Cubic spline interpolation of functions with high gradients in boundary layers. Comput. Math. Math. Phys. **57** (1), 7–25 (2017)
5. Bakhvalov, N.S.: The optimization of methods of solving boundary value problems with a boundary layer. USSR Comput. Math. Math. Phys. **9**, 139–166 (1969)
6. Boor, C. de.: Practical Guide to Splines. Springer-Verlag, New York (1985)
7. Demko, S.: Inverses of band matrices and local convergence of spline projections. SIAM J. Numer. Anal. **14**, 616–619 (1977)
8. Blatov, I.A., Strygin, V.V.: Fourth order accuracy collocation method for singularly perturbed boundary value problems. Siberian Math. J. **34** (1), 10–24 (1993)

On Exact Penalty Operators and Penalization Methods for Elliptic Unilateral Problems with Piecewise Smooth Obstacles

Rafail Z. Dautov

Abstract The aim of this paper is twofold: firstly, to prove that for piecewise smooth obstacle, the elliptic variational inequality of the first kind corresponding to a unilateral problem can be reformulated as a variational inequality of the second kind with a continuous functional and, secondly, to obtain accuracy estimates of penalization methods for such obstacle problems for a wide class of penalty functions. The accuracy estimates obtained in the current paper are of the same order as known estimates for smooth obstacles.

1 Introduction

Many phenomena in physics, biology, and finance can be described by partial differential equations that display a priori unknown interfaces or boundaries. Such problems are called free boundary problems. One of the simplest and the most important free boundary problems is the obstacle problem, in which, at least formally, a function u solves a partial differential equation on the set where it is strictly greater than a given function ψ, and equals this function elsewhere.

The prototype of problems that we will consider here is the linear obstacle problem with the obstacle $\psi \in H^1(\Omega)$. In this problem we are looking for a function u in $H^1(\Omega)$ which satisfy the following relations:

$$\begin{cases} -\Delta u - f \geq 0 \quad \text{in } \Omega, \quad u = u_D \quad \text{on } \partial\Omega, \\ u \geq \psi \quad \text{in } \Omega, \\ (-\Delta u - f)(u - \psi) = 0 \text{ in } \Omega. \end{cases}$$

R. Z. Dautov (✉)
Department of Numerical Mathematics, Kazan Federal University, Kazan, Russia

I. B. Badriev et al. (eds.), *Mesh Methods for Boundary-Value Problems and Applications*, Lecture Notes in Computational Science and Engineering 141,
https://doi.org/10.1007/978-3-030-87809-2_5

This problem is equivalent to the variational inequality of the first kind

$$u \in K: \quad \int_\Omega \nabla u \cdot \nabla (v-u)\, dx \geq \int_\Omega f(v-u)\, dx \quad \forall v \in K, \tag{1}$$

where $u_D \in H^1(\Omega)$,

$$K = \{v \in u_D + H_0^1(\Omega) : \ v \geq \psi \quad \text{on } \Omega\}.$$

Penalization techniques have been widely used in the study of obstacle problems (see e.g. [1–4]). The penalty methods are the basis of many well-known approximate methods for their solution and also used to demonstrate the existence and regularity of solutions. For problem (1) they consist of substituting the variational inequality by a family of nonlinear boundary value problems of the form[1]

$$\begin{cases} -\Delta u_\varepsilon + \beta_\varepsilon(x, u_\varepsilon - \psi) = f & \text{in } \Omega, \\ u_\varepsilon = u_D \quad \text{on } \partial\Omega, \end{cases}$$

where β_ε is a penalty function, $\varepsilon > 0$. The regularity of the solution of problem (1), as well as the accuracy of the penalty methods, are fairly well studied when

$$f \in L^2(\Omega), \quad \Delta\psi \in L^2(\Omega). \tag{2}$$

There are different choices of penalization. Notable examples of penalty functions are [2, p. 368], [4, p. 107]

$$\beta_\varepsilon(x,t) = -\frac{1}{\varepsilon}\, t^-, \quad \beta_\varepsilon(x,t) = (-\Delta\psi - f)^+ \theta_\varepsilon(t), \tag{3}$$

where $t^- = \max\{-t, 0\}$, $t^+ = \max\{t, 0\}$ are negative and positive parts of t, respectively; θ_ε is a sequence of Lipschitz functions which almost everywhere on R tends to θ as $\varepsilon \to 0$, where

$$\theta(t) = \begin{cases} -1, & t \leq 0, \\ 0, & t > 0. \end{cases} \tag{4}$$

Under conditions (2), (3) the following estimates are valid

$$\|u - u_\varepsilon\|_{L^\infty(\Omega)} \leq C\,\varepsilon, \quad \|u - u_\varepsilon\|_{H^1(\Omega)} \leq C\,\varepsilon^{1/2}, \tag{5}$$

where the constant C does not depend on ε.

---

[1] Often talk about the regularization method if the function $\beta_\varepsilon : \Omega \times R \to R$ uniformly over ε bounded and about the penalty method otherwise.

Many different penalty functions β_ε have been found over the years (see, for instance, [5–7]). Wide class of penalty functions guaranteeing estimates (5) are indicated in [8]. In this article, it is proved that under conditions (2) inequality (1) can be reformulated as the following variational inequality of the second kind: find $u \in H^1_D(\Omega)$ such that

$$\int_\Omega \nabla u \cdot \nabla(v-u)\,dx + j(v) - j(u) \geq \int_\Omega f(v-u)\,dx \quad \forall v \in H^1_D(\Omega), \tag{6}$$

where $H^1_D(\Omega) = u_D + H^1_0(\Omega)$,

$$j(v) = \int_\Omega g(x)\chi(v-\psi)\,dx, \quad g \in L^2(\Omega) : g \geq (-\Delta\psi - f)^+ \text{ in } \Omega,$$

and χ belongs to some set of continuous functions on R. Variational inequality (6) can be equivalently written as the inclusion

$$u \in H^1_D(\Omega) : \quad -\Delta u + \beta(u-\psi) \ni f \quad \text{in } \Omega, \tag{7}$$

where $\beta(v-\psi)$ is the subdifferential of $j(v)$.[2]

The operator $\beta : L^2(\Omega) \to L^2(\Omega)$ in (7) has been named an exact penalty operator. Penalty operator β_ε is obtained by its regularization. It is also proved in [8] that the second estimate in (5) can be improved to $O(\varepsilon^{3/4})$ in some cases.

This article is devoted to a generalization of the results [8] for the case of piecewise smooth obstacles, which includes an important subclass of polyhedra. We are not aware of any published work in which accuracy estimates of penalty (or regularization) method for problems with such obstacles were obtained.

The outline of this work is as follows. In Sect. 2, we formulate the original obstacle problem with a strongly monotone nonlinear operator. In Sect. 3, we reformulate it in the form of a variational inequality of the second kind and define the exact penalty operators. In the last Sect. 4, we define a penalty problem and obtain accuracy estimates.

We hope that the exact penalty operators obtained in this work will be useful also in studying the regularity of the solution of elliptic unilateral problems with piecewise smooth obstacles.

[2] In [8], the case $\psi = 0$ was considered. Under conditions (2) inequality (1) comes down to this case by shift $u \to u - \psi$ and $f \to f + \Delta\psi$.

2 Formulation of the Problem

The scalar product in R^n is denoted by $\xi \cdot \eta$ for all $\xi, \eta \in R^n$, $|\xi| = (\xi \cdot \xi)^{1/2}$, while $\langle \cdot, \cdot \rangle$ is generically used to indicate a duality pairing of the relevant function spaces.

2.1 *Functional Spaces*

Let $p \in [1, \infty]$. For an arbitrary domain $D \subset R^n$, $n \geq 1$, we use the traditional notations $L^p(D)$ for the Lebesgue space with the norm

$$\|v\|^p_{L^p(D)} = \int_D |v(x)|^p \, dx, \quad 1 \leq p < \infty,$$

$$\|v\|_{L^\infty(D)} = \operatorname*{ess\ sup}_{\Omega} |v|,$$

and $W^k_p(D)$ for the Sobolev space of order $k \geq 1$ of functions which weak derivatives of order $\leq k$ belong to $L^p(D)$:

$$W^k_p(D) = \{u \in L^p(D) : D^{|\alpha|} u \in L^p(D), \ |\alpha| \leq k\}.$$

The norms on it are denoted by $\| \cdot \|_{W^k_p(D)}$ and defined by the relation

$$\|v\|^p_{W^k_p(D)} = \sum_{|\alpha| \leq k} \|D^\alpha v\|^p_{L^p(D)}.$$

We use the notation $H^k(D) = W^k_2(D)$ and denote by $H^1_0(D)$ the Sobolev space of the functions that vanish on the boundary ∂D, endowed with the norm

$$\|u\|_{H^1_0(D)} = \left(\int_D |\nabla u|^2 \, \mathrm{d}x \right)^{1/2}.$$

Let $H^{-1}(D)$ be conjugate to $H^1_0(D)$ space,

$$\|f\|_{H^{-1}(D)} = \sup_{\eta \in H^1_0(D)} \frac{\langle f, \eta \rangle}{\|\eta\|_{H^1_0(D)}}.$$

Recall that $v^- \in H^1(\Omega)$ if $v \in H^1(\Omega)$, and $\nabla v^- = -\nabla v$ in $S = \{x \in \Omega : v(x) < 0\}$, $\nabla v^- = 0$ in $\Omega \setminus S$ (see e.g. [4, c. 50]).

In what follows, C will denote a positive constant that may vary from line to line.

2.2 Original Obstacle Problem

Let $\Omega \subset R^n$ be a bounded domain with Lipschitz boundary $\partial\Omega$, $n \geq 1$. Consider a sufficiently smooth vector field $(a_0, a) : \Omega \times R \times R^n \to R \times R^n$ and let A be an operator acting from $H^1(\Omega)$ into $H^{-1}(\Omega)$ defined by

$$Au = -\operatorname{div} a(x, u, \nabla u) + a_0(x, u, \nabla u).$$

For any $u \in H^1(\Omega)$ and $v \in H_0^1(\Omega)$ we have

$$\langle Au, v\rangle = \int_\Omega \big(a(x, u, \nabla u) \cdot \nabla v + a_0(x, u, \nabla u)v\big)\, dx.$$

Let ψ (the obstacle) and u_D (the boundary datum) be given functions in $H^1(\Omega)$ with $\psi \leq u_D$ a.e. on Ω, $H_D^1(\Omega) = u_D + H_0^1(\Omega)$. Let the convex set K in $H^1(\Omega)$ be defined by

$$K = \{v \in H_D^1(\Omega) :\ v \geq \psi \quad \text{a.e. on } \Omega\}.$$

Consider the following variational inequality of the first kind.

Problem (P_0) Find $u \in K$ such that

$$\langle Au, v - u\rangle \geq 0 \quad \forall\, v \in K. \tag{8}$$

2.3 Restrictions on the Operator A

We will assume that the vector field $a(x, s, \xi) = (a_1(x, s, \xi), \ldots, a_n(x, s, \xi))$ and the function $a_0(x, s, \xi)$ satisfy the following assumptions:

(H_1) $a_i \in W_\infty^1(\Omega) \times C^1(R) \times C^1(R^n)$, $\quad i = 0, \ldots, n$;
(H_2) for a.e. $x \in \Omega$ and for all $s \in R$ and $\xi \in R^n$

$$|a(x, s, \xi)| \leq C\big(|s| + |\xi|\big);$$
$$|a_0(x, s, \xi)| \leq C\big(|f(x)| + |s| + |\xi|\big), \quad f \in L^2(\Omega);$$

(H_3) for a.e. $x \in \Omega$ and for all $s, t \in R$ and $\xi, \eta \in R^n$

$$(a(x, s, \xi) - a(x, t, \eta)) \cdot (\xi - \eta) + (a_0(x, s, \xi) - a_0(x, t, \eta))(s - t) \geq \alpha\, |\xi - \eta|^2,$$

where $\alpha = \text{const} > 0$.

The existence and uniqueness of a solution u to the problem (P_0) under conditions (H_1)–(H_3) are well known [2, p. 247, Theorem 8.2].

Note that in the case of linear functions

$$a(x, s, \xi) = A(x)\xi,$$
$$a_0(x, s, \xi) = a_0(x)\, s - f(x).$$

conditions (H_1)–(H_3) will be satisfied if

$$A \in W^1_\infty(\Omega; R^{n\times n}), \quad a_0(x) \in L^\infty(\Omega), \quad f \in L^2(\Omega);$$
$$A(x)\xi \cdot \xi \geq \alpha\, |\xi|^2, \quad a_0(x) \geq 0 \quad \text{a.e. in } \Omega.$$

Condition (H_3) immediately implies that

$$\langle Au - Av, u - v\rangle \geq \alpha\, \|u - v\|^2_{H^1_0(\Omega)}. \tag{9}$$

for all $u, v \in H^1_D(\Omega)$. Additionally,

$$\langle Au - Av, (u - v)^-\rangle \leq -\alpha\, \|(u - v)^-\|^2_{H^1_0(\Omega)} \tag{10}$$

for all $u, v \in H^1(\Omega)$ such that $(u - v)^- \in H^1_0(\Omega)$. Indeed, put $a(u) = a(x, u, \nabla u)$, $a_0(u) = a_0(x, u, \nabla u)$. Then

$$\langle Au - Av, (u - v)^-\rangle = - \int\limits_{\{x\in\Omega:\, u(x)<v(x)\}} \Big((a(u) - a(v)) \cdot \nabla(u - v)$$
$$+(a_0(u) - a_0(v))(u - v)\Big)\, dx$$
$$\leq -\alpha \int\limits_{\{x\in\Omega:\, u(x)<v(x)\}} |\nabla(u - v)|^2\, dx = -\alpha\, \|(u - v)^-\|^2_{H^1_0(\Omega)}.$$

2.4 Restrictions on the Obstacle

Let us formulate additional conditions on the obstacle. Let $\{\Omega_1, \Omega_2, \ldots, \Omega_m\}$, $m \geq 2$, be a partition of Ω such that each subdomain Ω_i has a Lipschitz boundary $\partial\Omega_i$,

$$\Omega_i \cap \Omega_j = \emptyset \text{ for } i \neq j, \quad \bigcup_{i=1}^m \overline{\Omega_i} = \overline{\Omega}.$$

We will assume that the obstacle is piecewise smooth:

(H_4) $\psi \in C^{0,1}(\overline{\Omega})$, $\quad \psi|_{\overline{\Omega_i}} \in C^{1,1}(\overline{\Omega_i})$, $\quad 1 \le i \le m$.

Let $\{\Gamma_1, \Gamma_2, \dots, \Gamma_M\}$ be the set of all common parts of boundaries $\partial\Omega_i$ and $\partial\Omega_j$ of the neighboring subdomains Ω_i and Ω_j, $1 \le i \ne j \le m$. On each $\Gamma_k = \partial\Omega_i \cap \partial\Omega_j$ we define the function

$$h_k(x) = a(x, \psi, \nabla\psi) \cdot n_i(x)|_{\overline{\Omega_i}} + a(x, \psi, \nabla\psi) \cdot n_j(x)|_{\overline{\Omega_j}},$$

where $n_i(x)$ is the outward normal unit vector at the point $x \in \partial\Omega_i$. Put $\Gamma = \bigcup_{k=1}^{M} \Gamma_k$ and define

$$g \in L^\infty(\Omega): \quad g|_{\Omega_i} = A\psi|_{\Omega_i}, \quad 1 \le i \le m;$$
$$h \in L^\infty(\Gamma): \quad h|_{\Gamma_k} = h_k, \quad 1 \le k \le M.$$

Conditions (H_1) and (H_4) imply that functions g and h are well defined. According to the definition of these functions, we have

$$\langle A\psi, v\rangle = \int_\Omega g\, v\, dx + \int_\Gamma h\, v\, dx \quad \forall v \in H_0^1(\Omega).$$

Note that if the obstacle is smooth, say $\psi \in C^{1,1}(\overline{\Omega})$, then $h = 0$.

We define the functional $(A\psi)^\pm \in H^{-1}(\Omega)$ by the equality

$$\langle (A\psi)^\pm, v\rangle = \int_\Omega g^\pm\, v\, dx + \int_\Gamma h^\pm\, v\, dx \quad \forall v \in H_0^1(\Omega). \tag{11}$$

Since $H_0^1(\Omega) \subset L^2(\Gamma)$, it is easy to see that

$$\|(A\psi)^\pm\|_{H^{-1}(\Omega)} \le C\left(\|g^\pm\|_{L^2(\Omega)} + \|h^\pm\|_{L^2(\Gamma)}\right).$$

3 Equivalent Inequality Without Constraints

Let j be the indicator function of the convex set K

$$j(v) = \begin{cases} 0, & v \in K, \\ +\infty, & v \notin K. \end{cases} \tag{12}$$

The original variational inequality (8) can be represented in the form of an inequality without constraints.

Problem (P) Find $u \in H^1_D(\Omega)$ such that[3]

$$\langle Au, v - u\rangle + j(v) - j(u) \geq 0 \quad \forall v \in H^1_D(\Omega). \tag{13}$$

In this inequality, we replace the functional j with a functional having better properties. To this end, we introduce the class $J(\psi)$ of convex lower semicontinuous (l.s.c.) functionals $j : H^1_D(\Omega) \to R$ such that

(a) $j(v) = 0$ for functions $v \in K$;
(b) $j(v) \geq \langle (A\psi)^+, (v - \psi)^- \rangle$ for all $v \in H^1_D(\Omega)$.

The following theorem is the first of our main results.

Theorem 1 *Let the assumptions* (H_1)–(H_4) *be satisfied and let* $j \in J(\psi)$. *Then problems* (P_0) *and* (P) *are equivalent.*

Proof Let u be a solution to Problem (P). Since the solutions to problems (P_0) and (P) are unique, it suffices to prove that u is also a solution to (P_0).

Let us prove that $u \in K$. By choosing $v = u + (u - \psi)^- = \psi + (u - \psi)^+ \in K$ in inequality (13) and by taking into account properties (a) and (b) of the functional j, we obtain the inequalities

$$\langle Au, (u - \psi)^- \rangle \geq j(u) \geq \langle (A\psi)^+, (u - \psi)^- \rangle.$$

Hence, taking into account estimate (10), we have

$$\begin{aligned} -\alpha \, \|(u - v)^-\|^2_{H^1_0(\Omega)} &\geq \langle Au - A\psi, (u - \psi)^- \rangle \\ &\geq \langle (A\psi)^+ - A\psi, (u - \psi)^- \rangle = \langle (A\psi)^-, (u - \psi)^- \rangle \geq 0. \end{aligned}$$

Therefore, $(u - \psi)^- = 0$, i.e. $u \in K$. Choosing $v \in K$ in the inequality (13), we see that u is a solution to the original problem (P_0). □

3.1 Exact Penalty Operators

The above-introduced class $J(\psi)$ is quite large. Note that the indicator function (12) of the convex set K also belongs to $J(\psi)$. For the considered class of operators A and the set K we can select in $J(\psi)$ a subset of convex and continuous functionals.

[3] If j is a convex l.s.c. functional, then conditions (H_1)–(H_3) are sufficient for the existence and uniqueness of a solution of problem (P) [2, p. 251, Thm. 8.5].

To do this we define a function $\theta(t)$ on R such that

$$\theta \in C(-\infty, 0] \text{ is non-decreasing}, \quad \theta(t) \begin{cases} = 0, & t > 0, \\ \leq -1, & t \leq 0, \\ \geq a\,t + b, & |t| \text{ large}, \end{cases} \tag{14}$$

where $a \geq 0$, $b \in R$. Note that the maximal among functions (14) is defined in (4). We also define function $\Theta \in C^{0,1}_{\text{loc}}(R)$ and two function G, H on Ω:

$$\Theta(t) = \int_0^t \theta(t)\,dt,$$

$$G \in L^\infty(\Omega): \quad G \geq g^+ \text{ in } \Omega,$$

$$H \in L^\infty(\Gamma): \quad H \geq h^+ \text{ on } \Gamma.$$

We denote by $J_L(\psi)$ the set of functionals $H^1(\Omega) \to R$ of the form

$$j(v) = \int_\Omega G(x)\,\Theta(v(x) - \psi(x))\,dx + \int_\Gamma H(x)\,\Theta(v(x) - \psi(x))\,dx. \tag{15}$$

From the definition of function Θ it follows that j is continuous functionals (Lipschitz continuous if $a = 0$ in (14)). Since $\Theta(t) = 0$, if $t \geq 0$ and $\Theta(t) \geq t^-$ in R, then properties (a) and (b) are satisfied and $J_L(\psi) \subset J(\psi)$.

The function $\Theta(t)$ has no derivative only for $t = 0$, so its subdifferential is easy to calculate. It is equal to

$$\partial\Theta = \begin{cases} 0, & t > 0, \\ [\theta(0),\, 0], & t = 0, \\ \theta(t), & t < 0. \end{cases}$$

The functional j defined in (15) is the sum of the two convex continuous functionals on the whole space $H^1(\Omega)$. Therefore, according to the Moreau-Rockafellar theorem, his subdifferential can be calculated as the sum of subdifferentials of the functionals on the right hand side of (15).

We define in $H^1(\Omega)$ the multivalued operator β by the equality

$$\langle \beta(u), v \rangle = \int_\Omega G\,\partial\Theta(u) v\,dx + \int_\Gamma H\,\partial\Theta(u) v\,dx, \quad v \in H^1(\Omega).$$

Then $\partial j(u) = \beta(u - \psi)$ and problem (P) will be reduced to the inclusion

$$u \in H^1_D(\Omega): \quad Au + \beta(u - \psi) \ni 0.$$

The operator β will be called the exact penalty operator.

4 The Penalty Problem

We approximate the non-differentiable functional $j \in J_L(\psi)$ by a sequence j_ε of differentiable ones.

To this end, we approximate the function $\theta(t)$ (or $\partial\Theta$) by continuous functions $\theta_\varepsilon(t)$ nondecreasing on R so as to ensure that $\theta_\varepsilon(t) = \theta(t)$ for $t \le -\varepsilon_1$ and $t \ge \varepsilon_2$, where $\varepsilon_1,\ \varepsilon_2 \ge 0$, and $\varepsilon = \max(\varepsilon_1, \varepsilon_2)$. We set

$$\Theta_\varepsilon(t) = \int_0^t \theta_\varepsilon(t)\,dt,$$

$$j_\varepsilon(v) = \int_\Omega G\,\Theta_\varepsilon(v-\psi)\,dx + \int_\Gamma H\,\Theta_\varepsilon(v-\psi)\,dx \tag{16}$$

$$\langle \beta_\varepsilon(v), w\rangle = \int_\Omega G\,\theta_\varepsilon(v-\psi)w\,dx + \int_\Gamma H\,\theta_\varepsilon(v-\psi)w\,dx,$$

where $v,\, w \in H^1(\Omega)$. Note that Θ_ε is a small perturbation of Θ, since

$$|\Theta(t) - \Theta_\varepsilon(t)| \le |\theta(-\varepsilon_1)|\,(\varepsilon_1+\varepsilon_2), \quad t \in R. \tag{17}$$

The penalty problem is defined by

Problem (P_ε) Find $u_\varepsilon \in H^1_D(\Omega)$ such that

$$\langle Au_\varepsilon, v-u_\varepsilon\rangle + j_\varepsilon(v) - j_\varepsilon(u_\varepsilon) \ge 0 \quad \forall\, v \in H^1_D(\Omega). \tag{18}$$

This problem is equivalent to the equation

$$u_\varepsilon \in H^1_D(\Omega): \quad Au_\varepsilon + \beta_\varepsilon(u_\varepsilon - \psi) = 0,$$

since functional j_ε is differentiable and $j'_\varepsilon = \beta_\varepsilon$.

The following theorem is the second of our main results.

Theorem 2 *Let the assumptions* (H_1)–(H_4) *be satisfied, and* u *and* u_ε *are solutions to the problems* (P_0) *and* (P_ε), *respectively. Then*

$$\|u-u_\varepsilon\|^2_{H^1_0(\Omega)} \le C(\varepsilon_1)\,(\varepsilon_1+\varepsilon_2), \tag{19}$$

where

$$C(\varepsilon_1) = 2\,|\theta(-\varepsilon_1)|\Big(\int_\Omega G\,dx + \int_\Gamma H\,dx\Big).$$

Proof According to Theorem 1, u is also a solution to inequality (13). We take $v = u_\varepsilon$ in (13) and $v = u$ in (18), and add the two resulting inequalities to obtain

$$\langle Au - Au_\varepsilon, u - u_\varepsilon \rangle \le \big(j(u_\varepsilon) - j_\varepsilon(u_\varepsilon)\big) + \big(j_\varepsilon(u) - j(u)\big). \tag{20}$$

From the definitions (15), (16) of the functionals j, j_ε and the estimate (17) immediately follows

$$\big|j(v) - j_\varepsilon(v)\big| \le 0.5\, C(\varepsilon_1)\,(\varepsilon_1 + \varepsilon_2) \quad \forall\, v \in H^1(\Omega). \tag{21}$$

We use (21) to estimate the right-hand side of (20) from above and (9) to estimate the left-hand side from below. As a result, we get (19). □

4.1 Examples of Penalty Functions

The penalty function is defined by the functional parameters G, H, θ and θ_ε. We indicate the following two ways of choosing G and H:

$$(i)\quad G = \|g^+\|_{L^\infty(\Omega)}, \qquad H = \|h^+\|_{L^\infty(\Gamma)};$$
$$(ii)\quad G = g^+, \qquad H = h^+.$$

The method (i) is computationally simpler than (ii). The functions θ and θ_ε are more important. They can be chosen independently, or they can be consistent. Let's look at some examples.

Example 1 First, we select the θ function and then we regularize it to obtain θ_ε. Let θ be maximal among functions (14), i.e.,

$$\theta(t) = \begin{cases} -1, & t \le 0, \\ 0, & t > 0. \end{cases}$$

We define the function θ_ε by setting $\varepsilon_1 = 0$, $\varepsilon_2 = \varepsilon$:

$$\theta_\varepsilon(t) = \begin{cases} -1, & t \le 0, \\ t/\varepsilon - 1, & 0 \le t < \varepsilon, \\ 0, & t > \varepsilon. \end{cases}$$

In this case $\theta(-\varepsilon_1) = -1$, and estimate (19) takes the form

$$\|u - u_\varepsilon\|_{H_0^1(\Omega)} \le C\, \varepsilon^{1/2}. \tag{22}$$

Example 2 If θ_ε was originally selected and $\theta_\varepsilon(t) = 0$ if $t > 0$, say, in the classic way $\theta_\varepsilon(t) = -t^-/\varepsilon$, we can choose

$$\theta(t) = \begin{cases} \min\{\theta_\varepsilon(t), -1\}, & t \le 0, \\ 0, & t > 0, \end{cases} \quad \text{i.e.} \quad \theta(t) = \begin{cases} t/\varepsilon, & t \le -\varepsilon, \\ -1, & -\varepsilon \le t \le 0, \\ 0, & t > 0. \end{cases}$$

This function satisfies all conditions (14) and θ_ε is its regularization, $\varepsilon_1 = \varepsilon, \varepsilon_2 = 0$. In this case also $\theta(-\varepsilon_1) = -1$, and estimate (19) takes the form (22).

The same can be done with another choice of function θ_ε, known from publications.

Acknowledgment This paper has been supported by the Kazan Federal University Strategic Academic Leadership Program.

References

1. Brezis, H. and Stampacchia, G.: Sur la régularité de la solution d'inéquations elliptiques. Bull. Soc. Math. France. **96**, 153–180 (1968)
2. Lions, J.-L.: Quelques methodes de resolution des problemes aux limites non lineaires. Dunod, Gauthier-Villars, Paris (1969)
3. Bensoussan, A., Lions, J.-L.: Applications des inéquations variationnelles en contrôle stochastique, Dunod, Paris (1978)
4. Kinderlehrer, D., Stampacchia, G.: An Introduction to Variational Inequalities and their Applications, vol. 31 of Classics in Applied Mathematics. Society for Industrial and Applied Mathematics, Philadelphia (1980)
5. Nochetto, R.H.: Sharp L^∞-error estimates for semilinear elliptic problems with free boundaries. Numer. Math. **54**, 243–255 (1988)
6. Tran G., Schaeffer H., Feldman, W., Osher, S.: An L1 Penalty Method for General Obstacle Problems. SIAM Journal on Applied Mathematics. **75**, No. 4, 1424–1444 (2015)
7. S. Wang, X. Yang, A power penalty method for linear complementarity problems, Oper. Res. Lett. **36**, 211–214 (2008)
8. Dautov, R.Z.: On exact penalty operators for elliptic variational inequalities with an obstacle inside the domain. Differ. Equ. **31**, No. 6, 942–951 (1995)

Accurate Simulation of Guided Waves in Optical Fibers Using Finite Element Method Combined with Exact Non-reflecting Boundary Condition

Rafail Z. Dautov and Evgenii M. Karchevskii

Abstract We present an analysis of numerical results illustrating the potentials of a new method for calculating guided waves in optical fibers and dispersion curves of corresponding eigenvalues. The earlier proposed finite element method is based on a special exact non-reflecting boundary condition and mathematically justified. For linear Lagrangian elements, the analysis demonstrates that the speed of convergence of the presented algorithm is quadratic, which corresponds to previously obtained theoretical estimates.

1 Introduction

Problems of guided waves in optical fibers arise in mathematical and numerical modeling of light propagation in photonic circuits (see, for example, [1]). Mesh methods are used extensively to solve these important applied problems (see, for example, [2–6]). A finite element method for calculating guided waves in optical fibers and dispersion curves of corresponding eigenvalues was proposed in [7]. To this end, the vector electromagnetic problem for eigenwaves (particular solutions of the homogeneous Maxwel Equations of a special form), originally formulated on the plane, was reduced to a convenient for numerical solution linear parametric eigenvalue problem posed in a circle [7]. It was achieved using a specially tailored exact non-reflecting boundary condition.

A theoretical study of the accuracy of the approach proposed in [7] is done in [9]. The study of properties of the dispersion curves and the solvability of the obtained problem is based on the spectral theory of compact self-adjoint operators. Error estimates for approximating eigenvalues and eigenfunctions are derived. This

R. Z. Dautov (✉)
Department of Numerical Mathematics, Kazan Federal University, Kazan, Russia

E. M. Karchevskii
Department of Applied Mathematics, Kazan Federal University, Kazan, Russia

I. B. Badriev et al. (eds.), *Mesh Methods for Boundary-Value Problems and Applications*, Lecture Notes in Computational Science and Engineering 141,
https://doi.org/10.1007/978-3-030-87809-2_6

strong mathematical justification of the proposed numerical method is done on the base of general results of the abstract theory developed previously in article [8] to make the analysis of the continuous problem and the corresponding error analysis of the discrete problem. Earlier, we used a similar approach to solve the problem for surface eigenwaves of weakly guiding optical fibers [9].

The purpose of the current work is to investigate numerically the accuracy of the proposed in [7] and theoretically justified in [9] computational scheme. We present numerical results for waveguides of classical cross-sections (circular, square, and rectangular) as well as for waveguides of more complicated forms. Analysis of numerical results for linear Lagrangian elements demonstrates that the speed of convergence of the proposed algorithm is quadratic (with respect to the eigenvalues), which correspond to the theoretical estimates derived in [9] for elements of arbitrary order.

2 Theoretical Background

As usual (see, e.g., [10, 11]), we assume that the fiber is infinitely extended along its axis and is perfectly cylindrical. The refractive index n of the fiber is a real-valued function of only the transverse variable $x = (x_1, x_2) \in \mathrm{R}^2$. The core of the waveguide, i.e., the domain Ω_i, in the plane (x_1, x_2) is bounded, contains the origin, but is not necessarily simply connected. Within the cladding $\Omega_e = \mathrm{R}^2 \backslash \overline{\Omega}_i$, we have $n = n_\infty = \text{const} > 0$. We also suppose that

$$\inf_{x \in \mathrm{R}^2} n(x) \geq n_\infty, \quad n_+ = \sup_{x \in \mathrm{R}^2} n(x) > n_\infty. \tag{1}$$

A variational statement of the problem is given in [12] and looks as follows: find all values of $(\beta, k) \in \Lambda$ together with corresponding nonzero vectors $\mathrm{H} \in V^3(\mathrm{R}^2)$ such that for all $\mathrm{H}' \in V^3(\mathrm{R}^2)$ the next equality holds:

$$\int_{\mathrm{R}^2} \left(\frac{1}{n^2} \mathrm{rot}_\beta \mathrm{H} \cdot \overline{\mathrm{rot}_\beta \mathrm{H}'} + \frac{1}{n_\infty^2} \mathrm{div}_\beta \mathrm{H}\, \overline{\mathrm{div}_\beta \mathrm{H}'} \right) dx = k^2 \int_{\mathrm{R}^2} \mathrm{H} \cdot \overline{\mathrm{H}'}\, dx. \tag{2}$$

Here, β and k are propagation constants and wavenumbers of guided waves,

$$\Lambda = \{(\beta, k) : \beta / n_+ < k < \beta / n_\infty,\ \beta > 0\}. \tag{3}$$

The Sobolev space $V^3(\mathrm{R}^2)$ and the differential operators rot_β and div_β are introduced in [12].

Another formulation of the problem (2) based on an exact nonreflecting boundary condition derived from the analytical representation for solutions of equation (2) in the domain $\Omega_\infty = \mathrm{R}^2 \setminus \overline{\Omega}$,

$$\mathrm{H}_p(x) = \sum_{n=-\infty}^{\infty} \frac{K_n(pr)}{K_n(pR)} a_n(\mathrm{H})\, e^{in\varphi}, \quad a_n(\mathrm{H}) = \frac{1}{2\pi} \int_0^{2\pi} \mathrm{H}\big|_{r=R}\, e^{-in\varphi}\, d\varphi, \tag{4}$$

was proposed in paper [7]. Here, (r, φ) are the polar coordinates of x, $K_n(r)$ is the modified Bessel function of the second kind of order n. By Ω we denote the computational domain, namely, a circle such that $\Omega_i \subseteq \Omega$.

The original problem (2) was reduced in [7] to the equivalent linear eigenvalue problem of the form

$$\mathcal{A}(p)\mathcal{H} = \beta^2 \mathcal{B}(p)\mathcal{H} \tag{5}$$

in the circle Ω. Here, the eigenvector $\mathcal{H} = (\mathrm{H}_1, \mathrm{H}_2)$ represents the first two components of the magnetic field intensity vector H, the eigenvalue is β^2, the parameter p is the transverse wave number

$$p = (\beta^2 - k^2 n_\infty)^{1/2}, \tag{6}$$

the operators $\mathcal{A}(p)$ and $\mathcal{B}(p)$ are nonlocal and self-adjoint, $\mathcal{B}(p)$ is compact.

Let

$$K = \left\{(\beta, p) : \beta > 0,\ 0 < p < \sqrt{1 - (n_\infty/n_+)^2 \beta}\right\}. \tag{7}$$

If $(\beta, k) \in \Lambda$, then p is real and positive, and formula (6) defines the one-to-one correspondence between the sets Λ and K.

A Lagrangian finite element method for solution of equation (5) was proposed in [7]. Numerical calculation of operators of the discrete analog of problem (5) is quite economical [7]. Corresponding theoretical estimates were obtained in [13] for elements of arbitrary order m. It was proved in [13] that the speed of convergence of the proposed algorithm with respect to the eigenvalues and eigenvectors has order $2m$ and m, respectively. Particularly, for linear Lagrangian elements it is quadratic for eigenvalues.

3 Numerical Results

In this section, we illustrate the theoretical results [9] on the speed of convergence proposed in [7] finite element method by numerical experiments with linear Lagrangian elements.

3.1 Circular Waveguide

The exact solution of problem (2) obtained by the separation of variables method is well known for circular waveguides with constant refractive index (see for instance [10, 11]). Therefore, as a first numerical example, we took the circular waveguide with radius 1. The radius of the domain Ω was taken 1.5 and $n_+ = \sqrt{2}$, $n_\infty = 1$ (see an example of the triangulation of the computational domain Ω in Fig. 1).

The left panel of Fig. 2 shows the first seven dispersion curves $\beta = \beta(p)$ of the reduced problem (5) calculated using a mesh with $N_h = 2493$ nodes and with the number of Fourier harmonics $N = 10$. The solid lines is the exact solution

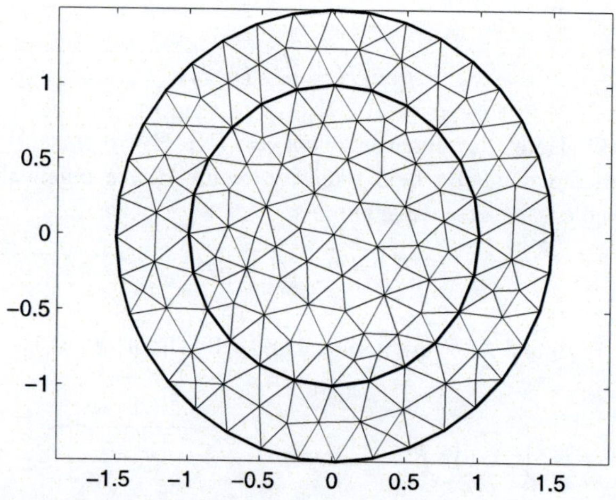

Fig. 1 Triangulation of the computational domain Ω for a circular waveguide. Here, $R = 1.5$ and $N_h = 146$

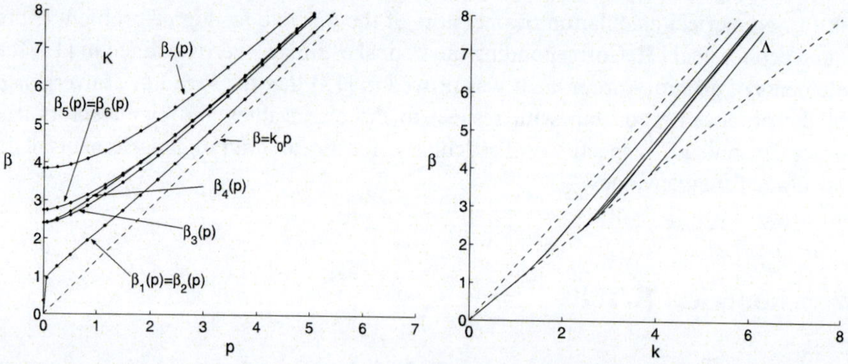

Fig. 2 The first seven dispersion curves $\beta = \beta(p)$ (left panel) and $\beta = \beta(k)$ (right panel) for a circular waveguide

Table 1 Circular waveguide: dependence of $e = h^{-2}|\beta_4 - \beta_4^h|/|\beta_4|$ on the parameters h and N for $p = 1$

$N \setminus N_h(n_\Gamma)$	45(16)	330(52)	1125(92)	2881(152)
1	0.640	0.748	0.631	0.668
3	0.641	0.748	0.631	0.668
5	0.641	0.748	0.631	0.668
7	0.641	0.748	0.631	0.668
15	0.642	0.748	0.631	0.668

and the dots is the computed solution. The dashed line is the boundary $\beta = k_0 p$ of the domain K. All the dispersion curves are above of this line. The right panel of this figure presents the first seven dispersion curves $\beta = \beta(k)$ of the original problem (2). The dashed lines are the boundaries $\beta = kn_+$ and $\beta = kn_\infty$ of the domain Λ.

Now we present the results of the analysis of the speed of convergence of the proposed algorithm with respect to number N_h of mesh points and the number N of Fourier harmonics. For given $p = 1$, we calculated approximate solutions β^h and compared them with exact solutions β. The numerical results are presented in Table 1 for β_4^h. Observing this table, we conclude that it is enough to take $N = 1$ or $N = 2$, then we have $|\beta_4 - \beta_4^h|/|\beta_4| \approx 0.7h^2$. Here, N_Γ is the number of nodes on the boundary Γ of the domain Ω.

3.2 *Square Waveguide*

The next example is a square dielectric waveguide. We choose it since results of physical experiments are known for this optical fiber (see [7] and references therein). The side of the square is 1, the radius of the circular computational domain Ω is 1.5, $n_+ = \sqrt{2.08}$, $n_\infty = 1$ (see an example of the triangulation of Ω in Fig. 3).

The left panel of Fig. 4 shows the first four dispersion curves $\beta = \beta(p)$ of the reduced problem (5) calculated using a mesh with $N_h = 2500$ nodes and with the number of Fourier harmonics $N = 10$. The bottom curve corresponds to the multiple eigenvalue $\beta_1(p) = \beta_2(p)$. The two other curves are intersecting. The experimental data are marked by dots and match well with numerical solutions. The right panel of this figure presents the corresponding solutions $\beta = \beta(k)$ of problem (2).

Now we present the results of the analysis of the speed of convergence of the proposed algorithm with respect to the parameters N_h and N for given $p = 1$. Any exact solution for square waveguide does not known. Therefore as an "exact solution" we use the approximate solution computed on the mesh with $N_h = 6000$ ($n_\Gamma = 212$). The numerical results are presented in Table 2 for β_3^h. Observing this table, we conclude that it is enough to take $N = 3$, then we have $|\beta_3 - \beta_3^h|/|\beta_3| \approx 1.6h^2$.

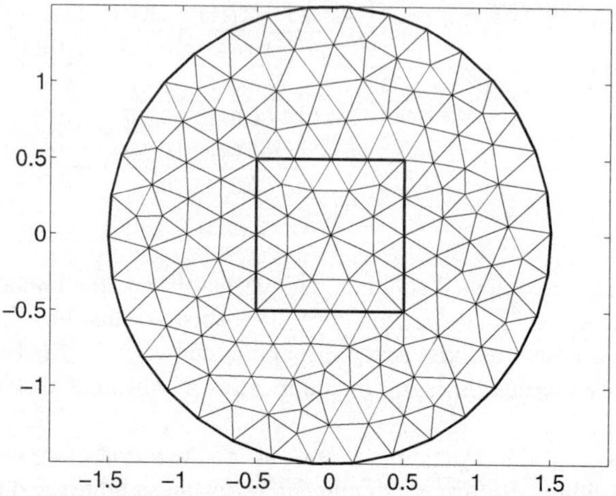

Fig. 3 Triangulation of the computational domain Ω for a square waveguide. Here, $R = 1.5$ and $N_h = 151$

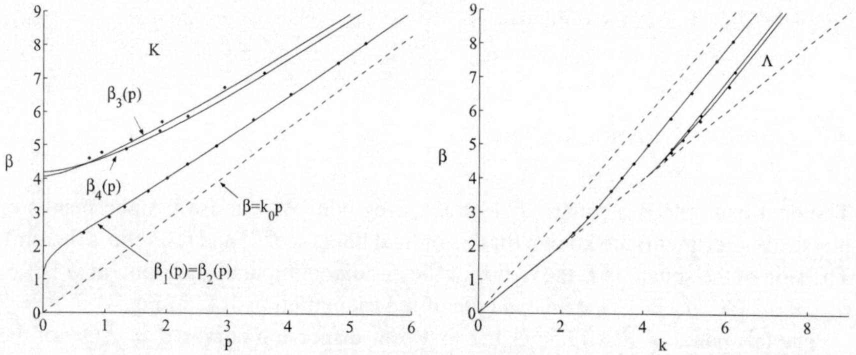

Fig. 4 The first four dispersion curves $\beta = \beta(p)$ (left panel) and $\beta = \beta(k)$ (right panel) for a square waveguide. Here, $\beta_1(k) = \beta_2(k)$

Table 2 Square waveguide: dependence of $e = h^{-2}|\beta_3 - \beta_3^h|/|\beta_3|$ on the parameters h and N for $p = 1$

$N \setminus N_h(n_\Gamma)$	31(16)	341(50)	1012(92)
1	2.26	1.60	1.61
3	2.27	1.61	1.64
5	2.27	1.61	1.64
7	2.27	1.61	1.64
15	2.27	1.61	1.64

3.3 Rectangular Waveguide

Another example with known results of physical experiments is a rectangular dielectric waveguide (see [7] and references therein). The sides of the rectangle are 1.5 and 1, the radius of the circle Ω is 1.5, $n_+ = \sqrt{2.08}$, $n_\infty = 1$ (see an example of the triangulation of the computational domain Ω in Fig. 5).

The left panel of Fig. 6 shows the first four dispersion curves $\beta = \beta(p)$ of the reduced problem (5) calculated using a mesh with $N_h = 2179$ nodes and with the number of Fourier harmonics $N = 10$. The experimental data are marked by dots and again match well with numerical solutions. The right panel of this figure presents the corresponding dispersion curves $\beta = \beta(k)$ for the original problem (2).

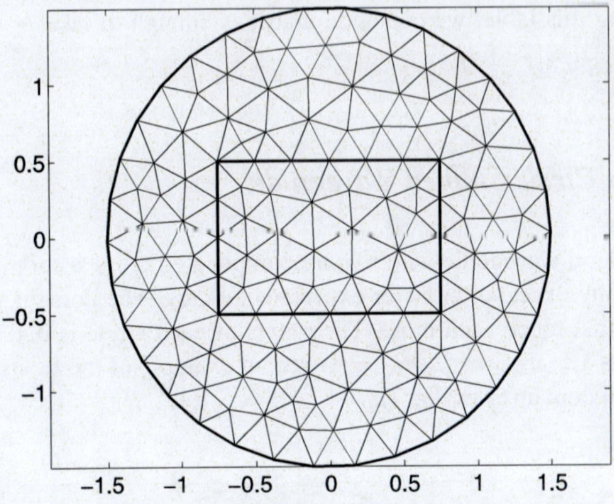

Fig. 5 Triangulation of the domain Ω for a rectangular waveguide. Here, $R = 1.5$, $N_h = 148$

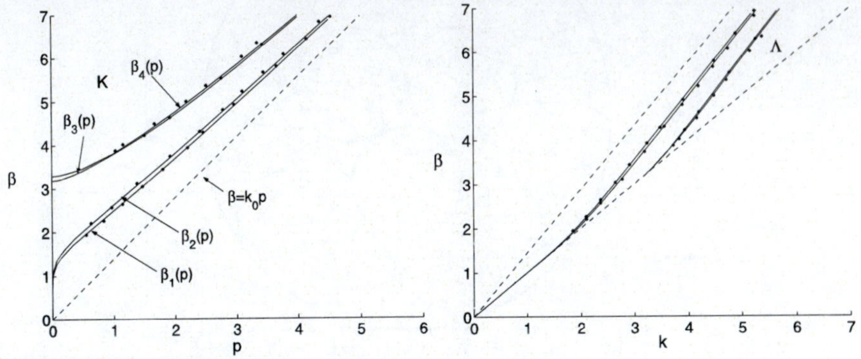

Fig. 6 The first four dispersion curves $\beta = \beta(p)$ (left panel) and $\beta = \beta(k)$ (right panel) for a rectangular waveguide

Table 3 Rectangular waveguide: dependence of $e = h^{-2}|\beta_3 - \beta_3^h|/|\beta_3|$ on the parameters h and N for $p = 1$

$N \setminus N_h(n_\Gamma)$	40(17)	304(50)	1016(92)
1	1.36	0.987	0.327
3	1.39	1.19	1.04
5	1.39	1.19	1.05
7	1.39	1.19	1.05
15	1.39	1.19	1.05

For this case, we also present the results of the convergence analysis of the proposed algorithm with respect to the parameters N_h and N for given $p = 1$. As for the previous example, any exact solution does not known. Therefore again, as an "exact solution" we use the approximate solution computed on the mesh with $N_h = 6015$ ($n_\Gamma = 212$). The numerical results are presented in Table 3 for β_3^h. Observing this table, we conclude that it is enough to take $N = 5$, then we have $|\beta_3 - \beta_3^h|/|\beta_3| \approx 1.1h^2$.

3.4 Three Circle Shaped Waveguide

Let us consider a waveguide with a more complicated cross section, for which we do not have any exact solutions or experimental data. The domain Ω_i consists of three circles that touch each other. The radius of each circle is 0.4. The radius of the circle Ω is 1.5, $n_+ = \sqrt{2}$, $n_\infty = 1$ (see an example of the triangulation of the computational domain Ω in Fig. 7).

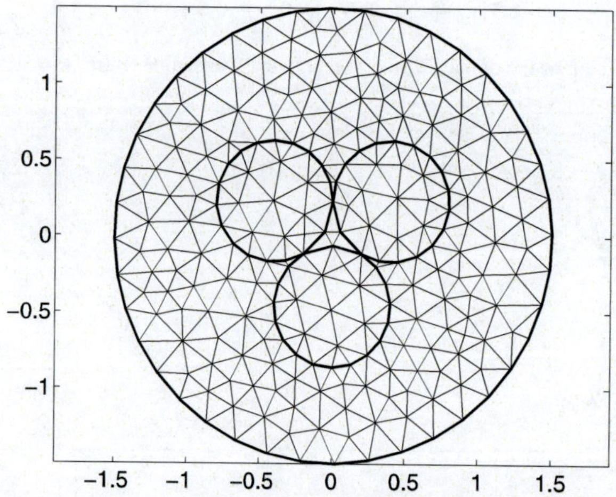

Fig. 7 Triangulation of the domain Ω for a three circle shaped waveguide. The radius of each circle is 0.4, $R = 1.5$, $N_h = 243$

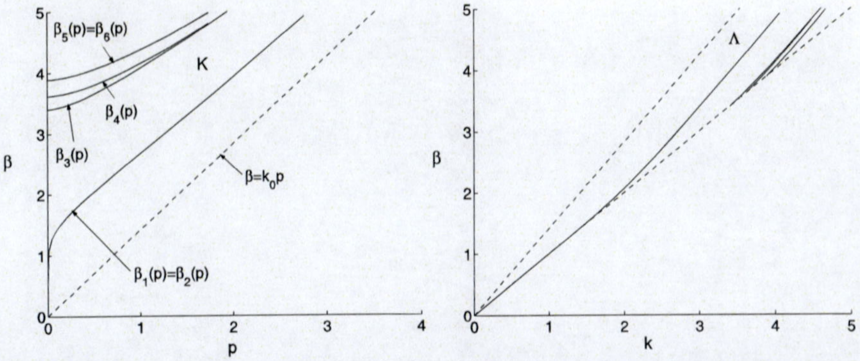

Fig. 8 The first six dispersion curves $\beta = \beta(p)$ (left panel) and $\beta = \beta(k)$ (right panel) for a three circle shaped waveguide

Table 4 Three circle shaped waveguide: dependence of $e = h^{-2}|\beta_4 - \beta_4^h|/|\beta_4|$ on the parameters h and N for $p = 1$

$N \setminus N_h(n_\Gamma)$	78(16)	335(50)	1093(90)
1	0.5	23.3	92.5
3	0.619	1.67	1.56
5	0.62	1.67	1.57
7	0.62	1.67	1.57
15	0.62	1.67	1.57

The left panel of Fig. 8 shows the first six dispersion curves $\beta = \beta(p)$ of the reduced problem (5) calculated using a mesh with $N_h = 2226$ nodes and with the number of Fourier harmonics $N = 10$. The first and the fifth curves shown in the figure correspond to the multiple eigenvalues $\beta_1(p) = \beta_2(p)$ and $\beta_5(p) = \beta_6(p)$, respectively. The right panel presents the corresponding solutions $\beta = \beta(k)$ of the original problem (2).

Now we again present the results of the convergence analysis of the proposed algorithm with respect to the parameters N_h and N for given $p = 1$. Clearly, any exact solution does not known, hence as an "exact solution" we use the approximate solution computed on the mesh with $N_h = 6006$ ($n_\Gamma = 216$). The numerical results are presented in Table 4 for β_4^h. Observing this table, we conclude that it is enough to take $N = 5$, then we have $|\beta_4 - \beta_4^h|/|\beta_4| \approx 1.6h^2$.

Figure 9, for given $p = 0.2$, presents isolines of absolute values of the eigenfunctions ($|H| = (H \cdot H)^{1/2}$) corresponding different eigenvalues β.

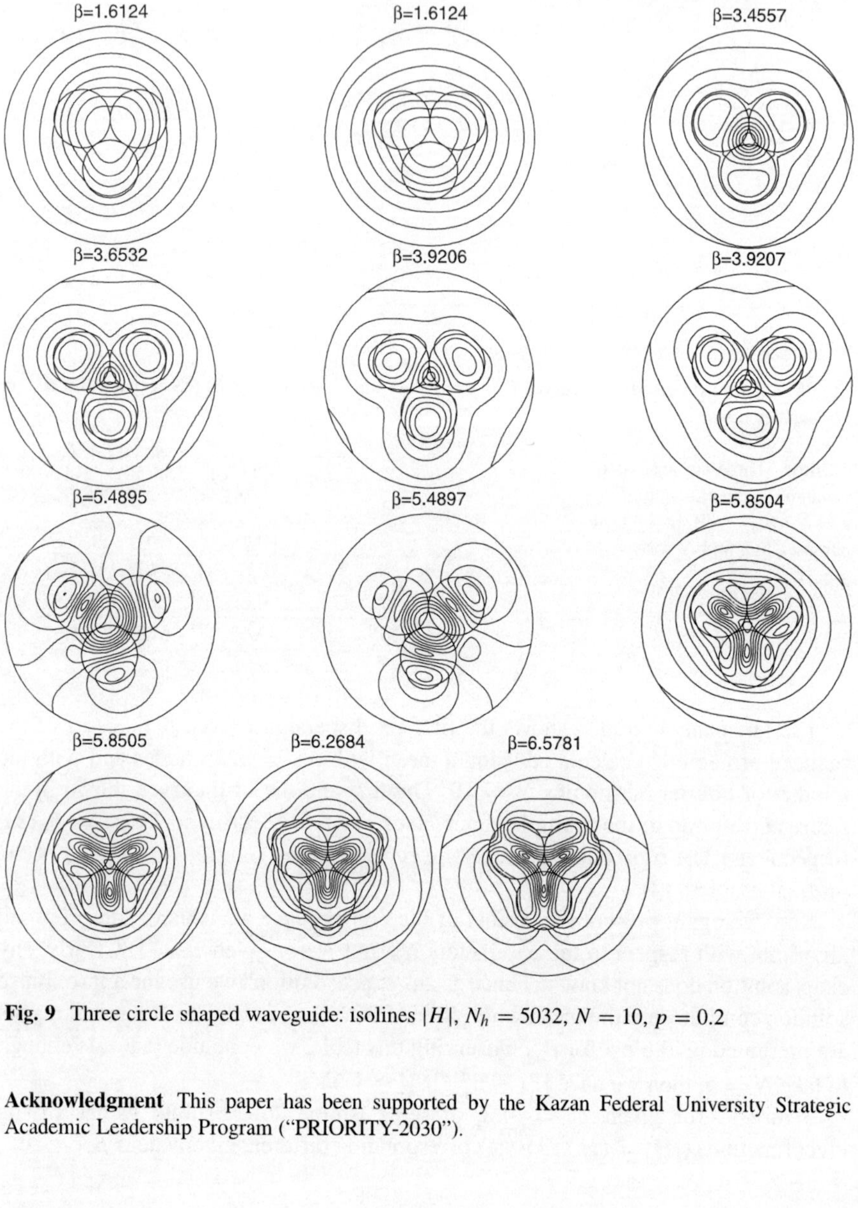

Fig. 9 Three circle shaped waveguide: isolines $|H|$, $N_h = 5032$, $N = 10$, $p = 0.2$

Acknowledgment This paper has been supported by the Kazan Federal University Strategic Academic Leadership Program ("PRIORITY-2030").

References

1. Obayya S.: Computational Photonics, John Wiley and Sons, UK (2011)
2. Hussein, R.A., Hameed, M.F.O., El-Azab, J., Abdelaziz, W.S., Obayya, S.S.A.: Analysis of ultra-high birefringent fully-anisotropic photonic crystal fiber. Opt. Quant. Electron. **47**, 2993–3007 (2015)

3. Pintus, P.: Accurate vectorial finite element mode solver for magneto-optic and anisotropic waveguides. Optics Express. **22**, 15737–15756 (2014)
4. Kowalczyk, P.: Analysis of microstructured optical fibers using compact macromodels. Optics Express. **19**, 19354–19364 (2011)
5. Monfared, Y.E., Javan, A.R.M., Kashani, A.R.M.: Confinement loss in hexagonal lattice photonic crystal fibers. Optik. **124**, 7049–7052 (2013)
6. Horikis, P.: Dielectric waveguides of arbitrary cross sectional shape. Appl. Math. Modelling. **37**, 5080–5091 (2013)
7. Dautov, R.Z., Karchevskii, E.M.: A numerical method for finding dispersion curves and guided waves of optical waveguides. Comput. Math. Math. Phys. **45**, 2119–2134 (2005)
8. Dautov, R.Z., Karchevskii, E.M.: Error estimates for a Galerkin method with perturbations for spectral problems of the theory of dielectric waveguides. Lobachevskii J. Math. **37**, 610–625 (2016)
9. Dautov, R.Z., Karchevskii, E.M.: Numerical modeling of optical fibers using the finite element method and an exact non-reflecting boundary condition. Comput. Methods Appl. Math. **18**, 581–602 (2018)
10. Marcuse, D.: Theory of Dielectric Optical Waveguide, Academic Press, New York (1974)
11. Snyder, A.W., Love, J.: Optical Waveguide Theory, Chapman and Hall, London (1983)
12. Bamberger, A., Bonnet, A.S.: Mathematical analysis of the guided modes of an optical fiber. SIAM J. Math. Anal. **21**, 1487–1510 (1990)
13. Dautov, R.Z., Karchevskii, E.M.: Accurate Full-Vectorial Finite Element Method Combined with Exact Non-Reflecting Boundary Condition for Computing Guided Waves in Optical Fibers. Comput. Methods Appl. Math. article number: 000010151520200162 (2021). https://doi.org/10.1515/cmam-2020-0162.

Simulation of Dynamic Response at Resonant Vibrations of a Plate with a Viscoelastic Damping Coating

Vyacheslav A. Firsov, Victor M. Shishkin, and Ruslan K. Gazizullin

Abstract A technique for determining the dynamic response at resonant vibrations of a rectangular plate with a soft viscoelastic damping coating is proposed. This technique is based on finite element approximations and linear physical equations of a viscoelastic solid. The material of the plate and damping layer is isotropic. It is believed that the plate with the damping layer is deformed according to the classical Kirchhoff–Love hypotheses. A special rectangular finite element with a low-modulus damping coating has been developed to model the inertial, stiff and damping properties of the marked plate. The system of resolving equations for plate vibrations in the resonance zone is obtained. Numerical experiments on approbation and estimation of reliability of the offered technique and the developed finite element are carried out.

1 Introduction

The value of acceptable vibration of any structure of a particular purpose is determined by its impact on the strength characteristics of the structure and its elements, on the performance, health, and well-being of people somehow associated with them, the operation of the equipment installed on it, etc. In terms of strength characteristics, one of the most dangerous modes of dynamic deformation of structures is a resonance, implemented in the structure when the frequencies of its natural vibrations coincide with the frequency of external cyclic impact. At such mode of loading, as it is known, amplitude values of dynamic stress-strain state parameters increase manifold. Their correct and reliable theoretical determination

V. A. Firsov · R. K. Gazizullin (✉)
Kazan National Research Technical University named after A. N. Tupolev – KAI, Kazan, Russia

V. M. Shishkin
Vyatka State University, Kirov, Russia

Kazan Federal University, Kazan, Russia

I. B. Badriev et al. (eds.), *Mesh Methods for Boundary-Value Problems and Applications*, Lecture Notes in Computational Science and Engineering 141,
https://doi.org/10.1007/978-3-030-87809-2_7

with the accuracy necessary for practical purposes requires proper consideration in the calculated ratios of damping properties of structural materials caused by internal friction. To date, extensive scientific literature has been devoted to the methods of their determination and construction to describe the corresponding mathematical models.

Modern thin-walled structures have a sufficiently dense spectrum of natural frequencies and can operate in a wide frequency range of disturbing forces. These factors make it difficult to use traditional methods of resonance tuning out and the use of various types of damping devices. This is especially true for aircraft structures and devices, where the use of such methods and devices is almost impossible. Hence, the ability of the structure to dampen dangerous resonance vibrations itself preventing the occurrence of significant displacements and overloads becomes critical. However, it should be noted that the majority of structural materials (metals, their alloys, and composites), along with their high strength and rigidity, have a very low damping capacity [1], and for many structures the main reason for energy dissipation at resonance is friction in the junctions of their individual elements (structural damping). It is important to note that the latter is a difficult prediction factor. Therefore, in order to increase damping parameters and reduce the dynamic intensity of thin-walled structures, their elements are often manufactured as two-layer structures consisting of a rigid hard layer and a relatively low rigid coating with high damping properties. Such elements are now widely used in aircraft and shipbuilding, automotive, civil and industrial buildings, in the design of devices to reduce their overloads in which various elastomers, mastics, and polymer compounds are used as damping coatings [2].

2 Rectangular Finite Element with Viscoelastic Damping Coating

The element consists of two layers (Fig. 1): a rigid isotropic layer 1 and a soft damping layer 2 (Fig 1a). The element is under the action of a surface dynamic load $q(x, y, t)$. Since the viscoelastic damping layer is soft, it can be assumed that the element is deformed within the classical Kirchhoff–Love hypotheses. The nodes of

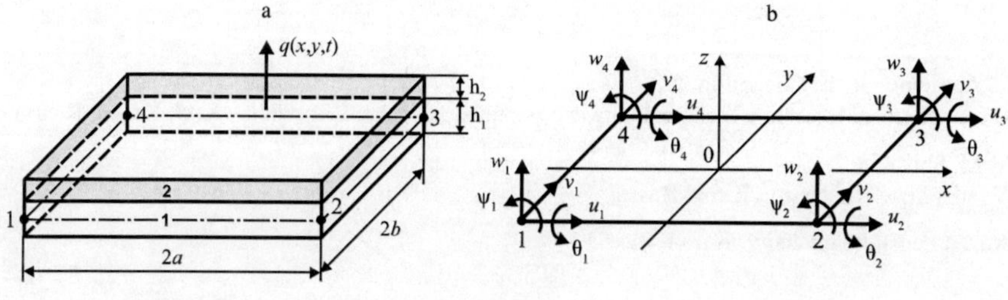

Fig. 1 Rectangular finite element of a two-layer plate

the element are located on the middle surface of layer 1. Each node $i\,(i = 1, 2, 3, 4)$ has five degrees of freedom (Fig. 1b): deflection w_i; displacements u_i, v_i in the plane $0xy$ and angles θ_i, ψ_i of rotation about the axes $0x$, $0y$, respectively. Let us introduce vectors

$$\mathbf{u} = \begin{Bmatrix} u_1 \\ u_2 \\ u_3 \\ u_4 \end{Bmatrix}, \quad \mathbf{v} = \begin{Bmatrix} v_1 \\ v_2 \\ v_3 \\ v_4 \end{Bmatrix}, \quad \mathbf{w} = \begin{Bmatrix} w_1 \\ w_2 \\ w_3 \\ w_4 \end{Bmatrix}, \quad \boldsymbol{\theta} = \begin{Bmatrix} \theta_1 \\ \theta_2 \\ \theta_3 \\ \theta_4 \end{Bmatrix}, \quad \boldsymbol{\psi} = \begin{Bmatrix} \psi_1 \\ \psi_2 \\ \psi_3 \\ \psi_4 \end{Bmatrix}.$$

The displacements u, v of an arbitrary point of the middle surface of an element are approximated by the expressions

$$u = \mathbf{S}\mathbf{u}, \quad v = \mathbf{S}\mathbf{v}, \tag{1}$$

where $\mathbf{S}$ is the row matrix of basis functions H_i $(i = 1, 2, 3, 4)$ depending on the dimensionless coordinates $\xi = x/a$ and $\eta = y/b$ of the element:

$$H_1 = (1-\xi)(1-\eta)/4\,; \quad H_2 = (1+\xi)(1-\eta)/4\,;$$
$$H_3 = (1+\xi)(1+\eta)/4\,; \quad H_4 = (1-\xi)(1+\eta)/4\,.$$

Dependencies (1) can be represented as a single matrix expression

$$\begin{Bmatrix} u \\ v \end{Bmatrix} = \mathbf{H}\mathbf{r}_\alpha, \tag{2}$$

where

$$\mathbf{H} = \left[\begin{array}{cc|cc|cc|cc} H_1 & 0 & H_2 & 0 & H_3 & 0 & H_4 & 0 \\ \hline 0 & H_1 & 0 & H_2 & 0 & H_3 & 0 & H_4 \end{array}\right]; \quad \mathbf{r}_\alpha = \{u_1\ v_1\ u_2\ v_2\ u_3\ v_3\ u_4\ v_4\}.$$

To reproduce the bending state of the plate, we define the deflection w in the form

$$w = \mathbf{f}^T\mathbf{c},$$
$$\mathbf{f} = \{1\ x\ y\ x^2\ y^2\ xy\ x^2y\ xy^2\ x^3\ y^3\ x^3y\ xy^3\}, \tag{3}$$
$$\mathbf{c} = \{c_0\ c_1\ c_2\ c_3\ c_4\ c_5\ c_6\ c_7\ c_8\ c_9\ c_{10}\ c_{11}\}.$$

The angles of rotation of the cross-sections of the plate with coordinates x and y in accordance with the accepted hypotheses are determined by the expressions

$$\theta = \frac{\partial w}{\partial y} = \frac{\partial\,\mathbf{f}^T}{\partial y}\,\mathbf{c}, \quad \psi = \frac{\partial w}{\partial x} = \frac{\partial\,\mathbf{f}^T}{\partial x}\,\mathbf{c}. \tag{4}$$

Element nodes have coordinates $x_1 = -a,\ y_1 = -b,\ x_2 = a,\ y_2 = -b,\ x_3 = a,\ y_3 = b,\ x_4 = -a,\ y_4 = b$. Substituting these coordinates into expressions (3) and (4), we obtain a system of 12 linear algebraic equations

$$\mathbf{Fc} = \mathbf{r}_\beta \tag{5}$$

with a matrix $\mathbf{F}$ depending on the coordinates of the element's nodes and the right-hand side $\mathbf{r}_\beta = \{w_1\ \psi_1\ \theta_1\ w_2\ \psi_2\ \theta_2\ w_3\ \psi_3\ \theta_3\ w_4\ \psi_4\ \theta_4\}$. After finding the vector $\mathbf{c}$ from system (5) and its substitution in approximation (2), we come to the expression

$$w = \mathbf{f}^T \mathbf{F}^{-1} \mathbf{r}_\beta,$$

From here, we can find the basis functions that determine the relationship between the deflection w and the components of the finite element vector $\mathbf{r}_\beta$:

$$N_j = \mathbf{f}^T \mathbf{F}_j \ \ (j = 1, 2, \ldots, 12).$$

Here $\mathbf{F}_j$ are the j-th columns of the matrix $\mathbf{F}^{-1}$ inverse to the matrix $\mathbf{F}$.

However, it should be noted that the procedure for analytical inversion of the matrix $\mathbf{F}$ using traditional (manual) technologies is practically unrealistic. The solution to the problem can be found in the application of symbolic calculation mode of the mathematical package MATLAB [3], which makes it possible to quickly find the functions N_j:

$$N_1 = (2 - 3\xi - 3\eta + 4\xi\eta + \xi^3 + \eta^3 - \xi^3\eta - \xi\eta^3)/8,$$
$$N_2 = a(1 - \xi - \eta - \xi^2 + \xi\eta + \xi^2\eta + \xi^3 - \xi^3\eta)/8,$$
$$N_3 = b(1 - \xi - \eta - \eta^2 + \xi\eta + \xi\eta^2 + \eta^3 - \xi\eta^3)/8,$$
$$N_4 = (2 + 3\xi - 3\eta - 4\xi\eta - \xi^3 + \eta^3 + \xi^3\eta + \xi\eta^3)/8,$$
$$N_5 = a(-1 - \xi + \eta + \xi^2 + \xi\eta - \xi^2\eta + \xi^3 - \xi^3\eta)/8,$$
$$N_6 = b(1 + \xi - \eta - \eta^2 - \xi\eta - \xi\eta^2 + \eta^3 + \xi\eta^3)/8,$$
$$N_7 = (2 + 3\xi + 3\eta + 4\xi\eta - \xi^3 - \eta^3 - \xi^3\eta - \xi\eta^3)/8,$$
$$N_8 = a(-1 - \xi - \eta + \xi^2 - \xi\eta + \xi^2\eta + \xi^3 + \xi^3\eta)/8,$$
$$N_9 = b(-1 - \xi - \eta + \eta^2 - \xi\eta + \xi\eta^2 + \eta^3 + \xi\eta^3)/8,$$
$$N_{10} = (2 - 3\xi + 3\eta - 4\xi\eta + \xi^3 - \eta^3 + \xi^3\eta + \xi\eta^3)/8,$$
$$N_{11} = a(1 - \xi + \eta - \xi^2 - \xi\eta - \xi^2\eta + \xi^3 + \xi^3\eta)/8,$$
$$N_{12} = b(-1 + \xi - \eta + \eta^2 + \xi\eta - \xi\eta^2 + \eta^3 - \xi\eta^3)/8$$

Having functions N_j, the deflection w can be represented as

$$w = \mathbf{N}\mathbf{r}_\beta = [\,N_1 | N_2 | \ldots | N_{12}\,]\,\mathbf{r}_\beta \tag{6}$$

According to the Kirchhoff–Love hypotheses, it can be assumed that each layer of the plate is in a plane stress state with normal stresses σ_x, σ_y, and shear stress τ_{xy}. These stresses correspond to strain ε_x, ε_y, and shear angle γ_{xy} determined by geometric relationships

$$\varepsilon_x = \frac{\partial}{\partial x}\left(u - z\frac{\partial w}{\partial x}\right) = \frac{\partial u}{\partial x} - z\frac{\partial^2 w}{\partial x^2}, \quad \varepsilon_y = \frac{\partial}{\partial y}\left(v - z\frac{\partial w}{\partial y}\right) = \frac{\partial v}{\partial y} - z\frac{\partial^2 w}{\partial y^2},$$

$$\gamma_{xy} = \frac{\partial}{\partial x}\left(v - z\frac{\partial w}{\partial y}\right) + \frac{\partial}{\partial y}\left(u - z\frac{\partial w}{\partial x}\right) = \frac{\partial v}{\partial x} + \frac{\partial u}{\partial y} - 2z\frac{\partial^2 w}{\partial x \partial y}.$$

The z coordinate for layer 1 varies within the $-h_1/2 \le z \le h_1/2$ limits, for layer 2 within the $h_1/2 \le z \le h_1/2 + h_2$ limits. It is convenient to write the presented dependencies in dimensionless coordinates ξ and η of the element:

$$\varepsilon_x = \frac{1}{a}\frac{\partial u}{\partial \xi} - z\frac{1}{a^2}\frac{\partial^2 w}{\partial \xi^2}; \quad \varepsilon_y = \frac{1}{b}\frac{\partial v}{\partial \eta} - z\frac{1}{b^2}\frac{\partial^2 w}{\partial \eta^2};$$
$$\gamma_{xy} = \frac{1}{a}\frac{\partial v}{\partial \xi} + \frac{1}{b}\frac{\partial u}{\partial \eta} - \frac{2z}{ab}\frac{\partial^2 w}{\partial \xi \partial \eta}. \tag{7}$$

Let us introduce the vector $\boldsymbol{\varepsilon} = \{\varepsilon_x\ \varepsilon_y\ \gamma_{xy}\}$ and differentiating operators

$$\mathbf{A}_\alpha^T = \left[\begin{array}{c|c|c} \frac{1}{a}\frac{\partial}{\partial \xi} & 0 & \frac{1}{b}\frac{\partial}{\partial \eta} \\ \hline 0 & \frac{1}{b}\frac{\partial}{\partial \eta} & \frac{1}{a}\frac{\partial}{\partial \xi} \end{array}\right], \quad \mathbf{A}_\beta^T = \left[\begin{array}{c|c|c} \frac{1}{a^2}\frac{\partial^2}{\partial \xi^2} & \frac{1}{b^2}\frac{\partial^2}{\partial \eta^2} & \frac{2}{ab}\frac{\partial^2}{\partial \xi \partial \eta} \end{array}\right].$$

Taking this into account, dependencies (7) can be represented as a single matrix expression

$$\boldsymbol{\varepsilon} = \mathbf{A}_\alpha \begin{Bmatrix} u \\ v \end{Bmatrix} - z\mathbf{A}_\beta w. \tag{8}$$

Substituting then representations (8) and (2) into (6), we obtain the connection between strains and nodal displacements of the finite element:

$$\boldsymbol{\varepsilon} = \mathbf{A}_\alpha \mathbf{H}\mathbf{r}_\alpha - z\mathbf{A}_\beta \mathbf{N}\mathbf{r}_\beta.$$

The resulting expression can be represented as

$$\boldsymbol{\varepsilon} = \mathbf{B}_\alpha \mathbf{r}_\alpha - z\mathbf{B}_\beta \mathbf{r}_\beta, \tag{9}$$

where

$$\mathbf{B}_\alpha = \mathbf{A}_\alpha \mathbf{H} = \left[\mathbf{B}_{\alpha,1} \middle| \mathbf{B}_{\alpha,2} \middle| \mathbf{B}_{\alpha,3} \middle| \mathbf{B}_{\alpha,4} \right], \quad \mathbf{B}_\beta = \mathbf{A}_\beta \mathbf{N} = \left[\mathbf{B}_{\beta,1} \middle| \mathbf{B}_{\beta,2} \middle| \ldots \middle| \mathbf{B}_{\beta,12} \right].$$

Blocks $\mathbf{B}_{\alpha,i}$ ($i = 1, 2, 3, 4$) and $\mathbf{B}_{\beta,j}$ ($j = 1, 2, \ldots, 12$) are defined by expressions

$$\mathbf{B}^T_{\alpha,i} = \left[\begin{array}{c|c|c} \frac{1}{a}\frac{\partial H_i}{\partial \xi} & 0 & \frac{1}{b}\frac{\partial H_i}{\partial \eta} \\ \hline 0 & \frac{1}{b}\frac{\partial H_i}{\partial \eta} & \frac{1}{a}\frac{\partial H_i}{\partial \xi} \end{array} \right], \quad \mathbf{B}^T_{\beta,j} = \left[\frac{1}{a^2}\frac{\partial^2 N_j}{\partial \xi^2} \middle| \frac{1}{b^2}\frac{\partial^2 N_j}{\partial \eta^2} \middle| \frac{2}{ab}\frac{\partial^2 N_j}{\partial xi \partial \eta} \right]. \tag{10}$$

The material of the rigid and damping layers of the plate is considered isotropic. To take into account the elastic and damping properties of the material, linear physical dependences can be used

$$\boldsymbol{\sigma}_k = \mathbf{D}_k \boldsymbol{\varepsilon} + \mathbf{D}_{g,k} \dot{\boldsymbol{\varepsilon}}, \tag{11}$$

representing a generalization of the well-known Kelvin–Voigt model [4, 5] for the case of a complex stress state of the material. Here $\boldsymbol{\sigma}_k = \{\sigma_x \; \sigma_y \; \tau_{xy}\}_k$ $(k = 1, \; 2)$ are stresses in k-th layer of the plate; D_k, $D_{g,k}$ are the stiffness matrix and the damping matrix of the material of this layer, respectively. For an isotropic viscoelastic material in a plane stressed state, the matrices D_k and $D_{g,k}$ will be as follows:

$$\mathbf{D}_k = \left[\begin{array}{c|c|c} E_k/(1-\nu_k^2) & E_k \nu_k/(1-\nu_k^2) & 0 \\ \hline E_k \nu_k/(1-\nu_k^2) & E_k/(1-\nu_k^2) & 0 \\ \hline 0 & 0 & G_k \end{array} \right];$$

$$\mathbf{D}_{g,k} = \frac{1}{\pi \omega} \left[\begin{array}{c|c|c} E_k \delta_{\varepsilon,k}/(1-\nu_k^2) & E_k \delta_{\varepsilon,k} \nu_k/(1-\nu_k^2) & 0 \\ \hline E_k \delta_{\varepsilon,k} \nu_k/(1-\nu_k^2) & E_k \delta_{\varepsilon,k}/(1-\nu_k^2) & 0 \\ \hline 0 & 0 & \delta_{\gamma,k} G_k \end{array} \right].$$

Here E_k, G_k, $\delta_{\varepsilon,k}$, $\delta_{\gamma,k}$ are elastic moduli and logarithmic decrements of vibrations of the layers' material, respectively, under tension-compression and shear; ν_k are Poisson's ratios; ω is the circular frequency of material deformation. Taking into account (9), dependences (11) are obtained as follows:

$$\boldsymbol{\sigma}_k = \mathbf{D}_k(\mathbf{B}_\alpha \mathbf{r}_\alpha - z\mathbf{B}_\beta \mathbf{r}_\beta) + \mathbf{D}_{g,k}(\mathbf{B}_\alpha \dot{\mathbf{r}}_\alpha - z\mathbf{B}_\beta \dot{\mathbf{r}}_\beta). \tag{12}$$

The first summand in (12) represents the elastic part of the stresses, which linearly depends on the nodal displacements $\mathbf{r}_\alpha$ and $\mathbf{r}_\beta$ of the element, the second summand is the inelastic part arising from the damping properties of the material and linearly depending on the nodal velocities $\dot{\mathbf{r}}_\alpha$ and $\dot{\mathbf{r}}_\beta$.

Let us write down the virtual work of the elastic part of the stresses in the rigid layer of the plate on the virtual strain $\delta\boldsymbol{\varepsilon}$ of this layer:

$$\delta A_1 = -\int_{h_1/2}^{h_1/2}\int_{-a}^{a}\int_{-b}^{b} \delta\boldsymbol{\varepsilon}^T \mathbf{D}_1(\mathbf{B}_\alpha \mathbf{r}_\alpha - z\mathbf{B}_\beta \mathbf{r}_\beta)\, dxdydz.$$

Substituting here relation (9), we obtain

$$\delta A_1 = -\int_{h_1/2}^{h_1/2}\int_{-a}^{a}\int_{-b}^{b} (\delta \mathbf{r}_\alpha^T \mathbf{B}_\alpha^T - z\,\delta \mathbf{r}_\beta^T \mathbf{B}_\beta^T)\mathbf{D}_1(\mathbf{B}_\alpha \mathbf{r}_\alpha - z\mathbf{B}_\beta \mathbf{r}_\beta)\, dxdydz.$$

After integration over the z coordinate, the last expression takes the form

$$\delta A_1 = -h_1 \delta \mathbf{r}_\alpha^T \int_{-a}^{a}\int_{-b}^{b} \mathbf{B}_\alpha^T \mathbf{D}_1 \mathbf{B}_\alpha dxdy\, \mathbf{r}_\alpha - \frac{h_1^3}{12}\delta \mathbf{r}_\beta^T \int_{-a}^{a}\int_{-b}^{b} B_\beta^T \mathbf{D}_1 \mathbf{B}_\beta dxdy\, \mathbf{r}_\beta. \quad (13)$$

Let's introduce a vector $\mathbf{r}^{(e)} = [\mathbf{r}_\alpha\ \mathbf{r}_\beta]$ containing all nodal displacements of the finite element. Taking into account this vector, expression (13) can be reduced to the form

$$\delta A_1 = -\delta\left(\mathbf{r}^{(e)}\right)^T \mathbf{K}_1 \mathbf{r}^{(e)}$$

where $\mathbf{K}_1$ is the block-diagonal matrix representing the contribution of layer 1 to the stiffness matrix of the finite element:

$$\mathbf{K}_1 = \left[\begin{array}{c|c} \mathbf{K}_{\alpha\alpha,1} & \mathbf{0} \\ \hline \mathbf{0} & \mathbf{K}_{\beta\beta,1} \end{array}\right].$$

Blocks $\mathbf{K}_{\alpha\alpha,1}$ and $\mathbf{K}_{\beta\beta,1}$ defined by expressions

$$\mathbf{K}_{\alpha\alpha,1} = h_1 \int_{-a}^{a}\int_{-b}^{b} \mathbf{B}_\alpha^T \mathbf{D}_1 \mathbf{B}_\alpha dxdy = h_1 ab \int_{-1}^{1}\int_{-1}^{1} \mathbf{B}_\alpha^T \mathbf{D}_1 \mathbf{B}_\alpha d\xi d\eta, \quad (14)$$

$$\mathbf{K}_{\beta\beta,1} = \frac{h_1^3}{12} \int_{-a}^{a}\int_{-b}^{b} \mathbf{B}_\beta^T \mathbf{D}_1 \mathbf{B}_\beta dxdy = \frac{h_1^3}{12} ab \int_{-1}^{1}\int_{-1}^{1} \mathbf{B}_\beta^T \mathbf{D}_1 \mathbf{B}_\beta d\xi d\eta. \quad (15)$$

From expressions (10) follows that the product $\mathbf{B}_\alpha^T \mathbf{D}_1 \mathbf{B}_\alpha$ quadratically depends on the dimensionless coordinates ξ and η of the element. In this case, the Gaussian formula [6] with two points in each coordinate direction can be used to calculate the integral in (14) accurately.

$$\int_{-1}^{1}\int_{-1}^{1} \mathbf{B}_\alpha^T \mathbf{D}_1 \mathbf{B}_\alpha d\xi d\eta = \sum_{m=1}^{2}\sum_{n=1}^{2} \mathbf{B}_\alpha^T(\xi_m, \eta_n)\mathbf{D}_1 \mathbf{B}_\alpha(\xi_m, \eta_n) Q_m P_n, \tag{16}$$

where $\xi_1 = \eta_1 = -0.57735$ and $\xi_2 = \eta_2 = 0.57735$ are coordinates of Gaussian points; $Q_1 = Q_2 = P_1 = P_2 = 1$ are weight factors. In the product $\mathbf{B}_\beta^T \mathbf{D}_1 \mathbf{B}_\beta$, the largest sum of the degrees of coordinates ξ and η, as again follows from (10), is equal to four. Therefore, to accurately calculate the integral in (15), it is necessary to take the Gaussian quadrature with three points along each of the element coordinates ξ and η:

$$\int_{-1}^{1}\int_{-1}^{1} \mathbf{B}_\beta^T \mathbf{D}_1 \mathbf{B}_\beta d\xi d\eta = \sum_{m=1}^{3}\sum_{n=1}^{3} \mathbf{B}_\beta^T(\xi_m, \eta_n)\mathbf{D}_1 \mathbf{B}_\beta(\xi_m, \eta_n) Q_m P_n, \tag{17}$$

$\xi_1 = \eta_1 = -0.77460$; $\xi_2 = \eta_2 = 0$; $\xi_3 = \eta_3 = 0.77460$; $Q_2 = P_2 = 0.88888$; $Q_1 = Q_3 = P_1 = P_3 = 0.55555$.

Similarly, the contribution to the stiffness matrix of the element of the second (damping) layer is obtained:

$$\mathbf{K}_2 = \left[\begin{array}{c|c} \mathbf{K}_{\alpha\alpha,2} & \mathbf{K}_{\alpha\beta,2} \\ \hline \mathbf{K}_{\alpha\beta,2}^T & \mathbf{K}_{\beta\beta,2} \end{array}\right];$$

$$\mathbf{K}_{\alpha\alpha,2} = h_2 ab \int_{-1}^{1}\int_{-1}^{1} \mathbf{B}_\alpha^T \mathbf{D}_2 \mathbf{B}_\alpha d\xi d\eta; \tag{18}$$

$$\mathbf{K}_{\alpha\beta,2} = -\frac{1}{2} h_2^2 \left(\frac{h_1}{h_2} + 1\right) ab \int_{-1}^{1}\int_{-1}^{1} \mathbf{B}_\alpha^T \mathbf{D}_2 \mathbf{B}_\beta d\xi d\eta; \tag{19}$$

$$\mathbf{K}_{\beta\beta,2} = \frac{h_2^3}{12}\left(3\frac{h_1^2}{h_2^2} + 6\frac{h_1}{h_2} + 4\right) ab \int_{-1}^{1}\int_{-1}^{1} \mathbf{B}_\beta^T \mathbf{D}_2 \mathbf{B}_\beta d\xi d\eta. \tag{20}$$

The integrals in expressions (18) and (19) are found at the same Gaussian points and weight factors as in formula (16). The integral in (20) is calculated by the formula (17) with the replacement of the matrix $\mathbf{D}_1$ in it by the matrix $\mathbf{D}_2$.

The complete stiffness matrix of a finite element is obtained by summing the contributions of the rigid and damping layers:

$$\mathbf{K}^{(e)} = \left[\begin{array}{c|c} \mathbf{K}_{\alpha\alpha,1} & \mathbf{0} \\ \hline \mathbf{0} & \mathbf{K}_{\beta\beta,1} \end{array}\right] + \left[\begin{array}{c|c} \mathbf{K}_{\alpha\alpha,2} & \mathbf{K}_{\alpha\beta,2} \\ \hline \mathbf{K}^{T}_{\alpha\beta,2} & \mathbf{K}_{\beta\beta,2} \end{array}\right].$$

In a similar form one can record a finite element damping matrix:

$$\mathbf{C}^{(e)} = \left[\begin{array}{c|c} \mathbf{C}_{\alpha\alpha,1} & \mathbf{0} \\ \hline \mathbf{0} & \mathbf{C}_{\beta\beta,1} \end{array}\right] + \left[\begin{array}{c|c} \mathbf{C}_{\alpha\alpha,2} & \mathbf{C}_{\alpha\beta,2} \\ \hline \mathbf{C}^{T}_{\alpha\beta,2} & \mathbf{C}_{\beta\beta,2} \end{array}\right].$$

The matrix $\mathbf{C}^{(e)}$ blocks are found according to the same formulas as the corresponding matrix $\mathbf{K}^{(e)}$ blocks with the replacement of stiffness matrices $\mathbf{D}_k$ $(k = 1, 2)$ in them by damping matrices $\mathbf{D}_{g,k}$ of the material of layers 1, 2 of the element.

Let us proceed to the construction of the finite element mass matrix $\mathbf{M}^{(e)}$. When building this matrix, we can consider that the volume forces of inertia are caused only by accelerations $\ddot{w}$ in the direction of the $0z$ axis of the element. Let us write down the virtual work of these forces on virtual displacements δw of the element:

$$\delta A = -h\rho \int_{-a}^{a}\int_{-b}^{b} \delta w\, \ddot{w}\, dx dy. \tag{21}$$

Here $h = h_1 + h_2$ and $\rho = (\rho_1 h_1 + \rho_2 h_2)/h$ are the thickness and average density of the element, respectively. To represent the displacements w when determining the inertial forces, we can take a bilinear approximation similar to representations (1): $w = \mathbf{S}\mathbf{w}$. Substituting this approximation into (21), we obtain

$$\delta A = -h\rho\, \delta \mathbf{w}^T \int_{-a}^{a}\int_{-b}^{b} \mathbf{S}^T \mathbf{S}\, dx dy\, \ddot{\mathbf{w}}.$$

The last expression can be represented as

$$\delta A = -\delta \mathbf{w}^T \mathbf{M}_w^{(e)} \ddot{\mathbf{w}},$$

where

$$\mathbf{M}_w^{(e)} = h\rho \int_{-a}^{a}\int_{-b}^{b} \mathbf{S}^T \mathbf{S}\, dx dy = h\rho\, ab \int_{-1}^{1}\int_{-1}^{1} \mathbf{S}^T \mathbf{S}\, d\xi d\eta\,.$$

The integral in the resulting expression is calculated using the Gauss formula with two points in each coordinate direction:

$$\int_{-1}^{1}\int_{-1}^{1} \mathbf{S}^T \mathbf{S}\, d\xi d\eta = \sum_{m=1}^{2}\sum_{n=1}^{2} \mathbf{S}^T(\xi_m,\ \eta_n)\mathbf{S}(\xi_m,\ \eta_n) Q_m P_n. \tag{22}$$

It should be noted that the resulting matrix $\mathbf{M}_w^{(e)}$ is constructed concerning the nodal displacements $w_i (i = 1, 2, 3, 4)$ of the finite element and has dimensions of 4×4, and the total mass matrix $\mathbf{M}^{(e)}$ of the element should have dimensions of 20×20 (in accordance with the number of nodal displacements in vector). To form the required matrix, one can use the procedure

$$\mathbf{M}^{(e)} = \mathbf{L}^T \mathbf{M}_w^{(e)} \mathbf{L},$$

where $\mathbf{L}$ is a control matrix of 4×20 size :

$$L = \begin{bmatrix} 0 & 0 & 0 & 0 & 0 & 0 & 0 & 0 & 1 & 0 & 0 & 0 & 0 & 0 & 0 & 0 & 0 & 0 & 0 & 0 \\ 0 & 0 & 0 & 0 & 0 & 0 & 0 & 0 & 0 & 0 & 0 & 1 & 0 & 0 & 0 & 0 & 0 & 0 & 0 & 0 \\ 0 & 0 & 0 & 0 & 0 & 0 & 0 & 0 & 0 & 0 & 0 & 0 & 0 & 0 & 1 & 0 & 0 & 0 & 0 & 0 \\ 0 & 0 & 0 & 0 & 0 & 0 & 0 & 0 & 0 & 0 & 0 & 0 & 0 & 0 & 0 & 0 & 0 & 1 & 0 & 0 \end{bmatrix}.$$

It remains to form the vector of external nodal forces (load vector) of the finite element, which can be obtained from the expression for the virtual work of the surface load $q(x, y, t)$:

$$\delta A = \int_{-a}^{a}\int_{-b}^{b} \delta w\, q(x, y, t)\, dxdy. \tag{23}$$

We represent the surface load and deflection w in the form

$$q(x, y, t) = \mathbf{Sq}(t), \quad w = \mathbf{Sw}.$$

Here $\mathbf{q}(t)$ is the vector of $q(x, y, t)$ values at the nodes of the finite element. Substituting these representations into expression (23), we obtain

$$\delta A = \delta\mathbf{w} \int_{-a}^{a}\int_{-b}^{b} \mathbf{S}^T \mathbf{S}\, dxdy\, \mathbf{q}(t).$$

Hence, the components of external nodal forces in the direction of displacements w_i of the element are obtained

$$\mathbf{P}_w^{(e)}(t) = \int_{-a}^{a}\int_{-b}^{b} \mathbf{S}^T\mathbf{S}\,dx\,dy\,\mathbf{q}(t) = ab\int_{-1}^{1}\int_{-1}^{1} \mathbf{S}^T\mathbf{S}\,d\xi\,d\eta\,\mathbf{q}(t).$$

The integral in the last expression is calculated by the formula (22). To form the load vector of the finite element relative to the nodal displacements $\mathbf{r}^{(e)}$, one can use the previous control matrix $\mathbf{L}$:

$$\mathbf{P}^{(e)}(t) = \mathbf{L}^T\mathbf{P}_w^{(e)}(t).$$

3 Formation of Solving Equations System

To obtain the motion equations of the finite element model of the plate, one can use the Lagrange–d'Alembert principle

$$-\delta\mathbf{r}^T\mathbf{M}\ddot{\mathbf{r}} - \delta\mathbf{r}^T\mathbf{C}\dot{\mathbf{r}} - \delta\mathbf{r}^T\mathbf{K}\mathbf{r} + \delta\mathbf{r}^T\mathbf{P}(t) = 0, \tag{24}$$

where $\mathbf{M},\ \mathbf{C},\ \mathbf{K},\ \mathbf{r},\ \mathbf{P}(t)$ are mass matrix, damping matrix, stiffness matrix, nodal displacement vector, and external nodal force vector of the noted model respectively. As vector $\delta\mathbf{r}$ components are independent and not equal to zero, then from (24) follows the system of equations of motion of the plate

$$\mathbf{M}\ddot{\mathbf{r}} + \mathbf{C}\dot{\mathbf{r}} + \mathbf{K}\mathbf{r} = \mathbf{P}(t). \tag{25}$$

Let us consider resonant vibrations under the action of a load $\mathbf{P}(t) = \mathbf{P}_0e^{ipt}$ with amplitude $\mathbf{P}_0$ and frequency $p = \omega_j$, where ω_j is one of the free vibrations frequencies of the plate. In this case, the vibrations of the plate occur in mode $\mathbf{F}_j$ corresponding to the frequency ω_j:

$$\mathbf{r} = s_j(t)\mathbf{F}_j, \tag{26}$$

where $s_j(t)$ is the generalized coordinate. Substituting (26) into the system (25) and then applying the procedure of the Bubnov–Galerkin method, we come to one equation for the coordinate $s_j(t)$:

$$m_j\ddot{s}_j + c_j\dot{s}_j + k_js_j = p_{0,j}e^{ipt} \tag{27}$$

with modal parameters

$$m_j = \mathbf{F}_j^T\mathbf{M}\mathbf{F}_j,\ \ c_j = \mathbf{F}_j^T\mathbf{C}\mathbf{F}_j,\ \ k_j = \mathbf{F}_j^T\mathbf{K}\mathbf{F}_j,\ \ p_{0,j} = \mathbf{F}_j^T\mathbf{P}_0.$$

The solution to Eq. (27) will be sought in the form

$$s_j(t) = s_{0,j}e^{i(pt-\phi_j)}, \quad (28)$$

where ϕ_j is the phase shift of the coordinate $s_j(t)$ relative to the load vector $\mathbf{P}_0 e^{ipt}$. Substituting (28) into Eq. (27) and then canceling the common factor e^{ipt}, we arrive at the system of resolving equations

$$\left[\begin{array}{c|c} k_j - p^2 m_j & pc_j \\ \hline pc_j & -k_j + p^2 m_j \end{array}\right]\begin{Bmatrix} s_{a,j} \\ s_{b,j} \end{Bmatrix} = \begin{Bmatrix} p_{0,j} \\ 0 \end{Bmatrix}, \quad (29)$$

where $s_{a,j} = s_{0,j}\cos\phi_j$, $s_{b,j} = s_{0,j}\sin\phi_j$. From system (29) components $s_{a,j}$ and $s_{b,j}$ are found. After that one can determine the amplitude $s_{0,j}$ and tgϕ_j: $s_{0,j} = \sqrt{s_{a,j}^2 + s_{b,j}^2}$; tg$\phi_j = s_{b,j}/s_{a,j}$.

4 Determination of Stress Amplitudes in Finite Elements Under Resonant Vibrations of a Plate

Since the modulus of elasticity of the rigid layer of the plate is much higher than the modulus of the damping layer, and the strains of the layers are of the same order, the strength of the plate is mainly determined by the rigid layer. The basis for determining the stress amplitudes in a given layer can be the previous dependence (12) (the index k, which means the layer number, is hereinafter omitted):

$$\boldsymbol{\sigma} = \{\sigma_x \ \sigma_y \ \tau_{xy}\} = \mathbf{D}(\mathbf{B}_\alpha \mathbf{r}_\alpha - z\mathbf{B}_\beta \mathbf{r}_\beta) + \mathbf{D}_g(\mathbf{B}_\alpha \dot{\mathbf{r}}_\alpha - z\mathbf{B}_\beta \dot{\mathbf{r}}_\beta).$$

Let us introduce the notation

$$\mathbf{D}_B = \mathbf{D}\left[\mathbf{B}_\alpha \middle| -z\,\mathbf{B}_\alpha\right], \ \mathbf{D}_{g,B} = \mathbf{D}_g\left[\mathbf{B}_\alpha \middle| -z\,\mathbf{B}_\alpha\right].$$

Taking into account these designations, the stresses $\boldsymbol{\sigma}$ can be calculated directly through the vector of nodal displacements $\mathbf{r}^{(e)} = \{\mathbf{r}_\alpha \ \mathbf{r}_\beta\}$ and the vector of nodal velocities $\dot{\mathbf{r}}^{(e)} = \{\dot{\mathbf{r}}_\alpha \ \dot{\mathbf{r}}_\beta\}$ of the finite element:

$$\boldsymbol{\sigma} = \mathbf{D}_B\,\mathbf{r}^{(e)} + \mathbf{D}_{g,B}\,\dot{\mathbf{r}}^{(e)}. \quad (30)$$

At the steady resonant vibrations of the plate, expression (30) can be represented in the complex form:

$$\mathbf{diag}[e^{i(pt-\gamma)}]\boldsymbol{\sigma}_0^{(e)} = (\mathbf{D}_B + ip\mathbf{D}_{g,B})\mathbf{diag}[e^{i(pt-\varphi)}]\mathbf{r}_0^{(e)}.$$

Here i is an imaginary unit; $\boldsymbol{\sigma}_0^{(e)}$ and $\mathbf{r}_0^{(e)}$ are vectors containing amplitudes of stresses and amplitudes of finite element nodal displacements, respectively; $\mathbf{diag}[e^{i(pt-\gamma)}]$ and $\mathbf{diag}[e^{i(pt-\varphi)}]$ are diagonal matrices with $e^{i(pt-\gamma_j)}$ and $e^{i(pt-\varphi_r)}$ elements; γ_j and φ_r are phase shifts of vector $\boldsymbol{\sigma}_0^{(e)}$ and $r_0^{(e)}$ components relative to the load vector of the finite element model of the plate. After simple transformations and reduction of the common factor e^{ipt}, we come to the expression

$$\boldsymbol{\sigma}_a + i\,\boldsymbol{\sigma}_b = (\mathbf{D}_B + i\,p\mathbf{D}_{g,B})(\mathbf{r}_a^{(e)} + i\mathbf{r}_b^{(e)}), \tag{31}$$

where the notation is introduced:

$$\boldsymbol{\sigma}_a = \mathbf{diag}[\cos\gamma]\boldsymbol{\sigma}_0^{(e)}; \qquad \boldsymbol{\sigma}_b = \mathbf{diag}[\sin\gamma]\,\boldsymbol{\sigma}_0^{(e)};$$

$$\mathbf{r}_a^{(e)} = \mathbf{diag}[\cos\varphi]\,\mathbf{r}_0^{(e)}; \qquad \mathbf{r}_b^{(e)} = \mathbf{diag}[\sin\varphi]\,\mathbf{r}_0^{(e)}.$$

The vectors $\boldsymbol{\sigma}_a$ and $\boldsymbol{\sigma}_b$ contain, respectively, in-phase and ortho-phase (shifted by $\pi/2$) with respect to the load components of the stress amplitudes. From expression (31) it follows:

$$\boldsymbol{\sigma}_a = \mathbf{D}_B\mathbf{r}_a^{(e)} - p\mathbf{D}_{g,B}\mathbf{r}_b^{(e)}; \quad \boldsymbol{\sigma}_b = \mathbf{D}_B\mathbf{r}_b^{(e)} + p\mathbf{D}_{g,B}\mathbf{r}_a^{(e)}.$$

This makes it possible to determine the stress amplitudes $\sigma_0^{(e)}$ and phase shifts γ_j using the formulas

$$\sigma_{0,j} = \sqrt{\sigma_{a,j}^2 + \sigma_{b,j}^2}, \quad \mathrm{tg}\,\gamma_j = \sigma_{b,j}/\sigma_{a,j}\,.$$

5 Numerical Experiments

A rectangular plate with dimensions of 960 × 580 mm, consisting of a rigid layer 1 and a low-module damping layer 2, hinged along all edges is considered. The material of the rigid layer is aluminium alloy D16AT, the material of the damping layer is mastic ADEM-NSh. The thickness of the plate layers: $h_1 = 1,8$ mm; $h_2 = 0,4$ mm. Characteristics of D16AT alloy: Young module $E = 7.2 \cdot 10^{10}$ N/m^2; Poisson's ratios $\nu = 0.3$; logarithmic decrement of vibrations $\delta = 0.0054$; density $\rho = 2700$ kg/m^3. ADEM-NSh mastic characteristics: $E = 5.4 \cdot 10^9$ N/m^2; $\nu = 0.28$; $\delta = 0.75$; $\rho = 1150$ kg/m^3. The resonant surface load $q(t) = q_0 \cos pt$ with amplitude $q_0 = 64.5$ N/m^2 and frequency $p = \omega_1 = 114.16$ s^{-1}, where ω_1 is the frequency of the main tone of the plate's free oscillations, found by inverse iterations [7, 8]. The finite element model of the plate consists of 60 identical elements (10 elements in the long side direction and 6 elements in the short side direction).

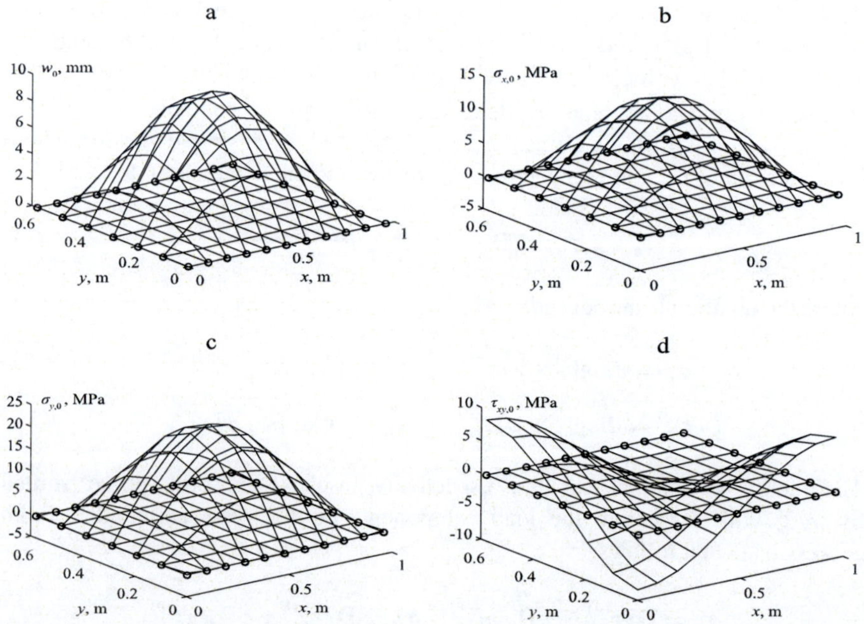

Fig. 2 Deflection amplitudes w_0 (**a**) and stress amplitudes $\sigma_{x,0}$ (**b**), $\sigma_{y,0}$ (**c**) and $\tau_{xy,0}$ (**d**) on the lower surface of the plate at resonance at the frequency $p = \omega_1$

Figure 2 shows the amplitudes of deflections w_0, the amplitudes of normal stresses $\sigma_{x,0}$ and $\sigma_{y,0}$, as well as the amplitudes of the shear stresses $\tau_{xy,0}$ on the lower surface of the plate at resonance. The results presented are in qualitative agreement with the concept of the operation of the plate under the given conditions of its loading and fixing.

The strongest criterion for evaluating the reliability of the obtained results can be the fulfillment of the condition of energy balance at resonance. This condition is the equality of scattered energy ΔW in the volume of the finite element model of the plate for one cycle of oscillations and of the full work A of external nodular forces $\mathbf{P}(t)$ during the same cycle: $\Delta W = A$. For determination of ΔW we can use the damping matrix $\mathbf{C}$ and the amplitude node displacements vector $\mathbf{r}_0$ [9]:

$$\Delta W = \pi p \mathbf{r}_0^T \mathbf{C} \mathbf{r}_0. \tag{32}$$

Full work A is the summation of external nodal forces $P_i(t)$ on one cycle of oscillations is:

$$A = \sum_k \int_0^{2\pi/p} P_k(t)\, \dot{r}_k(t)\, dt.$$

Here $P_k(t) = P_{0,k} \cos pt$, $r_k(t) = r_{0,k} \cos(pt - \varphi_k)$. After integrating the product $P_k(t)\,\dot{r}_k(t)$ we get

$$A = \pi \sum_k P_{0,k} \cdot r_{b,k} = \pi \mathbf{P}_0^T \mathbf{r}_b. \tag{33}$$

Calculations carried out according to formulas (32) and (33) for the considered two-layer plate confirm with high accuracy the fulfillment of the energy balance condition: $\Delta W = 0.44399\ \mathrm{N \cdot m}$; $A = 0.44399\ \mathrm{N \cdot m}$.

Acknowledgment This paper has been supported by the Russian Science Foundation (Project No. 19-19-00058) (Sect. 1,2) and by the Kazan Federal University Strategic Academic Leadership Program ("PRIORITY-2030") (Sect. 3–5).

References

1. Pisarenko, G.S., Yakovlev, A.P., Matveev, V.V.: Vibropogloshchayushchie svojstva konstrukcionnyh materialov : Spravochnik [Vibration-damping properties of structural materials : Handbook]. Naukova Dumka, Kiev (1971) [in Russian]
2. Chernyshev, V.M.: Dempfirovanie kolebanij mekhanicheskih sistem pokrytiyami iz polimernyh materialov [Damping of vibrations of mechanical systems with coatings made of polymer materials]. Nauka, Moscow (2004) [in Russian]
3. Mathews, J.H., Fink, K.D.: Numerical Methods Using MATLAB. Prentice Hall, Upper Saddle River (1999)
4. Postnikov, V.S.: Vnutrennee trenie v metallah [Internal friction in metals]. Metallurgy, Moscow (1969) [in Russian]
5. Khilchevsky, V.V., Dubenets, V.G.: Rasseyanie energii pri kolebaniyah tonkostennyh elementov konstrukcij [Energy dissipation during vibrations of thin-walled structural elements]. Vishcha school, Kiev (1977) [in Russian]
6. Zienkiewicz, O.C., Morgan, K.: Finite Elements and Approximation. John Wiley & Sons Inc., New York (1983)
7. Clough, R.W., Penzien, J.: Dynamics of Structures. McGraw Hill Inc., New York (1993)
8. Paimushin, V.N., Shishkin, V.M.: Modeling the elastic and damping properties of the multilayered torsion bar-blade structure of rotors of light helicopters of the new generation 2. Finite-element approximation of blades and a model of coupling of the torsion bar with the blades. Mech. Compos. Mater **51**(6), 771–788 (2015)
9. Paimushin, V.N., Firsov, V.A., Shishkin, V.M.: Modeling a dynamic response at resonant vibrations of an elongated plate with an integral damping coating. PNRPU Mech. Bull. **2020**(1), 74–86 (2020)

The Limit Theorem on the Trajectories Distribution

Farit G. Gabbasov, Aleksandr V. Gerasimov, Vyacheslav T. Dubrovin, R. M. Askhatov, and Maria S. Fadeeva

Abstract In mathematical modeling wide range of mathematical methods and concepts are used, in particular, ergodic theory which studies the statistical properties of motions in measure spaces (dynamic systems). The paper considers a dynamic system generated by measure space transformations. The central limit theorem with a convergence rate estimate for the trajectories distribution of a finite-dimensional torus is extended into transformations that are not ergodic.

1 Introduction

In mathematical modeling (for example, in studies of the deformation and motion of bodies, filtration, low-temperature plasma, etc. [1–8]) it becomes necessary to use dynamic systems generated by a wide range of measure spaces transformations. Analysis of the statistical properties of such systems is a topical problem not only for mathematicians but also for specialists from other industries. The results obtained in this research are an extension of dynamic system studies presented in [9–11]. They considered the dynamic systems generated by measure space transformations (automorphisms, endomorphisms). The history of the issue is as follows. In 1964 V.P. Leonov [12] proved the following central limit theorem.

Let Σ_k is a k-dimensional torus, $mes(.)$ is an invariant measure on it. Aside from the algebraic properties of Σ_k, $mes(.)$ can be identified with the Lebesgue measure defined on the hypercube $\Sigma_k = \{t : t = (t_1, \cdots, t_k), 0 \leq t_1 \leq 1, \cdots, 0 \leq t_k \leq 1\}$

F. G. Gabbasov
Kazan State University of Architecture and Engineering, Kazan, Russia
e-mail: gabbasov@kgasu.ru

A. V. Gerasimov
Kazan National Research Technological University, Kazan, Russia

V. T. Dubrovin (✉) · R. M. Askhatov · M. S. Fadeeva
Kazan (Volga Region) Federal University, Kazan, Russia
e-mail: Vyacheslav.Dubrovin@kpfu.ru

I. B. Badriev et al. (eds.), *Mesh Methods for Boundary-Value Problems and Applications*, Lecture Notes in Computational Science and Engineering 141,
https://doi.org/10.1007/978-3-030-87809-2_8

of the k-dimensional Euclidean space R^k. It is known that any measure-preserving algebraic torus endomorphism is defined using a nondegenerate integer matrix W by the convention $Tt = tW, t \in \Sigma_k$. Let W be an integer matrix. Among the roots of its characteristic polynomial, there is no one of unity. In this case the endomorphism T is ergodic. Suppose that the real-valued periodic function $f(t)$ with a period of 1 for each argument satisfies the following conditions $\int\limits_{\Omega_k} f(t)dt = 0, \int\limits_{\Omega_k} f^2(t)dt < \infty$ and there is given a restriction on its integral modulus of continuity: for some $A, \epsilon > 0$

$$\max_{1\le i\le k} \sup_{0\le h\le\delta} \int |f(t_1, \cdots, t_{i-1}, t_i + h, t_{i+1}, \cdots, t_k) - f(t_1, \cdots, t_k)|^2 dt \le A|\ln \delta|^{-2-\varepsilon}$$

Next, if $\sigma^2 = \lim\limits_{n\to\infty} \int\limits_{\Omega_k} \frac{1}{n}\left(\sum\limits_{k=1}^{n} f(tW^k)\right)^2 dt > 0,$

then $\lim\limits_{n\to\infty} mes\left\{t : t \in \Omega_k, \frac{1}{\sigma\sqrt{n}} \sum\limits_{m=1}^{n} f(tW^m) \le x\right\} = \frac{1}{\sqrt{2\pi}} \int\limits_{-\infty}^{x} e^{-u^2}du$. After this, it is reasonable that the question of studying the convergence rate in this limit relation arises. Articles of authors of this work were devoted to this. The results were obtained for the case of matrices W when all roots of their characteristic polynomials are greater than unity in modulus. Such matrices stretch Euclidean space in all directions. Exactly this matrix property makes it possible to obtain the property of the correlation exponential decay from simple geometric considerations and to apply the theory of summation of weakly dependent random variables to the studying problem. There were obtained the convergence rate estimates of the following orders $O(\ln n/n^{1/4})$, $O(1/n^{1/2-\varepsilon})$, $O(\ln n/n^{1/2})$. The estimates are given in chronological order. The results of these works can be extended to a certain class of other transformations. This work is devoted to this.

2 Statement and Proof of Results

Let define the transformation $t' = tW$of the Euclidean space R^k using a certain set of functions: $t'_1 = \phi_1(t_1, \ldots, t_k), \ \ldots, \ t'_k = \phi_k(t_1, \ldots, t_k)$. Let the following conditions are fulfilled:

1. $\{tW\} = \{\{t\}\ W\}$ where is the sign of fractional unit.
2. Functions $\phi_i, i = 1, \ldots, k$ and their partial derivatives are bounded in absolute value by some constant, and Jacobian of W is equal to a constant value J. Besides, $|J| > 1$.
3. There is a number $\delta > 0$, for which $\|tW - t'W\| \ge (1+\delta)\ \|t - t'\|, \ t, t' \in R^k$, where $\|\|$ is the designation of a vector length in R^k.

4. There is a number $\rho \in (0, 1)$, such that sup $\|dsW\| / \|ds\| \le (1-\rho)|J|$, where $ds = (dt_1, \ldots, dt_k)$ if the arc differential-vector of any smooth curve in R^k (by smooth curve it is meant a curve with a tangent at any point), and sup is taken along any such curve.
5. $|f(t) - f(t')| \le A\|t - t'\|^{\alpha}, t, t' \in \Omega_k$, where $\|t\| = t_1^2 + \ldots + t_k^2$.

Using the transformation W let define the transformation of torus T which can be identified with the unit hypercube in R^k according to the rule: $Tt = \{tW\}, t \in \Omega_k$.

Theorem 1 *Under the preceding foregoing rules, the following relation is the case:* $F_n(x) = mes\left\{t : t \in \Omega_k, \frac{1}{\sigma\sqrt{n}} \sum_{m=1}^{n} f(T^m t) \le x\right\} = \Phi(x) + O(n^{\varepsilon-0.5})$, *where* $\sigma^2 = \lim_{n\to\infty} \int_{-\infty}^{\infty} x^2 dF_n(x)$, $\Phi(x) = \frac{1}{\sqrt{2\pi}} \int_{-\infty}^{x} e^{-u^2} du$, *and* $f(t)$*is a real-valued periodic (with a period of 1 for each argument) Lebesgue square-integrable function that satisfies the Lipschitz condition 5 and relation* $\int_{\Omega_k} f(t)dt = 0$, *and* ε *is an arbitrarily small positive number.*

Proof Condition 1) is necessary for $T^d t = \{tW^d\}$, $d = 1, 2, \ldots$ Due to conditions 3 and 4 the following lemmas can be proved. □

Lemma 1 *Let* $g(t)$ *satisfies the Lipschitz condition 5, and* $h(t)$ *is a square-integrable on* R^k *function, such that* $B^2 = \int_{\Omega_k} h^2(t)dt < \infty$, $h(t + r) = h(t)$ *for any integer vector r. Next, if conditions 1–4 are fulfilled for the transformation W, then the following equality is correct:* $\int_{\Omega_k} g(t)h(T^q t)dt = \int_{\Omega_k} g(t)dt \int_{\Omega_k} h(t)dt + O(|J|^{-q\varepsilon} + \theta^{\alpha q})$ *where* $\varepsilon > 0$ *is a sufficiently small fixed number,* $\theta = 1/(1 + \delta)$, δ *is from condition 3), the constant in the symbol "O" depends only on the constant A in Lipschitz condition, on the root B of the integral, on the matrix W and the dimension k.*

Proof Let us write $\int_{\Omega_k} g(t)h(tW^q)dt = |J|^{-q} \int_{\Omega_k W^q} g(tW^{-q})h(t)dt$. The transformation W carries a unit cube into a certain area $\Omega_k W$. Here and elsewhere $R^{(k)}$ is a lattice of integer points from $\Omega_k W^q$. We represent the integration domain $\Omega_k W^q$ as $\Omega_k W^q = \Delta_0 \bigcup (\bigcup_r \Delta_r)$ where the backbone of Δ_0 is a union of the lattice $R^{(k)}$ parallelepipeds having at least one common point with the boundary of area $\Omega_k W^q$, and Δ_r runs through all remaining lattice parallelepipeds. According this

$$\int_{\Omega_k W^q} g(tW^{-q})h(t)dt = \int_{\Delta_0} g(tW^{-q})h(t)dt + \sum_r \int_{\Delta_r} g(tW^{-q})h(t)dt.$$

Further, because of periodicity $\int_{\Delta_r} g(tW^{-q})h(t)dt = \int_{\Omega_k} g((t + r)W^{-q})h(t)dt$, where $r = (r_1, \ldots, r_k)$ is a vector. It is necessary to shift Δ_r by this vector. Then, it holds the position of unit cube Ω_k at the origin. Due to condition 3) we have $\left\|t - t'\right\| \ge (1 + \delta)\left\|tW^{-1} - t'W\right\|$, $\left\|tW^{-1} - t'W\right\| \le \frac{1}{1+\delta}\left\|t - t'\right\|$ Repeating q times yields that $\left\|tW^{-q} - t'W^{-q}\right\| \leq \frac{1}{(1+\delta)^q}\left\|t - t'\right\|$ and $\left\|tW^{-q}\right\| \leq \frac{1}{(1+\delta)^q}\|t\|$ for all t from Ω_k. Therefore, $\sup_{t\in\Omega_k}\left\|tW^{-q}\right\| \leq \sup_{t\in\Omega_k}\|t\|\,\theta^{-q} = \sqrt{k}\theta^{-q}, \theta = 1/(1+\delta)$,

whence it follows that $\left|\int_{\Omega_k} g((t+r)W^{-q})h(t)dt - g(rW^{-q})\int_{\Omega_k} h(t)dt\right| = O(\theta^{-\alpha q})$ or $\left|\int_{\Delta_r} g((t)W^{-q})h(t)dt - g(rW^{-q})\int_{\Omega_k} h(t)dt\right| = O(\theta^{-\alpha q})$.

Further, since the volume of domain $\Omega_k W^q$ is equal to $|J|^q$, and the curve bounding this domain is smooth, then it follows that $P_q \leq |J|^q(1-\rho)^q$ from the condition 4) restricting the increase of perimeter P_q of the area $\Omega_k W^q$ in connecting with an increase of q. Hence, there is some sufficiently small positive number ε_0 such that $mes(\Delta_0) = O(P_q) = O(|J|^{q-q\varepsilon_0})$. Using it and Lipschitz condition 5) there is a result $\int_{\Delta_0} g(tW^{-q})h(t)dt = O(|J|^{q-q\varepsilon_0})$. After that, taking into account that the number of integer points in $R^{(k)}$ is $N_q = O(|J|^q - |J|^{q-q\varepsilon_0})$, let substitute the estimates that we obtained and get $\int_{\Omega_k W^q} g(tW^{-q})h(t)dt = \int_{\Omega_k} h(t)dt(\sum_r g(rW^{-q}) + O(|J|^{q-q\varepsilon_0} + |J|^q\theta^{\alpha q})$ By mean-value theorem there is a point $x_r \in \Delta_r W^{-q}$ that $g(x_r) = |J|^q \int_{\Delta_r W^{-q}} g(t)dt$. Since $g(x_r) = g(rW^{-q}) + O(\theta^\alpha)$, then $\int_{\Omega_k W^q} g(tW^{-q})h(t)dt =$ $|J|^q \int_{\Omega_k} h(t)dt \int_{\bigcup_r \Delta_r W^{-q}} g(t)dt + O(|J|^{q-q\varepsilon_0} + |J|^q\theta^{\alpha q})$. In the obtained expression let replace $\int_{\bigcup_r \Delta_r W^{-q}} g(t)dt$ to $\int_{\Omega_k} g(t)dt$. The replacement error will go into the remainder, since $\int_{\bigcup_r \Delta_r W^{-q}} g(t)dt = \int_{\Omega_k} g(t)dt + O(|J|^{-\varepsilon_0 q})$. Here we used the following estimate $mes(\Omega_k / \bigcup_r \Delta_r W^{-q}) = |J|^{-\varepsilon_0 q}$. The next formula is the result of replacement: $\int_{\Omega_k} g(t)h(T^q t)dt = |J|^q \int_{\Omega_k} g(t)dt \int_{\Omega_k} h(t)dt + O(|J|^{q-q\varepsilon_0} + |J|^q\theta^{\alpha q})$ which leads to the completion of the proof. □

Lemma 2 *There are numbers $k_1 \leq k_2 \leq \ldots \leq k_\nu$. In that case, if $k_{m+1} - k_m \geq \lambda$, then* $\left|\int_{\Omega_k}\prod_{i=1}^{\nu} f(tW^{k_i})dt - \int_{\Omega_k}\prod_{i=1}^{m} f(tW^{k_i})dt \int_{\Omega_k}\prod_{i=m+1}^{\nu} f(tW^{k_i})dt\right| \leq$ $(const)^\nu\theta_1^\lambda$ *where $0 < \theta_1 < 1$ and depends on A and α from condition 5, as well as on δ from condition 3) and the Jacobian $|J|$ of the transformation W.*

Proof Let us write $\int_{\Omega_k}\prod_{i=1}^{\nu} f(tW^{k_i})dt = |J|^{-k_m - g}\int_{\Omega_k W^{k_m+g}}\prod_{i=1}^{\nu} f(tW^{k_i - k_m - q})dt$.

Here $g = [2\lambda/3]$. As before, $R^{(k)}$ is a lattice of integer points from $\Omega_k W^{k_m+g}$, Δ_0 is a union of the lattice $R^{(k)}$ parallelepipeds having at least one common point with the boundary of area $\Omega_k W^{k_m+g}$, and Δ_r runs through all remaining lattice parallelepipeds. Then,

$$|J|^{-k_m-g} \int\limits_{\Omega_k W^{k_m+g}} \prod_{i=1}^{\nu} f(tW^{k_i-k_m-q})dt \tag{1}$$

$$= |J|^{-k_m-g} \int\limits_{\Delta_0} \prod_{i=1}^{\nu} f(tW^{k_i-k_m-q})dt + |J|^{-k_m-g} \sum_r \int\limits_{\Delta_r} \prod_{i=1}^{\nu} f(tW^{k_i-k_m-q})dt =$$

$$= |J|^{-k_m-g} \sum_r \int\limits_{\Delta_r} \prod_{i=1}^{m} f((t+r)W^{k_i-k_m-q}) \times \prod_{i=m+1}^{\nu} f(tW^{k_i-k_m-q})dt$$

$$+ O(|J|^{-\varepsilon_0(k_m+g)})$$

where ε_0 is a sufficiently small fixed positive number. Due to condition 3) the function $f_2(t) = \prod_{i=1}^{m} f((t+r)W^{k_i-k_m-q})$ satisfies the Lipschitz condition with constants $A \to (2A+1)^{\nu}, \alpha \to \alpha$. Actually, $f_2(t+h) = \prod_{i=1}^{m} f((t+r)W^{k_i-k_m-q} + hW^{k_i-k_m-q}) = \prod_{i=1}^{m} (f((t+r)W^{k_i-k_m-q}) + \rho_i A \| hW^{k_i-k_m-q} \|^{\alpha})$, (2) where $|\rho_i| \le 1$, $i = 1, \ldots, m$ However, if $|x_i| \le A$, $|\varepsilon_i| \le \varepsilon$, then $\left| \prod_{i=1}^{m} (x_i + \varepsilon_i) - \prod_{i=1}^{m} x_i \right| \le \varepsilon(1+2A)^m$, that in application to right-hand side of (2) gives $|f_2(t+h) - f_2(t)| \le 2(1+2A)^m$. Let apply Lemma 1 to the integral under the sum sign in (1), replacing in the estimate of the lemma A by $(2A+1)^{\nu}$ and setting $g(t) = \prod_{i=1}^{m} f((t+r)W^{k_i-k_m-q})$, $h(t) = \prod_{i=m+1}^{\nu} f(tW^{k_i-k_m-q})$, $q = k_{m+1} - k_m - g$. Since $k_{m+1} - k_m \ge [\lambda/3]$, then $q \ge [\lambda/3]$ by the definition of the number g, and by Lemma 1 $\int\limits_{\Omega_k} g(t)h(T^q t)dt = \int\limits_{\Omega_k} g(t)dt \int\limits_{\Omega_k} h(t)dt + O((1+2A)^{2\nu}(|J|^{-\lambda\varepsilon_0/3} + \theta^{\alpha\lambda/3})$.

Next, we substitute the result expression into (1):

$$\int\limits_{\Omega_k} \prod_{i=1}^{\nu} f(tW^{k_i})dt = \tag{2}$$

$$= |J|^{-k_m-g} \sum_r (\int\limits_{\Omega_{kr}} \prod_{i=1}^{m} f((t+r)W^{k_i-k_m-q})dt \times \int\limits_{\Omega_k} \prod_{i=m+1}^{\nu} f(tW^{k_i-k_m-q})dt +$$

$$+ O((1+2A)^{2\nu}(|J|^{-\lambda\varepsilon_0/3} + \theta^{\alpha\lambda/3}))) + O(|J|^{-\varepsilon_0(k_m+g)}) =$$

$$= |J|^{-k_m-g} \sum_r \int_{\Delta_{rr}} \prod_{i=1}^{m} f(tW^{k_i-k_m-q})dt \times \int_{\Omega_k} \prod_{i=m+1}^{\nu} f(tW^{k_i-k_m-q})dt +$$

$$+(1+|J|^{-\varepsilon_0}) \cdot O((1+2A)^{2\nu}(|J|^{-\lambda\varepsilon_0/3}+\theta^{\alpha\lambda/3})) + O(|J|^{-\varepsilon_0(k_m+g)})$$

Further, since

$$\sum_r \int_{\Delta_{rr}} \prod_{i=1}^{m} f(tW^{k_i-k_m-q})dt = \sum_r \frac{1}{mes\,\Delta_r W^{-(k_m+g)}} \int_{\Delta_r W^{-(k_m+g)}{}_r} \prod_{i=1}^{m} f(tW^{k_i})dt = \tag{3}$$

$$= |J|^{k_m+g} \int_{\bigcup_r \Delta_r W^{-(k_m+g)}{}_r} \prod_{i=1}^{m} f(tW^{k_i})dt$$

$$= |J|^{k_m+g} (\int_{\Omega_{kr}} \prod_{i=1}^{m} f(tW^{k_i})dt + O(|J|^{-\varepsilon_0(k_m+g)})),$$

then the next will follow from (3)

$$\int_{\Omega_k} \prod_{i=1}^{\nu} f(tW^{k_i})dt =$$

$$= \int_{\Omega_{kr}} \prod_{i=1}^{m} f(tW^{k_i})dt \times \int_{\Omega_k} \prod_{i=m+1}^{\nu} f(tW^{k_i})dt +$$

$$+(1+|J|^{-\varepsilon_0}) \cdot O((1+2A)^{2\nu}(|J|^{-\lambda\varepsilon_0/3}+\theta^{\alpha\lambda/3})) + O(|J|^{-\varepsilon_0(k_m+g)}) =$$

$$= \int_{\Omega_{kr}} \prod_{i=1}^{m} f(tW^{k_i})dt \times \int_{\Omega_k} \prod_{i=m+1}^{\nu} f(tW^{k_i})dt + O(c^\nu \theta^\lambda)$$

Lemma 2 is proved. □

Lemma 3 *The following relation is the case:*

$$\frac{1}{\sigma\sqrt{n}} \sum_{m=1}^{n} (f(tW^m) - \int_{\Omega_k} f(tW^m)dt) = \frac{1}{\sigma\sqrt{n}} \sum_{m=1}^{n} (f(tW^m) + O(1/\sqrt{n}).$$

Proof Since the transformation $T^d t = \{tW^d\}$, $d = 1, 2, \ldots$ does not remain the Lebesgue measure on Ω_k, then $\int_{\Omega_k} f(tW)dt \neq 0$, except that $\int_{\Omega_k} f(t)dt = 0$.

But it is easy to get around the difficulty that appears in this case. It can be done by considering functions $f(tW^m) - \int\limits_{\Omega_k} f(tW^m)dt, m = 0, 1, 2, \ldots.$ instead of functions $f(tW^m)$ in all calculations. Let us show that

$$\int\limits_{\Omega_k} f(tW^q)dt = O(e^{-aq}), \tag{4}$$

where $a > 0$ is a constant. Indeed, just as at the beginning of the proof of Lemma 1, we write $\int\limits_{\Omega_k} f(tW^q)dt = |J|^{-q} \int\limits_{\Omega_k W^q} f(t)dt = \left|J^{-q}\right| (\int\limits_{\Delta_0} f(t)dt + \sum\limits_r \int\limits_{\Delta_r} f(t)dt.$ Further, since $\int\limits_{\Delta_r} f(t)dt = \int\limits_{\Omega_k} f(t+r)dt = \int\limits_{\Omega_k} f(t)dt = 0, \int\limits_{\Delta_0} f(t)dt = O(|J|^{q-q\varepsilon_0}$, then the validity of (4) becomes obviously and the statement of Lemma 3 follows from this. Lemmas 1 and 2 show a weak dependence of terms in the sum $S_n = \sum_{j=1}^n f(T^j t)$. We denote the υth semi-invariant of this sum by $\chi_\nu(n)$, i.e. $\chi_\nu(n)=\frac{d^\nu}{dz^\nu} \ln \int_{\Omega_k} \exp(z \sum_{k=1}^n f(tW^k))dt\,|_{z=0}$. □

Lemma 4 *The estimate $\chi_\nu(n) = O(n)$. is the case for fixed ν, $2 \le \nu < \omega$, where ω is a sufficiently large positive number.*

Proof This lemma is proved using Lemmas 1 and 2 as in [13]. □

Lemma 5 *The estimate $\int_{\Omega_k} (\sum_{k=1}^m f(tW^k))^{2\nu} dt \le K^{2\nu}\nu!(M+1)^\nu(1 + \theta^K(M+1)^\nu)$, is the case for fixed ν, $1 \le \nu < \omega$, where $1 < \theta < 1, M = [m/K]$, $K-$ is any number from the interval (1, m/3). Here [a] is the designation of the integer part of the number a. The lemma is proved in the same way as in the article [1–9].*

Proof The proof of the theorem uses probabilistic terminology and tools. The theorem is proved using methods of summation of weakly dependent random variables based on the idea of S.N. Bernstein. He proposed to split the sums of weakly dependent random variables into long and short partial sums. The result of such kind separation is the fact that long sums become almost independent, and the contribution of short sums is small to the total distribution. Let us denote $\xi_j = f(T^j)$. The sum of the distribution of which is the problem of this studying is $S_n = \sum_{j=1}^n \xi_j$. Let Q and N be natural numbers increasing together with n. They fulfill the condition $|n - p(Q+N)| \le p$. The sum S_n is dividing as follows: $S_n = \sqrt{Q}\left(z_p + z_p^0\right) = \sqrt{Q}\sum_{j=1}^p y_j + \sqrt{Q}\sum_{j=1}^{p+1} y_j^0$, where $y_j = (1/\sqrt{Q})\sum_{r=1}^Q \xi_{(j-1)(Q+N)+r}$, $y_j^0 = (1/\sqrt{Q})\sum_{r=1}^N \xi_{jQ+(j-1)N+r}$, . $y_{p+1}^0 = \sum_{r=1}^n \xi_{p(Q+N)+r}$. Let, further, $\hat{z}_p = \sum_{j=1}^p \hat{y}_j$, where $\hat{y}_1, \ldots, \hat{y}_p$ are quantities satisfying the following properties: $mes\left\{t : t \in \Omega_k, \hat{y}_k < x\right\} = mes\{t : t \in \Omega_k, y_k < x\}$ $\int_{\Omega_k} \exp(i\hat{z}_p/\sqrt{p})dt = \prod_{j=1}^p \int_{\Omega_k} \exp(i\hat{y}_j/\sqrt{p})dt$. The second property shows that the quantities $\hat{y}_1, \ldots, \hat{y}_p$ are, as it were, independent random variables (in terminology of probability theory).

$$\sigma^2(Q) = \int_{\Omega_k} \left(\sum_{j=1}^{Q} \xi_j/\sqrt{Q}\right)^2 dt,$$

$$F_p(x) = mes\left(t : t \in \Omega_k, z_p \le x\sigma(Q)\sqrt{p}\right),$$

We denote $\hat{F}_p(x) = mes\left(t : t \in \Omega_k \hat{z}_p \le x\sigma(Q)\sqrt{p}\right),$

$$f_p(l) = \int_{\Omega_k} \exp(ilz_p/(\sigma(Q)\sqrt{p}))dt,$$

$$\hat{f}_p(l) = \int_{\Omega_k} \exp(il\hat{z}_p/(\sigma(Q)\sqrt{p}))dt.$$

Letters C_j, ω_j will denote positive constants which are independent of p, Q, N. Using Lemmas 1 and 2 it is possible to obtain the following

$$\left|f_p(l) - \hat{f}_p(l)\right| \le C_1\sqrt{p/Q}\exp(-C_2N) \quad (5)$$

We assume that $N = n^{1/\omega_1}\ \omega_1 = \omega^{0.75}$. Then, (5) will be written as follows:

$$\left|f_p(t) - \hat{f}_p(t)\right| \le C_2\sqrt{p/Q}n^{-\omega_1} \quad (6)$$

Next, we bring to use the function $G_{\nu p}(x) = \Phi(x) + \sum_{j=1}^{\nu} P_k(-\Phi)/p^{j/2}$, where

$$P_k(-\Phi) = \sum_{q=1}^{k} \frac{(-1)^{k+2q}}{q!} \sum_{\substack{k_1,\ldots,k_q \\ k_i \ge 3 \\ k_1+\ldots+k_q=k+2q}} \frac{\lambda_{k_1}\ldots\lambda_{k_q}}{k_1!\ldots k_q!}\Phi^{(k+2q)}(x)\ \Phi^{(r)}(x) =$$

$= (1/\sqrt{2\pi})(-1)^{r-1}H_{r-1}(x)e^{-x^2/2}$, $H(x)$ are Chebyshev-Hermite polynomials, $\lambda_r = \chi_r/\sigma^r(Q)$, χ_r is the rth semi-invariant of y_1. It is obvious that

$$\left|F_p(x) - G_{\nu p}(x)\right| \le \left|F_p(x) - \hat{F}_p(x)\right| + \left|\hat{F}_p(x) - G_{\nu p}(x)\right|. \quad (7)$$

Further,

$$\left|F_p(x) - \hat{F}_p(x)\right| \le 1/\sqrt{2}L(F_p, \hat{F}_p) + \max_x \left|\hat{F}_p(x) - \hat{F}_p(x + \Delta x)\right|, \quad (8)$$

where $L(F_p, \hat{F}_p)$-is the distance between distribution functions in the Levy metric; $|\Delta x| = L(F_p, \hat{F}_p)/\sqrt{2}$. It is known that

$L(F_p, \hat{F}_p) \le 1/\pi \int_0^U \left|f_p(t) - \hat{f}_p(t)\right| dt/t + 2e\ln U/U\ ,\ U > e.$

Choosing $U = n^{\omega_2}$ and applying (6), the following will be received

$$L(F_p, \hat{F}_p) \le C_4 p \ln n/n^{\omega_2} \quad (9)$$

The next follows from (3)–(5)

$$\left|F_p(x)-\hat{F}_p(x)\right| \le C_5\left(\sqrt{p/Q}n^{-\omega_3}+p\ln n/n^{\omega_2}\right) \tag{10}$$

Let us estimate $\left|\hat{F}_p(x)-G_{\nu p}(x)\right|$, using the inequality $\left|\hat{f}_p(l)-g_{\nu p}(l)\right| \le c(\nu)e^{-l^2/2}(|l|^{\nu+3}+|l|^{3(\nu+1)})/T_{\nu p}^{\nu+1}$, $|l| \le T_{\nu p}=\sqrt{p}\sigma^3(Q)/(8(\nu+3)r_{\nu+3}^{3/(\nu+3)}(Q)$, for the characteristic function $\hat{f}_p(l)$.

Here, $g_{\nu p}(l)=e^{-l^2/2}(1+\sum_{j=1}^{\nu}P_j(il)p^{-j/2})$ is the Fourier-Stieltjes transform of the function $G_{\nu p}(x) = \Phi(x)+\sum_{j=1}^{\nu}P_k(-\Phi)/p^{j/2}$. We use the Esseen inequality

$$\left|\hat{F}_p(x)-G_{\nu p}(x)\right| \le 24II/(\pi T)+1/\pi\int_{-T}^{T}\left|\hat{f}_p(l)-g_{\nu p}(l)\right|/l\,dl, \tag{11}$$

where $H>0$ is a constant. □

Next, it will be necessary to define the following.

Lemma 6 *If equality $F_n(x)=\Phi(x)+O(n^{-\alpha+\varepsilon}/(1+|x|^{\gamma}))$ holds for some α, $0<\alpha\le 0,5$, and some γ, $\gamma>1$, then there exist numbers $\delta,\beta,\tau>0$, $0<\varepsilon<\alpha/2$, such that $\max\limits_{\tau\le|l|\le\beta n^{\alpha-\varepsilon}}|f_n(l)|\le 1-\delta$ for sufficiently large n.*

Proof The lemma is proved using the proximity of the characteristic functions $f_n(l)$ and the normal distribution, as in [13] □

We choose $T=\varepsilon_0 Q^{\alpha-\varepsilon}T_{\nu p}$, $\varepsilon_0>0$ and estimate the integral $I=\int_{-T}^{T}\left|\hat{f}_p(l)-g_{\nu p}(l)\right|/l\,dl$. The estimate is carried out in the same way as in [9] using Lemma 6. Currently, $I=O(1/T_{\nu p}^{\nu+1}+1/p^{(\nu+1)/2}+1/T)$. From this estimate and from (7), (10), (11) the next will be obtain $\left|F_p(x)-G_{\nu p}(x)\right| = O(1/T_{\nu p}^{\nu+1}+1/(Q^{\alpha-\varepsilon}T_{\nu p})+\sqrt{p/Q}\ln T_{\nu p}n^{-\omega_3}+1/p^{(\nu+1)/2})$. Let us replace $F_p(x)$ with the distribution function $mes\left\{t:t\in\Omega_k,\sum_{i=1}^{n}f(T^it)/(\sigma(Q)\sqrt{pQ})<x\right\}$ and estimate the result error as in [9] using Markov inequality and estimate from Lemma 5.

Polynomials $P_k(-\Phi)$ which are part of $G_{\nu p}(x)$ will be estimated using Lemma 4. So, $G_{\nu p}(x) = \Phi(x)+O(\sum_{j=1}^{\nu}e^{-x^2/2}x^{3j}Q^{-j/2})$. Further, we choose $\nu=\omega^{1/3}$, $p=[n^{(1-2\alpha)/(2(1-\alpha))}]$ and Q from the condition $|n-(Q+N)|\le p$. This condition also implies $\sqrt{n/(pQ)}=1+O(N/Q)$. Considering all of this and the fact that $\sigma^2(Q)=\sigma^2+O(1/Q)$, it will be as in [9] and [13]:

$$F_n(x)=\Phi(x)+O(n^{\frac{1}{\omega_4}-\frac{1}{4(1-\alpha)}}/(1+|x|^{\gamma})),\ \omega_4=\sqrt[4]{\omega},\ \gamma=\omega^{1/12}. \tag{12}$$

Lemma 5 is not used for $\alpha = 0$. If $\alpha = 0$, then (12) transforms into $F_n(x) = \Phi(x) + O(n^{\frac{1}{\omega_4}-\frac{1}{4}}/(1+|x|^{\gamma}))$. Now let us take $\alpha = 1/\omega_4 - 1/4$ and get $F_n(x) = \Phi(x)+O(n^{\frac{3}{\omega_4}-\frac{1}{3}}/(1+|x|^{\gamma}))$. Continuing this process by successive approximations the statement of our theorem will be obtained (it will be for sufficiently large ω).

3 Conclusion

The obtained estimates for the convergence rate in the central limit theorem require the fulfillment of condition (1)–(4). Subsequently, adding conditions and improving the method of proof, it is possible to perfect this estimate. There is also the possibility of proving limit theorems with large deviation and multidimensional theorems.

Acknowledgment The work was supported by the Russian Science Foundation (project 16-11-10299).

References

1. Badriev I.B., Zadvornov O.A. : Analysis of the stationary filtration problem with a multivalued law in the presence of a point source. Differential Equations. 41(7), 915–922 (2005)
2. Badriev I.B., Fanyuk B.Y. : Iterative methods for solving seepage problems in multilayer beds in the presence of a point source. Lobachevskii Journal of Mathematics. 33(4),386–399 (2012)
3. Badriev I.B., Banderov V.V., Makarov M.V. : Mathematical simulation of the problem of the pre-critical sandwich plate bending in geometrically nonlinear one dimensional formulation. IOP Conference Series: Materials Science and Engineering. 208(1), art. No. 012002. (2017)
4. Badriev I.B., Makarov M.V., Paimuhin V.N. : Longitudinal and transverse bending by a cylindrical shape of the sandwich plate stiffened in the end sections by rigid bodies. IOP Conference Series: Materials Science and Engineering. 158(1),art. No. 012011. (2016)
5. Badriev I.B., Makarov M.V., Paimuhin V.N. : Geometrically nonlinear problem of longitudinal and transverse bending of a sandwich plate with transversally soft core. Lobachevskii Journal of Mathematics. 39(3),448–457. (2018)
6. Badriev I.B., Zheltukhin V.S., Chebakova V.Y. : On the solution of some nonlinear boundary value and initial boundary value problems.In: Materials of the XXII International Symposium "Dynamic and Technological Problems of Mechanics of Structures and Continuous Media" named after A.G. Gorshkov Moscow Aviation Institute (National Research University), pp. 31–33. LLC "TRP", Moscow (2016)
7. Chebakova V.Y.: Modeling of radio-frequency capacitive discharge under atmospheric pressure in argon. Lobachevskii Journal of Mathematics. **38**(6),1165–1178. (2017)
8. Badriev I.B., Chebakova V.Y., Zheltukhin V.S.: Capacitive coupled RF discharge: modelling at the local statement of the problem. Journal of Physics: Conference Series. 789(1),art. No. 012004. (2017)
9. Dubrovin V.T. : Central limit theorem for endomorphisms of Euclidean space. Uchenye zapiski Kazanskogo universiteta. Seria: Fiziko-matematicheskie nauki. 153(1),195–210 (2011)

10. Dubrovin V.T., Gabbasov F.G., Chebakova V.J. : Multidimensional central limit theorem for sums of functions of the trajectories of endomorphisms. Lobachevskii Journal of Mathematics. 37(4), 409–417 (2016)
11. Dubrovin V.T., Chebakova V.Y., Fadeeva M.S., Gabbasov F.G. Investigation of the statistical properties of a dynamic system generated by number-theoretic endomorphisms. Journal of Physics: Conference Series.1158(2), art. No. 220035, (2019)
12. Some applications of higher semi-invariants to the theory of random processes. Moscow: Nauka.64 p.(1964)
13. Moskvin D.A.: On the metric theory of automorphisms of a two-dimensional torus. Bulletin of the Academy of Sciences of the USSR. Mathematical series. 45(1), 60–100.(1981)

Design of the Best Linear Classifier for Box-Constrained Data Sets

Zulfiya R. Gabidullina

Abstract We construct a binary linear classifier for n-dimensional data sets with the special box-constrained structure. Data sets with this structure arise naturally in many real-world areas. We apply a linear separability criterion proposed in J. Optim. Theory Appl. (2012, https://doi.org/10.1007/s10957-012-0155-x). The Minkowski difference of the two data sets allows us to reduce a two-class classification problem to the problem in more easy to solve form. The greatest benefit of this reduction is that it allows to compute the parameters of a linear binary classifier by way of exact formulas. For this reason, a proposed framework has low computational costs. We rigorously prove that the developed linear classification model provides the possibility of constructing the data separator (or pseudo-separator) which really has the best estimate of its thickness. There are studied both regular and singular cases of separability arising in the theory and practice of linear classification of data sets.

1 Introduction

The data classification problem has numerous applications in a wide range of data mining tools. Our study is motivated namely by applications in many real-world areas. Indeed, the necessity of classifying the box-constrained data sets can naturally arise, for instance, in the problems of credit scoring and medical disease diagnosis. A useful systematic survey of the existing literature related to the data classification problem is contained, for example, in [1–6] (see also the references therein).

Z. R. Gabidullina (✉)
Kazan Federal University, Kazan, Russia
e-mail: zulfiya.gabidullina@kpfu.ru

I. B. Badriev et al. (eds.), *Mesh Methods for Boundary-Value Problems and Applications*, Lecture Notes in Computational Science and Engineering 141,
https://doi.org/10.1007/978-3-030-87809-2_9

For linear binary data classifying, the model of a supervised machine learning type can be described as follows:

1. For the two given training data sets (sets of objects with some characteristics—feature values), construct a linear binary classifier which makes a classification decision based on the training instances.
2. For a new unlabeled test instance, decide a class membership, i.e. identify class label.

Thus, the data classification process consists of the two main phases: (1) training phase, (2) testing phase.

The rest of this paper is organized as follows. In Sect. 2, we propose the auxiliary procedure for counting the vertices of the box-constrained set, the exact algebraic formula of the Minkowski difference of two sets given by box constraints. In Sect. 3, we propose the explicit formulas for computing the features weights corresponding to the best linear separation margin of sets. In Sect. 4, we construct a linear binary classifier and present a linear binary classification model. In addition, we rigorously justify the assignment criterion. In Sect. 5, there are drawn some conclusions.

2 The Minkowski Difference of Two Sets with Box-Constrained Structure

In this paper, we discuss the problem of the binary data classification in the case of the box-constrained sets. A box constraint on $x \in \mathbb{R}^n$ is usually given by the bilateral linear inequalities which restrict all variables to be in some intervals. Our approach consists in a reduction of the two sets separation problem to the problem of separating the origin of the Euclidean space from the Minkowski difference of these sets. The main benefit of this reduction is that it allows to calculate all the parameters of a linear binary classifier with the help of exact formulas. For this reason, the computational costs of a proposed framework are low.

We first note that in [7] there was proposed the exact formula of the Minkowski difference for the convex polyhedra given by the constraints of the general form. This section is devoted to the explicit algebraic expression of the Minkowski difference for the case when the both operands under this operation are determined by specific box constraints. By the way, let us notice that the obtained Minkowski difference formulation can be useful for the data analysis algorithm presented in [8].

Let be given the two following different sets:

$$L = \{z \in \mathbb{R}^n : a_i \leq z_i \leq b_i, \forall i \in I\},$$

$$M = \{p \in \mathbb{R}^n : e_i \leq p_i \leq d_i, \forall i \in I\}, \quad (1)$$

where $I = \{1, 2, \cdots, n\}$, $z = (z_1, z_2, \ldots, z_n)$, $p = (p_1, p_2, \ldots, p_n)$. Let the upper and lower bounds on the variables a_i & b_i, $\forall i \in I$ be any real numbers such that $a_i < b_i$. We make the analogous assumption about e_i & d_i, $\forall i \in I$: $e_i < d_i$. By construction, the sets L, $M \subset \mathbb{R}^n$ are convex and compact.

Let us describe an auxiliary procedure which will be often employed below.

Procedure (Count of the Vertices of the Box-Constrained Data Set)
Require: n-dimension of the space.
Require: $m = 2^n$-the number of vertices of

$$D = \{x \in \mathbb{R}^n : g_i \leq x_i \leq f_i,\ i \in I\}.$$

Require: lower bounds $g = (g_1, g_2, \ldots, g_n)$ for n variables.
Require: upper bounds $f = (f_1, f_2, \ldots, f_n)$ for n variables, $g_i < f_i$, $\forall i \in I$.

1. Declare the n-dimensional vector $y = (y(1), y(2), \ldots, y(n))$.
   Execute the loops presented below, setting that $\bar{I}1, \bar{I}2, \ldots, \bar{I}n$ are the control variables for the n loops, respectively. For these variables, there are initial the following values: $g_1, g_2, \ldots, g_n$. Likewise, $f_1, f_2, \ldots, f_n$ represent the final magnitudes of them. Due to syntax of the procedure pseudocode, to update the loop control variables, we use $f_1 - g_1, f_2 - g_2, \ldots, f_n - g_n$, respectively.

```
2.  DO Ī1 = g1, f1, f1 − g1
3.     y(1) = Ī1
4.     DO Ī2 = g2, f2, f2 − g2
5.        y(2) = Ī2
6.     ........................
7.          DO Īn = gn, fn, fn − gn
8.             y(n) = Īn
```

Having got all the coordinates of the vector y, we can print them.

```
9.             PRINT(y)
10.         END
11.    ........................
12.    END
13. END
```

As a result of executing the above described procedure, one obtains a complete list of all 2^n vertices of some n-dimensional box-constrained set D. The fact is demonstrated in the following elementary example.

Example 1 (Vertices of the Unit Cube in $\mathbb{R}^3$) Clearly, the unit cube, consisting of the vectors $x \in \mathbb{R}^3$ satisfying the constraint $-1 \leq x_i \leq 1$, $\forall i = \overline{1, 3}$, has the following eight vertices:

$$(-1, -1, -1), \quad (-1, -1, 1), \quad (-1, 1, -1), \quad (-1, 1, 1),$$
$$(1, -1, -1), \quad (1, -1, 1), \quad (1, 1, -1), \quad (1, 1, 1).$$

In general, the previous example illustrates how the above described procedure has established the order of combining the lower and upper bounds used in a linear bilateral inequality system describing some concrete box-constrained set.

For definiteness as well as simplicity of representing the further results, we utilize the already mentioned procedure for counting the vertices of sets L and M. In what follows, the first and second columns correspond to the full vertex collections of L and M, respectively. We introduce notations for the vertices of sets L and M: v_i^L and v_i^M, $i = 1, 2, \ldots, m$, respectively.

$$
\begin{array}{ll}
(a_1, a_2, \ldots, a_{n-2}, a_{n-1}, a_n), & (e_1, e_2, \ldots, e_{n-2}, e_{n-1}, e_n), \\
(a_1, a_2, \ldots, a_{n-2}, a_{n-1}, b_n), & (e_1, e_2, \ldots, e_{n-2}, e_{n-1}, d_n), \\
(a_1, a_2, \ldots, a_{n-2}, b_{n-1}, a_n), & (e_1, e_2, \ldots, e_{n-2}, d_{n-1}, e_n), \\
(a_1, a_2, \ldots, a_{n-2}, b_{n-1}, b_n), & (e_1, e_2, \ldots, e_{n-2}, d_{n-1}, d_n), \\
\ldots\ldots\ldots\ldots\ldots\ldots\ldots\ldots & \ldots\ldots\ldots\ldots\ldots\ldots\ldots\ldots \\
(a_1, b_2, \ldots, b_{n-2}, a_{n-1}, a_n), & (e_1, d_2, \ldots, d_{n-2}, e_{n-1}, e_n), \\
(a_1, b_2, \ldots, b_{n-2}, a_{n-1}, b_n), & (e_1, d_2, \ldots, d_{n-2}, e_{n-1}, d_n), \\
(a_1, b_2, \ldots, b_{n-2}, b_{n-1}, a_n), & (e_1, d_2, \ldots, d_{n-2}, d_{n-1}, e_n), \\
(a_1, b_2, \ldots, b_{n-2}, b_{n-1}, b_n), & (e_1, d_2, \ldots, d_{n-2}, d_{n-1}, d_n), \\
\ldots\ldots\ldots\ldots\ldots\ldots\ldots\ldots & \ldots\ldots\ldots\ldots\ldots\ldots\ldots\ldots \\
(b_1, a_2, \ldots, a_{n-2}, a_{n-1}, a_n), & (d_1, e_2, \ldots, e_{n-2}, e_{n-1}, e_n), \\
(b_1, a_2, \ldots, a_{n-2}, a_{n-1}, b_n), & (d_1, e_2, \ldots, e_{n-2}, e_{n-1}, d_n), \\
(b_1, a_2, \ldots, a_{n-2}, b_{n-1}, a_n), & (d_1, e_2, \ldots, e_{n-2}, d_{n-1}, e_n), \\
(b_1, a_2, \ldots, a_{n-2}, b_{n-1}, b_n), & (d_1, e_2, \ldots, e_{n-2}, d_{n-1}, d_n), \\
\ldots\ldots\ldots\ldots\ldots\ldots\ldots\ldots & \ldots\ldots\ldots\ldots\ldots\ldots\ldots\ldots \\
(b_1, b_2, \ldots, b_{n-2}, a_{n-1}, a_n), & (d_1, d_2, \ldots, d_{n-2}, e_{n-1}, e_n), \\
(b_1, b_2, \ldots, b_{n-2}, a_{n-1}, b_n), & (d_1, d_2, \ldots, d_{n-2}, e_{n-1}, d_n), \\
(b_1, b_2, \ldots, b_{n-2}, b_{n-1}, a_n), & (d_1, d_2, \ldots, d_{n-2}, d_{n-1}, e_n), \\
(b_1, b_2, \ldots, b_{n-2}, b_{n-1}, b_n), & (d_1, d_2, \ldots, d_{n-2}, d_{n-1}, d_n).
\end{array}
$$

The Minkowski difference of some two sets L and M is

$$L - M := \{a - b : \ a \in L, \ b \in M\}.$$

For $L,\ M \subset \mathbb{R}^n$, it is not hard to prove the following equality:

$$\min_{d \in L-M} \langle c, d\rangle = \min_{a \in L} \langle c, a\rangle - \max_{b \in M} \langle c, b\rangle \quad \forall c \in \mathbb{R}^n. \tag{2}$$

This equality justifies our approach which consists in a reduction of the two sets separation problem to the program of separating the origin of $\mathbb{R}^n$ from the Minkowski difference $L - M$.

Let us remind below the two important theorems which will be very useful for the formulation and justification of an explicit formula of the Minkowski difference of two box-constrained sets. We are not able go further without defining the following auxiliary set:

$$\Phi = \{x \in X : f_k(x) \le b_k,\ k \in K\},\ \ K = \{1, 2, \cdots, r\}, \tag{3}$$

where $f_k(x),\ k \in K$ are arbitrary real-scaled quasi-convex functions defined on a convex set $X \subseteq \mathbb{R}^n$. Here, $b_k,\ \forall k \in K$ are some scalars. A function $f(x)$ is said to be a quasi-convex on a convex set X if and only if $[S^{\bar{d}}, f]_X^{Lo}$ is a convex set for all $\bar{d} \in \mathbb{R}^1$, where

$$[S^{\bar{d}}, f]_X^{Lo} := \{x \in X : f(x) \le \bar{d}\}.$$

Here, the set Φ is convex as an intersection of the convex sets $[S^{b_i}, f_k]_X^{Lo},\ k \in K$.

Theorem 1 (Minkowski Difference When the Two Sets Under the Operation Are Given by a Constraint System and an Abstract Constraint, Respectively [9], p. 716) *Let be given an arbitrary nonempty set* $\Psi \subseteq \mathbb{R}^n$, *the set* $\Phi \ne \emptyset$ *be defined by (3),* $X = \mathbb{R}^n$, *then* $\Phi - \Psi = \Phi_1$, *where*

$$\Phi_1 = \{x \in \mathbb{R}^n : f_k(x + y) \le b_k,\ k \in K,\ y \in \Psi\},$$

$$\Phi - \Psi = \{q \in \mathbb{R}^n : q = x - y,\ x \in \Phi,\ y \in \Psi\}.$$

Theorem 2 (Minkowski Difference for Both Polyhedra Given as the Convex Hull of a Finite Collection of Some Given Vectors [10], p. 552) *If* $A = conv\{z_i\}_{i \in I_1},\ I_1 = \{1,\ 2,\ \ldots,\ \bar{l}\},\ B = conv\{p_j\}_{j \in J_1},\ J_1 = \{1,\ 2,\ \ldots,\ \bar{m}\}$, *then it holds* $A - B = conv\{z_i - p_j\}_{i \in I_1, j \in J_1}$.

Our objective now is to prove the explicit formula, when the both sets under consideration have the box-constrained structure.

Theorem 3 (Minkowski Difference When Two Sets Under the Operation Are Given by Box Constraints) *Let be given the arbitrary nonempty sets* L *and* M *defined by (1), then* $L - M = \Phi_2$, *where*

$$\Phi_2 = \{x \in \mathbb{R}^n : a - d \le x \le b - e\}, \tag{4}$$

$$L - M = \{x \in \mathbb{R}^n : x = z - p,\ z \in L,\ p \in M\}.$$

Proof

Part 1. We will first verify the inclusion $L - M \subseteq \Phi_2$. According to Theorem 1, it holds immediately

$$L - M = \{x \in \mathbb{R}^n : a \le x + y \le b,\ e \le y \le d\}.$$

Without loss of generality, we choose some arbitrary $\bar{x} \in L - M$. For any $y \in M$, we then obtain

$$\begin{cases} a - y \le \bar{x} \le b - y \\ -d \le -y \le -e. \end{cases}$$

Consequently, there is obviously fulfilled the following inequality:

$$a - d \le a - y \le \bar{x} \le b - y \le b - e,$$

i.e. $\bar{x} \in \Phi_2$. Through the arbitrary choice of $\bar{x} \in L - M$, we conclude that $L - M \subseteq \Phi_2$.

Part 2. Now, we need only check that the opposite inclusion is valid too. Taking into account the construction of the set Φ_2, applying the procedure of counting its vertices, we can state that Φ_2 has the following vertices:

$$(a_1 - d_1,\ a_2 - d_2,\ \ldots,\ a_{n-2} - d_{n-2},\ a_{n-1} - d_{n-1},\ a_n - d_n),$$
$$(a_1 - d_1,\ a_2 - d_2,\ \ldots,\ a_{n-2} - d_{n-2},\ a_{n-1} - d_{n-1},\ b_n - e_n),$$
$$(a_1 - d_1,\ a_2 - d_2,\ \ldots,\ a_{n-2} - d_{n-2},\ b_{n-1} - e_{n-1},\ a_n - d_n),$$
$$(a_1 - d_1,\ a_2 - d_2,\ \ldots,\ a_{n-2} - d_{n-2},\ b_{n-1} - e_{n-1},\ b_n - e_n),$$

....

$$(a_1 - d_1,\ b_2 - e_2,\ \ldots,\ b_{n-2} - e_{n-2},\ a_{n-1} - d_{n-1},\ a_n - d_n),$$
$$(a_1 - d_1,\ b_2 - e_2,\ \ldots,\ b_{n-2} - e_{n-2},\ a_{n-1} - d_{n-1},\ b_n - e_n),$$
$$(a_1 - d_1,\ b_2 - e_2,\ \ldots,\ b_{n-2} - e_{n-2},\ b_{n-1} - e_{n-1},\ a_n - d_n),$$
$$(a_1 - d_1,\ b_2 - e_2,\ \ldots,\ b_{n-2} - e_{n-2},\ b_{n-1} - e_{n-1},\ b_n - e_n),$$

....

$$(b_1 - e_1,\ a_2 - d_2,\ \ldots,\ a_{n-2} - d_{n-2},\ a_{n-1} - d_{n-1},\ a_n - d_n),$$
$$(b_1 - e_1,\ a_2 - d_2,\ \ldots,\ a_{n-2} - d_{n-2},\ a_{n-1} - d_{n-1},\ b_n - e_n),$$

$$(b_1 - e_1,\quad a_2 - d_2,\quad \ldots,\quad a_{n-2} - d_{n-2},\quad b_{n-1} - e_{n-1},\quad a_n - d_n),$$
$$(b_1 - e_1,\quad a_2 - d_2,\quad \ldots,\quad a_{n-2} - d_{n-2},\quad b_{n-1} - e_{n-1},\quad b_n - e_n),$$

....

$$(b_1 - e_1,\quad b_2 - e_2,\quad \ldots,\quad b_{n-2} - e_{n-2},\quad a_{n-1} - d_{n-1},\quad a_n - d_n),$$
$$(b_1 - e_1,\quad b_2 - e_2,\quad \ldots,\quad b_{n-2} - e_{n-2},\quad a_{n-1} - d_{n-1},\quad b_n - e_n),$$
$$(b_1 - e_1,\quad b_2 - e_2,\quad \ldots,\quad b_{n-2} - e_{n-2},\quad b_{n-1} - e_{n-1},\quad a_n - d_n),$$
$$(b_1 - e_1,\quad b_2 - e_2,\quad \ldots,\quad b_{n-2} - e_{n-2},\quad b_{n-1} - e_{n-1},\quad b_n - e_n).$$

Let us note that the quantity of the vertices of Φ_2 is $m = 2^n$. Due to Theorem 2, it holds

$$L - M = conv\{v_i^L - v_j^M\}_{i,\, j\in\{1,\, 2,\, \ldots,m\}}.$$

Thus, the set $L - M$ represents the convex hull of the m^2 vectors. Obviously, the vertices of the set Φ_2 belong to this sequence of the vectors forming $L - M$. By virtue of $L - M$ being convex, it contains any convex combination of the vertices of Φ_2. Therefore, $L - M$ include Φ_2. This evidently means that the inclusion $\Phi_2 \subseteq L - M$ is valid. From the forward and backward inclusions justified earlier, we deduce then that $L - M = \Phi_2$. That is what we want to prove. □

3 Calculation of Normal Vectors for a Binary Linear Classifier

We further explore in depth the question of how to calculate a normal vector for a linear classifier in the regular and singular cases of the sets separability. Namely, we consider the events when the two considered box-constrained data sets are: (1) strongly linearly separable; (2) non-strongly linearly separable; (3) inseparable.

In the first case, the sets separation problem can be solved, for instance, applying the optimization methods presented in [11–17]. In the first event, there can also be utilized the framework which is very similar to the successive projections methods described, for instance, in [18, 19]. For the second and third cases see, for example, [20, 21]. There can be also used the methods of solving the maximin problem (see the details, for example, in [15, 22]). However, the specificity of the training sets structure allows one to solve the separation problem very easy using the exact formulas. Here, we rigorously justify that the vector of the features weights computed by the exact formulas really provides the best data sets separation (in a sense of optimality of the appropriate separation margin).

Here, we apply the concept of "linear separability", so we have to remind at least briefly the meaning of this term for concreteness. The two pattern sets A and B are said to be linearly separable, if and only if there exists some non-zero vector $c \in \mathbb{R}^n$ such that:

$$\max_{b \in B} \langle c, b \rangle \leq \min_{a \in A} \langle c, a \rangle. \tag{5}$$

To define the strong separability of the objects in concern, the inequality (5) should be fulfilled strictly.

Our first goal is to compute, applying the explicit formula, a normal vector of the supporting hyperplanes determining the best linear separator for the training data sets. According to Theorem 3, the set $L - M$ is also given by box constraints. Under our conditions, this is the reason why $L - M$ is a convex and compact set.

Let c^* be a solution of the problem

$$\max_{\|c\|=1} t_{L-M}(c), \tag{6}$$

where $t_{L-M}(c) = \inf_{x \in L-M} \langle c, x \rangle$. By virtue of continuity on the whole space, the linear function $\langle c, x \rangle$, $c \in \mathbb{R}^n$ furnishes its infimum on the compact set $L - M$. Therefore, it is fulfilled $\inf_{x \in L-M} \langle c, x \rangle = \min_{x \in L-M} \langle c, z \rangle$. Let x^* be the optimizer of the following problem

$$\min_{x \in L-M} \langle c, x \rangle \tag{7}$$

for some $c \in \mathbb{R}^n$. The above facts obviously implies that $x^* \in L - M$. Let us note that we will utilize further a linear separability criterion introduced in [22]. This criterion is closely connected with problems (6)–(7).

Lemma 1 (Auxiliary Fact) *If $x^* \in L - M$ is the solution of (7), then*

$$\langle c, x^* \rangle = \min_{i=1,2,\ldots,m} \langle c, v_i^{L-M} \rangle.$$

For the proof details of the similar lemma in the more general setting, we direct the interested reader to [23] (see Lemma 13, p. 40). From Lemma 1, it follows that $t_{L-M}(c) = \min_{i=1,2,\ldots,m} \langle c, v_i^{L-M} \rangle$.

Definition 1 (Separator) If for some $c \in \mathbb{R}^n$ one has $t_{L-M}(c) > 0$, then the set

$$S(c) := \Big\{ x \in \mathbb{R}^n : \gamma_M(c) \leq \langle c, x \rangle \leq \gamma_L(c) \Big\},$$

formed by parallel hyperplanes $\pi(c, \gamma_L(c))$ and $\pi(c, \gamma_M(c))$ is the strong separation margin for the sets L and M, where $\gamma_L(c) = \min_{z \in L} \langle c, z \rangle$, $\gamma_M(c) = \max_{p \in M} \langle c, p \rangle$. The set $S(c)$ is called **a separator** (for details, see [22], p.161).

By the way, taking into account the list of vertices of $L - M$, according to Lemma 1, we evidently obtain $\gamma_L(c) = \min_{i=\overline{1,m}} \langle c, v_i^L \rangle$, $\gamma_M(c) = \max_{=\overline{1,m}} \langle c, v_i^M \rangle$. Here, the hyperplanes $\pi(c, \gamma_L(c))$ and $\pi(c, \gamma_M(c))$ are supporting to L and M, respectively. The thickness of the separation margin $S(c)$ is determined by the distance between its boundaries.

For separable data sets, the idea of maximizing the distance between two supporting hyperplanes has achieved widespread use in practice and theory of SVM, too.

Definition 2 (Pseudo-Separator) If for some $c \in \mathbb{R}^n$ it holds $t_{L-M}(c) < 0$, then the set

$$P(c) := \left\{ x \in \mathbb{R}^n : \gamma_L(c) \leq \langle c, x \rangle \leq \gamma_M(c) \right\}$$

is called **a pseudo-separator**.

Recall that the term "pseudo-separator" was introduced for the first time in [22] (p. 161). The set $P(c)$ represents the margin of unseparated points of sets. In the theory of data classification, the similar margin is usually called the margin of misclassified points, or for short, the misclassification margin.

In the degenerate case of the linear separability, i.e. if for some $c \in \mathbb{R}^n \setminus \{\mathbf{0}\}$ it holds $t_{L-M}(c) = 0$, then $S(c) = P(c) = \pi(c, \gamma(c))$, where $\gamma(c) = \gamma_L(c) = \gamma_M(c)$. Therefore, the set $S(c) = P(c)$ is the degenerate separator, because this separator degenerates into a hyperplane. As it will be demonstrated below, the thickness of the degenerate separator equals zero.

By construction of $S(c)$ and $P(c)$, to calculate the thickness of the geometric margin one has to determine the distance from $\pi(c, \gamma_L(c))$ to $\pi(c, \gamma_M(c))$: $\rho\big(\pi(c, \gamma_L(c)), \pi(c, \gamma_M(c))\big)$.

Now, the following theorem will justify this fact.

Theorem 4 (Thickness of the Separator or Pseudo-Separator) *If $c \in \mathbb{R}^n \setminus \{\mathbf{0}\}$, then the thickness of geometric margin formed between the parallel supporting hyperplanes of the sets L and M ($\pi(c, \gamma_L(c))$ and $\pi(c, \gamma_M(c))$, respectively) is equal to*

$$\rho\left(\pi(c, \gamma_L(c)), \pi(c, \gamma_M(c))\right) = \left| t_{L-M}\left(\frac{c}{\|c\|}\right) \right|. \tag{8}$$

If c^* is a maximizer of problem (6); then, based on a linear separability criterion introduced in [22]), the data sets L and M can be characterized as

- strongly separable when $t_{L-M}(c^*) > 0$,
- non-strongly linearly separable when $t_{L-M}(c^*) = 0$,
- inseparable when $t_{L-M}(c^*) < 0$.

Moreover, the optimality of the thickness of geometric margin is provided by virtue of Theorem 4 and the formulation of problem (6). Indeed, if the sets L and M are strongly separable, then having solved the problem (6) one can maximize the separator thickness of the sets L and M. Obviously, the thickness of the degenerate separator will be equal to zero. If some sets are linearly inseparable, then the problem (6) is equivalent to the next problem

$$-\max_{c\in\mathbb{R}^n\setminus\{\mathbf{0}\}} t_{L-M}\Big(\frac{c}{\|c\|}\Big) = \min_{c\in\mathbb{R}^n\setminus\{\mathbf{0}\}} -t_{L-M}\Big(\frac{c}{\|c\|}\Big) = \min_{c\in\mathbb{R}^n\setminus\{\mathbf{0}\}} \Big|t_{L-M}\Big(\frac{c}{\|c\|}\Big)\Big|$$

which provides one that there will be found a pseudo-separator having a minimal thickness for the considered sets.

Let us remind the well-known definition of the term "projection" which will be very useful below in the study.

Definition 3 (Projection) The point $\mathbf{P}_D(\bar{p}) \in D$ is called a projection of the point $\bar{p} \in \mathbb{R}^n$ onto D if and only if it holds

$$\|\mathbf{P}_D(\bar{p}) - \bar{p}\| \leq \|x - \bar{p}\|, \ \forall x \in D.$$

Due to Definition 3, $\mathbf{P}_D(\bar{p})$ is the point of D nearest to $\bar{p}$ among other points of D. If $\bar{p} \in D$, then it is obviously fulfilled $\mathbf{P}_D(\bar{p}) = \bar{p}$. As is widely known, if D is a nonempty convex and closed set in $\mathbb{R}^n$, then any point $\bar{p} \in \mathbb{R}^n$ has a unique projection onto D. In the context of the sets linear separability, we are interested mostly in projecting the origin of Euclidean space onto the set given by the box constraints.

It is well known from convex analysis that if the origin of Euclidean space $\mathbf{0} = (0, 0, \ldots, 0)$ does not belong to some box-constrained set

$$D = \{x \in \mathbb{R}^n : g_i \leq x_i \leq f_i, \ i \in I\},$$

then the coordinates of its projection $w = (w_1, w_2, \ldots, w_n)$ onto the set D can be calculated by an exact formula. For projecting the origin, we represent further the reformulation of the universal rule (see, for instance, [24], p.196) developed for projecting any point of $\mathbb{R}^n$ onto some box-constrained set:

$$w_i = \begin{cases} g_i, & \text{if } g_i > 0; \\ f_i, & \text{if } f_i < 0; \\ 0, & \text{if } g_i \leq 0 \leq f_i; \ i \in I. \end{cases} \tag{9}$$

Using the exact formula (9), we can therefore calculate $\mathbf{P}_{L-M}(\mathbf{0})$, which denotes a projection of the origin of $\mathbb{R}^n$ onto $L - M$.

The next results immediately give the possibility of evaluating the thickness of the separator with help of $\mathbf{P}_{L-M}(\mathbf{0})$.

Lemma 2 (Relationship Between the Solution of (6) and Projection of the Origin of $\mathbb{R}^n$ onto $L - M$ [23], p. 40)

1. *If* $t_{L-M}(c^*) > 0$, $v = c^* t_{L-M}(c^*)$, *then it holds* $\mathbf{P}_{L-M}(\mathbf{0}) = v$,
2. *If* $t_{L-M}(c^*) \leq 0$, *then* $\mathbf{P}_{L-M}(\mathbf{0}) = \mathbf{0}$.

For L and M being strongly separable, the previous lemma yields that there is automatically fulfilled the following equality:

$$\frac{\mathbf{P}_{L-M}(\mathbf{0})}{\|\mathbf{P}_{L-M}(\mathbf{0})\|} = c^*.$$

For strongly separable sets, this means that the maximum separator thickness corresponds to the normal vector $\mathbf{P}_{L-M}(\mathbf{0}) / \|\mathbf{P}_{L-M}(\mathbf{0})\|$.

For the exceptional case when the origin of $\mathbb{R}^n$ belongs to some set D, we need first to introduce a new term "pseudo-projection" as follows.

Definition 4 (Pseudo-Projection) We shall call the point $\tilde{\mathbf{P}}_{bd(D)}(\bar{v}) \in D$ a pseudo-projection of the point $\bar{v} \in D$ onto $bd(D)$ if and only if it holds

$$\|\tilde{\mathbf{P}}_{bd(D)}(\bar{v}) - \bar{v}\| \leq \|x - \bar{v}\|, \ \forall x \in bd(D),$$

where $bd(D)$ stands for the boundary of D.

Clearly, $\tilde{\mathbf{P}}_{bd(D)}(\mathbf{0}) = \mathbf{0}$ if and only if $\mathbf{0} \in bd(D)$. Let us consider further the case where $(\mathbf{0} \in D)$ & $(\mathbf{0} \notin bd(D))$, i.e. $\mathbf{0} \in int(D)$, Here $int(\cdot)$ stands for interior of the set D.

In the event of dealing with the set D having the box-constrained structure, to compute $\tilde{\mathbf{P}}_{bd(D)}(\mathbf{0})$ we need to project the origin of $\mathbb{R}^n$ from inside onto the $2n$ facets of D. By assumption, it obviously holds $g_i < \mathbf{0} < f_i$, $\forall i \in I$. Each pair of the facets corresponding to some variable x_{i_1}, $i_1 \in I$ evidently has the same normal vector as the pair of hyperplanes $\pi(c, g_{i_1})$ and $\pi(c, f_{i_1})$: $c = (c_1, c_2, \ldots, c_n)$, where $(c_{i_1} = 1)$ & $(c_i = 0 \ \ \forall i \in I, i \neq i_1)$. Moreover, the afore-mentioned normal vector is orthogonal to the corresponding facets of D. This fact allows one for the purpose of pseudo-projecting to make use an exact formula, which is well known from convex analysis to project the origin of $\mathbb{R}^n$ onto the hyperplane. Specifically, we apply this formula n times by setting:

$$w_i = \begin{cases} g_i, & \text{if } |g_i| < f_i; \\ g_i \vee f_i, & \text{if } |g_i| = f_i; \\ f_i, & \text{if } |g_i| > f_i; \ \ i \in I. \end{cases} \tag{10}$$

After calculations of $w_i, \quad \forall i \in I$, we determine the index $i_2 \in I$ for which there will be attained the minimum among the values of $|w_i|, \quad i \in I$: $\min_{i \in I} |w_i| = |w_{i_2}|$. If there are several indices $i \in I$ for which $|w_i|$ is minimal, one can choose any of them. Then, it should be constructed $\hat{w} = (\hat{w}_1, \hat{w}_2, \ldots, \hat{w}_n)$ in such a way that $(\hat{w}_i = 0 \quad \forall i \in I, \ i \neq i_2)$ & $(\hat{w}_{i_2} = w_{i_2})$. Thus, $\tilde{\mathbf{P}}_{bd(D)}(\mathbf{0}) = \hat{w}$. By construction, it immediately holds $\min_{w \in bd(D)} \|w\| = \|\tilde{\mathbf{P}}_{bd(D)}(\mathbf{0})\|$.

With the help of Example 1 we may obviously observe that a pseudo-projection can be not uniquely determined. Indeed, there are six points (projections of the zero vector onto the six facets) satisfying to the definition of the pseudo-projection of the origin onto the boundary of the three-dimensional unit cube. All of these points are equivalent to each other in the sense of having the same Euclidean distance between the origin of $\mathbb{R}^n$ and each of the six points. Namely, this identity allows one to select and apply any one of these points as a pseudo-projection.

One of our goals here is to calculate a normal vector of a pseudo-separator for the case when $\mathbf{0} \in int(L - M)$. From above, it follows that as such a normal vector there can serve the following normalized one:

$$c^* = -\tilde{\mathbf{P}}_{bd(L-M)}(\mathbf{0}) / \|\tilde{\mathbf{P}}_{bd(L-M)}(\mathbf{0})\|.$$

By construction, the thickness of the pseudo-separator corresponding to this support vector will be minimal, since it equals $\|\tilde{\mathbf{P}}_{bd(L-M)}(\mathbf{0})\|$.

Let us consider now the next case, namely when the origin of $\mathbb{R}^n$ lies on the boundary of some box-constrained set D, or briefly, $\mathbf{0} \in bd(D)$. In this event, at least for one index $i \in I$ it evidently holds $g_i = 0 < f_i$ or $g_i < 0 = f_i$. If there are several such indices, then one can choose any of them, for instance $i_1 \in I$. We construct the classifier normal as follows

$$c_i = \begin{cases} 0, & \text{if } (i \in I) \ \& \ (i \neq i_1); \\ -1, & \text{if } (i = i_1) \ \& \ (g_i < 0 = f_i); \\ 1, & \text{if } (i = i_1) \ \& \ (g_i = 0 < f_i). \end{cases} \tag{11}$$

By construction, we obviously obtain the degenerate case of the linear separability between the origin of the space and the set D. These objects are non-strongly linearly separable, so for this reason the linear separation margin has the null thickness.

4 A Linear Classification Model

To perform the linear binary classification of data sets, we need to construct a linear classifier based on some training sets. The term "linear classifier" is well known and widely used in the theory and practice of data classification. A linear classifier is generally defined for solving binary classification problems. It may be obviously interpreted as a rule, which assigns a test instance to some particular class of data.

Let us first define a studied term "linear classifier".

Definition 5 (Linear Classifier) For some given $c \in \mathbb{R}^n \setminus \{\mathbf{0}\}$, we shall say that $f : \mathbb{R}^n \to \mathbb{R}^1$ is a linear classifier for the two data sets L and M if and only if it holds

$$f(x) = \langle c, x \rangle - \gamma(c), \text{ where } \gamma(c) = \frac{\gamma_L(c) + \gamma_M(c)}{2},$$

$$\gamma_L = \min_{i=\overline{1,m}} \langle c, v_i^L \rangle, \; \gamma_M = \max_{i=\overline{1,m}} \langle c, v_i^M \rangle.$$

For the data classes L and M, the hyperplane $\pi(c, \gamma(c))$ divides the separation margin (separator or pseudo-separator), corresponding to the normal vector c, in half. For the linear function $f(x)$, the level set

$$[S^0, f]_{\mathbb{R}^n} := \{x \in \mathbb{R}^n : f(x) = 0\}$$

obviously coincides with the hyperplane $\pi(c, \gamma(c))$. This hyperplane also divides the Euclidean space into two closed linear subspaces. One of these subspaces includes the class L and another one contains the class M.

Procedure (Construction of a Linear Classifier)

1. Using (4), construct the Minkowski difference $L - M$.
2. Project the origin of $\mathbb{R}^n$ onto $L - M$ by (9).
   If $\|\mathbf{P}_{L-M}(\mathbf{0})\| \neq 0$, then calculate $c = \mathbf{P}_{L-M}(\mathbf{0})/\|\mathbf{P}_{L-M}(\mathbf{0})\|$.
   Else determine the pseudo-projection $\tilde{\mathbf{P}}_{bd(L-M)}(\mathbf{0})$ by (10).
   If $\|\tilde{\mathbf{P}}_{bd(L-M)}(\mathbf{0})\| \neq 0$, then calculate $c = -\tilde{\mathbf{P}}_{bd(L-M)}(\mathbf{0})/\|\tilde{\mathbf{P}}_{bd(L-M)}(\mathbf{0})\|$.
   Else use the formula (11) for calculating the normal vector c.
3. Applying Definition 5, construct $f(x)$.

The goal now is to build up **a linear binary classification model**. It may be determined as follows:

1. Construct a linear classifier $f(x)$.
2. Use an assignment criterion in order to decide class membership.

2.1 Assign a new unlabeled instance $x \in \mathbb{R}^n$ to the class L, if $f(x) > 0$.
2.2 Assign a new unlabeled point $x \in \mathbb{R}^n$ to the class M, if $f(x) < 0$.
2.3 If for $x \in \mathbb{R}^n$ it holds $f(x) = 0$, then this new object can equivalently be considered as a member of both the first and second class. For unambiguity, the unlabeled instance may be assigned to only one of the classes.

It is time to justify the assignment criterion utilized above in the linear classification model. We first explore the inseparable case. Due to the applied linear separability criterion, it evidently holds $\gamma_L < \gamma_M \Rightarrow \gamma_L < \gamma(c) < \gamma_M$. By definition, $\langle c, z\rangle \geq \gamma_L, \forall z \in L$ and $\langle c, p\rangle \leq \gamma_M, \forall p \in M$. When for some new instance $x \in \mathbb{R}^n$ we observe the positive value of the linear classifier, this new object should be assigned to L, since it holds $0 < f(x) < \langle c, x\rangle - \gamma_L$. If for the unlabeled object the value of $f(x)$ is negative, then we decide that this instance belongs to M, because $\langle c, x\rangle - \gamma_M < f(x) < 0$.

Investigate now the case when the training classes are strongly separable. According to the linear separability criterion, it is obviously fulfilled

$$\gamma_M < \gamma_L \Rightarrow \gamma_M < \gamma(c) < \gamma_L.$$

This implies that $\langle c, z\rangle \geq \gamma_L > \gamma(c), \forall z \in L$ and $\langle c, p\rangle \leq \gamma_M < \gamma(c), \forall p \in M$. Consequently, $(f(z) > 0, \forall z \in L)\,\&\,(f(p) < 0, \forall p \in M)$. If for some new object $x \in \mathbb{R}^n$ there is observed the positive value of the linear classifier, then we assign this new instance to L. When for the unlabeled instance the value of $f(x)$ is negative, then we conclude that this object belongs to M.

Consider now the case of non-strongly linearly separable training data sets. Owing to the linear separability criterion, one has

$$\gamma_M \leq \gamma_L \Rightarrow \gamma_M \leq \gamma(c) \leq \gamma_L.$$

This immediately yields $\langle c, z\rangle \geq \gamma_L \geq \gamma(c), \forall z \in L$ and $\langle c, p\rangle \leq \gamma_M \leq \gamma(c), \forall p \in M$. Therefore, $(f(z) \geq 0, \forall z \in L)\,\&\,(f(p) \leq 0, \forall p \in M)$. If for some new object $x \in \mathbb{R}^n$ the linear classifier has the nonnegative value, then we decide that this new point is a member of the first class. When for the unlabeled point the value of $f(x)$ is nonpositive, then we conclude that this object falls into the second class.

The principal, for the purposes of practical and theoretical applications, questions that immediately arise are what a linear classifier can be identified as the best one and what it means for the data classification model. To exactly answer these basic questions, we may solve, for instance, the problem (6) and verify the objective function optimal value: $t_{L-M}(c^*)$. Is the function value positive, equal to zero, or negative? The analysis shows what type of linear separability takes place for the separated box-constrained data sets. This also allows to detect with which one of the situations we will face. The cones of generalized support vectors (see, for instance, [9]) are empty, and may be some of them are not. Due to special box-constrained structure of the data sets being classified, there is no need to utilize the complex optimization framework for calculating the classifier weights of the features. For this purpose, we propose to apply the exact formulas. This makes classifying new test data to be fast. To perform a classification, all we need to do is calculate the linear classifier value and analyze its sign. This sign is crucial for the assignment criterion which is verified for deciding the class membership of new test objects. We evaluate the treated linear classifier as the best among those of others, since its normal vector corresponds to the optimal separation margin (separator or pseudo-separator) between the training data sets. By construction, the thickness of the separator is maximal in the separable case. For the inseparable event, the thickness of the pseudo-separator is minimal.

5 Conclusions

In the paper, we reduce the binary problem of sets separating to the problem of separating the origin of the Euclidean space from the Minkowski difference of sets being separated. This approach requires the presence of the exact algebraic expression for the Minkowski difference of box-constrained sets. Finally, we note that the realization of the presented linear binary classification model has low computational costs, since it is carried out by the exact formulas. There is no any need in applying the complex optimization framework. We propose the explicit formulas for computing the normal vector as well as threshold of the linear classifier. We provide the full justification of the assignment criterion for testing the unlabeled instances. The treated linear classifier is estimated as the best one due to optimality of the thickness of the appropriate separation margin between the training data sets.

Acknowledgment This paper has been supported by the Kazan Federal University Strategic Academic Leadership Program ("PRIORITY-2030").

References

1. Agarwal, Charu C.: Data Classification: Algorithms and Applications. Chapman & Hall/CRC, New York (2014)
2. Kesavaraj, G., Sukumaran, S.: A study on classification techniques in data mining, 2013 Fourth International Conference on Computing, Communications and Networking Technologies (ICCCNT), Tiruchengode, pp. 1–7 (2013) https://doi.org/10.1109/ICCCNT.2013.6726842.
3. Takeda, A., Mitsugi, H., and Kanamori, T.: A unified classification model based on robust optimization. Neural Comput. 25(3), 759–804 (2013)
4. Heling Jiang, An Yang, Fengyun Yan and Hong Miao: Research on pattern analysis and data classification methodology for data mining and knowledge discovery. International Journal of Hybrid Information Technology 9(3), 179–188 (2016)
5. Xanthopoulos P., Pardalos P.M., Trafalis T.B.: Linear Discriminant Analysis in Robust Data Mining. Springer (2013)
6. Kotsiantis, S.B.: Supervised Machine Learning: A Review of Classification Techniques. Informatica 31, 249–268 (2007)
7. Gabidullina, Z.R.: The Minkowski difference for convex polyhedra and some its applications. arXiv:1903.03590 (2019) https://arxiv.org/abs/1903.035902019
8. Gabidullina, Z.R.: An algorithm for data analysis via polyhedral optimization, Constructive Nonsmooth Analysis and Related Topics (Dedicated to the Memory of V.F. Demyanov), CNSA 2017, Proceedings. IEEE, 2017. - Vol., Is.. - Art. 7973957 (2017) http://ieeexplore.ieee.org/abstract/document/7973957/
9. Gabidullina, Z.R.: Necessary and sufficient conditions for emptiness of the cones of generalized support vectors. Optim. Lett. 9(4), 693–729 (2015)
10. Gabidullina, Z.R.: A theorem on strict separability of convex polyhedra and its applications in optimization. J. Optim. Theory Appl. 148(3), 550–570 (2011)
11. Gilbert, E.G., Johnson, D.W., Keerthi, S.S.: A Fast procedure for computing the distance between complex objects in three-dimensional space. IEEE J. of Robotics and Automation 4(2), 193–203 (1988)
12. Gabidullina, Z.R.: Adaptive conditional gradient method. J. Optim. Theory Appl. 183(3), 1077–1098 (2019)

13. Gabidullina, Z.R.: Solving of a projection problem for convex polyhedra given by a system of linear constraints, Constructive Nonsmooth Analysis and Related Topics (Dedicated to the Memory of V.F. Demyanov), CNSA 2017, Proceedings. IEEE, Vol., Is.. - Art. 7973958 (2017) http://ieeexplore.ieee.org/abstract/document/7973958/
14. Mitchell, V.F., Dem'yanov, V.F., and Malozemov, V.N.: Finding the point of polyhedron Closest to origin. SIAM J. Contr. 12, 19–26 (1974)
15. Dem'yanov V.F. and Malozemov V.N.: Introduction to Minimax. Dover Publications (1990)
16. Wolfe, P.: Finding the nearest point in a polytope. Math. Program. 11, 128–149 (1976)
17. Cameron, S.: Enhancing GJK: Computing minimum and penetration distances between convex polyhedra. In: Proceedings of International Conference on Robotics and Automation, 3112–3117, (1997)
18. Bregman, L.M.: The Relaxation method for Finding the Common point of Convex Sets and Its Applications to the Solution of Problems in Convex Programming. USSR Comput. Math.& Math. Phys. 7, 200–217 (1967)
19. Combettes, P.L. and Trussell, J.: Method of Successive Projections for Finding a Common Point of Sets in Metric Spaces. J. Optim. Theory Appl., 67(3), 487–507(1990)
20. G. van den Bergen: A fast and robust GJK implementation for collision detection of convex objects. Technical report, Department of Mathematics and Computing Science, Eindhoven University of Technology (1999)
21. Bennett, K.P., Mangasarian, O.L.: Robust linear programming discrimination of two linearly inseparable sets. Optim. Methods Softw. 1, 23–34 (1992)
22. Gabidullina, Z.R.: A Linear separability criterion for sets of euclidean space. J. Optim. Theory Appl. 158(1), 145–171 (2013)
23. Gabidullina, Z.R.: The problem of projecting the origin of Euclidean space onto the convex polyhedron. Lobachevskii J. Math. 39(1), 35–45 (2018)
24. Vasil'ev, F.P.: Numerical Methods for Solving Extremum Problems. Nauka, Moscow (1980)

Patient-Specific Bone Organ Modeling Using CT Based FEM

Oleg Gerasimov, Nikita Kharin, Evgeny Statsenko, Dmitri Mukhin, Dmitri Berezhnoi, and Oskar Sachenkov

Abstract The article presents an image-based method for numerical modeling of the inhomogeneous structures. Such an approach allows taking into account the material anisotropy during the integration by using the weight function. The implementation of this method includes the hypothesis of the correlation between the image data values and elastic properties of the material. A modeling of the specimen was based on the use of an eight-node 3D finite element of the continuous medium with bilinear approximation. In the work the distal part of the rat femur was modeled, the displacement field was calculated and the stress-strain state was locally averaged. In order to assess the reliability of the volumetric averaged stress state, the estimation of the strain energy error was performed.

1 Introduction

Numerical modeling became the most commonly used technique in various fields of scientific research at the moment. The practical application of the image data is a promising direction to assessing the behavior of heterogeneous structures under the external influence [1, 2]. The image-based modeling allows simulating the mechanical properties of the multi-connected porous materials [3–5]. Similar problems are especially relevant to the orthopedic clinical practice [6–8]. The obtaining information about the patient bone properties from the personal image data has a significant influence on the upcoming treatment quality at the diagnostic stage.

O. Gerasimov · N. Kharin · D. Mukhin · D. Berezhnoy · O. Sachenkov (✉)
Institute of Mathematics and Mechanics, Kazan Federal University, Kazan, Republic of Tatarstan, Russia

E. Statsenko
Institute of Geology and Petroleum Technologies, Kazan Federal University, Kazan, Republic of Tatarstan, Russia

I. B. Badriev et al. (eds.), *Mesh Methods for Boundary-Value Problems and Applications*, Lecture Notes in Computational Science and Engineering 141,
https://doi.org/10.1007/978-3-030-87809-2_10

There are several approaches to modeling mechanical properties based on the image data. Firstly, this includes an approximation of the inhomogeneity by the construction of the mean intercept length distribution and its approximation by the least square method [9–12]. In this case, physical relationships are formulated in terms of the tensor of elastic constants and the fabric tensor. A second approach includes the reduction of the material anisotropy to the orthotropy by determining constants from numerical experiments [13–15]. In the work, a third method to estimating the behavior of the inhomogeneous structures is considered.

The main imaging approach for these medical cases is performing a computed tomography, which involves creating a digital prototype of the investigated domain. Data of such a procedure are a 3D integer array, that contains values characterizing the permeability of the material in the microelement of volume. These values can be interpreted according to the quantitative Hounsfield scale of X-ray density. Thus, a digital prototype represents a structure of the porous medium element as a set of the elementary micro volumes, in each of them the percentage of the bone fraction is determined. According to these data and using some approximate method, a discrete mechanical model of the inhomogeneous medium element can be constructed [16–19].

The highest calculation accuracy can be reached in the case of modeling each microvolume of the continuous medium as one finite element [20–25]. However, this approach is resource intensive in problems of discrete modeling, postprocessor analysis, and, especially, at the processor computing stage. Therefore, it seems appropriate to increase the size of the finite elements. That allows considering each microvolume belonging to the element domain as a neighborhood of the integration node of the local stiffness matrix. Nevertheless, the question of determining a quantity of the integration points in each direction remains unclear. A small density of computed tomography data in the integration domain can lead to low calculation accuracy. The middle rectangle method can be used as the simplest integration method. However, the use of a large number of integration nodes, on the one hand, increases accuracy of the integration within the finite element, but on the other hand may reduce the flexibility of the whole model because of a small number of finite elements.

The purpose of the work is to implement a static calculation method for the elements of a porous structure based on the 3D isoparametric bilinear finite element of a continuous medium, built on its digital prototype revealed according to computed tomography data.

2 Materials and Methods

2.1 The Main Relations of the Finite Element Method

A well-known technique for constructing an eight-node 3D isoparametric finite element of a continuous medium with a bilinear approximation of the geometry and displacement field in the local coordinates was used [26]. Within the isoparametric concept for a geometry approximation (radius-vector $\{r\}$ of the point) and initial displacements (displacement vector $\{\theta\}$ of the point) the same system of functions is used:

$$\mathbf{r} = \{r\} = \begin{Bmatrix} x \\ y \\ z \end{Bmatrix} = \sum_{n=1}^{8} \begin{Bmatrix} x_n \\ y_n \\ z_n \end{Bmatrix} N_n(\xi, \eta), \tag{1}$$

$$\boldsymbol{\theta} = \{\theta\} = \begin{Bmatrix} u \\ v \\ w \end{Bmatrix} = \sum_{n=1}^{8} \begin{Bmatrix} u_n \\ v_n \\ w_n \end{Bmatrix} N_n(\xi, \eta, \zeta), \tag{2}$$

where $N_n(\xi, \eta, \zeta) = \frac{1}{8}(1+\xi_n\xi)(1+\eta_n\eta)(1+\zeta_n\zeta)$—known linear shape functions; ξ_n, η_n, ζ_n—local node coordinates of the finite element; *u, v, w*—displacement vector projections to the orts *x, y, z* of the global coordinate system. Relations (2) can be written in a matrix form:

$$\{\theta\} = [N]\{\theta^e\} \tag{3}$$

where $[N]$—matrix of the approximating functions, $\{\theta^e\}$—vector of the node displacements.

Components of the linear $\varepsilon_{xx}, \varepsilon_{yy}, \varepsilon_{zz}$ and shear $\gamma_{xy}, \gamma_{yz}, \gamma_{xz}$ strains describe a medium deformation. These components are represented as a reduced strain vector $\{\varepsilon\}$ and expressed in terms of displacements (2) by the well-known Cauchy relations [26]:

$$\begin{gathered} \varepsilon_{xx} = \frac{\partial u}{\partial x}, \qquad \varepsilon_{yy} = \frac{\partial v}{\partial y}, \qquad \varepsilon_{zz} = \frac{\partial w}{\partial z}, \\ \gamma_{xy} = \frac{\partial u}{\partial y} + \frac{\partial v}{\partial x}, \qquad \gamma_{yz} = \frac{\partial v}{\partial z} + \frac{\partial w}{\partial y}, \qquad \gamma_{xy} = \frac{\partial u}{\partial z} + \frac{\partial w}{\partial x}, \end{gathered} \tag{4}$$

which, in turn, can also be written in a matrix form

$$\{\varepsilon\} = [L]\{\theta\}, \tag{5}$$

A Cauchy stress tensor represents the stress state in the form of components of normal $\sigma_{xx}, \sigma_{yy}, \sigma_{zz}$ and shear $\tau_{xy}, \tau_{yz}, \tau_{xz}$ stresses, which, as well as the small-strain tensor can be written in a reduced stress vector $\{\sigma\}$ form. A Hooke's law, that relates the reduced stress and strain vectors represents in the form

$$\{\sigma\} = [D]\{\varepsilon\}, \tag{6}$$

where $[D]$—matrix of elastic constants of the homogeneous body, which can be represented as

$$[D] = \begin{bmatrix} \lambda + 2\mu & \lambda & \lambda & 0 & 0 & 0 \\ \lambda & \lambda + 2\mu & \lambda & 0 & 0 & 0 \\ \lambda & \lambda & \lambda + 2\mu & 0 & 0 & 0 \\ 0 & 0 & 0 & \mu & 0 & 0 \\ 0 & 0 & 0 & 0 & \mu & 0 \\ 0 & 0 & 0 & 0 & 0 & \mu \end{bmatrix}. \tag{7}$$

In this case, $[D]$ is an elasticity tensor for the isotropic material; λ and μ—Lame constants defined in terms of Young's modulus E and Poisson's ratio ν as

$$\mu = \frac{E}{2(1+\nu)}, \qquad \lambda = \frac{2\mu\nu}{1-2\nu}. \tag{8}$$

The stiffness matrix of the finite element is calculated according to the well-known relations [26]:

$$[K^e] = \int_V [B(\mathbf{r})]^T [D][B(\mathbf{r})] dV, \tag{9}$$

where $[B(\mathbf{r})]$—matrix connecting a reduced vector of small strains and vector of node displacements:

$$\{\varepsilon\} = [B(\mathbf{r})]\{\theta^e\}. \tag{10}$$

In local coordinates, relation (9) can be written in the form

$$[K^e] = \int_{-1}^{1}\int_{-1}^{1}\int_{-1}^{1} [B(\xi,\eta,\zeta)]^T [D][B(\xi,\eta,\zeta)]|J(\xi,\eta,\zeta)| d\xi d\eta d\zeta, \tag{11}$$

where $|J(\xi,\eta,\zeta)|$ is a determinant of the Jacobian matrix of the coordinate transformation.

2.2 Digital Prototype

A digital prototyping involves mapping a dataset to some element of the calculation area. Such a dataset is determined by a 3D structure, that contains the color intensity of the microvolumes composing the computational domain. These intensity values obtained by the computed tomography reflect X-ray density according to the Hounsfield scale [27, 28]. The process of creating a digital prototype implies dividing the investigated sample into a large number of virtual microcubes (voxels) the size of $\Delta x \times \Delta y \times \Delta z$ and with center coordinates at x_k, y_k, z_k. The linear sizes Δx, Δy, Δz generally are equal to each other and are defined by the resolution of a computer tomograph.

The values of computed tomography data corresponding to the voxel are binarized over the defined threshold, which can be calculated, for example, by the Otsu method. Such a procedure allows separating dense bone structure from the substance in the pores: values above the threshold determine the bone tissue, below—the pore.

2.3 Integration Based on Computed Tomography Data

The Gauss integration method is used in the case of calculating the local stiffness matrix of the isotropic continuous medium. The porous medium force to use the middle rectangle method. A finite element of the porous continuous medium, as well as for the isotropic case, represents a convex hexagon with single-curved four-node lateral surfaces. Integration points in the element are geometry coordinates of voxels from a model digital prototype.

Let us introduce into consideration the space of the continuous material Ω (Fig. 1a), the discrete space of computed tomography data Ω' (Fig. 1b) and the space of the finite element mesh Ω^e (Fig. 1c). Thus, it is possible to define some weight function $\omega(\mathbf{r})$, which values can be characterized by the space point based

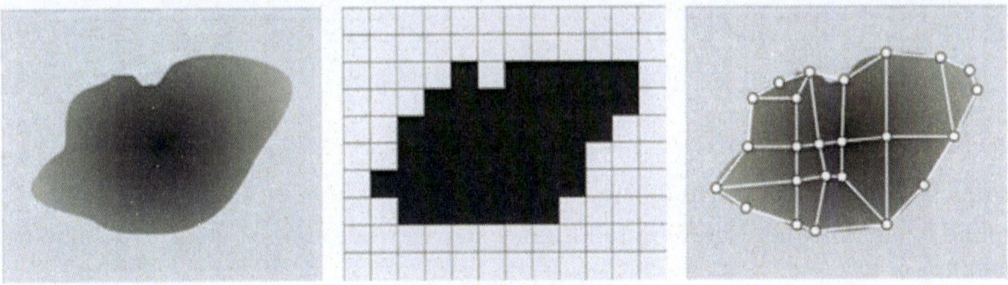

a b c

Fig. 1 Spaces under consideration: (**a**) space of the continuous medium Ω; (**b**) discrete space of computed tomography data Ω'; (**c**) space of the finite elements Ω^e

on computed tomography data. A relation connecting the elementary volume of the continuous space Ω and discrete space Ω' can be written in the form

$$dV' = \omega(\mathbf{r})dV. \tag{12}$$

In this case, passing from the integral over the space Ω to the integration over the discrete space Ω', subject to the formula (12), digital form of the integral (11) can be written as

$$[K^e] = \sum_{i=1}^{I}\sum_{j=1}^{J}\sum_{k=1}^{K}[B(\xi_i, \eta_j, \zeta_k)]^T[D][B(\xi_i, \eta_j, \zeta_k)]\cdot \\ \cdot |J(\xi_i, \eta_j, \zeta_k)|\omega(\xi_i, \eta_j, \zeta_k)d\xi d\eta d\zeta \tag{13}$$

where ξ_i, η_j, ζ_k—local coordinates of the integration points; $\Delta\xi$, $\Delta\eta$, $\Delta\zeta$—step size in three directions in the local coordinates; $\omega(\xi_i, \eta_j, \zeta_k)$—weights of the quadrature formula determined by a value of computed tomography data at the integration point; *I, J, K*—number of quadrature points along each local coordinate inside the finite element.

2.4 Stress-Strain State

In order to determine the stress-strain state, the following approach for the local averaging over the finite element volume is introduced [29–31]. In the case of a standard stress calculation procedure [32, 33], the finite element method uses the relation (6), where the deformations were expressed by the previously introduced formula (10).

Let $\tilde{\sigma}$ be an obtained from the finite element calculation arbitrary component of the stresses. This component is defined by the relation (6). An approximation of the corresponding smoothed value σ^0 over its node values within each finite element is introduced:

$$\sigma^0 = \sum_{n=1}^{R} N_i\, \alpha_n^0 = \{N\}^T\{\alpha^0\}, \tag{14}$$

where *R*—number of the finite element nodes. Approximation coefficients $\{\alpha^0\}$ are found from the condition for a minimum square deviation of σ^0 from $\tilde{\sigma}$ defined by relation

$$\Phi = \int_{V^e}\left(\sigma^0 - \tilde{\sigma}\right)^2 dV^e \to min. \tag{15}$$

Substituting expression for the smoothed value (14) into the minimum square condition (15) the following system of linear equations relative to the node values of the function σ^0 is derived:

$$[d]\left\{\alpha^0\right\} = \{b\}, \tag{16}$$

where

$$[d] = \int_{V^e} \{N\}\{N\}^T dV^e,\ \{b\} = \int_{V_e} \{N\}\tilde{\sigma} dV^e. \tag{17}$$

A similar procedure is applied to each component of the stress vector. The next step is an integration of the found functions of the volumetric stress approximation for each finite element. As in the method defined by the Eq. (13) the stress integration is performed with a weight corresponding to the computed tomography data at the integration point:

$$\{\bar{\sigma}\} = \frac{1}{V^e} \int_{V^e} \left\{\sigma^0\right\} dV^{e\prime} \tag{18}$$

where $\{\bar{\sigma}\}$—stress vector averaged over the finite element volume based on its computed tomography data.

2.5 *Error Estimation*

The usual continuity assumption used in displacement based finite element formulations results in a continuous displacement field from element to element, but a discontinuous stress field [34, 35]. To obtain more acceptable stresses, the averaging of the nodal stresses is done. Then, returning to the element level, the stresses at each node of the element are processes to yield:

$$\left\{\Delta\sigma_n^i\right\} = \left\{\sigma_n^a\right\} - \left\{\sigma_n^i\right\}, \tag{19}$$

where $\{\Delta\sigma_n^i\}$—stress error vector at node n of finite element i, $\{\sigma_n^a\}$—averaged stress vector at node n obtained by:

$$\left\{\sigma_n^a\right\} = \frac{\sum_{i=1}^{N_e^n} \left\{\sigma_n^i\right\}}{N_e^n}, \tag{20}$$

N_e^n—number of finite elements connecting to node n, $\left\{\sigma_n^i\right\}$—stress vector of node n of finite element i.

Then, for each element, the energy error is calculated as:

$$\bar{E}_i = \frac{1}{2} \int_{V^e} \{\Delta\sigma\}^T [D]^{-1} \{\Delta\sigma\} dV^{e\prime}, \tag{21}$$

where $\bar{E}_i$—energy error for finite element i, V^e—the volume of the finite element, $[D]$—elasticity tensor, $\{\Delta\sigma\}$ – stress error vector at points as needed, which is evaluated from all $\{\Delta\sigma_n\}$ of this finite element.

At the last stage of error estimation, the energy error can be normalized against the strain energy:

$$\check{\bar{E}}_i = 100 \cdot \left(\frac{\bar{E}_i}{\bar{U}_i + \bar{E}_i} \right)^{\frac{1}{2}}, \tag{22}$$

where $\check{\bar{E}}_i$—percentage error in energy norm, $\bar{U}_i$—strain energy of the element determined by the similar integration over the discrete space Ω':

$$\bar{U}_i = \int_{V^e} \{\bar{\sigma}\}^T \{\bar{\varepsilon}\} dV^{e\prime} \tag{23}$$

Thus, having estimated the energy error as a percentage, it is not difficult to determine the areas with the most reliable results of stresses and strains. The analysis of the stress-strain state in the future will be carried out definitely in these domains.

3 Results and Discussion

3.1 Meshing

The model problem is based on computed tomography data of a rat's femur. In this case, a distal part was used (Fig. 2a). Since the geometry of a specimen is non-trivial, the special approach to approximate the form by a finite element mesh was proposed. Computed tomography of the object is embedded into the regular rectangle mesh, which completely covers the sample geometry. At the preprocessor stage, a bone fraction of each finite element based on binarized computed tomography data is calculated. Then, the elements containing less than 5% of the material are removed from the mesh. Therefore, the remaining finite-element model has an optimized geometry to the computed tomography data of the investigated object (Fig. 2b, c).

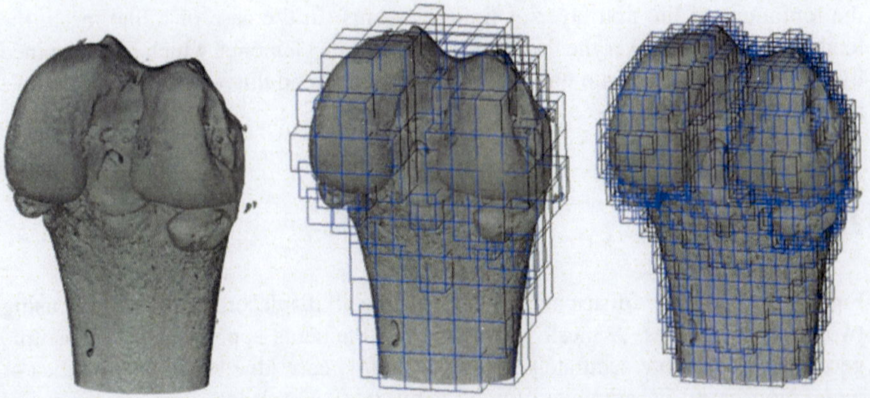

Fig. 2 Geometry approximation: (**a**) CT data view (Avizo); (**b**) and (**c**) superimposed finite-element mesh (>5% of bone)

3.2 *Scanning*

Scanning was performed using the micro-/nanofocus X-ray control system for computed tomography and 2D inspection *Phoenix V | tome | X S240* in the laboratory of X-ray computed tomography of the Institute of Geology and Petroleum Technology of Kazan (Volga region) Federal University. The system is equipped with two X-ray tubes: a microfocus with a maximum accelerating voltage of 240 kV with a power of 320 W and a nanofocus with a maximum accelerating voltage of 180 kV with a power of 15 W. *Datos | x reconstruction* software was used for primary data processing and creation of a three-dimensional (voxel) model of the sample based on X-ray images (projections). The sample fixed in the holder was placed on the rotating table of the X-ray computed tomography camera at the optimal distance from the X-ray source. The survey was conducted at an accelerating voltage of 90–100 kV and a current of 140–150 mA. The study area size was $8.73029 \times 7.62519 \times 10.71947$ mm, the number of voxels in the direction of the corresponding coordinate axes Ox, Oy, and Oz—$790 \times 690 \times 970$, the voxel size—$11.051^3 \mu m$.

3.3 *Modeling*

A numerical experiment was carried out for uniaxial compression. The nodes of the lower face, closest to the diaphysis, were fixed in displacements along the direction of Cartesian coordinate axes. A uniformly distributed load was applied to the upper edge of the distal area. In the case of a coarse mesh, the load was distributed over the top nodes of the first layer of finite elements. In the case of a fine mesh, the load was distributed over the first two layers of finite elements, which is determined by the relief surface of the upper section. Young's modulus was 2 GPa, Poisson's ratio—0.3, the applied force—30 N.

3.4 Solution

Figure 3 shows the distribution of longitudinal displacements obtained using two types of meshes. A mesh with numerous elements approximates the sample geometry much more accurately. However, in this case, due to the peculiarities of integration based on computed tomography data, significant surges in the nodal values for hollow elements are observed. This defines a large variation in the maximum displacement values.

The stress-strain state was locally averaged over the nodal values of each finite element based on the data of its computed tomography (Fig. 4). The results of solving the test problems showed that the mesh refinement has a greater effect on the solution convergence than the increase in the integration points. Due to this fact, an increase in the number of finite elements makes it possible to more accurately determine the concentrators in the stress-strain field without significant losses in the integration accuracy.

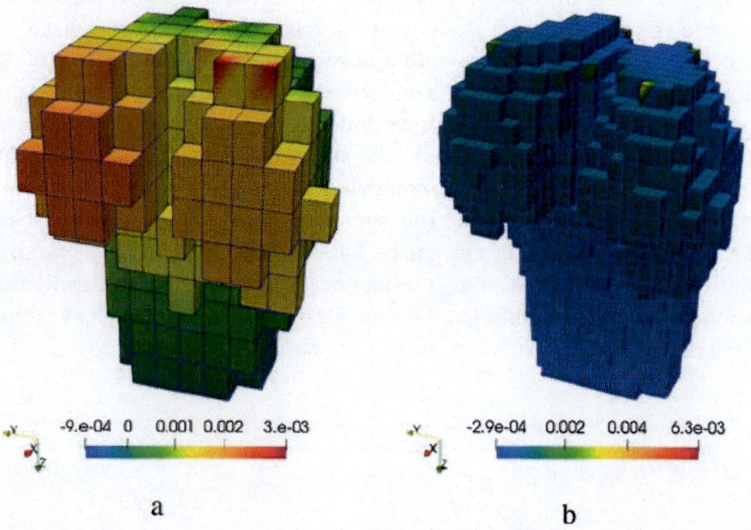

Fig. 3 Displacement field of the rat distal femur along the z-axis (mm): (**a**) 375 elements; (**b**) 2411 elements

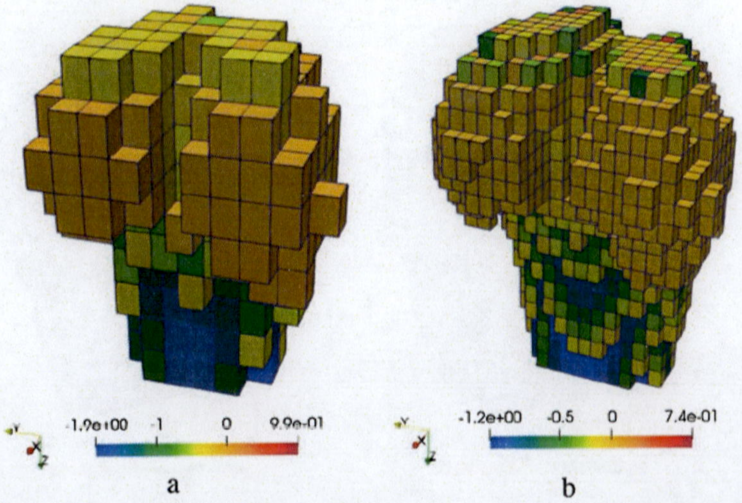

Fig. 4 Stress field of the rat distal femur along the z-axis (MPa): (**a**) 375 elements; (**b**) 2411 elements

3.5 Error Estimation

A comparison of the results presented for two finite element meshes requires an assessment of the obtained solution acceptability. For this purpose, a local calculation of the energy error was made for each finite element separately, based on the nodal stresses averaged over the entire mesh. The value obtained in this way was normalized against the strain energy calculated by a similar method of integration over the discrete space of computed tomography. Figure 5 shows the distribution field of the normalized energy error in percent.

The obtained results of the assessment make it possible to determine the areas with the smallest values of the energy error. Thus, for further comparative analysis of the stress state, a cross section from the blue range in the diaphysis area was selected (Fig. 6).

Figure 6 shows the stress field in the direction of the longitudinal z-axis. In a model with many finite elements, the compression stress is 45% less and the tensile stress is 25% less. The difference in values can be determined by the increased stiffness of the structure in case of a coarser mesh. Various approximation of the object geometry can also have an effect. This is directly related to the nature of filling finite elements with computed tomography data since integration is determined by non-zero values. The general view of the stress distribution in the study area is similar for both models. The calculation time for a model with a coarse mesh is 11 min, and for a model with a more accurate approximation, it is 13 min. Taking into an account the results obtained and the calculation time, it should be concluded that thickening of the grid in the areas with the highest energy error is preferred.

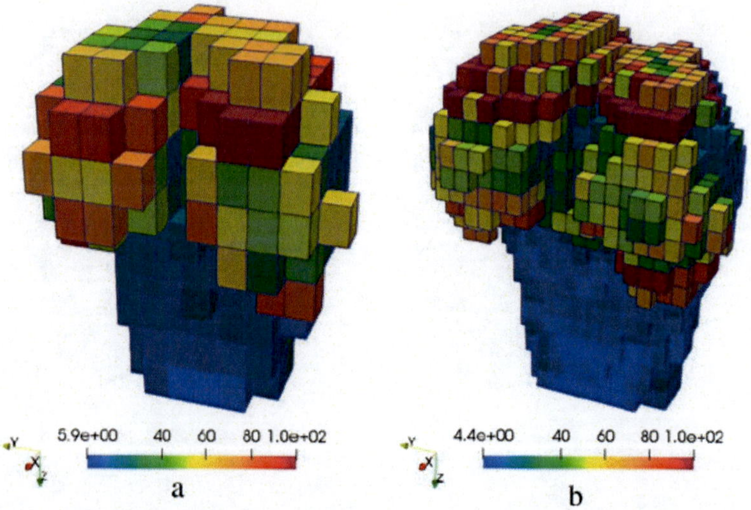

Fig. 5 Normalized energy error field of the rat distal femur (%): (**a**) 375 elements; (**b**) 2411 elements

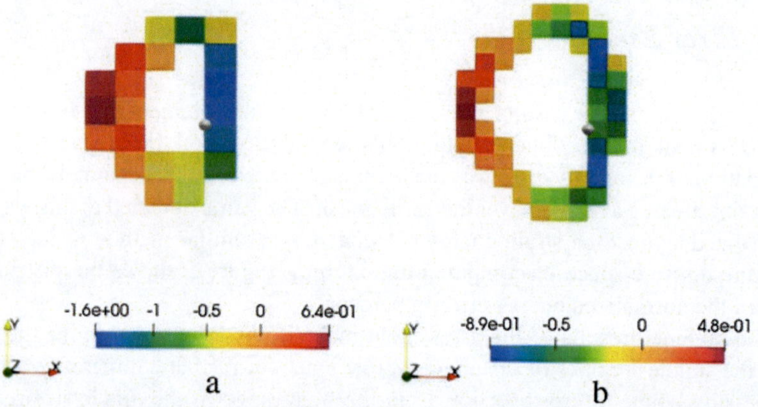

Fig. 6 Average stress state in the cross section of the rat distal femur along the z-axis, z = 8.575576 (MPa): (**a**) 375 elements; (**b**) 2411 elements

4 Conclusion

The article presents one of the possible approaches to describing the deformation processes of inhomogeneous media under the influence of external loads. The method consists of modeling the structure of the investigated area based on its digital prototype. Such an approach allows estimating the behavior of porous objects based on the material optical density. For this purpose, a discrete space over the object image was introduced. The main modeling tool, in this case, is determined by the integration of a local stiffness matrix with a weight function, the values of which are determined by computed tomography data. Thus, the numerical model allows taking into an account the volumetric distribution of the material properties over the volume of a solid finite element.

To estimate the influence of computed tomography data on the convergence of the proposed numerical technique, the test problems were solved. The obtained results were used to determine the relation between the number of integration points inside the calculation domain and the degree of model approximation based on area meshing. According to the analysis of the test problem solution, two numerical models of the rat distal femur with different degrees of approximation were constructed. In this case, a special approach to approximation of a regular rectangular finite element mesh was applied to the sample geometry. Such a method allows avoiding difficulties in the meshing of the nontrivial volumes by removing the empty elements based on a percentage of the material fraction. The possibilities of this method are restricted by the minimum size of the finite element of the initial mesh providing the absence of separate elements.

In the process of solving, a locally averaged stress-strain state of the bone sample was obtained. Calculations were made based on a similar method of integration by discrete space of computed tomography. In order to estimate and compare the obtained results, the calculation of energy error was made for each element separately. In this case, the energy error was normalized against the strain energy, which made it possible to determine acceptable areas for analysis with a small percentage of the error.

The paper presents a comparative analysis of the model problem solution for two types of finite element meshes. The obtained results reflect the influence of the accuracy of the numerical model approximation and also illustrate the dependence of displacements and stress-strain state on the material structure. It was found that an increased degree of approximation allows better determination of stress concentrators and provides more accurate results compared to a coarse mesh consisting of larger finite elements. The presented numerical approach justified the possibility of calculating objects with a porous structure of individual origin.

In the future, it is suggested to consider the ways of finite element mesh thickening for better approximation at the areas of energy error concentration. The studies can be extended using a different type of computed tomography data processing as well as an improved formula for numerical integration. It is also of interest to use other types of finite elements. The data obtained using such an approach can be used to assess the strength of the material at quasi-brittle fracture under static loading of inhomogeneous structures.

Acknowledgement This work was supported by the Russian Science Foundation (RSF Grant No. 18-75-10027).

References

1. Marwa, F., Wajih, E.Y., Philippe, L., Mohsen, M.: Improved USCT of Paired Bones Using Wavelet-based Image Processing. IJIGSP. **10** (9), 1–9 (2018). https://doi.org/10.5815/ijigsp.2018.09.01
2. Mithun, K.P.K., Mohammad, M.R.: Metal Artifact Reduction from Computed Tomography (CT) Images using Directional Restoration Filter. IJITCS. **6**(6), 47–54 (2014). https://doi.org/10.5815/ijitcs.2014.06.07
3. Alberich-Bayarri, A., Marti-Bonmati, L., Sanz-Requena, R., Belloch, E., Moratal, D.: In vivo trabecular bone morphologic and mechanical relationship using high-resolution 3-T MRI. Am. J. Roentgenol. **191**(3), 721–726 (2008). https://doi.org/10.2214/AJR.07.3528
4. Kayumov, R.A.: Structure of nonlinear elastic relationships for the highly anisotropic layer of a nonthin shell. Mech. Compos. Mater. **35**(5), 409–418 (1999). https://doi.org/10.1007/BF02329327
5. Kharin, N., Vorob'yev, O., Bolshakov, P., Sachenkov, O.: Determination of the orthotropic parameters of a representative sample by computed tomography. J. Phys. Conf. Ser. **1158**(3):032012 (2019).
6. Kichenko, A.A., Tverier, V.M., Nyashin, Y.I., Zaborskikh, A.A.: Experimental determination of the fabric tensor for cancellous bone tissue Russ. J. Biomech. **15**(4), 66–81 (2011).
7. Maquer, G., Musy, S.N., Wandel, J., Gross, T., Zysset, P.K.: Bone Volume Fraction and Fabric Anisotropy Are Better Determinants of Trabecular Bone Stiffness Than Other Morphological Variables. J. Bone Miner. Res. **30**(6), 1000–1008 (2015). https://doi.org/10.1002/jbmr.2437
8. Kichenko, A.A., Tverier, V.M., Nyashin, Y.I., Simanovskaya, E.Y., Elovikova, A.N.: Formation and elaboration of the classical theory of bone tissue structure description. Russ. J. Biomech. **12**(1), 66–85 (2008).
9. Zaytseva, T., Fedianin, A., Baltin, M., Eremeev, A, Baltina, T.: The neuromuscular apparatus of the calf muscles of the rat with restriction of motor function. Eur. J. Clin. Investig. **50**, 97 (2020).
10. Sachenkov, O., Hasanov, R., Andreev, P., Konoplev, Y.: Determination of muscle effort at the proximal femur rotation osteotomy. IOP Conf. Ser. Mater. Sci. Eng. **158**(1):012079 (2016).
11. Kayumov, R.A., Muhamedova, I.Z., Tazyukov, B.F., Shakirzjanov, F.R.: Parameter determination of hereditary models of deformation of composite materials based on identification method. J. Phys. Conf. Ser. **973**(1):012006 (2018). https://doi.org/10.1088/1742-6596/973/1/012006
12. Yaikova, V.V., Gerasimov, O.V., Fedyanin, A.O., Zaytsev, M.A., Baltin, M.E., Baltina, T.V., Sachenkov, O.A.: Automation of bone tissue histology. Front. Phys. **91** (2019) https://doi.org/10.3389/fphy.2019.00091
13. Ridwan-Pramana, A., Marcian, P., Borak, L., Narra, N., Forouzanfar, T., Wolff, J.: Finite element analysis of 6 large PMMA skull reconstructions: A multi-criteria evaluation approach. PLoS ONE. **12**:e0179325 (2017). https://doi.org/10.1371/journal.pone.0179325
14. Ruess. M., Tal, D., Trabelsi. N., Yosibash. Z., Rank. E.: The finite cell method for bone simulations: verification and validation. Biomech. Model. Mechanobiol. **11**(3–4), 425–437 (2012). https://doi.org/10.1007/s10237-011-0322-2
15. Chikova, T.N., Kichenko, A.A., Tverier, V.M., Nyashin, Y.I.: Biomechanical modelling of trabecular bone tissue in remodelling equilibrium. Russ. J. Biomech. **22**(3), 245–253 (2018). https://doi.org/10.15593/RJBiomeh/2018.3.01
16. Kharin, N.V., Vorobyev, O.V., Berezhnoi, D.V., Sachenkov, O.A.: Construction of a representative model based on computed tomography. PNRPU Mech. Bull. **3**, 95–102 (2018). https://doi.org/10.15593/perm.mech/2018.3.10
17. Sachenkov, O.A., Gerasimov, O.V., Koroleva, E.V., Mukhin, D.A., Yaikova, V.V., Akhtyamov, I.F., Shakirova, F.V., Korobeynikova, D.A., Khan, H.Ch.: Building the inhomogeneous finite element model by the data of computed tomography. Russ. J. Biomech. **22**(3), 291–303 (2018). https://doi.org/10.15593/RJBiomeh/2018.3.05

18. Legrain, G., Cartraud, P., Perreard, I., Mos, N.: An x-fem and level set computational approach for image-based modelling: application to homogenization. Int. J. Numer. Methods Eng. **86**(7), 915–934 (2011). https://doi.org/10.1002/nme.3085
19. Giovannelli, L., Rodenas, J.J., Navarro-Jimenez, J.M., Tur, M.: Direct medical image-based Finite Element modelling for patient-specific simulation of future implants. Finite Elem. Anal. Des. **136**, 37–47 (2017). https://doi.org/10.1016/j.finel.2017.07.010
20. Prez, M., Vendittoli, P.-A., Lavigne, M., Nuo, N.: Bone remodeling in the resurfaced femoral head: effect of cement mantle thickness and interface characteristic. Med. Eng. Phys. **36**(2), 185–195 (2014). https://doi.org/10.1016/j.medengphy.2013.10.013
21. Marcian, P., Wolff, J., Horakova, L., Kaiser, J., Zikmund, T., Borak, L.: Micro finite element analysis of dental implants under different loading conditions. Comput. Biol. Med. **96**, 157–165 (2018). https://doi.org/10.1016/j.compbiomed.2018.03.012
22. Kichenko, A.: Cancellous bone tissue remodelling: Mathematical modelling. Russ. J. Biomech. **23**(3), 284–304 (2019). https://doi.org/10.15593/RJBiomech/2019.3.02
23. Tveito, A., Jager, K.H., Kuchta, M., Mardal, K.-A., Rognes, M.E.: A cell-based framework for numerical modeling of electrical conduction in cardiac tissue. Front. Phys. **5**, 48 (2017). https://doi.org/10.3389/fphy.2017.00048
24. Nadal, E., Rodenas, J.J., Albelda, J., Tur, M., Tarancon, J.E., Fuenmayor, F.J.: Efficient finite element methodology based on cartesian grids: application to structural shape optimization. Abstr. Appl. Anal. 953786 (2013). https://doi.org/10.1155/2013/953786
25. Marco, O., Sevilla, R., Zhang, Y., Rodenas, J.J., Tur, M.: Exact 3d boundary representation in finite element analysis based on Cartesian grids independent of the geometry. Int. J. Numer. Methods Eng. **103**(6), 445–468 (2015). https://doi.org/10.1002/nme.4914
26. Zienkiewicz, O.C., Zhu, J.Z.: A simple error estimator and adaptive procedure for practical engineering analysis. Int. J. Numer. Methods Eng. **24**(2), 337–357 (1987). https://doi.org/10.1002/nme.1620240206
27. Carniel, T.A., Klahr, B., Fancello, E.A.: On multiscale boundary conditions in the computational homogenization of an RVE of tendon fascicles. J. Mech. Behav. Biomed. **91**, 131–138 (2019). https://doi.org/10.1016/j.jmbbm.2018.12.003
28. Gerasimov, O.V., Berezhnoi, D.V., Bolshakov, P.V., Statsenko, E.O., Sachenkov, O.A.: Mechanical model of a heterogeneous continuum based on numerical-digital algorithm processing computer tomography data. Russ. J. Biomech. **23**(1), 87–97 (2019).
29. Natali, A.N., Pavan, P.G., Ruggero, A.L.: Evaluation of stress induced in peri-implant bone tissue by misfit in multi-implant prosthesis, Dent. Mater. Off. Publ. Acad. Dent. Mater. **22**(4), 388–395 (2006). https://doi.org/10.1016/j.dental.2005.08.001
30. Sachenkov, O.A., Hasanov, R.F., Andreev, P.S., Konoplev, Yu.G.: Numerical study of stress-strain state of pelvis at the proximal femur rotation osteotomy. Russ. J. Biomech. **20**(3), 220–232 (2016). https://doi.org/10.15593/RJBiomech/2016.3.06
31. Schenk, D., Mathis, A., Lippuner, K., Zysset, Ph.: In vivo repeatability of homogenized finite element analysis based on multiple HR-pQCT sections for assessment of distal radius and tibia strength. Bone. 115575 (2020). https://doi.org/10.1016/j.bone.2020.115575
32. Grassi, L., Schileo, E., Taddei, F., Zani, L., Juszczyk, M., Cristofolini, L.: Accuracy of finite element predictions in sideways load configurations for the proximal human femur. J. Biomech. **45**(2), 394–399 (2012). https://doi.org/10.1016/j.jbiomech.2011.10.019
33. Marcian, P., Florian, Z., Horakova, L., Kaiser, J., Borak, L.: Microstructural finite-element analysis of influence of bone density and histomorphometric parameters on mechanical behavior of mandibular cancellous bone structure. Solid State Phenom. **258**, 362–365 (2017). https://doi.org/10.4028/www.scientific.net/SSP.258.362
34. Mithun, K.P.K., Gauhar, A., Mohammad, M.R., A.S.M. Delowar, H.: Automatically Gradient Threshold Estimation of Anisotropic Diffusion for Meyer's Watershed Algorithm Based Optimal Segmentation. IJIGSP. **6**(12), 26–31 (2014). https://doi.org/10.5815/ijigsp.2014.12.04
35. Hettich, G., Schierjott, R.A., Ramm, H., Graichen, H., Jansson, V., Rudert, M., Traina, F., Grupp, T.M.: Method for quantitative assessment of acetabular bone defects. J. Orthop. Res. **37**(1), 181–189 (2018). https://doi.org/10.1002/jor.24165

Parallel Algorithm for Solving Problem of Electromagnetic Wave Diffraction by a Tooth-Shaped Plate

Dinara Giniyatova, Dmitrii Tumakov, and Angelina Markina

Abstract In the present work the problem of plane electromagnetic wave diffraction by a thin metal tooth-shaped plate is considered. A numerical algorithm is developed using the method of moments with OpenMP and NVIDIA CUDA technologies implementation. The results of numerical modeling of a plane wave diffraction by the symmetrical four-tooth-shaped thin metallic plate is shown. A comparative analysis of the performance for CPU and GPU is carried out. It is shown that the method of moments implementation by graphical processor provides a sufficient gain in the performance.

1 Introduction

The problems of diffraction of electromagnetic waves arise in the study of various kinds of complex electrodynamic systems. For their analysis, it is necessary to use strict analytical methods of applied electrodynamics [1–3] or approximate numerical methods [4–6]. To date, the following methods are widely used in specialized software: the moment method (MoM), the finite element method (FEM) [7], and the finite difference method in the time domain (FDTD) [8]. All these methods lead to the need to solve complex systems of linear algebraic equations, the order of which directly depends on the desired degree of accuracy of solving the problem. The use of effective numerical methods and new computer technologies make it possible to solve similar problems within an acceptable time. A promising technology, from the point of view of calculation time, is a parallel computing on a graphics processor (NVIDIA CUDA) [9–12]. In the present paper, we consider a parallel algorithm for solving the diffraction problem by the method of moments on CUDA.

D. Giniyatova · D. Tumakov (✉) · A. Markina
Institute of Computational Mathematics and Information Technologies (Kazan Federal University), Kazan, Russia
e-mail: dtumakov@kpfu.ru

I. B. Badriev et al. (eds.), *Mesh Methods for Boundary-Value Problems and Applications*, Lecture Notes in Computational Science and Engineering 141,
https://doi.org/10.1007/978-3-030-87809-2_11

As is known [13, 14] the problem of diffraction of electromagnetic waves on a perfectly conducting surface can be described by the operator equation for the current surface. The operator can be a linear integral or integro-differential operator, and integration is carried out over the entire diffraction surface. To solve such equations, the method of moments is used. Harrington and his monograph "Field Computation by Moment Methods" described the method of moments most fully [15]; the current state of the method of moments in electrodynamics problems is described in the monographs by Sadiku [16] and Gibson [14]. In addition, a number of works are devoted to the mathematical justification of this method and the convergence of the approximate solution to the exact one [17]. As already noted above, the separation of the diffraction surface into small finite regions leads to the construction and further numerical solution of systems of linear algebraic equations of a very high order. On the other hand, this class of tasks lends itself well to parallelization, and the architecture of the graphic processor (GPU) is well optimized for parallel data processing.

For plates with complex geometry, the triangulation of the area is carried out by any triangulation method [18]. For some surfaces that have a number of features, such as comb-shaped plates [19], the triangulation algorithm can be accelerated. In the present paper, the fast triangulation of such a plate are considered by using OpenMP.

2 Diffraction Problem Statement

We consider the problem of electromagnetic field diffraction on a perfectly conducting thin plate of an arbitrary shape (see, for example, [20]). Let $\Omega \subset R^2$ be a bounded domain with a piece-wise-smooth boundary Γ consisting of a finite number of arcs of the class C^∞ converging at non-zero angles. The problem of diffraction of an external electromagnetic field $\boldsymbol{E}^0, \boldsymbol{H}^0$ on a perfectly conducting plate Ω, located in free space with a wave number k, $k^2 = \omega^2 \epsilon \mu$, consists in the determining scattered electromagnetic field

$$\boldsymbol{E}, \boldsymbol{H} \in C^2(R^3 \setminus \overline{\Omega}) \bigcap_{\delta>0} C(\overline{R}^3_+ \setminus \Gamma_\delta) \bigcap_{\delta>0} C(\overline{R}^3_- \setminus \Gamma_\delta) \tag{1}$$

satisfying homogeneous Maxwell equations:

$$\begin{aligned} \operatorname{Rot} \boldsymbol{H} &= -ik\boldsymbol{E}, \\ \operatorname{Rot} \boldsymbol{E} &= ik\boldsymbol{H}, \quad \mathbf{x} \in R^3 \setminus \overline{\Omega} \end{aligned} \tag{2}$$

boundary conditions for tangent components of the electric field on the plate surface:

$$\boldsymbol{E}_\tau|_\Omega = -\boldsymbol{E}^0_\tau|_\Omega \tag{3}$$

conditions of finite energy in any limited amount of space:

$$\boldsymbol{E}, \boldsymbol{H} \in L^2_{loc}(R^3) \tag{4}$$

and conditions at infinity:

$$\frac{\partial}{\partial r}\begin{pmatrix}\boldsymbol{E}\\ \boldsymbol{H}\end{pmatrix} - ik\begin{pmatrix}\boldsymbol{E}\\ \boldsymbol{H}\end{pmatrix} = o(r^{-1}), \quad \begin{pmatrix}\boldsymbol{E}\\ \boldsymbol{H}\end{pmatrix} = O(r^{-1}), \quad r : |\mathbf{x}| \to \infty. \tag{5}$$

For the full field, $\boldsymbol{E}^{tot} = \boldsymbol{E}^0 + \boldsymbol{E}, \quad \boldsymbol{H}^{tot} = \boldsymbol{H}^0 + \boldsymbol{H}$. We assume that all sources of the incident field are outside of the plate $\overline{\Omega}$ so that for some $\delta > 0$

$$\boldsymbol{E}^0 \in C^{\infty}(\Omega_\delta), \quad \Omega_\delta = \{\mathbf{x} : |\mathbf{x} - \mathbf{y}| < \delta, \mathbf{y} \in \Omega\} \tag{6}$$

whence it follows that

$$\boldsymbol{E}^0_\tau|_\Omega \in C^{\infty}(\overline{\Omega}). \tag{7}$$

Often, either a plane wave or an electric or magnetic dipole located outside of $\overline{\Omega}$ is considered as an incident field. In this case, conditions (6) and (7) are satisfied. The field $\boldsymbol{E}^0$, $\boldsymbol{H}^0$ is a solution to the system of Maxwell equations in free space without a plate.

One of the approaches to solving the problem (1)–(7) is to reduce it to an integrodifferential equation on a plate [13]. This method is often called the surface current method.

Now let S be the open surface of a perfectly conducting plate with the unit normal $\mathbf{n}$. By $\boldsymbol{E}^i$ we denote the electric field defined to be the field due to a source in the absence of a plate. It induces a surface currents $\boldsymbol{J}$ on S. Since S is an open surface, we consider $\boldsymbol{J}$ as the sum of the surface currents on opposite sides of S and, therefore, the normal component $\boldsymbol{J}$ should vanish on boundaries of S. The scattered electric field $\boldsymbol{E}^s$ can be computed by the formula [14]

$$\boldsymbol{E}^s = -i\omega\mathbf{A} - \nabla\Phi, \tag{8}$$

where $\boldsymbol{A}$ and Φ are the vector and scalar potentials, respectively. It is known [14] that the potentials are related to the excitatory current through the Green's function. In free space, the following formulas are valid

$$\mathbf{A}(\mathbf{r}) = \mu \int\limits_S \mathbf{J}(\mathbf{r}')G(\mathbf{r}, \mathbf{r}')dS', \tag{9}$$

$$\Phi(\mathbf{r}) = \frac{1}{\varepsilon} \int\limits_S \sigma G(\mathbf{r}, \mathbf{r}')dS', \tag{10}$$

where Green's function defined as

$$G(\mathbf{r}, \mathbf{r}') = \frac{e^{-ik|\mathbf{r}-\mathbf{r}'|}}{4\pi|\mathbf{r} - \mathbf{r}'|},$$

$k = \omega\sqrt{\mu\varepsilon} = 2\pi/\lambda$ (λ is a wavelength) and $|\mathbf{r} - \mathbf{r}'|$ is the distance between the arbitrarily located observation point $\mathbf{r}$ and the source point $\mathbf{r}'$ on S. The surface charge density σ is related to the surface divergence of current through the equation of continuity

$$\nabla_s \cdot \boldsymbol{J} = -i\omega\sigma. \tag{11}$$

The boundary condition for the electric field in the case of the perfectly conducting surface is

$$\mathbf{n} \times (\boldsymbol{E}^i + \boldsymbol{E}^s) = 0, \tag{12}$$

whence, using (8) we obtain the integro-differential equation with respect to $\mathbf{J}$

$$-\boldsymbol{E}^i_{tan} = (-i\omega\boldsymbol{A} - \nabla\Phi)_{tan}, \quad \mathbf{r} \in S. \tag{13}$$

Together with (9)–(11), Eq. (13) is the so-called the *electric field integral equation* (EFIE). Sometimes in the literature, Eq. (13) is called the equation in terms of mixed potentials (Mixed Potential Integral Equation). Nevertheless, hereinafter, we will use the term EFIE, implying Eq. (13), taking into account (9)–(11).

3 The Method of Moments

The method of moments (MoM) (see [14, 15]) is one of the most popular and powerful methods for electrodynamic modeling. Typically it is used for the analysis of electrically small flat structures made of metal, dielectric inclusions are allowed. Often it is applied to calculate surface currents on plane metal or dielectric structures when emitted in free space. The main practical advantage of the method of moments is that it is necessary to discretize (cover by patches) only metal component of the modeling structure, since the current distribution on metal surfaces is considered as an unknown quantity. Note that in other methods the main unknowns are usually electric/ magnetic fields, which are presented in all solution space. As a result, the "planar" mesh in MoM is much simpler and smaller than the equivalent "volume" mesh required for FEM and FDTD modeling. In fact, the method of moments is a way to solve Maxwell's equations written in the integral form (EFIE, MFIE) in the frequency domain.

We describe the basic idea of the method of moments. As it has been discussed earlier, the method of moments is a general technique for solving operator problem in the common form

$$Lf = g, \tag{14}$$

where L is a linear operator, g is the known source function or excitation, f is an unknown function to be determined. In our case, L is an integro-differential operator, f is an unknown current function $\boldsymbol{J}$, and g is a known excitation source (incident field $\boldsymbol{E}^i$) . We approximate the function f as the series expansion of N basis functions f_n with unknown weight coefficients α_n yet to be determined:

$$f \approx \sum_{n=1}^{N} \alpha_n f_n. \tag{15}$$

When (15) is substituted in (14) using the linearity of the operator L one obtains

$$\sum_{n=1}^{N} \alpha_n L(f_n) \approx g. \tag{16}$$

The basis functions f_n are chosen to correctly model the expected properties of the unknown function f and could be scalars or vector depending on considered problem. Next, both sides of (16) are multiplied by known testing or weighing function and the result integrated over a spatial area. The described procedure is called inner product or moment between the basis functions $f_n(r')$ and the test functions $g_m(r)$ and defined as:

$$\langle g_m, f_n \rangle = \int_{g_m} g_m(\mathbf{r}) \cdot \int_{f_n} f_n(\mathbf{r}'), \quad m = \overline{1, N}, \tag{17}$$

where the presented integrals can be linear, surface, or volume depending on the type of basis and test functions.

We require that the scalar product of each test function with the residual function be zero, then

$$\sum_{n=1}^{N} \alpha_n \langle g_m, L(f_n) \rangle = \langle g_m, g \rangle, \quad m = \overline{1, N}. \tag{18}$$

Equation (18) represents a system of linear algebraic equations for unknown coefficients α_n and in matrix form can be written as $\mathbf{Z}\,\mathbf{a} = \mathbf{b}$, where

$$\mathbf{Z} = \begin{pmatrix} \langle g_1, L(f_1)\rangle & \langle g_1, L(f_2)\rangle & \dots & \langle g_1, L(f_N)\rangle \\ \langle g_2, L(f_1)\rangle & \langle g_2, L(f_2)\rangle & \dots & \langle g_2, L(f_N)\rangle \\ \dots & \dots & & \\ \langle g_N, L(f_1)\rangle & \langle g_N, L(f_2)\rangle & \dots & \langle g_N, L(f_N)\rangle \end{pmatrix},$$

$$\mathbf{b} = \begin{pmatrix} \langle g_1, g\rangle \\ \langle g_2, g\rangle \\ \dots \\ \langle g_N, g\rangle \end{pmatrix}, \quad \mathbf{a} = \begin{pmatrix} \alpha_1 \\ \alpha_2 \\ \dots \\ \alpha_N \end{pmatrix}.$$

SLAE (18) could be solved using various numerical methods, Gaussian elimination or similar techniques like a version of LU-decomposition. Note that in the method of moments the resulting matrix of the system is compact but completely filled, in contrast, for example, to the methods based on differential equations whose matrices are huge and sparse.

Solving (18), we determine the unknown coefficients α_n by which the desired function f is reconstructed. Thus, $f = \langle \bar{\mathbf{f}}, \mathbf{Z}^{-1}\mathbf{b}\rangle, \quad \bar{\mathbf{f}} = (f_1, f_2, \dots, f_N)^T$. This completes the procedure of the method of moments.

3.1 *Basis and Testing Functions*

The basis and testing function could be arbitrary. However, to provide an efficient solution the basis function should be selected such that the relatively small number of functions will guarantee a well approximation. The testing functions should provide a reliable measure of discrepancy between two side of (16).

One of the most popular basis functions used in calculating the surface current are the so-called RWG functions proposed in [21]. They are conveniently used to search for an approximate EFIE solution when the surface of a perfectly conducting body is divided into elementary triangular patches. We will use standard terms, such as a face, to denote the surface of an elementary triangular patch, an edge (boundary edge) to indicate one of its sides, and a vertex to indicate the vertices of a triangle.

First of all, we note that each basis RWG function is associated with one inner edge and vanishes everywhere on S, except for a pair of triangles adjacent to this edge. Figure 1 shows two such triangles, T_n^+ and T_n^-, adjacent to the n-th edge. Points belonging to the triangle T_n^+ can be described both in global coordinates by the radius vector $\mathbf{r}$, and in local coordinates using the radius vector ρ_n^+ defined relative to the free vertex of the triangle T_n^+. A similar remark is also true for the triangle T_n^- with the only difference being that the vector ρ_n^- is directed from the point belonging to the triangle to the free vertex T_n^-. The choice of "positive" and "negative" triangles is arbitrary, given that for the entire cycle of calculating the surface current, it will not change.

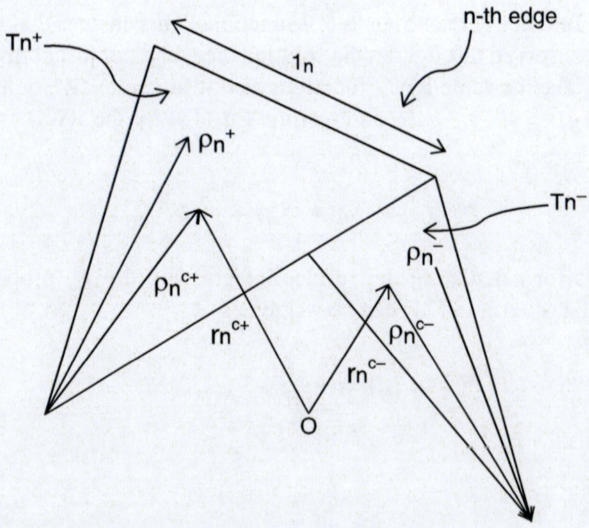

Fig. 1 Triangle pair and RWG parameters associated with inner edge n

Basis function associated with the n-th inner edge defined as:

$$f_n(\mathbf{r}) = \begin{cases} \dfrac{l_n}{2A_n^+}\rho_n^+, & \mathbf{r} \in T_n^+, \\ \dfrac{l_n}{2A_n^-}\rho_n^-, & \mathbf{r} \in T_n^-, \\ 0, & \text{otherwise}, \end{cases} \tag{19}$$

where l_n is the length of the n-th edge, A_n^+ and A_n^- are the areas of the triangles T_n^+ and T_n^-, respectively. The properties of RWG functions are described in detail in [21]. Following the method of moments, we represent the surface current everywhere on S in the form of an approximate formula

$$\boldsymbol{J} \approx \sum_{n=1}^{N} \alpha_n f_n(\mathbf{r}), \tag{20}$$

where N is the number of inner edges.

The next step in the method of moments is the testing procedure or multiplying the original equation by testing functions. Generally speaking, for the testing procedure, it is permissible to use any functions. However, their choice of a specific problem is crucial. One of the most effective methods is *The Galerkin method*, when the same basis functions are chosen as test functions. This ensures that the boundary conditions are observed throughout the solution area, and not just at discrete points. Therefore, we take the same RWG functions as test functions. We define the scalar product as $\langle f, g\rangle = \int_S f \cdot g\, dS$ and test the Eq. (13) by the RWG functions. We obtain

$$\langle \boldsymbol{E}^i, f_m\rangle = i\omega\langle \boldsymbol{A}, f_m\rangle + \langle \nabla\Phi, f_m\rangle. \tag{21}$$

Using methods for calculating the surface integral and the f_m property at the S boundaries, the last term in (21) can be written as

$$\langle \nabla\Phi, f_m\rangle = -\int\limits_S \Phi\, \nabla_s \cdot f_m dS. \tag{22}$$

Then, using

$$\nabla_s \cdot f_n = \begin{cases} \dfrac{l_n}{A_n^+}, & \mathbf{r} \in T_n^+, \\ -\dfrac{l_n}{A_n^-}, & \mathbf{r} \in T_n^-, \\ 0, & \text{otherwise}, \end{cases}$$

the integral in (22) can be approximated as follows

$$\int\limits_S \Phi\nabla_s \cdot f_m dS = l_m \left(\frac{1}{A_m^+} \int\limits_{T_m^+} \Phi dS - \frac{1}{A_m^-} \int\limits_{T_m^-} \Phi dS \right)$$
$$\cong l_m[\Phi(\mathbf{r}_m^{c+}) - \Phi(\mathbf{r}_m^{c-})]. \tag{23}$$

In (23), the average value of Φ for each triangle was replaced by the value Φ in the center of mass of the triangles. Using similar arguments, we can approximate the terms in (21) containing the vector potential and the incident field. We show this by the example of the term $\langle \boldsymbol{E}^i, f_m\rangle$:

$$\langle \boldsymbol{E}^i, f_m\rangle = \int\limits_S \boldsymbol{E}^i \cdot f_m dS = \frac{l_m}{2} \left(\frac{1}{A_m^+} \int\limits_{T_m^+} \boldsymbol{E}^i \cdot \rho_m^+ dS + \frac{1}{A_m^-} \int\limits_{T_m^-} \boldsymbol{E}^i \cdot \rho_m^- dS \right)$$
$$\cong \frac{l_m}{2} \left(\boldsymbol{E}^i(\mathbf{r}_m^{c+})\rho_m^{c+} + \boldsymbol{E}^i(\mathbf{r}_m^{c-})\rho_m^{c-} \right). \tag{24}$$

Thus, applying the testing procedure for EFIE by RWG functions, with (22)–(24), we obtain an equation

$$i\omega l_m\left[\boldsymbol{A}(\mathbf{r}_m^{c+})\frac{\rho_m^{c+}}{2}+\boldsymbol{A}(\mathbf{r}_m^{c-})\frac{\rho_m^{c-}}{2}\right]+l_m[\Phi(\mathbf{r}_m^{c+})-\Phi(\mathbf{r}_m^{c-})]$$
$$=l_m\left[\boldsymbol{E}^i(\mathbf{r}_m^{c+})\frac{\rho_m^{c+}}{2}+\boldsymbol{E}^i(\mathbf{r}_m^{c-})\frac{\rho_m^{c-}}{2}\right] \quad (25)$$

that is valid for each inner edge, $m=\overline{1,N}$.

3.2 *Obtaining the System of Linear Algebraic Equations*

After inserting the expansion for the surface current (20) into Eq. (25), we obtain a system of linear algebraic equations (SLAE) of size $N \times N$, which can be represented as

$$\mathbf{Z}\,\mathbf{I}=\mathbf{V}, \quad (26)$$

where $\mathbf{Z}=\lfloor Z_{mn}\rfloor$ is the $N \times N$ matrix, $\mathbf{I}=\lfloor \alpha_n\rfloor$ is the column of unknown coefficients, $\mathbf{V}=[V_m]$ is the column of the known right-hand side. The elements of the matrix $\mathbf{Z}$ and the column $\mathbf{V}$ are determined by the following formulas:

$$Z_{mn}=l_m\left[i\omega\left(A_{mn}^{+}\cdot\frac{\rho_m^{c+}}{2}+A_{mn}^{-}\cdot\frac{\rho_m^{c-}}{2}\right)+\Phi_{mn}^{-}-\Phi_{mn}^{+}\right], \quad (27)$$

$$V_m=l_m\left(\mathbf{E}_m^{+}\cdot\frac{\rho_m^{c+}}{2}+\mathbf{E}_m^{-}\cdot\frac{\rho_m^{c-}}{2}\right), \quad (28)$$

where

$$A_{mn}^{\pm}=\frac{\mu}{4\pi}\int\limits_S f_n(\mathbf{r}')\frac{e^{-ik|\mathbf{r}_m^{c\pm}-\mathbf{r}'|}}{|\mathbf{r}_m^{c\pm}-\mathbf{r}'|}dS', \quad (29)$$

$$\Phi_{mn}^{\pm}=-\frac{1}{4\pi\epsilon i\omega}\int\limits_S \nabla_s' f_n(\mathbf{r}')\frac{e^{-ik|\mathbf{r}_m^{c\pm}-\mathbf{r}'|}}{|\mathbf{r}_m^{c\pm}-\mathbf{r}'|}dS', \quad (30)$$

$$E_m^{\pm}=\boldsymbol{E}^i(\mathbf{r}_m^{c\pm}). \quad (31)$$

After defining the elements of the moment matrix **Z** and vector **V**, we can solve the resulting system (26) with respect to the vector of unknown coefficients α_n by one of the well-known methods for solving SLAEs.

4 Algorithm for Solving the Problem

The numerical solution of the problem can be conditionally divided into three main stages. At the first stage, we build a triangular grid of the plate surface and the array of RWG elements. At the second stage, we compute the elements of the moment matrix and derive the final SLAE; at the third stage, we solve the SLAE and build the required function.

A numerical method for solving the problem of diffraction by a rectangular metal plate proposed in [22]. In the case of tooth-shaped plates, at the first stage, we divide the original area into a set of rectangles with common boundaries. In the case of the four-tooth-shaped plate shown in Fig. 2 on the left, the original area is divided into seven rectangles: P_0,..,P_6.

For each rectangle we build a triangular mesh and generate an array of RWG elements. The boundary points of the triangulation for adjacent rectangles should be selected so that these points are common. To speed up the first stage of the program the OpenMP technology is used. In this case, the rectangular areas are processed in parallel by the corresponding threads (processor cores). After the threads are finished, we get an array of RWG elements that are generated separately for each rectangle. The array does not contain elements that consist of triangles located on the common boundaries of rectangular regions.

Next, in the sequential part of the program, we construct the boundary RWG elements and add them to the previously generated array (see an example of such

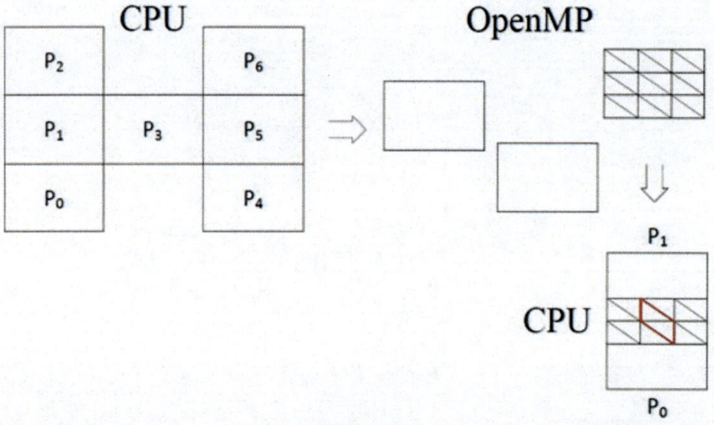

Fig. 2 RWG-mesh implementation scheme for tooth-shaped plate

element given for the boundary between the rectangles P_0 and P_1 on the bottom right of Fig. 2). Thus, after the first stage of the algorithm, the original tooth-shaped area is completely covered by the RWG mesh.

The next part is the calculation of the moment matrix elements by the formulas (27) and (28). This is the most laborious and time-consuming stage. However, each element of the moment matrix could be calculated independently. Consequently, the second stage of computations is easy to parallelize. For these purposes we use a graphics processor. An array of RWG elements is copied from RAM to video card memory; then each thread computes the corresponding matrix element.

The SLAE solving is carried out on the CPU by the Gaussian method after copying the moment matrix and the right-hand side from the GPU. Then the current vector on the plate is approximated using the formula (20).

5 Numerical Results

The calculation program is written in the programming language C using OpenMP and C/CUDA, provided by NVIDIA, which implements support for the CUDA API for compiling code that runs on a GPU. When launching the main computing core, which is responsible for calculating the elements of the moment matrix, a two-dimensional grid of blocks and two-dimensional blocks were used. This approach is convenient since the moment matrix is represented in memory as a two-dimensional data array.

For calculations, we used a personal computer with the Intel Core i3-5005U processor (2 GHz), RAM is 4 GB with the graphics accelerator GeForce 920M.

The case of normal incidence of an electromagnetic wave with a wavelength λ is considered. The calculations are performed for a metal perfectly conducting plate with the following parameters: width—0.9λ, height—0.675λ, tooth width—0.3λ, tooth depth—0.225λ.

Figure 3 shows the distributions of the absolute values of the current component $\boldsymbol{J}_x$ on the surface of a symmetrical four-tooth-shaped plate along the lines parallel to the axes Ox and Oy. The grid covering the tooth-shaped plate consists of 252 triangles, which corresponds to 345 RWG elements.

Distribution graph $|\boldsymbol{J}_x|$ (on the left in Fig. 3) is plotted along the lines $y = 0.3375\lambda$ and $y = 0.5625\lambda$. This component of the vector J is normal to the boundaries, and its values vanish at the edges. The values of the blue graph, corresponding to line B, also take zero values in the place where there is no metal.

The right part of Fig. 3 shows the graphs of the $|\boldsymbol{J}_x|$ distribution along the lines $x = 0.15\lambda$ and $y = 0.45\lambda$. In this case, the current component $|\boldsymbol{J}_x|$ is tangent to the boundaries and its values tend to infinity at the edges of the plate. $|\boldsymbol{J}_x|$ takes zero values outside the metal (see blue line).

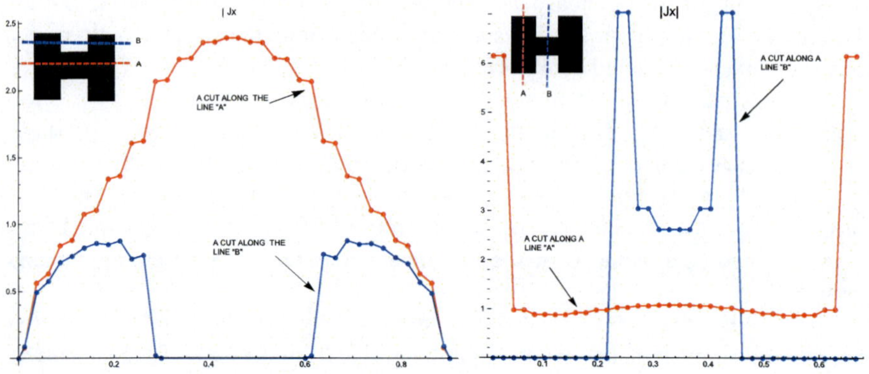

Fig. 3 Distribution of the current on various lines

Table 1 Time required for computing moment matrix on CPU and GPU

Number of RWG elements	Time on CPU (s)	Time on GPU (s)	Acceleration
225	2.80	0.71	3.93
587	15.19	1.40	10.84
1115	52.83	3.43	15.40
1443	95.63	5.95	16.06
1813	159.84	9.92	16.08

Table 1 summarizes data on the number of RWG elements (depends on the degree of area discretization) and the computation time required on CPU and GPU to generate the moment matrix for a different number of these elements.

Maximum 16x acceleration of GPU performance over CPU is achieved on a large number of RWG elements.

6 Conclusions

In this paper, the parallel algorithm for solving the problem of electromagnetic wave diffraction on tooth-shaped plates is proposed. The problem is reduced to the electric field integral equation (EFIE) and solved using the method of moments. For basis functions and testing procedure RWG functions are used. Two main stages in numerical algorithm—the construction of the RWG-mesh and the calculation of the moment matrix elements are discussed.

For the parallel implementation of the first stage, the OpenMP technology is used, and the elements of the moment matrix are calculated on the GPU using the CUDA technology. An almost 16-fold the acceleration of calculations on the video card is obtained. The numerical results of the algorithm for a symmetric four-tooth shaped metal plate are obtained. They show good correspondence with the results of previous work. Thus, the proposed parallel algorithm can be applied to the plane wave diffraction problems by the screens of complex shape with rectangular boundaries. Also, the proposed algorithm can be used to speed up the design of tooth-shaped antennas [23, 24].

Acknowledgments This paper has been supported by the Kazan Federal University Strategic Academic Leadership Program ("PRIORITY-2030").

References

1. Nagasaka, T., Kobayashi, K.: Wiener-Hopf analysis of the plane wave diffraction by a thin material strip. IEICE Transactions on Electronics **E100-C**, 11–19 (2017)
2. Luz, E., Granot, E., Malomed, B.A.: Analytical boundary-based method for diffraction calculations. Journal of Optics (2019) https://doi.org/10.1088/2040-8986/ab60c1
3. Nethercote, M.A., Assier, R.C., Abrahams, I.D.: Analytical methods for perfect wedge diffraction: A review. Wave Motion (2020) https://doi.org/10.1016/j.wavemoti.2019.102479
4. Tumakov, D.N., Tukhvatova, A.R.: Diffraction of an electromagnetic wave by gaps between plates. Lobachevskii J. Math. (2012) https://doi.org/10.1134/S1995080212040051.
5. Tumakov, D.N.:Iterative method for solving the problem of scattering of an electromagnetic wave by a partially shielded conducting sphere. App. Math. Sci. **8**, 5887–5898 (2014) https://doi.org/10.12988/ams.2014.48657
6. Selezov, I.T., Kryvonos, Y.G., Gandzha, I.S.: Some analytical and numerical methods in the theory of wave propagation and diffraction. Wave Prop. Diff. Springer, Singapore (2018) https://doi.org/10.1007/978-981-10-4923-1_1
7. Shibata, K., Kobayashi, M.: Difference between the method of moments and the finite element method for estimation of complex permittivity in liquids using a coaxial probe. Int. Symp. Electromagnetic Comp. (2019) https://doi.org/10.1109/EMCEurope.2019.8871604
8. Taflove, A., Hagness, S.: Computational Electrodynamics: the Finite-Difference Time-Domain Method, 2nd ed. Artech House, Norwood (2000)
9. Molostov, I.P., Scherbinin, V.V.: Application of NVIDIA CUDA technology for numerical simulation of electromagnetic pulses propagation. Izv. Altai State University **85**, 39–43 (2015)
10. Zhang, Y., Mei, X., Lin, H.: OpenMP-CUDA accelerated moment method for homogeneous dielectric objects. IEEE Antennas and Prop. Society International Symp. (2014) https://doi.org/10.1109/APS.2014.6905143
11. Li, J., Zhang, Z. et al.: GPU accelerated non-illuminated graphical electromagnetic computing method with high accuracy. Optik-International J. Light and Electron Optics **142**, 523–528 (2017)
12. Peng, S., Nie, Z.P.: Acceleration of the method of moments calculations by using graphics processing units. Anten. Prop. IEEE Trans. **56**, 2130–2133 (2008)
13. Balanis, C.A.: Antenna Theory: Analysis and Design, 4th ed. John Wiley & Sons, NY (2016)
14. Gibson, W.C.: The Method of Moments in Electromagnetics. Taylor & Francis Group, Abingdon (2008)
15. Harrington, R.F.: Field computation by moment methods. Anten. Prop. IEEE Trans. **4**, 229–231 (1993)
16. Sadiku, M.N.: Elements of Electromagnetics. Oxford University Press, NY (2001)
17. Mittra, R., Klein, C.A.: Stability and convergence of moment method solutions. In: Mittra R. (eds) Numerical and Asymptotic Techniques in Electromagnetics. Topics in Applied Physics. Springer, Berlin (1975)
18. De Loera, J.A., Rambau, J., Santos, F.: Triangulations: Structures for Algorithms and Applications. Algorithms and Computation in Mathematics. Springer, Berlin (2010)

19. Markina, A.G., Pleshchinskii, N.B., Tumakov, D.N.: On electrical characteristics of tooth-shaped microstrip antennas. Young Researchers in Electrical and Electronic Engineering (EIConRus). IEEE Conference of Russian, IEEE: 179–183 (2017) https://doi.org/10.1109/EIConRus.2017.7910523
20. Smirnov, Y.G., Medvedik, M.Y. et al: The solution of the problem of electromagnetic wave diffraction on screens of complex shape. Izv.Vuzov. Volga region. Physics and Mathematics **4**, 59–72 (2012)
21. Rao, M.S., Wilton, R.D., Glisson, A.W.: Electromagnetic scattering by surfaces of arbitrary shape. Anten. Prop. IEEE Prop. **AP-30**, 409–418 (1982)
22. Giniyatova, D., Tumakov, D., Markina A.: Solving Problem of Electromagnetic Wave Diffraction by a Metal Plate Using CUDA. In Proceedings of 2020 IEEE East-West Design & Test Symposium (EWDTS), 324–329 (2020) https://doi.org/10.1109/EWDTS50664.2020.9224674
23. Tumakov, D.N., Markina, A.G., Badriev, I.B.: Fast method for designing a well-matched symmetrical four-tooth-shaped microstrip antenna for Wi-Fi applications. J. Physics Conference Series (2019) https://doi.org/10.1088/1742-6596/1158/4/042029
24. Markina. A., Tumakov, D.: Designing the four-tooth-shaped microstrip antenna for Wi-Fi applications. 2019 1st International Conference on Control Systems, Mathematical Modelling, Automation and Energy Efficiency (SUMMA) (2019) https://doi.org/10.1109/SUMMA48161.2019.8947603

On Convergence of Explicit Differential Scheme for Solving One Parabolic Equation with Double Degeneration and Nonlocal Space Operator

Ludmila Glazyrina, Olga Glazyrina, and Maria Pavlova

Abstract We consider the initial-boundary value problem for nonlinear parabolic equations. This type of equation can be classified as a parabolic equation with double degeneration: degeneration can be present in space operator, and a nonlinear function which is under the derivative sign with respect to the variable t, may not be separated from zero. The space operator of the considered equation nonlinearly depends on the sought function, its gradient, and the non-local (integral) solution characteristic. This problem has an applied nature. Such equations appear, for example, in modeling the process of bacteria population spreading. In the present paper, we propose and investigate the explicit differential scheme. A priori estimates are obtained, and the convergence of the constructed algorithm is proved. The current work is a continuation of the research begun in our previous works, where the convergence of the explicit difference scheme in the case when nonlinearity is present only in the spatial operator have been investigated, for a problem with double degeneration, an approximate method has been studied. That method was constructed with the use of semidiscretization with respect to a variable t and the finite element method in the space variable with lowering nonlocality to the lower layer, the existence of an approximate solution and the convergence of the constructed algorithms was proved.

L. Glazyrina · O. Glazyrina (✉) · M. Pavlova
Institute of Computational Mathematics and Information Technologies, Kazan Federal University, Kazan, Russia
e-mail: glazyrina-ludmila@ya.ru; glazyrina-olga@ya.ru

I. B. Badriev et al. (eds.), *Mesh Methods for Boundary-Value Problems and Applications*, Lecture Notes in Computational Science and Engineering 141,
https://doi.org/10.1007/978-3-030-87809-2_12

1 Statement of the Problem

Let the Ω be bounded domain in the space R^n, Γ is its boundary, Ω, $Q_T = \Omega \times (0, T)$. In the domain Q_T consider the initial-boundary value problem

$$\frac{\partial \varphi(u)}{\partial t} - \sum_{i=1}^{n} \frac{\partial}{\partial x_i}\big(k_i(x, u, \nabla u, Bu)\big) = f, \quad x \in \Omega, \ t \in (0, T), \tag{1}$$

$$u(x, 0) = u_0(x) \quad x \in \Omega, \qquad u(x, t) = 0, \quad x \in \Gamma, \ t \in [0, T]. \tag{2}$$

Here k_i, u_0 are known functions, B is an operator of the form

$$Bu(t) = \int\limits_{\Omega'} g(x, u(x, t))\, dx\,, \tag{3}$$

g is a given function, Ω' is a domain that is contained in Ω or coincides with it.

We assume that function $\varphi(\xi)$ is an absolutely continuous, strongly increasing function and it satisfies the following inequalities for arbitrary $\xi \in R^1$,

$$b_0 \mid \xi \mid^{\alpha} - b_1 \le \Phi(\xi) \equiv \int\limits_0^{\xi} \varphi'(t) t dt \le b_2 \mid \xi \mid^{\alpha} + b_3, \quad \alpha > 1, \tag{4}$$

$$\mid \varphi(\xi) \mid \le b_i \mid \xi \mid^{\alpha - 1} + b_5, \tag{5}$$

$$(\varphi'(\xi)\xi)' \ge 0, \tag{6}$$

here b_{ij} are constants such that following inequalities are correct

$$b_{0i} > 0, \ b_{1i} \ge 0, \ b_{2i} > 0, \ b_{3i} \ge 0, \ b_{4i} > 0, \ b_{5i} \ge 0\,, i = 1, 2,$$

functions $k_i(x, \xi_0, \xi, \nu)$, $i = 1, \ldots, n$, are continuous with respect to ξ_0, ν and ξ, measurable with respect to x and for arbitrary $x \in \Omega$, $\xi_0, \nu \in R$, $\xi^1, \xi^2, \xi \in R^n$ satisfy the following conditions

$$\mid k_i(x, \xi_0, \xi, \nu) \mid \le d_0 \sum_{j=1}^{n} \mid \xi_j \mid^{p-1} + d_1\,, \quad d_0 > 0, \quad d_1 \ge 0\,, \quad p > 1\,, \tag{7}$$

$$\sum_{i=1}^{n} k_i(x, \xi_0, \xi, \nu)\xi_i \ge d_2 \sum_{i=1}^{n} \mid \xi_i \mid^{p} - d_3, \quad d_2 > 0, \quad d_3 \ge 0, \tag{8}$$

$$\sum_{i=1}^{n} \big(k_i(x, \xi_0, \xi^1, \nu) - k_i(x, \xi_0, \xi^2, \nu)\big)(\xi_i^1 - \xi_i^2) \ge 0. \tag{9}$$

Let's note that condition (7) implies that the operator L, acting from $\overset{\circ}{W}{}_p^1(\Omega)$ into $W_{p'}^{-1}(\Omega)$, where $p' = \dfrac{p}{p-1}$, is bounded. The conditions (8) and (9) provide, respectively, the coercivity and monotonicity with respect to the gradient of the operator L.

We assume that the function $g(x, \xi)$, defining the operator B, is continuous with respect to ξ, measurable with respect to x and satisfies the following condition

$$|g(x,\xi)| \le g_0(x) + |\xi|^s \qquad \text{for almost all } x \in \Omega, \tag{10}$$

where g_0 is a function integrable over Ω, $s \ge 0$.

Space operators with non-localities of the form (3) arise, for example, in the mathematical describing the diffusion of bacteria population when it is assumed that the propagation speed at a point is specified by the global state of the environment (e.g., see [1–3]).

Definition 1 A function $u \in L_p(0, T; \overset{\circ}{W}{}_p^1(\Omega)) \bigcap L_\infty(0, T; L_\alpha(\Omega))$ such that

$$u(x,0) = u_0(x) \quad \text{almost everywhere in } \Omega, \qquad \frac{\partial \varphi(u)}{\partial t} \in L_{p'}(0, T; W_{p'}^{-1}(\Omega)),$$

will be called a generalized solution of problem (1) and (2), if for any function v from the space $L_p(0, T; \overset{\circ}{W}{}_p^1(\Omega))$ the integral identity holds

$$\int_0^T \left\langle \frac{\partial \varphi(u)}{\partial t}, v \right\rangle dt + \int_0^T \int_\Omega \sum_{i=1}^n k_i\big(x, u, \nabla u, Bu\big) \frac{\partial v}{\partial x_i}\, dx dt = \int_0^T \langle f, v \rangle dt, \tag{11}$$

here $\langle g, v\rangle$ is the value of a functional g from $W_{p'}^{-1}(\Omega)$ on element v from $\overset{\circ}{W}{}_p^1(\Omega)$. When obtaining the results presented in this article, we use the technique from the papers [4–6]. The current work is a continuation of the research begun in the works [7, 8].

2 Auxiliary Results and Notation

In what follows, we will assume that the domain Ω is a n-dimensional parallelepiped: $\overline{\Omega} = \{x \in R_n : 0 \le x_i \le l_i,\ i = 1, 2, \ldots, n.\}$. On Ω construct a uniform mesh $\bar{\omega}_h$ with a mesh step h_i in the i-th direction, $\mathbf{h} = (h_1, \ldots, h_n)$, $h =$

$\min\limits_{1\le i\le n} h_i$. We will assume that there is a constant c such that $\overline{h} \le ch$, $\overline{h} = \max\limits_{1\le i\le n} h_i$. We denote

$$\overline{\omega}_h = \left\{x = (x_1, \dots, x_n) \in \overline{\Omega} : \ x_i = jh_i, \ j = 0, \dots, N_i, \ N_i = \frac{l_i}{h_i}\right\},$$

$$\gamma_h = \bar{\omega}_h \cap \Gamma, \quad \omega_h = \overline{\omega}_h \backslash \gamma_h.$$

On $[0, T]$ we construct a uniform mesh with a step τ:

$$\overline{\omega}_\tau = \left\{t \in [0, T] : \ t = j\tau, \ j = 0, \dots, M, \ M = \frac{T}{\tau}\right\}, \qquad \omega_\tau = \overline{\omega}_\tau \backslash \{0\}.$$

We denote by H the set of mesh functions defined on $\overline{\omega}$, $\overset{\circ}{H}$ are the functions from H, that equal zero on γ. Let further r is the n-dimensional vector with coordinates $r_i = \pm 1$, $\nabla_r y(x) = (\partial_{r_1} y(x), \partial_{r_2} y(x), \dots, \partial_{r_n} y(x))$,

$$\partial_{r_i} y(x) = \begin{cases} y_{x_i}(x), \ r_i = +1, \\ y_{\bar{x}_i}(x), \ r_i = -1. \end{cases}$$

Let us denote by $H_r(x)$ a mesh cell $\overline{\omega}$,, which contains all the mesh points participating in the notation of operator $\nabla_r y(x)$, ω_r is the set of points $x \in \overline{\omega}$, at which the operator $\nabla_r y(x)$ is defined. In the space of mesh functions $\overset{\circ}{H}$ introduce the following norms and scalar products

$$(y, v)_r = \sum_{x\in\omega_r} \tilde{H}_r \, y(x) \, v(x), \qquad [y, v] = (1/2^n) \sum_r (y, v)_r,$$

$$\| \, y \, \|_p = [| \, y \, |^p, 1]^{1/p}, \qquad \| \, y \, \|_{+p}^p = (1/2^n) \sum_r \sum_{i=1}^n (| \, \partial_{r_i} y \, |^p, 1)_r,$$

$$\| \, y \, \|_{-p'} = \sup_{v\neq 0} \frac{[y, v]}{\| \, v \, \|_{+p}},$$

here $\tilde{H}_r = \text{mes} \ H_r(x)$.

For mesh functions, we define piecewise constant extensions x and t each

$$\Pi_r z(x) = \{z(x'), x' \in \omega_r, x \in H_r(x')\},$$
$$\Pi^- w(t') = \{w(t), t = k\tau, (k-1)\tau < t' \le k\tau\},$$
$$\Pi^+ w(t') = \{w(t), t = k\tau, k\tau \le t' < (k+1)\tau\},$$
$$\Pi_r^+ w = \Pi^+ \Pi_r w, \quad \Pi_r^- w = \Pi^- \Pi_r w.$$

Lemma 1 (See [4]) *If $\varphi(\xi)$ is an absolutely continuous increasing function, then the following inequality holds*

$$(\varphi(\xi) - \varphi(\eta))\xi \geq \Phi(\xi) - \Phi(\eta), \quad \forall \xi, \eta \in R^1. \tag{12}$$

Lemma 2 (See [4]) *Let $\alpha \geq 2$, function φ satisfies the condition (4) and besides*

$$\varphi'(\xi) \geq b_6 \mid \xi \mid^{\alpha-2}, \qquad b_6 > 0. \tag{13}$$

Then for any constant $\theta > 1$ there is $\bar{c} = const > 0$, such that for any $\xi, \eta \in R^1$ the inequality holds

$$(\varphi(\xi) - \varphi(\eta))(\theta\xi - (\theta - 1)\eta) \geq \Phi(\xi) - \Phi(\eta) + \bar{c} \mid \xi - \eta \mid^{\alpha}. \tag{14}$$

Lemma 3 (See [4]) *Let $\varphi(\xi)$ be an absolutely continuous, monotonically increasing function satisfying the conditions (4)–(6). Then for any function v such that*

$$v \in L_p(0, T; \overset{\circ}{W}{}^1_p(\Omega)) \bigcap L_\infty(0, T; L_\alpha(\Omega)), \tag{15}$$

$$\frac{\partial \varphi(v)}{\partial t} \in L_{p'}(0, T; W^{-1}_{p'}(\Omega)), \tag{16}$$

$$v(x, 0) \in \overset{\circ}{W}{}^1_p(\Omega) \bigcap L_\alpha(\Omega), \tag{17}$$

the following equality holds

$$\int_0^T \langle \frac{\partial \varphi(v)}{\partial t}, v \rangle \, dt = \lim_{\lambda \to 0} \frac{1}{\lambda} \int_{T-\lambda}^{T} \int_{\Omega} \Phi(v(t)) \, dx \, dt - \int_{\Omega} \Phi(v(0)) \, dx. \tag{18}$$

It is easy to check the validity of the following lemma.

Lemma 4 (See [4]) *For any $y \in \overset{\circ}{H}$ the inequality holds*

$$\| y \|_{+p} \leq \lambda_\alpha \| y \|_\alpha, \tag{19}$$

where $\lambda_\alpha = \dfrac{c \sqrt[p]{n}}{h^{1+n(p-\alpha)/\alpha p}}$, *if* $p \geq \alpha$ *and* $\lambda_\alpha = \dfrac{c \sqrt[p]{n}}{h}$, *if* $1 < p < \alpha$.

3 Construction and Investigation of an Explicit Difference Scheme

For the problem (1) and (2), consider the explicit difference scheme

$$\varphi_t(y) + Ay(x,t) = f_{h\tau}(x,t), \quad x \in \omega_h,\ t \in \overline{\omega}_\tau \backslash \{T\}, \tag{20}$$

$$y(x,0) = y_0(x), \qquad y\,|_{\gamma_h} = 0.$$

Here A is a difference operator acting from $\overset{\circ}{H}$ to $\overset{\circ}{H}$, defined by the relation

$$[Ay, w] = \frac{1}{2^n}\sum_r \sum_{i=1}^n (a_i(x,y)k_i(x, \nabla_r y, B_h y), \partial_{ri} w)_r,$$

where $B_h y(t) = B(2^{-n}\sum\limits_r \Pi_r y(t))$, y_0 a difference analog of u_0 such that

$$\Pi_r y_0 \to u_0 \quad \text{in} \quad L_\alpha(\Omega), \tag{21}$$

$f_{h\tau}$ is a mesh function, that is an approximation of the original equation right side, which we define as follows

$$[f_{h\tau}, v] = \frac{1}{2^n}\sum_r \sum_{i=0}^n (f^r_{h\tau,i}, \partial_{r_i} v)_r \quad \forall v \in \overset{\circ}{H},$$

where

$$\partial_{r_0} v \equiv v, \quad f^r_{h\tau,i}(t) = \frac{1}{\tau\ \mathrm{mes}\big(H_r(x)\big)} \int_t^{t+\tau} \int_{H_r(x)} f_i(\xi,\eta)\, d\xi d\eta.$$

Conditions (7)–(8) on the coefficients k_i provide continuity, boundedness:

$$\| Ay \|_{-p'} \le c_0 \| y \|_{+p}^{p-1} + \bar{c}_0, \tag{22}$$

the coercivity of the operator A :

$$[Ay, y] \ge d_2 \| y \|_{+p}^{p} - d_3, \tag{23}$$

with constants $d_2 > 0$, $d_3 \ge 0$, $c_0 > 0$, $\bar{c}_0 \ge 0$, independent on $\bar{h}$ and τ. The unique solvability of the difference scheme (20) follows from the condition that the function φ is strictly monotonic.

Lemma 5 *Let $\alpha \geq 2$, function φ satisfies the conditions (4)–(5) and besides*

$$u_0 \in L_\alpha(\Omega), \quad f \in L_q(0, T; W_{p'}^{-1}(\Omega)), \ q = \max\{\alpha', p'\}.$$

Then for any

$$\tau \leq \begin{cases} c\,\dfrac{h^\alpha}{2^\alpha n^{\alpha/p}}, & 1 < p < \alpha, \\ c\,\dfrac{h^{p+n(p-\alpha)/\alpha}}{2^p n}, & p \geq \alpha, \end{cases} \tag{24}$$

for the solution of the difference scheme (20) the following a priori estimates hold

$$\sum_{t=0}^{t'} \tau \parallel y \parallel_{+p}^{p} \leq const, \tag{25}$$

$$\max_{t' \in \bar{\omega}_\tau} \parallel y(t') \parallel_\alpha^\alpha \leq const, \tag{26}$$

$$\sum_{t=0}^{t'} \tau^\alpha \parallel y_t \parallel_\alpha^\alpha \leq const \qquad \forall t' \in \bar{\omega}_\tau, \tag{27}$$

$$\frac{1}{k\tau} \sum_{t=0}^{T-k\tau} \tau[\varphi(y(t+k\tau)) - \varphi(y(t)), y(t+k\tau) - y(t)] \leq const \tag{28}$$

$$\forall k \in \{1, \ldots, N\}.$$

Proof Multiply both sides of (20) scalar in H by $\tau(\theta\hat{y} - (\theta - 1)y)$, where the constant $\theta > 1$. As a result, we get

$$\tau[\varphi_t(y), \theta\hat{y} - (\theta - 1)y] + \tau[Ay, \theta\hat{y} - (\theta - 1)y] = \tau[f_{h\tau}, \theta\hat{y} - (\theta - 1)y]$$

or

$$\begin{aligned} &\tau[\varphi_t(y), \theta\hat{y} - (\theta - 1)y] + \tau[Ay, y] = \\ &= \tau[f_{h\tau}, y] + \tau^2\theta[f_{h\tau}, y_t] - \tau^2\theta[Ay, y_t]. \end{aligned} \tag{29}$$

Using Lemma 2, we estimate the first summand in the left-hand side of the Eq. (29)

$$\tau[\varphi_t(y), \theta\hat{y} - (\theta - 1)y] \geq [\Phi(\hat{y}) - \Phi(y), 1] + \bar{c}\tau^\alpha \parallel y_t \parallel_\alpha^\alpha . \tag{30}$$

To estimate the first two summands on the right-hand side of (29) we use HÃűlder inequality, ε—inequality and a difference analog of the Friedrichs inequality, as a result we have

$$\tau[f_{h\tau}, y] \leq \frac{1}{\varepsilon_1^{p'} p'} \tau \sum_{j=0}^{n} \| f_{h\tau,j} \|_{p'}^{p'} + \frac{\varepsilon_1^p}{p} (1 + c_\Omega) \tau \| y \|_{+p}^p, \quad (31)$$

$$\tau^2[f_{h\tau}, y_t] \leq \frac{1}{\varepsilon_2^{\alpha'} \alpha'} \tau \sum_{j=0}^{n} \| f_{h\tau,j} \|_{p'}^{\alpha'} + \frac{\varepsilon_2^\alpha \tau^{\alpha+1}}{\alpha} (\| y_t \|_{+p}^\alpha + \| y_t \|_p^\alpha) \leq \quad (32)$$

$$\leq \frac{1}{\varepsilon_2^{\alpha'} \alpha'} \tau \sum_{j=0}^{n} \| f_{h\tau,j} \|_{p'}^{\alpha'} + \frac{\varepsilon_2^\alpha \tau^{\alpha+1}}{\alpha} (1 + c_\Omega) \lambda_\alpha^\alpha \| y_t \|_\alpha^\alpha + c_1 \tau,$$

here c_Ω is the constant from the difference analog of the Friedrichs inequality. From (22) follows that

$$\tau^2 \theta [Ay, y_t] \leq \tau^2 \theta (c_0 \| y \|_{+p}^{p-1} + \bar{c}_0) \| y_t \|_{+p} \equiv I + \tau^2 \theta \bar{c}_0 \| y_t \|_{+p} . \quad (33)$$

Further, using (30)–(33) and the coercivity of the operator A, from (29) is easy to obtain

$$[\Phi(\hat{y}) - \Phi(y), 1] + \bar{c} \tau^\alpha \| y_t \|_\alpha^\alpha + d_2 \tau \| y \|_{+p}^p - d_3 \tau \leq$$

$$\leq \frac{1}{\varepsilon_1^{p'} p'} \tau \sum_{j=0}^{n} \| f_{h\tau,j} \|_{p'}^{p'} + \frac{\varepsilon_1^p}{p} (1 + c_\Omega) \tau \| y \|_{+p}^p +$$

$$+ \frac{1}{\varepsilon_2^{\alpha'} \alpha'} \tau \sum_{j=0}^{n} \| f_{h\tau,j} \|_{p'}^{\alpha'} + \frac{\varepsilon_2^\alpha \tau^{\alpha+1}}{\alpha} (2 + c_\Omega) \lambda_\alpha^\alpha \| y_t \|_\alpha^\alpha + I + c_1 \tau. \quad (34)$$

Let $p \geq \alpha$. We estimate I using HÃűlder's inequality and Lemma 2, as a result we obtain

$$I \leq \tau^2 c_0 \theta \| y \|_{+p}^{p/\alpha'} \| y \|_{+p}^{(p-\alpha)/\alpha} \lambda_\alpha \| y_t \|_\alpha \leq$$

$$\leq \tau^2 c_0 \theta \| y \|_{+p}^{p/\alpha'} \lambda_\alpha^{p/\alpha} \| y \|_\alpha^{(p-\alpha)/\alpha} \| y_t \|_\alpha \leq$$

$$\leq \frac{\tau \varepsilon_3^{\alpha'}}{\alpha'} \| y \|_{+p}^p + \frac{\tau^{\alpha+1} (c_0 \theta)^\alpha \lambda_\alpha^p}{\alpha \varepsilon_3^\alpha} \| y \|_\alpha^{p-\alpha} \| y_t \|_\alpha^\alpha . \quad (35)$$

Substituting (35) into (34) and summing the resulting inequalities over t from 0 to $t' \in \bar{\omega}_\tau$, we will have

$$[\Phi(y(t')), 1] + \left(M_2 - \frac{\varepsilon_1^p}{p}(1+_\Omega^p) - \frac{\varepsilon_3^{\alpha'}}{\alpha'} \right) \sum_{t=0}^{t'} \tau \parallel y \parallel_{+p}^p +$$

$$+ \sum_{t=0}^{t'} \left(\bar{c} - \tau \frac{\varepsilon_2^\alpha}{\alpha}(2 + c_\Omega)\lambda_\alpha^\alpha - (c_0\theta)^\alpha \frac{\tau \lambda_\alpha^p}{\alpha \varepsilon_3^p} \parallel y(t) \parallel_\alpha^{p-\alpha} \right) \tau^\alpha \parallel y_t \parallel_\alpha^\alpha \leq$$

$$\leq \frac{1}{\varepsilon_1^{p'} p'} \sum_{t=0}^{t'} \tau \sum_{j=0}^{n} \parallel f_{h\tau,j}(t) \parallel_{p'}^{p'} +$$

$$+ \frac{1}{\varepsilon_2^{\alpha'} \alpha'} \sum_{t=0}^{t'} \tau \sum_{j=0}^{n} \parallel f_{h\tau,j}(t) \parallel_{p'}^{\alpha'} + [\Phi(y(0)), 1] + c_3. \tag{36}$$

First, let us prove that (36) implies the estimate

$$\parallel y(t') \parallel_\alpha^\alpha \leq c \left(\sum_{t=0}^{T} \tau \sum_{j=0}^{n} \parallel f_{h\tau,j}(t) \parallel_{p'}^{p'} + \sum_{t=0}^{T} \tau \sum_{j=0}^{n} \parallel f_{h\tau,j}(t) \parallel_{p'}^{\alpha'} + \right.$$

$$\left. + [\Phi(y(0)), 1] + 1 \right) = m^\alpha \qquad \forall t' \in \bar{\omega}_\tau, \tag{37}$$

where c, m are constants independent of $\bar{h}$ and τ. For $t' = 0$ estimate (37) holds. We assume that (37) is valid for all $t' \leq t_1;\ t', t_1 \in \omega_\tau$. Let us prove that (37) holds for $t' = t_1 + \tau$. To do this, write inequality (36) for $t' = t_1 + \tau$, considering, that $\parallel y(t) \parallel_\alpha^\alpha \leq m^\alpha \quad \forall t \leq t_1$,

$$[\Phi(y(t_1 + \tau)), 1] + \left(d_2 - \frac{\varepsilon_1^p}{p}(1 + c_\Omega^p) - \frac{\varepsilon_3^{\alpha'}}{\alpha'} \right) \sum_{t=0}^{t_1} \tau \parallel y \parallel_{+p}^p +$$

$$+ \left(\bar{c} - \tau \frac{\varepsilon_2^\alpha}{\alpha}(2 + c_\Omega)\lambda_\alpha^\alpha - (c_0\theta)^\alpha \frac{\tau \lambda_\alpha^p}{\alpha \varepsilon_3^p} m^{p-\alpha} \right) \sum_{t=0}^{t_1} \tau^\alpha \parallel y_t \parallel_\alpha^\alpha \leq$$

$$\leq \frac{1}{\varepsilon_1^{p'} p'} \sum_{t=0}^{t_1} \tau \sum_{j=0}^{n} \parallel f_{h\tau,j}(t) \parallel_{p'}^{p'} +$$

$$+ \frac{1}{\varepsilon_2^{\alpha'} \alpha'} \sum_{t=0}^{t_1} \tau \sum_{j=0}^{n} \parallel f_{h\tau,j}(t) \parallel_{p'}^{\alpha'} + [\Phi(y(0)), 1] + c_3. \tag{38}$$

Choosing $\varepsilon_1, \varepsilon_2, \varepsilon_3, \bar{h}$ and τ so that

$$d_2 - \frac{\varepsilon_1^p}{p}(1 + c_\Omega^p) - \frac{\varepsilon_3^{\alpha'}}{\alpha'} \geq \delta_1 > 0,$$

$$\bar{c} - \tau \frac{\varepsilon_2^\alpha}{\alpha}(2 + c_\Omega)\lambda_\alpha^\alpha - (c_0\theta)^\alpha \frac{\tau \lambda_\alpha^p}{\alpha \varepsilon_3^p} m^{p-\alpha} \geq \delta_2 > 0\,, \tag{39}$$

and using the condition (4), of (38) is easy to obtain (37) for $t' = t_1 + \tau$. Therefore, the estimate (37) will be valid for any $t' \in \bar{\omega}_\tau$. From (36) and (37) the estimates (25)–(27) follow. Note that the constant c in (24) is chosen so that the inequality (39) holds.

Similarly to the way above, it is easy to verify the validity of estimates(25)–(27) in the case $1 < p < \alpha$.

Let us further prove the validity of the estimate (28).To do this, we sum both sides (20) over t from $\bar{t}$ to $\bar{t} + (k-1)\tau$, then multiply the resulting equality scalarly in H by $\tau(y(\bar{t} + k\tau) - y(\bar{t}))$ and again sum over $\bar{t}$ from 0 to $T - k\tau$, as a result we will have

$$\frac{1}{k\tau} \sum_{\bar{t}=0}^{T-k\tau} \tau[\varphi(y(\bar{t} + k\tau)) - \varphi(y(\bar{t})), y(\bar{t} + k\tau) - y(\bar{t})] =$$

$$= -\frac{1}{k} \sum_{\bar{t}=0}^{T-k\tau} \sum_{t=\bar{t}}^{\bar{t}+(k-1)\tau} \tau[Ay(t), y(\bar{t} + k\tau) - y(\bar{t})] +$$

$$+\frac{1}{k} \sum_{\bar{t}=0}^{T-k\tau} \sum_{t=\bar{t}}^{\bar{t}+(k-1)\tau} \tau[f, y(\bar{t} + k\tau) - y(\bar{t})]. \tag{40}$$

Using the boundedness property of the operator A, HÃűlder's inequalities and (33), from (40) it is easy to obtain

$$\frac{1}{k\tau} \sum_{\bar{t}=0}^{T-k\tau} \tau[\varphi(y(\bar{t} + k\tau)) - \varphi(y(\bar{t})), y(\bar{t} + k\tau) - y(\bar{t})] \leq$$

$$\leq c_1 \sum_{\bar{t}=0}^{T-k\tau} \tau \parallel y(\bar{t}) \parallel_{+p}^p + \frac{2}{p'} \sum_{t=0}^{T} \tau \sum_{j=0}^{n} \parallel f_{h\tau,j}(t) \parallel_{p'}^{p'} .$$

From the last inequality and (25) it follows (28). The lemma is proved. □

The a priori estimates (25) and (26) imply the boundedness of the set $\{\Pi_r^\pm y\}$ in the spaces $L_p(Q_T)$ and $L_\infty(0, T; L_2(\Omega)$, as well as the boundedness of the set $\{\Pi_r^\pm \partial_{r_i} y\}$ in the space $L_p(Q_T)$. Due to the weak compactness of bounded sets in reflexive spaces and the *-weak compactness of bounded sets in $L_\infty(0, T; L_\alpha(\Omega))$ subsequences $\{\mathbf{h}^{(m)}\}_{m=1}^{\infty}$, $\{\tau_m\}_{m=1}^{\infty}$ [1] and the element u, which belongs to $L_p(0, T; \overset{\circ}{W}{}_p^1(\Omega)) \bigcap L_\infty(0, T; L_2(\Omega))$, such that for $\mathbf{h}^{(m)}, \tau_m \to 0$, exists there

$$\Pi_r^{\pm} y \rightharpoonup u \text{ in } L_p(Q_T), \tag{41}$$

$$\Pi_r^{\pm} \partial_{r_i} y \rightharpoonup \frac{\partial u}{\partial x_i} \text{ in } L_p(Q_T), \tag{42}$$

$$\Pi_r^{\pm} y \to u \text{ *-weak in } L_\infty(0, T; L_\alpha(\Omega)). \tag{43}$$

Using the estimates (26), (27), (29) and the mesh analog of the compactness theorem (see [4], lemma 9), it is easy to confirm the existence of subsequences $\{\mathbf{h}^{(m)}\}_{m=1}^{\infty}$, $\{\tau_m\}_{m=1}^{\infty}$, for which, along with (41)–(43) the limit relation of the form below holds

$$\Pi_r^{\pm} y \to u \text{ almost everywhere in } Q_T. \tag{44}$$

Further, the condition (7) and the estimate (25) imply the boundedness in $L_{p'}(Q_T)$ of the set $\{\Pi_r^{\pm} k_i(x, y, \nabla_r y, B_h y)\}$ for any $i \in \{1, 2, \dots, n\}$. Therefore, there are $K_i \in L_{p'}(Q_T)$ and sequences $\{\mathbf{h}^{(m)}\}_{m=1}^{\infty}$, $\{\tau_m\}_{m=1}^{\infty}$ such that

$$\Pi_r^{\pm} k_i(x, y, \nabla_r y, B_h y) \rightharpoonup K_i \text{ in } L_{p'}(Q_T). \tag{45}$$

For $s \le \alpha$ from (26), (44) and Lebesgue's theorem on passage to the limit, it is easy to show that

$$\Pi^{\pm} B(y) \to Bu \qquad \text{in } L_1(0, T). \tag{46}$$

Theorem 1 *Let the functions φ, k_i satisfy conditions (7)–(9) and (13), $\alpha \ge 2$ and the inequality (24) holds. Let, in addition, for $\tau, \bar{h} \to 0$*

$$\tau \lambda_\alpha^p \to 0, \quad \textit{if} \quad p \ge \alpha, \qquad \tau \lambda_\alpha^\alpha \to 0, \quad \textit{if} \quad 1 < p < \alpha. \tag{47}$$

Then for any function $f \in L_q(0, T; W_{p'}^{-1}(\Omega))$, where $q = \max\{\alpha', p'\}$, and the function $u_0, \in L_\alpha(\Omega) \bigcap \overset{\circ}{W}{}_p^1(\Omega)$ subsequence of piecewise constant extensions of the solution to the difference scheme (20), defined by the relations (41)–(46), converges to a generalized solution of the problem (1)–(2).

[1] In what follows, for the selected subsequences we will keep the notation of the sequences themselves.

Proof of this theorem is close to the proof of Lemma 3 from ([7]). Therefore, we present here only fragments of reasoning different from Lemma 3.

Let's scalarly multiply the difference scheme (20) by τz, where z —the drift of the function $\bar{z}$ from $C^\infty(0, T; C_0^\infty(\Omega))$, $\bar{z}(x, T) = 0$ and sum over t from 0 to $T - \tau$. As a result, we get

$$\sum_{t=0}^{T-\tau} \tau[\varphi_t, z] + \sum_{t=0}^{T-\tau} \tau[Ay, z] = \sum_{t=0}^{T-\tau} \tau[f_{h\tau}, z].$$

We transform the first summand by using the formula for summation by parts. We write the resulting equality using piecewise constant extensions in the form of the integral identity

$$\frac{1}{2^n}\sum_r \Big\{ -\int_0^T \int_\Omega \Pi_r^- \varphi(y) \Pi_r^-(z_{\bar{t}}) dx dt +$$

$$+ \sum_{i=1}^n \int_0^T \int_\Omega \Pi_r^+ k_i(x, y, \nabla_r y, B_h y) \Pi_r^+ \partial_{r_i} z dx dt \Big\} =$$

$$= \frac{1}{2^n} \sum_r \sum_{i=1}^n \int_0^T \int_\Omega \Pi_r^+ f_{h\tau, i} \Pi_r^+ \partial_{r_i} z dx dt. \tag{48}$$

In the equality (48), we pass to the limit as $\tau, h \to 0$. As a result, we will have

$$-\int_0^T \int_\Omega \varphi(u) \frac{\partial \bar{z}}{\partial t} dx dt - \int_\Omega \varphi(u_0) \bar{z}(x, 0) dx +$$

$$+ \sum_{i=1}^n \int_0^T \int_\Omega K_i \frac{\partial \bar{z}}{\partial x_i} dx dt = \int_0^T \langle f, \bar{z} \rangle dt. \tag{49}$$

Following ([7], lemma 3), from (49) it is easy to obtain that

$$\int_0^T \langle \frac{\partial \varphi(u)}{\partial t}, \bar{z} \rangle \, dt + \sum_{i=1}^n \int_0^T \int_\Omega K_i \frac{\partial \bar{z}}{\partial x_i} dx dt =$$

$$= \int_0^T \langle f, \bar{z} \rangle \, dt \qquad \forall \bar{z} \in L_p(0, T; \overset{\circ}{W}{}_p^1(\Omega)) \tag{50}$$

and, besides, $u(x,0) = u_0(x)$ almost everywhere in Ω. Let us prove further that

$$\sum_{i=1}^{n} \int_0^T \int_\Omega K_i \frac{\partial \bar{z}}{\partial x_i} dxdt = \sum_{i=1}^{n} \int_0^T \int_\Omega k_i(x, u, \nabla u, Bu) \frac{\partial \bar{z}}{\partial x_i} dxdt \tag{51}$$

for any function $\bar{z}$ from $L_p(0, T; \overset{\circ}{W}{}^1_p(\Omega))$. To do this, we consider the following inequality

$$[\varphi(\hat{y}) - \varphi(y), \hat{y}] + \sum_{i=1}^{n} \tau[(k_i(x, y, \nabla y, B_h y) - k_i(x, \nabla \hat{v}, B_h y)), \partial_{r_i}(y - \hat{v})] \geq$$

$$\geq [\Phi(\hat{y}) - \Phi(y), 1], \tag{52}$$

where y is the solution of the difference scheme (20), $v(x,t)$ is the drift of the function $\bar{v}(x,t) \in C^\infty(0, T; C_0^\infty(\Omega))$ to the points of the mesh $\bar{\omega}_\tau \times \bar{\omega}$. The validity of (52) follows from (9) and the Lemma 1. Considering that the function y satisfies equality (20), we rewrite inequality (52) as follows

$$[f_{h\tau}, \hat{y}] + \tau[Ay, y_t] - \sum_{i=1}^{n} [k_i(x, y\nabla\hat{v}, B_h y), \partial_{r_i}(y - \hat{v})] -$$

$$- \sum_{i=1}^{n} [k_i(x, y\nabla y, B_h y), \partial_{r_i}\hat{v}] \geq \frac{1}{\tau}[\Phi(\hat{y}) - \Phi(y), 1].$$

Using the extension Π_r^+, we write the last inequality for all $t \in [0, T]$ and integrate the resulting inequality over the segment $[0, t']$, $t' \in [0, T]$. As a result we will have

$$J_1(t') = \frac{1}{2^n} \sum_r \int_0^{t'} \{\langle \Pi_r^+ f_{h\tau}, \Pi_r^+ y\rangle - \sum_{i=1}^{n} \int_\Omega \Pi_r^+ k_i(x, y, \nabla y, B_h y) \Pi_r^+ \partial_{r_i} \hat{v} dx -$$

$$- \sum_{i=1}^{n} \int_\Omega \Pi_r^+ k_i(x, \nabla\hat{v}, B_h y) \Pi_r^+ \partial_{r_i}(y - \hat{v}) dx\} dt + \tag{53}$$

$$+ \sum_{t=0}^{T-\tau} \tau^2 \mid [Ay, y_t] \mid \geq \frac{1}{2^n} \sum_r \frac{1}{\tau} \int_0^{t'} \int_\Omega \{\Phi(\Pi_r^+ \hat{y}) - \Phi(\Pi_r^+ y)\} dxdt.$$

Further, using the [7] methodology, when the condition (47) holds we establish the validity of the limit equality

$$\lim_{\tau,h\to 0}\sum_{t=0}^{T-\tau}\tau^2\big|[Ay,y_t]\big|=0. \tag{54}$$

Let us notice, that

$$\frac{1}{\tau}\int_0^{t'}\int_\Omega\{\Phi(\Pi_r^+\hat{y})-\Phi(\Pi_r^+y)\}dxdt=\frac{1}{\tau}\int_{t'}^{t'+\tau}\int_\Omega\Phi(\Pi_r^+y)dxdt-\int_\Omega\Phi(u_0(x))dx.$$

Let further t^* be a mesh point ω_τ, belonging to $(t',t'+\tau]$, $\mu(t')=(t'+\tau-t^*)/\tau$, Λ_τ—linear extension with respect to t. Using the convexity of the function Φ, we have

$$\frac{1}{\tau}\int_{t'}^{t'+\tau}\int_\Omega\Phi(\Pi_r^+y(t))dxdt=\frac{1}{\tau}\Bigg\{\int_{t^*}^{t'+\tau}\int_\Omega\Phi(\Pi_r^+y(t))dxdt$$

$$+\int_{t'}^{t^*}\int_\Omega\Phi(\Pi_r^+y(t))dxdt\Bigg\}=$$

$$=\mu(t')\int_\Omega\Phi(\Pi_r y(t^*))dx+(1-\mu(t'))\int_\Omega\Phi(\Pi_r y(t^*-\tau))dx= \tag{55}$$

$$=\int_\Omega\Big\{\mu(t')\Phi(\Pi_r y(t^*))dx+(1-\mu(t'))\Phi(\Pi_r y(t^*-\tau))\Big\}dx\ge$$

$$\ge\int_\Omega\Phi(\Pi_r(\mu(t')y(t^*)+(1-\mu(t'))y(t^*-\tau)))dx=\int_\Omega\Phi(\Lambda_\tau\Pi_r(y(t')))\,dx.$$

Let us prove further that

$$\Pi_r^+\big(k_i(x,y,\nabla_r\hat{v},B_hy)\big)\to k_i(x,u,\nabla\bar{v},Bu)\quad\text{in}\quad L_{p'}(Q_T). \tag{56}$$

We denote

$$J=\int_{Q_T}\big|\Pi_r^+\big(k_i(x,y,\nabla_r\hat{v},B_hy)\big)-k_i(x,u\nabla\bar{v},Bu)\big|^{p'}\,dx\,dt\,. \tag{57}$$

Limit relations (44) and (46), smoothness of the function v and continuity of $k_i(x, \xi, \eta, v)$ for each of the arguments allow us to assert that the integrand function in (57) tends to 0 as $h, \tau \to 0$ almost everywhere in Q_T. In addition, from the estimate (7) it follows that

$$\left|\Pi_r^+(k_i(x, y, \nabla_r \hat{v}, B_h y)) - k_i(x, u, \nabla \bar{v}, Bu)\right|^{p'} \le$$

$$\le \left(d_0 \sum_{i=1}^{n} \left\{|\partial_{r_i} \bar{v}|^{p-1} + |\frac{\partial \bar{v}}{\partial x_i}|^{p-1}\right\} + 2 d_1\right)^{p'}.$$

The right-hand side of the last inequality, due to the smoothness of v is a function integrable over Q_T, therefore, by the Lebesgue theorem on the passage to the limit $J \to 0$ for $\tau, \;\; h \to 0$, it means that (56) holds.

From the inequalities (53)–(55) it follows that

$$\overline{\lim_{\tau,h\to 0}} J_\tau(t') \ge \underline{\lim_{\tau,h\to 0}} \int_\Omega \Phi(\Lambda_\tau \Pi_r(y(t'))dx - \int_\Omega \Phi(u_0(x))dx. \tag{58}$$

From the relations (41)–(46) and (56) it follows that

$$\overline{\lim_{\tau,h\to 0}} J_\tau(t') = \lim_{\tau,h\to 0} J_\tau(t') = J(t') \equiv \int_0^{t'} \{\langle f, u\rangle -$$

$$- \sum_{i=1}^{n} \int_\Omega K_i \frac{\partial v}{\partial x_i} dx - \sum_{i=1}^{n} \int_\Omega k_i(x, u \nabla \bar{v}, Bu) \frac{\partial (u - \bar{v})}{\partial x_i} dx\} dt. \tag{59}$$

Considering (50), we will obtain

$$J(t') = \int_0^{t'} \{\langle \frac{\partial \varphi(u)}{\partial t}, u\rangle + \sum_{i=1}^{n} \int_\Omega (K_i - k_i(x, \nabla \bar{v}, Bu) \frac{\partial (u - \bar{v})}{\partial x_i} dx\} dt. \tag{60}$$

Substituting (59), (60) in the inequality (58) and integrating the result over t' from $T - \lambda$ to $T, \; \lambda = const > 0$, we will have

$$\int_{T-\lambda}^{T} J(t')dt' \ge \int_{T-\lambda}^{T} \underline{\lim_{\tau,h\to 0}} \int_\Omega \Phi(\Lambda_\tau \Pi_r(y(t'))dx dt' - \lambda \int_\Omega \Phi(u_0(x))dx. \tag{61}$$

The convexity of the function $\Phi(\xi)$ implies the weak lower semicontinuity on $L_\alpha(\Omega)$ of the functional $\int\limits_\Omega \Phi(w(x))dx$. Therefore

$$\int\limits_{T-\lambda}^{T} \lim_{\tau,h\to 0} \int\limits_\Omega \Phi(\Lambda_\tau \Pi_r(y(t'))dxdt' \geq \int\limits_{T-\lambda}^{T} \int\limits_\Omega \Phi(u(t'))dxdt'. \tag{62}$$

We transform the left-hand side of inequality (61) using the mean value theorem. The application of this theorem is admissible, since the function $J(t')$ is absolutely continuous with respect to t'. Considering (62), we will obtain

$$\lambda J(\bar{t}) = \int\limits_{T-\lambda}^{T} \int\limits_\Omega \Phi(u(t'))dxdt' - \lambda \int\limits_\Omega \Phi(u_0(x))dx,$$

here $\bar{t} \in [T-\lambda, T]$. We divide both sides of the last inequality by λ and pass to the limit as $\lambda \to 0$, as a result we get

$$\int\limits_0^T \langle \frac{\partial \varphi(u)}{\partial t}, u\rangle dt + \int\limits_0^T \int\limits_\Omega \sum_{i=1}^n (K_i - k_i(x, u, \nabla\bar{v}, Bu))\frac{\partial(u-\bar{v})}{\partial x_i} dxdt \geq$$

$$\geq \lim_{\lambda\to 0} \frac{1}{\lambda} \int\limits_{T-\lambda}^{T} \int\limits_\Omega \Phi(u(t'))dxdt' - \int\limits_\Omega \Phi(u_0(x))dx.$$

The last inequality and the 3 lemma imply

$$\int\limits_0^T \int\limits_0^T \int\limits_\Omega \sum_{i=1}^n (K_i - k_i(x, u, \nabla\bar{v}, Bu))\frac{\partial(u-\bar{v})}{\partial x_i} dxdt \geq 0. \tag{63}$$

Assuming in the inequality (63) first $\bar{v} = u + \lambda w$, and then $\bar{v} = u - \lambda w$, where $\lambda = const > 0$, w is an arbitrary function from $L_p(0, T; \overset{\circ}{W}{}_p^1(\Omega))$, it is easy to obtain equality (51). The theorem is proved. □

Acknowledgement This paper has been supported by the Kazan Federal University Strategic Academic Leadership Program ("PRIORITY-2030").

References

1. Chipot, M., Molinet, L.: Asymptotic behavior of some nonlocal diffusion problems. Applicable Analysis, **80**(3/4), 279–315 (2001)
2. Chipot, M., Lovat, B.: Existence and uniqueness results for a class of nonlocal elliptic problems, advances in quenching. Dyn. Contin. Discrete Impuls. Syst., Ser. A Math. Anal., **8**(1), 35–51 (2001)
3. Simon, L.: On quasilinear parabolic functional differential equation with discontinuous terms. Annales Univ. Shi. Budapest, **47**, 211 -229 (2004)
4. Alt, H.W., Luckhaus, S.: Quasilinear elliptic-parabolic differential equation. Math.Z., **183**(8), 311–341 (1983)
5. Xu, X.: Existence and convergence theorems for double nonlinear partial differential equations of elliptic-parabolic type. J. Math. Anal. Appl., **150**, 205-233 (1990)
6. Zemanm, J.: On existence of the weak solution for nonlinear diffusion equation. Appl. of mathematics, **36**(1), 9–20 (1991)
7. Glazyrina, O.V., Pavlova, M.F.: Issledovanie shodimosti javnoy raznostnoy shemy dlja parabolicheskogo uravnenija s nelokal'nym prostranstvennym operatorom. Uchenye Zapiski Kazanskogo Universiteta. Seriya Fiziko-Matematicheskie Nauki, **153**(1), 24–39 (2013) (in russian)
8. Glazyrina, L.L., Glazyrina, O.V., Pavlova, M.F.: On convergence of implicit finite element method scheme for a solution of a parabolic equation with double degeneration and nonlocal space operator. In: IOP Conference Series: Journal of Physics, vol.1158, is.2, Art. Id 022048 (2019)

The Program System for Design Optimization of Data Transmission Networks

Vadim M. Gostev

Abstract The paper aims at analyzing the problem of data transmission networks (DTN) design and the description of DTN design process using the design optimization system (DOS). Methods and technologies of structure and parameters synthesis of DTN on the DOS base are considered.

1 Introduction

Data transmission network (DTN) is the structural core of a geographically distributed wide-area computer network. DTN is a backbone communications mesh subnet that provides information exchange between servers and workstations. The basis of DTN is formed by communication nodes that interconnected by communication links. Nodes manage transmission processes of data streams on DTN and are usually implemented on the basis of high-performance multi-protocol routers.

The cost of a router depends on its performance (throughput) measured by the number of packets processed per unit of time, as well as on the quantity, types and speed of interfaces, types and versions of supported protocols. Communication links are created usually on the basis of dedicated (leased) channels of digital communication systems. The cost of channels renting mostly depends on their capacity and the distance between the connected points (nodes of data transmission network).

V. M. Gostev (✉)
Institute of Computational Mathematics and Information Technologies, Kazan Federal University, Kazan, Russia
e-mail: gvm@kpfu.ru

I. B. Badriev et al. (eds.), *Mesh Methods for Boundary-Value Problems and Applications*, Lecture Notes in Computational Science and Engineering 141,
https://doi.org/10.1007/978-3-030-87809-2_13

2 The Process of Data Transmission Networks Optimization

Among many problems that arise during the DTN life cycle, the problem of system (conceptual) design takes a special place. The quality of decisions that are made during design process largely determines the efficiency of the functioning of information systems created on the basis of a wide-area computer network.

The DTN design process includes the following stages. First, it is necessary to choose the topological structure of the DTN, that is to determine the number and location of communication nodes and also to determine for each node with which other nodes it will be directly connected by communication links. In the second stage, there is a need to select main routes of data packets transferring between each pair of nodes. The third stage is making a choice of routers throughput and communication links capacity.

The value of router throughput directly influences the choice of a specific router model from the range of available devices (by comparing their characteristics such as cost and throughput) for installation in each communication node. Communication link capacity influences on the characteristics of communication equipment installing in nodes and thus influences the cost of installation or renting of communication link.

The main criteria for estimating a data transmission network project are:

- expected time characteristics of data transmission—average and maximum delays of messages and packets in DTN; those delays determine the wide-area computer network response time (quality of service for DTN subscribers);
- throughput (overall performance) of DTN;
- cost characteristics, including capital costs of the communication nodes equipment, as well as operating costs (the cost of renting communication channels, cost of network administration).

The choosen values of parameters of routers and communication links must be conformed with each other and ensure the compatibility of all DTN components by protocols and interfaces. In addition, during the development of the DTN project, it may be necessary to provide a reserve for connecting of new subscribers to DTN and take into account the forecasted dynamics of the increase of external load as well as the phased development of the network parts.

The making of all design decisions is interdependent and requires taking into account a large number of factors. For example, the choice of DTN topology affects primarily the cost of the network. However, this choice, together with other design decisions, also affects the reliability of the network and the values of the time delays during the transmission of packets through links from sources to destinations. The choices of routers throughput and links capacity determine the cost of the network and packet delay. Without solving the routing problem (and hence the distribution of data streams over nodes and links), packet delays cannot be calculated. So project quality criteria usually conflict with each other. For example, it is difficult to find a compromise between the amount of funds allocated for the development of a DTN and such technical characteristics as reliability, average packet delay time and DTN throughput.

Thus, the problem of DTN design is a complicated multicriteria problem, which is characterized by a complex nature, inconsistency, and poor formalizability of the set of requirements for the designed object, the need to carefully consider numerous interrelated factors of various natures. A comprehensive solution of this problem requires the development of adequate mathematical models, computationally efficient methods and technologies that would allow taking into account the most important requirements for modern DTN (for the formation and optimization of the project).

3 The Data Transmission Networks Design Optimization System

The complexity of the DTN design problem does not allow formulating one common mathematical task (in general statement) that describes the whole problem. As noted, for example, in [1], "the DTN optimization problem is so complicated that there is no hope of solving it in general terms". In such circumstances, interactive human-machine technologies can help to cope with the complexity of this problem [2]. A number of such technologies are implemented in the data transmission network design optimization system (DTN DOS) [3]. The technologies provide versatile support for the activities of a human in the design process, i.e. during the searching, forming and estimating the effectiveness of solutions to the problem under our consideration.

DTN DOS provides the designer with a set of tools that help him to develop and to evaluate versions of design solutions. The design process (as a decision-making process) is based on a combination of abilities of human-designer (his ability to solve informal problems, his experience, knowledge, understanding of specific project situation) with the computing capabilities of a computer. That human-computer interaction allows solving complex mathematical problems of analysis, evaluation and optimization as components of the general DTN design problem.

The system allows solving the problems of structural-topological and parametric design of DTN, performing calculations and evaluating the parameters of projected networks based on the use of their models, comparing various design solutions and assessing their effectiveness, optimizing design solutions in terms of cost, reliability, performance and time delays. The system provides support for a multi-stage iterative man-machine design process with the ability to repeatedly perform individual stages and solve individual problems in order to correct, refine and optimize previously adopted design solutions, as well as the implementation of design methods with varying levels of complexity. At the same time, the system provides the designer with a user-friendly interface that meets modern requirements (graphical display of projected networks in a multi-window mode, tools for working with versions of the projected network, means for supporting technologies for automatic completion of partial solutions, means for automatically improving existing solutions).

The process of designing a data transmission network based on a design optimization system includes the following main stages:

1. structural and topological design (STD);
2. routes selection (RS)—the choice of routes for data transmission between communication nodes;
3. capacity selection (CS)—the choice of throughput capacities of nodes and communication links;
4. the estimation of quality and efficiency (EQE) of the designed DTN.

The design process is supported by a complex of functional and supporting subsystems of the DTN DOS.

Functional subsystems implement the solution of design problems (in automatic, manual and hybrid modes) at the appropriate design stages.

The subsystem of structural and topological design includes a set of programs that implement methods for optimizing the design of the topological structure of DTN and tools for supporting manual design (building the initial network topology, its further completion, improvement). Within the framework of the subsystem, algorithms for solving the following basic problems are implemented: construction of the initial topology, construction of a graph with a given degree of vertices, construction of a graph with the minimum total length of the shortest paths between all pairs of vertices under a cost constraint, algorithms for optimal completion and improvement of the topological DTN structures.

The route selection subsystem includes a set of programs for solving data flow routing problems: a flow deviation method to minimize the average latency, maximum delay minimization algorithms, the shortest path algorithms, various flow distribution methods aimed at increasing the DTN throughput.

The capacity selection subsystem includes a set of programs that implement methods for the optimal selection of communication links capacity and nodes throughput as well as tools for completing and improving DTN design versions.

Supporting subsystems maintain the operation of functional subsystems. The set of supporting subsystems forms the system environment (shell) of the design optimization system. The main functions of the system environment are design process management, data management, implementation of the man-machine interface. The design optimization system includes the following supporting subsystems: monitor (design process control subsystem), support subsystem for working with versions of design solutions, data management subsystem, graphic display subsystem, subsystem of means of interaction between the designer and the system.

4 The Estimating Quality and Efficiency of the Data Transmission Networks

The complexity of the problem under consideration, the large number of parameters, conditions and restrictions of various levels of detail do not allow describing and investigating the designing DTN in full enough within the framework of a unified mathematical model. To increase the efficiency of the design process, the integrated modeling technology is implemented in the design optimization system (DOS): in the process of forming and analyzing solution versions, both analytical and simulation models are used. On the basis of analytical models, the following functional subsystems of DOS are built: structural and topological design, routes selection, capacity selection.

The use of analytical models makes it possible to relatively quickly form a complete version of the DTN project using optimization algorithms of various complexity levels. However, as a rule the relatively high speed of these algorithms is achieved due to a number of simplifying assumptions. In particular, a continuous distribution of the lengths of transferring packets (and, therefore, a continuous distribution of the time processing of packets in the nodes and links), the independence of the packet lengths, the absolute reliability of the DTN elements, the independence of the packet delays in all transit elements of the route are assumed [4]. These assumptions reduce the accuracy of the design characteristics. In a such situation to evaluate whether this accuracy is within acceptable limits and to outline the boundaries of the scope of analytical models it is possible using simulation modeling [5, 6]. Simulation models can be used to substantiate and verify the developed analytical models of DTN elements, to evaluate the effectiveness of heuristic algorithms based on analytical models, and also to evaluate the effectiveness of DTN projects developed on the basis of analytical models. The use of simulation models is time-consuming to obtain results, however, it allows one to abandon some simplifications and to obtain more accurate estimates of the DTN characteristics versus analytical models.

The integrated using of analytical and simulation models forms the basis for the technology for estimating the quality and efficiency (EQE) of designed DTN. That technology implemented in the EQE subsystem of the DTN design optimization system.

The EQE subsystem provides a solution to the problems of calculating the characteristics of projected networks and criteria for estimating the quality of their functioning with a different character of the distribution of external load on the network. The criteria are average and maximum packet delays, estimates of the maximum throughput and cost of the projected network. The subsystem also provides a solution the problems of failures modeling in the operation of networks and estimating the quality of the functioning of networks in these conditions, what makes it possible to identify the "bottlenecks" of a particular version of the network.

DTN simulation models are built automatically by a special model generator included in the EQE subsystem. The generator input receives an information model of the DTN created by the designer at the previous stages (structural and topological design, routes selection, capacity selection). Analyzing the DTN parameters, the generator forms a model in the GPSS language. The model simulates those functions of communication nodes that reflect the key aspects of routing algorithms and the main functions of the network protocols: packet queuing, packet header analysis, packet processing, selection of the outgoing direction of packet transmission in accordance with the routing procedure. This takes into account the packet delay in the node during the time takes to complete these operations. At the process of automatic generation, libraries are used that contain models of individual functional modules of the communication nodes and communication links, as well as models of external data sources (servers and workstations). The resulting model is passed to the input of the GPSS interpreter.

The output characteristics obtained as a result of the simulation experiment are the average packet delay time, standard deviation from the average packet delay time, packet delay distribution functions, the number of packets processed by each communication node, the number of packets transmitted, average lengths of queues and average values waiting time in each communication node, the load of communication links and processors of the routers at every communication node. Analyzing the simulation results, designer can correct the current version of the DTN project (change the topology, routes, nodes throughput, links capacities, packet sizes) and repeat the simulation stage using the means of the EQE subsystem of DTN DOS.

5 Mathematical Models and Methods for Data Transmission Networks Design Optimization

For a comprehensive solving to the general problem of DTN design and for solving of particular problems of optimization of design solutions various methods and technologies are proposed.

Let the DTN topology be defined by the graph $G = (V,\ U)$, where the set of vertices $V = \{v_1, v_2, \ldots, v_n\}$ represents the set of DTN communication nodes, the set of edges $U = \{u_1, u_2, \ldots, u_m\}$ represents a set of communication links. We denote by c_k, $k = \overline{1, m}$ the capacities of communication links, by f_k, $k = \overline{1, m}$—flow rates through links, through γ—total DTN traffic.

In [4], an estimate of the average delay of messages on communication links via DTN is given:

$$T = \frac{1}{\gamma} \sum_{k=1}^{m} \frac{f_k}{c_k - f_k}, \tag{1}$$

which occurs when certain, sufficiently stringent, assumptions are met (Kleinrock regularization, see [4]).

The DTN design problem is formulated as follows. For the set of nodes $V = \{v_1, v_2, \ldots, v_n\}$ and for the external flows matrix $\Gamma = (\gamma_{ij})$ it is necessary to find: 1)the set of edges $U = \{u_1, u_2, \ldots, u_m\}$ describing the DTN links structure; 2) the links capacities $c = \{c_1, c_2, \ldots, c_m\}$; 3) the flows' distribution $f = \{f_1, f_2, \ldots, f_m\}$, corresponding to the matrix $\Gamma = (\gamma_{ij})$, which minimize the total cost of the DTN when the average delay T is limited and a given degree of graph $G = (V, U)$ connectivity is ensuring.

The software of DTN design optimization system includes:

1. some well-known, proven mathematical models, numerical methods and algorithms for solving problems of analysis, estimation and optimization DTN [6–9, 11–13];
2. a number of new mathematical models, methods and technologies for DTN design optimization [2, 3].

The relevance of the development of new methods and technologies is determined for the two following reasons.

Firstly, "old" methods do not take into account delays in communication nodes. This can be explained by the fact that communication links in "early" DTN had relatively low throughput, while processors in the nodes had a relatively higher productivity. And the communication links were the "bottlenecks" that had a decisive impact on the overall DTN throughput. In modern conditions, when high-speed digital links are used in DTN, delays in communication nodes can have the same order of magnitude as delays in links. Therefore, in the DTN design process, it is also necessary to ensure control of nodal delays.

Secondly, the well-known methods for optimizing the DTN design, in particular, methods for the optimal choice of links capacities are practically unsuitable for solving problems of DTN design with an asymmetric load on the network while exactly this case is typical for modern wide-area computer networks.

6 An Example of Integrated DTN Modeling

Among the characteristics of DTN the most important place is occupied by the values of packet delays. In [2], a method for estimating the average delay of packets is proposed. The method is taking into account the delays in the communication nodes. In this case to estimate the delays in the communication links well-known approaches based on the use of the M/M/1 and M/D/1 models are used. A similar approach is used to estimate packet delays in the communication nodes.

Based on the M/M/1 model, i.e. consider that the input stream of packets for processing by the processor of the communication node is Poisson with intensity λ, and the processing time is $1/q$ considered as a continuous random variable distributed exponentially with an average value, then the average packet delay in a separate node will be expressed as

$$\theta_M = \frac{1}{q - \lambda} = \frac{1}{q - \mu g}, \tag{2}$$

where $g = \lambda/\mu$ is the average value of the flow through the node in bit/s. Here, too, the flow admissibility condition must be met: $g < q/\mu$.

Since the processing of packets in the communication nodes usually consists in performing operations that are the same for all packets regardless of their size, we can assume that the processing time is constant for all packets and depends only on the processor's performance. Therefore, it is possible to consider a separate communication node as an M/D/1 system, at the input of which a Poisson flow with an intensity λ arrives, and the service time is constant, equal $1/q$ (therefore, the service rate is q packets per second). Then the average packet delay in a separate node can be calculated as

$$\theta_D = \frac{1}{2q} + \frac{1}{2(q - \lambda)} = \frac{1}{2q} + \frac{1}{2(q - \mu g)}, \tag{3}$$

Note that $\theta_D = (1 - \rho/2)\theta_M$, where $\rho = \mu g/q$ - the average load on the node, and for $0 \le \rho < 1$ takes place $0,5\theta_M < \theta_D \le \theta_M < 2\theta_D$.

To evaluate the proposed approach (to check the adequacy of the analytical models), a series of computational experiments were carried out on the basis of the DTN DOS to calculate the characteristics of the DTN in order to study the dependence of the average packet delay time on the nature of the external load on the network. For networks with a different number of communication nodes (from 4 to 50), the change in the volume and structure of information flows from DTN users was modeled. Cases of symmetric and asymmetric external load on DTN, various laws of distribution of intervals of the arrival of packets from external sources, as well as service time (packet processing) in nodes and efficiency were considered. The result of one of the typical experimental is as follows. Here, the dependencies of average delays on the total load on the DTN are obtained for a network of 10 nodes (packet size is 1500 bytes). For the projected DTN with a limit on the average packet delay of 6 ms at a total load (global flow) of 120 Mbit/s, a series of tests was carried out using the EQE subsystem. Four methods for assessing delays were used: analytical and simulation modeling of the control and efficiency using the M/M/1 and M/D/1 models. The experiments have shown that the proposed method for analytical evaluation of delays in DTN at nominal (design) external load gives results that differ from the results of simulation by 5–10%, and can well be used in the early design stages for a quick estimation of quality and efficiency functioning of the developed versions of DTN. For a more accurate assessment and forecast of the "behavior" of the network (especially when the rated load is exceeded), the means of the EQE subsystem can be used.

7 The DTN DOS as Basis of Electronic Scientific and Educational Complex (ESEC)

In order to improve the quality of e-learning the concept of formation and development of the system of electronic scientific and educational complexes (ESEC) was developed [10]. The main objective of such a system—to ensure the information integrity of the university's scientific and educational space and its integration into the global information space on the basis of a new generation of electronic scientific and educational resources. ESECs should provide comprehensive information support of the educational process, as well as the integration of research and educational activities. The structure of a typical ESEC includes: problem-oriented portal; digital library; tools for collaboration (collective project activities as part of studies, independent research work of students, joint research activities of teachers and students).

An example of the implementation of the concept of ESEC is an Electronic scientific-educational complex "Data Transmission Networks Design Optimization". The complex provides comprehensive support for research and educational activities in the field of computer networks—on classroom training sessions with the use of modern educational technologies (on lectures, seminars, etc.) and to self-teaching and research work of students. Information support of the educational process on the basis of ESEC made via the portal "Advanced Information and Communication Technologies". The portal provides access to a range of electronic teaching materials (work programs, lecture materials, manuals, links to electronic resources) for a lot of courses in "Advanced ICT" area.

For the implementation of the relevant technologies and the organization of educational and research work in the field of computer networks, the DTN DOS is included in ESEC "Data Transmission Networks Design Optimization". The basis for organizing of training sessions using the DTN DOS is the creation of problem situations, through which students are involved in the problem solving process. Here, the transition from the principle of mastering knowledge through repetition and memorization to the principle of mastering knowledge in the process of independent intellectual activity of students is realized. Working with the system, students can study the composition and structure of DTN, master the use of simulation methods, methods of optimizing design solutions and also study and apply in practice the principles of a systematic approach to the design of complex objects. The implementation of these works gives a significant teaching effect, since in parallel with the study of architecture and methods of designing computer networks, students deepen their knowledge and develop skills in the field of simulation, system and applied programming.

8 Conclusion

The program system for design optimization of data transmission networks allows solving the problems of structural-topological and parametric design of DTN. The system provides support for a multi-stage iterative man-machine design process with the implementation of design methods with varying levels of complexity. The software of DTN design optimization system includes both well-known and new mathematical models, methods and technologies for DTN design optimization.

An integrated approach to modeling allows within the framework of a unified DTN design process to use the advantages of analytical (at the stages of forming and adjusting the project) and simulation (at the stage of estimating the quality and efficiency of the resulting project) models. Along with direct application in the process of DTN design, this approach can be used to justify and verify the developed analytical models. An example of such use is the experimental substantiation of a heuristic method for estimating packet delays in DTN.

In addition, the DTN DOS serves as a technological basis for further development of the ESEC concept including methods for designing and creating hardware, software, information, and organizational support for ESEC, experimental assessment of the labor intensity of implementing the developed ESEC architecture and estimating the effectiveness of new educational technologies.

Acknowledgment This paper has been supported by the Kazan Federal University Strategic Academic Leadership Program ("PRIORITY-2030").

References

1. Davies D.W., Barber D.L.A., Price W.L., Solomonides C.M. (1979) Computer Networks and Their Protocols, Wiley
2. Gostev V.M., Khabibullin R.F. (1995) The SELENA teaching/research system for automated design of remote data processing networks. Journal of Mathematical Sciences, Vol.74, No.5, pp.1214–1218
3. Gostev V. (2012) The implementation of educational information environment of federal university based on innovative technologies. Innovative Information Technologies (I2T): Materials of International scientific-practical conference, Prague, pp.48–50
4. Kleinrock L. (1976) Queueing Systems, Volume II: Computer Applications. Wiley Interscience: New York.
5. Musaria K.M., Fawzi M.A., Uzunoglu N.K. (2012) Design and simulation of a data transmission network for industrial control system subject to reliability improvement, Intern. Conf. on Future Communication Networks ICFCN'12, pp.23–28
6. Krivchenkov A. (2017) Model of Wireless Data Network in GPSS Language. Reliability and Statistics in Transportation and Communication. RelStat. Lecture Notes in Networks and Systems, vol 36. Springer, Cham.
7. Gerla M., Kleinrock L. (1977) On the Topological Design of Distributed Computer Networks, IEEE Transactions On Communications, Vol.COM-25, No.1., pp.48–60
8. Zaychenko Y., Gasanov A., Hamidov G. (2016) New Generation Networks Performance Analysis and Optimisation, IEEE 10th International Conference on Application of information and Communication technologies, pp.594–598

9. Zaychenko Y., Zaychenko H., Hamidov G (2017) Structure optimization of new generation networks, Image and Signal Processing BioMedical Engineering and Informatics 10th International Congress, pp.1–5
10. Gostev V. (2016) Electronic Scientific-Educational Complex "Geoinformation Technologies and Systems", 16th International Multidisciplinary Scientific GeoConference SGEM2016, Vol.III, pp.809–816
11. Vishnevskii V.M., Leonov A.O., Levchenko N.I. (2007) Topological optimization of the large-scale data transmission networks, Automation and Remote Control, Vol. 68, No. 5, pp.760–772
12. Schwartz M. (1977) Computer-Communication Network Design and Analysis, Prentice-Hall, Englewood Cliffs, N.J.
13. Bertsekas, D.P., Gallaher, R.G. (1987) Data Networks, Prentice-Hall, Englewood Cliffs, N.J.

Mathematical Modeling of Transient Processes in Circular Channel with the Boiling of Refrigerant-113

Damir A. Gubaidullin and Boris A. Snigerev

Abstract The two-fluid model has been extensively used in modeling boiling flow of water, however, there are few equivalent studies of boiling flow of cryogenic liquids. In the present study, the two-fluid model was developed with by incorporating new closure correlations, then boiling flow of liquid refrigerant-113 in a vertical annulus tube was numerically simulated using the modified model. Comparison with experimental data shows that the modified model is satisfactorily improved in accuracy. This study demonstrated that the following parameters and models are important for accurate prediction: the lift force, the bubble diameter distribution and the active site density, among which, the active site density has the most significant effect.

1 Introduction

Boiling flows are frequently found in industry and engineering due to the large amount of heat that can be transferred within such flows with minimum temperature differences. In the nuclear industry, boiling affects in different ways the operation of almost all water-cooled nuclear reactors. Recently, the use of computational fluid dynamic (CFD) approaches to predict boiling flows is increasing and, in the nuclear area, numerical methods is being developed to solve thermal hydraulic safety issues such as establishing the critical heat flux, which is perhaps the major threat to the integrity of nuclear fuel rods [1, 2]. In recent years, the advances in computational technologies have allowed flow boiling simulations faster than before. Generally, according to the problem specifications, objectives and characteristics, there are some important models for numerical simulation of two-phase flow and boiling heat transfer in channels. Most often, numerical study of two-phase flow is based

D. A. Gubaidullin (✉) · B. A. Snigerev
Institute of Mechanics and Engineering FRC Kazan Scientific Center Russian Academy of Sciences, Kazan, Tatarstan, Russia
e-mail: Gubaidullin@imm.knc.ru; Snigerev@imm.knc.ru

I. B. Badriev et al. (eds.), *Mesh Methods for Boundary-Value Problems and Applications*, Lecture Notes in Computational Science and Engineering 141,
https://doi.org/10.1007/978-3-030-87809-2_14

on Eulerian- Eulerian mathematical approach. This approach considers the liquid phase as a continuum phase and the particles phase (vapor bubbles in boiling flow) as another continuum phase. Then, the conservation equations are solved by considering interphase forces and exchanged heat on a control volume for both of phases [3–7].

With the aim of predicting the boiling process, different wall boiling models have been incorporated in modern CFD codes. For two-fluid averaged models, these approaches are in the large majority based on the Rensselaer Polytechnic Institute (RPI) boiling model from Kurul and Podowski [8], where the heat flux from the wall is partitioned between the mechanisms responsible for the heat transfer process, these being single-phase convection, quenching and evaporation. In recent years, many authors have used more or less refined versions of the RPI boiling model to predict boiling flows [9, 10]. After departure from the heated wall, the bubbles join the bulk of the flow and the size distribution of these bubbles, polydispersed in general, governs interphase exchanges of mass, momentum and energy. Therefore, in models of these flows, knowledge of the average diameter of the bubbles is required in many closure relations, and additional models have been used to predict the average bubble diameter distribution. Initially, bubble size was derived from experimental data or empirical correlations of subcooling in the liquid phase. Some key wall boiling parameters, including the nucleation site density (N_w), bubble departure diameter (D_w) bubble departure frequency (f), are used to close the model. These parameters should be carefully identified due to the significant impact on the boiling physics as well as the local flow patterns of the two-phase flow [11, 12]. The closure models of N_w, D_w and f have been expressed by empirical correlations. Since the correlations were obtained empirically from experiments, their reliability is strongly dependent on the working conditions. In fact, for the refrigerants at low pressure the variations of liquid properties with subcooling and saturation temperature along the heated channel are relatively obvious. Taking refrigerant-113 for example, when liquid subcooling varies from 0 to 40 K, density, thermal conductivity, viscosity and specific heat at constant pressure increase by 6.7%, 13.5%, 57.1%, 4.7%, respectively. Moreover, the saturation temperature of refrigerant-113 at low pressure varies evidently along the heated channel (for about 5 K in a 3 m long vertical channel), which is closely correlative with the predictions of local flow characteristics in subcooled boiling flow. Therefore, temperature dependent properties and saturation temperature variation along the heated channel should be given extra considerations in order to accurately simulate the process of subcooled boiling flow.

In this paper, the accuracy of an Eulerian-Eulerian, two-fluid CFD model is evaluated over database of subcooled boiling flows of liquid refrigerant-113 from number of experiments are described in [13, 14]. The model applied the basic theories of mass, heat and momentum transfer, Reynolds stress turbulence model and a boiling model, derived using the heat flux partitioning approach. The database covers a large range of conditions in subcooled boiling flows of refrigerants in vertical annular channels. Overall, a satisfactory predictive accuracy is achieved for some quantities of interest, such as the void fraction and the turbulence and liquid temperature fields, but results are less satisfactory in other areas, more specifically for the mean velocity profiles close to the wall in annular channels. Agreement may be improved with advances in the treatment of large bubbles and bubble break-up and coalescence, as well as in improved modelling of the boiling region close to the wall, and more specifically the bubble departure diameter, the wall treatment and the contribution of bubbles to turbulence.

2 Mathematical Two-Fluid Model for Subcooled Boiling

In a two-phase mechanistic model, both the gas and liquid phases are treated as continua, and two sets of conservation equations governing the balance of mass, momentum and energy of each phase are solved. Consider the interfacial mass, momentum and energy transfer, the governing equations for the two-fluid model are given by Nigmatulin [1], Yeoh and Tu [11], Gubaidullin and Snigerev [15]:

$$\frac{\partial}{\partial t}(\rho_k\alpha_k) + \nabla\cdot(\rho_k\alpha_k\mathbf{u}_k) = \dot{m}_{lg} - \dot{m}_{gl},\quad \alpha_l + \alpha_g = 1,\quad k = l, g,$$

$$\begin{aligned}\frac{\partial}{\partial t}(\rho_k\alpha_k\mathbf{u}_k) + \nabla\cdot(\rho_k\alpha_k\mathbf{u}_k\,\mathbf{u}_k) &= -\alpha_k\nabla P + \alpha_k\rho_k\mathbf{g} + \dot{m}_{lg}\mathbf{u}_l - \dot{m}_{gl}\mathbf{u}_g + \\ &\quad + \nabla\cdot\left(\alpha_k\mu_{eff,k}\left[\nabla\mathbf{u}_k + (\nabla\mathbf{u}_k)^T\right]\right) + \mathbf{M}_{lg},\end{aligned} \tag{1}$$

$$\frac{\partial}{\partial t}(\rho_k\alpha_k\,h_k) + \nabla\cdot(\rho_k\alpha_k\mathbf{u}_k\,h_k) = \nabla\cdot[\alpha_k\lambda_k\,\nabla T_k] + \dot{m}_{lg}\,h_l - \dot{m}_{gl}\,h_g + Q_{lg}.$$

In the equation system (1) t is time, ϱ_l and ϱ_g are continuous and dispersed phase densities respectively, α_l, α_g are volumetric concentrations of gas and bearing phase, λ_k is heat conductivity coefficient of k phase, $\mathbf{g}$ is gravity acceleration vector, $\mathbf{u}_k = u_{1k}i + u_{2k}j + u_{3k}k$ is velocity vector k-phases, P is carrying phase pressure, $\mu_{eff,k}$ is effective dynamic the viscosity of the k phase, $\mathbf{M}_{lg}$ is vector of interphase interaction force, h_k is enthalpy of k phase, T_k is temperature of k phase, Q_{lg} is heat transfer between phases, $\dot{m}_{lg}$ is mass transfer rate between the phases of l and g. To simulate turbulence, the Reynolds stress transfer model is used, which includes the effective viscosity of the medium $\mu_{eff,l}$, determined by the ratio $\mu_{eff,l} = \mu_{lam,l} + \mu_{t,\,l} + \mu_{BI,l}$. To describe the additional dissipation of the kinetic energy of turbulence by the pulsation of the bubbles, the viscosity of $\mu_{BI,l}$ is entered [11, 16]. For calculation of $\mu_{eff,l}$ applies the Kolmogorov formula, and for $\mu_{BI,l}$ the ratio from [9] is applied

$$\mu_{t,\,l} = \frac{C_\mu\,\varrho_l\,k_l^2}{\varepsilon_l},\quad \mu_{BI,l} = C_{\mu b}\,\rho_l\,\alpha_l\,d_s\left|\mathbf{u}_g - \mathbf{u}_l\right|,\quad \mu_{\text{eff},\,g} = \frac{\varrho_g}{\varrho_l}\mu_{\text{eff},\,l}. \tag{2}$$

The total interphase interaction force $\mathbf{M}_{lg}$ plays a very important role when modeling multiphase flows [1, 17]. The inter-phase momentum exchange was accounted through the forces acting on the dispersed bubbles as:

$$M_{li} = -M_{gi} = -\left(M_{gi}^{D} + M_{gi}^{L} + M_{gi}^{TD} + M_{gi}^{W}\right), \tag{3}$$

where M^D, M^L, M^{TD}, M^W are inter phase momentum transfer contributions from drag, lift, turbulent dispersion and wall forces respectively. The volumetric source of the momentum exchange between the two phases due to the drag force exerted by the liquid is given by

$$\mathbf{M}^D = \frac{3}{4}\rho_l\alpha_l\alpha_g\frac{C_D}{d_b}|\mathbf{u}_g - \mathbf{u}_l|(\mathbf{u}_g - \mathbf{u}_l). \tag{4}$$

The drag coefficient C_D was calculated using the correlation of [8]. The lift force that is experienced by bubbles due to velocity gradients in continuous phase was estimated as:

$$\mathbf{M}^L = C_L\rho_l\alpha_g(\mathbf{u}_g - \mathbf{u}_l) \times \nabla \times \mathbf{u}_l. \tag{5}$$

The C_L was calculated using the model proposed by Wang et al. [7]. The liquid phase turbulence influences the vapor distribution and it was accounted by the turbulent dispersion force model of [16] as $\mathbf{M} = C_{TD}\rho_l k_l \nabla\alpha_g$ where, C_{TD} is turbulent dispersion coefficient which may vary from 0–10. The wall force acts opposite to the lift force and forces the bubbles to move towards pipe center. Since bulk of the liquid is below the saturation temperature, bubbles formed at the wall start condensing when they move inside. Similarly, evaporation can take place in the bulk of the liquid. This interphase mass transfer was accounted by including appropriate source and sink terms into the continuity equation. The rate of evaporation is given by

$$\dot{m}_{lg} = \frac{h_{lg}A_{lg}(T_l - T_g)}{h_{fg}}. \tag{6}$$

Further, $\dot{m}_{lg} + \dot{m}_{gl} = 0$. The interfacial heat transfer Q_{lg} is given by $Q_{lg} = h_{lg}A_{lg}(T_{sat} - T_l)$, where h_{lg} is heat transfer coefficient calculated as [18] $h_{lg} = Nu\lambda_l/d_b$ and Nu is Nusselt number given by $Nu = 2 + 0.6Re^{0.5}Pr^{0.3}$. The interfacial area is calculated as $A_{lg} = 6\alpha_g/d_b$. For present work, since bubble size was found to be almost constant, a constant mean bubble diameter was used in the simulations. An additional source term active only at the near wall cells was included to account for the vapor generation at the heated wall. It was calculated using wall heat flux partitioning model as $\Gamma_g = q_e A_s/h_{fg}V_{cv}$ where V_{cv} is control volume and A_s is heated wall area.

According to Kurul and Podowski [8] wall flux partitioning model, the heat flux between the heated wall and liquid is exchanged via three mechanisms

$$q = q_q + q_e + q_c, \tag{7}$$

where q_q, q_e, q_c are convective, evaporative and quenching heat respectively. At the heated walls, bubbles are formed due to vaporisation of liquid at the nucleation sites and the part of the wall heat used for this is called evaporation flux. Once bubbles reach critical bubble size they detach from the wall, cold liquid replaces the space occupied by bubbles and receives heat from the wall. This flux is called quenching heat flux. The rest of the area of the wall, that is not covered with the bubbles, is used for the single phase convective heat transfer. The quenching heat flux is given by

$$q_q = A_q h_q (T_w - T_l). \tag{8}$$

The area available for quenching heat transfer is $A_q = \pi N d_{dep}^2$. The quenching heat transfer coefficient was calculated as $h_q = 2\lambda_l f \sqrt{\tau/\pi k}$. The bubble waiting time is given as, $\tau = 0.8/f_{dep}$. The evaporation heat flux is given by

$$q_e = \pi/6 d_{dep}^3 \rho_l f_{dep} N h_{lg}, \tag{9}$$

where nucleation site density N was calculated as [19, 20] $N = N_{ref}[(T_w - T_{sat})]^{1.805}$ and bubble departure frequency was calculated as [21] $f_{dep} = (4g(\rho_l - \rho_g)3\,\rho_l\, d_{dep})^{0.5}$. The convective heat flux was calculated using following correlation

$$q_c = (1 - A_q)\frac{\rho_l C_{p,l} u^*}{T^+}(T_w - T_l), \tag{10}$$

where T^+ is the non-dimensionless temperature. The system of Eqs. (1)–(10) describe the hydrodynamics and heat and mass transfer in the movement of steam-gas-liquid mixture in thermal power engineering apparatuses of chemical technology.

The simulations were carried out for an axis-symmetric geometry. A velocity inlet condition was specified at the bottom and pressure boundary condition was imposed at the top outlet. The no slip velocity boundary condition was specified at the wall for the vapor and liquid phase. For vapor phase, some researchers have argued that free slip condition is best suited and thus effect of free slip boundary condition for vapor phase was also investigated. The $k - \epsilon$ turbulence model with bubble induced turbulence source term [16] was used to simulate turbulence in the continuous phase.

The solution algorithm PISO was used to solve pressure-velocity coupling. In this work, a combination of Gauss upwind, Gauss linear and Gauss limited Linear schemes were used for discretization of spatial derivatives. For the time derivative, first order accurate Euler implicit method was used. The set of discretized equations was solved by the generalised geometric-algebraic multi-grid (GAMG) solver with Diagonal incomplete-Cholesky (symmetric) smoother for pressure and the preconditioned bi-conjugate gradient (PBiCG) solver with Diagonal incomplete-LU (DILU) pre-conditioner for the rest of the variables. The existing two phase solver twoPhaseEulerFoam available for isothermal gas-liquid flows was further developed to account for phase change phenomenon.

3 Numerical Results

The verification of the presented mathematical model was carried out by comparing the results of calculations with experimental data. The data of experiments presented in [14] were chosen, in which the boiling ascending bubble flow of subcooled liquid refrigerant-113 at high pressure in a annulus tube with step heating (diameter of heated inner tube $D_1 = 15.8$ mm, insulated outer tube $D_2 = 42.02$ mm, height $H_1 = 3660$ mm, height of the heated section $H_2 = 2750$ mm) was studied. In the Table 1 the basic characteristics of gas-liquid flow at the inlet of the pipe are presented: G_l is mass flow rate of liquid at the inlet of the pipe, P is pressure, q_W is heat flux on the wall of the pipe, T_{sat} is saturation temperature of liquid, $T_{1,IN}$ is water temperature at the inlet, ΔT is degree of subcooling of liquid. Two-phase flow in a vertical pipe is assumed to be axisymmetric, therefore for numerical simulation the calculated the area consisting of a circular sector with a radius of $r_0 = D_2/2 = 21.0 \times 10^{-3}$ m, a length of L = 3.66 m and a solution angle of 40°. Numerical calculations were performed on finite volume grids consisting of $M_e = 64{,}000, 124{,}000, 264{,}000$ nodes of the computational grid. In the section of the plane $x_1 x_2$, the number of partitions by coordinates along the pipe axis and by length for different grids is $M_1 = 20 \times 200$, $M_2 = 40 \times 400$, $M_3 = 60 \times 600$, respectively. The following parameters of the liquid refrigerant-113 in subcooling temperature in range from 0 to 40 K are set: density change in scope $\rho_l = 1423 - 1526\,\text{kg/m}^3$, dynamic viscosity coefficient $\mu_l = 340.8 \cdot 10^{-5} - 535.8 \cdot 10^{-5}$ Pa · s, the latent heat of vaporization is $L_{lg} = 1.3 \cdot 10^6$ J/kg, constant pressure heat capacity for liquid and gas $C_{pl} = 978 - 932$ J/kg K, coefficients of heat conductivity of medias $k_l = 0.57 - 0.65$ W/m K. Confidence in the predictions of CFD codes relies on validation of their results against relevant experimental data. In this regard, it is important that models provide accurate predictions over many experiments, with parameter variations as wide as possible. Therefore, a database was built from 6 experiments from [14]. In this

Table 1 Characteristics of mode parameters of experiments [14]

	P MPa	G_l kg/m^2 s	q_W kW/m^2	T_{sat} C	$T_{2,IN}$ C
roy1	0.269	568.0	79.4	85	42.7
roy3	0.269	784.0	125.8	85	52.0

research work [14] the subcooled boiling of refrigerant R-113 in a vertical annulus channel of 3.66 m in length, 0.0158 m in inner diameter and 0.0422 m in outer diameter was testing. A laser Doppler velocimetry system allowed measurement of the velocity field and the turbulent fluctuations, with an optical probe used to obtain the void fraction and the bubble diameter. The liquid and vapour temperatures were measured with micro-thermocouples. Measurements were taken at 0.269 MPa and in the ranges 565–785 G_l kg/m^2 s for the mass flux, 79.4–125.9 q_W kW/m^2 for the heat flux and 42.1–50.2 C^0 for the inlet temperature. Comparisons for the experiment [14] are presented in Figs. 1 and 2.

In these, and subsequent figures, symbols are used for experimental data and lines for model predictions. In annular channels (Figs. 1 and 2), the radial position is nondimensionalized with the distance between the outer and inner radius and,

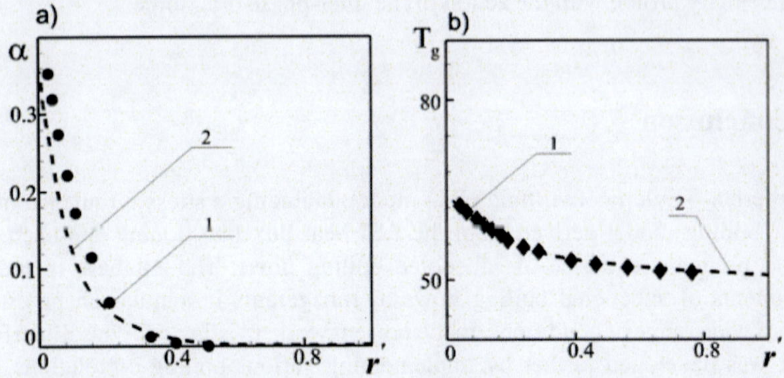

Fig. 1 Comparison of predicted physical variables against experimental data for case roy3 (1–symbols experiment [14], 2–line calculation: (**a**) void fraction α; (**b**) temperature T_g)

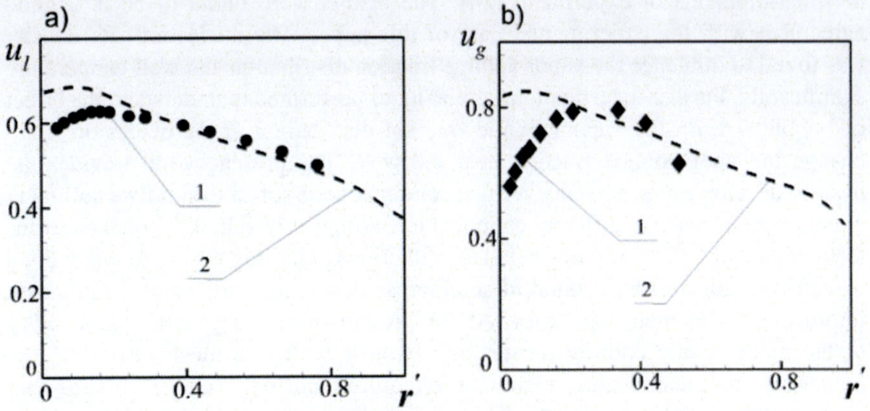

Fig. 2 Comparison of predicted physical variables against experimental data for case roy3 (1–symbols experiment [14], 2–line calculation: (**a**) velocity of liquid u_l; (**b**) velocity of gas u_g)

therefore, in the plots $r' = (r - R_i)/(R_0 - R_i) = 0.0$ identifies the inner wall, whereas $r' = (r - R_i)/(R_0 - R_i) = 1.0$ corresponds to the outer wall. Only the inner wall is heated in [14]. In the following, discussion of the results is presented for each physical quantity predicted. Even if the specific quantitative accuracy depends on the particular experiment, the void fraction and temperature profile is generally predicted with reasonable accuracy. In Fig. 1, temperature profile are well predicted for [14], the only discrepancy being a minor underestimation void fraction in the case of roy3. Compared with the experimental data, large discrepancy occurs at the near-wall region for the axial liquid and vapor velocity profiles in the two predicted results, and the discrepancy will be enhanced with high wall heat flux. As the wall heat flux increases, high void fraction profile will be found in the bubble boundary layer at the measurement plane, which makes the vapor bubbles move faster than the bubbles in lower wall heat flux, and eventually develops the high predicted axial liquid velocity profile with the action of the inter-phase drag force.

4 Conclusion

An Eulerian—Eulerian two-fluid CFD model, including a stress turbulence model $k - \epsilon$, boiling model derived from the RPI heat flux partitioning approach, was used to predict a database of subcooled boiling flows. The database includes 2 experiments of subcooled boiling flows of refrigerants in annular channels, and covers a wide range of conditions. In the present work, existing twoPhaseEulerFoam solver was developed further by implementing various boiling correlations, wall heat flux partitioning model and energy equation. Also, different inter-phase coupling forces and bubble turbulence terms were included in the code. The modified code was used to simulate the boiling and predictions were verified using the measurements of experiment [14]. The results were found to be in a good agreement with the experimental data of this paper. The nucleation site density was found to influence the vapor volume fraction distribution and wall temperature significantly. Further investigations are being to performed to understand the effect of bubble size on the vapor volume fraction distribution, in particular on sharp change in vapor volume fraction near the wall. The present work provides the basis to develop experimentally verified computer code solver to simulate boiling in complex geometries in different chemical technology. Overall, the model confirms the potential of CFD to provide detailed predictions of boiling flows and rather good agreement with data was found in some areas, but others still require significant improvements in model accuracy. At the present time, the general applicability of the model is not entirely satisfactory. Even if built in a mechanistic fashion, numerous empirical closure relations are required, not only for wall boiling, but also for the turbulence models. This clearly limits the overall models general applicability and, therefore, the development of more mechanistic closures is highly desirable.

Acknowledgement This research was supported by the Russian Foundation for Basic Research, project no. 19-01-000442.

References

1. R. I. Nigmatulin, *Foundations of mechanics of heterogeneous media*, Moscow, Science (1978)
2. J. G. Collier and J. G. Thome, *Convective Boiling and Condensation, third ed.*, Oxford University Press, Oxford (1994)
3. C. Erfeng, L. Yanzhong, C. Xianghua, CFD simulation of upward subcooled boiling flow of refrigerant-113 using the two-fluid model. Applied Thermal Engineering. **29**, 2508–2517 (2009)
4. E. Krepper, B. Koncar, Y. Egorov, CFD modelling and subcooled boiling – concept, validation and application to fuel assembly design. Nucl. Eng. Des. **237**, 716–731 (2007)
5. S. Vashisth, K. D. P. Nigam, Prediction of flow profiles and interfacial phenomena for two-phase flow in coiled tubes. Chem Eng Process Process Intensif. **48**, 452–463 (2009)
6. D. Bestion, Applicability of two-phase CFD to nuclear reactor thermalhydraulics and elaboration of best practice guidelines. Nucl. Eng. Des. **253**, 311–321 (2012)
7. S. M. Wang, J. Wen, Y. Li, Population balance modelling for subcooled boiling flow of liquid nitrogen in a vertical tube. International Journal of Heat and Mass Transfer. **60**, 632–645 (2013)
8. N. Kurul, M. Z. Podowski, On the modeling of multi-dimensional effects in boiling channels. *in: 27th National Heat Transfer Conference, Minneapolis*. (1991)
9. B. Koncar, E. Krepper, CFD simulation of convective flow boiling of refrigerant in a vertical annulus. Nuclear Engineering and Design. **238**, 693–706 (2008)
10. Z. Li, Y. Wu, J. Lu, Heat transfer to supercritical water in circular tubes with circumferentially non-uniform heating. Appl. Therm. Eng. **70**(1), 190–200 (2014)
11. G. H. Yeoh, J. Y. Tu, Numerical modelling of bubbly flows with and without heat and mass transfer. Applied Mathematical Modelling. **30**(10), 1067–1095 (2006)
12. S. C. P. Cheung, S. Vahaji, G. H. Yeoh, J. Y. Tu, Modeling subcooled flow boiling in vertical channels at low pressures - Part 1: Assessment of empirical correlations. International Journal of Heat and Mass Transfer. **75**, 736–753 (2014)
13. T. H. Lee, G. C. Park, D. J. Lee, Local flow characteristics of subcooled boiling flow of water in a vertical concentric annulus. Int. J. Multiphase Flow. **28**, 1351–1368 (2002)
14. R. P. Roy, S. Kang, J. A. Zarate, A. Laporta, Turbulent Subcooled Boiling Flow – Experiments and Simulations. ASME Journal of Heat Transfer. **124**(73), 72–93 (2002)
15. D. A. Gubaidullin, B. A. Snigerev, Numerical Simulation of the Turbulent Upward Flow of a Gas-Liquid Bubble Mixture in a Vertical Pipe: Comparison with Experimental Data. High Temperature. **56**(1), 61–69 (2018)
16. Y. Sato, M. Sadatomi, K. Sekoguchi, Momentum and heat transfer in two phase bubble flow-II. International Journal of Multiphase Flow. **7**(2), 179–190 (1981)
17. D. A. Gubaidullin, B. A. Snigerev, Numerical Simulations of Subcooled Boiling Flow in Vertical Pipe at High Pressure. Lobachevskii Journal of Mathematics. **40**(6), 745–750 (2019)
18. W. E. Ranz, W. R. Marshall, Evaporation from drops. Chemical Engineering Progress. **48**(3), 141–146 (1952)
19. H. K. Forster, N. Zuber, Dynamics of vapor bubbles and boiling heat transfer. AIChE J. **1**(4), 531–535 (1972)
20. M. Lemmert, M. Chawla, Influence of flow velocity on surface boiling heat transfer coefficient. Heat Transfer in Boiling. **2**, 237–247 (1977)
21. R. Cole, A photographic study of pool boiling in the region of the critical heat flux. AIChe Journal. **6**, 533–542 (1960)

Hybrid Methods for Network Equilibrium Problems

Igor Konnov and Olga Pinyagina

Abstract In the present paper, we propose a hybrid approach for network equilibrium problems. This approach combines the methods of conditional gradient and partial linearization. To apply the hybrid method, the whole set of origin-destination pairs is arbitrarily divided into two parts, for one of them the subproblem of direction finding is solved by the conditional gradient method, for the other, the partial linearization method is used. We propose two variants of the hybrid method with inexact direction finding and adaptive step-size choice.

1 Introduction

Network equilibrium problems are widely used in different areas such as telecommunication and transportation networks (for example, see [2, 3, 11–13]). These problems have the simple feasible simplex-like sets and special decomposable structure, therefore one can use modifications of the conditional gradient or partial linearization methods (CGM and PLM for short), where the subproblem of finding a descent direction can easily be solved without any iterative procedure [5, 9] or even inexactly [7]. In addition, the adaptive step-size choice can also be applied (see [6]).

In the present paper, we propose a hybrid approach for the network equilibrium problems, which combines CGM and PLM in the common iterative procedure. To apply the hybrid method, we split the set of origin-destination pairs into two subsets, for one of them the subproblem of direction finding is solved by CGM, for the other,

I. Konnov
Department of System Analysis and Information Technologies, Kazan (Volga Region) Federal University, Kazan, Russia
e-mail: konn-igor@yandex.ru

O. Pinyagina (✉)
Department of Data Mining and Operation Research, Kazan (Volga Region) Federal University, Kazan, Russia
e-mail: Olga.Piniaguina@kpfu.ru

I. B. Badriev et al. (eds.), *Mesh Methods for Boundary-Value Problems and Applications*, Lecture Notes in Computational Science and Engineering 141,
https://doi.org/10.1007/978-3-030-87809-2_15

PLM is used. We propose two variants of the hybrid method, the first one with inexact direction finding and the second one with adaptive step-size choice.

2 Preliminaries

Let us remind the general schemes of CGM and PLM. CGM was originally proposed by M. Frank and Ph. Wolfe in [4] for quadratic programming problems and further developed in [8]. Let $f : R^n \to R$ be a smooth function, D be a convex closed bounded set in R^n. We consider the following constrained optimization problem

$$\min_{x \in D} \longrightarrow f(x) \tag{1}$$

and the auxiliary linearized problem

$$\min_{y \in D} \longrightarrow \langle f'(x), y \rangle . \tag{2}$$

Under the given assumptions, both problems (1) and (2) have solutions, these solutions are nonunique in general. We denote by $Z(x)$ the set of solutions to problem (2).

At the kth iteration of CGM, $k = 0, 1, \dots$, we have a point $x^k \in D$. We solve problem (2) with $x = x^k$ and find a point $z^k \in Z(x^k)$. If $x^k \in Z(x^k)$, the necessary optimality condition for problem (1) holds, and it is a solution if f is convex. Otherwise, we determine the descent direction $d^k = z^k - x^k$, choose the step-size $\lambda_k \in [0, 1]$, and take the next iterate $x^{k+1} = x^k + \lambda_k d^k$. The step-size can be found with using a suitable exact or inexact line-search (or even without line-search) approach.

The partial linearization approach was proposed in [10] for optimization problems and developed in [13] for variational inequalities (VI for short). This approach has advantages when the objective function can be decomposed into two parts, one of them is suitable for linearization, and the other is sufficiently simple. We will consider the following optimization problem

$$\min_{x \in D} \longrightarrow \mu(x) \tag{3}$$

where the objective function $\mu : R^n \to R$ is the sum of two functions $\mu(x) = f(x) + h(x)$, the first of them f is smooth, the second one h is convex, and the feasible domain D is a convex closed set in R^n.

We now describe PLM for problem (3). Let us given a point $x^k \in D$ at the kth iteration, $k = 1, 2, \dots$. Find $z^k \in D$ as a solution to the auxiliary problem

$$\min_{x \in D} \longrightarrow \langle f'(x^k), x \rangle + h(x),$$

set $d^k = z^k - x^k$ and define the next iterate $x^{k+1} = x^k + \lambda_k d^k$, where the step λ_k can be found using a suitable exact or inexact procedure for the one-dimensional minimization problem. The above-stated method converges to a stationary point of the problem provided that the feasible set D is bounded.

When solving the network equilibrium problem by CGM or PLM methods, the subproblem of finding a descent direction needs no iterative procedure [5]. The special versions of CGM and PLM with inexact solution to the direction finding subproblem were proposed in [6] and [7], respectively. In the present paper, we propose a hybrid approach, combining these two methods in the general iterative process.

3 Network Equilibrium Problems

Let us recall the formulation of the network equilibrium problem, which was originally given in [3]. Usually, it describes a model of traffic or information data flows.

Let V be a set of network nodes, A be a set of directed arcs (links). In addition, a set W of origin-destination (O/D) pairs (i, j), $i, j \in V$ is given. For each O/D-pair $w \in W$ a set of paths P_w is given; each path presents a simple chain of arcs starting at the origin node and ending at the destination node of O/D-pair. We denote by x_p a variable flow value passing along path p, for all $p \in P_w$, $w \in W$.

Additionally, for each O/D-pair $w \in W$ a demand variable y_w is given. It presents a flow outgoing from the origin and ingoing to the destination. We assume these variables to be bounded from below and above with boundaries $0 \le \hat{\gamma}_w < \check{\gamma}_w$.

The network equilibrium problem is to find a distribution of the required demands for all O/D pairs among the sets of paths by using a certain (equilibrium) criterion.

The feasible set has the form:

$$U = \left\{ (x, y) \;\middle|\; \sum_{p \in P_w} x_p = y_w, x_p \ge 0, p \in P_w, y_w \in [\hat{\gamma}_w, \check{\gamma}_w], w \in W \right\}.$$

The following incidence matrix with elements

$$\alpha_{pa} = \begin{cases} 1, & \text{if arc } a \text{ belongs to path } p; \\ 0, & \text{otherwise} \end{cases}$$

gives the correspondence of paths and arcs.

Then the arc flow value is calculated as the sum of the corresponding path flows, for each arc $a \in A$:

$$f_a = \sum_{w \in W} \sum_{p \in P_w} \alpha_{pa} x_p. \tag{4}$$

Let a continuous cost function c_a be given for each $a \in A$; it can depend on all the arc flows in general. In addition, a so-called disutility continuous function h_w is given for each O/D pair $w \in W$. In the general case, the disutility functions can depend on the whole demand vector y.

The path cost function is defined as follows:

$$g_p(x) = \sum_{a \in A} \alpha_{pa} c_a(f),$$

for each path p, where f is the vector of arc flows $f_a, a \in A$.

We denote by G and H the vectors with the components g_p, $p \in P_w$, $w \in W$, and h_w, $w \in W$, respectively.

For finding an equilibrium state of this network, one can solve the following variational inequality: find an element $(x^*, y^*) \in U$ such that

$$\langle G(x^*), x - x^* \rangle - \langle H(y^*), y - y^* \rangle \geq 0 \quad \forall (x, y) \in U\,. \tag{5}$$

In what follows, we assume that each arc cost function c_a depends on f_a only, $\forall a \in A$, each disutility function h_w depends on y_w only, $\forall w \in W$. Then the mappings G and H are potential, and there exist the functions

$$\mu_a(f_a) = \int_0^{f_a} c_a(t)dt \quad \forall a \in A\,, \quad \sigma_w(y_w) = \int_0^{y_w} h_w(t)dt \quad \forall w \in W\,.$$

Note that VI (5) presents the optimality condition for the following optimization problem:

$$\min_{u \in U} \longrightarrow \psi(u), \tag{6}$$

where $u = (x, y)$,

$$\psi(x, y) = \left\{ \sum_{a \in A} \mu_a(f_a) - \sum_{w \in W} \sigma_w(y_w) \right\},$$

$f_a, \forall a \in A$ are defined in (4). We denote by ψ^* the optimal value of the goal function in problem (6). Therefore, each solution to problem (6) solves problem (5). The reverse assertion is also true, if, for example, the mappings G and $-H$ are monotone.

4 Hybrid Methods for Network Equilibrium Problems

Auxiliary problem (2) of the ordinary CGM has the following sense for the network equilibrium problem. At the kth iteration ($k = 0, 1, \dots$) of the main process, we have the vector of path flows x^k and demands y^k. We calculate the values of cost and disutility functions $g_p(x^k)$, $h_w(y^k)$ for all $p \in P_w$, $w \in W$. The problem is to find a vector $(\bar{x}^k, \bar{y}^k) \in U$, which is a solution to the auxiliary linearized VI:

$$\sum_{w\in W}\left[\sum_{p\in P_w} g_p(x^k)(x_p - \bar{x}_p^k) - h_w(y^k)(y_w - \bar{y}_w^k)\right] \geq 0, \ \forall (x, y) \in U, \qquad (7)$$

or the equivalent optimization problem

$$\min_{(x,y)\in U} \longrightarrow \sum_{w\in W}\left[\sum_{p\in P_w} g_p(x^k)x_p - h_w(y^k)y_w\right]. \qquad (8)$$

We see that problems (7) or (8) can be decomposed into a set of independent problems for each O/D pair. They can be solved using the simple algorithm from [9]. We remind the scheme of this algorithm for solving such a separate problem.

Algorithm A
For certain O/D pair $w \in W$, we calculate a set of shortest paths $\bar{P}_w^k$ with cost values $g_p(x^k)$. Let $\tilde{\lambda}_w = g_p(x^k)$, $\forall p \in \bar{P}_w^k$. Then the following three cases are possible.

(1) If $h_w(y^k) < \tilde{\lambda}_w$, then set $\bar{y}_w^k = \hat{y}_w$.
(2) If $h_w(y^k) > \tilde{\lambda}_w$, then set $\bar{y}_w^k = \check{y}_w$.
(3) Otherwise we have $h_w(y^k) = \tilde{\lambda}_w$, then choose any feasible demand $\bar{y}_w^k \in [\hat{y}_w, \check{y}_w]$.

Distribute the demand value $\bar{y}_w^k$ among paths $p \in \bar{P}_w^k$ (it is possible to associate the whole demand with one path). Set $\bar{x}_p^k = 0 \ \forall p \in P_w \backslash \bar{P}_w^k$.

In [5], the authors applied the partial linearization approach to the network equilibrium problem with elastic demand. Let us remind its general scheme. We suppose that $h_w(y) = h_w(y_w)$, h_w are monotonically decreasing functions, $\forall w \in W$.

The auxiliary direction finding problem is described as follows. At the kth iteration ($k = 0, 1, \dots$) of the main process, we have the vector of path flows x^k. We calculate the values of cost functions $g_p(x^k)$, for all $p \in P_w$, $w \in W$. We intend to find a vector $(\bar{x}^k, \bar{y}^k) \in W$, which is a solution to the auxiliary linearized VI:

$$\sum_{w\in W}\left[\sum_{p\in P_w} g_p(x^k)(x_p - \bar{x}_p^k) - h_w(\bar{y}^k)(y_w - \bar{y}_w^k)\right] \geq 0 \quad \forall (x, y) \in U, \qquad (9)$$

or to the equivalent optimization problem

$$\min_{(x,y)\in U} \longrightarrow \sum_{w\in W} \left[\sum_{p\in P_w} g_p(x^k)x_p - \sigma_w(y_w) \right], \tag{10}$$

where $h_w(y_w) = \sigma'_w(y_w)$. The above stated problems (9) or (10) can also be decomposed into a family of independent problems for each O/D pair. Hence the algorithm has the following simple scheme (see [5]).

Algorithm B
For certain O/D pair $w \in W$, we calculate the set of shortest paths $\bar{P}_w^k$ with costs values $g_p(x^k)$. Let $\tilde{\lambda}_w = g_p(x^k)$, $\forall p \in \bar{P}_w^k$. Hence the following three cases are possible.

(1) If $h_w(\hat{\gamma}_w) \leq \tilde{\lambda}_w$, then set $\bar{y}_w^k = \hat{\gamma}_w$.
(2) If $h_w(\check{\gamma}_w) \geq \tilde{\lambda}_w$, then set $\bar{y}_w^k = \check{\gamma}_w$.
(3) Otherwise we have $h_w(\check{\gamma}_w) < \tilde{\lambda}_w < h_w(\hat{\gamma}_w)$, then find the value of demand $\bar{y}_w^k \in [\hat{\gamma}_w, \check{\gamma}_w]$ such that $h_w(\bar{y}_w^k) = \tilde{\lambda}_w$.

Distribute the demand value $\bar{y}_w^k$ among paths $p \in \bar{P}_w^k$ (it is possible to associate the whole demand with one path). Set $\bar{x}_p^k = 0\ \forall p \in P_w \setminus \bar{P}_w^k$.

Now we intend to combine these two methods in a general iterative scheme.

4.1 A Hybrid Methods with Inexact Direction Finding

Let us arbitrarily split the set W into two subsets W_1 and W_2. In what follows, we assume that $h_w(y) = h_w(y_w)$, h_w are monotonically decreasing functions, $\forall w \in W_2$.

In solving the direction finding subproblem, for W_1 we use CGM, and for W_2 we apply PLM. Then at the kth iteration ($k = 0, 1, \ldots$) of the main process we can formulate the following auxiliary problem of direction search in the form of VI

$$\begin{aligned} &\sum_{w\in W_1} \left[\sum_{p\in P_w} g_p(x^k)(x_p - \bar{x}_p^k) - h_w(y^k)(y_w - \bar{y}_w^k) \right] + \\ &\sum_{w\in W_2} \left[\sum_{p\in P_w} g_p(x^k)(x_p - \bar{x}_p^k) - h_w(\bar{y}^k)(y_w - \bar{y}_w^k) \right] \geq 0 \quad \forall (x,y) \in U, \end{aligned} \tag{11}$$

or the equivalent optimization problem

$$\min_{(x,y)\in U} \longrightarrow \sum_{w\in W_1}\left[\sum_{p\in P_w} g_p(x^k)x_p - h_w(y^k)y_w\right]$$

$$+\sum_{w\in W_2}\left[\sum_{p\in P_w} g_p(x^k)x_p - \sigma_w(y_w)\right]. \quad (12)$$

In what follows, we will use the approach from [6, 7], in which the special versions of PLM and CGM with *inexact* solution to the direction finding subproblem were proposed.

At the kth iteration, we successively apply Algorithm A for OD-pairs $w \in W_1$ and Algorithm B for OD-pairs $w \in W_2$ and try to find a *sufficiently good* direction, which satisfies the condition

$$\sum_{w\in W^k}\left[\sum_{p\in P_w} g_p(x^k)(x_p^k - \bar{x}_p^k) - h_w(y^k)(y_w^k - \bar{y}_w^k)\right] \geq \delta. \quad (13)$$

Here $W^k \subset W$, $\delta > 0$ is a given tolerance. We vanish the missing components of the vector $(\bar{x}^k, \bar{y}^k)$:

$$\bar{x}_p^k = 0,\ p \in P_w,\ w \in W\backslash W^k, \quad \bar{y}_p^k = 0,\ w \in W\backslash W^k.$$

Therefore, we formulate the hybrid method for network equilibrium problems with inexact solution to the direction finding subproblem.

Hybrid Method 1 for Network Equilibrium Problems (HM1)

Step 0. Choose an initial point $u^0 \in U$, a sequence of tolerances $\{\delta_l\} \searrow 0$, $l = 1, 2, \ldots$, and numbers $\theta \in (0, 1)$, $\beta \in (0, 1)$. Set $l = 1$.

Step 1. Set $k = 0$, $v^k = u^{l-1}$, $(x^k, y^k) = v^k$.

Step 2. Sequentially using Algorithm A for $w \in W_1$ and Algorithm B for $w \in W_2$, find a set of O/D pairs $W^k \subset W$ and the vector $(\bar{x}^k, \bar{y}^k) \in U$ such that condition (13) holds. If it is not possible, then set $u^l = v^k$, $l = l + 1$ and go to Step 1.

Step 3. Set $\bar{v}^k = (\bar{x}^k, \bar{y}^k)$, $d^k = \bar{v}^k - v^k$. Find the smallest nonnegative number m that it holds

$$\psi(v^k + \theta^m d^k) - \psi(v^k) \leq \beta\theta^m \langle \psi'(v^k), d^k\rangle. \quad (14)$$

Set $\lambda_k = \theta^m$, $v^{k+1} = v^k + \lambda_k d^k$, $k = k + 1$ and go to Step 2.

The proposed algorithm has a two-level scheme. On the inner level, the objective function is minimized within a fixed tolerance value and on the outer level, this tolerance is decreased.

Now we substantiate convergence properties of the proposed modification, following [7].

It is easy to see that the line-search procedure at Step 3 of HM1 is finite. In fact, suppose that the line-search procedure is infinite, then (14) will never be fulfilled. We obtain

$$\theta^{-m}\left[\psi(v^k + \theta^m d^k) - \psi(v^k)\right] > \beta\langle\psi'(v^k), d^k\rangle,$$

for all $m \to \infty$. Taking the limit as $m \to \infty$ we have $\langle\psi'(v^k), d^k\rangle \geq \beta\langle\psi'(v^k), d^k\rangle$, therefore $\langle\psi'(v^k), d^k\rangle \geq 0$. We obtain a contradiction to (13).

Now we prove the finiteness of the inner process of the algorithm in Steps 2–3 (see also Lemma 1 from [7]).

Lemma 1 *The inner iterative process (Steps 2–3) of HM1 is finite.*

Proof Let us suppose that the sequence $\{v^k\}$ is infinite at stage l. Due to (14) we have $\psi^* \leq \psi(v^k)$, $\psi(v^{k+1}) \leq \psi(v^k) - \beta\delta_l\lambda_k$, therefore $\lim\limits_{k\to\infty} \lambda_k = 0$. Note that the sequences $\{v^k\}$ and $\{\bar{v}^k\}$ are bounded, then so is the sequence $\{d^k\}$. We can take subsequences $\{v^{k_s}\}$ converging to a certain point $\bar{v}$ and $\{d^{k_s}\}$ converging to a certain point $\bar{d}$ as $s \to \infty$. In view of (13) we have

$$\langle\psi'(\bar{v}), \bar{d}\rangle = \lim_{s\to\infty}\langle\psi'(v^{k_s}), d^{k_s}\rangle \leq -\delta_l. \tag{15}$$

At the same time, (14) is not fulfilled for the value of step-size λ_k/θ. We have for $k = k_s$

$$(\lambda_{k_s}/\theta)^{-1}\left[\psi(v^{k_s} + (\lambda_k/\theta)d^{k_s}) - \psi(v^{k_s})\right] > \beta\langle\psi'(v^{k_s}), d^{k_s}\rangle.$$

Taking the limit as $s \to \infty$, we obtain

$$\langle\psi'(\bar{v}), \bar{d}\rangle = \lim_{s\to\infty}(\lambda_{k_s}/\theta)^{-1}\left[\psi(v^{k_s} + (\lambda_k/\theta)d^{k_s}) - \psi(v^{k_s})\right] \geq \beta\langle\psi'(\bar{v}), \bar{d}\rangle,$$

hence, $(1-\beta)\langle\psi'(\bar{v}), \bar{d}\rangle \geq 0$ that contradicts (15).

Now we are ready to prove the convergence properties of HM1 (see also Theorem 1 from [7]).

Theorem 1 *The sequence $\{u^l\}$ generated by HM1 has limit points, all of them are solutions to VI (5). If the function ψ is convex, they are also solutions to optimization problem (6).*

Within the proof of this theorem, we denote by $\{\bar{u}^l\}$ the sequence of exact solutions to problem (11) or (12). Note that by construction the sequences $\{u^l\}$ and $\{\bar{u}^l\}$ are bounded and have limit points. In addition, $\psi(u^*) \leq \psi(u^{l+1}) \leq \psi(u^l)$, therefore the limit $\lim_{l\to\infty} \psi(u^l) = \psi'$ exists. Take any limit point $u' = (x', y')$ of the sequence $\{u^l\}$, denote by $\{u^{l_s}\}$ a subsequence converging to this point. Take any limit point $\bar{u} = (\bar{x}, \bar{y})$ of the sequence $\{\bar{u}^l\}$, denote by $\{\bar{u}^{l_s}\}$ a subsequence converging to this point. By construction due to the condition of return from the inner level of the method to the outer level at step 2 for all $l > 0$ we have

$$\langle G(x^l), x^l - \bar{x}^l\rangle - \langle H(y^l), y^l - \bar{y}^l\rangle < \delta_l. \tag{16}$$

On the other hand, for a point $\bar{u}^l = (\bar{x}^l, \bar{y}^l)$ we get for all $l > 0$

$$\begin{aligned}\langle G(x^l), x - \bar{x}^l\rangle &- \sum_{w\in W_1} h_w(y^l)(y_w - \bar{y}^l_w)\\ &- \sum_{w\in W_2} h_w(\bar{y}^l)(y_w - \bar{y}^l_w) \geq 0 \ \forall (x, y) \in U.\end{aligned} \tag{17}$$

Since the functions h_w $\forall w \in W$ are non-increasing, from (16) we obtain

$$\langle G(x^l), x^l - \bar{x}^l\rangle - \sum_{w\in W_1} h_w(y^l)(y^l_w - \bar{y}^l_w) - \sum_{w\in W_2} h_w(\bar{y}^l)(y^l_w - \bar{y}^l_w) < \delta_l.$$

Summing the two above inequalities and taking the limit as $s \to \infty$, we have

$$\begin{aligned}\lim_{s\to\infty} \langle G(x^{l_s}), x - x^{l_s}\rangle &- \sum_{w\in W_1} h_w(y^{l_s})(y_w - y^{l_s}_w) - \sum_{w\in W_2} h_w(\bar{y}^{l_s})(y_w - y^{l_s}_w) =\\ \langle G(x'), x - x'\rangle &- \sum_{w\in W_1} h_w(y)(y_w - y'_w) - \sum_{w\in W_2} h_w(\bar{y})(y_w - y'_w) \geq 0,\\ &\forall (x, y) \in U.\end{aligned} \tag{18}$$

Now let us show that $h_w(\bar{y}_w) = h_w(y'_w)$ $\forall w \in W_2$. Assume the contrary, let for at least one $\bar{w} \in W_2$ $h_{\bar{w}}(\bar{y}_{\bar{w}}) \neq h_{\bar{w}}(y'_{\bar{w}})$, then evidently

$$\langle h_{\bar{w}}(\bar{y}_{\bar{w}}) - h_{\bar{w}}(y'_{\bar{w}}), \bar{y}_{\bar{w}} - y'_{\bar{w}}\rangle < 0. \tag{19}$$

Then, on the one hand, taking the limit as $s \to \infty$ in (16) and taking into account (19), we have:

$$\langle G(x'), x' - \bar{x}\rangle - \sum_{w\in W_1} h_w(y')(y'_w - \bar{y}_w) - \sum_{w\in W_2} h_w(\bar{y})(y'_w - \bar{y}_w) < 0.$$

On the other hand, taking the limit as $s \to \infty$ in (17) and setting $y = y'$, we obtain

$$\langle G(x'), x' - \bar{x} \rangle - \sum_{w \in W_1} h_w(y')(y'_w - \bar{y}_w) - \sum_{w \in W_2} h_w(\bar{y})(y'_w - \bar{y}_w) \geq 0.$$

A contradiction is obtained, therefore $h_w(\bar{y}_w) = h_w({y'}_w)$ $\forall w \in W_2$ and from (18) we have

$$\langle G(x'), x - x' \rangle - \langle H(y'), y - y' \rangle \geq 0, \ \forall (x, y) \in U.$$

Hence, the point $u' = (x', y')$ solves VI (5). In addition, if ψ is convex, then this point also solves optimization problem (6), as desired.

4.2 A Hybrid Method with Adaptive Step-Size

In this section, we propose a variant of the hybrid method, which does not require any iterative step-size line-search. The main idea of this approach [6] is a given majorant step-size sequence converging to zero. In accordance with this majorant, we take the next decreased value of the step-size only when the current iterate does not give a sufficient descent, which is estimated with the help of an Armijo-type condition.

Hybrid Method 2 for Network Equilibrium Problems (HM2)

Step 0. Choose an initial point $u^0 \in U$, sequences $\{\delta_l\} \searrow 0$, $\{\tau_p\} \to 0$, $\tau_p \in (0, 1)$, and number $\beta \in (0, 1)$. Set $l = 1$.

Step 1. Set $k = 1$, $p = 0$, $v^k = u^{l-1}$, $(x^k, y^k) = v^k$, choose an initial step-size $\lambda_0 \in (0, \tau_0)$.

Step 2. Sequentially using Algorithm A for $w \in W_1$ and Algorithm B for $w \in W_2$, find a set of O/D pairs $W^k \subset W$ and the vector $(\bar{x}^k, \bar{y}^k) \in U$ such that condition (13) holds. If it is not possible, then set $u^l = v^k$, $l = l + 1$ and go to Step 1.

Step 3. Set $\bar{v}^k = (\bar{x}^k, \bar{y}^k)$, $d^k = \bar{v}^k - v^k$, $v^{k+1} = v^k + \lambda_k d^k$.
If

$$\psi(v^{k+1}) - \psi(v^k) \leq \beta \lambda_k \langle \psi'(v^k), d^k \rangle, \tag{20}$$

then set $\lambda_{k+1} \in [\lambda_k, \tau_p]$. Otherwise set ${\lambda'}_{k+1} = \min\{\lambda_k, \tau_{p+1}\}$, $p = p + 1$, take $\lambda_{k+1} \in (0, {\lambda'}_{k+1})$. Set $k = k + 1$ and go to Step 2.

First of all, let us prove the finiteness of the inner iterative process.

Lemma 2 *The inner iterative process (Steps 2–3) of HM2 is finite.*

Proof Let us suppose that the sequence $\{v^k\}$ of HM2 is infinite at stage l. We note that the sequences $\{v^k\}$ and $\{\bar{v}^k\}$ are contained in a bounded set U, hence they have limit points. So is the sequence $\{d^k\}$. Then the two cases are possible.

Case 1. The Number of Changes of Index p is Finite Then we have $\lambda_k \geq \bar{\lambda} > 0$ for numbers k large enough. Therefore we get from condition (20)

$$\psi(v^{k+1}) \leq \psi(v^k) + \beta\lambda_k\langle\psi'(v^k), d^k\rangle \leq \psi(v^k) + \beta\bar{\lambda}\langle\psi'(v^k), d^k\rangle$$

for k large enough. Since $\psi(v^k) \geq \psi^* > -\infty$ for all k, we obtain

$$\lim_{k\to\infty} \langle\psi'(v^k), d^k\rangle = 0.$$

On the other hand, due to (13) $\langle\psi'(v^k), d^k\rangle \leq -\delta_l$, therefore

$$\lim_{k\to\infty} \langle\psi'(v^k), d^k\rangle \leq -\delta_l < 0.$$

We obtain a contradiction, hence the number of changes of index p cannot be finite.

Case 2. The Number of Changes of Index p Is Infinite In this case, there exists an infinite sequence of indices $\{k_s\}$ such that $v^{k_s+1} = v^{k_s} + \lambda_{k_s} d^{k_s}$ and condition (20) is violated:

$$\psi(v^{k_s+1}) - \psi(v^{k_s}) > \beta\lambda_{k_s}\langle\psi'(v^{k_s}), d^{k_s}\rangle .$$

At the same time, $\lambda_{k_s} \in (0, \tau_p]$, $\lambda_{k_s+1} \in (0, \tau_{p+1}]$ and $\lim\limits_{p\to\infty} \tau_p = 0$. Hence, $\lim\limits_{s\to\infty} \lambda_{k_s} = 0$. Proceeding to the limit as $s \to \infty$ in the correlation

$$(\lambda_{k_s})^{-1}\psi(v^{k_s+1}) - \psi(v^{k_s}) > \beta\langle\psi'(v^{k_s}), d^{k_s}\rangle ,$$

we obtain $\langle\psi'(v'), d'\rangle(1-\beta) \geq 0$, where v' and d' are the corresponding limit points. As we noted above, from (13) we have $\langle\psi'(v^k), d^k\rangle \leq -\delta_l$, therefore $\langle\psi'(v'), d'\rangle\rangle \leq -\delta_l < 0$. We also obtain a contradiction, hence this case is impossible, and the sequence $\{v^k\}$ of HM2 cannot be infinite at stage l, as desired.

Theorem 2 *The sequence $\{u^l\}$ generated by HM2 has limit points, all of them are solutions to VI (5). If the function ψ is convex, they are also solutions to optimization problem (6).*

Proof follows the lines of Theorem 1.

5 Numerical Tests

Combining CGM and PLM within a general scheme, we can obtain more flexible implementations, which take into account the peculiarities of the initial problem. In the following example, the hybrid method is more efficient by computational time and the iterations numbers in most cases.

Let us consider a known network structure (Fig. 1) [1] in which all arcs are assumed to be bypass (i.e., each arc cost functions c_a depends on the arc flow f_a only, for all $a \in A$).

The cost functions are $c_a(f_a) = 1 + 4f_a, \forall a \in A$. The disutility functions are $h_w(y_w) = 30 - 0.5y_w, \forall w \in W$. The lower boundaries of demand $\hat{\gamma}_w$ are (2, 1, 1, 1, 1), and all the upper boundaries $\check{\gamma}_w, w \in W$ are equal to 50.

We compared HM1 and the corresponding pure versions of CGM and PLM with the Armijo-type line-search and inexact direction finding. The stop criterion has the form $\langle \psi'(u^l), u^l - z^l \rangle \leq \Delta$ for given $\Delta > 0$, where z^l is a solution to the problem

$$\min_{z \in U} \longrightarrow \langle \psi'(u^l), z \rangle .$$

The parameters of methods are $\theta = 0.5$, $\beta = 0.5$, $\delta_0 = 1$, $\delta_{l+1} = 0.5\delta_l$. Applying HM1, we use CGM for O/D pairs with odd numbers and PLM for even numbers .

The calculation results for different accuracy values are presented in Table 1.

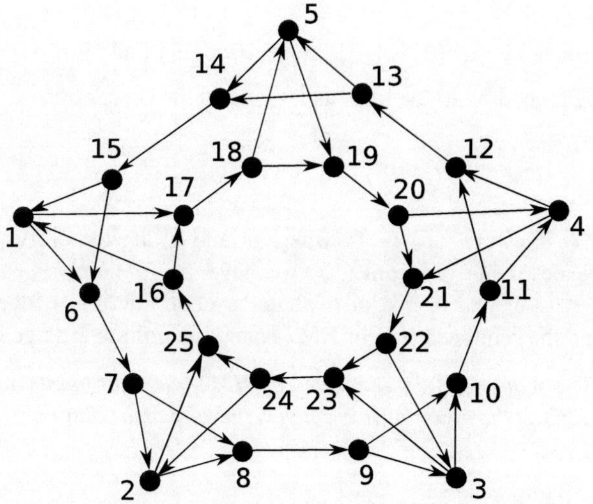

Fig. 1 Network of 25 nodes, 5 O/D pairs (1–4), (2–5), (3–1), (4–2), (5–3)

Table 1 Example 1, numbers of iterations and calculation time

	CGM		*PLM*		*HM1*	
Δ	Iterations	Time (ms)	Iterations	Time (ms)	Iterations	Time (ms)
0.1	2953	78	2578	78	2474	63
0.05	5853	187	5001	218	5032	156
0.01	39,005	1263	38,030	1108	19,979	608

6 Conclusion

In the present paper, a hybrid approach for the network equilibrium problems was proposed. This approach combines the methods of conditional gradient and partial linearization. The set of origin-destination pairs should be arbitrarily divided into two parts, and the subproblem of finding the direction is solved by CGM for the first subset of indices and by PLM for the second one. This approach allows us to take into account the specifics of problems under consideration and it is promising for further investigations.

Acknowledgement This work was supported by Russian Foundation for Basic Research, project No 19-01-00431.

References

1. Bertsekas, D.P., Gafni, E.M.: Projection methods for variational inequalities with application to the traffic assignment problem. In: Nondifferential and Variational Techniques in Optimization, pp. 139–159. Springer, Berlin, Heidelberg (1982)
2. Dafermos, S.: Traffic equilibrium and variational inequalities. Transp. Sci. **14** (1), 42–54 (1980)
3. Dafermos, S.: The general multimodal network equilibrium problem with elastic demand. Networks. **12** (1) 57–72 (1982)
4. Frank, M., Wolfe, Ph.: An algorithm for quadratic programming. Naval Research Logistics Quarterly. **3** (12), 95–110 (1956)
5. Konnov, I., Pinyagina, O.: Partial linearization method for network equilibrium problems with elastic demands. In: Kochetov Y., Khachay M., Beresnev V., Nurminski E., Pardalos P. (eds) Discrete Optimization and Operations Research. DOOR 2016. Lecture Notes in Computer Science, Vol. 9869, pp. 418–429. Springer, Cham (2016)
6. Konnov, I.V.: Simplified versions of the conditional gradient method. Optimization. **67** (12), 2275–2290 (2018)
7. Konnov, I., Laitinen, E., Pinyagina, O.: Inexact partial linearization methods for network equilibrium problems. J. Appl. Ind. Math. **14**, 92–103 (2020)
8. Levitin, E.S., Polyak, B.T.: Constrained minimization methods, USSR Computational Mathematics and Mathematical Physics. **6** (5), 1–50 (1966)
9. Magnanti, T.L.: Models and algorithms for predicting urban traffic equilibria. In: Transportation Planning Models, pp. 153–185. North–Holland, Amsterdam (1984)
10. Mine, H., Fukushima, M.: A minimization method for the sum of a convex function and a continuously differentiable function. J. Optim. Theor. Appl. **33**, 9–23 (1981)

11. Nagurney, A.: Network Economics: A Variational Inequality Approach. Kluwer, Dordrecht (1999)
12. Patriksson, M.: Nonlinear Programming and Variational Inequality Problems: a Unified Approach. Kluwer, Dordrecht (1999)
13. Patriksson, M.: The Traffic Assignment Problem: Models and Methods. Dover, Mineola, NY (2015)

Numerical Simulation of Water-Oil Inflow into the Producing Well from Non-uniform Oil Reservoir

Vladimir M. Konyukhov, Ivan V. Konyukhov, and Leysan R. Ilyasova

Abstract Mathematical and numerical models of non-stationary mass transfer in a water-oil mixture flowing into the bottom hole of a production well from a layered-heterogeneous oil reservoir are developed. The model consists of two group of differential equations. The first of them simulates a one-dimensional dispersed oil-water flow with discrete oil droplets included in a continuous water phase, and the second one—two-dimensional two-phase isothermal filtration governing by Darcy's law with taking into account the compressibility of phases and a porous medium. To solve system of equations the finite difference schema is developed. The mass transfer equations in the bottom hole and in the reservoir are approximated upstream by implicit difference equations. The general system of nonlinear algebraic equations is solved iteratively with the use of the original method to calculate the pressure in the reservoir and Newtonian linearization. The developed numerical model is implemented in computer software that allows to carry out the numerical experiments with simultaneous visualization of the results of calculations. The influence of the reservoir structure and its uncovering conditions by the well on the characteristics of the process in the bottom hole of the well and the transition time of mass transfer processes to a quasi-stationary hydrodynamic regime are estimated.

V. M. Konyukhov · L. R. Ilyasova
Kazan Federal University, Kazan, Russian Federation
e-mail: vladimir.konyukhov@kpfu.ru

I. V. Konyukhov (✉)
Innopolis University, Innopolis, Russian Federation
e-mail: i.konyukhov@innopolis.ru

I. B. Badriev et al. (eds.), *Mesh Methods for Boundary-Value Problems and Applications*, Lecture Notes in Computational Science and Engineering 141,
https://doi.org/10.1007/978-3-030-87809-2_16

1 Introduction

Natural productive oil reservoirs, as a rule, have a layered-heterogeneous structure. Filtration and capacitance and geometric characteristics of their layers may differ significantly from each other [1–3]. The reservoir structure determines the intensity and composition of filtration flows entering the bottom hole of the producing wells.

The processes at the bottom hole of the well, uncovering the reservoir, and in the reservoir itself are closely interrelated [4, 5]. They are determined not only by the reservoir structure and properties of the porous medium, but also by the heterogeneity of the two-phase mixture that includes oil and water entering the bottom hole.

In turn, the physical and chemical properties (density, viscosity, etc.) of these phases are significantly different. In addition, at low speeds of the mixture, due to the difference in phase densities, the effects of gravitational separation of water and oil can occur in the two-phase flow. As a result, the water content gradually decreases in the upper part of the bottom-hole of the well and increases in its lower part. Water sedimentation and oil floating-up lead to changes in the effective properties of the mixture—its density and viscosity (for example, the effective viscosity of the mixture can rise almost tenfold with an increase in the water content from 10 to 60% [7, 9]). The pressure distribution along the well becomes piece-wise linear with a clearly defined breakpoint corresponding to the water-oil interface.

The length of the bottom-hole section of the well, the conditions for uncovering a layered-inhomogeneous reservoir, the intensity and composition of filtration flows, as well as the difference in viscosity and density of phases affect the mass transfer and hydrodynamic processes in this area. Therefore, the water content of the mixture entering directly into the lifting pipes of a low-yield well can differ within a few hours from its average integral values, which are usually set as boundary conditions at the inlet of these pipes.

Such a delay is especially important to take into account in the calculating of transient processes when the well is put into operation after the repair of a submersible pumping unit located near the bottom hole of the producing well. In such a situation, the use of integral values in boundary conditions can result in significant overstating of the water content at the pump inlet.

Because of this, the calculated working characteristics of centrifugal pump and electric motor, which significantly depend on the composition and flow rate of the mixture at the pump inlet (as well as the duration of their non-stationary operation stage when the average integral oil content in the bottom hole has not yet been reached) can be determinated with very large errors. As a result, these design conditions may be unacceptable for normal operation of the pump unit, both due to overloading or underloading of the motor and its insufficient cooling by the mixture flow, as well as lead to subsequent incorrect impact from the surface control station on the working regime of the unit (for example, an emergency shutdown of the motor or the wrong change the frequency of its current) [13].

The aim of this paper is to numerically study the features of non-stationary mass transfer in the water-oil flow entering the bottom-hole section of a production well from the layered-heterogeneous oil reservoir, to estimate the influence of its inhomogeneity and uncovering conditions on the characteristics of these processes and the time of their transition to a quasi-stationary hydrodynamic regime.

2 Mathematical Model

Let us consider the situation when the well has been filled with water before its commissioning into operation. After turning the submersible pumping unit on, the oil-water mixture begins to flow into the bottom hole from the reservoir. Figure 1 schematically shows a vertical cross-section of a symmetrical planar-radial layered-heterogeneous reservoir uncovered by a producing well located on its left boundary. The reservoir consists of layers of different thickness $H_l = \gamma_l - \gamma_{l-1}$, absolute permeability K_l and dynamic porosity m_l, $l = 1, 2, \ldots, N$. The total thickness of the reservoir $H = H_1 + H_2 + \ldots + H_N$.

The coordinate axes Or and Oz are directed respectively along its roof and the vertical axis of the producing well, the side surface of which is located on the left boundary of the formation at $r = r_0$. The origin (0, 0) is located in the upper-left corner of the reservoir roof. Surfaces $z = H$ and $z = 0$ of its bottom and roof are impermeable. The boundaries γ_l (l = 1,2, ...,N − 1) of the layers are permeable, so that they are hydrodynamically connected. The well can uncover all layers or only some of them. The "permeable" (γ_L, γ_R) and "impenetrable" (Γ_L, Γ_R) parts of the left and right lateral boundaries of the reservoir at $r = r_0$ and $r = R_r$ uncovered by perforation are shown in the figure as dashed and solid lines, respectively. To describe the movement of a two-phase flow with incompressible continuous water and discrete oil phases at the bottom hole and in the reservoir, a mathematical model is developed that can be written as follows:

$$\frac{\partial \varphi}{\partial \tau} + \upsilon_2 \frac{\partial \varphi}{\partial z} = q_2 - \varphi \frac{\partial \upsilon_2}{\partial z}; \quad 0 < z < H, \quad t > 0; \tag{1}$$

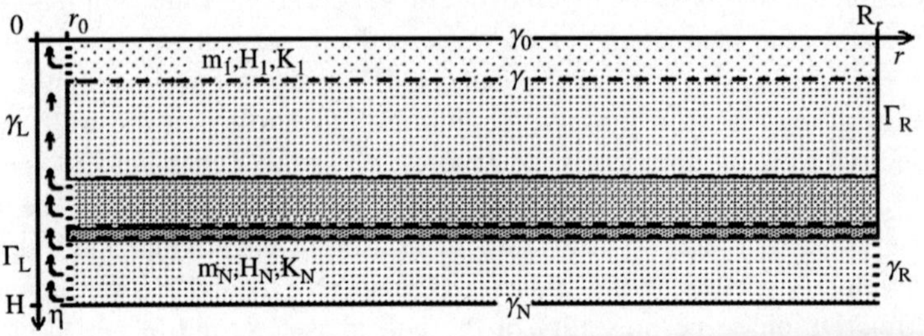

Fig. 1 Vertical cross-section of the oil reservoir and the bottom-hole part of the producing well

$$G(z,\tau) = \frac{S_0}{g}\left[\frac{\partial P}{\partial \tau}(z,\tau) - \frac{\partial P}{\partial \tau}(0,\tau)\right] + \tag{2}$$

$$+S_0\rho_1^* \int_0^z q_1(z,\tau)\,dz + S_0\rho_2^* \int_0^z q_2(z,\tau)\,dz;$$

$$-\frac{\partial P}{\partial z} = g\left(\rho_1^*(1-\varphi) + \rho_2^*\varphi\right); \tag{3}$$

$$G_1 = S_0\rho_1^*(1-\varphi)\,\upsilon_1;\quad G_2 = G - G_1;\quad \upsilon_1 = \frac{1-\beta}{1-\varphi}\upsilon;\quad \upsilon_2 = C_\varsigma \upsilon + \nu_2; \tag{4}$$

$$\upsilon = \frac{G/S_0}{\rho_1^*(1-\beta)+\rho_2^*\beta};\quad \beta = C_\varsigma\varphi + \frac{\varphi\nu_2}{\upsilon};\quad \nu_2 = g\frac{a^2\left(\rho_1^*-\rho_2^*\right)}{3\varsigma(\varphi)\mu_1}\cdot\frac{\mu_1+\mu_2}{\mu_1+1.5\mu_2}; \tag{5}$$

$$\alpha_T\frac{\partial P}{\partial \tau} + \frac{1}{r}\frac{\partial(rV_r)}{\partial r} + \frac{\partial V_z}{\partial z} = 0;\quad V_r = -KK^*\frac{\partial P}{\partial r};\quad V_z = -KK^*\frac{\partial P}{\partial z}; \tag{6}$$

$$m\frac{\partial S}{\partial \tau} + \alpha_{T1}^* S\frac{\partial P}{\partial \tau} + \frac{1}{r}\frac{\partial\left(rV_{1,r}\right)}{\partial r} + \frac{\partial V_{1,z}}{\partial z} = 0;\quad V_{1,r} = f\,V_r;\quad V_{2,z} = f\,V_z; \tag{7}$$

$$\alpha_{T1}^* = \alpha_{Tm} + m\alpha_{T1};\quad \alpha_{T2}^* = \alpha_{Tm} + m\alpha_{T2};\quad \alpha_T = \alpha_{T1}^* S + \alpha_{T2}^*(1-S); \tag{8}$$

$$K_1^* = \begin{cases} 0 &, \quad 0 \le S \le S_*; \\ \left((S-S_*)/S_*\right)^3, & S_* \le S \le 1; \end{cases} \tag{9}$$

$$K_2^* = \begin{cases} \left((S^*-S)/(S^*-S_*)\right)^3, & 0 \le S \le S^*; \\ 0 &, \quad S^* \le S \le 1; \end{cases} \quad K^* = K_1^*/\mu_1 + K_2^*/\mu_2.$$

Here P is the pressure; ρ_i^*, υ_i , G_i, μ_i, φ_i and β_i are the density, the velocity, the mass flow rate (debit), the dynamic viscosity, the actual and volumetric consumed concentration of i-th phase averaged over cross section S_0 of the well (sub-indexes 1 and 2 denote the characteristics of water and oil, respectively); τ is the time; $q_i(z,\tau) = 2/r_o V_{r,i}(r_0, z, \tau)$ is the density of the mass filtration flow of the i-th phase from the reservoir through the bottom-hole surface of the well; υ_2 is the drift velocity of the oil drops, a is their radius; C_ς is the Zuber-Findlay coefficient [10, 11]; g is the gravitational acceleration; S is the water saturation of the mixture in the reservoir; V_r, V_z and $V_{1,r}$, $V_{1,z}$ are projections of the mixture and water phase filtration velocity vectors on the and axes Oz and Or; μ_i and K_i^* are the viscosity and the relative phase permeability of the i-th phase; K and m are the absolute permeability and the dynamic porosity; $f = K_1^*/(\mu_1 K^*)$ is the fraction of water in the total filtration flow (Buckley-Leverett function); α_T and $\alpha_{T,i}$ are volumetric elasticity coefficients of the porous medium and i-th phase; S_* is the irreducible water saturation; S^* is the limiting water saturation.

Equations (1), (2) and (3) describing the continuity of the oil phase, the integral mass balance of the mixture, the conservation of momentum, and additional relations (4) and (5) are developed on the base of general multiphase flow equations (see, e.g., [7, 8] in the framework of the drift flow model [10, 11] without taking into account inertia forces. They simulate a one-dimensional dispersed oil-water flow with discrete oil droplets included in a continuous water phase.

Equations (5)–(9) describe two-dimensional two-phase isothermal filtration governing by Darcy's law without taking into account capillary effects and gravity, as well as taking into an account the compressibility of phases and a porous medium [1, 2, 4–6].

To close the mathematical model (1)–(9), we must set the initial and boundary conditions. Let there be no movement in the well and the oil reservoir at the initial time $\tau = 0$, and the well is filled with water with a hydrostatic pressure distribution, so that

$$\upsilon_1(0, z) = \upsilon_2(0, z) = \upsilon(0, z) = 0; \tag{10}$$

$$q_1(0, z) = q_2(0, z) = 0; \quad G_1(0, z) = G_2(0, z) = 0;$$

$$\varphi(0, z) = \beta(0, z) = 0; \quad \rho(0, z) = \rho_1^*; \quad P(0, z) = P_H^0 + g\rho_1^* z, 0 \le z \le H, \tag{11}$$

where $P(0, 0) = P_H(\tau = 0) = P_H^0$ is the bottom hole pressure in the well at the level $z = 0$ of the reservoir roof at time $\tau = 0$. The function $P_H(\tau) = P(\tau, 0) = P_H^0 \cdot \Psi(\tau)$ in the considered problem is a given function of time τ and it simulates the pressure drop at the bottom hole of the well after turning the electric motor of the submersible pumping unit on.

At the boundaries γ_m of layers where the absolute permeability K has a discontinuity of the first kind, the conjugation conditions $[P] = 0$, $[V_z] = 0$, $[f] = 0$ take place at $z = \gamma_m$, $m = \overline{1, N-1}$.

The initial conditions

$$(r, z, 0) = S^0(r, z), \quad P(r, z, 0) = P_H^0 + g\rho_1^* z, \quad 0 \le z \le H \tag{12}$$

determine the state of the reservoir before turning the electric motor on, when the pressure in the reservoir has the same hydrostatic distribution as in the bottom hole part of the well. At the right boundary $r = R_r$ of the reservoir at $\tau > 0$, the pressure $P(R_r, z, \tau)$ is determined by the initial hydrostatic distribution. At the left boundary $P(r_0, z, \tau)$ it is calculated by solving the Eq. (3) at the bottom hole part of the well at $0 \le z \le H$.

It should be noted that this article presents only some basic relationships which define the characteristics of two-phase flows in the well and oil reservoir. A complete set of special constitutive relations to close the equations is too large and it can be found in our publications [4, 5, 12, 13].

3 Numerical Model and Algorithm

The system of Eqs. (1)–(10) is nonlinear and is solved numerically by the finite difference method using iterative algorithms. To solve problems (6) and (7), we introduce a grid D_h in the reservoir area $D = \{z \in [0, H], \ r \in [r_o, R_r]\}$ with a constant step $h_r = (R_r - r_0)/N_r$ along the Or axis and a variable step along the Oz axis, where N_r is the number of grid points along the Or axis. At the same time, in each layer the step is constant and equal to $h_n = H_n/N_n$, $n = 1..N$, where N_n is the number of grid points in the layer. With such a construction of the grid, we have $h_\kappa = h_n$ at $\kappa = 1 + \sum_{l=1}^{n-1} N_l \ldots \sum_{l=1}^{n} N_l$, $n = 1..N$. To improve the approximation of the flows $\mathbf{V}$ and $\mathbf{V_i}$ in Eqs. (6) and (7) we use an additional grid $\bar{D}_{i,\kappa}$ shifted by half-step along Or and Oz axes so that the exterior boundaries of its unit cells are placed at the boundary of the filtration area D and at the boundaries γ_n of layers. The total number of grid points is $N_r \cdot N_z$, where $N_z = \sum_{n=1}^{N} N_n$. We also denote by τ_j the points of grid along the time axis $O\tau$ with constant step h_τ. Conservative finite-difference equations approximating the system (6) and (7) for $i = 1..N_r$, $\kappa = 1..N_z$ with order $O(h_\tau + h_\kappa^2 + h_r^2)$, can be written as:

$$\Lambda\,[V]_{i.\kappa}^{j+1} = r_{i-1/2} h_r h_\kappa \alpha_{T i,\kappa} \cdot \left(P_{i.\kappa}^{j+1} - P_{i,\kappa}\right)\Big/h_\tau; \tag{13}$$

$$\Lambda\,[V_1]_{i.\kappa}^{j+1} = m r_{i-1/2} \frac{h_r h_\kappa}{h_\tau} \left\{ \left(J_{i,\kappa}^{j+1} - J_{i,\kappa}^{j}\right) + \frac{\alpha_{T1}^*}{m} \left(P_{i,\kappa}^{j+1} - P_{i,\kappa}^{j}\right) J_{i,\kappa}^{j,j+1} \right\}; \tag{14}$$

$$J_{i,\kappa}^{j,j+1} = \begin{cases} J_{i,\kappa}^{j}, & P_{i,\kappa}^{j+1} \le P_{i,\kappa}^{j}; \\ J_{i,\kappa}^{j+1}, & P_{i,\kappa}^{j+1} > P_{i,\kappa}^{j}; \end{cases} \qquad J_{i,\kappa} = \frac{1}{r_{i-1/2} h_r h_\kappa} \int_{D_{i,\kappa}} S r dr dz; \tag{15}$$

$$V_{i+1/2,\kappa}^{j+1} = h_\kappa \zeta_i \left(K K^*\right)_{i+1/2,\kappa}^{j} \left(P_{i+1,\kappa}^{j+1} - P_{i,\kappa}^{j+1}\right); \quad V_{1,i+1/2,\kappa}^{j+1} = f_{i+1/2,\kappa}^{j}\, V_{i+1/2,\kappa}^{j+1}; \tag{16}$$

$$V_{i,\kappa+1/2}^{j+1} = A_{i,\kappa+1/2} \left(P_{i,\kappa+1}^{j+1} - P_{i,\kappa}^{j+1}\right); \quad V_{1,i,\kappa+1/2}^{j+1} = f_{i,\kappa+1/2}^{j}\, V_{i,\kappa+1/2}^{j+1}; \tag{17}$$

$$\zeta_i = \begin{cases} \ln^{-1}\left(h_r/(2r_0)\right), & i - 1/2 = 1/2; \\ \ln^{-1}\left((2i+1)/(2i-1)\right), & 1 \le i < i_0; \\ i, & i > i_0; \end{cases} ;$$

$$A_{i,\kappa+1/2} = 2 r_{i-1/2} h_r \left(\frac{h_\kappa}{K_{i,\kappa} K_{i,\kappa}^*} + \frac{h_{\kappa+1}}{K_{i,\kappa+1} K_{i,\kappa+1}^*} \right),$$

where $\Lambda[V]_{i,\kappa} = V_{i+1/2,\kappa} + V_{i-1/2,\kappa} + V_{i,\kappa+1/2} + V_{i,\kappa-1/2}$; ζ_i is the correction coefficients [4, 5] that take into account the logarithmic character of pressure distribution in the vicinity of the well at the approximation of flows $V^{j+1}_{i+1/2,\kappa}$; $i_0 h_r$ is the radius of the vicinity in which the solution has the logarithmic singularity. The values $K^*_{i+1/2,\kappa}$ are calculated using water saturation $S^j_{i+1/2,\kappa} = 0.5\left(J^j_{i+1,\kappa} + J^j_{i,\kappa}\right)$.

The pressure field $P^{j+1}_{i,\kappa}$ is calculated from a system of implicit finite-difference equations obtained from (13) by substituting the total flows $V^{j+1}_{i,\kappa+1/2}$, $V^{j+1}_{i,\kappa+1/2}$ (16) and (17) defined for the water saturation values $S^j_{i+1/2,\kappa}$ and $S^j_{i,\kappa+1/2}$ from the previous time point τ_j. The transport equation (14) is used to calculate the average integral values $J^{j+1}_{i,\kappa}$ (15) of water saturation in the unit cells.

The values of the water saturation $S^{j+1}_{i+1/2,\kappa}$ and $S^{j+1}_{i,\kappa+1/2}$ are defined with the use of the fractional-linear interpolation through integral values $J^{j+1}_{i,\kappa}$ taking account the flow direction. For example, if $V^{j+1}_{i+1/2,\kappa} < 0$, so that the liquid moves oppositely to the direction of the Or axis from cell $D_{i+1,\kappa}$ to cell $D_{i,\kappa}$ then

$$S_{i+1/2,\kappa} = \begin{cases} S^*_\kappa\,, & S^*_\kappa - \varepsilon^* \le J_{i,\kappa}, \\ F\,, \quad F \in \left[J_{i+1,\kappa}, J_{i,\kappa}\right], & S_{*,\kappa} + \varepsilon_* \le J_{i,\kappa} < S^*_\kappa - \varepsilon^*, \\ J_{i,\kappa}, \quad F \notin \left[J_{i+1,\kappa}, J_{i,\kappa}\right], & S_{*,\kappa} + \varepsilon_* \le J_{i,\kappa} < S^*_\kappa - \varepsilon^*, \\ S_{*,\kappa}\,, & J_{i,\kappa} \le S^*_\kappa + \varepsilon_*, \end{cases} \tag{18}$$

$$F = \begin{cases} 0.5\left(J_{i-1,\kappa} + J_{i,\kappa}\right) J_{i,\kappa} \big/ J_{i-1,\kappa}\,, & J_{i-1,\kappa} \ge J_{i,\kappa}, \\ 0.5\left(1 + J_{i,\kappa} - \left(1 - J_{i,\kappa}\right)^2 \big/ \left(1 - J_{i-1,\kappa}\right)\right), & J_{i-1,\kappa} < J_{i,\kappa}. \end{cases}$$

Here ε^* is a small value that does not exceed the expected error in calculating of $S_{i,\kappa}$; ε_* is the parameter of schema that is no more than two-thirds of the amplitude of the saturation jump. Conditions (18) characterize the finiteness of the velocity of the oil displacement front by water (see [4]).

Water flows (16), (17) through the boundaries of elementary cells are calculated from the values of the function $S^{j+1}_{i+1/2,\kappa}$ and $S^{j+1}_{i,\kappa+1/2}$ at the points of the grid $\bar{D}$. After calculating the pressure and water saturation, the mass flows $q^{j+1}_{1,\kappa}$, $q^{j+1}_{2,\kappa}$ of both water and oil entering the well through its side face are determined.

This allows us to proceed the numerical solution of differential equations (1)–(5). These equations are solved using the implicit finite-difference scheme of the first approximation order $O(h_\tau + h_z)$ at points (τ_j, z_κ) of the same grid D_h. The oil transfer equation (1) is approximated upstream by an implicit difference schema [14]. Thus:

$$\frac{\varphi^{j+1}_\kappa - \varphi^j_\kappa}{h_\tau} + \upsilon^{j+1}_{2,\kappa} \frac{\varphi^{j+1}_\kappa - \varphi^{j+1}_{\kappa-1}}{h_\kappa} = q^{j+1}_{2,\kappa} - \varphi^{j+1}_\kappa \frac{\upsilon^{j+1}_{2,\kappa} - \upsilon^{j+1}_{2,\kappa-1}}{h_\kappa}; \tag{19}$$

$$P^{j+1}_\kappa = P^{j+1}_{\kappa-1} - h_\kappa g \rho^{j+1}_\kappa \tag{20}$$

$$G_{\kappa}^{j+1} = G_{\kappa-1}^{j+1} + \frac{S_0}{g}\left[\frac{P_{\kappa}^{j+1} - P_{\kappa}^{j}}{h_\tau} - \frac{P_{\kappa-1}^{j+1} - P_{\kappa-1}^{j}}{h_\tau}\right] + \tag{21}$$

$$+h_\kappa S_0 \rho_1^* \frac{q_{1,\kappa}^{j+1} + q_{1,\kappa-1}^{j+1}}{2} + h_\kappa S_0 \rho_2^* \frac{q_{2,\kappa}^{j+1} + q_{2,\kappa-1}^{j+1}}{2};$$

$$\rho_{\kappa}^{j+1} = \varphi_{\kappa}^{j+1}\left(\rho_2^* - \rho_1^*\right) + \rho_1^*; \quad \beta_{\kappa}^{j+1} =_{\varsigma,\kappa}^{j+1} \varphi_{\kappa}^{j+1} + \varphi_{\kappa}^{j+1} \nu_{2,\kappa}^{j+1} / \upsilon_{\kappa}^{j+1}; \tag{22}$$

$$\upsilon_{2,\kappa}^{j+1} = C_{\zeta,\kappa}^{j+1} \upsilon_{\kappa}^{j+1} + \nu_{2,\kappa}^{j+1}; \quad \upsilon_{\kappa}^{j+1} = \frac{G_{\kappa}^{j+1}}{S_0 \cdot (\rho_1^*(1 - \beta_{\kappa}^{j+1}) + \rho_2^* \beta_{\kappa}^{j+1})}; \tag{23}$$

$$\upsilon_{1,\kappa,i}^{j+1} = \frac{1 - \beta_{\kappa}^{j+1}}{1 - \varphi_{\kappa}^{j+1}} \upsilon_{\kappa}^{j+1}; \quad \nu_{2,\kappa}^{j+1} = g \frac{a^2\left(\rho_1^* - \rho_2^*\right)}{3\varsigma_{\kappa}^{j+1} \mu_1} \cdot \frac{1 + \chi}{1 + 1.5\chi}; \quad \chi = \mu_2/\mu_1; \tag{24}$$

$$G_{1,\kappa}^{j+1} = S_0 \rho_1^* \left(1 - \varphi_{\kappa}^{j+1}\right) \upsilon_{1,\kappa}^{j+1}; \quad G_{2,\kappa}^{j+1} = G_{\kappa}^{j+1} - G_{1,\kappa}^{j+1}; \tag{25}$$

$$\varsigma_{\kappa}^{j+1} = \left(1 + \varphi_{\kappa}^{j+1}(4 + 5\chi)/(1 + \chi)\right) \Big/ \left(1 + 0.5\varphi_{\kappa}^{j+1}\left(1 + 4\chi + 5\chi^2\right)/(1 + \chi)^2\right).$$

The developed general numerical model of interrelated processes in the well and reservoir is a nonlinear system of algebraic equations. This problem is solved by iterative methods. The method [4, 5] with a convergence rate of about 3–5 iterations is used to compute reservoir pressure and water saturation. The system of Eqs. (19)–(19) governing the processes in the producing well is solved by the Newton linearization method. The numerical model and algorithms are implemented in computer software that allows carrying out the multi-parametric numerical experiments to study the filtration processes in the reservoir and mass transfer at the bottom hole of the well with simultaneous visualization o the results of computations.

4 Results

The main results of our research can be formulated as follows.

Theoretical Results To compare the effectiveness of iterative methods, computational experiments were carried out for various cases of reservoir structure that determine the character of dependencies of mass inflows and of water and oil into the well. Based on the analysis of their results, the high convergence rate of the Newton method for solving the problem in the well and the iterative method for calculating pressure and water saturation in the reservoir (3–5 iterations) is shown.

Studies of the convergence and stability of the difference schema using the Courant criterion allow us to draw the following conclusions:

1. the solution of the developed difference schema is stable and converges to itself;
2. the Courant criterion allows us to calculate the values of a variable time step, which provide the possibility to compute the numerical solution in accordance with this condition of its stability at each time moment. Simultaneously, the counting time is significantly reduced (tenfold) in comparison with the calculations with fixed time steps over the entire time range of the problem solution;
3. the practical multi-variant calculations can be performed with sufficiently large steps of the spatial grid using the Courant criterion to determine the time step of the difference schema.

Practical Results Based on the analysis of the results of calculations for different variants of the development of the reservoir layers at the bore-hole zone and the inflow of oil-water mixture to the well, taking into account the influence of the flow structure, when $C_\zeta = C_\zeta\ (\varphi, \upsilon, \mu)$, it is shown that that

1. the spatial distributions of such basic characteristics of two-phase flow in the bottom-hole zone of well as pressure, actual and consumed concentration, flow rate, density and velocity of two-phase mixture, the velocity of individual phases and the drift rate of oil drops are determined by the type of functions $q_1\ (z, \tau)$, $q_2\ (z, \tau)$ and are usually nonlinear;
2. the time to form the quasi-stationary composition of two-phase mixture inflowing to the well pipe from its bottom hole zone can reach several tens of minutes and is determined by length of this zone, structure of two-phase flow, the uncovering conditions and filtration-capacitive characteristics of the oil reservoir. This effect must be taken into account in calculation of the transient hydrodynamic processes that occur during the commissioning of the non-operating producing well into operation.

5 Conclusions

1. The mathematical model, algorithms, and software are developed for calculating interrelated non-stationary hydrodynamic and mass transfer processes during filtration of a two-phase oil-water mixture in a layered-nonuniform oil reservoir and its movement in the bottom-hole zone of a producing well.
2. On the basis of computational experiments, the convergence and stability of the numerical solution of the problem, as well as the features of non-stationary processes in the reservoir and oil well during their transition on the steady state. The duration of these processes is also estimated.
3. High performance of computations with the use of the developed software is shown.

4. The obtained results are used in a cyber-physical system for modeling and forecasting field technological processes of oil production, as well as training the specialists working in oil industry.

Acknowledgement This paper has been supported by the Kazan Federal University Strategic Academic Leadership Program ("PRIORITY-2030").

References

1. Barenblatt, G.I., Yentov, V.M., Ryzhik ,V.M.: Motion of Liquids and Gases in Nature Seams. Nedra, Moscow (1984) (in Russian)
2. Bear, J.: Dynamics of Fluids in Porous. Media.Dover Publications Inc. New York City (1988)
3. Ertekin, T., Abou-Kassem, J.H., King G.R.: Basic Applied Reservoir Simulation. SPE Textbook Series, Richardson, Texas. V. 7 (2001)
4. Chekalin, A.N., Konyukhov, V.M., Kosterin, A.V.: Two-phase Multicomponent Filtration in Oil Reservoirs of Complex Structure. Kazan State University, Kazan (2009) (in Russian)
5. Diyashev, R.N., Khisamov, R.S., Konyukhov, V.M., Chekalin, A. N.: Forced Fluid Extraction From Reservoirs with Double Porosity Saturated with non-Newtonian Oil. Publishing House FEN of Academy of Sciences of Republic of Tatarstan, Kazan (2012) (in Russian)
6. Wu, Y-S., Zhang, K., Ding, C., Pruess, K., Elmroth, E., Bodvarsson, G.S.: An efficient parallel-computing method for modeling nonisothermal multiphase flow and multicomponent transport in porous and fractured media. Adv. Wat. Resor. **25, 3** 243–261 (2002)
7. Salamatin, A.N.: Mathematical Model of Disperse Flows. Publishing of Kazan State University, Kazan (1987) (in Russian)
8. Pudovkin, M.A. Salamatin, A.N., Chugunov, V.A.: Thermal Processes in Producing Wells. Publishing House of Kazan University, Kazan (1977) (in Russian)
9. Lucas, G.P., Penagiotopulos, N.: Oil Volume Fraction and Velocity Profiles in Vertical, Bubbly Oil-In-Water Flows Flow. Measurement and Instrumentation **20** 127–135 (2009)
10. Bratland O.: Pipe Flow 2: Multiphase Flow Assurance (Available at drbratland.com). (2010)
11. Wallis, G.B. One-Dimensional Two-Phase Flow. McGraw-Hill Book Company, New York (1969)
12. Konyukhov, V.M., Chekalin, A.N., Konyukhov, I.V.: Computer modeling of heat and mass transfer in the unified complex "oil reservoir - system of wells". J. Fundam. Appl. Sci. **9 (1S)**, 1508–1523 (2017)
13. Konyukhov, I.V., Konyukhov, V.M.: Cyber-physical system for control the heat and mass transfer in the oil reservoir and producing pumping well. Cybernetics and Physics. **8 (3)**, 137–142 (2019)
14. Roache, P.J.: Computational fluid dynamics. Albuquerque, Hermosa Publs, VII+ (1976)

On Error Control at Numerical Solution of Forth Order Elliptic Equations with Strongly Discontinuous Reaction Coefficient

Vadim G. Korneev

Abstract The paper studies errors of numerical solutions to the equation $\Delta\Delta u + \kappa^2 u = f$ by classical, i.e., C^1-conform, and mixed Ciarlet-Raviart finite element methods. We focus on the case of the element wise constant coefficient κ^2, which chaotically varies between finite elements in the sufficiently wide range, and present the guaranteed, robust, and computable a posteriori error bounds. For the classical FEM's, our bounds are robust for $\kappa^2 \in [0, ch^{-4}]$, where $c =$ const and h is the maximal size of finite elements. One of their good properties is that at $\kappa \equiv$ const their coefficients coincide with ones in the not improvable in the order of the accuracy a posteriori bound, obtained earlier especially for this case (Korneev, 2017). In case of a jumpihg κ the coefficients are only insignificantly worse than those at $\kappa \equiv$ const while computation of the bounds does not require equilibration. The a posteriori error bound for the mixed Ciarlet-Raviart method incorporates, as a part of the estimator, the a posteriori error bound for the primal problem.

1 Introduction

We consider the problem

$$\begin{aligned} \Delta\Delta u + \sigma u &= f(x) \qquad \text{in}\,\Omega\,, \\ u = \partial u/\partial \boldsymbol{\nu} &= 0 \qquad \text{on}\,\partial\Omega\,, \end{aligned} \tag{1.1}$$

with $f \in L^2(\Omega)$ and the first boundary condition and $\boldsymbol{\nu}$ being the internal normal to the boundary. It is assumed that Ω is a polygon covered by the family of the quasiuniform triangulations $\mathcal{T}_h$, defined for any $h > 0$ and each containing

V. G. Korneev (✉)
St. Petersburg State University, St. Petersburg, Russia
e-mail: Vad.Korneev2011@yandex.ru

I. B. Badriev et al. (eds.), *Mesh Methods for Boundary-Value Problems and Applications*, Lecture Notes in Computational Science and Engineering 141,
https://doi.org/10.1007/978-3-030-87809-2_17

compatible triangles τ_r, $r = 1, 2, \ldots, \mathcal{R}_h$ of "size" $h > 0$ with $\overline{\Omega} = \cup_{r=1}^{\mathcal{R}_h} \overline{\tau}_r$. It is essential, that $\sigma = \kappa^2$ is assumed to be element wise constant, i.e.,

$$\sigma = \sigma_r = \text{const} \quad \text{for} \quad x \in \tau_r, \quad r = 1, 2 \ldots, \mathcal{R}_h \,, \tag{1.2}$$

and satisfies only one restriction $\sigma \leq \lambda$, with $\lambda^{-1} = O(h^4)$ in many important cases.

Variety of numerical techniques, including conforming, nonconforming, mixed, discontinuous Galerkin, and other types of numerical methods were studied in the respect of a posteriori error bounds at their application to the biharmonic and the thin plate in bending equations, as well as to the singularly perturbed equation $\varepsilon^2 \Delta^2 u - \Delta u = f$, see, e.g., [1–15]. However, less attention were paid to the popular in applications problem of the thin plate in bending, resting upon Winkler's foundation, a particular example of which is (1.1). This is especially true for the case when the subgrade modulus is discontinuous and has significant jumps. Aposteriory error bounds for C^1-conforming finite element solutions of the problem (1.1) with the element wise constant σ, varying arbitrarily between finite elements in the segment $[0, ch^{-4}]$, $c = \text{const}$, were derived quite recently in [16]. The present paper improves the results of this work for C^1-conforming finite element methods and attempts to expand them to the mixed Ciarlet-Raviart method.

It is necessary to note that a posteriori bounds for the 2nd-order reaction-diffusion equations with the discontinuous reaction coefficient gained some attention in the literature. However, it was restricted to the case of the subdomain wise or finite element wise constant reaction coefficient that varies "mildly" between neighbouring elements. A typical variant of such change is found in [17, 18], where it was mainly motivated by the derivation of the a posteriori error bound alongside with the sort of almost inverse bound, termed the bound of local efficiency. Results of the present paper for C^1-conforming finite element methods are literally (with obvious changes) expanded to the 2nd-order reaction-diffusion equations and show that the range of admissible jumps for obtaining robust a posteriori error bounds is much wider, than it could be expected. At that, the price for widening this range is an insignificant worsening of the coefficients in front of the typical norms in the right parts of the a posteriori bounds.

Among the widely spread in practice a posteriori error bounds, there are two types of them, namely the residual based bounds and the bounds based on the use of the equilibrated flaxes/stresses/ stress resultants. A drawback of the first type is that the derivation of the bounds heavily leans upon the approximation bounds, and the coefficients in the former bounds strongly depend on the constants in the latter. An example of such a bound for the approximate solution of the equation $\Delta\Delta u = f$ is found, e.g., in [13]. At the implementation of the second type bounds, for each numerical solution, e.g., of the second order elliptic equation, one finds a single testing flax. It is evaluated by some *equilibrated* flax recovery procedure, which depends not only on the mesh, but also on the problem. This constricts the universality of the bounds. Besides, the equilibration increases the diffusion type error, and the value of the damage is seen only from the inverse like bound, the constants in which usually are not too optimistic. Thus, bounds of both types, which have been intensively developing in several last decades, have their own restrictions and their real accuracy needs to be checked by practice.

Consistent bounds of [19–21] belong to the most old class of a posteriori error majorants and possess two important properties: (1) the order of accuracy of a consistent bound is the same as for the corresponding sharp a priori bound; (2) for the proof of the property (1), as well as for the calculation of the bound in the practice, it is sufficient to use any testing fields of the flaxes/stresses/stress resultants, which possess respective approximation properties, without resorting to the flax equilibration procedure. In result, such bounds are more universal and their calculation is simplified, because for providing the sharpness one can use any testing flax, having the approximation properties not worse than those of the numerical flax. Recovery procedures, providing such properties, are easy to find in the literature for the reason that they were thoroughly studied and widely tested in the residual type error estimators, see [22–25] and references there. In this paper, the technique, used in [19–21], is adjusted for obtaining the a posteriori error bounds for a more complicated class of problems and numerical methods.

The notations $\|\cdot\|_k$, $|\cdot|_k$ will stand for the norms and quasi-norms in Sobolev's spaces $H^k(\Omega) = W_2^k(\Omega)$ with the agreement that $|\cdot|_0 = \|\cdot\|_0 = \|\cdot\|$. Additionally we introduce the spaces $H_0^2(\Omega) := \{v \in H^2(\Omega) : v = \partial v/\partial \boldsymbol{\nu} = 0 \text{ on } \partial\Omega\}$, $H_0^2(\Omega, \Delta\Delta) = \{v \in H_0^2(\Omega) : \Delta\Delta v \in L^2(\Omega)\}$, and $\mathbf{L}^2(\Omega) = \left(L^2(\Omega)\right)^4$. In relation with the problem (1.1), it is helpful to introduce the subspace $\mathbf{M}(\Omega) = \{\mathbf{m} = \{m_{k,l}\}_{k,l=1}^2 \in \mathbf{L}^2(\Omega) : m_{1,2} = m_{2,1}\}$ of vector-functions $\mathbf{m}$ and the operators $\mathcal{D} : H^2(\Omega) \to \mathbf{M}(\Omega)$ and $\mathcal{D}^* : \mathbf{M}(\Omega, \mathcal{D}^*) \to L^2(\Omega)$, defined as

$$\mathcal{D}v = \left\{ \frac{\partial^2 v}{\partial x_k \partial x_l} \right\}_{k,l=1}^2, \qquad \mathcal{D}^*\mathbf{m} = \sum_{k,l=1}^2 \frac{\partial^2 m_{k,l}}{\partial x_k \partial x_l},$$

where $\mathbf{M}(\Omega, \mathcal{D}^*) = \{\mathbf{m} \in \mathbf{M}(\Omega) : \mathcal{D}^*\mathbf{m} \in L^2(\Omega)\}$. If (1.1) is viewed as the thin plate bending problem (at the cylindrical stiffness equal to unity and Poisson coefficient equal to zero), vector-functions $\mathbf{m} = \mathcal{D}u$ have the meaning of components of the bending and twisting moments acting in the plate. For this reason and for brevity they are called *moments*. Where it does not cause confusion, for norms $\|\cdot\|_{\mathbf{L}^2(\cdot)}$ in the spaces $\mathbf{L}^2(\cdot)$ we use the notation $\|\cdot\|$, so that $\|\mathbf{m}\|$ will stand for $\|\mathbf{m}\|_{\mathbf{L}^2(\Omega)}$.

2 C^1-Conform Finite Element Method for Primal Problem

We assume that the finite element assemblage $\mathcal{K}_h$ is defined on Ω and induces the space $\mathcal{V}_h(\Omega) = \mathcal{V}_h^p(\Omega) := \{\phi_h \in C^1(\Omega) : \phi_h\big|_{\tau_r} \in \mathcal{P}_p,\ p \geq 3,\ r = 1, 2, \ldots, \mathcal{R}_h\}$ and its subspace $\mathcal{V}_{h,0}(\Omega) = \{\phi_h \in \mathcal{V}_h(\Omega) : \phi_h = \partial\phi_h/\partial\boldsymbol{\nu} = 0 \text{ on } \partial\Omega\}$, which is used for solution of the primal formulation of the problem. Here $\mathcal{P}_p$ is the space of polynomials of the degrees $\le p$. For convenience, we define the norm in $H^2(\Omega)$ as $\|\cdot\|_2^2 = \|\cdot\|_0^2 + |\cdot|_2^2$.

Let $a(w, v) = (\mathcal{D}w, \mathcal{D}v)$, $V(\Omega)$ be the Hilbert space of functions with the scalar product $[w, v] = a(w, v) + (\sigma w, v)$ and the norm $|\!|\!|v|\!|\!| = [v, v]^{1/2}$, and $V_0(\Omega) = \{v \in V(\Omega) : v = \partial v/\partial \boldsymbol{\nu} = 0 \text{ on } \partial\Omega\}$. The weak form of the problem (1.1) reads: find $u \in V_0(\Omega)$ such that

$$a(u, v) + (\sigma u, v) = (f, v)\,, \quad \forall\, v \in V_0(\Omega)\,. \tag{2.1}$$

The finite element solution from the space $\mathcal{V}_{h,0}(\Omega)$, denoted $u_{\rm fem}$, satisfies the integral identity

$$a(u_{\rm fem}, v) + (\sigma u_{\rm fem}, v) = (f, v)\,, \quad \forall\, v \in \mathcal{V}_{h,0}(\Omega)\,. \tag{2.2}$$

The proof of a posteriori error majorans is based on the fundamental properties of the finite element method reflected in the approximation and inverse estimates and on the adequate estimate of the value $\mu_{\rm fem}^{-1} = \sup \|\mathcal{D}e_{\rm fem}\|/\|\kappa e_{\rm fem}\|$. In case of $\sigma = \text{const}$ and, at some cost, in case of piece wise constant σ, it can be replaced by the much simpler estimate $\mu_\circ^{-1} = \sup \|\mathcal{D}e_\circ\|/\|e_\circ\|$, where $e_\circ = u_\circ - u$, $u_\circ$ is the finite element function, which minimizes the norm $\|\mathcal{D}(\phi - u)\|$ on the space of functions $\phi \in \mathcal{V}_{h,0}(\Omega)$. With the use of the latter inequality, the a posteriori error bounds for the C^1-conform finite element method solutions to (2.2) were derived in [16]. At that, the Aubin-Nitsche trick [26, 27] was one of the ingredients of the proof and implied that the boundary $\partial\Omega$ satisfies the condition for H^4-solvability of the problem at $\sigma \equiv 0$, $\forall f \in L^2(\Omega)$. Clearly, this condition restricts the range of problems, to which a posteriory bounds can be efficiently applied. In order to obtain the a posteriory error bounds applicable to a wider range of problems, in the current paper we avoid relying on the H^4-solvability condition, but use some additional properties of the finite element approximations. We assume the existence of the projection operator $\mathfrak{I}_h : H^2(\Omega) \to \mathcal{V}_h(\Omega)$, in particular quasinterpolation operator, which has the following properties:

- (i) if $v \in \mathcal{V}_h(\Omega)$, then $\mathfrak{I}_h v = v$;
- (ii) $(v - \mathfrak{I}_h v) \in H_0^2(\Omega)$, if $v|_{\partial\Omega} \in \mathcal{V}_{\rm tr}(\partial\Omega) := [\mathcal{V}_h(\Omega)]\big|_{\partial\Omega}$;
- (iii) $\forall v \in H^s(\Omega)$, $\|v - \mathfrak{I}_h v\|_{t,\Omega} \le c(t,s) h^{s-t} \|v\|_{s,\Omega}$ for $t = 0, 1, 2$, and $s \ge 2$, or $s = 2, 3, \dots, l \le p+1$, if $v \in H_0^l(\Omega) := H_0^2(\Omega) H^l(\Omega)$;
- (iv) $|\mathfrak{I}_h v|_{2,\Omega} \le \breve{c} |\, v\,|_{2,\Omega}$ and $\|\mathfrak{I}_h v\|_{2,\Omega} \le \hat{c} \|\, v\,\|_{2,\Omega}$ $\forall v \in H^2(\Omega)$.

  Above, $c(s,t)$, $\breve{c}$ and $\hat{c}$ are positive constants, depending on the constants in the quasiuniformity conditions for finite element assemblage.

Theorem 1 *Let $u \in H_0^2(\Omega, \Delta\Delta)$, $u_{\rm fem}$ be the finite element solution, $u_{\rm fem} \in \mathcal{V}_{h,0}(\Omega)$, and $\mathbf{m} \in \mathbf{M}(\Omega, \mathcal{D}^*)$. Let, additionally, the linear operator $\mathfrak{I}_h$ with* *the properties* **i**)–**iv**) *exist. Then for any* σ *and* σ_*, *satisfying the inequalities* $0 \leq \sigma \leq \sigma_* = 1/(c(0,2)h^2)^2$, *the bound*

$$|||u_{\text{fem}} - u|||^2 \leq \Theta \mathcal{M}(\sigma_*, u_{\text{fem}}, \mathbf{m}), \tag{2.3}$$

$$\mathcal{M}(\sigma_*, \phi, \mathbf{m}) = \|\mathcal{D}\phi + \mathbf{m}\|^2_{\mathbf{L}^2(\Omega)} + \frac{1}{\sigma_*}\|f - \sigma\phi - \mathcal{D}^*\mathbf{m}\|^2_{L^2(\Omega)},$$

holds with

$$\Theta = \frac{2\widetilde{c}^2\sigma_* - (\widetilde{c}^2 - 1)\sigma_{\min}}{\sigma_* + \sigma_{\min}}, \tag{2.4}$$

$\widetilde{c} = \breve{c} + 2c_{2,0}c(0,2)$ *and the constant* $c_{2,0}$ *from the inverse inequality (2.8), see below.*

Proof For $e = e_{\text{fem}} := u_{\text{fem}} - u$ and $\forall w \in \mathcal{V}_{h,0}(\Omega)$, by using the Galerkin property, integrating by parts and by applying the Cauchy inequalities

$$(\phi, \psi) \leq \|\phi\|\,\|\psi\| \quad \text{and } (\phi_1, \psi_1) + (\phi_2, \psi_2) \leq$$

$$\leq [\|\phi_1\|^2 + \sigma_*^{-1}\|\phi_2\|^2]^{1/2}\,[\|\psi_1\|^2 + \sigma_*\|\psi_2\|^2]^{1/2},$$

we get

$$|||e|||^2 = (\mathcal{D}e, \mathcal{D}e) + (\sigma e, e) = (\mathcal{D}e, \mathcal{D}(e+w)) + (\sigma e, e+w) =$$

$$(\mathcal{D}v - \mathbf{m}, \mathcal{D}e) - (\mathcal{D}u - \mathbf{m}, \mathcal{D}e + w) + (\sigma e, e + w) =$$

$$(\mathcal{D}v - \mathbf{m}, \mathcal{D}e + w) - (f - \sigma v - \mathcal{D}^*\mathbf{m}, e + w) \leq$$

$$\|\mathcal{D}v - \mathbf{m}\|\|\mathcal{D}e + w\| + \|f - \sigma v - \mathcal{D}^*\mathbf{m}\|\,\|e + w\| \leq$$

$$\leq \left\{\|\mathcal{D}v - \mathbf{m}\|^2 + \tfrac{1}{\sigma_*}\|f - \sigma v - \mathcal{D}^*\mathbf{m}\|^2\right\}^{1/2}\left\{\|\mathcal{D}e + w\|^2 + \sigma_*\|e + w\|^2\right\}^{1/2}. \tag{2.5}$$

Let π_h be the operator of L^2 orthogonal projection on the space $\mathcal{V}_{h,0}(\Omega)$. Having set $w = -\pi_h e_{\text{fem}}$, we see that $e + w = (I - \pi_h)e =: e_0$ and, therefore, for every $\beta \in [0, 1]$ we can write

$$\|\mathcal{D}(e+w)\|^2 + \sigma_*\|e+w\|^2 = \|\mathcal{D}e_0\|^2 + \sigma_*\|e_0\|^2 \leq \|\mathcal{D}e_0\|^2 +$$

$$+\Big[\sigma_* - \sigma_{\min}\Big(1 + \tfrac{\beta}{\sigma_{\max}}(\sigma_* - \sigma_{\max})\Big)\Big]\|e_0\|^2 + \sigma_{\min}\Big(1 + \tfrac{\beta}{\sigma_{\max}}(\sigma_* - \sigma_{\max})\Big)\|e_0\|^2 =$$

$$= \|\mathcal{D}e_0\|^2 + B_1\|e_0\|^2 + B_2\sigma_{\min}\|e_0\|^2, \tag{2.6}$$

where the notations $B_1,\ B_2$ are obvious.

As it follows from the properties of the operators $\mathfrak{I}_h$ and π_h, for $\phi - \pi_h\phi$ with any $\phi \in H_0^2(\Omega)$, there are valid the bounds

$$\|\phi - \pi_h\phi\|_0 \le \|\phi\|_0 ,$$
$$\|\phi - \pi_h\phi\|_0 \le c(0,2)h^2\|\mathcal{D}\phi\|_0 , \tag{2.7}$$
$$\|\mathcal{D}(\phi - \pi_h\phi)\|_0 \le \widetilde{c}\|\mathcal{D}\phi\|_0 ,$$

in which the constant $\widetilde{c}$ depends only on two constants from conditions of quasiuniformity of triangulation. The proof is needed only for the last bound, and it follows from the relations:

$$\|\mathcal{D}(\phi - \pi_h\phi)\|_0 \le \|\mathcal{D}(\phi - \mathfrak{I}_h\phi)\|_0 + \|\mathcal{D}(\mathfrak{I}_h\phi - \pi_h\phi)\|_0 \le \breve{c}\|\mathcal{D}\phi\|_0 +$$
$$c_{2,0}h^{-2}\|\mathfrak{I}_h\phi - \pi_h\phi\|_0 \le \breve{c}\|\mathcal{D}\phi\|_0 + c_{2,0}h^{-2}\Big[\|\mathfrak{I}_h\phi - \phi\|_0 + \|\phi - \pi_h\phi\|_0\Big] \le$$
$$\le \Big(\breve{c} + 2c_{2,0}c(0,2)\Big)\|\mathcal{D}\phi\|_0 ,$$

where $c_{2,0}$ is the constant in the inverse inequality

$$\|\mathcal{D}(\mathfrak{I}_h\phi - \pi_h\phi)\|_0 \le c_{2,0}h^{-2}\|\mathfrak{I}_h\phi - \pi\phi\|_0 . \tag{2.8}$$

Thus, we conclude that in (2.7)

$$\widetilde{c} = \breve{c} + 2c_{2,0}c(0,2) .$$

Now, application of the second and third inequalities (2.7) to the first and second terms in the right part of (2.6) yields

$$\begin{aligned}\|\mathcal{D}e_0\|^2 + \sigma_*\|e_0\|^2 &\le (1+\sigma_*^{-1}B_1)\|\mathcal{D}e_0\|^2 + \sigma_{\min}B_2\|e_{\mathrm{fem}}\| \le \\ &\le \widetilde{c}^2(1+\sigma_*^{-1}B_1)\|\mathcal{D}e_{\mathrm{fem}}\|^2 + B_2\|\kappa e_{\mathrm{fem}}\| .\end{aligned} \tag{2.9}$$

The values of $B_1,\ B_2$ depend on the choice of β, and it is not diffiult to notice that, if there exists the solution of the equation

$$\widetilde{c}^2(1+\sigma_*^{-1}B_1) = B_2 , \tag{2.10}$$

then it is the optimal value of β. Indeed, there is the solution

$$\beta = \widetilde{c}^2\sigma_{\max}\frac{(2-\widetilde{c}^{-2})\sigma_* - \sigma_{\min}}{(\sigma_* + \sigma_{\min})(\sigma_* - \sigma_{\max})}.$$

Substituting this value of β in (2.9) and furder in (2.5) at $w = \pi_h e_{\text{fem}}$, we conclude the proof. We have two expressions for Θ, defined by the left and the right parts of (2.10). Substituting the found β in any of them, we get the needed expression (2.4) for Θ. □

Remark 1 We minimized the second multiplier in (2.5) by taking minimal coefficient $1/\sigma_*$ from admissible and minimizing Θ by means of the corresponding choice of parameter β. A more general situation of $\sigma_* \in [\sigma_{\max}, \lambda]$, $\lambda = 1/(c(0,2)h^2)^2$ can be treated in the same way and is resolved by

$$\beta = \sigma_{\max} \frac{\tilde{c}^2\lambda(1 - \tilde{c}^{-2}) + \tilde{c}^2(\sigma_* - \sigma_{\min})}{(\sigma_* + \sigma_{\min})(\sigma_* - \sigma_{\max})}$$

and

$$\Theta = 1 + \frac{\tilde{c}^2\lambda(1 - \tilde{c}^{-2}) + \tilde{c}^2(\sigma_* - \sigma_{\min})}{(\sigma_* + \sigma_{\min})}. \tag{2.11}$$

We see that minimal Θ of (2.11) is reached at maximal σ_*, i.e., at $\sigma_* = \lambda = 1/(c(0,2)h^2)^2]$, and, therefore, is the same as in (2.4). At definition of Θ and σ_* we can take into consideration the real values of the norms in the right part of (2.3). Let us denote these norms as $\mathcal{N}_1$ and $\mathcal{N}_2$. If the values $\mathcal{N}_1$ and $\lambda^{-1}\mathcal{N}_2$ differ considerably, this can improve efficiency.

For providing high accuracy it is important to pick up the testing vector-function **m** with the components as close as possible to their exact values. It is usually done with the use of the respective *recovery procedures*, in particular, the same as used for the derivation of the residual type a posteriori error bounds. As it is noted in the book [23] and several papers, flax recovery procedures demonstrated very high efficiency at the use for evaluation of a posteriori error bounds for the finite element solutions of the 2nd order elliptic equations. If attended for evaluation of (2.3), on the basis of the finite element solution u_{fem}, they produce **m** as an element of some appropriate finite element subspace $\widetilde{\boldsymbol{M}}(\Omega) \subset \mathbf{M}(\Omega, \mathcal{D}^*)$. The most popular in the practice is the one called *averaging procedure* exemplified in [13, 28]. Another efficient and optimal in the computational cost procedure for finding **m** is the least squares procedure. In it, moments $m_{k,l}$ are defined as $L^2(\Omega)$ orthogonal projections of the derivatives $\partial^2 u_{\text{fem}}/\partial x_k \partial x_l$ upon the corresponding subspaces $\widetilde{M}_{k,l}(\Omega)$ of the space $\widetilde{\boldsymbol{M}}(\Omega)$, e.g., $\widetilde{M}_{k,l}(\Omega) = \mathcal{V}_h^{p_m}(\Omega)$ with appropriate $p_m \geq p - 2$.

3 Ciarlet-Raviart Mixed Metod

The weak mixed Ciarlet-Raviart type formulation of the problem (1.1) reads, cf. [29]: find the vector-function $\mathbf{w} = (v, u)^\top \in H^1(\Omega) \times \mathring{H}^1(\Omega)$, $\mathring{H}^1(\Omega) = \{\phi \in H^1(\Omega) : \phi = 0 \text{ on } \partial\Omega\}$, satisfying the system of the integral identities

$$\begin{aligned} (v, q) - \langle \nabla u, \nabla q \rangle &= 0\,, && \forall q \in H^1(\Omega)\,, \\ \langle \nabla v, \nabla g \rangle + (\sigma u, g) &= (f, g)\,, && \forall g \in \mathring{H}^1(\Omega)\,, \end{aligned} \tag{3.1}$$

where $\langle \cdot, \cdot \rangle$ is the scalar product of vector-functions $\langle \mathbf{z}, \mathbf{y} \rangle = (z_1, y_1) + (z_2, y_2)$ for $\mathbf{y} = (y_1, y_2)^\top$ and $\mathbf{z} = (z_1, z_2)^\top$.

In general, for solving equations (3.1) two finite element assemblages, denoted $\mathcal{K}_{h,u}$ and $\mathcal{K}_{h,v}$, are employed, which induce respectively the space $\mathsf{U}_h(\Omega) = \{\phi_h \in C(\Omega) : \phi_h\big|_{\tau_r} \in \mathcal{P}_{p_u},\ p_u \geq 2,\ r = 1, 2, \ldots, \mathcal{R}_h\}$ and the space $\mathsf{V}_h(\Omega) = \{\psi_h \in C(\Omega) : \psi_h\big|_{\tau_r} \in \mathcal{P}_{p_v},\ p_v \geq 2,\ r = 1, 2, \ldots, \mathcal{R}_h\}$. The mixed finite element solution $\mathbf{w}_h = (v_h, u_h)^\top \in \mathsf{V}_h(\Omega) \times \mathring{\mathsf{U}}_h(\Omega)$, where $\mathring{\mathsf{U}}_h(\Omega) = \{\phi_h \in \mathsf{U}_h(\Omega) : \phi_h = 0 \text{ on } \partial\Omega\}$, satisfies the system of equations

$$\begin{aligned} (v_h, q_h) - \langle \nabla u_h, \nabla q_h \rangle &= 0\,, && \forall q_h \in \mathsf{V}_h(\Omega)\,, \\ \langle \nabla v_h, \nabla g_h \rangle + (\sigma u_h, g_h) &= (f, g_h)\,, && \forall g \in \mathring{\mathsf{U}}_h(\Omega)\,. \end{aligned} \tag{3.2}$$

The error of the finite element solution, denoted

$$\mathbf{e}_{\text{fem}} = (e_v, e_u)^\top \qquad e_u = u_h - u\,, \qquad e_v = v_h + \Delta u\,,$$

obviously, satisfies the integral identities

$$\begin{aligned} (e_v, q_h) - \langle \nabla e_u, \nabla q_h \rangle &= 0\,, && \forall q_h \in \mathsf{V}_h(\Omega)\,, \\ \langle \nabla e_v, \nabla g_h \rangle + (\sigma e_u, g_h) &= 0\,, && \forall g_h \in \mathring{\mathsf{U}}_h(\Omega)\,, \end{aligned} \tag{3.3}$$

Turning to the a posteriori error bound for the Ciarlet-Raviart mixed method, we note that usual approximation properties corresponding to the degrees of polynomials on finite elements are assumed for the involved in the estimation finite element spaces:

($\mathcal{A}_1$) For any $w \in \mathring{H}^l(\Omega) := \mathring{H}^1(\Omega) \cap H^l(\Omega)$ the space $\mathring{\mathsf{U}}_h(\Omega)$ provides such an approximation $\widetilde{w} = G_{h,u} w$ that at $k = 0, 1$ and $1 \leq l \leq p_u + 1$ we have

$$|\widetilde{w} - w|_k \leq c_{k,l} h^{l-k} |w|_l\,, \qquad c_{k,l} = \text{const}\,, \tag{3.4}$$

where $G_{h,u} : \overset{\circ}{H}{}^1(\Omega) \mapsto \mathring{\mathsf{U}}_h(\Omega)$ is a linear operator. Similar approximation estimates hold for $w \in H^l(\Omega)$, $\widetilde{w} \in \mathsf{V}_h(\Omega)$, $k = 0, 1$, and $1 \leq l \leq p_v + 1$.

($\mathcal{A}_2$) For any $w \in H_0^l(\Omega) := H^l(\Omega) \cap H_0^2(\Omega)$, $l \geq 2$ the space $\mathcal{V}_{h,0}(\Omega)$ provides an approximation $\widetilde{w} = Q_h w$ that at $k = 0, 1, 2$ and $2 \leq l \leq p + 1$ we have

$$|\widetilde{w} - w|_k \leq c_{k,l} h^{l-k} |w|_l \,, \qquad c_{k,l} = \text{const} \,, \tag{3.5}$$

where $Q_h : H_0^2(\Omega) \mapsto \mathcal{V}_{h,0}(\Omega)$ is a linear operator.

For the error of the solution by the mixed finite element method, we use the norm

$$|\!|\!| \mathbf{e}_{\text{fem}} |\!|\!| = \frac{1}{\sqrt{2}} \Big\{ \|e_v\|^2 + \|\Delta_h e_u\|^2 + 2\|\kappa e_u\|^2 \Big\}^{1/2} . \tag{3.6}$$

For any $\phi \in \mathcal{V}_{h,0}(\Omega)$ and $e_\phi = \phi - u$ we introduce λ_ϕ as the value, saisfying the inequality

$$\|e_\phi\|^2 \leq \lambda_\phi^{-1} \|\mathcal{D} e_\phi\|^2 . \tag{3.7}$$

Lemma 1 *Let the assumptions ($\mathcal{A}_\alpha$), $\alpha = 1, 2$, be fulfilled and $\mathbf{w}_h = (v_h, u_h)^\top \in \mathsf{V}_h(\Omega) \times \mathring{\mathsf{U}}_h(\Omega)$ be the solution to the system (3.2), $\tilde{u}$ be any function from $\mathcal{V}_h(\Omega)$, and $\mathbf{m}$ be any vector-function belonging to $\mathbf{M}(\Omega, \mathcal{D}^*)$. Then at σ and σ_* satisfying the inequalities $0 \leq \sigma \leq \sigma_* \leq \lambda_{\tilde{u}}$ the a posteriori error bound*

$$\begin{aligned} |\!|\!| \mathbf{e}_{\text{fem}} |\!|\!|^2 \leq \|\Delta_h(u_h - \tilde{u})\|^2 + \|v_h - \Delta_h \tilde{u}\|^2 + 2\|\kappa(u_h - \tilde{u})\|^2 + \\ +2\Theta \mathcal{M}(\sigma_*, \tilde{u}, \mathbf{m}) \,, \end{aligned} \tag{3.8}$$

holds with $\Theta = 1 + (\sigma_* - \sigma_{\min})/\lambda_{\tilde{u}}$.

Proof At any $\tilde{u} \in \mathcal{V}_{h,0}(\Omega)$, two first summands in the figure brackets of (3.6) we transform to the form

$$\begin{aligned} \|\Delta_h e_u\|^2 + \|\kappa e_u\|^2 = (\Delta_h(u_h - \tilde{u}), \Delta_h e_u) + (\sigma(u_h - \tilde{u}), e_u) + \\ +(\Delta(\tilde{u} - u), e_u) + (\sigma(\tilde{u} - u), e_u) \,, \end{aligned} \tag{3.9}$$

and in a similar way transform the rest terms:

$$\begin{aligned} \|e_v\|^2 + \|\kappa e_u\|^2 = ((v_h - \Delta\tilde{u}), e_v) + (\sigma(u_h - \tilde{u}), e_u) + \\ +(\Delta(\tilde{u} - u), e_v) + (\sigma(\tilde{u} - u), e_u) \,. \end{aligned} \tag{3.10}$$

The function $\tilde{u}$ can be considered as an approximation of the problem (1.1) and, therefore, for $\|\Delta\tilde{e}\|^2 + \|\kappa\tilde{e}\|^2$ similar to the mentioned in Remark 1 bound (2.3) can be used:

$$\|\Delta\tilde{e}\|^2 + \|\kappa\tilde{e}\|^2 \leq \Theta\mathcal{M}(\sigma_*, \tilde{u}, \mathbf{m}),$$
$$\mathcal{M}(\sigma_*, \tilde{u}, \mathbf{m}) = \|\mathcal{D}\tilde{u} + \mathbf{m}\|^2_{\mathbf{L}^2(\Omega)} + \frac{1}{\sigma_*}\|f - \sigma\tilde{u} - \mathcal{D}^*\mathbf{m}\|^2_{L_2(\Omega)}, \tag{3.11}$$

where $\Theta = 1 + (\sigma_* - \sigma_{\min})/\lambda_{\tilde{u}}$. The proof of it follows the path close to the path of the proof of (2.3) with the difference, caused by the fact that $\tilde{u}$ does not possess the Galerkin property. This fact causes also the difference in Θ. Combining (3.9), (3.10) and (3.6), the use of Cauchy inequality and the bound (3.11), result in the inequality

$$|\!|\!|\mathbf{e}_{\mathrm{fem}}|\!|\!|^2 \leq \{|\!|\!|u_h - \tilde{u}|\!|\!|^2 + \|v_h - \Delta_h\tilde{u}\|^2 + 2\|\kappa(u_h - \tilde{u})\|^2 + 2\Theta\mathcal{M}(\sigma_*, \tilde{u}, \mathbf{m})\}^{1/2}\times$$
$$\times\frac{1}{\sqrt{2}}\{\|\Delta_h e_u\|^2 + \|e_v\|^2 + 2\|\kappa e_u\|^2\}^{1/2}, \tag{3.12}$$

from which the bound (3.8) follows. □

In general the value $\lambda_{\tilde{u}}^{-1}$ is bounded by the constant c_F^{-1} from the Friedreichs type inequality $c_F\|\phi\|^2 \leq \|\Delta\phi\|^2$, $\forall\phi \in \mathring{H}^1(\Omega)$.

The function $\tilde{u}$ for the use in (3.8) can be defined at least in two ways: by the least squares projection of u_h upon the space $\mathcal{V}_{h,0}(\Omega)$, i.e. $\tilde{u} = \pi_h u_h$, or as $\tilde{u} = E_h u_h$, where $E_h : \mathring{\mathrm{U}}_h(\Omega) \mapsto \mathcal{V}_{h,0}(\Omega)$ is the recovery operator, based on averaging and described in [13, 28]. The testing moments $\mathbf{m}$ for the use in (3.8) can be defined on the basis the function $\tilde{u}$ by means of exactly the same recovery procedures as was mentioned for moments $\mathbf{m}$ in (2.3).

Conclusion Guaranteed, reliable and computable a posteriori error bounds for solutions of the problem (1.1) with constant coefficients by C^1-conform finite element methods were obtained in [15, 19, 21]. An additional feature of the problem, which is taken into consideration in [16] and here and seems new even for the studies of the conform methods for the 2nd-order elliptic equations, is related to the reaction coefficient. It is assumed to be finite element wise constant and changing chaotically between finite elements in a wide range. In this paper, we suggested another way of the derivation of these a posteriori error bounds, admitting to improve their coefficients. Desides, we removed the requirement to the boundary, arising from the elliptic regularity condition on the subsidiary problem in Ω, which appear in the Aubin–Nitsche trick. The a posteriori error bound for the mixed Ciarlet-Raviart method with relatively easily calculated constants was also presented, which, however, assumes sharpening in the future research.

For simplicity, we restricted consideration to the polygonal domain $\Omega \subset R^2$, which is covered by the quasi-uniform (regular) triangulation. However, the results can be expanded to arbitrary sufficiently smooth domains. This is for the reason that the techniques for constructing curvilinear C^0 and C^n, $n \geq 2$, finite elements in [30, 31] and [31–33], respectively, allow one to create the finite element assemblages, which exactly represent Ω by means of the special curvilinear finite elements, used along curvilinear parts of the boundary. These finite element assemblages satisfy the generalized conditions of quasiuniformity, see, e.g., [34, Section 3.2], and induce the finite element spaces of classes C^0 and C^1, which provide the same orders of a priori bounds of approximation and convergence.

References

1. R. Verfürth: *A review of a posteriori error estimation and adaptive mesh-refinement techniques*. Wiley-Teubner, Chichester (1995)
2. A. Charbonneau, K. Dossou, and R. Pierre: A residual-based a posteriori error estimator for the Ciarlet- Raviart formulation of the first biharmonic problem. *Numer. Meth. Part. Different. Eq.* **13**, 93–111 (1997)
3. P. Neittaanmaki and S. I. Repin: A posteriori error estimates for boundary-value problems related to the biharmonic operator. *East-West J. Numer. Math.* **2**, 157–178 (2001)
4. S. Adjerid: A posteriori error estimates for fourth-order elliptic problems. *Comput. Meth. Appl. Mech. Eng.* **191**, 2539–2559 (2002)
5. Th. Gratsch, and K.-J. Bathe: A posteriori error estimation techniques in practical finite element analysis. *Comput. Structur.* **83**, 235–265 (2005)
6. K. Liu: A Gradient Recovery-based a Posteriori Error Estimators for the Ciarlet–Raviart Formulation of the Second Biharmonic Equations. *Appl. Math. Sci.* **1**, 997–1007 (2007)
7. da Veiga Beirao, J. Niiranen, and R. L. Stenberg: A posteriori error estimates for the Morley plate bending element. *Numer. Math.* **106**, 165–179 (2007)
8. X. Feng and H.Wu: A posteriori error estimates for finite element approximations of the Cahn-Hilliard equation and Hele-Shaw flow. *J. Comput. Math.* **26**, 767–796 (2008)
9. M. Wang and S. Zhang: A posteriori estimators of nonconforming finite element method for fourth order elliptic perturbation problems. *J. Comp. Math.* **26**, 554–577 (2008)
10. S. C. Brenner, T. Gudi, and L.-Y. Sung: An a posteriori error estimator for a quadratic C^0 interior penalty method for the biharmonic problem. *IMA J. Numer. Analys.* **30** (3), 777–798 (2010). https://doi.org/10.1093/imanum/drn057
11. P. Hansbo and M. G. Larson: A posteriori error estimates for continuous/discontinuous Galerkin approximations of the Kirchhoff–Love plate. *Comput. Meth. Appl. Mechan. Eng.* **200**(47–48), 3289–3295 (2011). https://doi.org/10.1016/j.cma.2011.07.007
12. E. H. Georgoulis, P. Houston and J. Virtanen: An a posteriori error indicator for discontinuous Galerkin approximations of fourth-order elliptic problems. *IMA J. Numer. Analys.* **31**, 281–298 (2011)
13. T. Gudi: Residual-based a posteriori error estimator for the mixed finite element approximation of biharmonic equation. *Numer. Meth. Part. Different. Eq.* **27**, 315–328 (2011)
14. S.H. Du, R. Lin, Z.M. Zhang: Robust residual-based a posteriori error estimators for mixed finite element methods for fourth order elliptic singularly perturbed problems. *arXiv:1609.04506v1 [math.NA]* 15 Sep. 2016, 1–21.
15. V.G. Korneev, On the accuracy of a posteriori functional error majorants for approximate solutions of elliptic equations: *Doklady Mathemat.* **96** (1), 380–383 (2017)
16. V. G. Korneev: A note on a posteriori error bounds for numerical solutions of elliptic equations with piece wise constant reaction coefficient having large jumps. *Computational Mathematics and Mathematical Physics* **60** (11), 1754–1760 (2020)

17. M. Ainsworth, T. Vejchodský: Robust error bounds for finite element approximation of reaction-diffusion problems with non-constant reaction coefficient in arbitrary space dimension. *Comput. Meth. Appl. Mech. Engineer.* **281**, 184–199 (2014)
18. M. Ainsworth, T. Vejchodský: A simple approach to reliable and robust a posteriori error estimation for singularly perturbed problems. *Comput. Methods Appl. Mech. Engrg.* **353**, 373–390 (2019). https://doi.org/10.1016/j.cma.2019.05.014
19. V. G. Korneev: On a renewed approach to a posteriori error bounds for approximate solutions of reaction-diffusion equations. In: Th. Apel, Y. Langer, A. Meyer and O. Steinbach (eds.) *Advanced Finite Element Methods with Applications*, 207–228, Springer (2019)
20. V. G. Korneev. On Error Control in the Numerical Solution of Reaction–Diffusion Equation . *Computational Mathematics and Mathematical Physics* **59** (1), 1–18 (2019)
21. V. Korneev, V. Kostylev: Some a posteriori error bounds for numerical solutions of plate in bending problems. *Lobachevskii J. Math.* **39** (7), 904–915 (2018)
22. J. Xu, Z. Zhang: Analysis of recovery type a posteriori error estimators for mildly structured grids. *Math. Comput.* **73** (247), 1139–1152 (2003)
23. M. Ainsworth, T. Oden: *A posteriori estimation in finite element analysis*. John Wiley & Sons, Inc., New York, (2000)
24. Z. Zhang: Ultracovergence of the patch recovery technique. *Math. Comput.* **65** (216), 1431–1437 (1996)
25. O. C. Zienkiewicz, J. Z. Zhu: The superconvergence patch recovery (SPR) and adaptive finite element refinement. *Comput. Meth. Appl. Mech. Engineer.* **101**, 207–224 (1992)
26. J.-P. Aubin: *Approximation of elliptic boundary-value problems*. Wiley-Interscience (1972).
27. J. Nitsche: Zur Konvergenz von Naherungsverfahren bezuglich verschiedener Normen. *Numer. Math.* **15** (3), 224–228 (1970)
28. S. C. Brenner and L.-Y. Sung: C^0 interior penalty methods for fourth order elliptic boundary value problems on polygonal domains. , *J. Sci. Comput.* 22/23, 83–118 (2005).
29. P. G. Ciarlet: *The finite element method for elliptic problems*. North-Holland, Amsterdam, (1978).
30. V. G. Korneev: O postroenii variaciono-raznostnynh shem vysokih poriadrov tochnosti *Vestnik Leningradskogo universiteta* **25** (19), 28–40 (1970) [in Russian]
31. V. G. Korneev: *High-Order Accuracy Finite Element Schemes*. Publ. House of Leningr. State Univ., Leningrad, (1977) [in Russian]
32. V. G. Korneev: *Exact boundary approximation at numerical solution of high order elliptic equations*. Publ. House of Leningr. State Univ., Leningrad, 1991, 83 p. [in Russian]
33. V. G. Korneev, K. A. Khusanov: Curvilinear finite elements of class C^1 with singular coordinate functions. *Different. Eq.* **22** (12), 2144–2157 (1986).
34. V. G. Korneev and U. Langer: *Dirichlet-Dirichlet Domain Decomposition Methods for Elliptic Problems: h and hp Finite Element Discretizations*. World Scientific, London, (2015)

On the Solvability of a One-Dimensional Problem of Filtration Consolidation with a Limiting Gradient

Check for updates

Alexander V. Kosterin, Maria F. Pavlova, and Elena V. Rung

Abstract It is considered that one-dimensional initial-boundary value problem models the process of joint motion of a viscoelastic porous medium and a liquid saturating the medium. In the filtration theory, this process is called filtration consolidation. From a mathematical viewpoint, the model under study is a system of nonlinear partial differential equations with respect to the pressure and displacement of liquid in the pores. Herewith, the degeneration of the spatial operator is allowed in the equation for pressure. The definition for a generalized solution is introduced. It is proved that under certain assumptions for the solution smoothness the generalized statement of the problem is equivalent to the original statement. Using the semi-discretization method in combination with the Galerkin method and the monotonicity method, the generalized solvability of the problem is established.

1 Introduction

Filtration consolidation is the process of interaction between the deformation of porous medium (skeleton) and the filtration of liquid saturating the medium under the influence of external forces. Herewith, if the pores of the medium are not completely occupied by liquid, then it is unsaturated filtration consolidation; otherwise, it is saturated filtration consolidation.

The problem of saturated-unsaturated filtration consolidation was studied, for example, in [1–4]. In the above articles, existence theorems for generalized solutions of several initial-boundary value problems of saturated-unsaturated filtration consolidation were proved.

The foundations of the theory of saturated filtration consolidation were laid in such works as [5–8]. In the works, mathematical models of filtration consolidation were built, and studies of the models from the standpoint of continuum mechanics

A. V. Kosterin · M. F. Pavlova · E. V. Rung (✉)
Kazan Federal University, Kazan, Russia
e-mail: Alexander.Kosterin@kpfu.ru

I. B. Badriev et al. (eds.), *Mesh Methods for Boundary-Value Problems and Applications*, Lecture Notes in Computational Science and Engineering 141,
https://doi.org/10.1007/978-3-030-87809-2_18

were carried out. A rigorous mathematical analysis of problems of saturated filtration consolidation was carried out in [9]. The present article aims to continue those studies, namely, here we intend to prove the generalized solvability of one problem of saturated filtration consolidation with a limiting gradient.

2 Problem Statement

Let us consider a one-dimensional problem of consolidation of a saturated porous medium on the interval $0 \le x \le L$. We adopt a model of the filtration consolidation process [10], which includes:

- force balance equation

$$-\frac{\partial \sigma^f}{\partial x} + \frac{\partial p}{\partial x} = f(x,t), \tag{1}$$

- rheological relationship for a porous skeleton in the form of the Kelvin-Voigt law [11]

$$\sigma^f = \varepsilon_x + \frac{\partial \varepsilon_x}{\partial t}, \tag{2}$$

- equation of phases joint deformation (consolidation equation)

$$\frac{\partial q}{\partial x} + \frac{\partial \varepsilon_x}{\partial t} = 0, \tag{3}$$

- nonlinear filtration law

$$q = -g\left(\left|\frac{\partial p}{\partial x}\right|\right)\frac{\partial p}{\partial x}. \tag{4}$$

Here $p(x,t)$ is liquid pressure in the pores, $u(x,t)$ is motion of the skeleton particles, σ^f is effective stress in the skeleton [8], $\varepsilon_x = \dfrac{\partial u}{\partial x}$ is deformation component, q is filtration rate.

Substituting relations (2), (4) into Eqs. (1), (3), we obtain the following system of equations for the unknown functions $u(x,t)$, $p(x,t)$:

$$-\frac{\partial}{\partial x}\left(\frac{\partial u}{\partial x} + \frac{\partial^2 u}{\partial x \partial t}\right) + \frac{\partial p}{\partial x} = f(x,t), \tag{5}$$

$$\frac{\partial^2 u}{\partial x \partial t} - \frac{\partial}{\partial x}\left(g\left(\left|\frac{\partial p}{\partial x}\right|\right)\frac{\partial p}{\partial x}\right) = 0. \tag{6}$$

We assume that for $t \in (0, T]$ the following boundary conditions are satisfied

$$u(0,t) = 0, \quad \frac{\partial u}{\partial x}(L,t) + \frac{\partial^2 u}{\partial x \partial t}(L,t) = 0, \tag{7}$$

$$p(0,t) = p(L,t) = 0. \tag{8}$$

The initial conditions are given as

$$u(x,0) = u_0(x), \quad p(x,0) = p_0(x). \tag{9}$$

In what follows, we assume that the functions $g(\xi)$, $f(x,t)$ satisfy the following conditions:

A$_1$. $g(\xi)\xi$, $\xi \geq 0$ is an absolutely continuous in ξ, nonnegative, nondecreasing function and there exist $\xi_0 \geq 0$, η, $\mu > 0$, such that at $\xi \geq \xi_0$ the following inequality holds

$$\eta(\xi - \xi_0) \leq g(|\xi|)\xi \leq \mu(\xi - \xi_0). \tag{10}$$

A$_2$. The function $f(x,t)$ is continuous at $(x,t) \in Q_T$, where $Q_T = [0, L] \times [0, T]$.

Note that the class of problems under consideration is rather wide and includes, in particular, filtration consolidation problems with a limiting gradient, when $g(\xi) \equiv 0$ at $\xi \leq \beta$, $\beta > 0$.

3 Defining a Generalized Solution

Let $\overset{\circ}{V}$ be the closure of smooth functions equal to zero at $x = 0$ in the norm of the space $W_2^{(1)}(0, L)$, and let $\overset{\circ}{V}_1$ be the closure of smooth functions equal to zero on the boundary of the interval $[0, L]$, in the norm of the same space.

Definition 1 By a generalized solution to problem (5)–(9), we imply functions (u, p), for which the following conditions hold:

$$u \in W_2^{(1)}(0, T; \overset{\circ}{V}), \quad p \in L_2(0, T; \overset{\circ}{V}_1),$$

$$u(x,0) = u_0(x), \quad p(x,0) = p_0(x) \quad \text{almost everywhere on } (0, L),$$

and for any functions $v \in W_2^{(1)}(0, T; \overset{\circ}{V})$, $z \in L_2(0, T; \overset{\circ}{V}_1)$ the following equality is true:

$$\int_0^T \int_0^L \left\{ \left(\frac{\partial u}{\partial x} + \frac{\partial^2 u}{\partial x \partial t} \right) \frac{\partial^2 v}{\partial x \partial t} - p \frac{\partial^2 v}{\partial x \partial t} + \frac{\partial^2 u}{\partial x \partial t} z \right.$$

$$\left. + g\left(\left| \frac{\partial p}{\partial x} \right| \right) \frac{\partial p}{\partial x} \frac{\partial z}{\partial x} \right\} dxdt = \int_0^T \int_0^L f(x,t) \frac{\partial v}{\partial t} dxdt. \qquad (11)$$

Theorem 1 *Let u, p be a solution to problem (5)–(9) satisfying the following conditions:*

$$u \in W_2^{(1)}(0, T; \overset{\circ}{V}), \quad p \in L_2(0, T; \overset{\circ}{V}_1),$$

then u, p are a generalized solution to the problem.

And vice versa, if u, p is a generalized solution to problem (5)–(9) such that

$$u(x,t), p(x,t) \in C^{(2)}(0, L) \qquad \forall t \in (0, T), \qquad (12)$$

then the functions u, p satisfy relations (5)–(9).

Proof Let u, p be a solution to problem (5)–(9). It is required to establish that u, p satisfy equality (11).

Let $v \in W_2^{(1)}(0, T; \overset{\circ}{V})$. We multiply the equality (5) by the function $\frac{\partial v}{\partial t}$ and integrate the resulting equality over x from 0 to L, t from 0 to T. As a result, we obtain

$$-\int_0^T \int_0^L \frac{\partial}{\partial x} \left(\frac{\partial u}{\partial x} + \frac{\partial^2 u}{\partial x \partial t} - p \right) \frac{\partial v}{\partial t} dxdt = \int_0^T \int_0^L f(x,t) \frac{\partial v}{\partial t} dxdt. \qquad (13)$$

Using the formula for integration by parts, we transform the left-hand side of equality (13):

$$-\int_0^T \int_0^L \frac{\partial}{\partial x} \left(\frac{\partial u}{\partial x} + \frac{\partial^2 u}{\partial x \partial t} - p \right) \frac{\partial v}{\partial t} dxdt = \int_0^T \int_0^L \left(\frac{\partial u}{\partial x} + \frac{\partial^2 u}{\partial x \partial t} - p \right) \frac{\partial^2 v}{\partial x \partial t} dxdt$$

$$-\int_0^T \left(\frac{\partial u}{\partial x}(L,t) + \frac{\partial^2 u}{\partial x \partial t}(L,t) - p(L,t) \right) \frac{\partial v}{\partial t}(L,t)\, dt \qquad (14)$$

Based on (13), (14) and boundary conditions (7), (8), the following equality holds true

$$\int_0^T\int_0^L \left(\frac{\partial u}{\partial x} + \frac{\partial^2 u}{\partial x \partial t} - p\right) \frac{\partial^2 v}{\partial x \partial t}\, dxdt = \int_0^T\int_0^L f(x,t)\frac{\partial v}{\partial t}\, dxdt. \tag{15}$$

Now we multiply equality (6) by the function $z \in L_2(0, T; \overset{\circ}{V}_1)$ and integrate the resulting equality over x from 0 to L, t from 0 to T. We apply the formula of integration by parts and, as a result, we obtain

$$\int_0^T\int_0^L \left\{\frac{\partial^2 u}{\partial x \partial t} z + g\left(\left|\frac{\partial p}{\partial x}\right|\right)\frac{\partial p}{\partial x}\frac{\partial z}{\partial x}\right\} dxdt = 0. \tag{16}$$

It follows from relations (15) and (16) that the functions u, p satisfy equality (11).

Let us prove the second part of the statement of Theorem 1. Let u, p be a generalized solution to problem (5)–(9) satisfying condition (12). Let us use the formula of integration by parts to transform equality (11). We obtain

$$-\int_0^T\int_0^L \frac{\partial}{\partial x}\left(\frac{\partial u}{\partial x} + \frac{\partial^2 u}{\partial x \partial t} - p\right)\frac{\partial v}{\partial t}\, dxdt$$

$$+\int_0^T\int_0^L \left\{\frac{\partial^2 u}{\partial x \partial t} - \frac{\partial}{\partial x}\left(g\left(\left|\frac{\partial p}{\partial x}\right|\right)\frac{\partial p}{\partial x}\right)\right\} z\, dxdt$$

$$+\int_0^T \left(\frac{\partial u}{\partial x}(L,t) + \frac{\partial^2 u}{\partial x \partial t}(L,t)\right)\frac{\partial v}{\partial t}(L,t)\, dt = \int_0^T\int_0^L f\frac{\partial v}{\partial t}\, dxdt. \tag{17}$$

Assuming in Eq. (17) that $\dfrac{\partial v}{\partial t} = \phi$, $z = 0$, where ϕ—is an arbitrary function from $C_0^\infty(Q_T)$ and taking into account the density $C_0^\infty(Q_T)$ in $L_2(Q_T)$, it is easy to show that (5) follows from (17). If in relation (17) we put $v = 0$, $z = \phi$, then arguing in a similar manner, we obtain that (6) follows from (17).

Let us now prove that boundary conditions (7) are satisfied. Since (5), (6) are proved, it follows from (17) that the following equality holds true for any function $v \in W_2^{(1)}(0, T; \overset{\circ}{V})$

$$\int_0^T \left(\frac{\partial u}{\partial x}(L,t) + \frac{\partial^2 u}{\partial x \partial t}(L,t)\right)\frac{\partial v}{\partial t}(L,t)\, dt = 0,$$

whence, due to arbitrariness of v, the validity of the following boundary condition is proved

$$\frac{\partial u}{\partial x}(L,t)+\frac{\partial^2 u}{\partial x \partial t}(L,t)=0.$$

Theorem 1 is proved. □

4 Existence Theorem

Theorem 2 *For any* $u_0 \in \overset{\circ}{V}$, $p_0 \in \overset{\circ}{V}_1$ *there exists a generalized solution to problem (5)–(9).*

Proof Let us use the semi-discretization method in combination with the Galerkin method. Let

$$\bar{\omega}_\tau = \{t = k\tau,\ 0 \le k \le M,\ M\tau = T\}$$

be a grid on the interval $[0, T]$, $\omega_\tau = \bar{\omega}_\tau \setminus \{0\}$.

Let $\{\varphi_i\}$ and $\{\psi_i\}$ be full systems of basis functions in the spaces $\overset{\circ}{V}$ and $\overset{\circ}{V}_1$ respectively. Let also V^n and V_1^n be finite-dimensional spaces spanned by the system of functions $\{\varphi_i\}_{i=1}^n$ and $\{\psi_i\}_{i=1}^n$ respectively.

Definition 2 By the approximate solution to problem (5)–(9), constructed by the method of semi-discretization in combination with the Galerkin method, we imply the functions $(\hat{u}^n(t), \hat{p}^n(t))$ for which the following conditions hold:

$$\hat{u}^n(t) \in V^n, \quad \hat{p}^n(t) \in V_1^n \qquad \forall t \in \omega_\tau,$$

$$u^n(x,0) = u_0(x), \quad p^n(x,0) = p_0(x) \quad \text{almost everywhere on } (0, L),$$

and for any functions $v^n \in V^n$, $z^n \in V_1^n$ the following equality is true

$$\int_0^L \left\{\left(\frac{\partial \hat{u}^n}{\partial x}+\frac{\partial u_t^n}{\partial x}\right)\frac{\partial v_t^n}{\partial x} - \hat{p}^n\,\frac{\partial v_t^n}{\partial x} + \frac{\partial u_t^n}{\partial x}\,\hat{z}^n \right.$$

$$\left. + g\left(\left|\frac{\partial \hat{p}^n}{\partial x}\right|\right)\frac{\partial \hat{p}^n}{\partial x}\,\frac{\partial \hat{z}^n}{\partial x}\right\}dx = \int_0^L \hat{f}(x,t)\,v_t^n dx. \tag{18}$$

Here $\hat{v} = v(t+\tau)$, $v_t = \dfrac{\hat{v}-v}{\tau}$. □

Lemma 1 *The Galerkin system (18) has at least one solution.* □

Proof Obviously, it suffices to establish the existence of $\hat{p}^n$, $\hat{u}^n$ satisfying (18), under the assumption that p^n, u^n are known.

Since the choice of the functions v^n, z^n is arbitrary, the Galerkin system (18) is equivalent to the following system

$$\int_0^L \left\{ \left(\frac{\partial \hat{u}^n}{\partial x} + \frac{\partial u_t^n}{\partial x} \right) \frac{\partial v_t^n}{\partial x} - \hat{p}^n \frac{\partial v_t^n}{\partial x} \right\} dx = \int_0^L \hat{f}(x,t)\, v_t^n dx,$$

$$\int_0^L \left\{ \frac{\partial u_t^n}{\partial x} \hat{z}^n + g\left(\left| \frac{\partial \hat{p}^n}{\partial x} \right| \right) \frac{\partial \hat{p}^n}{\partial x} \frac{\partial \hat{z}^n}{\partial x} \right\} dx = 0.$$

Approximate solutions are sought by the Galerkin method in the form

$$\hat{u}^n = \sum_{k=1}^n \zeta_k^{(1)} \varphi_k, \qquad \hat{p}^n = \sum_{k=1}^n \zeta_k^{(2)} \psi_k.$$

The unknown coefficients $\zeta_k^{(i)}$, $k = \overline{1,n}$, $i = 1, 2$ are determined by the following system of equations:

$$\int_0^L \left\{ \left(\frac{\partial \hat{u}^n}{\partial x} + \frac{\partial u_t^n}{\partial x} \right) \frac{\partial (\varphi_k)_t}{\partial x} - \hat{p}^n \frac{\partial (\varphi_k)_t}{\partial x} \right\} dx = \int_0^L \hat{f}(x,t)\, (\varphi_k)_t dx, \tag{19}$$

$$\int_0^L \left\{ \frac{\partial u_t^n}{\partial x} \hat{\psi}_k + g\left(\left| \frac{\partial \hat{p}^n}{\partial x} \right| \right) \frac{\partial \hat{p}^n}{\partial x} \frac{\partial \hat{\psi}_k}{\partial x} \right\} dx = 0. \tag{20}$$

Let $\mathbf{H} : R^{2n} \to R^{2n}$ be a nonlinear operator such that the equation

$$\mathbf{H}(\zeta) = 0$$

is equivalent to system (19)–(20). Let us make sure that R^{2n} contains a sphere centered at zero of finite radius, on which

$$(\mathbf{H}(\zeta), \zeta)_{R^{2n}} \geq 0.$$

We have

$$(\mathbf{H}(\zeta),\zeta)_{R^{2n}}=\int\limits_0^L\left(\frac{\partial\hat{u}^n}{\partial x}+\frac{\partial u_t^n}{\partial x}\right)\frac{\partial u_t^n}{\partial x}dx$$

$$+\int\limits_0^L g\left(\left|\frac{\partial\hat{p}^n}{\partial x}\right|\right)\left(\frac{\partial\hat{p}^n}{\partial x}\right)^2dx-\int\limits_0^L\hat{f}(x,t)\,u_t^n dx. \qquad (21)$$

The first term on the right-hand side of equality (21) can be transformed to the form:

$$\int\limits_0^L\left(\frac{\partial\hat{u}^n}{\partial x}+\frac{\partial u_t^n}{\partial x}\right)\frac{\partial u_t^n}{\partial x}dx=\left(\frac{1}{\tau}+\frac{1}{\tau^2}\right)\|\,\hat{u}^n\,\|_1^2$$

$$-\left(\frac{1}{\tau}+\frac{2}{\tau^2}\right)\int\limits_0^L\frac{\partial\hat{u}^n}{\partial x}\frac{\partial u^n}{\partial x}dx+\frac{1}{\tau^2}\|\,u^n\,\|_1^2\,. \qquad (22)$$

Here $\|\,v\,\|_1^2=\int\limits_0^L\left(\frac{\partial v}{\partial x}\right)^2dx.$

Using the Cauchy-Bunyakovsky inequality

$$(x,y)\ \le\ \delta||x||^2+\frac{1}{4\delta}||y||^2, \qquad (23)$$

from (22) it is easy to obtain the following estimate

$$\int\limits_0^L\left(\frac{\partial\hat{u}^n}{\partial x}+\frac{\partial u_t^n}{\partial x}\right)\frac{\partial u_t^n}{\partial x}dx\ge\left(\frac{1}{\tau}+\frac{1}{\tau^2}-\delta\right)\|\,\hat{u}^n\,\|_1^2$$

$$+\left(\frac{1}{\tau^2}-\frac{1}{4\delta}\left(\frac{1}{\tau}+\frac{2}{\tau^2}\right)^2\right)\|\,u^n\,\|_1^2\,.$$

□

Using (23) and inequality (10), for the second term in equality (21) we have

$$\int\limits_0^L g\left(\left|\frac{\partial\hat{p}^n}{\partial x}\right|\right)\left(\frac{\partial\hat{p}^n}{\partial x}\right)^2dx\ge(\eta-\delta)\,\|\,\hat{p}^n\,\|_1^2-\frac{\eta^2\xi_0^2L^2}{4\delta}.$$

To estimate the last term of (21), we use the boundedness of the function $f(\xi)$, inequality (23), and the Friedrichs inequality. As a result, we obtain

$$\left| \int_0^L \hat{f}(x,t) \cdot u_t^n dx \right| \leq \delta \parallel \hat{u}^n \parallel_1^2 + \frac{C^2 C_F^2 L^2}{4\delta\tau^2} + \frac{CC_F}{\tau} \parallel u^n \parallel_1^2,$$

here C_F is a constant of the Friedrichs inequality, C is a constant such that

$$|f(\xi, \zeta)| \leq C \qquad \forall \xi \in [0, L], \quad \forall \zeta \in [0, T].$$

Substituting the estimates obtained in (21), we have

$$(\mathbf{H}(\zeta), \zeta)_{R^{2n}} \geq \overline{K}(\delta) \left(\parallel \hat{u}^n \parallel_1^2 + \parallel \hat{p}^n \parallel_1^2 \right) - \overline{R}(\delta)\,, \tag{24}$$

where

$$\overline{K}(\delta) = \min \left\{ \left(\frac{1}{\tau} + \frac{1}{\tau^2} - 2\delta \right), \eta - \delta \right\},$$

$$\overline{R}(\delta) = \left(\frac{CC_F}{\tau} - \frac{1}{\tau^2} + \frac{1}{4\delta} \left(\frac{1}{\tau} + \frac{2}{\tau^2} \right)^2 \right) \parallel u^n \parallel_1^2 + \frac{C^2 C_F^2 L^2}{4\delta\tau^2} + \frac{\eta^2 \xi_0^2 L^2}{4\delta}.$$

Let δ^* be a constant such that for all $0 < \delta \leq \delta^*$ the following inequality holds

$$\overline{K}(\delta) \ \geq \ \beta = const > 0\,,$$

and $S \subset R^{2n}$ be a sphere centered at zero at which the right-hand side of inequality (24) is non-negative. Then, by the topological lemma ([12], p. 66), there is at least one solution to the Galerkin system inside this sphere. The proof of Lemma 1 is complete. □

Lemma 2 *For the approximate solution (18), the following a priori estimates are valid*

$$\max_{t'} \|u^n(t')\|_1^2 \leq C, \qquad \sum_{t=0}^{t'} \tau \|p^n(t)\|_1^2 \leq C, \tag{25}$$

$$\sum_{t=0}^{t'-\tau} \tau \left\| (u^n)_t \right\|_1^2 \leq C, \qquad \sum_{t=0}^{t'} \tau \left\| \frac{\partial u_t^n}{\partial x} \right\|_{L_2(0,L)}^2 \leq C, \tag{26}$$

$$\sum_{t=0}^{t'-\tau} \tau \left\| g\left(\left| \frac{\partial \hat{p}^n}{\partial x} \right| \right) \frac{\partial \hat{p}^n}{\partial x} \right\|_{L_2(0,L)}^2 \leq C. \tag{27}$$

Proof Let us assume in (18) $v^n = u^n$, $z^n = p^n$ and obtain

$$\int_0^L \left\{ \left(\frac{\partial \hat{u}^n}{\partial x} + \frac{\partial u_t^n}{\partial x} \right) \frac{\partial u_t^n}{\partial x} + g\left(\left| \frac{\partial \hat{p}^n}{\partial x} \right| \right) \left(\frac{\partial \hat{p}^n}{\partial x} \right)^2 \right\} dx = \int_0^L \hat{f}(x,t)\, u_t^n dx. \quad (28)$$

Note that

$$\frac{\partial \hat{u}^n}{\partial x}\frac{\partial u_t^n}{\partial x} = \frac{\partial \hat{u}^n}{\partial x}\frac{\partial}{\partial x}\left(\frac{\hat{u}^n - u^n}{\tau} \right) = \frac{\partial \hat{u}^n}{\partial x}\frac{1}{\tau}\left(\frac{\partial \hat{u}^n}{\partial x} - \frac{\partial u^n}{\partial x} \right)$$
$$= \frac{1}{2}\left(\frac{\partial \hat{u}^n}{\partial x} \right)^2 - \frac{1}{2}\left(\frac{\partial u^n}{\partial x} \right)^2 + \frac{\tau^2}{2}\left(\frac{\partial u_t^n}{\partial x} \right)^2. \quad (29)$$

We substitute equality (29) into (28), multiply by τ and sum the resulting relation over t from 0 to $t' - \tau$ and obtain

$$\frac{1}{2}\|u^n(t')\|_1^2 - \frac{1}{2}\|u^n(0)\|_1^2 + \frac{1}{2}\sum_{t=0}^{t'-\tau} \tau^2 \|u_t^n(t)\|_1^2 + \sum_{t=0}^{t'-\tau} \tau \left\| \frac{\partial u_t^n}{\partial x} \right\|_{L_2(0,L)}^2$$
$$+ \sum_{t=0}^{t'-\tau} \tau \int_0^L g\left(\left| \frac{\partial \hat{p}^n}{\partial x} \right| \right) \left(\frac{\partial \hat{p}^n}{\partial x} \right)^2 dx = \sum_{t=0}^{t'-\tau} \tau \int_0^L \hat{f}(x,t) \cdot u_t^n dx. \quad (30)$$

□

From (30), taking into account inequality (10), we have a priori estimates (25)–(26). Also, considering that

$$\left\| g\left(\left| \frac{\partial \hat{p}^n}{\partial x} \right|^2 \right) \frac{\partial \hat{p}^n}{\partial x} \right\|_{L_2(0,L)}^2 \le \left\| \frac{\partial \hat{p}^n}{\partial x} \right\|_{L_2(0,L)}^2,$$

we have estimate (27). The proof of Lemma 2 is complete. □

Lemma 3 *There exist function*

$$u \in W_2^{(1)}(0,T; \overset{\circ}{V}), \qquad p \in L_2(0,T; \overset{\circ}{V}_1)$$

and sequences $\{\tau\}$, $\{n\}$ *such that at* $\tau \to 0$, $n \to \infty$

$$\Pi^+ u^n \rightharpoonup u, \quad \Pi^+ u_t^n \rightharpoonup \frac{\partial u}{\partial t} \quad in\ L_2(0,T; \overset{\circ}{V}), \quad (31)$$

$$\frac{\partial \Pi^+ u_t^n}{\partial x} \rightharpoonup \frac{\partial^2 u}{\partial x \partial t} \quad in\ L_2(0,T; L_2(0,L)), \quad (32)$$

$$\Pi^+ p^n \rightharpoonup p \ \ in\ L_2(0,T; \overset{\circ}{V}_1). \quad (33)$$

Here $\Pi^+ z$ *is piecewise-constant filling of* z:

$$\Pi^+ z(t) = \{z(k\tau) : k\tau \le t < (k+1)\tau\}.$$

Proof The validity of statements (31)–(33) follows from a priori estimates (25)–(26) and the weak compactness of bounded sets in a reflexive Banach space. The proof of Lemma 3 is complete. □

Lemma 4 *Functions* u, p *satisfying relations (31)–(33) are a generalized solution to problem (5)–(9).* □

Proof Let the functions u, p satisfy relations (31)–(33), it is required to prove that u, p satisfy identity (11). To do this, in (18) we put

$$v^n(x,t) = \frac{1}{\tau}\int_t^{t+\tau} \tilde{v}^n(x,\xi)d\xi, \quad z^n(x,t) = \frac{1}{\tau}\int_t^{t+\tau} \tilde{z}^n(x,\xi)d\xi,$$

where $\tilde{v}^n$, $\tilde{z}^n$ are functions from $C^\infty(0,T;\overset{\circ}{V}{}^n)$ and $C^\infty(0,T;\overset{\circ}{V}{}_1^n)$ respectively, such that $\tilde{v}^n(x,T) = \tilde{z}^n(x,T) = 0$. We multiply (18) by τ, sum over t from 0 to $T-\tau$. The result, using the filling operator Π^+, can be written in the form

$$\int_0^T\int_0^L \left\{ \left(\frac{\partial \Pi^+ \hat{u}^n}{\partial x} + \frac{\partial \Pi^+ u_t^n}{\partial x} \right) \frac{\partial \Pi^+ v_t^n}{\partial x} - \Pi^+ \hat{p}^n \frac{\partial \Pi^+ v_t^n}{\partial x} + \frac{\partial \Pi^+ u_t^n}{\partial x} \Pi^+ \hat{z}^n \right.$$

$$\left. + g\left(\left| \frac{\partial \Pi^+ \hat{p}^n}{\partial x} \right| \right) \frac{\partial \Pi^+ \hat{p}^n}{\partial x} \frac{\partial \Pi^+ \hat{z}^n}{\partial x} \right\} dxdt = \int_0^T\int_0^L \hat{f}(x,t)\, \Pi^+ v_t^n dxdt. \tag{34}$$

From the boundedness of g and estimate (27) it follows that there exists a function χ from the space $L_2(0,T;L_2(0,L))$ such that

$$g\left(\left| \frac{\partial \Pi^+ \hat{p}^n}{\partial x} \right| \right) \frac{\partial \Pi^+ \hat{p}^n}{\partial x} \rightharpoonup \chi \quad \text{in} \quad L_2(0,T;L_2(0,L)). \tag{35}$$

Taking into account (31)–(33) and (35) in equality (34), we pass to the limit in $\tau \to 0$ and $n \to \infty$ and obtain

$$\int_0^T\int_0^L \left\{ \left(\frac{\partial u}{\partial x} + \frac{\partial^2 u}{\partial x \partial t} \right) \frac{\partial^2 v}{\partial x \partial t} - p \frac{\partial^2 v}{\partial x \partial t} + \frac{\partial^2 u}{\partial x \partial t} z + \chi \frac{\partial z}{\partial x} \right\} dxdt$$

$$= \int_0^T\int_0^L f(x,t) \frac{\partial v}{\partial t} dxdt. \tag{36}$$

Let us prove that $\chi = g\left(\left|\frac{\partial p}{\partial x}\right|\right)\frac{\partial p}{\partial x}$. To do this, we use the monotonicity method. We write down the apparent inequality

$$\sum_{t=0}^{T-\tau}\tau\int_0^L\left(\frac{\partial u^n}{\partial x}-\frac{\partial v^n}{\partial x}\right)_t\left(\frac{\partial \hat{u}^n}{\partial x}-\frac{\partial \hat{v}^n}{\partial x}\right)dx \geq \frac{1}{2}\|u^n(T)-v^n(T)\|_1^2\frac{\partial^2 u}{\partial x\partial t}z$$

$$-\frac{1}{2}\|u^n(0)-v^n(0)\|_1^2 \geq -\frac{1}{2}\|u_0-v^n(x,0)\|_1^2,$$

where v^n is an arbitrary smooth function $v \in C^\infty(0,T;\overset{\circ}{V}{}^n)$. From this inequality and the monotonicity of the function $g(\xi)$ it follows that

$$\sum_{t=0}^{T-\tau}\tau\int_0^L\left(\frac{\partial u^n}{\partial x}-\frac{\partial v^n}{\partial x}\right)_t\left(\frac{\partial \hat{u}^n}{\partial x}-\frac{\partial \hat{v}^n}{\partial x}\right)dx$$

$$+\sum_{t=0}^{T-\tau}\tau\int_0^L\left\{g\left(\left|\frac{\partial \hat{p}^n}{\partial x}\right|\right)\frac{\partial \hat{p}^n}{\partial x}-g\left(\left|\frac{\partial \hat{z}^n}{\partial x}\right|\right)\frac{\partial \hat{z}^n}{\partial x}\right\}\frac{\partial\left(\hat{p}^n-\hat{z}^n\right)}{\partial x}dx$$

$$\geq -\frac{1}{2}\|u_0-v^n(x,0)\|_1^2.$$

The last relation is equivalent to the following integral inequality

$$I=\int_0^T\int_0^L\left(\frac{\partial \Pi^+u_t^n}{\partial x}-\frac{\partial \Pi^+v_t^n}{\partial x}\right)\frac{\partial \Pi^+\left(\hat{u}^n-\hat{v}^n\right)}{\partial x}dxdt$$

$$+\int_0^T\int_0^L g\left(\left|\frac{\partial \Pi^+\hat{p}^n}{\partial x}\right|\right)\frac{\partial \Pi^+\hat{p}^n}{\partial x}\frac{\partial \Pi^+\left(\hat{p}^n-\hat{z}^n\right)}{\partial x}dxdt$$

$$-\int_0^T\int_0^L g\left(\left|\frac{\partial \Pi^+\hat{z}^n}{\partial x}\right|\right)\frac{\partial \Pi^+\hat{z}^n}{\partial x}\frac{\partial \Pi^+\left(\hat{p}^n-\hat{z}^n\right)}{\partial x}dxdt$$

$$\geq -\frac{1}{2}\|u_0-v^n(x,0)\|_1^2.$$

We represent I as the sum $I = I_1 + I_2$, where

$$I_1 = \int_0^T \int_0^L \left\{ \frac{\partial \Pi^+ u_t^n}{\partial x} \frac{\partial \Pi^+ (\hat{u}^n - \hat{v}^n)}{\partial x} + g\left(\left|\frac{\partial \Pi^+ \hat{p}^n}{\partial x}\right|\right) \frac{\partial \Pi^+ \hat{p}^n}{\partial x} \frac{\partial \Pi^+ (\hat{p}^n - \hat{z}^n)}{\partial x} \right\} dxdt,$$

$$I_2 = -\int_0^T \int_0^L \left\{ \frac{\partial \Pi^+ v_t^n}{\partial x} \frac{\partial \Pi^+ (\hat{u}^n - \hat{v}^n)}{\partial x} + g\left(\left|\frac{\partial \Pi^+ \hat{z}^n}{\partial x}\right|\right) \frac{\partial \Pi^+ \hat{z}^n}{\partial x} \frac{\partial \Pi^+ (\hat{p}^n - \hat{z}^n)}{\partial x} \right\} dxdt.$$

To transform the first relation I_1, we use equality (34) at $v^n = u^n - v^n$, $p^n = p^n - z^n$ and obtain

$$I_1 = \int_0^T \int_0^L \left\{ - \frac{\partial \Pi^+ u_t^n}{\partial x} \frac{\partial \Pi^+ (u_t^n - v_t^n)}{\partial x} - \Pi^+ \hat{p}^n \frac{\partial \Pi^+ v_t^n}{\partial x} + \frac{\partial \Pi^+ u_t^n}{\partial x} \Pi^+ \hat{z}^n - \frac{\partial \Pi^+ u_t^n}{\partial x} \frac{\partial \Pi^+ v^n}{\partial x} + \frac{\partial \Pi^+ u^n}{\partial x} \frac{\partial \Pi^+ v_t^n}{\partial x} + \hat{f}(x,t)\, \Pi^+ \left(u^n - v^n\right)_t \right\} dxdt. \tag{37}$$

In (37), we make the passage to the limit as $\tau \to 0$, $n \to \infty$, taking into account (31)–(33) and (35). As a result, we obtain

$$I_1 \to \int_0^T \int_0^L \left\{ - \frac{\partial^2 u}{\partial x \partial t} \frac{\partial^2 (u - v)}{\partial x \partial t} - p \frac{\partial^2 v}{\partial x \partial t} + \frac{\partial^2 u}{\partial x \partial t} z - \frac{\partial^2 u}{\partial x \partial t} \frac{\partial v}{\partial x} + \frac{\partial u}{\partial x} \frac{\partial^2 v}{\partial x \partial t} + f(x,t) \frac{\partial (u - v)}{\partial t} \right\} dxdt. \tag{38}$$

Using equality (36), the right-hand side of relation (38) takes the following form

$$I_1 \to \int_0^T \int_0^L \left\{ \frac{\partial^2 u}{\partial x \partial t} \frac{\partial (u - v)}{\partial x} + \chi \frac{\partial (p - z)}{\partial x} \right\} dxdt.$$

Apparently, from (31)–(33) and (35) for $\tau \to 0$, $n \to \infty$ we obtain

$$I_2 \to -\int_0^T\int_0^L \left\{ \frac{\partial^2 v}{\partial x \partial t}\frac{\partial (u-v)}{\partial x} + g\left(\left|\frac{\partial z}{\partial x}\right|^2\right)\frac{\partial z}{\partial x}\frac{\partial (p-z)}{\partial x}\right\} dxdt.$$

Thus, it follows from the definition of I that

$$\int_0^T\int_0^L \left\{ \frac{\partial^2 (u-v)}{\partial x \partial t}\frac{\partial (u-v)}{\partial x} + \left(\chi - g\left(\left|\frac{\partial z}{\partial x}\right|^2\right)\frac{\partial z}{\partial x}\right)\frac{\partial (p-z)}{\partial x}\right\} dxdt$$

$$\geq -\frac{1}{2}\|u_0 - v(x,0)\|_1^2. \tag{39}$$

In (39), we choose $v = u + \lambda w$, $z = p + \lambda q$, where $\lambda = \text{const} > 0$, and w, q are arbitrary functions from $C^\infty(0,T; C^\infty(0,L))$, where $w(x,0) = 0$ for $x \in (0,L)$. As a result, we obtain

$$\lambda \int_0^T\int_0^L \left(\chi - g\left(\left|\frac{\partial (p+\lambda q)}{\partial x}\right|^2\right)\frac{\partial (p+\lambda q)}{\partial x}\right)\frac{\partial q}{\partial x} dxdt$$

$$+\lambda^2 \int_0^T\int_0^L \frac{\partial^2 w}{\partial x \partial t}\frac{\partial w}{\partial x} dxdt \geq -\frac{\lambda}{2}\|w(x,0)\|_1^2 = 0. \tag{40}$$

We divide inequality (40) by λ and pass to the limit as $\lambda \to 0$, we obtain

$$\int_0^T\int_0^L \left(\chi - g\left(\left|\frac{\partial p}{\partial x}\right|^2\right)\frac{\partial p}{\partial x}\right)\frac{\partial q}{\partial x} dxdt \geq 0.$$

Since q is an arbitrary function, the inequality holds at $q = v$ and $q = -v$, where $v \in L_2(0,T; W_2^1(0,L))$ is an arbitrary function; therefore, we have

$$\chi = g\left(\left|\frac{\partial p}{\partial x}\right|^2\right)\frac{\partial p}{\partial x}.$$

The proof of Lemma 4 is complete. □

The assertion of Theorem 2 follows from Lemmas 1–4. □

Acknowledgement This paper has been supported by the Kazan Federal University Strategic Academic Leadership Program ("PRIORITY-2030").

References

1. Akhtareev, A.A., Dautov, R.Z.: Mixed variable method for modeling saturated-unsaturated flows. Uchen. Zap. Kazan Gos. Univ. Fiz-Mat. Nauki. **149**(4), 58–72 (2007)
2. Pavlova, M.F., Rung, E.V.: On the solvability of the problem of saturated-unsaturated filtration consolidation. Diff. Eq. **48**(7), 990–1004 (2012)
3. Pavlova, M.F., Rung, E.V.: On solvability of a elliptic-parabolic problem of nonlinear filtration theory. IOP Conference Series: Materials Science and Engineering (2016). https://doi.org/10.1088/1757-899X/158/1/012076
4. Pavlova, M.F., Rung, E.V.: On the existence of a generalized solution of the saturated-unsaturated filtration problem. Diff. Eq. **54**(3), 352–362 (2018)
5. Zareckij, Y.K.: Theory of soil consolidation. Nauka, Moscow (1967)
6. Nikolaevskij, V.N.: Mechanics of porous and fractured media. Nedra, Moscow (1984)
7. Biot, M.A.: The mechanics of deformation and acoustic propagation in porous media. J. Appl. Phys. **33**(4), 1482–1498 (1962)
8. Egorov, A.G., Kosterin, A.V., Skvorcov, E.V.: Consolidation and acoustic waves in saturated porous media. KSU Publishing House, Kazan (1990)
9. Drobotenko, M.I., Kosterin, A.V.: Regularization of the problem of filtration consolidation of an elastic porous medium. Izv. Vyssh. Uchebn. Zaved. **4**, 18–22 (1984)
10. Kadyrov, F.M., Kosterin, A.V., Skvorcov E.V.: Plane problem of filtration consolidation for an elastic half-space with discontinuous initial conditions . Prikl. mekh. i tekhn. fiz. **57**(6), 1–7 (2016)
11. Barenblatt, G.I., Entov, V.M., Ryzhik, V.M.: Movement of liquids and gases in natural reservoir. Nedra, Moscow (1984)
12. Lions, J.-L.: Quelques méthodes de résolution des problémes aux limites nonlinéaires. Dunod, Paris (1969)

Mathematical Modeling (Faedo–Galerkin Method, Solution Existence Theorem) of Nonlinear Dynamics for MEMS/NEMS Devices Elements (Micropolar Theory) in the Rectangular Shells form in Plane, Taking into Account the Temperature and Deformation Fields Connection

Ekaterina Yu. Krylova, Irina V. Papkova, Anton V. Krysko, and Vadim A. Krysko

Abstract In this work, a mathematical model of the elements nonlinear dynamics for NEMS/MEMS devices is constructed in the form of two flexible rectangular shallow isotropic Kirchhoff-Love shells, taking into account their contact interaction, the connectivity of the deformation and temperature fields. Size-dependent effects are taken into account according to the micropolar moment theory of elasticity. The sought equations in displacements are obtained from the Hamilton-Ostrogradsky energy principle. The existence theorem for a solution is proved. It is proved that an approximate solution in the second boundary value problem, which determines the condition of thermomechanical evolution for two rectangular shallow isotropic shells, can be found by the Bubnov–Galerkin method. This theorem is new, and in the future it allows to construct algorithms for solving the dynamics of a coupled problem for MEMS/NEMS devices elements in the form of rectangular flexible elastic shells in terms of their contact interaction.

E. Yu. Krylova
Saratov State University, Saratov, Russia
e-mail: kat.krylova@bk.ru

I. V. Papkova (✉) · A. V. Krysko · V. A. Krysko
Yuri Gagarin State Technical University Of Saratov, Saratov, Russia
e-mail: ikravzova@mail.ru; anton.krysko@gmail.com; tak@sun.ru

I. B. Badriev et al. (eds.), *Mesh Methods for Boundary-Value Problems and Applications*, Lecture Notes in Computational Science and Engineering 141,
https://doi.org/10.1007/978-3-030-87809-2_19

1 Introduction

At the moment, the application field of NEMS sensors and devices is very wide. Unique properties of NEMS devices have predetermined their use in physics, chemistry, biology, medicine, criminology, military and consumer technology, navigation and control systems. Interesting properties of NEMS devices usually arise from their active part behavior (including dynamic), which can represent various nano-objects types, such as nano-rods, nanotubes, nano-beams, nanoplates and nanoshells, and their various combinations. It is important to note that NEMS dissipates very little energy and this makes them extremely sensitive to external influences, especially thermal effects and noise fields. Nano-mechanics is based on theories that can account for scale effects at the nano-scale level. Their good overview is given in [1]. One of such theories is the currently actively developed micropolar (asymmetric, moment) theory [2–14]. The question of the temperature effects influence on the nano-plates behavior is also considered in works [15, 16]. An important issue is the existence of solutions for the nonlinear differential equation systems that describe the mechanical structures behavior. For this, it is necessary to prove theorems on the solutions existence. In the case of solving a nonlinear stationary problem, approximate solutions are constructed using the Faedo–Galerkin method. In the case of non-stationary problems, solutions are constructed using the Faedo–Galerkin method, then a priori estimates of the energy type inequalities are established for them [17–20]. In the works of these authors, theorems on the solution existence for classical mathematical models were proved. In this paper, we consider evidence of the solution existence for mathematical models based on the micropolar theory.

2 Formulation of the Problem

Let us consider the related problem of thermoelasticity, which determines the thermomechanical evaluation conditions for shallow micropolar homogeneous isotropic shells in the Kirchhoff-Love hypotheses framework, taking into account the three-dimensional equation of heat conduction and contact interaction (1)–(4)

Differential equations describing the shell element motion, obtained on the basis of the Hamilton-Ostrogradsky variational principle, taking into account the micropolar theory, Karman's theory and Cantor's contact interaction theory, have the form:

$$
\begin{gathered}
\frac{\partial N_{xx}^{k}}{\partial x}+\frac{\partial T^{k}}{\partial y}+\frac{1}{2}\frac{\partial^{2} Y_{yz}^{k}}{\partial y^{2}}+\frac{1}{2}\frac{\partial^{2} Y_{xz}^{k}}{\partial x \partial y}=\rho h^{k}\frac{\partial^{2} u^{k}}{\partial t^{2}},\\
\frac{\partial N_{yy}^{k}}{\partial y}+\frac{\partial T^{k}}{\partial x}-\frac{1}{2}\frac{\partial^{2} Y_{xz}^{k}}{\partial x^{2}}-\frac{1}{2}\frac{\partial^{2} Y_{yz}^{k}}{\partial x \partial y}=\rho h^{k}\frac{\partial^{2} v^{k}}{\partial t^{2}},\\
\frac{\partial^{2} M_{xx}^{k}}{\partial x^{2}}+\frac{\partial^{2} M_{yy}^{k}}{\partial y^{2}}+2\frac{\partial^{2} H^{k}}{\partial x \partial y}+\frac{\partial}{\partial x}\left(N_{x}^{k}\frac{\partial w^{k}}{\partial x}\right)+\frac{\partial}{\partial y}\left(N_{y}^{k}\frac{\partial w^{k}}{\partial y}\right)+2\frac{\partial T^{k}}{\partial x}\frac{\partial w^{k}}{\partial y}+\\
+2\frac{\partial T^{k}}{\partial y}\frac{\partial w^{k}}{\partial x}+4T\frac{\partial^{2} w^{k}}{\partial x \partial y}-k_{y}^{k}N_{yy}^{k}-k_{x}^{k}N_{xx}^{k}-\frac{\partial^{2} Y_{xx}^{k}}{\partial x \partial y}+\frac{\partial^{2} Y_{yy}^{k}}{\partial y \partial x}+\frac{\partial^{2} Y_{xy}^{k}}{\partial x^{2}}-\\
-\frac{\partial^{2} Y_{xy}^{k}}{\partial y^{2}}+2q^{k}\mp 2K\left(w^{1}-w^{2}-\delta\right)\psi=\rho h^{k}\frac{\partial^{2} w^{k}}{\partial t^{2}}+\varepsilon\rho h^{k}\frac{\partial w^{k}}{\partial t}.
\end{gathered} \tag{1}
$$

The three-dimensional heat equation in a coupled setting can be written as follows:

$$
\frac{C_{0}}{T_{0}}\frac{\partial T^{k}}{\partial t}-\frac{\lambda}{T_{0}}\left(\frac{\partial^{2} T^{k}}{\partial x^{2}}+\frac{\partial^{2} T^{k}}{\partial y^{2}}+\frac{\partial^{2} T^{k}}{\partial z^{2}}\right)+\frac{E\alpha_{t}}{1-v}\left(\frac{\partial \varepsilon_{xx}^{k}}{\partial t}+\frac{\partial \varepsilon_{yy}^{k}}{\partial t}\right)=\frac{1}{T_{0}}g_{t}^{k}. \tag{2}
$$

The boundary and initial conditions are chosen in the following form:

$$
\begin{gathered}
u^{k}\,|_{\Gamma}=0, \quad \frac{\partial u^{k}}{\partial x}\,|_{\Gamma}=0, \quad \frac{\partial u^{k}}{\partial y}\,|_{\Gamma}=0,\\
v^{k}\,|_{\Gamma}=0, \quad \frac{\partial v^{k}}{\partial x}\,|_{\Gamma}=0, \quad \frac{\partial v^{k}}{\partial y}\,|_{\Gamma}=0,\\
w^{k}\,|_{\Gamma}=0, \quad \frac{\partial w^{k}}{\partial x}\,|_{\Gamma}=0, \quad \frac{\partial w^{k}}{\partial y}\,|_{\Gamma}=0,\\
T^{k}\,|_{S}=0.
\end{gathered} \tag{3}
$$

$$
\begin{gathered}
w^{k}\Big|_{t=t_{0}}=\phi_{w}^{k}(x, y), \quad \frac{\partial w^{k}}{\partial t}\Big|_{t=t_{0}}=\psi_{w}^{k}(x, y)\\
u^{k}\Big|_{t=t_{0}}=\phi_{u}^{k}(x, y), \quad \frac{\partial u^{k}}{\partial t}\Big|_{t=t_{0}}=\psi_{u}^{k}(x, y)\\
v^{k}\Big|_{t=t_{0}}=\phi_{v}^{k}(x, y), \quad \frac{\partial v^{k}}{\partial t}\Big|_{t=t_{0}}=\psi_{v}^{k}(x, y)\\
T^{k}\Big|_{t=t_{0}}=\phi_{t}^{k}(x, y, z)
\end{gathered} \tag{4}
$$

In the boundary value problem (1)–(4), the following notations are used.

$$\Omega = \Omega_k, \quad D = D_k,$$

where k-structure layer number; δ-gap between shells.

Ω-rectangle (in plane O_{xy}) with bound $\partial\Omega$:

$$\Omega = (0, a) \times (0, b), \quad \overline{\Omega} = [0, a] \times [0, b], \quad \partial\Omega = \overline{\Omega} \setminus \Omega;$$

D-parallelepiped in space O_{xyz} with boundary plane ∂D:

$$D = (a, b) \times \left(-\frac{h}{2}, \frac{h}{2}\right), \quad \overline{D} = [a, b] \times \left[-\frac{h}{2}, \frac{h}{2}\right], \quad \partial D = \overline{D} \setminus D;$$

$$Q_1 = \Omega \times (t_0, t_1); \quad Q_2 = D \times (t_0, t_1); \quad \Gamma = \partial\Omega \times [t_0, t_1]; \quad S = \partial D \times [t_0, t_1];$$

$[t_0, t_1]$– observation time span of shell evolution, $t \in [t_0, t_1]$. Classic strains and moments, as well as higher order strains and moments:

$$\left(N^k_{xx}, N^k_{yy}, T^k\right) = \int_{-h^k}^{h^k} \left(\sigma^k_{xx}, \sigma^k_{yy}, \sigma^k_{xy}\right) dz,$$

$$\left(M^k_{xx}, M^k_{yy}, H^k\right) = \int_{-h^k}^{h^k} \left(\sigma^k_{xx}, \sigma^k_{yy}, \sigma^k_{xy}\right) zdz,$$

$$Y^k_{xx} = \int_{-h^k}^{h^k} m^k_{xx} dz, \quad Y^k_{xy} = \int_{-h^k}^{h^k} m^k_{xy} dz, \quad Y^k_{zx} = \int_{-h^k}^{h^k} m^k_{zx} dz, \quad x \rightleftarrows y.$$

Nonzero stress tensor components:

$$\sigma^k_{xx} = \frac{E}{1 - \nu^2}\left[\varepsilon^k_{xx} + \nu\varepsilon^k_{yy}\right], \quad x \rightleftarrows y, \quad \sigma^k_{xy} = \frac{E}{(1 + \nu)}\varepsilon^k_{xy}.$$

Higher-order nonzero moments tensor components [21]:

$$\left(m^k_{xx}, m^k_{xy}, m^k_{zx}\right) = \frac{El^2}{1 + \nu}\left(\chi^k_{xx}, \chi^k_{xy}, \chi^k_{zx}\right), \quad x \rightleftarrows y$$

Deformation tensor components taking into account temperature effects:

$$\begin{aligned}
\varepsilon_{xx}^{k} &= \frac{\partial u^{k}}{\partial x} + \frac{1}{2}\left(\frac{\partial w^{k}}{\partial x}\right)^{2} - k_{x}^{k} w^{k} - z\frac{\partial^{2} w^{k}}{\partial x^{2}} + \alpha_{t} T^{k}(x, y, z); \\
\varepsilon_{yy}^{k} &= \frac{\partial v^{k}}{\partial y} + \frac{1}{2}\left(\frac{\partial w^{k}}{\partial y}\right)^{2} - k_{y}^{k} w^{k} - z\frac{\partial^{2} w^{k}}{\partial y^{2}} + \alpha_{t} T^{k}(x, y, z); \\
\varepsilon_{xy}^{k} &= \frac{1}{2}\left(\frac{\partial u^{k}}{\partial y} + \frac{\partial v^{k}}{\partial x}\right) + \frac{\partial w^{k}}{\partial x}\frac{\partial w^{k}}{\partial y} - z\frac{\partial^{2} w^{k}}{\partial x \partial y}.
\end{aligned} \tag{5}$$

Torsional bending tensor nonzero components written for the case when the displacement and rotation fields are not independent:

$$\begin{aligned}
&\chi_{xx}^{k} = \frac{\partial^{2} w^{k}}{\partial x \partial y}; \quad \chi_{yy}^{k} = -\frac{\partial^{2} w^{k}}{\partial y \partial x}; \quad \chi_{xy}^{k} = \frac{1}{2}\left(\frac{\partial^{2} w^{k}}{\partial y^{2}} - \frac{\partial^{2} w^{k}}{\partial x^{2}}\right); \\
&\chi_{xz}^{k} = \frac{1}{4}\left(\frac{\partial^{2} v^{k}}{\partial x^{2}} - \frac{\partial^{2} u^{k}}{\partial x \partial y}\right); \quad \chi_{yz}^{k} = \frac{1}{4}\left(\frac{\partial^{2} v^{k}}{\partial y \partial x} - \frac{\partial^{2} u^{k}}{\partial y^{2}}\right).
\end{aligned} \tag{6}$$

$w^k(x, y, t)$, $u^k(x, y, t)$, $v^k(x, y, t)$—the required deflection and displacement functions defined on the area $\overline{Q_1} = \overline{\Omega} \times [t_0, t_1]$; $T^k(x, y, z, t)$—the required functions that determine the temperature field in the region $\overline{Q_2} = \overline{D} \times [t_0, t_1]$; $h > 0$—constant shell thickness; $\rho > 0$—constant density of the shell material; $E > 0$—Young's modulus; $0 < \nu < 0.5$—Poisson's ratio; $l > 0$—additional material length parameter associated with the bend-torsion tensor; $\varepsilon > 0$—constant damping factor; α—thermal expansion factor; $T_0 > 0$—initial shell temperature; $C_0 > 0$—specific heat; $\lambda > 0$—thermal conductivity factor; $g_t^k(x, y, z, t)$—known functions defined on area Q_2 and determining the bulk density of internal heat sources; $q^k(x, y, t)$—known function of the transverse load intensity on the shell, defined on the region; Q_1; k_x^k, k_y^k—initial curvatures of the shells middle surfaces.

The notation of all the main functional spaces, norms and scalar products correspond to those adopted in the works [22, 23]: $L^2(A)$—Lebesgue space of square-integrable functions; $|\cdot|_A$—norm in Hilbert space $L^2(A)$, and $(\cdot, \cdot)_A$—dot product in this space; $H_0^1(A)$—the subspace of the Sobolev space $H^1(A)$, in which the dense set is the set of all functions of their $C^1(\overline{A})$ equal to zero near the boundary ∂A; $H_0^2(A)$—the subspace of space $H^2(A)$, in which the dense set is the set of all functions of their $C^2(\overline{A})$ equal to zero near the boundary ∂A.

Theorem 1 *Let the boundary contour $\partial\Omega$ have smoothness sufficient for the used embedding theorems and the following conditions are satisfied: $q^k \in L^2(Q_1)$, $g_t^k \in L^2(Q_2)$, $\phi_w^k, \phi_u^k, \phi_v^k \in H_0^2(\Omega)$, $\psi_w^k, \psi_u^k, \psi_v^k \in L^2(\Omega)$, $\phi_t^k \in L^2(D)$. Then:*

(1) there is at least one solution $\left\{\tilde{u}^k, \tilde{v}^k, \tilde{w}^k, \tilde{T}^k\right\}$ *for task (1)–(4), wherein*

$$\tilde{u}^k, \tilde{v}^k, \tilde{w}^k \in L^2\left(t_0, t_1, H_0^2(\Omega)\right);$$
$$\frac{\partial \tilde{w}^k}{\partial t}, \frac{\partial \tilde{u}^k}{\partial t}, \frac{\partial \tilde{v}^k}{\partial t} \in L^2\left(t_0, t_1, L^2(\Omega)\right) \tag{7}$$
$$\tilde{T}^k \in L^2\left(t_0, t_1; H_0^1(D)\right) k = \overline{1,2}$$

(2) an approximate solution to problem (1)–(4) can be found by the Bubnov–Galerkin method, while the entire set of approximate solutions is weakly compact in spaces corresponding to conditions (7), and its limit points determine the generalized solution of problem (1)–(4) ;

(3) the phase space V of the mechanical system defined by the boundary value problem (1)–(4), with generalized solutions from spaces (7), is an infinite-dimensional functional space of the following form $V = \left(L^2(\Omega)\right)^6 \times T$, *where* T *–the configuration space of such a system,* $T = \left(H_0^2(\Omega)\right)^6 \times (H_0(D))^2$, *while for almost all* $t \in [t_0, t_1]$,

$$\left\{\frac{\partial \tilde{w}^k}{\partial t}, \frac{\partial \tilde{u}^k}{\partial t}, \frac{\partial \tilde{v}^k}{\partial t}, \tilde{w}^k, \tilde{u}^k, \tilde{v}^k, \tilde{T}^k\right\} \in V.$$

3 The Main Stages of the Theorem Proof

3.1 Construction of an Approximate Solution

Construction of an approximate solution to problem (1)–(4). Let functions sequences $\{\chi_{wk}^n\}, \{\chi_{vk}^n\}, \{\chi_{uk}^n\}$ define a basis in the space $H_0^2(\Omega)$ orthonormalized with respect to the norm of $L^2(\Omega)$, and the sequence $\{\chi_{Tk}^n\}$ define the same orthonormal basis in the space $H_0^1(\Omega)$. Following the Bubnov–Galerkin method, an approximate solution $\{w^{kn}, u^{kn}, v^{kn}, T^{kn}\}$ of problem (1)–(4) will be sought in the following finite expansions (sums) form:

$$w^{kn} = \sum_{l=1}^{n_{wk}} g_{wkl}(t)\chi_{wkl}(x, y), \quad v^{kn} = \sum_{l=1}^{n_{vk}} g_{vkl}(t)\chi_{vkl}(x, y),$$
$$u^{nk} = \sum_{l=1}^{n_{uk}} g_{ukl}(t)\chi_{ukl}(x, y), \quad T^{kn} = \sum_{l=1}^{n_{Tk}} g_{Tkl}(t)\chi_{Tkl}(x, y) \tag{8}$$

Expansion coefficients are determined as solutions of the following Cauchy problem for an ordinary differential equations system:

$$\begin{aligned}
&\rho\left(\frac{\partial^2 u^{kn}}{\partial t^2}, \chi^k_{u_{luk}}\right)_D + \left(\sigma^{kn}_{xx}, \frac{\partial \chi^k_{u_{luk}}}{\partial x}\right)_D + \left(\sigma^{kn}_{xy}, \frac{\partial \chi^k_{u_{luk}}}{\partial y}\right)_D + \\
&+\frac{1}{2}\left(-m^{kn}_{yz}, \frac{\partial^2 \chi^k_{u_{luk}}}{\partial x^2}\right)_D + \frac{1}{2}\left(-m^{kn}_{xz}, \frac{\partial^2 \chi^k_{u_{luk}}}{\partial x \partial y}\right)_D = 0, \\
&\rho\left(\frac{\partial^2 v^{kn}}{\partial t^2}, \chi^k_{v_{lvk}}\right)_D + \left(\sigma^{kn}_{yy}, \frac{\partial \chi^k_{v_{lvk}}}{\partial y}\right)_D + \left(\sigma^{kn}_{xy}, \frac{\partial \chi^k_{v_{lvk}}}{\partial x}\right)_D + \\
&+\frac{1}{2}\left(m^{kn}_{xz}, \frac{\partial^2 \chi^k_{v_{lvk}}}{\partial x^2}\right)_D + \frac{1}{2}\left(m^{kn}_{yz}, \frac{\partial^2 \chi^k_{v_{lvk}}}{\partial x \partial y}\right)_D = 0, \\
&\rho\left(\frac{\partial^2 w^{kn}}{\partial t^2}, \chi^k_{w_{lwk}}\right)_D + \varepsilon\rho\left(\frac{\partial w^{kn}}{\partial t}, \chi^k_{w_{lwk}}\right)_D + \left(-z\sigma^{kn}_{xx}, \frac{\partial^2 \chi^k_{w_{lwk}}}{\partial x^2}\right)_D + \\
&+\left(-z\sigma^{kn}_{yy}, \frac{\partial^2 \chi^k_{w_{lwk}}}{\partial y^2}\right)_D + k^k_y\left(-\sigma^{kn}_{yy}, \chi^k_{w_{lwk}}\right)_D + k^k_x\left(-\sigma^{kn}_{xx}, \chi^k_{w_{lwk}}\right)_D + \\
&+2\left(-z\sigma^{kn}_{xy}, \frac{\partial^2 \chi^k_{w_{lwk}}}{\partial x \partial y}\right)_D + \left(\sigma^{kn}_{xx}\frac{\partial w^{kn}}{\partial x}, \frac{\partial \chi^k_{w_{lwk}}}{\partial x}\right)_D + \left(\sigma^{kn}_{yy}\frac{\partial w^{kn}}{\partial y}, \frac{\partial \chi^k_{w_{lwk}}}{\partial y}\right)_D + \\
&+2\left(\sigma^{kn}_{xy}\frac{\partial w^{kn}}{\partial x}, \frac{\partial \chi^k_{w_{lwk}}}{\partial y}\right)_D + 2\left(\sigma^{kn}_{xy}\frac{\partial w^{kn}}{\partial y}, \frac{\partial \chi^k_{w_{lwk}}}{\partial x}\right)_D + \left(m^{kn}_{xx}, \frac{\partial^2 \chi^k_{w_{lwk}}}{\partial x \partial y}\right)_D + \\
&+\left(-m^{kn}_{yy}, \frac{\partial^2 \chi^k_{w_{lwk}}}{\partial y \partial x}\right)_D + \left(-m^{kn}_{xy}, \frac{\partial^2 \chi^k_{w_{lwk}}}{\partial x^2}\right)_D + \left(m^{kn}_{xy}, \frac{\partial^2 \chi^k_{w_{lwk}}}{\partial y^2}\right)_D + \\
&+\left(\mp 2K\left(w^{1n} - w^{2n} - \delta\right)\psi, \chi^k_{w_{lwk}}\right)_\Omega = \left(q^k, \chi^k_{w_{lwk}}\right)_\Omega, \\
&C_0\left(\frac{\partial T^{nk}}{\partial t}, \chi^n_{T_{lTk}}\right)_D + \lambda\left(\frac{\partial T^{nk}}{\partial x}\frac{\partial \chi^n_{T_{lTk}}}{\partial x}\right)_D + \lambda\left(\frac{\partial T^{nk}}{\partial y}, \frac{\partial \chi^n_{T_{lTk}}}{\partial y}\right)_D + \\
&+\lambda\left(\frac{\partial T^{nk}}{\partial z}, \frac{\partial \chi^n_{T_{lTk}}}{\partial z}\right)_D + \frac{E\alpha_t}{1-\upsilon}\left[\left(\frac{\partial \varepsilon^k_{xx}}{\partial t}, \chi^n_{T_{lTk}}\right)_D + \left(\frac{\partial \varepsilon^k_{yy}}{\partial t}, \chi^n_{T_{lTk}}\right)_D\right] \\
&= \left(g_t, \chi^n_{T_{lTk}}\right)_D.
\end{aligned} \tag{9}$$

In (9) $l_{uk}=\overline{1,n_{uk}}, l_{vk}=\overline{0,n_{vk}}, l_{wk}=\overline{1,n_{wk}}, l_{Tk}=\overline{1,n_{Tk}}$. The initial conditions will take the form:

$$\begin{aligned}
&w^{kn}(x,y,t_0)=\phi_{wk}^n, \quad \phi_{wk}^n=\sum_{l=1}^{n_{wk}} a_{wkl}\chi_{wkl}, \quad \phi_{wk}^n\to\phi_{wk} \quad \text{from } H_0^2(\Omega),\\
&\frac{\partial w^{kn}}{\partial t}(x,y,t_0)=\psi_{wk}^n, \quad \psi_{wk}^n=\sum_{l=1}^{n_{wk}} b_{wkl}\chi_{wkl}, \quad \psi_{wk}^n\to\psi_{wk} \quad \text{from } H_0^1(\Omega),\\
&u^{nk}(x,y,t_0)=\phi_{uk}^n, \quad \phi_{uk}^n=\sum_{l=1}^{n_{uk}} a_{ukl}\chi_{ukl}, \quad \phi_{uk}^n\to\phi_{uk} \quad \text{from } H_0^2(\Omega),\\
&\frac{\partial u^{nk}}{\partial t}(x,y,t_0)=\psi_{uk}^n, \quad \psi_{uk}^n=\sum_{l=1}^{n_{uk}} b_{ukl}\chi_{ukl}, \quad \psi_{uk}^n\to\phi_{uk} \quad \text{from } H_0^1(\Omega),\\
&v^{nk}(x,y,t_0)=\phi_{vk}^n, \quad \phi_{vk}^n=\sum_{l=1}^{n_{vk}} a_{vkl}\chi_{vkl}, \quad \phi_{vk}^n\to\phi_{vk} \quad \text{from } H_0^2(\Omega),\\
&\frac{\partial v^{nk}}{\partial t}(x,y,t_0)=\psi_{vk}^n, \quad \psi_{vk}^n=\sum_{l=1}^{n_{vk}} b_{vlk}\chi_{vkl}, \quad \psi_{vk}^n\to\psi_{vk} \quad \text{from } H_0^1(\Omega),\\
&T^{nk}(x,y,z,t)=\phi_{Tk}^n, \quad \phi_{Tk}^n=\sum_{l=1}^{n_{Tk}} a_{Tkl}\chi_{Tkl}, \quad \phi_{Tk}^n\to\phi_{Tk} \quad \text{from } H_0^1(D)
\end{aligned} \tag{10}$$

or

$$\begin{aligned}
&g_{wkl_{wk}}(t_0)=a_{wkl_{wk}}, \quad \frac{\partial g_{wkl_{wk}}}{\partial t}(t_0)=b_{wkl_{wk}},\\
&g_{ukl_{uk}}(t_0)=a_{ukl_{uk}}, \quad \frac{\partial g_{ukl_{uk}}}{\partial t}(t_0)=b_{ukl_{uk}},\\
&g_{vkl_{vk}}(t_0)=a_{vkl_{vk}}, \quad \frac{\partial g_{vkl_{vk}}}{\partial t}(t_0)=b_{vkl_{vk}}, \quad g_{Tkl_{Tk}}(t_0)=a_{Tkl_{Tk}}
\end{aligned}$$

In this case, the "arrows"Ì in (10) indicate the convergence according to the corresponding norms. The solution of the Cauchy problem (9)–(10) on some interval $[t_0,t_n]$, $t_n\le t_1$ follows from the Schauder Yu theorem [17, 24]. In general, this corresponds to the Peano's theorem proof on the solution existence to the Cauchy problem for an ordinary differential equations system. For the shell theory problem a detailed proof of such a theorem is given in the works of Kirichenko V.F. [19, 20]. It is taken into account that the functions $\int_{t_0}^{t_n} \iint_{\Omega} q^k(x, y, t)\, \chi_{w_{lw}}^{k}\, dx dx dt$ belong to the space $H^1\left((t_0, t_1)\right)$ and, therefore, belong to the space $C\left([t_0, t_1]\right)$ [25–27]. And also from the indicated Cauchy problem solvability it follows that $g_{wkl_{wk}}, g_{ukl_{uk}}, g_{vkl_{vk}}, g_{Tkl_{Tk}} \in H^2\left((t_0, t_1)\right)$.

3.2 The Priori Estimates

Let us obtain a priori estimates for the approximate solution constructed by the Bubnov–Galerkin method. We multiply the equations of the system by $\frac{dg_{ukl_{uk}}}{dt}$, $\frac{dg_{vkl_{vk}}}{dt}$, $\frac{dg_{wkl_{wk}}}{dt}$, $g_{Tkl_{Tk}}$, respectively, then sum the result:

$$\frac{1}{2}\frac{\partial}{\partial t}\left\{\rho\left|\frac{\partial u^{kn}}{\partial t}\right|_D^2 + \rho\left|\frac{\partial v^{kn}}{\partial t}\right|_D^2 + \rho\left|\frac{\partial w^{kn}}{\partial t}\right|_D^2 + \frac{E}{1-\nu^2}\left|\varepsilon_{xx}^{nk}\right|_D^2 + \frac{E}{1-\nu^2}\left|\varepsilon_{yy}^{nk}\right|_D^2 + \right.$$
$$+2\nu\left(\varepsilon_{xx}^{nk}, \varepsilon_{yy}^{nk}\right)_D + \frac{2E}{1+\nu}\left|\varepsilon_{xy}^{nk}\right|_D^2 + \left|m_{xx}^{nk}\right|_D^2 + \left|m_{yy}^{nk}\right|_D^2 + \left|m_{xy}^{nk}\right|_D^2 + \left|m_{xz}^{nk}\right|_D^2 +$$
$$\left. + \left|m_{yz}^{nk}\right|_D^2 + \frac{C_0}{T_0}\left|T^{kn}\right|_D^2\right\} + \rho\varepsilon\left|\frac{\partial w^{kn}}{\partial t}\right|_D^2 + \lambda\left(\left|\frac{\partial T^{kn}}{\partial x}\right|_D^2 + \left|\frac{\partial T^{kn}}{\partial y}\right|_D^2 + \left|\frac{\partial T^{kn}}{\partial z}\right|_D^2\right) +$$
$$+\left(\mp K\psi\left(w^{1n} - w^{2n} - \delta\right), \frac{\partial w^{kn}}{\partial t}\right)_D + \frac{E\alpha_t}{1-\nu}\left[\left(\frac{\partial \varepsilon_{xx}^{nk}}{\partial t}, T^{nk}\right)_D + \right.$$
$$\left. + \left(\frac{\partial \varepsilon_{yy}^{nk}}{\partial t}, T^{nk}\right)_D\right] - \frac{E\alpha_t}{1-\nu}\left[\left(T^{nk}, \frac{\partial \varepsilon_{xx}^{nk}}{\partial t}\right)_D + \left(T^{nk}, \frac{\partial \varepsilon_{yy}^{nk}}{\partial t}\right)_D\right] =$$
$$= \left(g_t, T^{nk}\right)_D + \left(q^k, \frac{\partial w^{kn}}{\partial t}\right)_\Omega \tag{11}$$

In expression (11), we write the components of the stress tensor through the components of the strain tensor (5), and the components of the moment stress tensor through the components of the bending and torsion tensor (6). Integrating the result over the segment $[t_0, t] \in [t_0, t_n]$, we obtain the following inequality:

$$\frac{1}{2}\left\{\rho\left|\frac{\partial u^{kn}}{\partial t}\right|_D^2+\rho\left|\frac{\partial v^{kn}}{\partial t}\right|_D^2+\rho\left|\frac{\partial w^{kn}}{\partial t}\right|_D^2+\frac{E}{1-\nu^2}\left|\varepsilon_{xx}^{nk}\right|_D^2+\frac{E}{1-\nu^2}\left|\varepsilon_{yy}^{nk}\right|_D^2+\right.$$
$$+2\nu\left(\varepsilon_{xx}^{nk},\varepsilon_{yy}^{nk}\right)_D+\frac{2E}{1+\nu}\left|\varepsilon_{xy}^{nk}\right|_D^2+\left|m_{xx}^{nk}\right|_D^2+\left|m_{yy}^{nk}\right|_D^2+\left|m_{xy}^{nk}\right|_D^2+\left|m_{xz}^{nk}\right|_D^2+$$
$$\left.+\left|m_{yz}^{nk}\right|_D^2+\frac{C_0}{T_0}\left|T^{kn}\right|_D^2\right\}+\rho\varepsilon\int_{t_0}^{t_1}\left|\frac{\partial w^{kn}}{\partial t}\right|_D^2 d\tau+\lambda\int_{t_0}^{t_1}\left(\left|\frac{\partial T^{kn}}{\partial x}\right|_D^2+\right.$$
$$\left.+\left|\frac{\partial T^{kn}}{\partial y}\right|_D^2+\left|\frac{\partial T^{kn}}{\partial z}\right|_D^2\right)d\tau+\int_{t_0}^{t_1}\left(\mp K\psi\left(w^{1n}-w^{2n}-\delta\right),\frac{\partial w^{kn}}{\partial t}\right)_D d\tau+$$
$$+\frac{E\alpha_t}{1-\nu}\int_{t_0}^{t_1}\left[\left(\frac{\partial\varepsilon_{xx}^k}{\partial t},T^{nk}\right)_D+\left(\frac{\partial\varepsilon_{yy}^k}{\partial t},T^{nk}\right)_D\right]d\tau-\frac{E\alpha_t}{1-\nu}\int_{t_0}^{t_1}\left[\left(T^{nk},\frac{\partial\varepsilon_{xx}^k}{\partial t}\right)_D+\right.$$
$$\left.+\left(T^{nk},\frac{\partial\varepsilon_{yy}^k}{\partial t}\right)_D\right]d\tau=\frac{1}{2}\left\{\rho\left|\frac{\partial u^{kn}(t_0)}{\partial t}\right|_D^2+\rho\left|\frac{\partial v^{kn}(t_0)}{\partial t}\right|_D^2+\rho\left|\frac{\partial w^{kn}(t_0)}{\partial t}\right|_D^2+\right.$$
$$+\frac{E}{1-\nu^2}\left|\varepsilon_{xx}^{nk}(t_0)\right|_D^2+\frac{E}{1-\nu^2}\left|\varepsilon_{yy}^{nk}(t_0)\right|_D^2+2\nu\left(\varepsilon_{xx}^{nk}(t_0),\varepsilon_{yy}^{nk}(t_0)\right)_D+$$
$$+\frac{2E}{1+\nu}\left|\varepsilon_{xy}^{nk}(t_0)\right|_D^2+\left|m_{xx}^{nk}(t_0)\right|_D^2+\left|m_{yy}^{nk}(t_0)\right|_D^2+\left|m_{xy}^{nk}(t_0)\right|_D^2+\left|m_{xz}^{nk}(t_0)\right|_D^2+$$
$$\left.+\left|m_{yz}^{nk}(t_0)\right|_D^2+\frac{C_0}{T_0}\left|\phi_t^{nk}\right|_D^2\right\}+\int_{t_0}^{t_1}\left(g_t,T^{nk}\right)_D d\tau+\int_{t_0}^{t_1}\left(q,\frac{\partial w^{kn}}{\partial t}\right)_\Omega d\tau. \quad (12)$$

By virtue of (10), the sequences $\{\phi_{wk}^n\}$, $\{\psi_{wk}^n\}$, $\{\phi_{uk}^n\}$, $\{\psi_{uk}^n\}$, $\{\phi_{vk}^n\}$, $\{\psi_{vk}^n\}$, $\{\phi_{Tk}^n\}$ are convergent in the norms of the indicated spaces and, therefore, bounded with respect to such norms. Taking into account this fact, theorem conditions and inequality (13), expression (12) takes the form (14).

$$|ab|\le\frac{1}{2}a^2+\frac{1}{2}b^2\quad\forall a,b\in R \quad (13)$$

$$\frac{1}{2}\left\{\rho\left|\frac{\partial u^{kn}}{\partial t}\right|_D^2+\rho\left|\frac{\partial v^{kn}}{\partial t}\right|_D^2+\rho\left|\frac{\partial w^{kn}}{\partial t}\right|_D^2+\frac{E}{1-\nu}\left|\varepsilon_{xx}^{nk}\right|_D^2+\frac{E}{1+\nu}\left|\varepsilon_{yy}^{nk}\right|_D^2+\right.$$
$$+\frac{2E}{1+\nu}\left|\varepsilon_{xy}^{nk}\right|_D^2+\left|m_{xx}^{nk}\right|_D^2+\left|m_{yy}^{nk}\right|_D^2+\left|m_{xy}^{nk}\right|_D^2+\left|m_{xz}^{nk}\right|_D^2+\left|m_{yz}^{nk}\right|_D^2+$$
$$\left.+\frac{C_0}{T_0}\left|T^{kn}\right|_D^2+\left|\mp K\psi w^{kn}\right|_D^2\right\}+\rho\varepsilon\int_{t_0}^{t_1}\left|\frac{\partial w^{kn}}{\partial t}\right|_D^2 d\tau+$$
$$+\frac{\lambda}{T_0}\int_{t_0}^{t_1}\left(\left|\frac{\partial T^{kn}}{\partial x}\right|_D^2+\left|\frac{\partial T^{kn}}{\partial y}\right|_D^2+\left|\frac{\partial T^{kn}}{\partial z}\right|_D^2\right)d\tau\le C(t_1)+\int_{t_0}^{t_1}\left(q,\frac{\partial w^{kn}}{\partial t}\right)_\Omega d\tau+$$
$$+\int_{t_0}^{t_1}\left(g_t,T^{nk}\right)_D d\tau+\int_{t_0}^{t_1}\left(\mp K\psi\left(w^{jn}+\delta\right),\frac{\partial w^{kn}}{\partial t}\right)_D d\tau, \tag{14}$$

where $j=1$ for $k=2$ and $j=2$ for $k=1$.

$\int_{t_0}^{t_1}\left(q^k,\frac{\partial w^{kn}}{\partial t}\right)_\Omega dt\le\frac{1}{2}\int_{t_0}^{t_1}\left|q^k\right|_\Omega^2 d\tau+\frac{1}{2}\int_{t_0}^{t_1}\left|\frac{\partial w^{kn}}{\partial t}\right|_\Omega^2 dt\le C(t_1)$, here $C(t_1)$— positive constant depending only on the length of the segment $[t_0,t_1]$.

$$\int_{t_0}^{t_1}\left(g_t,T^{nk}\right)_D dt\le\int_{t_0}^{t_1}h\left|g_t\right|_\Omega^2 dt+\int_{t_0}^{t_1}h\left|T^{nk}\right|_\Omega^2 dt\le C(t_1)$$

$$\int_{t_0}^{t_1}\left(\mp K\psi\left(w^{jn}+\delta\right),\frac{\partial w^{kn}}{\partial t}\right)_D=\int_{t_0}^{t_1}\int_D\mp K\psi\left(w^{jn}+\delta\right)\frac{\partial w^{kn}}{\partial t}dDdt\le$$
$$\le\frac{1}{2}\int_{t_0}^{t_1}\left|\mp K\psi h\left(w^{jn}+\delta\right)\right|_\Omega^2+\frac{1}{2}\int_{t_0}^{t_1}\left|h\frac{\partial w^{kn}}{\partial t}\right|_\Omega^2\le C(t_1)$$

Thus, the availability of a priori estimates

$$\left|\frac{\partial w^{nk}}{\partial t}\right|_\Omega^2\le C,\quad\left|\frac{\partial u^{nk}}{\partial t}\right|_\Omega^2\le C,\quad\left|\frac{\partial v^{nk}}{\partial t}\right|_\Omega^2\le C,$$
$$\left|\frac{\partial^2 w^{nk}}{\partial x^2}\right|_\Omega^2\le C,\quad\left|\frac{\partial^2 u^{nk}}{\partial x^2}\right|_\Omega^2\le C,\quad\left|\frac{\partial^2 v^{nk}}{\partial x^2}\right|_\Omega^2\le C,$$
$$\left|\frac{\partial^2 w^{nk}}{\partial y^2}\right|_\Omega^2\le C,\quad\left|\frac{\partial^2 u^{nk}}{\partial y^2}\right|_\Omega^2\le C,\quad\left|\frac{\partial^2 v^{nk}}{\partial y^2}\right|_\Omega^2\le C,$$
$$\left|\frac{\partial^2 w^{nk}}{\partial x\partial y}\right|_\Omega^2\le C,\quad\left|\frac{\partial^2 u^{nk}}{\partial x\partial y}\right|_\Omega^2\le C,\quad\left|\frac{\partial^2 v^{nk}}{\partial x\partial y}\right|_\Omega^2\le C,$$
$$\int_{t_0}^{t_1}\left|\frac{\partial T^{nk}}{\partial x}\right|_D^2 dt+\int_{t_0}^{t_1}\left|\frac{\partial T^{nk}}{\partial y}\right|_D^2 dt+\int_{t_0}^{t_1}\left|\frac{\partial T^{nk}}{\partial z}\right|_D^2 dt,\quad\left|T^{nk}\right|_D^2\le C\tau\le C.$$

The priori estimates presence given by the methods proposed in [27] allows us to extend the solution of the Cauchy problem (9)–(10) to the entire interval $[t_0, t_1]$. And also to conclude that the sets of approximate solutions to problem $\{w^{nk}, u^{nk}, v^{nk}, T^{nk}\}$ (8) obtained by the Bubnov–Galerkin method are weakly compact in the spaces from conditions (7). The sets $\{w^{nk}, u^{nk}, v^{nk}, T^{nk}\}$ satisfy the conditions:

$$\begin{gathered}
\text{The sets} \quad \left\{w^{nk}\right\}, \left\{u^{nk}\right\}, \left\{v^{nk}\right\} \quad \text{are limited in} \quad L^2\left(t_0, t_1, H_0^2(\Omega)\right); \\
\text{The sets} \quad \left\{\frac{\partial w^{nk}}{\partial t}\right\}, \left\{\frac{\partial u^{nk}}{\partial t}\right\}, \left\{\frac{\partial v^{nk}}{\partial t}\right\} \quad \text{are limited in} \quad L^2\left(t_0, t_1, H_0^1(\Omega)\right); \\
\text{The set} \quad \left\{T^{nk}\right\} \quad \text{is limited in} \quad L^2\left(t_0, t_1, H_0^1(D)\right).
\end{gathered} \tag{15}$$

3.3 *Transition to Limit*

All spaces from conditions (15) are Hilbert spaces and, therefore, all bounded sets from (15) are weakly compact in the corresponding spaces [28]. Thus, from the sequences $\{w^{nk}, u^{nk}, v^{nk}, T^{nk}\}$ we can distinguish weakly converging subsequences such that:

$$\begin{gathered}
\left\{w^{nk}\right\} \to \tilde{w}^{nk} \quad \text{– weakly in} \quad L^2\left(t_0, t_1, H_0^2(\Omega)\right); \\
\left\{\frac{\partial w^{nk}}{\partial t}\right\} \to \frac{\partial \tilde{w}^{nk}}{\partial t} \quad \text{– weakly in} \quad L^2\left(t_0, t_1, H_0^1(\Omega)\right); \\
\left\{u^{nk}\right\} \to \tilde{u}^{nk} \quad \text{– weakly in} \quad L^2\left(t_0, t_1, H_0^2(\Omega)\right); \\
\left\{\frac{\partial u^{nk}}{\partial t}\right\} \to \frac{\partial \tilde{u}^{nk}}{\partial t} \quad \text{– weakly in} \quad L^2\left(t_0, t_1, H_0^1(\Omega)\right); \\
\left\{v^{nk}\right\} \to \tilde{v}^{nk} \quad \text{– weakly in} \quad L^2\left(t_0, t_1, H_0^2(\Omega)\right); \\
\left\{\frac{\partial v^{nk}}{\partial t}\right\} \to \frac{\partial \tilde{v}^{nk}}{\partial t} \quad \text{– weakly in} \quad L^2\left(t_0, t_1, H_0^1(\Omega)\right); \\
\left\{T^{nk}\right\} \to \tilde{T}^{nk} \quad \text{– weakly in} \quad L^2\left(t_0, t_1, H_0^1(D)\right).
\end{gathered} \tag{16}$$

In this case, the limiting properties of the generalized derivatives are taken into account [29]. In [17] a well-known proving method that functions $\left\{\tilde{w}^{nk}, \tilde{u}^{nk}, \tilde{v}^{nk}, \tilde{T}^{nk}\right\}$ are a generalized solution to problem (1)–(4) is given.

To conclude the theorem proof, note that the configuration space of the mechanical system (in the considered shell form) defined by the boundary value problem (1)–(4) with generalized solutions from spaces (7) is the infinite-dimensional functional space V of the form $\left(L^2(\Omega)\right)^6 \times \left(H_0^2(\Omega)\right)^6 \times (H_0(D))^2$ (from point 3 of the theorem), since elements $\left\{\frac{\partial \tilde{w}^k}{\partial t}, \frac{\partial \tilde{u}^k}{\partial t}, \frac{\partial \tilde{v}^k}{\partial t}, \tilde{w}^k, \tilde{u}^k, \tilde{v}^k, \tilde{T}^k\right\}$ from this space specify the mechanical system position with a mathematical model in the form of a boundary value problem (1)–(4).

Acknowledgments Work was supported by the Russian Science Foundation project 19-19-00215.

References

1. Thai, H.-T., Vo, T.P., Nguyen, T.-K., Kim, S.-E.: A review of continuum mechanics models for size-dependent analysis of beams and plates. Composite Structures. **177**, 196–21 (2017)
2. Eremeev, V.I., Zubov, L.M.: Mechanics of elastic shells. Nauka, Moscow, (2008)
3. Neff, P.: A geometrically exact planar Cosserat shell-model with microstructure: existence of minimizers for zero Cosserat couple modulus. Math. Models Methods Appl. Sci. **17**, Is. 3, 363–392 (2007)
4. Birsan, M.: On Saint-Venant's principle in the theory of Cosserat elastic shells. Int. J. Eng. Sci. **45**, Is. 2–8, 187–198 (2007)
5. Sarkisyan, S.O., Alvadgyan, S.I.: Models of static deformation of anisotropic micropolar elastic thin beam and singularities of their strength-hardness characteristics. Problems of Atomic Science and Technology **4**, 196–204 (2011)
6. Krylova, E. Yu., Papkova, I. V., Yakovleva, T. V., Krysko, V. A.: Theory of vibrations of carbon nanotubes like flexible micropolar mesh cylindrical shells taking into account shift. Izv.Saratov Univ. (N. S.), Ser. Math. Mech. Inform. **19**, iss. 3, 305–316 (2019)
7. Altenbach, J., Altenbach, H., Eremeyev, V.: On generalized Cosserat-type theories of plates and shells: a short review and bibliography. Archive of Applied Mechanics. **80 (1)**, 73–92 (2010)
8. Sargsyan, S.H., Zhamakochyan, K.A.: Applied theory of micropolar elastic thin plates with constrained rotation and the finite element method. Materials Physics and Mechanics. **35(1)**, 145–154 (2018)
9. Sargsyan, S.H. General theory of micropolar elastic thin shells. Physical Mesomechanics. **15(1–2)**, 69–79 (2012)
10. Altenbach, H., Eremeyev, V.A.: Cosserat-type shells CISM international centre for mechanical sciences. Courses and Lectures. **541**, 131–178 (2013)
11. Sargsyan, S. H., Sargsyan, A. H.: General dynamic theory of micropolar elastic thin plates with free rotation and special features of their natural oscillations. Acoustical Physics. **57(4)**, 473–481 (2011)
12. Ansari, R., Shakouri, A.H., Bazdid-Vahdati, M., Norouzzadeh, A., Rouhi, H.: A nonclassical finite element approach for the nonlinear analysis of micropolar plates. J. Comput. Nonlinear Dynam. **12(1)**, 011019 (2017)
13. Gharahi, A., Schiavone, P.: Uniqueness of solution for plane deformations of a micropolar elastic solid with surface effects. Continuum Mechanics and Thermodynamics. 1–14 (2019)
14. Sargsyan, A.H., Sargsyan, S.H.: Dynamic model of micropolar elastic thin plates with independent fields of displacements and rotations. Journal of Sound and Vibration. **333(18)**, 4354–437 (2014)

15. Fazelzadeh,S.A., Rahmani, S., Ghavanloo, E., Marzocca, P.: Thermoelastic vibration of doubly-curved nano-composite shells reinforced by graphene nanoplatelets. Journal of Thermal Stresses. **42(1)**, 1–17 (2019)
16. Rahimi, Z., Sumelka, W., Ahmadi, S. R., Baleanu, D. Study and control of thermoelastic damping of in-plane vibration of the functionally graded nano-plate. Journal of Vibration and Control. (2019) doi: 10.1177/1077546319861009
17. Vorovich, I.I., Lebedev, L.P.: On the solvability of non-linear shallow shell equilibrium problems. Journal of Applied Mathematics and Mechanics. **52 (5)**, 636–641 (1988)
18. Khludnev, A.M.: Problem on the contact of two elastic plates. Journal of Applied Mathematics and Mechanics. textbf47 (1), 110–115 (1983)
19. Awrejcewicz, J., Krysko, V.A., Sopenko, A.A., Kirichenko, A.V., Krysko, A.V.: Mathematical modelling of physically/geometrically non-linear micro-shells with account of coupling of temperature and deformation fields. Chaos, Solitons and Fractals. **104**, 635–654 (2017)
20. Kirichenko, V.F., Kirichenko, A.V., Krysko, V. A, Awrejcewicz, J., Krysko, A.V.: On the non-classical mathematical models of coupled problems of thermo-elasticity for multi-layer shallow shells with initial imperfections. Internationsl Jornal of non-liniar mechanics. **74**, 51–72 (2015)
21. Yang, F., Chong,A. C. M., Lam, D. C. C., Tong, P.: Couple stress based strain gradient theory for elasticity. Int. J. Solids Struct. **39**, 2731–2743 (2002)
22. Bolotin, V.V., Novichkov, YU.N.: Mekhanika mnogoslojnyh konstrukcij. Mashinostroenie, Moscow, (1980)
23. Vasidzu, K.: Variacionnye metody v teorii uprugosti i plastichnosti. Mir, Moscow, (1987)
24. Lyusternik, L. A., Sobolev, V. I.: Elementy funkcional'nogo analiza. Nauka, Moscow, (1965)
25. Ladyzhenskaya, O.A.: Kraevye zadachi matematicheskoj fiziki. Nauka, Moscow, (1973)
26. Mihajlov, V.P.: Differencial'nye uravneniya v chastnyh proizvodnyh. Nauka, Moscow, (1983)
27. Rektoris, K.: Variacionnye metody v matematicheskoj fizike i tekhnike. Mir, Moscow, (1985)
28. Trenogin, V.A.: Funkcional'nyj analiz. Nauka, Moscow, (1980)
29. Mihlin, S.G.: Linejnye uravneniya v chastnyh proizvodnyh. Vysshaya shkola, Moscow, (1977)

Component-Based Software Model for Numerical Simulation of Constrained Oscillations of Liquid Drops and Layers

Igor Kuzmin and Leonid Tonkov

Abstract The study of microhydrodynamic processes have not only practical significance, but also have a wide field for theoretical approaches and numerical investigation. The article deals with a numerical investigation of constrained oscillation of a liquid drop on a substrate, which harmonically oscillates, and oscillation of the liquid layer located on the surface of a bending plate. Forced vibrations of the cantilevered plate are excited by the piezoelectric element. The mathematical model is based on a system of Navier–Stokes equations for an immiscible incompressible two-phase mixture. The problem of numerical simulation of the interaction between a deformed solid and a fluid layer is a Fluid-Structure Interaction problem and requires a solution of both the elastodynamic and the hydrodynamics equations. The partitioned approach to solving fluid-interaction problems is one of the most common. Its allows solving each of the physical problems independently, using specific numerical schemes and a proprietary parallelism model. The elastodynamic problem taking into account geometric and physical nonlinearity is solved by the finite element method. The proposed mathematical models allow us to study the dynamics of the free surface of small liquid volumes and the processes of redistribution of a liquid layer on a flexible vibrating base.

1 Introduction

Understanding multiphase flow at low Weber numbers is of considerable importance in a variety of environmental, industrial, and engineering applications such as atomization of the fuel, contaminant cleanup, fluid absorption, and separation in porous media and many others. However, accurate numerical simulation of such flows is a tricky computational problem when interfacial tension effects become dominant.

I. Kuzmin · L. Tonkov (✉)
Udmurt State University, Izhevsk, Russia
e-mail: letonkov@mail.ru

I. B. Badriev et al. (eds.), *Mesh Methods for Boundary-Value Problems and Applications*, Lecture Notes in Computational Science and Engineering 141,
https://doi.org/10.1007/978-3-030-87809-2_20

Mesh-based numerical methods are conventionally considered, as the preferred approach for most applications, however, is the need for an algorithm to determine the shape of interface boundary and its evolution with time.

One of the widespread approaches to solve the investigating problem is representing a bulk as an immiscible incompressible two phase mixture described by Navier–Stokes equations with the dynamic equilibrium condition at the interface and subsequent application algorithm, that represents the interface implicitly by marking the fluids on both sides of the interface, using a scalar indicator function such as a volume fraction (Volume-Of-Fluid method) [1].

The main advantage of this approach is that it does not require complicated interface tracking algorithms. This is important for modeling two-phase flows through complex geometries with large interface motions and interactions. The surface tension force and the contact angle effect arise from the calculation of interface normal vector $n_s = \nabla\alpha/\|\nabla\alpha\|$ and curvature $K = \nabla \cdot n_s$.

The prediction of a liquid droplet natural frequencies and free surface shapes under constrained oscillations are extensively studied by analytical [2], numerical and experimental [3] methods. Consider these problems as the convenient testing tool of verification and validation numerical methods and algorithms for capillary simulation of the flows with a free surface.

It is of interest to investigate the interaction of the liquid with elastic bodies when bending vibrations are caused. Usually, for investigating the instability in liquid drops or layers, rigid substrate is used, which vibrates with the same amplitude along the entire contact area. The vibrations of bodies such as beams are bending vibrations with distributed amplitude. At high frequencies of the bending vibrations of beams, the length of the bending waves in them is comparable to the sizes of the region of the contact with a liquid layer and distributed vibrations can appreciably influence the liquid behavior. In our previous studies [4, 5], we investigated the interaction of a thin plate that performs bending vibrations and liquids at the interphase boundary.

2 Mathematical Model and Numerical Method

2.1 Liquid Dynamic Equations

The equations of motion for an isothermal, immiscible incompressible two-phase mixture flow of Newtonian fluids can be written using a single-fluid continuum approach as follows:

$$\frac{\partial \rho \mathbf{v}}{\partial t} + \nabla \cdot (\rho \mathbf{v}\mathbf{v}) = -\nabla p + \nabla \cdot \boldsymbol{\tau} + \mathbf{f}_{\mathrm{sv}},$$
$$\nabla \mathbf{v} = 0, \tag{1}$$

where $\mathbf{v}$ is velocity vector, total pressure p is the sum of dynamic and hydrostatic pressure, $\boldsymbol{\tau} = \mu(\nabla \mathbf{v} + \nabla \mathbf{v}^T)$ is viscous stress tensor, $\mathbf{f}_{sv}$ is surface tension force per unit volume. The density and viscosity are defined by

$$\rho = \alpha \rho_l + (1 - \alpha)\rho_g, \quad \rho = \alpha \mu_l + (1 - \alpha)\mu_g, \tag{2}$$

where subscripts «l» and «g» denotes liquid ($\alpha = 1$) and gas ($\alpha = 0$) phase respectively. The scalar indicator function α is evolved with an advection equation of the conservative form:

$$\frac{\partial \alpha}{\partial t} + \nabla \cdot (\alpha \mathbf{v}) = 0. \tag{3}$$

Volume-Of-Fluid method (VOF), defined by Eqs. (1)–(3) is mass conservative, computationally efficient and flexible for treating complex interface shapes. Therefore, the VOF-method is a popular and powerful tool for the direct numerical simulation of immiscible two-phase flow.

2.2 Advection of Indicator Function

By its definition, the indicator function has the form of a step function in the continuum limit, while numerical approximation of convective terms in Eqs. (1), and (3) leads to smear function jump. Let us distinguish among the other two general approaches to deal with this problem. One of them is using a low-dissipative scheme with Van-Leer limiter for the approximation of convective terms, the other is an introduction of artificial compression term.

The last approach leads to the following form of advection equation (3):

$$\frac{\partial \alpha}{\partial t} + \nabla \cdot (\alpha \mathbf{v}) + \nabla \cdot (\alpha(1 - \alpha)\mathbf{v}_r) = 0, \tag{4}$$

where $\mathbf{v}_r$ is a compression velocity, the value of which is based on the scaled maximum fluid velocity magnitude in the transition region.

2.3 Calculation of Interface Curvature and Normal Vector

The solution of Eq. (4) updates the indicator function in such a way that the interface remains as sharp as possible. The reverse side of this is cumulative errors of capillary forces due to the unstable calculation of the normal vector and interface curvature.

For more accurate and stable calculation of the normal in cells near the interface, we first use smoothing of the indicator function procedure. This is numerically done using the following relationship:

$$\alpha_s^{i+1} = C_{\mathrm{fc}} \left\langle \langle \alpha_s^i \rangle_{\mathrm{f}} \right\rangle_{\mathrm{c}} + (1 - C_{\mathrm{fc}})\alpha_s^i, \quad \alpha_s^0 = \alpha, \quad i = 0, 1, \ldots, N, \tag{5}$$

where the first operator $\langle\cdot\rangle_{\mathrm{f}}$ means that the field values interpolated from the cell centers to the face centers and the second operator $\langle\cdot\rangle_{\mathrm{c}}$ means that the field values at cell centers calculated by averaging values at face centers. A value of $C_{\mathrm{fc}} = 0.5$ and $N = 2$ is used in present simulations.

The smoothed indicator function α_s is then used to obtain the interface normal vectors $\mathbf{n}_s = \nabla\alpha_s/|\nabla\alpha_s|$ at cell centers. The next step is to calculate interface curvature $K = \nabla\cdot\mathbf{n}_s$ in accordance with the control volume method, the divergence of the vector function is calculated as follows:

$$\nabla\cdot\mathbf{n}_s = \frac{1}{V_i}\sum_{f\in S_i}\left[\frac{\nabla\alpha_s}{|\nabla\alpha_s|}\right]_f\cdot\mathbf{S}_f,$$

where for each grid block i, V_i is its volume, S_i is set of its faces, $\mathbf{S}_f$ is the outward vector area of the face.

Direct calculation of gradient $\nabla\alpha_s$ with subsequent normalization leads to nonzero vectors $\mathbf{n}_s$ outside the transition region. To deal with this problem, an extra filtration procedure is used for dummy face flux $\psi = \nabla\alpha_s\cdot\mathbf{S}$. This filtering will explicitly set the dummy fluxes ψ to zero when their magnitude is of the order of the numerical errors. The filtered flux reads:

$$\bar{\psi} = \psi - \max(\min(\psi, \psi_*), -\psi_*), \tag{6}$$

where ψ_* is a threshold value below which flux $\bar{\psi}$ is set to zero. The threshold value is chosen as $\phi_* = C_\phi|\mathbf{S}_f|\overline{|\nabla\alpha_s|_f}$, where $\overline{|\nabla\alpha_s|_f}$ is the average gradient magnitude over all faces where they are non-zero. The filtering coefficient should be chosen sufficiently small. In our simulations, we use $C_\psi = 0.01 \div 0.03$.

Once the interface curvature is computed, we smooth the calculated value in the direction normal to the interface, similar to that suggested in [6].

2.4 Equation of Motion for Elastodynamics of the Plate

The equations of motion of elastodynamic problem in the Lagrangian formulation, in the general case, take the form:

$$\begin{gathered}\rho_s\frac{\partial^2\mathbf{u}}{\partial t^2} = \nabla\cdot(\mathbf{F}\cdot\mathbf{S}) + \rho_s\mathbf{f}\\ \mathbf{u}(t_0) = \mathbf{u}^0, \quad \dot{\mathbf{u}}(t_0) = \dot{\mathbf{u}}^0,\\ \mathbf{n}_s\cdot(\mathbf{F}\cdot\mathbf{S}) = p,\end{gathered} \tag{7}$$

where **u**—displacement vector, ρ_s, ∇—density and divergence operator in the reference configuration, f, p—vector of mass forces and pressure, **F**—deformation gradient, $\mathbf{S} = \mathbf{F}^{-1} \cdot \sigma \cdot \mathbf{F}^{-T} \det \mathbf{F}$—symmetrical stress tensor of Piola–Kirchhoff, σ—Cauchy stress tensor.

The elastodynamic problem taking into account geometric and physical nonlinearity is solved by the finite element method. The integration of elastodynamic problem equations is performed by the explicit scheme takes into account the dissipative properties of the system [7].

The taking into account the influence of the liquid mass distribution on the plate's vibrations is based on the weak coupling algorithm. In this case, the coupling of solutions between the two problems is performed at the interface boundary between fluid and structure. The considered mathematical model makes it possible to reproduce the characteristic features of the liquid layer distribution on the plate surface.

3 Programming Model

The distributed programming model is based on the ZeorC Ice [10] middleware, within the client-server model. The FEStudioFSI client application is connected with other applications and implements the logic of the entire program.

In the case of a strong coupling approach, the client are synchronized data transfer, performed parameter adjustments, and checked the convergence of the iterative process of solution coupling on the interface boundary.

In the considered model, servers are applications that solved individual physical problems. The client stores information about the proxy objects of the servers (Fig. 1), each of which, being an Ice-object, provides a unique external name, and hides all low-level details of the process of data exchange with the corresponding server.

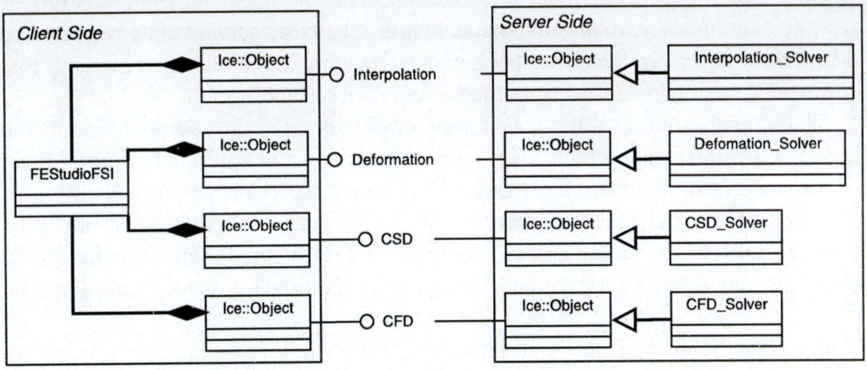

Fig. 1 The distributed model for coupling independent applications

Instances of the CFD_Solver server responsible for solving the fluid dynamics problem are implemented using the OpenFOAM library. The solution to the elastodynamics problem is carried out by the CSD_Solver server, implemented within the FEStudio [11] package.

The Interpolation_Solver server is responsible for interpolating data and transferring it between servers that provide a solution to physical problems. The Defomation_Solver server performs the mesh deformation required to solve the fluid dynamics problem.

The described mechanism for building an application on the Ice infrastructure allows us, by replacing objects, to obtain a distributed object-oriented program that allows us to solve a specific FSI problem. This approach provides the greatest flexibility and allows you to combine independent applications that only need to use the appropriate API.

ZeroC Ice is object-oriented application software. It provides the means for developing object-oriented distributed applications. Clients are active entities that request certain services from the server. Servers are passive entities that provide services in response to client requests. The Ice programming model is based on the concept of an Ice-object. It is an abstraction that can respond to client requests, run on a single or multiple servers.

Each Ice-object has a unique identifier and a set of interfaces—facets. To call an Ice-object, the client needs to use a proxy. Proxy is a client-side local address space agent of an Ice-object. The proxy code for a specific programming language is generated by the Slice compiler, which is a standalone tool of Ice-workflow. A proxy encapsulates the information required to invoke an Ice object: the server's physical address, object ID, and optionally a facet ID.

Remote call of methods that are implemented on the server-side is done through the generated proxies. Besides, the client provides the consistency of information common to the interacting servers, for example, the displacement vector and the pressure at the interface boundary.

On the server-side, the behavior of Ice objects is implemented using servants. A servant is an instance of an implemented class. The basic servant code is generated by the Slice compiler, the developer is required to implement the class methods that correspond to the operations from the Ice-object interface.

When a call comes in, the Ice runtime environment at the server side finds the servant corresponding to the callable object and delegates call handling to it. Each of the servers implements the methods necessary for the distributed solution of the FSI problem: calculating displacements (CSD_Server servant), determining the pressure field (CFD_Server servant), transferring the obtained solutions to the client, obtaining new data on displacement and pressure required for the next solution step, etc.

Figure 2 shows the distributed model of the FEStudioFSI application. It is a set of several Ice-objects corresponding to OpenFOAM servers, FEStudio servers, and FEStudioFSI clients.

The Ice environment is supported both synchronous and asynchronous calling models. In the latter case, the client, calling the object using a proxy, along with the

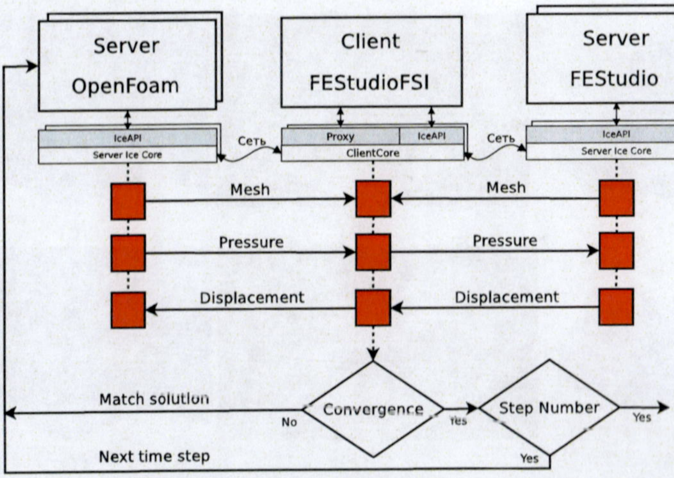

Fig. 2 Distributed model for FSI problem

usual parameters of the operation, sends the proxy a special callback object. After calling the proxy, the control is immediately returned to the client. When the call to the object completes, the client-side Ice runtime invokes the callback method, passing it the results of the call or exception.

4 Results and Discussion

First, we consider a three-dimensional droplet of volume 87 μl positioned on a cylindrical substrate with radius $R = 4$mm that oscillates along vertical axe O_z due to harmonic force, produced by a piezoelectric transducer. The feature of the process is the droplet pinning on the substrate with a cone cavity with cone-angle $\beta = 140°$. In this case, we carried out both an experimental study and numerical simulation. The experiments were conducted with the use of a facility a detailed description of which is presented in [4]. In the experiment, zonal mode (4.0) (Fig. 3 b, e) and tesseral mode (3,1) (Fig. 4 b, e) was obtained in the excitation frequency range from 38 to 45 Hz. In the numerical experiments, the value of the substrate oscillation frequency was 40 Hz.

Computational block-structured grid was generated by rotating a 2-D flat grid around the axe of symmetry to become a three-dimensional grid containing 1,752,500 hexagonal cells. It should be noted that in the numerical experiment it is necessary to initially introduce small asymmetry in the forcing vibrations of the substrate to achieve the non-axisymmetric (tesseral) mode of the drop oscillations. Both the experimental and numerical drops experienced similar free surface shape

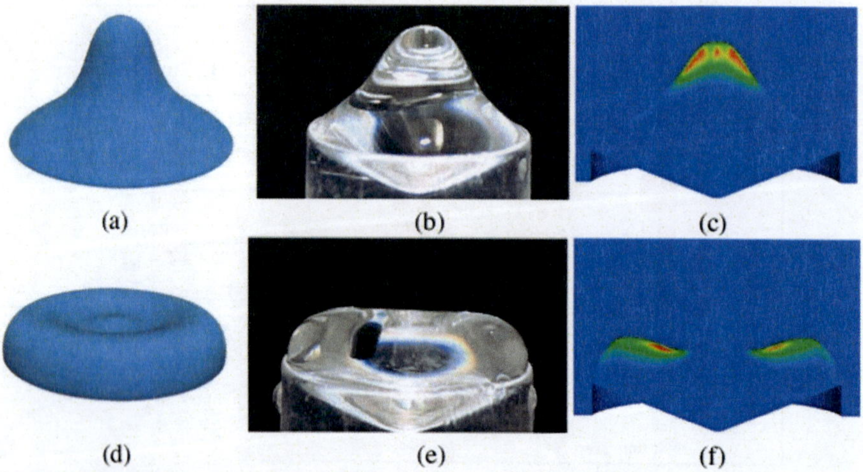

Fig. 3 Zonal oscillation mode (4,0). Calculated (**a**), (**d**) and observed (**b**), (**e**) free surface shape of the drop jointly with magnitude of the Umov–Poynting vector field (**c**), (**e**); when the phase of the oscillation (**a**)–(**c**) $\phi = 0$, (**d**)–(**f**) $\phi = \pi$

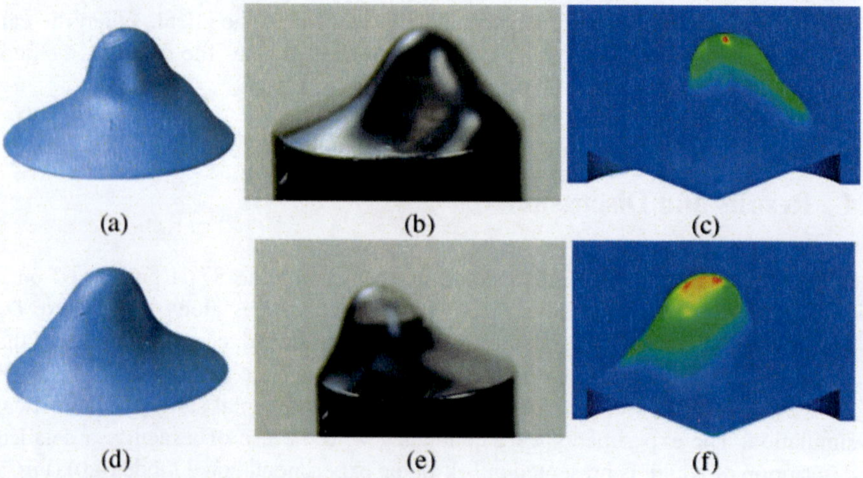

Fig. 4 Tesseral oscillation mode (3,1). Calculated (**a**), (**d**) and observed (**b**), (**e**) free surface shape of the drop jointly with magnitude of the Umov–Poynting vector field (**c**), (**e**); when the phase of the oscillation (**a**)–(**c**) $\phi = 0$, (d)–(f) $\phi = \pi$

(Figs. 3 and 4) and close values of maximum and minimum drop heights. Drop height was measured from the top cross-section of the substrate.

For a more thorough analysis of the numerical solution, the Umov–Poynting vector field was constructed. The Umov–Poynting vector $\mathbf{v}(p + \rho\mathbf{v}\mathbf{v}/2)$ describes total energy flux in liquid. Figures 3c, f and 4c, f shows the magnitude of the energy flux in the corresponding phase of the oscillation. One can see that, for both zonal

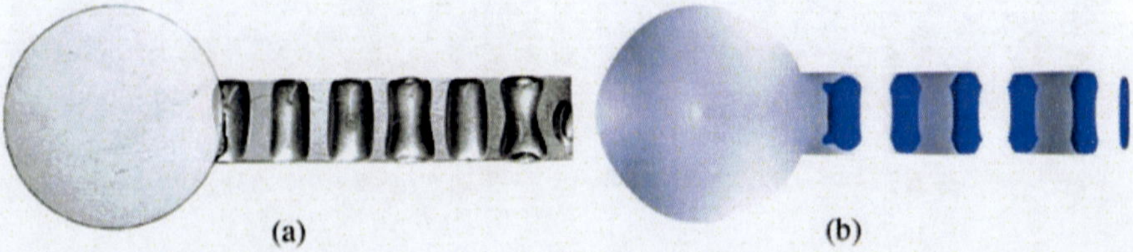

Fig. 5 The droplet of vacuum oil on the vibrating plate in the experiment described in [9] (**a**) and calculated results (**b**)

and tesseral modes, the most intense energy flow occurs at the top part of the drop near the interface surface.

Despite the pinning of the drop, the low-frequency eigenforms obtained in the experiment and reproduced by the numerical simulation are close to those shown in [8]. The developed numerical scheme allows to obtain a detailed structure of microflows in an oscillating drop and contribution of different mechanisms to the transition from one mode to another.

The taking into account the influence of the liquid mass distribution on the plate's vibrations is based on the weak coupling algorithm. In this case, the coupling of solutions between the two problems is performed at the interface boundary between fluid and structure. The considered algorithms for implicit coupling were used to numerical simulation of the physical experimental investigation of the interaction of the vibrating console plate with a layer of viscous liquid deposited on its surface [4]. Forced vibrations of a plate with a frequency of 4.5 kHz are excited by a piezoelectric element, with a cantilevered plate.

Figure 5a shows the result of the experiment [9] performed for the vacuum oil with and Fig. 5b shows the result of numerical simulation. At the excitation of vibrations, viscous liquids applied as a thin layer on the plate surface initially flow to the plate surface areas with the antinodes of vibrations taking a convex form.

The coupled solution of the problems is carried out on hexahedral non-matching meshes with a size of 1,300,000 cells for the fluid dynamics problem and 23,000 cells for the elastodynamic problem. The point-concentrated force is applied at the center of the piezoelectric element. It is important to note that the vibrations of a thin plate in the form of the superposition of longitudinal (see Fig. 6)) and transverse waves allow obtaining stable droplet patterns (see Fig. 6, $t = 0.16$) which cannot be formed on a rigid substrate.

Compared with the experiment in numerical simulation, the destruction of a thin liquid film between droplets formed at antinodes occurs more slowly. This is a feature of the numerical solution of the advection equation of the indicator function near the wall.

The study showed that the topological features of the distribution of the fluid are determined by the peculiarities of the bending vibrations of the plate. The comparison of the results of numerical simulation with the experimental data allows

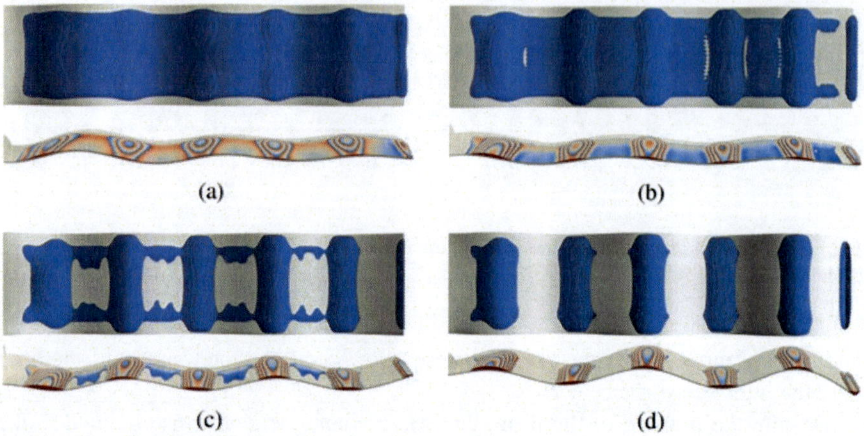

Fig. 6 The distribution of liquid over the surface of the plate and the longitudinal bending of the plate at different times in numerical simulation: **(a)**–**(d)** $t = 0.04$ s, $t = 0.08$ s, $t = 0.12$ s, $t = 0.16$ s

us to conclude that the numerical methods and algorithms used to describe the processes of interaction between the liquid layer and the vibrating plate quite accurately.

Acknowledgments The study of the first author was funded by the Udmurt State University, project no. 2020-04-16. The study of the second author was funded by the Ministry of Science and Higher Education of the Russian Federation, project no. 0827-2020-0010.

References

1. Hirt, C.W., Nichols, B.D.: Volume of fluid (VOF) method for the dynamics of free boundaries. Journal of Computational Physics. vol 39, **1**, 201–225 (1981)
2. Lubimov, D.V., Lubimova, T.P., Shklyaev, S.V. Behavior of a drop on an oscillating solid plate. Physics of Fluids. (2006) https://doi.org/10.1063/1.2137358
3. Park, C.-S., Kim, H., Lim, H.-C. Study of internal flow and evaporation characteristics inside a water droplet on a vertically vibrating hydrophobic surface. Experimental Thermal and Fluid Science. (2016) https://doi.org/10.1016/j.expthermflusci.2016.05.018
4. Aleksandrov, V.A., Kopysov, S.P., Tonkov, L.E. Vortex Flows in the Liquid Layer and Droplets on a Vibrating Flexible Plate. Microgravity Science and Technology. (2018) https://doi.org/10.1007/s12217-017-9579-0
5. Aleksandrov, V.A., Kopysov, S.P., Tonkov, L.E. Jet formation at interaction of a vibrating plate with liquid. IOP Conference Series: Materials Science and Engineering. (2017) https://doi.org/10.1088/1757-899X/208/1/012001
6. Shams, M., Raeini, A.Q., Martin, J.B., Branko, B. A numerical model of two-phase flow at the micro-scale using the volume-of-fluid method. Journal of Computational Physics. (2018) https://doi.org/10.1016/j.jcp.2017.12.027

7. Chang, S.-Y. A new family of explicit methods for linear structural dynamics. Computers & Structures. (2018) https://doi.org/10.1016/j.compstruc.2010.03.002
8. Chang, C.-T., Bostwick, J.B., Steen, P.H., Daniel, S. Substrate constraint modifies the Rayleigh spectrum of vibrating sessile drops. Physical Review E. (2013) https://doi.org/10.1103/PhysRevE.88.023015
9. Aleksandrov, V.A. Interaction of a vibrating rod and liquid at the interface. Chemical Physics and Mesoscopics. vol 15, **1**, 116–216 (2013)
10. Henning, M. A new approach to object-oriented middleware. Internet Computing, IEEE. vol 8, **1**, 66–75 (2004)
11. Kopysov, S.P., Kuzmin, I.M., Nedozhogin, N.S., Novikov, A.K., Rychkov, N.V., Sagdeeva, Y.A., Tonkov, L.E. Parallel implementation of finite element algorithms on graphics accelerators in the FEStudio software package. Computer research and modeling. vol 6, **1**, 79–97, (2014)

Modeling of Long-Term Strength of a Rod Under Creep Conditions and Finite Deformations

Evgenii B. Kuznetsov and Sergey S. Leonov

Abstract The paper investigates the process of creep and long-term strength of a long metal rod of circular cross-section, taking into account the finite deformations. Deformation of the rod describes by nonlinear transport equations, and the creep process describes by equations of the kinetic creep theory. An analytical solution to the problem is given for the case of constant stress.

1 General Model

We use the finite deformation model [1, 2] to describe the medium's motion, in which differential transport equations specify reversible and irreversible deformations. In the spatial case, the constitutive kinematic relations of the model in the Euler variables have the form:

$$d_{ij} = \frac{1}{2}\left(\frac{\partial u_i}{\partial x_j} + \frac{\partial u_j}{\partial x_i} - \frac{\partial u_k}{\partial x_i}\frac{\partial u_k}{\partial x_j}\right) =$$
$$= e_{ij} + p_{ij} - \frac{1}{2}e_{ik}e_{kj} - e_{ik}p_{kj} - p_{ik}e_{kj} + e_{ik}p_{ks}e_{sj},$$
$$\frac{de_{ij}}{dt} = \varepsilon_{ij} - \gamma_{ij} + r_{ik}e_{kj} - e_{ik}r_{kj} - \frac{1}{2}[(\varepsilon_{ik} - \gamma_{ik} + z_{ik})e_{kj} +$$
$$+ e_{ik}(\varepsilon_{kj} - \gamma_{kj} + z_{kj})],$$

E. B. Kuznetsov · S. S. Leonov (✉)
Moscow Aviation Institute, Moscow, Russia
e-mail: kuznetsov@mai.ru; powerandglory@yandex.rus

I. B. Badriev et al. (eds.), *Mesh Methods for Boundary-Value Problems and Applications*, Lecture Notes in Computational Science and Engineering 141,
https://doi.org/10.1007/978-3-030-87809-2_21

$$\frac{dp_{ij}}{dt} = \gamma_{ij} - p_{ik}r_{kj} + r_{ik}p_{kj} - p_{ik}\gamma_{kj} - \gamma_{ki}p_{kj},$$

$$\varepsilon_{ij} = \frac{1}{2}(\frac{\partial v_i}{\partial x_j} + \frac{\partial v_j}{\partial x_i}),\ w_{ij} = \frac{1}{2}(v_{i,j} - v_{j,i}),\ v_i = \frac{du_i}{dt} = \frac{\partial u_i}{\partial t} + \frac{\partial u_i}{\partial x_j}v_j, \tag{1}$$

$$r_{ij} = w_{ij} + z_{ij}(e_{ks}, \varepsilon_{ks}),$$

$$z_{ij} = A^{-1}[(\varepsilon_{im}e_{mj} - e_{im}\varepsilon_{mj})B^2 + B(\varepsilon_{im}e_{mn}e_{nj} - e_{im}e_{mn}\varepsilon_{nj}) + \\ + e_{im}\varepsilon_{mn}e_{nk}e_{kj} - e_{im}e_{mn}\varepsilon_{nk}e_{kj}],$$

$$A = 8 - 8E_1 + 3E_1^2 - E_2 - \frac{1}{3}E_1^3 + \frac{1}{3}E_3,\ B = 2 - E_1,\ E_1 = e_{kk},\\ E_2 = e_{ij}e_{ji},\ E_3 = e_{ij}e_{jk}e_{ki}.$$

All indices in relations (1) vary from 1 to 3, u_i and v_i are components of vectors of displacements and velocities of medium points, d_{ij} are components of the total deformation tensor (Almansi deformations), e_{ij} are linear components of the tensor of reversible deformations (elastic deformations), p_{ij} are components of the tensor of irreversible deformations (deformations of creep or plasticity), r_{ij} are components of rotation tensor, ε_{ij} and γ_{ij} are the components of the total and irreversible deformations rates, t is time.

As in the classical theory, the stresses in a medium are entirely determined by reversible deformations

$$\sigma_{ij} = -p_0\delta_{ij} + \frac{\partial W}{\partial e_{ij}}(\delta_{ij} - e_{ij}),$$

$$W = -2\mu I_1 - \mu I_2 + bI_1^2 + (b - \mu)I_1I_2 - \chi I_1^3 + \ldots \tag{2}$$

$$I_1 = e_{kk} - \tfrac{1}{2}e_{ks}e_{sk},\ I_2 = e_{ks}e_{sk} - e_{ks}e_{st}e_{tk} + \tfrac{1}{4}e_{ks}e_{st}e_{tn}e_{nk}.$$

In relations (2) σ_{ij} are components of the Euler–Cauchy stress tensor, p_0 is additional hydrostatic pressure, W is the elastic potential; μ is a module of shear of the investigated material; b, χ are elastic modules of higher order.

2 Tension of the Circular Cross-Section Rod

Consider a metal circular cross-section rod, the length of which is many times greater than its cross-section. An constant tensile force applied to the rod sets the stress σ_0 at the initial time moment. The rod is under creep conditions. In this one-dimensional case, the indices for the corresponding functions can be omitted. Then the first equation of system (1) takes the form

$$d = \frac{\partial u}{\partial x} - \frac{1}{2}\left(\frac{\partial u}{\partial x}\right)^2 = e - \frac{1}{2}e^2 + (1 - e)^2 p. \tag{3}$$

In relation (3), e is a linear part of the elastic strain tensor, $e - \frac{1}{2}e^2$ are the reversible components of the Almansi total strain tensor d. Introducing the notation $z = \frac{\partial u}{\partial x}$, we obtain the quadratic equation

$$z^2 - 2z + B = 0, \tag{4}$$

where $B = 2e - e^2 + 2(1-e)^2 p$.

Equation (4) has following roots

$$\left(\frac{\partial u}{\partial x}\right)_1 = 1 + \sqrt{1-B}, \quad \left(\frac{\partial u}{\partial x}\right)_2 = 1 - \sqrt{1-B}.$$

According to (1), the rate of total deformations can be calculated by the formula:

$$\varepsilon = \frac{d}{dt}\frac{\partial u}{\partial x} = \mp \frac{1-e}{\sqrt{1-B}}\left((1-2p)\frac{de}{dt} + (1-e)\frac{dp}{dt}\right). \tag{5}$$

In this case, the equation of elastic strains transfer (the second equation of system (1)) takes the form

$$\frac{de}{dt} = (1-e)(\varepsilon - \gamma^c), \tag{6}$$

where γ^c is a creep strain rate. We exclude from Eq. (6) the total strain rate ε using Eq. (5). Then, taking into account that the creep strain transfer equation p^c, the third equation of system (1), has the form $\frac{dp^c}{dt} = (1-2p^c)\gamma^c$, and Eq. (6) after simplification is transformed to the following

$$\frac{de}{dt} = -(1-e)\gamma^c.$$

The creep of the rod we modeled by the Yu.N. Rabotnov's kinetic equations [3], that take the form of the following system of equations:

$$\begin{aligned} \frac{de}{dt} &= -(1-e)\gamma^c, \\ \frac{dp^c}{dt} &= (1-2p^c)\gamma^c, \\ \frac{d\omega}{dt} &= \frac{B}{A}\gamma^c. \end{aligned} \tag{7}$$

Here

$$\gamma^c = \frac{A\sigma^n}{\omega^\alpha (1-\omega^{\alpha+1})^m},$$

$$\sigma = \sigma_0 + (1 - e)\frac{dW}{de},$$

$$W = -2\mu I_1 - \mu I_2 + bI_1^2 + (b - \mu)I_1 I_2 - \chi I_1^3 + \ldots$$

$$I_1 = e - e^2/2, \quad I_2 = e^2 - e^3 + e^4/4,$$

$$\frac{dW}{de} = -2\mu I_1' - \mu I_2' + 2bI_1 I_1' + (b - \mu)(I_1' I_2 + I_1 I_2') - 3\chi I_1^2 I_1' + \ldots$$

$$I_1' = 1 - e, \quad I_2' = 2e - 3e^2 + e^3, \quad ()' = \frac{d}{de},$$

d is a component of the total strain tensor (Almansi strain), e is a linear component of the reversible deformations (elastic deformations) tensor, the p^c is a component of the irreversible creep strain tensor, the γ^c is a component of the creep strain rate, ω is a damage parameter of the rod material, t is time, W is an elastic potential, μ is the considered material module of shear, b, χ is the higher order elastic modules, σ is a stress in the rod, σ_0 is initial stress in the rod, A, B, n, m, α are parameters of creep. As in the classical theory, we supposed that the rod's stresses are completely determined by reversible deformations. At the initial time moment $t = 0$, the rod is undeformed, and the applied stress in a short time τ increases from 0 to σ_0, and the initial conditions will be homogeneous. However, the τ interval is much shorter than the creep time t_* of the material so that the initial conditions can be taken as:

$$e(0) = \frac{\sigma_0}{E}, \quad p^c(0) = 0, \quad \omega(0) = 0 \tag{8}$$

and the system of equations (7) should be solved under the initial conditions (8) until the time $t = t_*$, at which the damage parameter takes the value $\omega(t_*) = 1$. It determines the failure of the rod. Under conditions (8), E is the elastic modulus of the rod material.

2.1 Analitical Solution of the Problem (7)–(8)

Finding the solution to the problem (7)–(8). Consider the case of constant stress $\sigma = \sigma_0 = \text{const}$. Then from the last equation of system (7) we find

$$\int_0^\omega \omega^\alpha (1 - \omega^{\alpha+1})^m d\omega = \int_0^t B\sigma_0^n dt.$$

We calculate the resulting integrals:

$$\frac{1}{(\alpha+1)(m+1)} - \frac{(1-\omega^{\alpha+1})^{m+1}}{(m+1)(\alpha+1)} = B \cdot \sigma_0^n \cdot t.$$

Resolving this relation to the damage parameter ω, we finally find

$$\omega = \left\{1 - \left[1 - (m+1)\cdot(\alpha+1)\cdot B \cdot \sigma_0^n \cdot t\right]^{\frac{1}{m+1}}\right\}^{\frac{1}{\alpha+1}}. \tag{9}$$

Dividing the second equation of system (7) by the third, we obtain the relation

$$\frac{dp^c}{1-2p^c} = \frac{A}{B}d\omega,$$

from which, using the initial conditions (8), we find an expression for the creep strain

$$p^c = \frac{1}{2}\cdot\left(1 - \mathrm{e}^{\frac{-2A}{B}\omega}\right). \tag{10}$$

Then we find the elastic deformation e. To do this, we divide the second equation of system (7) by the first:

$$\frac{dp^c}{de} = -\frac{1-2p^c}{1-e}.$$

Separating the variables and integrating the resulting relationship, using the initial conditions (8), we find an expression for the elastic strain

$$e = 1 + \frac{\sigma_0 - E}{E\sqrt{1-2p^c}}. \tag{11}$$

Supplementing expressions (9)–(11) with the value of the long-term strength of the structure

$$t^* = \frac{1}{(m+1)\cdot(\alpha+1)\cdot B \cdot \sigma_0}, \tag{12}$$

obtained from (9) for $\omega = 1$, we find a complete solution of problem (7)–(8) for $\sigma = \sigma_0 = \mathrm{const}$.

3 Conclusion

In this work, we condider the problem of finite deformations of a long metal rod with a circular cross-section under creep conditions. An analytical solution to problem (7) and (8) is obtained in the form (9)–(12). However, obtaining the analytical solutions and the further investigation of the problem (7) and (8) under varying stresses are complicated. This is due to the lack of adequate model parameters. Therefore, our further research will aim to construct a mathematical model of creep, taking into account finite deformations based on experimental data with identification of creep characteristics. When considering problem (7) and (8), we will use an effective method of numerical integration based on the solution continuation with respect to the best argument [4].

Acknowledgments This paper is based on research carried out with the financial support of the grant of the Russian Science Foundation (project 18-19-00474).

References

1. Burenin, A.A., Kovtanyuk L.V.: Large irreversible deformations and elastic aftereffect. Dalnauka, Vladivostok (2013). In Russian
2. Kovtanyuk, L.V., Lemza, A.O. (2019) Large irreversible creep deformations under conditions of local plastic flow. *Works of XII All-Russian Congress on Fundamental Problems of Theoretical and Applied Mechanics, Vol. 3* (pp. 321–323). Ufa: RITs BashSU. In Russian
3. Rabotnov, Yu.N.: Creep problems in structural members. Amsterdam/London, North-Holland Publishing Company. 1969.
4. Kuznetsov, E.B., Leonov, S.S.: Technique for selecting the functions of the constitutive equations of creep and long-term strength with one scalar damage parameter. Journal of Applied Mechanics and Technical Physics. **57**:2, 369–377 (2016)

Locally One-Dimensional Schemes for Quasilinear Parabolic Equations with Time Fractional Derivative

Alexander V. Lapin and Ksenija O. Levinskaya

Abstract An initial-boundary value problem for a quasilinear parabolic equation with Caputo-type fractional time derivative and mixed boundary conditions is considered. The coefficients of the elliptic part of the equation depend on the derivatives of the solution and satisfy the conditions providing strong monotonicity and Lipschitz-continuity of the corresponding operator. This operator can be split into the sum of locally one-dimensional operators of second order.

The problem is approximated by two locally one-dimensional (LOD) finite difference schemes. The stability of LOD schemes is proved and the accuracy estimates are given under the condition of sufficient smoothness of the input data and the solution of the differential problem. For the constructed nonlinear mesh problems, easily implementable preconditioned iterative methods are used. The results of numerical experiments confirming the theoretical conclusions are presented.

1 Introduction

Fractional calculus is an actual approach to modeling various phenomena in physics, mechanics, and control theory. In particular, PDEs with fractional time derivatives arise in mathematical modeling of anomalous diffusion and dynamic processes in materials with memory. Numerous articles are devoted to the approximation of boundary value problems for linear problems. Several papers consider implicit mesh schemes for nonlinear equations with fractional time derivatives and either with right-hand sides depending on the solution (see [1–3]) or with a nonlinear

A. V. Lapin
Sechenov University, Moscow, Russia
e-mail: avlapine@mail.ru

K. O. Levinskaya (✉)
Kazan Federal University, Kazan, Russia
e-mail: sisina.kseniya@yandex.ru

I. B. Badriev et al. (eds.), *Mesh Methods for Boundary-Value Problems and Applications*, Lecture Notes in Computational Science and Engineering 141,
https://doi.org/10.1007/978-3-030-87809-2_22

diffusion coefficient depending on the solution [4]. The article [5] proposes a new mathematical model for a viscoelastic-plastic process by formulating a temporary fractional equation containing an elliptic operator, which depends on the gradient of the solution. In [6], the authors construct and study several mesh schemes approximating the problem with a fractional time derivative and a quasilinear elliptic part. Two classes of effectively implemented mesh schemes for the evolutionary equations are well known, they are alternating direction implicit (ADI) schemes [7–9] and locally one dimensional (LOD) schemes [9–13]. These schemes have been thoroughly investigated for the parabolic equations with integer derivatives. In [14–16], on analysis of ADI schemes was performed for linear time-fractional equations. In this article, we develop the results [6], investigating theoretically and numerically two LOD schemes for quasilinear fractional time equations.

2 Differential Problem

Let $\Omega = (0, 1) \times (0, 1)$ be the unit square with the boundary $\partial\Omega = \Gamma_D \cup \Gamma_N$, $\text{mes}\,\Gamma_D \neq \emptyset$, and $Q = \Omega \times (0, T]$ be the cylinder with the boundary $\partial Q = \partial\Omega \times (0, T]$. Denote by $\Sigma_D = \Gamma_D \times (0, T]$ and $\Sigma_N = \Gamma_N \times (0, T]$ the parts of ∂Q and by n the unit vector of the outward normal to Σ_N.

We define a quasilinear elliptic differential operator and its co-normal derivative by the equalities

$$Lu = -\sum_{i=1}^{2} \frac{\partial}{\partial x_i} g_i\big(x, t, \frac{\partial u}{\partial x_i}\big), \quad \frac{\partial u}{\partial \nu_L} = \sum_{i=1}^{2} g_i\big(x, t, \frac{\partial u}{\partial x_i}\big) \cos(n, x_i).$$

Nonlinear coefficients $g_i(p) = g_i(x, t, p)$ are assumed to be continuous and satisfy the following assumptions for all $(x, t) \in \bar{Q}$ and $p, q \in \mathbb{R}$:

$$(g_i(p) - g_i(q))(p - q) \geqslant c_0(p - q)^2, \ c_0 > 0, \tag{1}$$

$$|g_i(p) - g_i(q)| \leqslant c_1|p - q|. \tag{2}$$

Next,

$$G(t) : (0, +\infty) \to \mathbb{R} \text{ is a continuous, positive, and strictly decreasing function, } \int_0^{+\infty} G(t)\, dt < \infty. \tag{3}$$

Define a Caputo-type fractional time derivative:

$$\mathcal{D}_t y(t) = \int_0^t G(t-s)\frac{\partial y}{\partial s}(s)\, ds. \tag{4}$$

We list several well-known fractional derivatives, which are the particular cases of (4) with the kernels satisfying the assumptions (3):

- the generalized Caputo fractional derivative with $G(t) = \dfrac{r(t)}{\Gamma(1-\alpha)\, t^{\alpha}}$, $0 < \alpha < 1$, $\Gamma(x)$ is gamma-function, a weighting function $r(t) \in C^2[0,T]$, $r(t) > 0$ and $r'(t) \leqslant 0$ for all $t \in [0,T]$ ($r(t) \equiv 1$ corresponds to the classical Caputo fractional derivative);
- multi-term fractional derivative with $G(t) = \sum_{k=1}^{s} \dfrac{c_k}{\Gamma(1-\alpha_k)\, t^{\alpha_k}}$, $0 < \alpha_1 < \ldots < \alpha_s < 1$, $c_k > 0$.

In this article, we construct and study a mesh approximation of the parabolic problem with the mixed boundary conditions

$$\begin{gathered} \mathcal{D}_t u + Lu = f \text{ in } Q, \\ u = 0 \text{ on } \Sigma_D, \quad \frac{\partial u}{\partial \nu_L} = q \text{ on } \Sigma_N, \\ u = 0 \text{ for } t = 0,\ x \in \Omega. \end{gathered} \tag{5}$$

Assuming the sufficient smoothness of all functions in the statement of the problem (5) and multiplying (5) with a smooth test function $v(x,t)$, which vanishes on the boundary Σ_D, after integrating over Q, we obtain a variational equation

$$\int_Q \mathcal{D}_t u\, v\, dxdt + \int_Q \sum_{i=1}^{2} g_i\Big(x,t,\frac{\partial u}{\partial x_i}\Big)\frac{\partial v}{\partial x_i}\, dxdt = \int_Q f\, v\, dxdt + \int_{\Sigma_N} q\, v\, d\Gamma dt. \tag{6}$$

Reciprocally, if a sufficiently smooth function u satisfies variational equality (6), we can prove, by following the standard procedure, that u is a solution of (5). So, below we will construct approximations of the problem (5) using the variational equality (6).

Let us make some remarks on the well-known results on the existence of a weak solution of the time fractional parabolic equations.

For the case of linear Dirichlet boundary value problem and classical Caputo derivative $\mathcal{D}_t$ in [17] the existence of a unique solution to the problem (5) from $H^{\alpha}(0,T;L^2(\Omega)) \cap L^2(0,T;H_0^1(\Omega) \cap H^2(\Omega))$ is proved and the corresponding a priori estimate through the $L^2(Q_T)$-norm of the right-hand side f is given.

In [18] the unique existence of a "very weak" solution from $B^{\alpha/2}(Q_T) = H^{\alpha/2}(0,T;L^2(\Omega)) \cap L^2(0,T;H^1(\Omega))$ is substantiated for (5) with mixed boundary conditions and classical Caputo derivative. The existence of a similar very weak solution for a quasilinear Dirichlet boundary value problem with classical Caputo derivative is proved in [6]. The result of [6] can be generalized for the case of mixed boundary value problem (5) and generalized Caputo time derivative. Nevertheless, we omit this proof and focus on studying the mesh approximations of the problem.

3 Approximation

Let $\omega_\tau = \{t_j = j\tau,\ j = 0, 1, \dots M;\ M\tau = T\}$ be a uniform mesh on the segment $[0, T]$ and $y^j = y(t_j)$ for a continuous function $y(t)$. The conventional $L1$-approximation of a first order fractional derivative of a continuous function $y(t),\ y(0) = 0$, at a mesh point $t_k \in \omega_t$ is defined by the equality

$$\mathcal{D}_t y(t_k) \approx \partial_t y(t_k) = d_1 y^k + \sum_{j=1}^{k-1} (d_{j+1} - d_j) y^{k-j},\ d_j = \frac{1}{\tau} \int\limits_{t_{k-j}}^{t_{k-j+1}} G(t_k - s)\, ds.$$

Due to (3) the coefficients satisfy the inequalities

$$d_1 > d_2 > \cdots > d_M > 0. \tag{7}$$

We construct a finite difference approximation of the elliptic part of the equation using bilinear finite elements (see [19]) and the composite trapezoidal quadrature formulas. Let T_h be a family of non-overlapping closed rectangles e (finite elements) with maximal side h. We suppose that T_h is a conforming and regular triangulation $\overline{\Omega} = \bigcup\limits_{e \in T_h} e$ of $\overline{\Omega}$ [19, p. 124] and T_h generates the triangulation ∂T_h of $\overline{\Gamma}_N$, i.e. $\overline{\Gamma}_N$ consists of an integer number of sides ∂e of elements $e \in T_h$. Let V_h be the space of continuous and piecewise bilinear functions (bilinear on each e) that vanish on the boundary Γ_D, and Q_h be the space of the piecewise linear functions on Γ_N (linear on each $\partial e \in \Gamma_N$), which are the traces on Γ_N of the functions from V_h. In what follows, to shorten the notation, we will omit the index h of mesh functions from the spaces V_h and Q_h.

We use quadrature formulas approximating the integrals $\int_e g(x)dx$ and $\int_{\partial e} g(x)\, d\Gamma$ of a continuous function $g(x)$:

$$S_e(g) = \frac{\text{meas}\,(e)}{4} \sum_{\alpha=1}^{4} g(x_\alpha), \quad S_{\partial e}(g) = \frac{\text{meas}\,(\partial e)}{2} \sum_{\alpha=1}^{2} g(x_\alpha),$$

where x_α are the vertices of e and ∂e, respectively, and the composite quadrature formulas

$$S(g) = \sum_{e \in T_h} S_e(g), \quad S_\Gamma(g) = \sum_{\partial e \in \partial T_h} S_{\partial e}(g),$$

approximating the integrals over the domain Ω and the boundary Γ_N.

On the space V_h, we define mesh analogs of L^2-norm and H^1-norm, and also H^1-seminorms:

$$\|v\|_0^2 = S(v^2), \quad \|v\|_1^2 = S(|\nabla v|^2), \quad v \in V_h, \quad \rho_i^2(v) = S\Big(\Big(\frac{\partial v}{\partial x_i}\Big)^2\Big), \; i = 1, 2.$$

Next, let $V_{h\tau}$ be the linear space of the mesh functions $y(t) : \omega_\tau \to V_h$, and $Q_{h\tau}$ be the linear space of the mesh functions $q(t) : \omega_\tau \to Q_h$. We use the notation $y^k = y(t_k)$ for a mesh function from $V_{h\tau}$ or $Q_{h\tau}$.

Using introduced spaces of the mesh functions and the quadrature formulas, we construct approximations of the terms on the left side of Eq. (6) and its right side:

$$a_i^k(y, v) = S\Big(g_i\big(x, t^k, \frac{\partial y}{\partial x_i}\big)\frac{\partial v}{\partial x_i}\Big) \text{ for } y, v \in V_h, \quad i = 1, 2, \quad k = 1, 2, \ldots, M;$$

$$F^k(v) = S\Big(f^k\, v\Big) + S_\Gamma\Big(q^k\, v\Big) \text{ for } f, v \in V_{h\tau}, \; q \in Q_{h\tau}, \quad k = 1, 2, \ldots, M.$$

In what follows we use a splitting of mesh function F into a sum of two functions:

$$F^k = \phi_1^k + \phi_2^k, \quad k = 1, 2, \ldots, M.$$

Now everything is ready for constructing approximations of the variational problem (6). We will consider two types of locally one-dimensional schemes. Namely, for a given $y^0 = 0$ we are looking for y^k for all $k = 1, 2, \ldots, M$ from one of the following two systems of nonlinear equations

$$\begin{aligned}
&\frac{1}{2}d_1 S\Big(w_1^k\, v\Big) + a_1^k(w_1^k, v) = S(\phi_1^k\, v) - \frac{1}{2}\sum_{j=1}^{k-1}(d_{j+1} - d_j)S(y^{k-j}\, v),\\
&\frac{1}{2}d_1 S\Big(w_2^k\, v\Big) + a_2^k(w_2^k, v) = S(\phi_2^k\, v) - \frac{1}{2}\sum_{j=1}^{k-1}(d_{j+1} - d_j)S(y^{k-j}\, v),\\
&y^k = \frac{1}{2}(w_1^k + w_2^k)
\end{aligned} \tag{8}$$

or

$$(d_1 - d_2)S\Big(w^k v\Big) + a_1^k(w^k, v) = S(\phi_1^k v) + (d_1 - d_2)S(y^{k-1} v),$$

$$d_1 S\Big(y^k v\Big) + a_2^k(y^k, v) = (d_1 - d_2)S(w^k v) + S(\phi_2^k v) - \sum_{j=2}^{k-1}(d_{j+1} - d_j)S(y^{k-j} v). \tag{9}$$

Lemma 1 *Problems* (8) *and* (9) *have unique solutions.*

Proof Due to assumptions (1) and (2) the forms $a_i^k : V_h \times V_h \to \mathbb{R}$ satisfy the properties of the monotonicity and Lipschitz-continuity:

$$\begin{aligned} a_i^k(y, v - y) - a_i^k(v, v - y) &\geqslant c_0\rho_i^2(y - v) \quad \forall y, v \in V_h, \\ a_i^k(y - v, w) &\leqslant c_1\rho_i(y - v)\rho_i(w) \quad \forall y, v, w \in V_h. \end{aligned} \tag{10}$$

Since $d_1 > 0$ and $d_1 - d_2 > 0$, then due to (10) every equation in (8) and (9) is an equation with a uniformly monotone and continuous operator, which immediately implies the existence of a unique solution. □

4 Stability and Accuracy Estimates

It is well-known that the locally one-dimensional schemes have only the so-called aggregate approximation. This forces us to obtain stability estimates containing on the right-hand side some norms of the sum $\phi_1^k + \phi_1^k$ and additions with a small parameter which depends on the mesh step τ. We prove the stability estimates in $L^2(\omega_t; L^2(\omega_x))$-norm, which is defined by the equality

$$\|v\|^2_{L^2(\omega_t; L^2(\omega_x))} = \|v\|^2_{L^2} = \sum_{k=1}^{M} \tau\|v^k\|_0^2.$$

The functions from the space V_h vanish on a set of the mesh nodes $\omega \cup \Gamma_D$, so, there exists a positive constant ξ such that

$$\|v\|_1^2 = \rho_1^2(v) + \rho_2^2(v) \geqslant \xi\|v\|_0^2 \quad \forall v \in V_h.$$

Moreover, a similar inequality holds for at least one of the seminorms, let it be true for ρ_1:

$$\rho_1^2(v) \geqslant \xi_0\|v\|_0^2 \ \ \forall v \in V_h, \ \ \xi_0 > 0. \tag{11}$$

Lemma 2 *Define the lower triangle Toeplitz matrix*

$$B = \begin{pmatrix} d_2 & 0 & 0 & 0 \cdots & 0 & 0 & 0 \\ 0 & d_2 & 0 & 0 \cdots & 0 & 0 & 0 \\ d_3 - d_2 & 0 & d_2 & 0 \cdots & 0 & 0 & 0 \\ \cdots & \cdots & \cdots & \cdots\cdots & \cdots & \cdots \\ d_M - d_{M-1} & d_{M-1} - d_{M-2} & d_{M-2} - d_{M-3} & \cdots \cdots & d_3 - d_2 & 0 & d_2 \end{pmatrix}$$

The matrix $0.5\big(B + B^T\big)$ *is positive definite, so, in particular, for any* $n \leqslant M$

$$\sum_{k=1}^{n} (B_n \bar{y})^k \, y^k \geqslant 0 \quad \forall \bar{y} = (y^1, y^2, \ldots, y^n). \tag{12}$$

Proof Since the matrix B is strongly diagonally dominant both in rows and in columns, then the matrix $0.5\big(B + B^T\big)$ is also strongly diagonally dominant. Next, it has positive diagonal and non-positive off-diagonal elements. Because of these properties $0.5\big(B + B^T\big)$ is M-matrix. Since at the same time it is Stieltjes matrix, then it is positive definite. Any diagonal submatrix of $0.5\big(B + B^T\big)$ is also positive definite, so, for any $n \leqslant M$ the inequality (2) is true. □

Using the matrix B, the approximation of time derivative at a point $t_k = k\tau \in \omega_\tau$ can be written as

$$\partial_t y(t_k) = (d_1 - d_2)(y^k - y^{k-1}) + \big(B_n \bar{y}\big)^k, \quad \bar{y} = (y^1, y^2, \ldots y^n),$$

where B_n is the diagonal submatrix of the first n rows and columns of the matrix B.

The main result on the stability of the mesh schemes is as follows:

Theorem 1 *Denote by* y_1^k *and* y_2^k *the solutions of the LOD schemes, corresponding to the right-hand sides* ϕ_{1i} *and* ϕ_{2i}*. Let* $y = y_1 - y_2$ *and* $\phi_i = \phi_{1i} - \phi_{2i}$*. Let the assumptions* (1) *and* (2) *be satisfied.*

For the solution of LOD scheme (8) *the following a priori estimate holds:*

$$\|y\|_{L^2}^2 \leqslant C \, \|\phi_1 + \phi_2\|_{L^2}^2 + \frac{C}{d_1 - d_2}\big(\|\phi_1\|_{L^2}^2 + \|\phi_2\|_{L^2}^2\big). \tag{13}$$

If, in addition, (11) *takes place, then for the solution of LOD scheme* (9) *the following estimate holds:*

$$\|y\|_{L^2}^2 \leqslant C \, \|\phi_1 + \phi_2\|_{L^2}^2 + \frac{C}{d_1 - d_2}\|\phi_2\|_{L^2}^2. \tag{14}$$

Above C *means a generic constant independent of mesh steps* h *and* τ.

Proof The proof of the first estimate coincides up with certain details and notation with the proof of Theorem 3 in [6]; therefore, we present only the proof of the second estimate.

For the difference of the solutions of problem (9) with input data ϕ_{1i}^k and ϕ_{2i}^k, the equations can be written in the following form:

$$(d_1 - d_2)S\Big((w^k - y^{k-1})\,v\Big) + a_1^k(w_1^k, v) - a_1^k(w_2^k, v) = S(\phi_1^k\, v),$$
$$(d_1 - d_2)S\Big(((y^k - w^k)v\Big) + d_2 S\Big(y^k\, v\Big) + a_2^k(y_1^k, v) - a_2^k(y_2^k, v) +$$
$$+\sum_{j=2}^{k-1}(d_{j+1} - d_j)S(y^{k-j}\, v) = S(\phi_2^k\, v).$$

Taking $v = w^k$ in the first equation and $v = y^k$ in the second, then using (10) and summing up the resulting inequalities, we get:

$$\frac{d_1 - d_2}{2}(\|y^k\|_0^2 - \|y^{k-1}\|_0^2 + \|w^k - y^{k-1}\|_0^2 + \|y^k - w^k\|_0^2) + d_2\|y^k\|_0^2 +$$
$$+\sum_{j=2}^{k-1}(d_{j+1} - d_j)S(y^{k-j}\, y^k) + c_0\,\rho_1^2(w^k) + c_0\,\rho_2^2(y^k) \leqslant (\phi_1^k,\, w^k) + (\phi_2^k, y^k). \tag{15}$$

Summation of (15) over k from 1 to n and using the initial value $y^0 = 0$ and the inequality

$$\sum_{k=1}^{n}\Big(d_2\|y^k\|_0^2 + \sum_{j=2}^{k-1}(d_{j+1} - d_j)S(y^{k-j}\, y^k)\Big) = S(\sum_{k=1}^{n}(B_n\bar{y})^k\, y^k) \geqslant 0$$

gives

$$\frac{d_1 - d_2}{2}\|y^n\|_0^2 + \frac{d_1 - d_2}{2}\sum_{k=1}^{n}(\|w^k - y^{k-1}\|_0^2 + \|y^k - w^k\|_0^2) +$$
$$+ c_0\,\rho_1^2(w^k) + c_0\,\rho_2^2(y^k) \leqslant \sum_{k=1}^{n}\Big((\phi_1^k,\, w^k) + (\phi_2^k, y^k)\Big).$$

To estimate the right-hand side we use the estimate (11), whence

$$S(\phi_1^k\, w^k) + S(\phi_2^k\, y^k) = S((\phi_1^k + \phi_2^k)\, w^k) + S(\phi_2^k(y^k - w^k)) \leqslant$$
$$\leqslant \frac{1}{4c_0\,\xi_0}\|\phi_1^k + \phi_2^k\|_0^2 + c_0\,\xi_0\|w^k\|_0^2 + \frac{1}{2(d_1 - d_2)}\|\phi_2^k\|_0^2 + \frac{d_1 - d_2}{2}\|y^k - w^k\|_0^2.$$

The last two inequalities lead to the estimate

$$\frac{d_1-d_2}{2}\|y^n\|_0^2+\frac{d_1-d_2}{2}\sum_{k=1}^{n}\|w^k-y^{k-1}\|_0^2+c_0\,\xi_0\sum_{k=1}^{n}\|y^k\|_0^2\leqslant$$

$$\leqslant\frac{1}{4c_0\,\xi_0}\sum_{k=1}^{n}\|\phi_1^k+\phi_2^k\|_0^2+\frac{1}{2(d_1-d_2)}\sum_{k=1}^{n}\|\phi_2^k\|_0^2,$$

therefore

$$c_0\,\xi_0\sum_{k=1}^{n}\|y^k\|_0^2\leqslant\frac{1}{4c_0\,\xi_0}\sum_{k=1}^{n}\|\phi_1^k+\phi_2^k\|_0^2+\frac{1}{2(d_1-d_2)}\sum_{k=1}^{n}\|\phi_2^k\|_0^2.$$

Due to the definition of $L^2(\omega_t;\,L^2(\omega_x))$-norm in V_h this inequality leads to the estimate (14). □

Using the proved stability estimates, we can derive the accuracy estimates under the assumption that the input data and the solution of the differential problem are sufficiently smooth. Note that, in the accuracy estimates, the most significant terms in the stability inequalities are $\frac{C_2}{d_1-d_2}(\|\phi_1\|_{L^2}^2+\|\phi_2\|_{L^2}^2)$ for the scheme (8) and $\frac{C_4}{d_1-d_2}\|\phi_2\|_{L^2}^2$ for the scheme (9). Since $\|\phi_i\|_{L^2}=O(1)$, then the smallness of these terms is ensured only by the smallness of $(d_1-d_2)^{-1}$.

In the case of the generalized Caputo derivative $d_1-d_2=O(\tau^{-\alpha})$. For the approximation of the time derivative it is known the estimate $\partial_t^\alpha\bar{u}^k=\mathcal{D}_t^\alpha u(t_k)+O(\tau^{2-\alpha})$ (cf., e.g., [20]), and the elliptic part is approximated with the order $O(h^2)$. So,

$$\|\phi_1^k+\phi_2^k\|_{L^2(\omega_x)}=O(\tau^{2-\alpha}+h^2)$$

for all time levels k. As a consequence, we can cite the following result (Theorem 4 in [6]):

Theorem 2 *Let the coefficients $g_i(x,t,p)$, $i=1,2$ and the right hand side of the Eq.* (5), *as well as its solution $u(x,t)$, be sufficiently smooth. Denote by y the solution of a mesh scheme and by u the mesh function which coincides with the solution of the differential problem* (5) *at the mesh points. Then for both LOD schemes approximating differential problem with generalized Caputo time-fractional derivative the following accuracy estimate is valid:*

$$\|y-u\|_{L^2(\omega_t;L^2(\omega_x))}=O(\tau^{\alpha/2}+h^2). \tag{16}$$

The accuracy estimates for the case of multi-term fractional derivative can be proved in similar way using the provided stability estimates and analyzing the approximation errors.

5 Numerical Results

In the experiments, we consider Dirichlet boundary value problem and classical Caputo time-fractional derivative. To implement the non-linear mesh schemes we used the stationary preconditioned iterative method. The mesh approximations of Laplace operator are used in constructing the preconditioners. The constructed iterative methods converge with the geometric rate of the convergence (not depending on τ and h) and they are easily implemented. For more details, we refer a reader to the authors' article [6].

The theoretically proved accuracy estimate for both considered LOD schemes in case of smooth input data and solutions is

$$\|y - u\|_{L^2} = C_t\tau^{\alpha/2} + C_x h^2 = O(\tau^{\alpha/2} + h^2), \tag{17}$$

with constants C_t and C_x independent on mesh steps h and τ.

We took equal functions $g_i(x, t, p) = g(p), i = 1, 2$, and $g(p)$ linearly depended on p at their "small" values and nonlinearly with their large values: $g(p) = \{p \quad \text{if} \quad p < \frac{1}{\xi^2}; \ \sqrt{p} - \frac{\xi - 1}{\xi^2} \quad \text{if} \quad p \geqslant \frac{1}{\xi^2}\}$. Here the parameter ξ is responsible for the non-linearity zone, namely, the more ξ, the larger this zone. We present the results for the parameters $\xi = 4, T = 1$ and exact solution $u = t\sin(\pi x_1)\sin(\pi x_2)$.

We verified the sharpness of the estimates (17) by numerically determining the constants in these estimates. When determining the values of C_x,, we selected a sufficiently fine mesh in time to minimize the approximation error in the time variable, and performed the calculations using a sequence of meshes in spatial variables. We used the same approach to check the asymptotic accuracy in time. Moreover, we check the order of accuracy for τ. To do this, for a sufficiently fine mesh of spatial variables, we perform calculations using a sequence of cells with doubling the number of nodes in time and calculate the number

$$\nu_t = \log \frac{E_\tau}{E_{\frac{\tau}{2}}} (\log 2)^{-1},$$

where $E_\tau = \|y - u\|_{L^2}$ is the norm of error in case of time step τ and $E_{\frac{\tau}{2}}$ has a similar meaning. This number corresponds to the accuracy estimate $\|y - u\|_{L^2} = O(\tau^{\nu_t})$. According to the theoretical results, we expected to find the approximate equality $\nu_t \approx \alpha/2$ for both LOD schemes. As can be seen from the results in Tables 1 and 2, the calculated results confirm all the theoretical estimates.

Table 1 Spatial variable accuracy[a]

	LOD scheme 1		LOD scheme 2	
h	$\|y-u\|_{L^2}$	C_x	$\|y-u\|_{L^2}$	C_x
1/50	0.0120850	30.212	0.0089578	22.394
1/100	0.0030218	30.218	0.0021326	21.326
1/200	0.0007546	30.184	0.0005337	21.348
1/250	0.0004829	30.181	0.0003387	21.169
1/400	0.0001886	30.176	0.0001309	20.948
1/500	0.0001207	30.162	0.0000854	21.359

[a] $\tau = 1/1000$, $\alpha = 0.5$

Table 2 Time variable accuracy[b]

α	τ	$\|y-u\|_{L^2}$	C_t	ν_t
LOD scheme 1				
0.3	1/50	0.0010397	0.0018	
	1/100	0.0009350	0.0018	0.1531
	1/200	0.0008415	0.0018	0.1520
0.5	1/50	0.0023039	0.0061	
	1/100	0.0018762	0.0059	0.2962
	1/200	0.0015273	0.0057	0.2968
0.7	1/50	0.0028741	0.0113	
	1/100	0.0021637	0.0108	0.4096
	1/200	0.0016289	0.0104	0.4096
0.9	1/50	0.0031106	0.0180	
	1/100	0.0022686	0.0180	0.4553
	1/200	0.0016675	0.0180	0.4441
LOD scheme 2				
0.3	1/50	0.0006759	0.0012	
	1/100	0.0005901	0.0011	0.1958
	1/200	0.0005148	0.0011	0.1969
0.5	1/50	0.0015498	0.0041	
	1/100	0.0012647	0.0039	0.2932
	1/200	0.0010317	0.0038	0.2937
0.7	1/50	0.0018603	0.0073	
	1/100	0.0013712	0.0068	0.4400
	1/200	0.0010123	0.0064	0.4378
0.9	1/50	0.0025547	0.0148	
	1/100	0.0016628	0.0132	0.6195
	1/200	0.0010830	0.0117	0.6212

[b] $h = 1/500$

6 Conclusions

Based on the proven and experimentally confirmed accuracy estimates, we can conclude that when applying LOD mesh schemes, we must take a significantly small time step to achieve the desired accuracy. This statement is all the more powerful the smaller the parameter $\alpha \in (0, 1)$. On the other hand, LOD schemes have several essential advantages, such as:

- the possibility of applying for the boundary value problems in the domains with curvilinear boundaries and with various types of boundary conditions;
- for nonlinear problems—effective implementation using one-step iterative methods with tridiagonal preconditions;
- massive parallelization.

Thus, LOD mesh schemes provide an effective tool for solving multidimensional quasilinear equations with a fractional time derivative and without mixed space derivatives.

Acknowledgments This work was supported by Russian Science Foundation, project No. 20-61-46017.

References

1. Jin, B., Li, B., Zhou, Z.: Numerical analysis of nonlinear subdiffusion equations. SIAM J. Numer. Anal. **56** (1), 1–23 (2018)
2. Li, D., Liao, H.-L., Sun, W., Wang, J., Zhang, J.: Analysis of $L1$-Galerkin FEMs for time-fractional nonlinear parabolic problems. Commun. Comput. Phys. **24** (1), 86–103 (2018)
3. Li, D., Zhang, J., Zhang, Z.: Unconditionally optimal error estimates of a linearized Galerkin method for nonlinear time fractional reaction-subdiffusion equations. J. Sci. Comput. **76** (2), 848–866 (2018)
4. Gerasimov, D.N., Kondratieva, V.A., Sinkevich, O.A.: An anomalous non-self-similar infiltration and fractional diffusion equation. Phys. D. **239** (16), 1593–1597 (2010)
5. Tatar, S., Tnaztepe, R., Zeki, M.: Numerical solutions of direct and inverse problems for a time fractional viscoelastoplastic equation. J. Eng. Mech. **143** (7) (2017)
6. Lapin, A., Levinskaya, K.: Numerical solution of a quasilinear parabolic equation with a fractional time derivative. Lobachevskii J. Math. **41** (12) (2020), accepted
7. Douglas Jr., J., Rachford Jr., H.H.: On the numerical solution of heat conduction problems in two and three space variables. Trans. Am. Math. Soc. **82**, 421–439 (1956)
8. Peaceman, D.W., Rachford Jr., H.H.: The numerical solution of parabolic and elliptic differential equations. J. Soc. Ind. Appl. Math. **3**, 28–41 (1955)
9. Samarskii, A.A., Nikolaev, E.S.: Numerical Methods for Grid Equations: Volume II Iterative Methods. Basel (1989)
10. D'Yakonov, E.G.: Difference schemes with splitting operator for multi-dimensional nonstationary problems. Zh. Vychisl. Mat. i Mat. Fiz. **2**, 549–568 (1962)
11. Marchuk, G.I.: Some applicatons of splitting-up methods to the solution of problems in mathematical physics. Aplikace Matematiky. **1**, 103–132 (1968)
12. Yanenko, N.N.: The Method of Fractional Steps: the Solution of Problems of Mathematical Physics in Several Variables. Springer, Berlin (1971)

13. Gordeziani, D.T., Meladze, D.V.: The simulation of the third boundary value problem for multidimensional parabolic equations in an arbitrary domain by one-dimensional equations. USSR Comp. Math. Math. Phys. **14** (1), 249–253 (1974)
14. Zhang, Y., Sun, Z., Zhao, X.: Compact alternating direction implicit scheme for the two dimensional fractional diffusion-wave equation. SIAM J. Numer. Anal. **50**, 1535–1555 (2012)
15. Chen, A., Li, C.: A novel compact ADI scheme for the time-fractional subdiffusion equation in two space dimensions. Int. J. Comput. Math. **93**, 889–914 (2016)
16. Gao, G.H., Sun, Z.Z.: Two alternating direction implicit difference schemes for two-dimensional distributed-order fractional diffusion equations. J. Sci. Comput. **66**, 1281–1312 (2016)
17. Gorenflo, R., Luchko, Yu., Yamamoto, M.: Time-fractional diffusion equation in the fractional Sobolev spaces. Fractional Calculus and Applied Analysis. **18** (3), 799–820 (2015)
18. Wang, J.-G., Ran, Y.-H., Yuan, Z.-B.: Uniqueness and numerical scheme for the Robin coefficient identification of the time-fractional diffusion equation. Comp. Math. Appl. **75** (11), 4107–4114 (2018)
19. Ciarlet, Ph.G.: The finite element method for elliptic problems. North Holland, Amsterdam (1978)
20. Langlands, T.A.M., Henry, B.I.: The accuracy and stability of an implicit solution method for the fractional diffusion equation. J. of Comput. Phys. **205** (2), 719–736 (2005)

Enhanced Step-Wise Approximation to Speech File in a Noisy Environment

R. Latypov and E. Stolov

Abstract Reducing the size of speech files for transmitting through a noisy channel with low capacity is an actual problem. The current compression methods for speech files can provide a very high degree of compression, but any change in the compressed file can lead to the impossibility of restoring content. We propose to leverage a step-wise version of the original speech file that requires 2 or 3 bits for a sample. The obtained file does not fit musical playback products but provides adequate speech file audibility even if there is a noise channel. In the paper, we consider two problems. One is details of the transformation of the standard speech file into its step-wise form, namely choice of thresholds for creation step-function; the other is a method for enhancing the perception of the speech file step-wise version by a human. A suboptimal algorithm for fast thresholds calculation is developed in transforming the original signal to a step-wise form. Enhancing the perception of speech is gained through the regression function, which has to be placed at the receiver point. It is shown that the regression on the voice of a speaker is useful for enhancing speech produced by other speakers.

1 Introduction

Data transfer over a noisy channel with low capacity is actual as before, especially when dealing with speech information. The standard sample frequency for telephone conversation is 8 kHz, but the speech saves its audibility even for 4 kHz sample frequency. Standard methods for sound compression can provide a very high degree of compression [1], but the methods are sensitive to small distortion in a compressed file. Such distortions are inevitable if a noise presents in the channel. Implementation of error correction coding [2] can not lead to a solution since the theory is valid for a particular error model. It means that some additional methods

R. Latypov (✉) · E. Stolov
Kazan Federal University, Kazan, Russia
e-mail: Roustam.Latypov@kpfu.ru; ystolov@list.ru

I. B. Badriev et al. (eds.), *Mesh Methods for Boundary-Value Problems and Applications*, Lecture Notes in Computational Science and Engineering 141,
https://doi.org/10.1007/978-3-030-87809-2_23

must be used in this situation. In this paper, we consider a noised single-channel intended for transmitting speech signals. Many authors investigated the problem. Mainly, they reduce the problem to the standard speech enhancing method. First of all, various versions of adaptive filters were proposed [3].

Later, distinct methods based on the special codebook [1] were applied. There are diversities in codebook structures. It can consist of templates of speech files, coefficients of linear prediction, and other features. The receiver looks for available templates in the acquired file and changes them for corresponding fragments of speech or uses them in a restoration procedure. As a rule, an additive noise signal model is supposed, and much attention is dedicated to methods for extraction parameters of the noise [6].

A more complicated mathematical model considering features of speech signals (spectrum, linear prediction, and others) is used in vocoder [4]. Such the device contains a speech synthesis system based on the parameters extracted from the received file. That works only if the speech signal parameters are transmitted correctly, and the quality of the reconstructed signal, damaged by the channel, requires additional research. A particular version of the problem under investigation is the declipping procedure [5]. Here, the media file is damaged because of technical problems. The methods enhancing the sound of the file are also based on mathematical models. However,t the peculiarity of the situation is the type of errors we are dealing with—there are intervals of samples having the same value.

It is not the case we have while transmitting a signal over a noisy channel. Recently, the main focus was directed to implementing the neural net for the reconstruction of speech. A net is trained on a clean signal and then performs corrections of the signal on its input. An overview of enhancing methods, including neural net regression, is presented in [7]. There are many approaches to data preparation for net input. It is known that the speech file saves a part of its audibility after changing the source signal with its step-wise version. The approach that is close to our paper's technique is leveraging the ideal binary mask (IBM) to noised speech signal [8]. After processing, the source signal was converted into a binary file in the following way. A fragment was changed to 0 or 1 depending on the prescribed value L_0 of signal-to-noise ratio (SNR) obtained for this fragment. If the calculated SNR is less than L_0, then the fragment consists of zeros, otherwise—of ones. This procedure provides a reducing level of noise in the signal using simple calculations. In [9], an enhancement speech procedure is developed for real babble noise reduction based on the binary transform. Such an approach is not suitable for music production but acceptable when the ratio intelligibility/bitrate is essential.

In our paper, we consider the speech file as a stream of bits transmitted over a noise channel and suppose that any bit can be inverted during transmission with the same probability P. No other properties of the noise are assumed. Let $Frag(t)$ be a fragment of speech file. For example, its step-wise version $Step(t)$ can have the form

$$Step(t) = \begin{cases} 0 & \text{if } |Frag(t)| < Thr_1, \\ 2 \cdot \mathbf{sign}(Frag(t)) & \text{if } |Frag(t)| \geq Thr_2. \\ \mathbf{sign}(Frag(t)) & \text{otherwise.} \end{cases} \quad (1)$$

Here $Thr_1, Thr_2, \quad Thr_1 < Thr_2$ are two thresholds defining the transform. Such a five-level version of the speech file can be recognized if the used sample frequency is about 4 kHz. The number of levels can be arbitrary. Some numbers are convenient for sample coding. In the case of a single threshold, one has three possible signal values, and two bits are enough for coding any sample. Using three thresholds, one has seven possible values for the signal and three bits for sample coding. The transfer of a step-wise file requires a small bitrate. For example, utilizing three bits per sample, we reduce the standard bitrate by five times. It will be shown that the produced signal comes out noise-resistant if a particular technique is implemented.

After the conversion speech file into its step-wise version, one observes significant degradation of perception quality. Trying to gain a suboptimal approximation of speech signal by its step-wise version, we have to apply different thresholds for each fragment, whereas the range of the produced step-function is the same. It leads to additional distortions of a signal, but the speech signal keeps its intelligibility.

Recently many approaches are suggested to solve the task. We show that utilizing a linear regression function solves this problem partly. Using the source file and its step-wise approximation, we create a linear regression employing a simple algorithm. The implementation of that technology enhances the perception of the step-wise version at the transmission point. The transformed signal is transmitted, and the trained regression is placed in front of the receiver. That is the way one gets an enhanced version of the step-wise signal. We discovered that the function trained for a given voice keeps its enhancing properties for other voices.

While investigating the problem, we always have to evaluate the quality of approximation of a source signal S by an approximation signal $Appr$. The evaluation is based on utilizing Signal-to-noise ratio (SNR) in the form

$$SNR = 10 \cdot \log_{10} \left(\frac{\sigma^2(S)}{\sigma^2(S - Coef \cdot Appr)F} \right). \quad (2)$$

where coefficient $Coef$ provides equal lengths (according to Euclid metrics) of both the signals.

2 Optimal Thresholds

The first problem we solve in this paper is thresholds for optimal coding of a signal. The main result relating to the topic belongs to Lloyd [10] when the signal is a stochastic process with known distribution. Here, we consider the situation when a regular function must be approximated with a step-function. We investigate the situation of two thresholds as far as other numbers of thresholds are analyzed similarly. The $Frag(t)$ is supposed to be a continuous function. A horizontal line given by the equation $Y = Thr_1$ meets the function at the points $(A1, Frag(A1))$, $(B1, Frag(B1))$, $(A2, Frag(A2))$, $(B2, Frag(B2))$, …(see Fig. 1). Another horizontal line defined by the equation $Y = Thr_2$ meets the function at the points $(C1, Frag(C1))$, $(D1, Frag(D1))$, $(C2, Frag(C2))$, $(D2, Frag(D2))$, …The approximation $Appr(t) = Coe \cdot Step(t)$ where $Step(t)$ is defined by (1) and the coefficient

$$Coe = ||L_2(Frag||/||L_2(Step)||. \tag{3}$$

The criterion for optimization is

$$||L_2(Frag - Appr)|| \to \min. \tag{4}$$

Instead of $Frag(t)$, we can raise the same problem for the function $Frag(K \cdot t)$ for any positive K. The main result of this section is the following Proposition.

Proposition 1 *Let $Frag(K \cdot t)$ be a function having to continue derivative on the interval $[0, E/K]$. The values Thr_1, Thr_2, where (4) takes its minimal values are independent of K.*

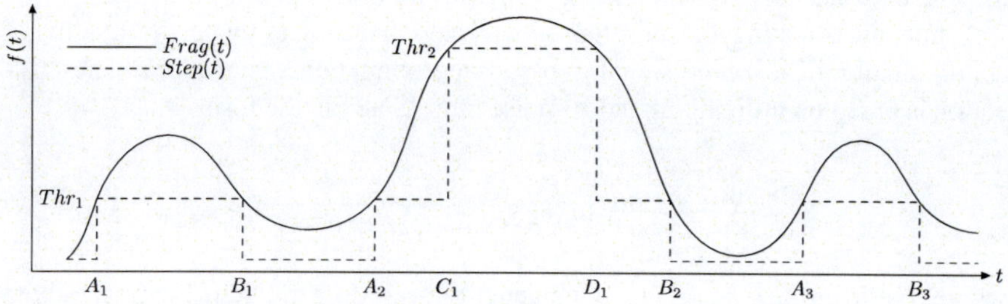

Fig. 1 Approximation of source signal (positive part)

Proof The function $Frag(t)$ is defined for $t \in [0, E]$. We restrict ourselves with the case of two thresholds $Thr_1 = Frag(A_1)$ and $Thr_2 = Frag(C_1)$ and start considering the situation for $K = 1$. The function

$$Appr(t) = \begin{cases} \mathbf{sign}(Step(t)) \cdot Coe, & \text{if } |Step(t)| = Thr_1, \\ 2 \cdot \mathbf{sign}(Step(t)) \cdot Coe, & \text{if } |Step(t)| = Thr_2, \\ 0 & \text{otherwise.} \end{cases}$$

It means that the current configuration is defined by all boundary points $\{U_i, V_i\}$ of the intervals where $Appr(t)$ is constant. For example, for the signal in Fig. 1, the mentioned intervals are $[0, A_1)$, $[A_1, B_1)$, $[B_1, C_1)$, $[C_1, D_1)$ $[D_1, A_2)$, $[A_2, B_2)$, $[B_2, E)$. The interval $[0, E]$ is a union of non overlapped intervals $Inter_i = [U_i, V_i)$ connected with the value of $Step(t)$ at points of those intervals. Let us collect all the intervals corresponding to the same values of $|Step(t)|$ 1, 2, and 0 into sets Φ, Ψ, and Ξ, respectively. In what follows, we use notation $Foo(\{U_i, V_i\})$ to emphasize the function's dependence on all intervals. Under the definition,

$$R(\{U_i, V_i\}) = ||L_2(Frag - Appr)||^2 = \sum_{Inter_i \in \Phi} \int_{Inter_i} (Frag(t) - Appr(t))^2 \, dt$$

$$+ \sum_{Inter_j \in \Psi} \int_{Inter_j} (Frag(t) - Appr(t))^2 \, dt + \sum_{Inter_z \in \Xi} \int_{Inter_z} Frag^2(t) \, dt$$

Removing the parentheses, we get

$$R(\{U_i, V_i\}) = ||L_2(Frag)||^2 + Coe^2 \sum_{Inter_i \in \Phi} \mathbf{size}(Inter_i)$$

$$-2 \cdot \int_{Inter_i} Frag(t) \cdot Appr(t) \, dt$$

$$+ (Coe \cdot 2)^2 \sum_{Inter_j \in \Psi} \mathbf{size}(Inter_j) - 2 \cdot \int_{Inter_i} Frag(t) \cdot Appr(t) \, dt.$$

We have

$$Coe^2 \left(\sum_{Inter_i \in \Phi} (\mathbf{size}(Inter_i) + 4 \sum_{Inter_j \in \Psi} \mathbf{size}(Inter_j) \right)$$

$$= ||L_2(Appr)||^2 = ||L_2(Frag)||^2,$$

so

$$R(\{U_i, V_i\}) = 2||L_2(Frag)||^2$$
$$-2 \cdot Coe\left(\sum_{Inter_i \in \Phi} \pm 1 \int\limits_{Inter_i} Frag(t)\,dt + \sum_{Inter_j \in \Psi} \pm 2 \int\limits_{Inter_j} Frag(t)\,dt\right), \tag{5}$$

and

$$Coe^2 = \frac{||L_2(Frag)||^2}{\sum_{Inter_i \in \Phi} \mathbf{size}(Inter_i) + 4\sum_{Inter_j \in \Psi} \mathbf{size}(Inter_j)}. \tag{6}$$

In other words

$$R(\{U_i, V_i\}) = 2||L_2(Frag)||^2 + G(\{U_i, V_i\}). \tag{7}$$

Here $G(\{U_i, V_i\})$ is a function depending on the set of all intervals and related to step-function values.

Now, consider the general case where we are dealing with $FragK(t) = Frag(K \cdot t)$, $t \in [0, E/K]$. We have $||L_2(FragK)||^2 = ||L_2(Frag)||^2/K$, $Thr'_1 = FragK(A'_1) = Frag(K \cdot A'_1)$, $Thr'_2 = FragK(C'_1) = Frag(K \cdot C'_1)$. If U, V are boundary points of any interval in set Φ or Ψ, then

$$\int\limits_U^V FragK(t)\,dt = \frac{1}{K}\int\limits_{KU}^{KV} Frag(t)\,dt,$$

$$CoeK = \frac{||L_2(Frag)||^2}{\sum_{Inter_i \in \Phi} K \cdot \mathbf{size}(Inter_i) + 4\sum_{Inter_j \in \Psi} K \cdot \mathbf{size}(Inter_j)},$$

and $K \cdot \mathbf{size}([U, V]) = K \cdot U - K \cdot V$. It means that for any K and given boundary points $\{U_i, V_i\}$ of the intervals

$$R(\{U'_i, V'_i\}) = 2 \cdot ||L_2(Frag)||^2/K + G(\{K \cdot U_i, K \cdot V_i\})/K \tag{8}$$

Thus searching for minimal value for $R(\{U'_i, V'_i\}$ we go to the same values of the optimal levels. □

Since the spectrum of $Frag(K \cdot t)$ depends on K, we can apply the same thresholds to construct step-function for different speech file fragments. The fast evaluation method for arbitrary fragment thresholds is developed in the next section.

Remark 1 Although $\min ||L_2(Frag - Appr)||$ depends on K (8), the quality of approximation in terms of SNR is independent of K,

$$SNR = \frac{||L_2(Frag)||^2}{||L_2(Frag - Appr)||^2} = \frac{||L_2(Frag)||^2}{2 \cdot ||L_2(Frag)||^2 + G(\{U_i, V_i\})}.$$

Remark 2 There is a special case where optimal values of SNR can be found theoretically, $Frag(t),\ t \in [0, \pi]$ (see Fig 2). For example, if we leverage a single threshold $Thr_1 = \sin(A)$, then we have a single interval $[A, \pi - A]$ where $Appr(t) = Coe \cdot Thr$. According to (7)

$$Dist^2 = ||L_2(\sin - Appr)||^2 = 2||L_2(\sin)||^2 + 2 \cdot Coe \cdot Thr_1 \int_A^{\pi - A} \sin(t)\, dt = \pi - 4 \cdot Coe \cdot \cos(A), \quad Coe \cdot Thr_1 = \sqrt{\frac{\pi}{2 \cdot (\pi - 2 \cdot A)}}. \tag{9}$$

In the case of two thresholds, $Thr_1 = \sin(A)$, $Thr_2 = \sin(C)$,

$$Dist^2 = \pi + 4Coe \cdot (Thr_1 \cdot (\cos(C) - \cos(A)) - Thr_2 \cdot \cos(C)), \quad Coe^2 = \frac{\pi}{2(2Thr_1^2 \cdot (C - A) + Thr_2^2 \cdot (\pi - 2C))}. \tag{10}$$

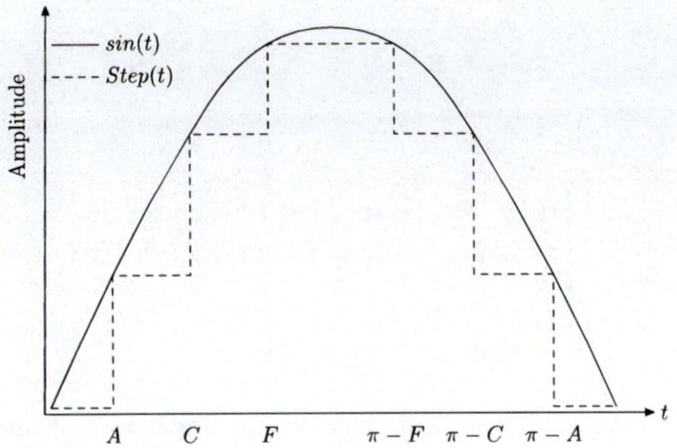

Fig. 2 Approximation of Sine function (three thresholds)

Table 1 Best values of thresholds and SNRs while approximation Sine with step-function depending on the number of thresholds

Number of thresholds	Thr_1	Thr_2	Thr_3	SNR (dB)
1	0.39	–	–	13
2	0.27	0.6	–	15
3	0.2	0.44	0.72	17.8

In the case of three thresholds, $Thr_1 = \sin(A)$, $Thr_2 = \sin(C)$, and $Thr_3 = \sin(F)$ we have

$$\begin{aligned} Dist^2 &= \pi + 4Coe \cdot (Thr_1 \cdot (\cos(B) - \cos(A)) + \\ Thr_2 &\cdot (\cos(C) - \cos(F)) - Thr_3 \cdot \cos(F)), \\ Coe^2 &= \frac{\pi}{2(2Thr_1^2 \cdot (C - A) + 2Thr_2^2 \cdot (F - B) + Thr_3^2 \cdot (\pi - 2F))}. \end{aligned} \tag{11}$$

Now, the best value of Thr_1, Thr_2, Thr_3 can be found by tabulation of (9), (10), and (11). All the results concerning the cases of one, two, and three thresholds are collected in Table 1.

3 Practical Evaluation of Optimal Thresholds for Speech Files Fragments

In this section, we consider the procedure to obtain three thresholds providing a suboptimal approximation of a signal $Source = a_0, a_1, \ldots$ by a step-sequence $Appr$ with a range in (12)

$$Range = \{-3, -2, -1, 0, 1, 2, 3\}, \tag{12}$$

and defined by

$$Appr[n] = \begin{cases} 0 & \text{if } |a_n| < Thr_1, \\ 1 \cdot \mathbf{sign}(a_n) & \text{if } |a_n| \geq Thr_1 \& |a_n| < Thr_2, \\ 3 \cdot \mathbf{sign}(a_n) & \text{if } |a_n| \geq Thr_3. \\ 2 \cdot \mathbf{sign}(a_n) & \text{otherwise}. \end{cases} \tag{13}$$

In practice, we are dealing with a discrete sequence, and any evaluation of optimal thresholds can be gained by applying the k-means procedure [11]. K-means produce centroids of clusters, and one has to leverage middles between two neighbor centroids as thresholds obtaining coding (13). The results related to the optimal thresholds for a fragment $Frag$ can be carried over obviously to the fragment $U \cdot Frag$ with arbitrary constant U.

3.1 Reduction to the Case of a Single Threshold

In a real situation of data transmission, threshold calculation for any fragment must be performed very fast. Hence, our goal is to drop off the resources needed for getting the results. We show that the situation with a few thresholds can be reduced to the case of a single threshold. For example, in three thresholds, the optimal values of Thr_1, Thr_2, Thr_3 for $U \cdot Frag(t)$ depend on U, but ratios Thr_2/Thr_1 and Thr_3/Thr_1 does not. If $Frag(t) = \sin(t)$, then the third row of Table 1 brings the ratios 2.2, 3.6 for these thresholds. The main idea of the reduction assumes that those ratios are constant for all fragments of speech files and close to obtained for sin function. In Table 2, the results of experimental calculations of SNRs and ratios are collected. The data are established as follows. File #1 is a part of the audiobook; files ##2–5 received by direct records from TV programs. All of them are saved with a sampling frequency of 44100 Hz. File #6 is a result of the downsampling of #1. In each file, 12 fragments of size 256 were selected. The k-means procedure evaluates optimal thresholds [12] the way explained above. For each file, all 12 ratios SNRs are collected, and the found medians are placed in Table 2. For each file, all 12 ratios are collected, and the found median is placed in Table 2. Now, while searching for evaluation of optimal thresholds, we have to evaluate just Thr_1 and set

$$Thr_2 = 3Thr_1,\ Thr_3 = 5Thr_1. \tag{14}$$

The results of calculation SNR with thresholds defined by (14) are placed in Table 3. Comparing SNRs in Tables 2 and 3 shows that they weakly depend on the implemented method, and the second approach fits collection statistics.

3.2 Evaluation of Suboptimal Thresholds Employing Linear Regression

At this point, we develop a method for fast evaluation of Thr_1. Usage k-means provides collection statistics for linear regression implementation to evaluate the

Table 2 Experimental median values of ratios of thresholds and SNRs in the case of three thresholds

Index	File	Thr_2/Thr_1	Thr_3/Thr_1	SNR (dB)
1	Professional Russian narrator	2.9	4.8	14.2
2	Russian female voice	3.0	5.0	13.9
3	Russian male voice	3.0	5.0	14.0
4	Tatar female voice	2.9	4.8	14.9
5	Tatar male voice	3.1	5.2	15.2
6	Sample frequency 4410 Hz	2.9	5.0	12.2

Table 3 Experimental median values of SNR with suboptimal threshold defined in (14) and various length *Len* of fragments

Index	File	Thr_2/Thr_1	Thr_3/Thr_1	SNR (dB)
1	Professional Russian narrator	14.0	13.2	13.1
2	Russian female voice	14.2	13.7	13.6
3	Russian male voice	14.1	13.5	13.0
4	Tatar female voice	15.2	14.8	14.2
5	Tatar male voice	14.2	13.9	13.5
6	Sample frequency 4410 Hz	12.6	12.3	12.0

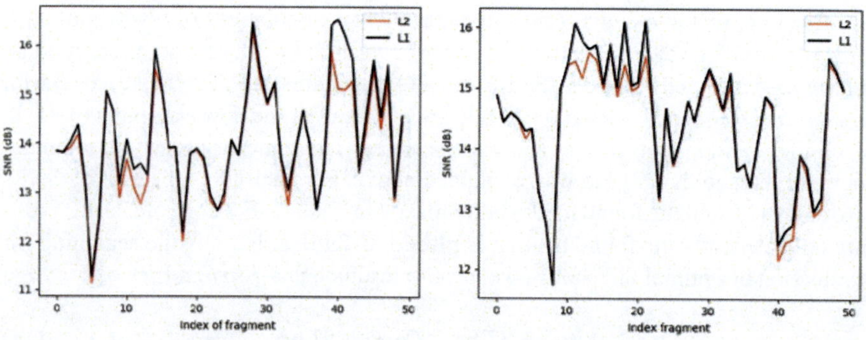

Fig. 3 SNR of approximation source signal by suboptimal step-wise function. Label 'L1' –SNR established by exhaustive search of thresholds using (14), label 'L2'—SNR through thresholds obtained by regression with coefficients calculated for file #3 in Table 3. (**a**) file #4, (**b**) file #5

suboptimal threshold. The above experiments suggest that a connection between the maximum of the fragment and the produced suboptimal threshold exists. Having statistics for a big set of fragments, one can obtain a simple formula to receive the threshold. Utilizing the fragment's maximal value as an argument in linear regression is a bad idea since that parameter is very variable in the speech file. This circumstance leads to insufficient quality of the approximation. We propose to leverage the fragment $Frag$ features: the maximal value Mx and the standard deviation Std to avoid this obstacle. If the threshold Thr relates to $Frag$, then Thr/Mx corresponds to Mx/Std in the training procedure. We leveraged the realization of the regression procedure through the package [12]. The obtained parameters of regression vary from one speaker to another. Nevertheless, the parameters established on the base of a speech file presented in our database produce acceptable SNR for speech files belonging to other speakers. In Fig. 3, the results of the two experiments are shown. Coefficients of regression are calculated for file #3 in Table 3. The target values in the training procedure are Thr_1Opt/Mx in (14). Here Thr_1Opt is found by exhaustive search of Thr_1 for gaining maximal SNR in approximation procedure, and Mx is the maximal value of fragment. The training procedure's argument is Mx/Std, where Std is the fragment's standard deviation. The total number of fragments used for training is 50, and the length of each fragment is 512 samples, sample frequency equals 4,4100 Hz. The established regression function is used for obtaining suboptimal values of Thr_1 for two other files. The graphs show that such a procedure brings the acceptable quality of the approximation. In what follows, we use the notation

$$Thr_1 = \mathbf{Thr}(File) \tag{15}$$

for the value of Thr_1 obtained through an exhaustive search of threshold in fragments of the file $File$ with the assumption (14). The approximation of the file $File$ was produced by (13) with the application of

$$Appr = \mathbf{ApprProc}(File, Thr_1). \tag{16}$$

4 Linear Regression to Enhance the Perception of the Step-Wise Signal

This section and below section show three thresholds in the approximation procedure, and files are written with 4,4100 Hz sampling frequency. Applying a fast procedure for conversion source file into its step-wise version, we receive a step-function with a range of values in $Range$, (12). The idea of enhancing the quality of speech perception by a human, presented by step-function, based on linear regression, is as follows. There is a one-to-one correspondence between samples of source signal and values of step-function. We construct a window of odd length M. The window slides over the values of the step-function. Any position of the window defines the input signal for the regression. Those are values of the step function inside the window. The target value for the net at the window's current position is the sample corresponding to the symbol in the middle of the window.

4.1 *Distribution of Input Signals for Regression*

Let $Seq = (s_0, s_1, \ldots, s_{M-1})$, $s_i \in Range$ (12) be a sequence of symbols inside the window of length M. We can think of this sequence as a record of integer for radix 7. That is the way to obtain the distribution of signal on inputs of regression. The graphs in Fig. 4a show that the input signals distribution has a special form, it has peaks at the same points, and those points are independent of source signals. Additional analysis establishes that points of local maximums correspond to constant intervals of kind $(s, s, \ldots, s)$, $s \in Range$, (12) of length M. The histogram structure is independent of M if M is no more than half of the constant interval's maximal lengths. Such a structure of the histogram is a feature

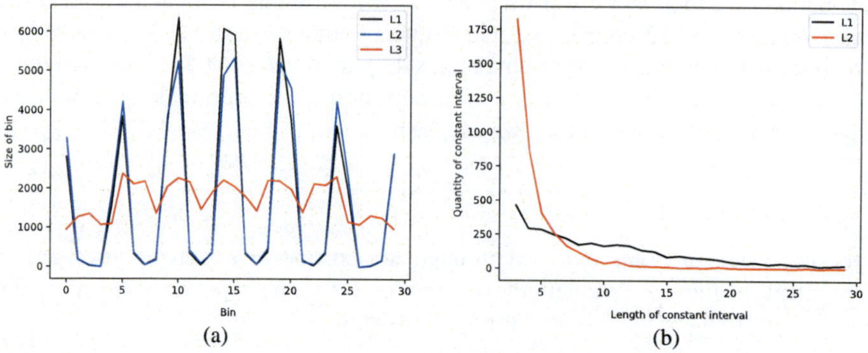

Fig. 4 Distribution of input signals and lengths of constant intervals. (**a**) Length of window = 5, label "L1"—file #2, label "L2"—file #4 in Table 3, label "L3"—noise file. (**b**) Distribution of lengths of constant intervals, label "L1"—file #1, label "L2"—file #6 in Table 3

of a speech file. For the random file, we see another picture. This phenomenon was mentioned in [13], and its origin is a theme of additional research.

The graphs in Fig. 4b show the problem we face when selecting the length M of the window. Let us have a constant interval of the length L, $L >> M$. While the window slides inside this interval, the regression's input will be the same for any window position, but the target values are different. That means that M and L must be comparable numbers; otherwise, the training procedure can not be successful. It follows from Fig. 4 that reducing sample frequency reduces the number of long constant intervals.

4.2 Linear Regression and Training Procedure

As before, we use the package [12] for the realization of our calculations. We chose $M = 57$ for the graphs (a) in Fig. 4, We can not assert that it is the best possible value, but it fits our requirements. The set of arguments for training is a list of fragments of length M, having the form $[(s_0, s_1, \ldots, s_{M-1}), (s_1, s_2, \ldots, s_M), \ldots]$ consisting of values of $Appr$, and the target values $Targs$ are the values from the source speech file, corresponding to the position integer part of $M/2$ of arguments. Let $RegSign$ be the signal produced by the regression. That is a particular case of the implementation of a finite impulse response filter. For comparing the effect of linear regression on the sound's perception, we compare SNRs of $Targs$ to $Appr$ and $Targs$ to $RegSign$. What more, we used the coefficients, established for one file, for construction $RegSign$, which is produced on the base of the $Appr$ function belonging to another speaker. The results are presented in Fig. 5. One can see that the SNRs gained by regression exceeds the SNRs gained by step-wise approximation for most fragments. The examples show that the regression coefficients can be

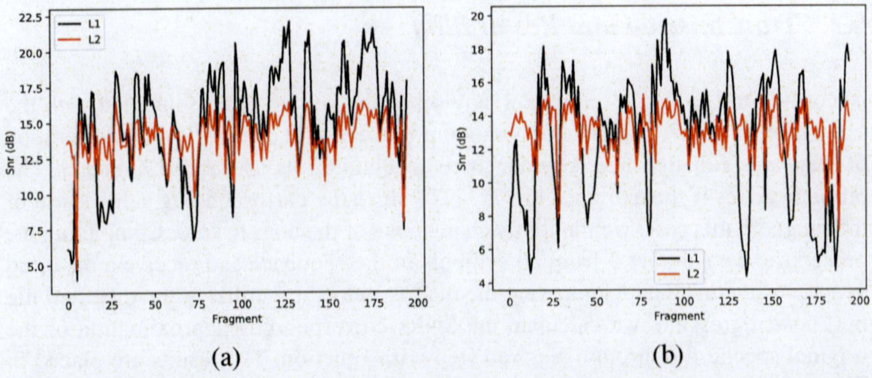

Fig. 5 Compare of SNRs gained by approximation of speech file by step-wise function and linear regression. **(a)** file #4 in Table 3, label "L1"—regression, "L2"—step-wise approximation. **(b)** "L1"—regression of step-wise function built on the base of file #3 utilizing the coefficients of regression established for file #4, "L2"—a step-wise approximation of file #3 in Table 3

established one time and can be leveraged to enhance step-wise approximations built by other speakers. In practice, the latter situation leads to a bit little difference in quality, but produced results are applicable in any case. The procedure production of coefficients of linear regression using $Appr$, $Targs$, and M denote as

$$Reg = \mathbf{RegFun}(Appr, Targs, M). \qquad (17)$$

5 Speech Transfer in a Noisy Environment

This section presents how our technique can be implemented for speech transfer via a noised channel. We investigate the model where any bit in a stream can be inverted with a constant probability P. There are no assumptions about the events' joint distribution.

5.1 *Preparation of Step-Wise File for Transmitting*

Let us suppose that the step-wise sequence that has to be transmitted has the form $s_0, s_1, \ldots, s_i \in Range$, (12). We add 3 to each symbol in the sequence and convert it into a new one with items in $\{0, 1, 2, 3, 4, 5, 6\}$. Then we change any symbol in the new sequence by its binary code using 3 bits for any item. That is the stream which must be transmitted via the channel. Conversion step-function to binary stream denote as

$$Stream = \mathbf{StreamFun}(Appr). \qquad (18)$$

5.2 Transmission and Restoration

According to the suggested model, any bit in the stream can be inverted with the same probability P, but the total number of bits does not change. For the restoration of step-wise function, we consider three sequential bits as a radix 2 integer. The problem arises if the extracted bits is '111' since the corresponding value is out of the range. In this case, we randomly change one of the ones to zero. Completing the procedure, we subtract 3 from all symbols in the sequence and receive a distorted version of the original. Comparing the quality gained by utilizing the distorted file by linear regression, we calculate the SNRs corresponding approximation of the original speech file through restored step-wise function. The results are placed in Fig. 6. One can see that enhancing by regression leads to significant improvement, whereas usage of constant coefficients of regression brings acceptable results. Thus, the regression coefficients calculated one time can be leveraged for enhancing other

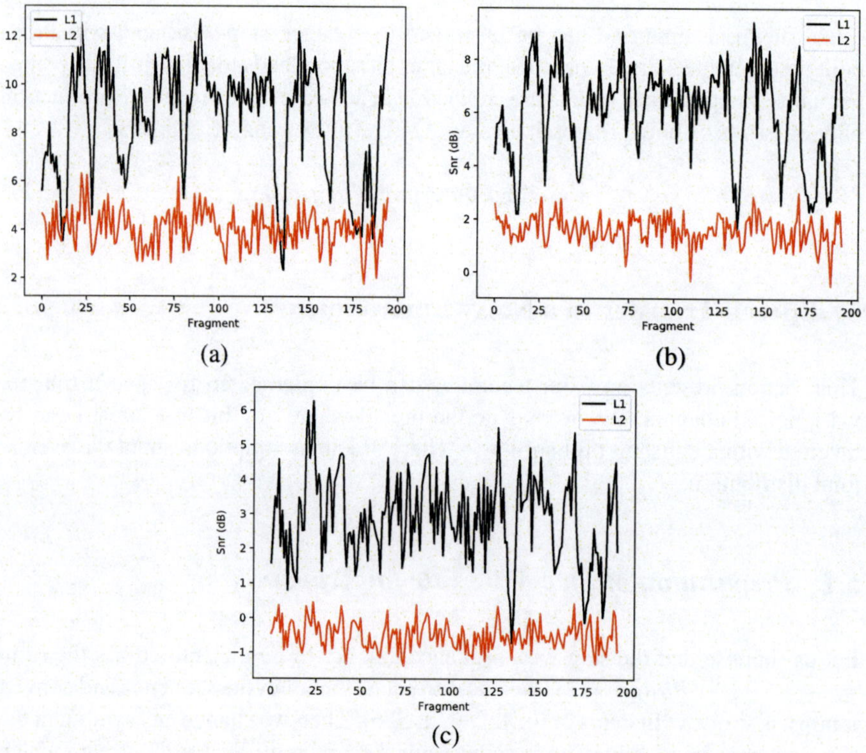

Fig. 6 Compare of SNRs restored and enhanced by linear regression distorted step-wise functions for various P. Used coefficients of the regression obtained based on file #3, and the speech file is #4 in Table 3. **(a)** file #4 in Table 3. Labels: "L1"—enhanced version, "L2"—direct restoration; **(a)** $P = 0.05$, **(b)** $P = 0.1$, **(c)** $P = 0.2$

Algorithm 1 Data transfer and restoration

Require: Length of window M, files $File_0, File_1, \ldots, File_N$

$Thr_1 \leftarrow \mathbf{Thr}(File_0)$ {Employ (15)}

$Appr_0 \leftarrow \mathbf{ApprProc}(File_0, Thr_1)$ { According to (16)}

$Reg_0 \leftarrow \mathbf{RegFun}(Appr_0, File_0, M)$ { According to (17)}

$Stream_0 = \mathbf{StreamFun}(Appr_0)$ {Employ (18)}

$Reg_1 \leftarrow Reg_0$ {Coefficients copied to receive the point of the channel without errors}

$\overline{Stream_0} \leftarrow Stream_0$ {Receiving distorted version}

for $i = 1$ to N **do**

    $Appr_i \leftarrow \mathbf{ApprProc}(File_i, Thr_1)$

    $Stream_i = \mathbf{StreamFun}(Appr_i)$

    $\overline{Stream_i} \leftarrow Stream_i$

end for

for $i = 0$ to N **do**

    $RegSign_i \leftarrow Reg_1, \overline{Stream_i}$ {Regression}

end for

speech files. The data transfer and its restoration, according to our technique, are presented in Algorithm 1.

Acknowledgement This paper has been supported by the Kazan Federal University Strategic Academic Leadership Program ("PRIORITY-2030").

References

1. Ramasubramanian, V., Doddala, H.: Ultra low bit-rate speech coding. Springer, Heidelberg (2015) https://doi.org/10.1007/978-1-4939-1341-1
2. Moon, T.K: Error correction coding. Wiley (2005)
3. Sambur, M.: Adaptive noise canceling for speech signals. IEEE Trans. on Acoustics, Speech, and Signal Processing, **26** 419–423 (1978) https://doi.org/10.1109/TASSP.1978.1163137
4. Morise, M., Yokomori, F., Ozave, K.: WORLD A vocoder-based high-quality speech synthesis system for real-time applications. IEICE Trans. Inf. & SYST.,**E99-D**, 1877–1884 (2016)
5. Gaultier, C., Kiti'c, S., Gribonval, R., Bertin, N.:Sparsity-based audio declipping methods: overview, new algorithms, and large-scale evaluation. Preprint, hal-02611226, 1–14 (2020)
6. Kavalekalam, M.S., Nielsen, J.K., Christensen, M.G., Boldt, J.B.: A study of noise PSD estimators for single channel speech enhancement. In: 2018 IEEE International Conference on Acoustics, Speech and Signal Processing (ICASSP), Calgary, AB, 5464–5468 (2018) https://doi.org/10.1109/ICASSP.2018.8461703
7. Hu, Y., Loizou, P.C.: Subjective evaluation and comparison of speech enhancement algorithms. Speech Commun, **49**, 588–601 (2007)
8. Wang, Y., Wand, D.:Towards scaling up classification-based speech separation. IEEE Trans. Audio, Speech, and Lang. Proc. **21** 1381–1389 (2013)
9. Saleem, N., Irfan, M., Chen, X., Ali, M.: Deep neural network based supervised speech enhancement in speech-babble noise. In: 2018 IEEE/ACIS 17th International Conference on Computer and Information Science (ICIS), Singapore, 871–874, (2018) https://doi.org/10.1109/ICIS.2018.8466542
10. Lloyd, S.P.:Least squares quantization in PCM. IEEE Trans. Inform. Theory, **IT-28** 129–136 (1982)

11. Girod, B.: Image and Video Compression. https://web.stanford.edu/class/ee398a/handouts/lectures/05-Quantization.pdf. Cited 29 Sep. 2020
12. Pedregosa, F. et al.: Scikit-learn: machine learning in Python. Journal of Machine Learning Research, **12**, 2825–2830 (2011).
13. Latypov R., Stolov E.:A new method towards speech files local features investigation. In: IEEE 62nd ELMAR Symposium, Zadar, Croatia (2020)

Simulation of Two-Phase Flow Toward a Horizontal Multistage Hydraulically Fractured Well Using Accelerated Explicit-Implicit Algorithms

Alexander B. Mazo and Marsel R. Khamidullin

Abstract Explicit-implicit algorithms for accelerating the three-dimensional two-phase flow calculation towards a horizontal well with a multistage hydraulic fracturing are presented. Acceleration is achieved by dividing the calculation domain to local zones where depending on the local Courant number an explicit or implicit scheme for the saturation transfer equation is applied.

1 Introduction

Mathematical models of fluid flow towards a horizontal well (HW) with multistage hydraulic fracturing (MHF) are based on common equations of two-phase (oil—water) flow [1], which contain a parabolic equation for pressure p and a hyperbolic equation of saturation s transfer. There are three main schemes for the numerical solution of the equations of traditional models of reservoir penetrated by a system of vertical wells.

The most common scheme is IMPES (Implicit Pressure, Explicit Saturation) [2–4] when pressure is calculated according to an implicit scheme, and saturation according to an explicit one. This scheme is conditionally stable; the grid step h and time step τ must satisfy the Courant–Friedrichs–Lewy (CFL) condition

$$\tau \ll \frac{h}{\max\left(f'\,|u|\right)} \quad \text{or } \mathrm{C} = \frac{\tau \max\left(f'\,|u|\right)}{h} \ll 1, \tag{1}$$

where u is the flow rate, $f' = df/ds$ is the derivative of the Buckley–Leverett function. The feature of the problem is that $|u|$ sharply increases near wells while condition (1) requires solving an explicit scheme for s with a small step τ which significantly slows down the numerical solution of the problem. There is a simple

A. B. Mazo · M. R. Khamidullin (✉)
Kazan Federal University, Kazan, Russia

I. B. Badriev et al. (eds.), *Mesh Methods for Boundary-Value Problems and Applications*, Lecture Notes in Computational Science and Engineering 141,
https://doi.org/10.1007/978-3-030-87809-2_24

way to speed up computations according to the IMPES scheme [5]. The calculation domain is divided into several zones where maximum $|u|$ and the corresponding time step τ which guarantees the fulfillment of condition (1) are defined. Thus, to get a solution for the time interval Δt in each of the zones a different number of time layers defined as $N_s = \Delta t/\tau$ is required.

Another common way to speed up computation is the Fully Implicit Method (FIM) [6], when p and s are calculated using a purely implicit scheme. In this case iterative procedures are used to find a solution to a nonlinear system at each time layer. Therefore the FIM scheme requires more computational resources; however unlike the IMPES scheme it is unconditionally stable.

The combination of the IMPES and FIM methods advantages is implemented in the Adaptive Implicit Method (AIM) [7–9]. The main idea of this method is to locally apply FIM or IMPES, depending on the computational efficiency according to the stability condition in the form of (1). The features of applying these schemes in detail were studied in [10].

Within the considered design schemes various methods of computational mathematics and methods of calculations organization are used to speed up the two-phase flow simulation.

In [11] an iterative method was proposed for the solution of combined problems for saturation and pressure which accelerates the calculation by 30% compared to FIM and in contrast to IMPES converges at a larger Courant number C. The GMRES (Generalized Minimal Residual) method [12] with AMG (Algebraic Multigrid) [13] as preconditioner is used to solve systems of linear algebraic equations. For spatial approximation the finite difference method and for time discretization the inverse Euler method is used. The work [14] is devoted to the design of optimal iterative methods for solving nonlinear flow in porous media at each time layer. It is shown that the nonlinear multigrid method is more efficient than Newton's method; moreover, this reduces the cost of RAM. The idea of the method is that before the linearization of equations by Newton's method a multilevel iterative method with a preconditioner is used. In [15] various algorithms for solving grid equations using adaptive grids with local refinement are considered.

Recently, to speed up the simulation of multiphase flow in the reservoir on detailed grids parallel computations on multiprocessors are used. The task is divided into several subtasks which are distributed among the processors. In [16] an algorithm for parallelizing the saturation calculation is presented. The FIM method is used in the areas near the wells and the IMPES method is used in the rest of the area. In [17] numerical algorithms were proposed for parallelizing the solution of two-phase flow problems on grids with local refined sections in which the AIM scheme is used to calculate the saturation. The paper [18] describes the features of the implementation of algorithms for solving this problem on heterogeneous computing systems.

In this article explicit-implicit computational acceleration schemes are used to solve a three-dimensional problem of two-phase flow near a horizontal well with a multistage hydraulic fracturing. A distinctive feature of this problem is that the flow rate $|u|$ increases significantly not only near the horizontal wellbore but also in the hydraulic fractures.

2 Problem Statement

To illustrate the effectiveness of the proposed numerical algorithms the following model problem with one hydraulic fracture is considered. Solution area D is a part of the reservoir in the form of a rectangular parallelepiped of height $2H$ with rounded edges. In the center of the domain there is a cylindrical cut γ—a well of radius r_w and length L (Fig. 1). The area D is vertically limited by the planes $z = \pm H$—the top and bottom of the formation. The lateral surface Γ, located at a distance l from the well γ is an external reservoir boundary. The hydraulic fracture is located orthogonal to the Oy axis directed along the wellbore. Fracture faces are a pair of rectangular planes F^+ and F^- with normals in the directions $n^\pm = \pm y$ having dimensions $2H \times 2h$ located at a distance $2\delta \ll H$ from each other (fracture opening). The fracture has permeability k^f and porosity m^f which are much higher than the absolute permeability k and porosity m of the formation. The reservoir is considered to be homogeneous. Capillary and gravitational forces are neglected.

In dimensionless form the equations for pressure and saturation in the domain D are following [1]

$$\beta\frac{\partial p}{\partial t} + \mathrm{div}\mathbf{u} = 0, \qquad \mathbf{u} = -\sigma\,\mathrm{grad}\,p. \tag{2}$$

$$m\frac{\partial s}{\partial t} + \mathrm{div}\left(f\left(s\right)\mathbf{u}\right) = 0, \tag{3}$$

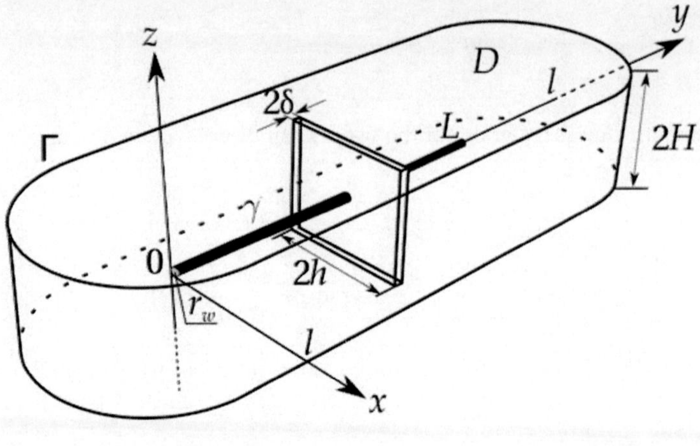

Fig. 1 HW scheme with single MHF fracture

$$f(s) = \frac{k_w(s)}{\sigma(s)}, \quad \sigma = k_w(s) + \mathrm{K}_\mu k_o(s),$$
$$k_w(s) = s^3, \quad k_o(s) = (1-s)^3. \tag{4}$$

Here $\beta \sim 10^{-3}$ is the elastic capacity; $f(s)$ is the Buckley-Leverett function; $k_w(s), k_o(s)$ are the relative phase permeabilities of water and oil respectively; $\sigma(s)$—transmissibility; K_μ is the ratio of water to oil viscosity.

Dimensionless initial conditions

$$t = 0, \ (x, y, z) \in D: \quad p = 1, \ s = 0 \tag{5}$$

means that the reservoir is saturated with oil at hydrostatic pressure. Top and bottom of the formation (coordinates are normalized to H) are impermeable

$$z = \pm 1: \quad u_n = -\sigma \frac{\partial p}{\partial n} = 0, \tag{6}$$

where n is the outward normal.

For $t > 0$ the waterflooding process is simulated at a constant pressure $p = 1$ and water saturation $s = 1$ on the contour Γ while the pressure $p = p_\gamma = 0$ is set at the well γ. The hydrodynamic interaction of the reservoir, the surface γ of the well and the surfaces $F^\pm$ of the fracture is expressed in the continuity of pressure and the normal velocity of the flow.

The equation for the dimensionless pressure p^f averaged over the fracture opening 2δ is written as follows:

$$\Delta_{xz} p^f + \frac{1}{2M} \sigma \left. \frac{\partial p}{\partial y} \right|_{F^-}^{F^+} = 0, \quad -h < x < h, \ -1 < z < 1, \quad M = \frac{k^f \delta}{kH},$$
$$x = \pm h, \ z = \pm 1: \quad \frac{\partial p^f}{\partial n} = 0; \qquad p^f = p_\gamma = 0 \ \text{ for } \ (x, y, z) \in \gamma. \tag{7}$$

The dimensionless equation for the water saturation s^f in the fracture looks like:

$$m^f \frac{\partial s^f}{\partial t} + \nabla_{xz} \left(f\left(s^f\right) \mathbf{u}^f \right) + \frac{2\delta}{H} \left(f(s) u_n \right) |_{F^-}^{F^+} = 0,$$
$$\mathbf{u}^f = -\frac{k^f}{k} \sigma\left(s^f\right) \operatorname{grad} p^f, \quad \frac{k^f}{k} \gg 1. \tag{8}$$

Note that the last term of this equation models inflow of water from the reservoir into the well and is calculated on the fracture faces F^+ and F^- from the reservoir side.

The mathematical model of two-phase flow towards an *output* HW with MHF in dimensionless form consists of Eqs. (2)–(6) in reservoir D and (7), (8) in fractures.

3 Explicit and Implicit Finite Volume Schemes

The numerical solution of the problem is based on the finite volume (FV) method [19]. For each finite volume V_i bounded by a set of faces Γ_i^j the average pressures, saturation and normal velocities across the faces are determined as:

$$P_i = \frac{1}{|V_i|}\int\limits_{V_i} p dV, \quad S_i = \frac{1}{|V_i|\ m_i}\int\limits_{V_i} msdV, \quad u_i^{j,n} = \frac{\sigma_i^j}{h_i}\left(P_i - \sum_j \alpha_i^j P_j\right). \tag{9}$$

Here m_i is the average porosity of a finite volume V_i, $u_i^{j,n}$ is the average flow rate through the faceΓ_i^j in the direction of the outward normal n, $|V_i|$ is the volume of V_i, α_i^j are the collocation coefficients [20, 21], h_i is the distance from the V_i center to the collocation point. Integration of Eqs. (2)–(6) in the domain D and (7) and (8) over the volume V_i leads to the following grid scheme.

For D it has the form

$$\beta_i \frac{\hat{P}_i - P_i}{\tau} + \sum_j u_i^{j,n}\left|\Gamma_i^j\right| = 0, \tag{10}$$

$$m_i\,|V_i|\,\frac{\hat{S}_i - S_i}{\tau} + \sum_j f_i^j u_i^{j,n}\left|\Gamma_i^j\right| = 0, \tag{11}$$

where $\left|\Gamma_i^j\right|$ is the face area. The notation $\hat{P}\,(t) = P\,(t+\tau)$ is used for the function on the upper time layer.

Equations in a fracture for a flat finite volume $V^f{}_i$ whose boundaries Γ_i^j are line segments, are written in a similar way. The pressure is determined by the equation

$$\begin{gathered}\sum_j \left(u^f\right)_i^{j,n}\left|\Gamma_i^j\right| + \frac{1}{2M}\left|V^f{}_i\right|\left[\frac{P_i^f - P^+}{h_i^+} + \frac{P_i^f - P^-}{h_i^-}\right] = 0,\\ \left(u^f\right)_i^{j,n} = \frac{\sigma\left(S^f\right)}{h_i}\left(P_i^f - \sum_j \alpha_i^j P_j^f\right),\end{gathered} \tag{12}$$

where $h_i^{\pm}$ are the distances from the center of the flat volume $V^f{}_i$ to the centers of the reservoir finite volumes adjacent to the faces of the fracture $F^{\pm}$ and $P^{\pm}$ are the average pressures in these finite volumes. The finite volume equation for water saturation in a fracture has the form

$$m_i^f \left|V_i^f\right| \frac{\hat{S}_i^f - S_i^f}{\tau} + \sum_j f_i^j \left(u^f\right)_i^{j,n} \left|\Gamma_i^j\right| + f\left(S^{\pm}\right) Q_w = 0,$$
$$Q_w = \frac{2\delta}{H}\left[\frac{P^+ - P_i^f}{h_i^+} + \frac{P^- - P_i^f}{h_i^-}\right], \tag{13}$$

where Q_w describes the inflow to the fracture element from the reservoir, $S^{\pm}$ are the average saturations in the reservoir FV adjacent to the fracture faces $F^{\pm}$, and ${S^f}_i^j$ is the water saturation at the Γ_i^j face. According to the "upwind" scheme its value is determined by the sign of the velocity $\left(u^f\right)_i^{j,n}$

$${S^f}_i^j = \begin{cases} S_i^f, & \left(u^f\right)_i^{j,n} > 0 \\ S_j^f, & \left(u^f\right)_i^{j,n} < 0. \end{cases} \tag{14}$$

Explicit for saturation IMPES scheme in Eqs. (11) and (13) is obtained if saturation at the current time layer is taken as the argument of the Buckley-Leverett function. If $\hat{S}$ is used on the upper time layer, then it is a completely implicit scheme for FIM.

4 Numerical Solution Algorithms

The solution of the problem for finding pressures $\hat{P}, \quad \hat{P}^f$ by Eqs. (2) and (7) and saturations $\hat{S}, \quad \hat{S}^f$ by Eqs. (11) and (13) on the upper time layer is performed in several stages

1. the pressures $\hat{P}_i, \quad \hat{P}_i^f$ are calculated according to Eqs. (10) and (12) in which the coefficients are calculated for the saturations $S_i, \quad S_i^f$ from the current time layer;
2. total flow rates are calculated through all faces of the finite volumes in the reservoir (9) and hydraulic fractures (12);
3. the saturations $\hat{S}_i$ in the reservoir and fractures S_i^f are calculated.

Stages 1–2 are the same for IMPES and FIM schemes, but stage 3 is different: in the IMPES scheme, the saturation is calculated using explicit formulas, and in the FIM scheme, an extra problem must be solved. Let's consider this issue in more detail.

For the FIM method, Eq. (11) in operator form can be written as a system of nonlinear equations for the problem in the reservoir:

$$A\hat{S} = F(S), \tag{15}$$

where

$$Ay = y + \frac{\tau}{|V|\,m}\sum_j f^j(y)\,u^{j,n}\left|\Gamma^j\right|, \quad F(S) = S. \tag{16}$$

The Newton method is used to solve system (15)

$$\dot{A}_k\left(y^{k+1} - y^k\right) + Ay^k = F, \tag{17}$$

where $k = 0, 1\ldots$—iteration index; $\dot{A}_k$ is a linear operator—the Gateaux derivative of the operator A at the point y^k

$$\dot{A}_k z = z + \frac{\tau}{|V|\,m}\sum_j f'^j\left(y^k\right)u^{j,n}\left|\Gamma^j\right|z. \tag{18}$$

The problem (17) means that at each iteration the linear system of equations must be solved

$$\dot{A}_k\xi^{k+1} = r, \quad r = F - Ay^k, \quad \xi^{k+1} = y^{k+1} - y^k. \tag{19}$$

The Newton iterative process (17) for Eq. (13) is constructed in a similar way. In this case instead of (16) and (18) we have

$$\begin{aligned} Ay &= y + \frac{\tau}{\left|V^f\right|m^f}\sum_j f(y)\left(u^f\right)^{j,n}\left|\Gamma^j\right| + f\left(y^{\pm}\right)Q_w, \\ \dot{A}_n z &= z + \frac{\tau}{\left|V^f\right|m^f}\left[\sum_j f'\left(y^k\right)\left(u^f\right)^{j,n}\left|\Gamma^j\right|z + f'\left(y^{\pm}\right)Q_w z\right], \\ F &= S^f. \end{aligned} \tag{20}$$

Here f' is the derivative of the function f.

The initial approximation y_0 in the iterative process (17) for both problems is specified as a function of saturation at the current time layer: $y^0 = S$. Calculations have shown that for such initial approximation the convergence of Newton's method is not always ensured. In this case several approximations to the solution $\hat{S}$ are preliminary performed using the two-layer iterative process

$$B\frac{S^{(k+1)} - S^{(k)}}{\lambda} + AS^{(k)} = F, \tag{21}$$

where k is the iteration index, B is the preconditioner, λ is the iteration step. For problem (15) and (16) B was chosen as the linear operator

$$By = y + \frac{\tau}{|V|\,m} \sum_{j} y u^{j,n} \left|\Gamma^{j}\right|, \tag{22}$$

and for problem (15) and (20) the operator:

$$By = y + \frac{\tau}{\left|V^{f}\right| m^{f}} \sum_{j} y \left(u^{f}\right)^{j,n} \left|\Gamma^{j}\right| + f\left(y^{\pm}\right) Q_{w}. \tag{23}$$

Note that preconditioners (22) and (23) differ from nonlinear operators A (16) and (20) only in that a linear function y is used instead of the nonlinear Buckley-Leverett function $f(y)$ (see (4)). Both of these functions are monotonic and coincide at the boundaries of saturation variation $y = 0$ and $y = 1$.

The numerical scheme was implemented with the operators A, B, $\dot{A}$ approximated on a grid of finite volumes with a significant refinement towards the well (see Fig. 3); minimum element size $\sqrt[3]{|V|}$ is 0.004, the maximum is 0.46, the total number of grid cells is $N = 10^5$; the time step was $\tau = 0.1$. Numerical experiments showed that for given parameters of the scheme the convergence of Newton's method it is sufficient to take one or two steps according to the method (21) at $\lambda = 0.6$. The residual r norm convergence is shown in Fig. 2; the numbers at the curves indicate the number of preliminary iterations by the method (21).

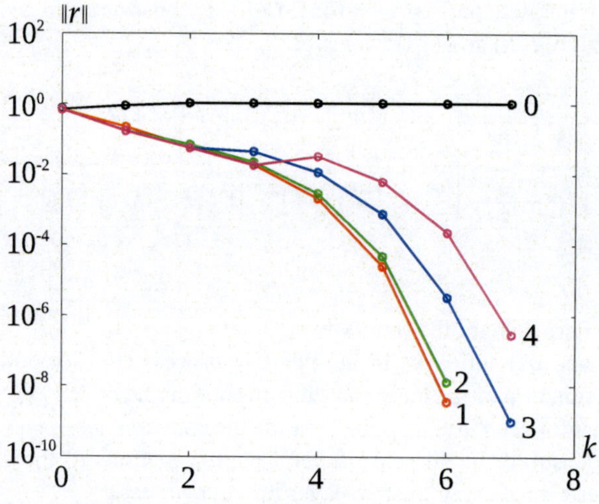

Fig. 2 Convergence of an iterative process for FIM at $\tau = 0.1$

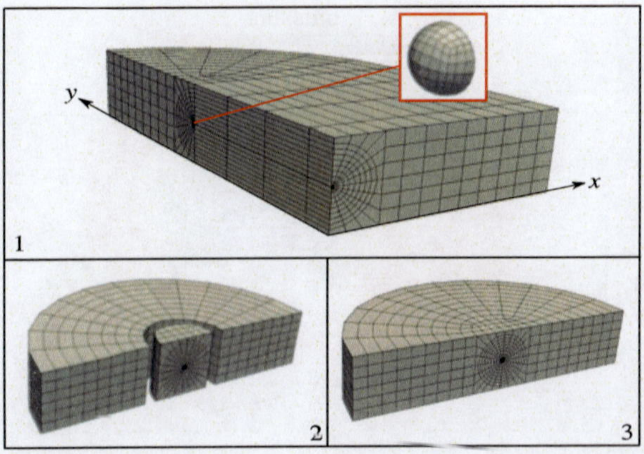

Fig. 3 Finite volumes grid

5 Model Problem

The proposed algorithms were tested by solving a model problem of flow in a homogeneous reservoir. The dimensionless parameters of the problem are as follows: permeability $k = 1$, porosity $m = 0.2$, formation thickness $2H = 2$, contour radius $l = 10$, well radius $r_w = 10^{-2}$, its length $L = 10$, fracture length $h = 3$ and its conductivity $M = 10$, viscosity ratio $K_\mu = 0.1$. Tests were run on Intel (R) Core (TM) i3 CPU 540 3.07 GHz personal computer. The systems of equations for pressure $P,\ P^f$ and saturation $S,\ S^f$ were solved by the iteratively stabilized bi-conjugate gradient method (BCGS Bi-Conjugate Gradient Squared [22]) with the AMG as preconditioner. Iterations were terminated when the residual norm decreased to values $||r|| \leq 10^{-6}$.

In Fig. 3 a grid of finite volumes covering domain D is presented. The features of the grid are as follows: refinement near the well, the wellbore ends are modeled in the form of hemispheres which are obtained as a result of stretching a structured cubic grid onto a sphere (Fig. 3, 1). In the remaining area a structured cylindrical finite volume mesh is constructed (Fig. 3, 2). Then the spherical and cylindrical meshes are combined using the transformation of coordinates according to the Laplace equation [23] (Fig. 3, 3). The mesh along the wellbore is constructed by extruding the 2D mesh in the Oxz plane.

In Fig. 4 the pressure distribution in the domain D at the moment $t = \tau$ is presented.

The streamlines and the pressure field in the sections of a homogeneous reservoir containing a horizontal well with a single hydraulic fracture are shown in Fig. 5. It is seen how the fluid flow from the contour is divided into two parts: one part enters the fractures the other directly into the well.

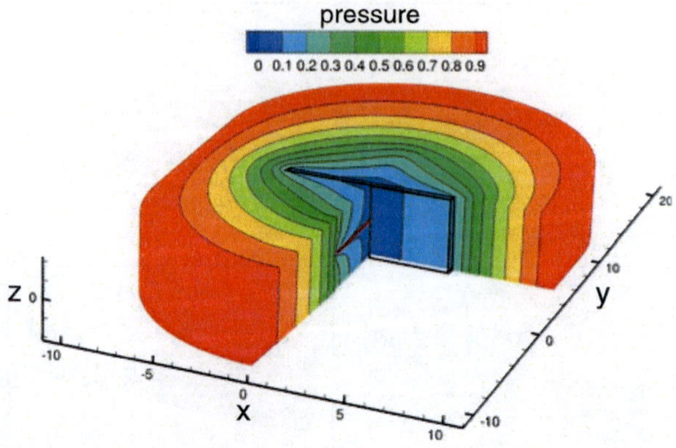

Fig. 4 Pressure distribution in the computational domain D at the moment $t = \tau$

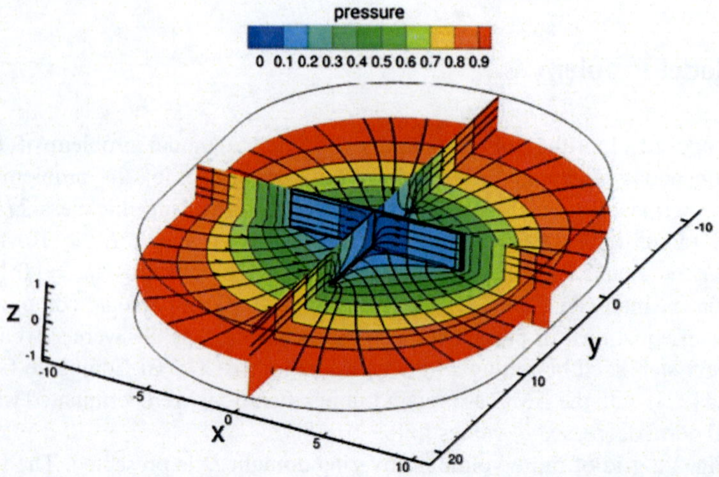

Fig. 5 Streamlines and pressure distribution in cross sections of the homogeneous reservoir containing HW with a single hydraulic fracture

In Fig. 6 the distribution of saturation in the reservoir at different time moments of waterflooding; moments t_1, t_2, t_3, t_4 correspond to the well product water cut 0, 0, 43, 98% is presented. It is seen that the presence of a fracture accelerates the movement of isosat towards the zones of reduced pressure (Fig. 4). It should be noted that the advance of isosat to the wellbore ends is noticeably slower than to the wellbore main part. This is because near the wellbore ends the flow structure has spherical symmetry while in the rest of the part it is radial. It is known [1] that the time of complete withdrawal of liquid with spherical symmetry of the flow occurs over a longer period of time than with radial symmetry.

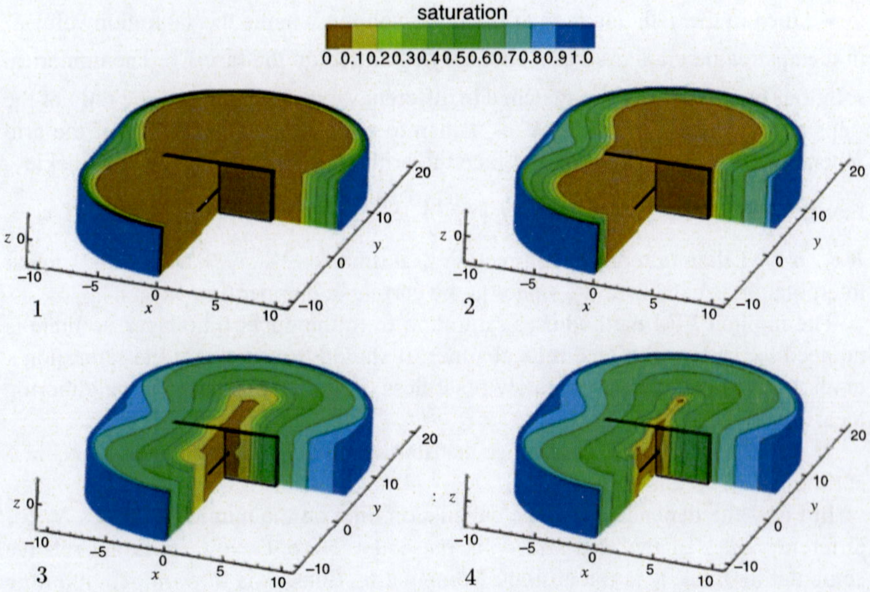

Fig. 6 Saturation distribution during waterflooding of a reservoir containing HW with a single hydraulic fracture (1) $t_1 = 0.25$; (2) $t_2 = 0.5$; (3) $t_3 = 0.75$; (4) $t_4 = 1.07$

6 Computational Efficiency of an Explicit-Implicit Scheme

For the considered model problem the IMPES scheme is stable for $\tau < 10^{-3}$. According to this scheme for the model problem with grid size $N = 10^5$ the water cut reaches 98% at moment $t = 1.07$. This time is used as a standard for estimating the acceleration according to the combined scheme using division of the domain D into local zones in some of which the IMPES scheme is applied and in the other FIM.

The division of the domain into local zones will be carried out by the modulus of velocity. For this the following algorithm is used: for a given saturation distribution, the pressure field in finite volumes and velocities on the faces is calculated. After that the admissible time step τ_i is determined which satisfies the condition (1) in each finite volume V_i. The division of the region occurs by grouping the finite volumes according to the values of τ_i. This method of separation leads to the fact that zones are identified in a small vicinity of the well, at its ends and near the hydraulic fracture. Local zones will be rebuilded for each step of calculating the pressure and velocities.

Using different steps in local zones to calculate saturation by explicit IMPES scheme requires the matching of numerical solutions on the faces separating the finite volumes. Let V_i and V_j be two finite volumes separated by the face Γ_i^j in which the time steps τ_i and τ_j are used. If the velocity u is directed from the cell V_i

to V_j, then in the j-th equation of the finite-volume scheme the saturation value S_i in the upstream cell is used as a boundary condition on the face Γ_i^j. The numerical solutions on the Γ_i^j face are matched in different ways depending on the ratio of the steps τ_i and τ_j. If $\tau_j = K\tau_i, K > 1$ then to ensure the conservatism of the grid scheme for the water flow on the face the arithmetic mean value of the Buckley-Leverett function $f^i = K^{-1} \sum_{k=1}^{K} f\left(S_i^{(k)}\right), S_i^{(k)} = S_i\left(t + k \cdot \tau_i\right)$ is used. If $\tau_i = K\tau_j, K > 1$ then to replace the function s_i at times $t + k \cdot \tau_j, k = 1..K-1$ linear interpolation is used over the values at the current and upper time layers.

The implicit FIM method uses saturation from the upper time layer, so there is no need to match solutions in local zones. It should be noted that the saturation s in all zones is calculated separately regardless of the methods used in neighboring local zones.

The Table 1 shows the percentage distribution of cells in local zones for $N_z = 5$ at times t_1, t_2, t_3 and t_4.

In Fig. 7 the dependence of the calculation time on the number of zones N_z for different values of the grid size N is presented. Here $T = t_c/t_0$ is the relative computation time, t_c is the absolute computation time, t_0 is the computation time according to the IMPES scheme with other identical settings for $N_z = 1$. It can be seen that the larger grid size N is the more significant acceleration gives increasing the number of local zones. For $N = 10^5$, the separation of the region speeds up the calculation by approximately three times for $N_z = 10$. A further increase in the number of local zones has a weak effect on acceleration.

Extra acceleration can be obtained using a completely implicit FIM scheme in some local zones. Since one iteration of the implicit scheme requires more computational costs than one step of the explicit scheme, then using the FIM is justified only in those local zones where N_s is many times greater than one (corresponds to one time step). Numerical experiments show that the use of the FIM method is justified for $N_s > 50$.

In Fig. 8 the relative calculated time for $N = 10^5$ with a different number of local zones is presented. It can be seen that the use of the FIM method in combination with local zones (AIM) accelerates the calculation by an order for $N_z = 5$ and 14 times for $N_z = 10$. A further increase in the number of zones does not give a gain in the calculation speed. Calculations showed that optimal zone count for AIM methods is equal to 6.

Table 1 Percentage distribution of cells in local zones for $N_z = 5$

Zone index	t_0	t_1	t_2	t_3	t_4
1	100	82	59	1	1
2	0	16	24	3	1
3	0	2	17	16	4
4	0	0	0	48	7
5	0	0	0	32	87

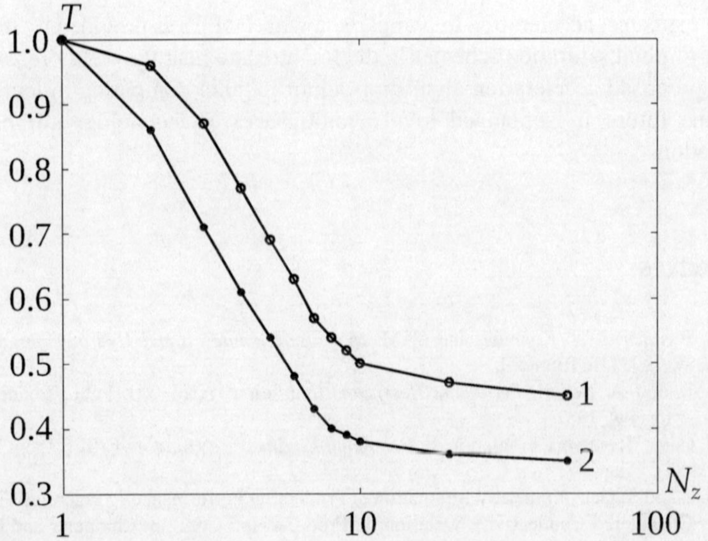

Fig. 7 Relative solution time T according to the IMPES scheme with a different number of cells ($1 - N = 10^4$, $2 - N = 10^5$)

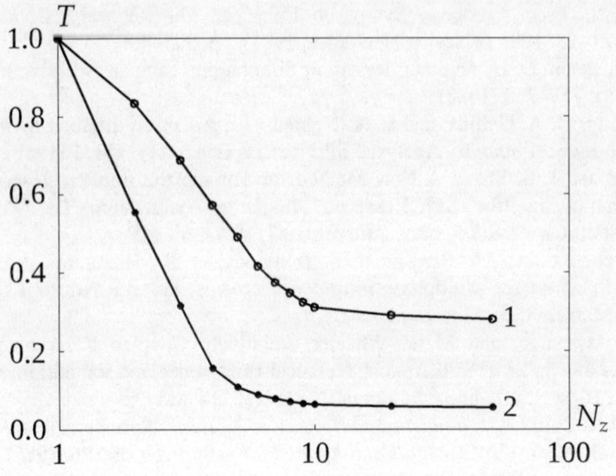

Fig. 8 Relative solution time T for IMPES (1) and AIM (2) methods for $N = 10^5$

7 Conclusion

Two phase flow in porous media containing a horizontal well with a single transverse hydraulic fracture in a three-dimensional formulation with a grid refinement near the wellbore calculation acceleration algorithms have been developed. Acceleration is achieved by automatically dividing the area into local zones with its own time step. The greatest acceleration in comparison with IMPES is provided by the use of explicit-implicit saturation schemes in the fracture and near the wellbore zones.

The proposed acceleration algorithms admit parallel computing in local zones, so in the future it is planned to use multiprocessor technology for hardware acceleration.

References

1. K. S. Basniev, I. N. Kochina, and V. M. Maksimov, *Underground Hydrodynamics* (Nedra, Moscow, 1993) [in Russian].
2. K. Aziz and A. Settari, *Petroleum Reservoir Simulation* (Appl. Sci. Publ., London, 1979; Nedra, Moscow, 1982).
3. K. H. Coats, Reservoir simulation, in *Petroleum Engineering Handbook* (SPE Press, Richardson, 1987), Chap. 48.
4. R. Shen and S. Gao: Numerical Simulation of Production Performance of Fractured Horizontal Wells Considered Conductivity Variation, in Proc. 3rd Int. Conf. on Computer and Electrical Engineering (IACSIT Press, Singapore, 2012), Vol. 53, No. 2.36.
5. A. I. Shangaraeva and D. V. Shevchenko: Speed up of the Oil Saturation Numerical Algorithm for the Plane-Parallel Filtration, Appl. Math. Sci. **9** , 7467–7474 (2015).
6. J. R. Appleyard, I. M. Cheshire, and R. K. Pollard: Special Techniques for Fully-Implicit Simulators, in Proc. European Symp. on Enhanced Oil Recovery, Bournemouth, U.K., September 21–23, 1981 (Elsevier, New York, 1981), pp. 395–408.
7. G. W. Thomas and D. H. Thurnau: Reservoir Simulation Using an Adaptive Implicit Method, SPE J. **23** (5), 759–768 (1983).
8. L. S.-K. Fung, D. A. Collins, and L. X. Nghiem: An Adaptive -Implicit Switching Criterion Based on Numerical Stability Analysis, SPE Reserv. Eng. 4 (1), 45–51 (1989).
9. J. W. Watts and J. S. Shaw: A New Method for Solving the Implicit Reservoir Simulation Matrix Equation, in Proc. SPE Reservoir Simulation Symposium, The Woodlands, USA, January 31-February 2, 2005. https://doi.org/10.2118/93068-MS
10. N. A. Marchenko, A. Kh. Pergament, S. B. Popov, et al.: Hierarchy of Explicit-Implicit Difference Schemes for Multiphase Filtration Problems, Preprint No. 97 (Keldysh Institute of Applied Mathematics, Moscow, 2008).
11. B. Lu, T. Alshaalan, and M. F. Wheeler: Iteratively Coupled Reservoir Simulation for Multiphase Flow, in Proc. SPE Annual Technical Conference and Exhibition, Anaheim, USA, November 11–14, (2007) https://doi.org/10.2118/110114-MS
12. J. Stoer and R. Bulirsch, *Introduction to Numerical Analysis* (Springer, New York, 2002).
13. W. Hackbusch, *Multi-Grid Methods and Applications* (Springer, Berlin, 1985).
14. M. C. Christensen, K. L. Eskildsen, A. P. Engsig-Karup, and M. Wakefield: Nonlinear Multigrid for Reservoir Simulation, SPE J. **21** (3), 0888–0898 (2016).
15. W. A. Mulder and R. H. J. Gmeling-Meyling: Numerical Simulation of Two-Phase Flow Using Locally Refined Grids in Three Space Dimensions, SPE Advanced Technology Series 1 (1) (1993) https://doi.org/10.2118/21230-PA
16. G. S. Shiralkar, G. C. Fleming, J. W. Watts, et al.: Development and Field Application of a High Performance, Unstructured Simulator with Parallel Capability, in Proc. SPE Reservoir Simulation Symposium, The Woodlands, USA, January 31-February 2, (2005) https://doi.org/10.2118/93080-MS
17. P. A. Mazurov and A. V. Tsepaev: Parallel Algorithms for Solving Two-Phase Flow Problems with Fine Grid Segments, Vychisl. Metody Programm. **7**, 251–258 (2006).

18. A. V. Tsepaev: Application of Heterogeneous Computing Systems for Solving the Problem of Fluid Flow by Domain Decomposition Methods, Vychisl. Metody Programm. **13**, 38–43 (2012).
19. M. R. Khamidullin: Numerical Simulation of One-Phase Flow to Multi-Stage Hydraulically Fractured Horizontal Well, Uchen. Zap. Kazan. Gos. Univ. Fiz. Mat. **158** (2), 287–301 (2016).
20. A. B. Mazo, K. A. Potashev, E. I. Kalinin, and D. V. Bulygin: Oil Reservoir Simulation with the Superelement Method, Mat. Model. **25** (8), 51–64 (2013).
21. K. D. Nikitin: Nonlinear Finite Volume Method for Two-Phase Flow in Porous Media, Mat. Model. **22** (11), 131–147 (2010).
22. A. Henk: Iterative Krylov Methods for Large Linear Systems (Cambridge Univ. Press, Cambridge, 2003).
23. C. A. J. Fletcher, Computational Techniques for Fluid Dynamics (Springer, Heidelberg, 1988; Mir, Moscow, 1991).

Dynamical Processes in the Space of φ-Distributions

Valery S. Mokeichev and Anatoly M. Sidorov

Abstract Many mathematical physics problems during modeling require the amplification of the partial differential equation's solution. The cause of this consists in the definition's insufficiency of classical or generalized solution as it can be seen from the problem's physical meaning. For example, the investigation of the solvability of the mathematical model of a fixed string's vibrations leads to the situation, when there is no solution even in the space of 2π-periodic generalized functions. It is necessary to expand the definition of the mathematical model's solution. In particular, the authors have introduced the definitions of the φ-distribution and φ-solution, so it is possible to describe the linear math models theory $P(t, x, D)u = f(t, x)$ and to prove the correspond processes dynamism. The process is dynamic, when there is a unique solution of the math model's Cauchy problem for all initial data. The main result described in this article is the existence of the Cauchy problem's unique solution in the φ-distributions space with values in the Banach space is proved.

1 Introduction

The concept of a φ-solution of some linear equations was introduced early. Namely, the generalized function [1] (the Schwartz distribution in foreign terminology) was called the φ-solution of the equation $Au = f$, where $\varphi = \{\varphi_p(x), p \in \mathbb{N}\}$, if $u = \sum_p u_p\varphi_p(x)$, and the series $\sum_p A(u_p\varphi_p(x))$ converges in some space to $f(x)$. In fact, this is the Fourier method for finding solutions to linear equations. Later it turned out that this concept is very convenient for finding solutions to linear boundary value problems for partial differential equations with deviations

V. S. Mokeichev · A. M. Sidorov (✉)
Institute of Computational Mathematics and Information Technologies, Kazan Federal University, Kazan, Russia
e-mail: Valery.Mokeychev@kpfu.ru; Anatoly.Sidorov@kpfu.ru

I. B. Badriev et al. (eds.), *Mesh Methods for Boundary-Value Problems and Applications*, Lecture Notes in Computational Science and Engineering 141,
https://doi.org/10.1007/978-3-030-87809-2_25

of arguments. However, the rigid binding of φ-solutions to the set of generalized functions led to the fact that a number of mathematical models turned out to be insoluble, which should not be in the opinion of J. Hadamard [2]. The undecidability of a mathematical model means that either one of the points is not taken into an account in the process of its construction, or the concept of a solution is poorly chosen. The concept of the generalized solution is insufficient for the solvability of a number of partial differential equations [4, 5].

In [3] the concept of a φ-solution was introduced without reference to the space of generalized functions. For this, the concept of φ-distribution was introduced on its basis, so called, the concept of φ-solution, and a theory of solvability of linear boundary value problems for partial differential equations with deviations of arguments was developed. In [6], the space of φ-distributions was studied in the case, when φ is a system of functions, the concepts of differentiability and integrability of φ-distributions were correctly introduced, and the expandability problem of generalized functions in series in a given system of functions was also studied.

2 Mathematical Model of the Problem

There are mathematical physics problems that require amplification of the partial differential equation's solution. It can be explained the definition's insufficiency of classical or generalized solution. For example, the mathematical model of a fixed string's vibrations on $[0, \pi]$ can be written in the form

$$u_{tt}^{(2)} - c^2 u_{xx}^{(2)} = f(t, x),\ (t, x) \in \mathbb{R} \times [0, \pi],$$
$$u(t, 0) = u(t, \pi) = 0,$$

where $u,\ u_t^{(1)},\ f(t, x) \in L^2(\mathbb{R} \times [0, \pi])$ are 2π-periodic function with respect to t.

It was proved in [3] that if c is Liouville number then there is no model's solution even in the space of 2π-periodic generalized functions. It is necessary to expand the definition of the math model's solution.

There we introduced the definitions of φ-distribution and φ-solution to describe the solution of the linear problem, to explain the linear problems solvability's theory without meaning of the equation's scalarity and type (e.g. elliptic, parabolic or hyperbolic). It allows us to solve some mathematical models which have no solutions in the generalized functions spaces [3]. We present the φ_B-distribution theory's objects so-called φ-distributions with values in Banach space B necessary for the further concept. This theory is detailed in the articles [10, 11].

Let K^n be the set of vectors $p = (p_1, \ldots, p_n)$ with integer coordinates values not necessarily the same as $\mathbb{Z}^n$ and $|p| = |p_1| + \ldots + |p_n|$. Let

$$\varphi = \{\varphi_p,\ p \in K^n\} \quad \text{and} \quad \varphi^* = \{\varphi_p^*,\ p \in K^n\}$$

be the biorthogonal elements systems with respect to scalar product $< \cdot, \cdot >$, i.e.

$$< \varphi_p, \varphi_q^* >= \begin{cases} 0, & p \neq q; \\ 1, & p = q. \end{cases}$$

We will assume that space B and system φ such that for all $m \in \mathbb{N}$ and for all $a_p \in B$ there is a sum

$$\sum_{|p| \leqslant m} a_p \varphi_p. \tag{1}$$

Denote by L_φ the set of elements written for some $m \in \mathbb{N}$ in the form (1). The set L_{φ^*} are defined the same way.

Definition 1 Linear mapping $u : L_{\varphi^*} \to B$ is φ_B—distribution.

Definition 2 The sequence $(u_m)_{m=1}^{\infty}$ of φ_B—distributions converges to u as element of φ_B—distribution, when $\lim\limits_{m \to \infty} \|u_m(\psi) - u(\psi)\| = 0$ for all $\psi \in L_{\varphi^*}$ in the norm $\| \cdot \|$ of space B.

Denote by D'_φ the set of all φ_B-distributions with two elements addition, multiplication by number and convergence operations given in definition (2).

Theorem 1 *D'_φ is a full space.*

Definition 3 Fourier coefficient of $u \in D'_\varphi$ by a system φ is $u_p = u(\varphi_p^*)$, $p \in K^n$. Fourier series of φ_B-distribution u by a system φ is $\sum\limits_p u_p \varphi_p$.

Theorem 2 *$u = v$ if and only if $u_p = v_p$ for all $p \in K^n$.*

Theorem 3 *There are only elements from D'_φ can be written by Fourier series by a system φ.*

Let $\Omega \subset \mathbb{R}^n$ be Lebesgue measurable set with non-zero measure,

$$\varphi = \{e^{(\mu + ip)x}, \ p \in \mathbb{Z}^n\}, \ x \in \Omega, \tag{2}$$

where $\mu = (\mu_1, \ldots, \mu_n) \in \mathbb{C}^n$ is an arbitrary vector. Then a system

$$\varphi^* = \{(2\pi)^{-n} \cdot e^{(-\overline{\mu} + ip)x}, \ p \in \mathbb{Z}^n\}$$

is biorthogonal to φ with respect to the scalar product

$$(f_1, f_2) = \int_0^{2\pi} f_1(x) \overline{f_2(x)} \, dx,$$

where dx is a Lebesque measure.

Let $|a, b| \subset \mathbb{R}$ is open, semi-open or closed set, I is a finite set of multi-indices $\alpha = (\alpha_1, \dots, \alpha_n)$ with non-negative components, $D_x^\alpha = D_1^{\alpha_1} \cdot \ldots \cdot D_n^{\alpha_n}$, where $D_t^j = \dfrac{\partial^j}{\partial t^j}$, $D_k = \dfrac{\partial}{\partial x^k}$, B is a Banach space, D'_φ is a space of φ_B-distributions and φ are the elements system (2). Let the mathematical model of a process $u(t, x)$ in D'-space be

$$\sum_{j=0}^{M} \sum_{\alpha \in I} c_{\alpha,j}(t, x) D_t^j D_x^\alpha u = f(t, x), \ t \in |a, b|, \ x \in \Omega, \tag{3}$$

where

$$c_{\alpha,j}(t, x) = \sum_{|q| \leqslant M_1} c_{\alpha,j,q}(t) e^{iq\cdot x}, \ q \in \mathbb{Z}^n$$

with $c_{\alpha,j,q}(t) : B \to B$ is a linear operator for all $t \in |a, b|$, $f : |a, b| \times B \to D'_\varphi$.

The process $u(t, x)$ is dynamic if it's every state is defined by the initial state for $t > t_0$, $t_0 \in |a, b|$, i.e.

$$D_t^j u(t_0, x) = g_j(x), \ x \in \Omega, \ j = 0, 1, \dots, M - 1. \tag{4}$$

So, the process is dynamic, when Cauchy problem (3) and (4) for it's mathematical model has a unique solution.

One of the methods for studying the Cauchy problem is the Fourier method. Among the works on this method, we note the works [6–9].

3 Fourier Series Expansion of the Problem's Solution

Let us define conditions for dynamic process $u(t, x)$ with math model (3). Denote by

$$u(t, x) = \sum_p u_p(t) e^{(\mu + ip)x},$$
$$f(t, x) = \sum_p f_p(t) e^{(\mu + ip)x},$$
$$g_j(x) = \sum_p g_{j,p} e^{(\mu + ip)x}$$

the Fourier series expansions by a system φ of φ_B-distributions $u(t, x)$, $f(t, x)$, $g_j(x)$ for $j = 0, 1, \dots, M - 1$. The Fourier coefficients $u_p(t)$, $f_p(t)$ almost for all $t \in |a, b|$ and $g_{j,p}$ are elements of space B.

Definition 4 The function $x(t) : |a, b| \to B$ is absolutely continuous function, if there is function $y(t) : |a, b| \to B$ with $\|y\| \in L^1_{loc}(|a, b|)$, and for all subsets $[a', b'] \subset |a, b|$

$$x(t) = \int_{a'}^{t} y(s)\, ds + x(a'),\ t \in [a', b'].$$

Definition 5 φ_B-distribution u is the solution of problem (3) and (4), if Fourier coefficients $u_p(t)$ are absolutely continuous, (4) holds in D'_φ and Eq. (3) holds almost everywhere in D'_φ.

Let us note that

$$D_t^j u = \sum_p u_p^{(j)}(t) e^{(\mu+ip)x},$$
$$D_t^j D_x^\alpha u = \sum_p u_p^{(j)}(t)(\mu + ip)^\alpha e^{(\mu+ip)x},$$

where the series on the right-hand side converge in D'_φ due to absolutely continuity of $u_p^{(j)}(t),\ j = 0, \dots, M-1$, and the series converge almost everywhere in D'_φ for $j = M$.

We are going to apply the Fourier series expansion in (3) and (4) using system φ. Then we have

$$\sum_p \sum_{j=0}^{M} \sum_{\alpha \in I} \sum_{|q| \leqslant M_1} c_{\alpha,j,q}(t)(\mu + i(p-q))^\alpha u_{p-q}^{(j)}(t) e^{(\mu+ip)x} = \sum_p f_p(t) e^{(\mu+ip)x}. \tag{5}$$

$$\sum_p u_p^{(j)}(t_0) e^{(\mu+ip)x} = \sum_p g_{j,p} e^{((\mu+ip)x}. \tag{6}$$

Due to (2), Eqs. (5) and (6) are equivalent to

$$\begin{aligned} &\sum_{\alpha \in I} c_{\alpha,M,0}(t)(\mu + ip)^\alpha u_p^{(M)}(t) + \\ &+ \sum_{j=0}^{M-1} \sum_{\alpha \in I} \sum_{|q| \leqslant M_1} c_{\alpha,j,q}(t)(\mu + i(p-q))^\alpha u_{p-q}^{(j)}(t) = \\ &= f_p(t),\ t \in |a, b|,\ p \in \mathbb{Z}^n, \end{aligned} \tag{7}$$

$$u_p^{(j)}(t_0) = g_{j,p},\ j = 0, 1, \dots, m-1,\ p \in \mathbb{Z}^n. \tag{8}$$

So, the Cauchy problem (3) and (4) is an infinite system of equations (7) with infinite initial conditions (8).

Denote by

$$Q_{p,j,q}(t) = \sum_{\alpha \in I} c_{\alpha,j,q}(t)(\mu + i(p-q))^{\alpha}, \ p \in \mathbb{Z}^n, \ j = 0, \dots, M, \ |q| \leqslant M_1.$$

Then $Q : B \to B$ are linear operators for all $t \in |a, b|$. Then we can rewrite (7) in the form

$$Q_{p,m,0}(t)u_p^{(M)}(t) + \sum_{j=0}^{M-1} \sum_{|q| \leqslant M_1} Q_{p,j,q}(t)u_{p-q}^{(j)}(t) = f_p(t). \tag{9}$$

Let operator $Q_{p,M,0}(t)$ is invertible for some $\mu \in \mathbb{C}^n$ and for all $p \in \mathbb{Z}^n$, $t \in |a, b|$. Then let's apply inverse operator $Q_{p,M,0}^{-1}(t)$ to (9):

$$u_p^{(M)}(t) + \sum_{j=0}^{M-1} \sum_{|q| \leqslant M_1} F_{p,j,q}(t)u_{p-q}^{(j)}(t) = h_p(t), \tag{10}$$

where $F_{p,j,q}(t) = Q_{p,M,0}^{-1}(t) \cdot Q_{p,j,q}(t)$, $h_p(t) = Q_{p,M,0}^{-1}(t) f_p(t)$. We will use matrices for Eq. (10). Denote by

$$\begin{aligned} W_p(t) &= (u_p(t), u_p^{(1)}(t), \dots, u_p^{(M-1)}(t))^T, \\ W_p^{(1)}(t) &= (u_p^{(1)}(t), u_p^{(2)}(t), \dots, u_p^{(M)}(t))^T, \\ V_p &= (g_{0,p}; g_{1,p}; \dots; g_{M-1,p})^T. \end{aligned}$$

Let $M \times M$ matrix $A_{p,q}(t)$ be in the form

$$A_{p,q}(t) = \begin{pmatrix} 0 & B_q & 0 & \dots & 0 \\ 0 & 0 & B_q & \dots & 0 \\ \vdots & \vdots & \vdots & \dots & \vdots \\ 0 & 0 & 0 & \dots & B_q \\ -F_{p,q,0}(t) & -F_{p,q,1}(t) & -F_{p,q,2}(t) & \dots & -F_{p,q,M-1}(t) \end{pmatrix},$$

where $B_q = \{0 \text{ for } q \neq 0; \ I \text{ for } q = 0\}$ with identity operator $I : B \to B$. Then we have (10) in matrix form

$$W_p^{(1)}(t) = \sum_{|q| \leqslant M_1} A_{p,q}(t) W_{p-q}(t) + H_p(t), \ p \in \mathbb{Z}^n, \tag{11}$$

where $H_p(t) = (0, 0, \dots, 0, h_p(t))^T$.

Let $\|\cdot\|_1$ be the norm in the space of linear bounded operators defined on $|a, b|$ with values in D'_{φ}, $\|\cdot\|_2$ be the norm of $M \times M$ matrices with elements in the form of linear bounded operators defined on $|a, b|$ with values in D'_φ and $\|\cdot\|_3$ be the norm in Banach space $B^M = B \times B \times \ldots \times B$.

Theorem 4 *Let operators $Q_{p,M,0}(t)$ are invertible for some $\mu \in \mathbb{C}^n$ and for all $t \in |a, b|$, $p \in \mathbb{Z}^n$. There is a function $C(t) \in L^1_{loc}(|a, b|)$ such that $\|F_{p,q,j}(t)\| \leqslant \leqslant C(t)$, $|q| \leqslant M_1$, $j = 0, 1, \ldots, M-1$, $H_p(t) \in L^1_{loc}(|a, b|)$. Then process $u(t, x)$ with mathematical model in the form (3) is dynamic in D'_φ with elements $\varphi = \{e^{(\mu+ip)x}, p \in \mathbb{Z}^n\}$.*

Proof We need to prove that Cauchy problem (3) and (4) has a unique solution in D'_φ for all $t \in |a, b|$, i.e. the system above with infinity equations number and infinity Cauchy initial conditions $W_p(t_0) = V_p$, $p \in \mathbb{Z}^n$ has a unique solution. We are going to prove it for any fixed $[a', b'] \subset |a, b|$. The system (11) is the same as the system

$$W_p(t) = \sum_{|q|\leqslant M_1} \int_{t_0}^{t} A_{p,q}(s) W_{p-q}(s)\, ds + \int_{t_0}^{t} H_p(s)\, ds + V_p, \ t_0, t \in [a', b']. \tag{12}$$

Define a sequence $(W_{p,m}(t))$, $t \in [a', b']$ for all $p \in \mathbb{Z}^n$ following way:

$$\begin{aligned} W_{p,0}(t) &= \int_{t_0}^{t} H_p(s)\, ds + V_p, \\ W_{p,m}(t) &= \sum_{|q|\leqslant M_1} \int_{t_0}^{t} A_{p,q}(s) W_{p-q,m-1}(s)\, ds + \\ &+ \int_{t_0}^{t} H_p(s)\, ds + V_p, \ m = 1, 2, \ldots \end{aligned} \tag{13}$$

Due to inequality $\|F_{p,q,j}(t)\| \leqslant C(t)$ there is a number T such that

$$\|A_{p,q}(t)\|_2 \leqslant T \cdot C(t) \text{ for all } p \in \mathbb{Z}^n, \ |q| \leqslant M_1, \ t \in [a', b'].$$

Let (b_p), $p \in \mathbb{Z}^n$ be a sequence of positive numbers with $b_{p-q} \leqslant T_1 b_p$ for $|q| \leqslant M_1$ and T with T_1 don't depend on p,

$$\sup_{t\in[a',b']} \|W_{p,1}(t) - W_{p,0}(t)\|_3 \leqslant b_p. \tag{14}$$

Denote by γ the number of $q \in \mathbb{Z}^n$ satisfying inequality $|q| \leqslant M_1$. Let us prove for some $m \in \mathbb{N}$ inequality (15) holds.

$$\|W_{p,m}(t) - W_{p,m-1}(t)\|_3 \leqslant \frac{(\gamma T T_1)^{m-1} b_p}{(m-1)!} \left| \int_{t_0}^{t} C(s)\, ds \right|^{m-1}. \tag{15}$$

For $m = 1$ it holds due to (14). Assume that for $m = k$ inequality

$$\|W_{p,k}(t) - W_{p,k-1}(t)\|_3 \leqslant \frac{(\gamma T T_1)^{k-1} b_p}{(k-1)!} \left| \int_{t_0}^{t} C(s)\, ds \right|^{k-1} \tag{16}$$

holds. Then we are going to prove that it holds for $m = k + 1$ too. Obviously, due to (13) and (16) we get

$$\begin{aligned}
&\|W_{p,k+1}(t) - W_{p,k}(t)\|_3 \leqslant \\
&\leqslant \sum_{|q| \leqslant M_1} \left| \int_{t_0}^{t} \|A_{p,q}(s)\|_2 \cdot \|W_{p-q,k}(s) - W_{p-q,k-1}(s)\|_3\, ds \right| \leqslant \\
&\leqslant \sum_{|q| \leqslant M_1} \frac{T(\gamma T T_1)^{k-1} b_{p-q}}{(k-1)!} \left| \int_{t_0}^{t} C(s) \left| \int_{t_0}^{s} C(y)\, dy \right|^{k-1} ds \right| \leqslant \\
&\leqslant \frac{\gamma T (\gamma T T_1)^{k-1} T_1 b_p}{(k-1)! \cdot k} \left| \int_{t_0}^{t} C(s)\, ds \right|^{k} = \frac{(\gamma T T_1)^{k} b_p}{(k)!} \left| \int_{t_0}^{t} C(s)\, ds \right|^{k}.
\end{aligned}$$

Due to (15) the series $\|W_{p,0}(t)\|_3 + \sum_{m=1}^{\infty} \|W_{p,m}(t) - W_{p,m-1}(t)\|_3$ uniformly converges on $[a', b']$ for all $p \in \mathbb{Z}^n$. In this way, the series $W_{p,0}(t) + \sum_{m=1}^{\infty} (W_{p,m}(t) - W_{p,m-1}(t))$ uniformly converges for all $t \in [a', b']$, and there is a function $W_p(t)$ such that for all $t \in [a', b']$ and all $p \in \mathbb{Z}^n$ there is $\lim_{m\to\infty} \|W_{p,m}(t) - W_p(t)\|_3 = 0$. If we apply the limit in (13), we get (12). So, the function $W_p(t),\ p \in \mathbb{Z}^n$ is a solution of (12).

Let us prove the solution's uniqueness. We assume that there is another solution $\widetilde{W}_p(t),\ p \in \mathbb{Z}^n$. Denote by $y_p(t) = W_p(t) - \widetilde{W}_p(t),\ p \in \mathbb{Z}^n$. Then

$$y_p(t) = \sum_{|q| \leqslant M_1} \int_{t_0}^{t} A_{p,q}(s) y_{p-q}(s)\, ds,$$

$$\|y_p(t)\|_3 \leqslant \gamma T \left| \int_{t_0}^{t} C(s) \|y_{p-q}(s)\|_3\, ds \right|.$$

The function $\|y_p(t)\|_3$ is a continuous function, so there are numbers $d_p > 0,\ p \in \mathbb{Z}^n$ and there is a number A that it doesn't depend on p such that $\|y_p(t)\|_3 \leqslant d_p,\ d_{p-q} \leqslant A d_p$ for all $|q| \leqslant M_1$. In addition, we have

$$\begin{aligned}
\|y_p(t)\|_3 &\leqslant \sum_{|q| \leqslant M_1} \left| \int_{t_0}^{t} \|A_{p,q}(s)\|_2 \cdot \|y_{p-q}(s)\|_3\, ds \right| \leqslant \\
&\leqslant \sum_{|q| \leqslant M_1} T d_{p-q} \left| \int_{t_0}^{t} C(s)\, ds \right| \leqslant \gamma A T d_p \left| \int_{t_0}^{t} C(s)\, ds \right|,
\end{aligned}$$

i.e.

$$\|y_p(t)\|_3 \leqslant \gamma A T d_p \left| \int_{t_0}^{t} C(s)\, ds \right|. \tag{17}$$

Let's apply inequality (17) and then we get

$$\|y_p(t)\|_3 \leqslant \sum_{|q| \leqslant M_1} \left| \int_{t_0}^{t} \|A_{p,q}(s)\|_2 \cdot \|y_{p-q}(s)\|_3\, ds \right| \leqslant$$

$$\leqslant \sum_{|q| \leqslant M_1} T \left| \int_{t_0}^{t} \gamma A T d_{p-q} \left| \int_{t_0}^{s} C(y)\, dy \right| ds \right| \leqslant \frac{(\gamma A T)^2 d_p}{2} \left| \int_{t_0}^{t} C(s)\, ds \right|^2.$$

If we continue this procedure, we get

$$\|y_p(t)\|_3 \leqslant \frac{(\gamma A T)^m d_p}{m!} \left| \int_{t_0}^{t} C(s)\, ds \right|^m, \quad m = 1, 2, \ldots$$

Note that $\|y_p(t)\|_3 \leqslant 0$ for $m \to \infty$. As a result, $y_p(t) \equiv 0$ and solution's uniqueness is proved.

Due to $W_p(t) = (u_p(t), u_p^{(1)}(t), \ldots, u_p^{(M-1)}(t))^T$, we conclude that φ_B-distribution $u(t,x) = \sum_p u_p(t) e^{(\mu + ip)x}$ is a solution of problem (3) and (4), because the Fourier coefficients are absolutely continuous, equalities (4) hold in D'_φ and equality (3) holds almost everywhere in D'_φ. □

This work continues the author's researches [12], where coefficients in mathematical model (3) don't depend on spacial variables.

In the process of the solving the mixed problem for hyperbolic and parabolic differential equations by the Fourier method [13–15], we find the φ-solution with a specially constructed sequence φ. If φ is a finite sequence, then in some cases the φ-solutions can be interpreted as the numerical solutions of the problem found by the Galerkin method [16].

Acknowledgments This paper has been supported by the Kazan Federal University Strategic Academic Leadership Program ("PRIORITY-2030").

References

1. Sobolev, S.L.: Methode nouvelle a resoudre le probleme de Cauchy pour les equations lineaires hyperboliques normales. Math. Sb. **43**(1), 39–72 (1936)
2. Hadamard, J.S.: Lectures on Cauchy's problem in linear partial differential equations. Dover Publ., New York (1952)

3. Mokeichev, V.S., Mokeichev, A.V.: A new approach to the theory of linear problems for systems of partial differential equations. I. Russian Math. **43**(1), 22–32 (1999)
4. Yegorov, Yu.V.: The linear differential equations of the main type. Nauka, Moscow (1984)
5. Hermander, L.: The analysis of linear partial differential operators. The distribution theory and Fourier analysis. Mir, Moscow (1986)
6. Lomov, I.S.: Loaded differential operators: convergence of spectral expansions. Differential Equations. **50**(8), 1070–1079 (2014)
7. Lomov, I.S.: The Convergence of Expansions in Eigenfunctions of a Differential Operator with Integral Boundary Conditions. Doklady Akademii nauk **481**(6), 599–604 (2018) DOI: 10.31857/S086956520002091-3
8. Lomov, I.S.: Uniform convergence of expansions in root functions of a differential operator with integral boundary conditions. Differential Equations. **55**(4), 471–482 (2019)
9. Dubinskiy, Yu.A.: The Cauchy problem in complex domain. MEI, Moscow (1996)
10. Mokeichev, V.S., Sidorov, A.M.: On an expansion in the series by given system of elements. Issled. Prikl. Mat. Inf. **25**, 163–167 (2004)
11. Mokeichev, V.S.: Metric, Banach and Hilbert spaces of φ_B-distributions. Russian Math. **62**(5), 64–70 (2018).
12. Mokeichev, V.S., Sidorov, A.M.: A dynamical process of several variables. Uchenye Zapiski Kazanskogo Universiteta. Seriya Fiziko-Matematicheskie Nauki **160**(4), 762–770 (2018)
13. Maurin, K.: Methods of Hilbert spaces. PWN, Warsaw (1967)
14. Bers, L., John, F., Schechter, M.: Partial differential equations. Interscience Publ., New York, London, Sydney (1964)
15. Vladimirov, V.S.: The mathematical physics equations. Nauka, Moscow (1985)
16. Tikhonov, A.N., Arsenin, V.Ya.: The methods for solving of ill-posed problems. Nauka, Moscow (1974)

An Approach to Synthesis of the Neuromorphic Functional Models for Analog Components and Blocks

Sergey Mosin

Abstract Numerical simulation of analog circuits and functional blocks (FB) is an important design stage of analog and mixed-signal integrated circuits as well as the state-of-the-art embedded systems. The application of adequate mathematical models for components and FB defines the quality of simulation and influences the time and cost characteristics of the up-to-date microelectronic devices development process. An approach to the automated synthesis of functional models for analog components and blocks using machine learning methods—neuromorphic functional models (NFM)—is proposed in the paper. The approach implements the possibility to use either analytically defined dependencies (model-based) or dependencies obtained during natural experimental measurements (data-driven) as the raw data for the NFM synthesis. The design flow including the description of the mathematical models for an analog circuit (MMC) applying the NFMs and further numerical simulation in accordance with the assigned type of the circuit analysis is presented. The results of experimental research for a model of the semiconductor diode D1N4934 and circuits of the voltage rectifiers on its base are showed. The obtained results demonstrate the high precision of the synthesized NFM and the high quality of simulation. The comparison of obtained results with results of simulation in the Cadence CAD tools based on a structural model of the diode is performed. The simulation errors consist of less than 1% of the input signals' amplitude.

1 Introduction

Mathematical numerical simulation is actively applied in computer-aided design (CAD) tools for a microelectronic design already several decades providing support of the development and implementation processes for the up-to-date microelectron-

S. Mosin (✉)
Kazan Federal University, Institute of Computational Mathematics and Information Technologies, Kazan, Russia
e-mail: smosin@ieee.org

I. B. Badriev et al. (eds.), *Mesh Methods for Boundary-Value Problems and Applications*, Lecture Notes in Computational Science and Engineering 141,
https://doi.org/10.1007/978-3-030-87809-2_26

ics devices [1]. Two main entities of the mathematical support in CAD tools are distinguished. Firstly, the mathematical models, which are used for the description of the developed circuits for a particular type of simulation, and, secondly, the mathematical methods ensuring the computation of investigated characteristics based on a mathematical model of a device according to selected type of analysis.

The truth tables and Boolean algebra equations are used, as rule, for description the models of digital devices. Hardware description languages (HDL) are widely used for behavior description of digital devices, for instance, VHDL (Very high speed integrated circuits Hardware Description Language) [2, 3], Verilog, SystemC [4, 5], etc. Methods of functional logic modeling are applied for computation of digital circuits' characteristics.

The models of analog circuits are produced in the form of differential, nonlinear and linear algebraic equations and systems of equations. Description of analog circuits is performed using SPICE-like scripts [6] or description languages of analog and mixed-signal devices such as VHDL-AMS [7], Verilog-A [8, 9], etc. The corresponding system of equations are generated for realization of a particular analysis type: in the static mode (DC—Direct Current), in the frequency domain (AC—Alternative Current) and in the time domain (Tran—Transient). Specific computation methods are used for each type of analog circuit analysis. The Gauss method or LU-decomposition method is applied for circuit simulation in the frequency domain, where the circuit model is represented by a system of linear algebraic equations. The Newton method or simple iteration method is applied for circuit simulation in the static mode, where the circuit model is represented by a system of nonlinear algebraic equations. The finite difference methods of different orders are applied for circuit simulation in the time domain, where the circuit model is represented by a system of differential equations.

Simulation of analog circuits is a more complex process in comparison with digital circuits' simulation due to the following reasons, firstly, from the computational point of view because the process has an iterative character, secondly, from the functional point of view because adequate mathematical models are required for each component used in the circuit for each type of analysis [10, 11].

The development of mathematical models for new components, as rule, requires an essential time cost and therefore restricts a quick introduction to the state-of-the-art elements and blocks in the design process [12]. There are different approaches to enhancing the analog models in order to provide required precision and fasten analog circuit simulation [13–15].

An approach to the synthesis of the neuromorphic functional models (NFM) for components oriented to numerical analysis of analog circuits is proposed in the paper. An advantage of the proposed approach is the ability to synthesis a NFM based on results of functional simulation (model-based) and results of natural investigation of the analog components and blocks (data-driven). A neuromorphic functional model synthesized in result can be used for analog circuit simulation as well as for the description entity of complex functional blocks before implementation in neural network hardware at the embedded systems design.

The paper is organized as the following. Section 2 describes the design flow of the mathematical model of an analog circuit with the application of the neuromorphic functional models for components and the features of such a circuit's numerical simulation. The design flow of automated synthesis of the NFM for active analog components is presented in Sect. 3. The results of experimental research are proposed in Sect. 4. Conclusion includes an efficiency analysis of the proposed approach and provides quantity estimation of the results of analog circuits' simulation using the neuromorphic functional models for some components.

2 Functional Models of Analog Components and Analog Circuits

The mathematical models of analog passive components are based on the Ohm's law for a resistor, the charge conservation law for a capacitor and the electromagnetic induction law for an inductance.

The mathematical models of analog active components, which also analytically reflect the relationship of the current flowing through the component in dependence on the applied voltage using a big number of internal parameters, have significantly greater complexity [10]. For instance, the mathematical model (MM) of a semiconductor diode is based on the Shockley's equation, MM of a bipolar transistor based on the Ebers-Moll model or the Gummel-Poon model. The main complexity of designing a MM of active component deals with adequate adjustment of internal parameters in order to guarantee correspondence of the model output characteristics to characteristics of a real (physical) component.

A functional model for active components that provides an approximation of the output characteristics over a given range of input parameters is proposed in the paper as an alternative to analytical structural models. An artificial neural network (ANN) trained to implement the following functional dependence is proposed as an approximator

$$\mathbf{Y}_{out} = f(\mathbf{X}_{in}) , \tag{1}$$

where $\mathbf{X}_{in}$ is a vector of input values, $\mathbf{Y}_{out}$ is an associative vector of output values.

The current flowing through the component ($I \in \mathbf{Y}_{out}$) is considered at the NFM as the output characteristic for the one-terminal components (Fig. 1), and the applied voltage to the component is the input characteristic ($V \in \mathbf{X}_{in}$).

The input and output currents ($I_{out}, I_{in} \in \mathbf{Y}_{out}$) are considered at the NFM as the output characteristics for the two-terminal components (Fig. 2), and the input current ($I_{in} \in \mathbf{X}_{in}$), the input voltage ($V_{in} \in \mathbf{X}_{in}$) and the output voltage ($V_{out} \in \mathbf{X}_{in}$) are considered as the input characteristics.

The mathematical model of the analog circuit (MMC) is described by applying the nodal potential method that uses the structural and/or functional models for the

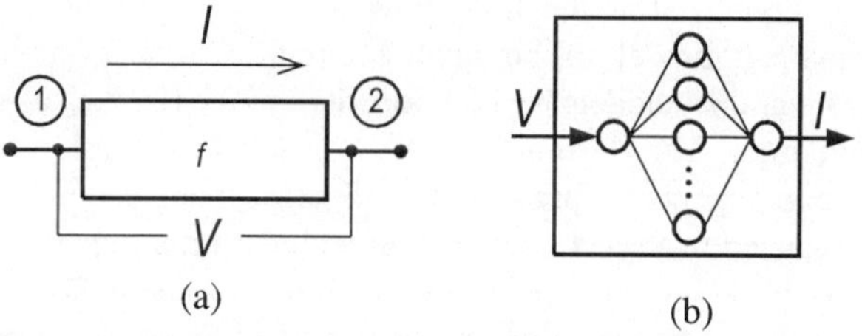

Fig. 1 One-terminal component: schematic (**a**), NFM (**b**)

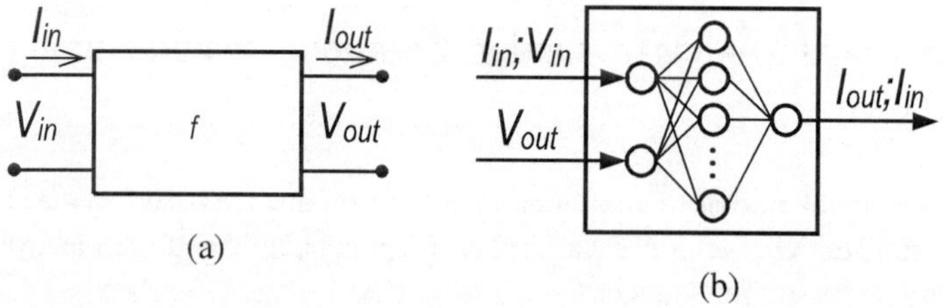

Fig. 2 Two-terminal component: schematic (**a**), NFM (**b**)

active components

$$\mathbf{YV} = \mathbf{I}\,, \tag{2}$$

where **Y** is a nodal conductance matrix; **V** is a vector of nodal potentials; **I** is a vector of nodal currents.

The values of the current returned by the neural network when an effective voltage is applied to its input are used during the description of the MMC. For example, the design flow for constructing the MMC and performing circuit analysis in the static mode includes the following steps:

1. Formation of vector **V**.
   (a) Initialization of the vector by the zero values.
   (b) For all independent voltage source V_{pn} connected between nodes p and n, where p corresponds to a node of positive polarity, and n to a node of negative polarity add the nominal voltage V_{pn} to the element of vector **V** with index p (if $p \neq 0$), subtract the nominal voltage V_{pn} from the element of vector **V** with index n (if $n \neq 0$).
2. Formation of vector **I**.
   (a) Initialization of the vector by the zero values.
   (b) Express the flowing current i_k for all passive and active components in the circuit. Add i_k to the element of vector **I** with the index corresponding to the circuit node into which the current flows. Subtract i_k from the element

of vector **I** with the index corresponding to a circuit node from which the current flows. The operations are not considered for node 0. The current values i_k are determined using either structural or functional models. For the case of using the functional model for a component m connected in the circuit between nodes i and j the trained neural network returns the value of current when the effective voltage is applied to the input of the ANN

$$i_m = f_m(\mathbf{V}(j) - \mathbf{V}(i)). \tag{3}$$

3. Formation of matrix **Y**.

   (a) Initialization of the matrix by the zero values.
   (b) For each component k in the circuit connected between nodes i and j add the corresponding conductance y_k with a positive sign to the elements of the matrix with the indexes (i, i) and (j, j) and with a negative sign to the elements with the indexes (i, j) and (j, i). If i or j is equal to 0, the conductance is not added to the matrix.

$$\mathbf{Y}(i,i) = \mathbf{Y}(i,i) + y_k, \mathbf{Y}(j,j) = \mathbf{Y}(j,j) + y_k,$$
$$\mathbf{Y}(i,j) = \mathbf{Y}(i,j) - y_k, \mathbf{Y}(j,i) = \mathbf{Y}(j,i) - y_k, \tag{4}$$
$$\forall i, j \in N, i \neq 0, j \neq 0,$$

   where N is the set of circuit nodes.

4. To solve the system of nonlinear algebraic equations for **V**, for example, by the Newton's method

$$\mathbf{Y}(\mathbf{V}^{(k)})\Delta\mathbf{V}^{(k)} = -\mathbf{I}(\mathbf{V}^{(k)}),$$
$$\mathbf{V}^{(k+1)} = \mathbf{V}^{(k)} - \mathbf{I}(\mathbf{V}^{(k)})/\mathbf{Y}(\mathbf{V}^{(k)}), \tag{5}$$
$$\Delta\mathbf{V}^{(k)} = \mathbf{V}^{(k+1)} - \mathbf{V}^{(k)}.$$

5. If $\Delta\mathbf{V}^{(k)} > \epsilon$ (5), where ϵ is a threshold value of the absolute error, then repeat steps 2–4 recalculating the currents **I** and conductance **Y** for obtained values $\mathbf{V}^{(k+1)}$. Otherwise, $\mathbf{V}^{(k+1)}$ is the resulting vector of the nodal potentials.

3 The Design Flow of Automated Synthesis of the Neuromorphic Functional Model for Analog Active Components

The artificial neural network represented by the two-layer perceptron providing an approximation of the following functional dependence (1) is used as a base of the neuromorphic functional model.

The design flow of automated synthesis of the NFM including the following steps is proposed (Fig. 3).

1. Generation of initial data can be performed by two ways, firstly, based on the results of modeling the analytical dependence of the current on the voltage (model-based) or, secondly, based on the results of measuring the characteristics during field testing of a component (data-driven). The raw data is represented as

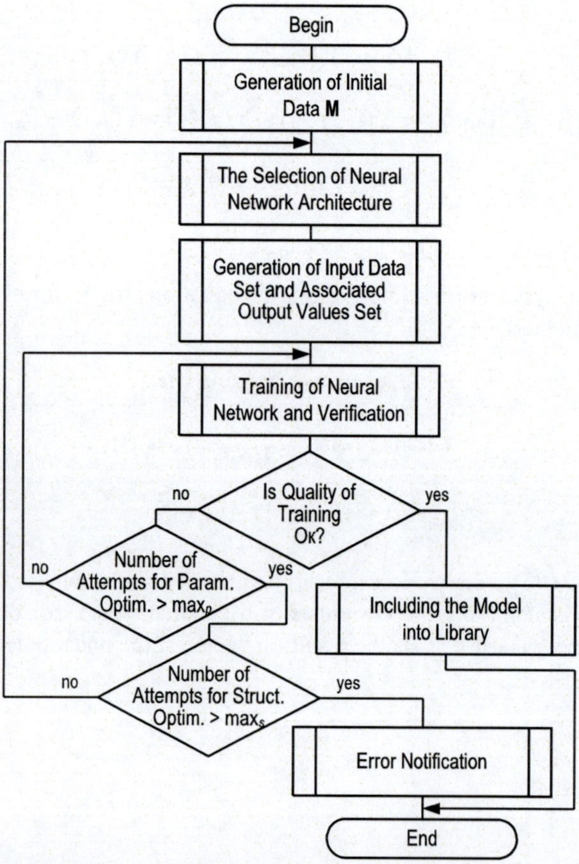

Fig. 3 Design flow of automated synthesis of the NFM

a tuples array of the following form

$$\mathbf{M} = \{m_n =< x_1^{(n)}, \ldots, x_{N_x}^{(n)}, y_1^{(n)}, \ldots, y_{N_y}^{(n)} >\},$$

$$x_i \in \mathbf{X}_{in}, y_k \in \mathbf{Y}_{out}, i = 1..N_x, k = 1..N_y,$$

where N_x is the number of the input parameters of the model, N_y is the number of the output parameters of the model, N_s is the number of discrete values of the functional dependence (1) in N_x-dimensional space of changes the input values

$$S_i^{(n)} \leq x_i^{(n)} \leq E_i^{(n)}, \forall i = 1..N_x, S_i^{(n)} = \min(x_i^{(n)}), E_i^{(n)} = \max(x_i^{(n)}).$$

2. The selection of the ANN architecture is focused on determining the number of layers, the number of neurons in each layer, and the type of activation function. A two-layer perceptron providing an approximation of the functional dependence is used as a base of the NFM. The number of neurons in the input layer is determined by the number of input parameters of the model and is equal to N_x. The number of neurons in the output layer is determined by the number of model output parameters (N_y). The number of neurons in the hidden layer (N) is determined empirically taking into account the corollary from the theorems of Arnold–Kolmogorov Hecht–Nielsen [16]

$$\frac{N_y k_{trn}}{1 + \log_2(k_{trn})} \leq N_w \leq N_y \left(\frac{k_{trn}}{N_x} + 1 \right) (N_x + N_y + 1) + N_y, \tag{6}$$

where N_x is the dimension of the input sequence; N_y is the dimension of the output sequence; N_y is the number of elements in the training set; N_w is the required number of synaptic connections.

Whence, the number of neurons of the hidden layer (N) of the two-layer perceptron will be equal to

$$N = \frac{N_w}{N_x + N_y}. \tag{7}$$

3. To generate the training ($\mathbf{M}_{trn}$) and the testing ($\mathbf{M}_{tst}$) subsets by a uniform sampling from the set of raw data $\mathbf{M}$

$$\mathbf{M}_{trn} \in \mathbf{M}, \mathbf{M}_{tst} \in \mathbf{M}, \mathbf{M}_{trn} \cap \mathbf{M}_{tst} = \emptyset,$$

$$k_{trn} = |\mathbf{M}_{trn}|, k_{tst} = |\mathbf{M}_{tst}|, k_{trn} > k_{tst},$$

where k_{tst} is the number of elements in the testing subset.

Both subsets have the same structure including the matrix (vector) of the input sets formed by the input parameters of the model x_i, $i = 1..N_x$ and the associated matrix (vector) of output values formed by the output parameters of the model y_k, $k = 1..N_y$.

4. The training of an ANN is performed using the subset $\mathbf{M}_{trn}$. The training process is stopped when either the training error becomes less or equal to the threshold value, or when the number of executed iterations exceeds the maximum available value. The quality of ANN training is tested using the subset $\mathbf{M}_{tst}$. If the quality of ANN training does not correspond to requirements, then the cycle of parametric synthesis is initiated. In this case, the re-training of the ANN architecture selected on step 2 with random re-assigning of the initial conditions is performed. If the required quality of an ANN training cannot be achieved during the limited number of parametric synthesis attempts ($\max_p$), then the cycle of the structural synthesis is initiated, which dealt with a return to step 2 and making changes in the ANN architecture (the number of neurons in the hidden layer, increasing the number of intermediate layers, etc.). If after $\max_s$ attempts the structural synthesis could not provide the required quality of ANN training, then the process is stopped with the generation of the corresponding notification.
5. The successfully trained ANN is stored in the library for further application during describing and simulating the electronic circuits.

4 Experimental Results

The model of the semiconductor diode D1N4934 was used as an object of experimental research. The raw data for the synthesis of the neuromorphic functional model (NFM) was generated at a simulation of the structural model of the D1N4934 diode in the CADENCE CAD tools. The set of raw data **M** includes 12001 tuples $m_n =< V_n, I_n >$ obtained as a result of performing the DC-analysis, where V_n is the effective voltage applied to the diode changing in the range from −6 to +6 V with step 0.01 V; I_n is the current flowing through the diode at applied corresponding voltage V_n; n=1..12001.

The training and testing subsets are generated from the set of raw data **M**. The training subset $\mathbf{M}_{trn}$ includes 1201 elements uniformly extracted from the **M** with step 10 starting from the element m_1. The rest 10800 elements not including into $\mathbf{M}_{trn}$ form the testing subset $\mathbf{M}_{tst}$.

The NFM based on the architecture of a two-layer perceptron with one neuron on the input, one neuron on the output and N neurons in the hidden layer. According to (6) and (7) N can take value in the range $54 \leq N \leq 1804$.

Table 1 The results of an ANN training

N	The number of epochs	Time, s	Performance error
54	48	0.21	1.69e−06
100	84	3.98	6.80e−08
1000	12	18.14	5.74e−14
1804	5	22.34	1.30e−18

Table 2 The estimation of a VAC approximation quality for the trained NFMs

N	The average approximation values
54	0.1391e+01
100	0.1475e+01
1000	0.5399e+07
1804	0.1986e+08

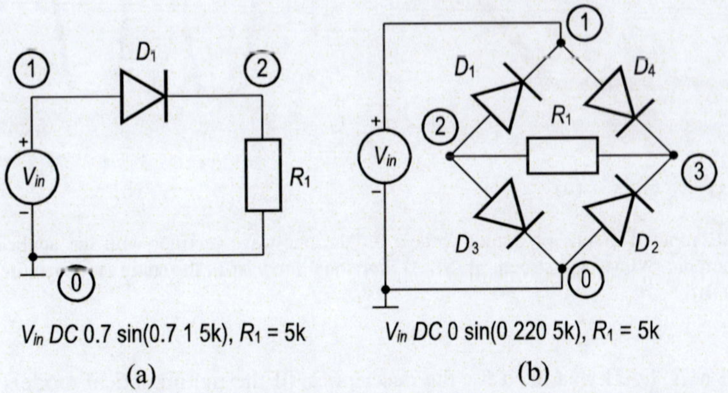

Fig. 4 The voltage rectifier circuit: one-wave rectifier (**a**), two-wave bridge rectifier (**b**)

The results of an ANN training for different number of neurons in the hidden layer (N) are represented in Table 1. The training has been performed in the tool of mathematical and engineering calculation MATLAB on the computational system with processor IntelCore i7-4770 CPU @3.4 GHz and RAM 8 GB.

The average approximation values of a volt-ampere characteristic (VAC) for the diode have been calculated for the trained NFMs using the testing set (see Table 2).

The obtained results demonstrate that the NFM with a minimal number of neurons in the hidden layer provides high precision comparable to the accuracy of the structural model of the diode. The curves of the diode's VAC for both models (structural and neuromorphic functional) are practically coincide. An effect of the ANN underfitting is observed at the use of 1000 and 1804 neurons in the hidden layer that leads to essential reducing the approximation quality.

The trained NFM with 54 neurons in the hidden layer is used for describing and simulation of the one-wave rectifier (Fig. 4a) and the two-wave bridge rectifier (Fig. 4b) circuits.

Table 3 The results of an ANN training

Error	One-wave rectifier		Two-wave rectifier	
	DC	Tran	DC	Tran
Max_Abs, V	0.0130	0.0037	0.0267	0.4988
Avg_Abs, V	0.0050	0.0019	0.0175	0.1254
RMSE, V	0.0058	0.0021	0.0149	0.1528

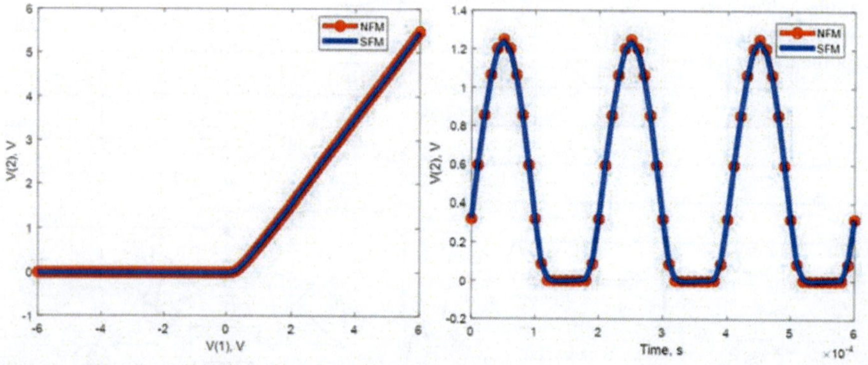

Fig. 5 Combined graphs of output voltage for the one-wave rectifier with the application the neuromorphic (NFM) and structural (SFM) functional models: in the static mode (**a**), in the time domain (**b**)

The MATLAB was used for the description of the mathematical models for the circuits of the one-wave rectifier and two-wave bridge rectifier using the NFM for the diode(s), as well as for further MMCs calculation. The obtained results were compared with results of simulating the corresponding rectifiers' circuits in the CADENCE CAD tools (Table 3). The DC-analysis with changing input voltage in the range −6 to +6 V with step 0.1 V and the Tran-analysis in the time domain during three periods of the input sine-wave voltage were performed.

Comparative analysis demonstrates the sufficient proximity of the simulation results using the NFM and the structural model of the diode. The absolute error and mean square root error for both circuits are less than 1% of the input signal amplitude. Combined graphs of the simulating results for the one-wave rectifier in the static mode and in the time domain are presented in Fig. 5, and for the two-wave bridge rectifier—in Fig. 6.

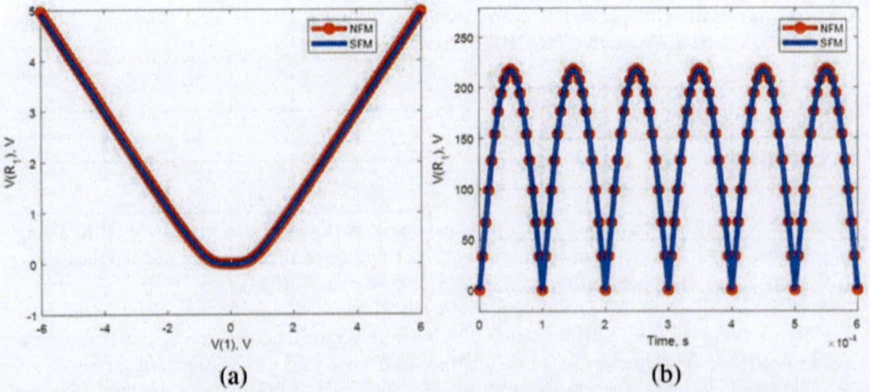

Fig. 6 Combined graphs of output voltage for the two-wave bridge rectifier with the application the neuromorphic (NFM) and structural (SFM) functional models: in the static mode (**a**), in the time domain (**b**)

5 Conclusion

The proposed approach to the synthesis of neuromorphic functional models for components and their use at the analog circuit design has demonstrated high efficiency. The synthesis of the NFM requires low computational cost and time consumption. Once trained NFM for components or functional blocks is stored in the library and can subsequently be used repeatedly in the description of MMC and device simulation. The proposed design flow for automated synthesis of the NFM is implemented in the form of CAD software in the MATLAB tool.

The synthesized NFMs provide sufficient circuit's simulation accuracy and can be used in the early stages of the design prior to the development and verification of structural models for the new components or functional blocks. The results of experimental studies for the model of the semiconductor diode D1N4934 and the circuits of voltage rectifiers based on the diode have ensured the adequacy of the obtained NFM. The average approximation error in comparison with the structural model does not exceed 1.5; the average absolute simulation error of the first circuit was less than 0.0131 V for the static mode and less than 0.00191 V for the time domain, and for the second circuit was less than 0.01751 and 0.12541 V, respectively, for simulating in the static mode and time domain. The resulting errors are less than 1% of the input voltage amplitude. The mean square root error of simulating the first circuit in the static mode and time domain was less than 0.00581 and 0.00211 V, respectively, and for the second circuit was less than 0.01491 and 0.15281 V, respectively.

Thus, the neuromorphic functional model of components and functional blocks can be used at the analog circuit design for the design time reduction. Moreover, the NFM can be used for the embedded system design with the implementation of some functionality in the neural network hardware.

Acknowledgement This paper has been supported by the Kazan Federal University Strategic Academic Leadership Program ("PRIORITY-2030").

References

1. Mosin, S. G.: The Features of Integrated Technologies Development in Area of ASIC Design. In: Proc. of 9th International Conference - The Experience of Designing and Applications of CAD Systems in Microelectronics, 292–295. Lviv-Polyana (2007)
2. Bassoli, M., Bianchi, V., De Munari, I.: A model-based design floating-point accumulator. Case of study: FPGA implementation of a support vector machine kernel function. Sensors (Switzerland) **20(5)**, paper No. 1362 (2020) https://doi.org/10.3390/s20051362
3. Baumeister, J., Finkbeiner, B., Schwenger, M., Torfah, H.: FPGA stream-monitoring of real-time properties. ACM Trans. on Embedded Comp. Sys., **18(5s)**, paper No. a88 (2019) https://doi.org/10.1145/3358220
4. Song, C., Wu, X., Tao, Y.: FPGA virtual platform based on SystemC and Verilog. IOP Conference Series: Materials Science and Engineering **768(7)**, paper No. 072001 (2020) https://doi.org/10.1088/1757-899X/768/7/072001
5. Pomante, L., Muttillo, V., Santic, M., Serri, P.: SystemC-based electronic system-level design space exploration environment for dedicated heterogeneous multi-processor systems. Microprocessors and Microsystems **72**, paper No. 102898 (2020)
6. Nagel, L. W., Pederson, D. O.: SPICE: Simulation Program With Integrated Circuit Emphasis. Univ. California, Berkeley, CA, USA (1973)
7. Trofimov, M., Mosin, S.: The realization of algorithmic description on VHDL-AMS. In: Proc. of International Conference Modern Problems of Radio Engineering, Telecommunications and Computer Science, 350–352. Lviv-Slavsko (2004)
8. Pecheux, F., Lallement, C., Vachoux, A.: VHDL-AMS and Verilog-AMS as alternative hardware description languages for efficient modeling of multidiscipline systems. IEEE Trans. Comput.-Aided Design Integr. Circuits Syst. **24(2)**, 204–225 (2005)
9. Brinson, M.E., Kuznetsov, V.: A new approach to compact semiconductor device modelling with Qucs Verilog-A analogue module synthesis. Int. J. of Numerical Modelling: Electronic Networks, Devices and Fields **29(6)**, 1070–1088 (2016)
10. Wei, Y., Doboli, A.: Systematic development of analog circuit structural macromodels through behavioral model decoupling. In Proc. of Design Automation Conference, paper No. 5.2, 5-7-62. Association for Computing Machinery, Anaheim California USA (2005)
11. Lora, M., Vinco, S., Fraccaroli, E., Quaglia, D., Fummi, F.: Analog models manipulation for effective integration in smart system virtual platforms. IEEE Trans. Comput.-Aided Design Integr. Circuits Syst. **37(2)**, 378–391 (2018)
12. Chang, Y., et al.: Compact Model for Tunnel Diode Body Contact SOI n-MOSFETs. IEEE Trans.on Electron Devices **66(1)**, 249–254 (2019)
13. Aridhi, H., Zaki, M. H., Tahar, S.: Enhancing Model Order Reduction for Nonlinear Analog Circuit Simulation. IEEE Trans. on Very Large Scale Integration (VLSI) Systems **24(3)**, 1036–1049 (2016)
14. Bond, B. N., et al.: Compact modeling of nonlinear analog circuits using system identification via semidefinite programming and incremental stability certification. IEEE Trans. Comput.-Aided Design Integr. Circuits Syst. **29(8)**, 1149–1162 (2010)
15. De Jonghe, D., Gielen, G.: Characterization of analog circuits using transfer function trajectories. IEEE Trans. Circuits Syst. I, Reg. Papers **59(8)**, 1796–1804 (2012)
16. Hecht-Nielsen, R.: Kolmogorov's mapping neural network existence theorem. In Proc. of IEEE First Annual Int. Conf. on Neural Networks, **3**, 11–13. San Diego (1987).

Scalability Pipelined Algorithm of the Conjugate Gradient Method on Heterogeneous Platforms

Nikita S. Nedozhogin, Sergey P. Kopysov, and Alexandr K. Novikov

Abstract This paper presents a parallelized iterative solver for large sparse linear systems implemented on a heterogeneous platform. Traditionally, these problems do not scale well on multi-CPU/multi-GPUs clusters. We consider the standard preconditioned Conjugate Gradient (PCG) algorithm, and as an alternative the pipelined variant, a formulation that is potentially better suited for hybrid CPU/GPU computing since it requires only one synchronization point per iteration, instead of two for standard CG. On heterogeneous cluster, the PCG iteration needs the vector entries generated by current GPU and other GPUs, so the communication between GPUs becomes a major performance bottleneck. In this paper, we study the implementation of the pipeline PCG on multi-CPU/multi-GPU platform. This paper presents an approach to reduce the communications between cluster compute nodes for these solvers. Additionally, computation and communication are overlapped to reduce the impact of data exchange. To achieve scalability, we adopt pipelined version of the conjugate gradient method with one synchronization point, the possibility of asynchronous calculations, load balancing between the CPU and GPU for parallel solving the large linear systems. The algorithm is implemented with the combined use of technologies: MPI, OpenMP and CUDA. We show that almost optimum speed up on 8-CPU/2GPU may be reached (relatively to a one GPU execution). The parallelized solver achieves a speedup of up to 5.49 times on 16 NVIDIA Tesla GPUs, as compared to a one GPU.

1 Introduction

Highly heterogeneous HPC platforms, where multicore processors are coupled with graphics processing units (GPUs), have been widely used in high performance computing as one approach to continuing performance improvement while managing

N. S. Nedozhogin (✉) · S. P. Kopysov · A. K. Novikov
Udmurt State University, Izhevsk, Russia

I. B. Badriev et al. (eds.), *Mesh Methods for Boundary-Value Problems and Applications*, Lecture Notes in Computational Science and Engineering 141,
https://doi.org/10.1007/978-3-030-87809-2_27

the new challenge of energy efficiency [10]. Although some software packages and programming languages could be used directly, the introduction of multicore processors in HPC resulted in redesign of some critical software packages and significant refactoring of some existing parallel applications. Hybrid CPU/GPU computing is one method of realizing performance gains independent of the iterative method used. With hybrid CPU/GPU computing, we focus on separating the computationally intensive portions of the program among several workers.

In [1] proposes a combination of a hybrid CPU-GPU and a pure GPU software implementation of a direct algorithm for solving shifted linear systems with a large number of complex shifts and multiple right-hand sides. This is implemented as a blocked highly parallel CPU-GPU hybrid algorithm; individual blocks are reduced by the CPU, and the necessary updates of the rest of the matrix are split among the cores of the CPU and the GPU. Thus, in [9] a hybrid method for solving systems of equations of Schur complement by preconditioned iterative methods from Krylov subspaces was built and implemented when used together the cores of central (CPU) and graphic processing units (GPU). The classical preconditioned conjugate gradient method(PCG) [6] was applied for the block ordered matrix and the separation of calculations in matrix operations between the CPU and one or more GPUs, when the system of equations in Schur complement was solved in parallel.

Distributed-memory implementation of Algebraic Recursive Multilevel Solver are presented in [7], that based on MPI and CUDA to adapt for heterogeneous CPU/GPU architectures. The tasks performed on the GPU are related to the preconditioning of each part of the distributed matrix (local preconditioning) which is handled in the distributed version by each MPI process. The solving step remains on the CPU.

Runtime systems with dynamical task scheduler were recently applied to PCG solver on heterogeneous multi-CPUs/multi-GPUs architectures using PARALUTION and StarPU libraries [8]. The authors considered the multiple advantages of heterogeneous architecture (Multi-CPUs/Multi-GPUs) to increase the performance of PCG solver by using StarPU runtime system.

In [3] numerical experiments are presented using heterogeneous computing hardware that show lower computing times and better speed-up for the pipelined variant of conjugate gradients [2]. To reduce the cost of global communication, in [12] have implemented a pipelined CG algorithm and properly fused some of the vector operations to reduce the addition overhead. In this paper, we consider an approach that reduces the cost of data exchanging between the CPU and GPU by reducing the number of synchronization points and pipelined computing when system of linear algebraic equations (SLAE) is solved on heterogeneous platforms.

We consider the pipelined variant, which is potentially better for heterogeneous multi-CPUs/multi-GPUs computing, since it requires only one synchronization point per iteration, instead of two for standard CG.

This paper presents a pipeline technique for conjugate gradient method and discusses its parallel implementation on multi-CPU/multi-GPU platform for solving large sparse linear systems.Hybrid parallel computing approaches are adopted to significantly improve performance of the solver. Specifically, we introduce a hybrid solution by fully utilizing multi-core nodes available through multi-threading techniques by means of OpenMP, and exploit an access to massively parallel hardware through GPU-offloading with CUDA, in which data are transfered to the GPU for processing. The combination of GPU-offloading and CPU-threading is explored through a hybrid CPU/GPU compute implementation.

In our work, we modify the basic CG algorithm to minimize the cost of collective communication. A modified but mathematically equal variant of the conjugate gradient algorithm is employed to reduce the cost of global communication. By using the modified algorithm, the three vector dot products in each iteration can be done simultaneously with only one nonblocking collective communication that can be further overlapped with other operations.

2 Pipelined Algorithm of the Conjugate Gradient Method

We consider now the pipelined version of the conjugate gradient method, which is mathematically equivalent to the classical form of the preconditioned CG method and has the same convergence rate.

In this algorithm, the modification of the vectors r_{j+1}, x_{j+1}, s_{j+1}, p_{j+1} and matrix-vector products provide the pipelined computations. The dot products (line 4) can be overlapped with the computation of the product by the preconditioner (line 2) and the matrix-vector product (line 3). However, the number of triads in the algorithm increases to eight, in contrast to three for the classic version and four in [2]. In this case, a parallel computation of triads and two dot products at the beginning of the iterative process and one synchronization point is possible.

The pipelined version CG presented in this work can be used with any preconditioner. There are two ways to organize computations in the preconditioned pipelined CG, which provide a compromise between scalability and the total number of operations. The first approach is that all computations are executed by GPU, and the CPU acts only as an intermediary for communications between GPUs within and without computational node. The second approach uses the CPU as another computing unit, i.e. a part of the matrix, which is similar for computations on the GPU, is also allocated for computations on the CPU. The article considers the intermediate result of these two approaches. On the one hand, the CPU mainly acts as a communication and control device. On the other hand, the CPU is also involved in computing of the matrix-vector product and summing of the dot products.

Thus, the CG pipeline scheme is characterized by a different order of computations, the presence of global communication, which can overlap with local computations, such as matrix-vector product and operations with a preconditioner, and the possibility of organizing asynchronous communications.

The two variants of the conjugate gradient method were compared: the classical scheme and the pipelined one. Table 1 presents the results of numerical experiments where the execution time of a sequential version of the classical CG and the CGwO pipelined scheme (Algorithm in Fig. 1) executed on the CPU and GPU are shown.

Table 1 Statistics of the test problems. Problem names, dimensions (N), number of nonzeros (nnz), device type (DT) and problem analysis in terms of the timing in seconds

Matrix	N	nnz	# iter.	DT	Time, s	
					CG	CGwO
`Plat362`	362	5786	991	M2090	6.88E-01	**3.07E-01**
				K40m	4.13E-01	3.12E-01
`1138_bus`	1138	4054	717	M2090	3.81E-01	**1.84E-01**
				K40m	5.31E-01	2.01E-01
				debug	6.82E-01	1.90E-01
`Muu`	7102	170134	12	M2090	2.64E-01	4.68E-03
				K40m	3.31E-01	**4.55E-03**
`Kuu`	7102	340200	378	M2090	4.31E-01	**1.31E-01**
				K40m	4.39E-01	1.35E-01
`Pres_Poisson`	14822	715804	661	M2090	6.72E-01	3.13E-01
				K40m	6.346E-01	**2.73E-01**
`Inline_1`	503712	36816342	5642	M2090	4.74E+01	5.17E+01
				K40m	**3.06E+01**	3.37E+01
`Fault_639`	638802	28614564	4444	M2090	3.83E+01	4.32E+01
				K40m	**2.44E+01**	2.77E+01
				debug	2.44E+01	2.77E+01
`thermal2`	1228045	8580313	2493	M2090	1.35E+01	1.82E+01
				K40m	**8.33E+00**	1.18E+01
`G3_circuit`	1585478	7660826	592	M2090	3.43E+00	4.32E+00
				K40m	**1.94E+00**	2.92E+00
`Quenn_4147`	4147110	399499284	8257	M2090	5.46E+02	5.78E+02
				K40m	**3.55E+02**	3.75E+02

Note that in the variants for the GPU, joint computation of all dot products of vectors in one kernel function was implemented, independently of each other. For this, when starting the CUDA kernel, the dimension of the `Grid` hierarchy of CUDA threads was set in two-dimensional form: 3 sets of blocks, each for performing computations on its own pair of vectors. This allowed us to reduce the number of exchanges between the CPU and GPU memory, combining all the resulting scalars in one communication.

Matrices from the SuiteSparse Matrix Collection [4] were used in the test computations. The right hand side vector was formed as a row-wise sum of matrix elements. Thus, the solution of the system $Ax = b$, dimension $N \times N$ (with the number of nonzero elements nnz) is a vector $x = (1, 1, \dots, 1)^T$.

For systems of equations of small dimension, the solution time on the CPU according to the classical CG scheme is significantly less than the GPU execution time for the same number of iterations (see Table 1). In tables, the fastest options in the line are in bold. For large systems, the costs of synchronization and forwarding between the CPU and GPU overlap with the speed of the GPU. In the pipelined version of CGwO, the computational execution costs on the GPU are reduced almost

Fig. 1 Algorithm 1: pipelined algorithm CGwO

$r = b - Ax;$
$u = M^{-1}r;$
$w = Au;$
$\gamma_1 = (r, u);\ \delta = (w, u);$
while $||r||_2/||b||_2 > \varepsilon$ **do**
  $m = M^{-1}w;$
  $n = Am;$
  if $(j = 0)$ **then**
    $\beta = 0;$
  end
  else
    $\beta = \gamma_1/\gamma_0\ ;$
  end
  $\alpha = \gamma_1/(\delta - \beta\gamma_1/\alpha)\ ;$
  $z = n + \beta z;\ w = w - \alpha z;\ s = w + \beta s;\ r = r - \alpha s;$
  $p = u + \beta p;\ x = x + \alpha p;\ q = m + \beta q;\ u = u + \alpha q;$
  $\gamma_0 = \gamma_1;$
  $\gamma_1 = (r, u);\ \delta = (w, u);$
end

threefold for all the considered systems of equations only due to the reduction of exchanges between the GPU and the CPU in the computation of dot products.

3 CG with the Combined Use of CPU and GPU

Let us consider the application of the Algorithm in Fig. 1 for the parallel solution of super-large systems of equations on computing nodes, each of which contains several CPUs and GPUs. To solve SLAEs on several GPUs, we construct a block pipelined algorithm for the conjugate gradient method. On heterogeneous platform, data exchange between different GPUs within the same computing node is carried out with OpenMP technology, and the exchange between different computing nodes is carried out by MPI technology.

For example, consider a node containing a central eight-core processor and two graphics accelerators. The number of OpenMP threads is selected by the number of available CPU cores. The first two OpenMP threads are responsible for exchanging data and running on two GPUs. Threads 2–6 provide computations on the CPU and can perform computations on a block of the SLAE matrix. The last thread provides data exchange with other computing nodes by MPI.

3.1 Matrix Partitioning

To divide the matrix A into blocks, we construct the graph $G_A(V, E)$, where $V = \{i\}$ is the set of vertices associated with the row index of the matrix (the number of vertices is equal to the number of rows of the matrix A); $E = \{(i, j)\}$ is the set of edges. Two vertices i and j are considered to be connected if the matrix A has a nonzero element with indices i and j. The resulting graph is divided into subgraphs whose number is d. For example, to split a graph, you can use the [8, 11] layer-by-layer partitioning algorithm, which reduces communication costs due to the need to exchange only with two neighboring computing nodes.

After that, each vertex of the graph is assigned its own GPU or CPU. On each computing unit, the vertices are divided into internal and boundary. The latter are connected with at least one vertex belonging to another subgraph.

After partitioning, each block A_k of the original matrix A contains the following submatrices:

- $A_k^{[i_k,i_k]}$—matrix associated with the internal vertices;
- $A_k^{[i_k,b_k]}$, $A_k^{[b_k,i_k]}$—matrices associated with the internal and boundary vertices;
- $A_k^{[b_k,b_l]}$—matrix associated with the boundary vertices of the k-th and l-th blocks.

Then the matrix A can be written in the following form:

$$A = \begin{pmatrix} A_1^{[i_1,i_1]} & A_1^{[i_1,b_1]} & \dots & 0 & 0 \\ A_1^{[b_1,i_1]} & A_1^{[b_1,b_1]} & \dots & 0 & A_1^{[b_1,b_d]} \\ \vdots & \vdots & \ddots & \vdots & \vdots \\ 0 & 0 & \dots & A_d^{[i_d,i_d]} & A_d^{[i_d,b_d]} \\ 0 & A_d^{[b_d,b_1]} & \dots & A_d^{[b_d,i_d]} & A_d^{[b_d,b_d]} \end{pmatrix}.$$

We divide the matrix-vector product $n = Am$ into two components by using the obtained partition:

$$n_k^b = A_k^{[b_k,i_k]} m_k^i + \sum_{l=1}^{l \le d} A_k^{[b_k,b_l]} m_l^b, \qquad n_k^i = A_k^{[i_k,i_k]} m_k^i + A_k^{[i_k,b_k]} n_k^b. \tag{1}$$

Here k corresponds to the computing device. The block representation of the vectors involved in the algorithm is inherited from the matrix partitioning. For example, the vector m has the form $m^T = \left(m_1^i, m_1^b, \dots, m_k^i, m_k^b, \dots, m_d^i, m_d^b\right)$. The implementation of the matrix-vector product reduces the cost of communication between blocks at each iteration of conjugate gradient method. To perform this operation, an exchange of vectors m_k^b is required, the size of which is less than the dimension of the initial vector m.

The partitioning of the preconditioner M is carried out in a similar way.

3.2 Block Pipelined Algorithm

The matrix blocks were mapped on the available CPU and GPU with the block partitioning of the matrix and vectors. The number and size of blocks let on to map the load in accordance with the performance of the computing units, including the allocation of several blocks to one.

Let us represent parallel block scheme of the method CGwO that is performed each k-th computing unit in the form of Algorithm 2. Two parallel branches of this algorithm are executed accordingly on the CPU and CPU/GPU. Operations performed in parallel are shown in one line of the algorithm. Vector operations on each computing unit occur in two stages, for internal and boundary nodes. The designations of the internal and boundary nodes for vectors are omitted, with the exception of the matrix-vector multiplication. Dot products are performed independently by each computing unit on its parts of vectors. The summation of intermediate scalars occurs in parallel threads responsible for communication, which is the synchronization point at each iteration of the algorithm.

In block CGwO, compared to Algorithm 1, the preconditioning step has been moved (line 5 to line 21). This is done in order to combine vector operations on the computing unit and the assembly of the vector parts of the right hand side to perform matrix-vector multiplication in preconditioning. The 13 line on the right uses the ternary operator: if $j = 0$, then $\beta = 0$, in other cases $\beta = \gamma_1/\gamma_0$. The subscript h is used for vectors that are stored only in CPU memory.

3.3 MPI+OpenMP+CUDA Programming Model

Numerical experiments on the Algorithm in Fig. 2 were carried out on heterogeneous platform with various configuration of computing nodes containing several CPUs and GPUs. In the general case, the parallel computing on several heterogeneous computing nodes containing one or more CPUs and several GPUs is implemented by the combination of several technologies: MPI, OpenMP and CUDA. In this article, the approach is to properly divide the computational workload between the CPU and the GPU, so that the CPU can aid the GPU in sharing the computational costs. Our programming strategy is based on implementation strategy, where a hybrid MPI+CUDA+OpenMP programming model is used to realize concurrent CPU+GPU computations. The principal concept for the strategy is to overlap computation with communication using OpenMP's nested parallelism capability to generate two independent groups of threads. The first thread group handles the CUDA, MPI communication and computation of the halo boundary points on the CPU using OpenMP threads. The second thread group computes the interior points on the CPU.

Let us consider the software organization of computations using as example some cluster, which includes two computing nodes (8-CPU cores and 2-GPUs).

Data: Matrix partitioning into blocks $A_k^{[i_k,i_k]}$, $A_k^{[i_k,b_k]}$, $A_k^{[b_k,i_k]}$, $A_k^{[b_k,b_l]}$.

1	$r = b$;	
2	$u = M^{-1}r$;	
	// Parallel algorithm branches	
	// (CPU $\vee$ GPU)$_k$	// CPU
3	$w_k^i = A_k^{[i_k,i_k]} \cdot u_k^i + A_k^{[i_k,b_k]} \cdot u_k^b$;	Assembly of the vectors u_k^b;
4	$w_k^b = A_k^{[b_k,b_k]} \cdot u_k^b + A_k^{[b_k,i_k]} \cdot u_k^i$;	$w_h^b = \sum_{l=1,l\neq k}^{l\leq d} A_k^{[b_k,b_l]} \cdot u_k^b$;
5		Copying w_h^b on the GPU$_k$;
6	$w_k^b = w_k^b + w_h^b$;	
7	$m = M^{-1}w$;	Assembly of the vectors m_k^b;
8	$\gamma_{1k} = (r_k, u_k)$; $\delta_k = (w_k, u_k)$;	Assembly $\delta = \sum_k \delta_k$; $\gamma_1 = \sum_k \gamma_{1k}$;
	while $\|\|r\|\|_2/\|\|b\|\|_2 > \varepsilon$ **do**	
9	$n_k^i = A_k^{[i_k,i_k]} \cdot m_k^i + A_k^{[i_k,b_k]} \cdot m_k^b$;	
10	$n_k^b = A_k^{[b_k,b_k]} \cdot m_k^b + A_k^{[b_k,i_k]} \cdot m_k^i$;	$n_h^b = \sum_{l=1,l\neq k}^{l\leq d} A_k^{[b_k,b_l]} \cdot m_k^b$;
11		Copying n_h^b on the GPU$_k$;
12	$n_k^b = n_k^b + n_h^b$;	
13	$z = n + \beta z$;	$\beta = ((j = 0)\,?\,0\,:\,\gamma_1/\gamma_0)$;
14	$w = w - \alpha z$;	$\alpha = \gamma_1/(\delta - \beta\gamma_1/\alpha)$;
15	$q = m + \beta q$;	
16	$s = w + \beta s$;	
17	$p = u + \beta p$;	Assembly of the vectors w_k^b;
18	$x = x + \alpha p$;	
19	$r = r - \alpha s$;	
20	$u = u + \alpha q$;	Assembly vectors m_k^b;
21	$m = M^{-1}w$;	
22	$\gamma_0 = \gamma_1$;	Assembly $\delta = \sum_k \delta_k$; $\gamma_1 = \sum_k \gamma_{1k}$;
23	$\gamma_{1k} = (r_k, u_k)$; $\delta_k = (w_k, u_k)$;	
	end	

Fig. 2 Algorithm 2: block algorithm CGwO performed on *k*-th device

Each computing node is associated with a parallel MPI process. In a parallel process, 9 parallel OpenMP threads are generated, which is one more than the available CPU cores. The eighth OpenMP thread is responsible for communications between different computing nodes (using MPI technology, vector assembly using the `Allgatherv` function, adding scalars `Allreduce`) and various GPUs. In the 2 Algorithm, the operations performed by this thread are presented to the right. Zero and first OpenMP threads are the host threads for one of the available GPU devices and are responsible for transfer data between the GPU/CPU (calls to asynchronous copying functions) and auxiliary computations. Each available GPU device (further considered as a computing unit) is associated with one of the parallel OpenMP threads, which is responsible for transferring data between the GPU and CPU (calls to asynchronous copy functions) and participates with the eighth treads in matrix-vector product on boundary vertices (lines 4, 9 right column). The remaining parallel threads (second to seventh) perform the calculations as a separate computing unit for their matrix block. The operations performed by computing units in the 2 Algorithm are shown on the left.

The preconditioning in lines 2, 7 and 21 implies the use of block matrix-vector multiplication of the form (1) considered above.

4 Numerical Experiments

The numerical experiments were performed on the cluster Uran of Supercomputer center IMM UB RAS, Yekaterinburg, Russia. Uran involves heterogeneous partitions with computing nodes (CNs), which differ by CPUs, GPUs, memory sizes and networks. The cluster partitions with the following characteristics were used:

- partition "debug": 4 CNs tesla [31–32,46–47] with two 8-cores CPU Intel Xeon E5-2660 (2.2 GHz), cache memory is 20 MB L3 cache, RAM is 96 GB and 8 GPU Tesla M2090 (6 GB per device), network is 1 Gb/s Ethernet.
- partition "tesla[21–30]": 10 CNs with two 6-cores CPU Intel Xeon X5675 (3.07 GHz), RAM is 192 GB, cache memory is 12 MB L3 cache and 8 GPU Tesla M2090 (6 GB per device), with network is Infiniband 20 Gb/s.
- partition "tesla[33–45]": 13 CNs with two 8-cores CPU Intel Xeon E5-2660 (2.2 GHz), cache memory is 20 MB L3 cache, RAM is 96 GB, and 8 GPU Tesla M2090 (6 GB per device), network is Infiniband 20 Gb/s.
- partition "tesla[48–52]": 5 CNs with two 8-cores CPU Intel Xeon E5-2650 (2.6 GHz), cache memory is 20 MB L3 cache, RAM is 64 GB and 3 GPU Tesla K40m (12 GB per device), network is Infiniband 20 Gb/s.

Clusters such as Uran are that CPU+GPU codes are not effective, if the performance difference between CPU and GPU is too big. In these cases, GPU-only code might be a better alternative. Luckily, it is capable of using both GPU-only and 1-CPU/2-GPU code. The numerical experiments were carried out on well-known datasets, which we will consider in more detail.

4.1 *Benchmarking with HPCG Matrices*

The High Performance Conjugate Gradient (HPCG) [5] is a benchmark program that solves a sparse linear system arising in solving a three-dimensional heat diffusion problem. HPCG intends to solve the linear system generated from the finite difference discretization of the Poisson equation: $-\Delta u = b$, with homogeneous Dirichlet boundary conditions applied along the boundary of a three-dimensional cubic domain Ω. Based on a semistructured mesh with equidistant mesh spaces in the x, y and z directions, respectively, the discretization employed in HPCG leads to a second-order accurate 27-point stencil.

The resulting sparse linear system has the following properties: A—sparse matrix with 27 nonzero entries per row for interior equations and 7 to 18 nonzero terms for boundary equations; A—symmetric, positive definite, nonsingular linear operator.

We generate a synthetic symmetric positive definite (SPD) matrix A using an array-of-pointers-style compressed sparse row format, an exact solution vector of all 1.0 values, a corresponding right-hand-side vector b, and initial guess for x of all 0.0 values. The sparsity pattern of the synthetic matrix is really a regular 27-point 3-dimensional stencil pattern.

We tested one CPU performance on the Uran cluster based on three typical data sizes, including:

1. $125 \times 125 \times 160$, $nnz = 66503662$;
2. $160 \times 160 \times 201$, $nnz = 137318884$;
3. $200 \times 200 \times 250$, $nnz = 267487792$;
4. $250 \times 250 \times 310$, $nnz = 519219712$;
5. $310 \times 310 \times 390$, $nnz = 1005862912$;
6. $390 \times 390 \times 485$, $nnz = 1986735009$;

In all variants, the pipelined version of CG converged in 43 iterations with ε equal to 10^{-6}. For example, the solving time of the first three data sizes on one OpenMP thread were 8.39, 17.41, 33.22 s, respectively. We were able to obtain the result only for the first two sizes: 1.33, 2.72 when using single GPU. The remaining data sizes is not placed in the memory of one graphics accelerator. These results allow us to estimate that the performance of one OpenMP thread is approximately 6.5 times lower than the performance of single GPU for the linear system solving by the conjugate gradient method.

Performance metrices are executed and compared through scalability studies and absolute runtime results. To estimate the computational performance and the impact of MPI and OpenMP communications, our numerical experiments were executed with different numbers of CPUs and GPUs. The results are shown in the Figs. 3 and 4. Each figure corresponds to the number of cluster nodes (n-CPU) involved in the computations and the number of graphics accelerators on each node (m-GPU).

Figure 5 shows the results by subdomains for the case when 2 GPUs are used per computational node. Here, the matrix size of the linear system is approximately doubled. It can be noted that, starting from a size equal to 10,000,000, there is a good scalability of the algorithm. The problem execution time remains practically unchanged by doubling the problem size and doubling the number of subdomains.

4.2 Benchmarking on the SuiteSparse Matrix Collection

The results of comparing two algorithms of the conjugate gradient method on SLAEs containing test matrices [4] are considered. The problems range from small matrices, used as counter-examples to hypotheses in sparse matrix research, to large test cases arising in large-scale computation.

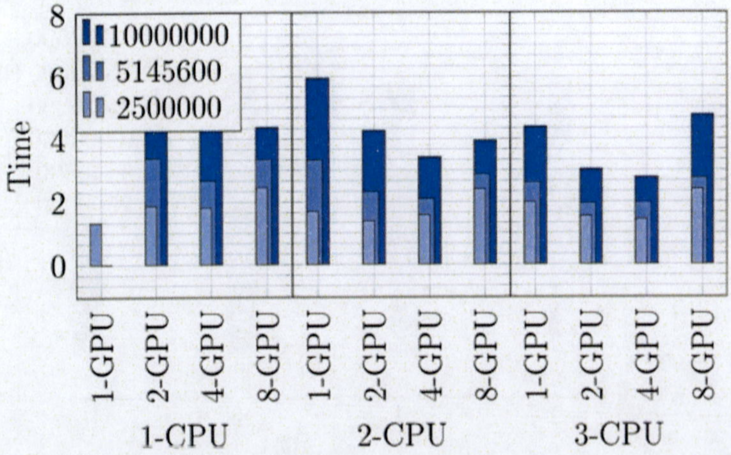

Fig. 3 Scalability of the block algorithms CGwO by CPU and GPU

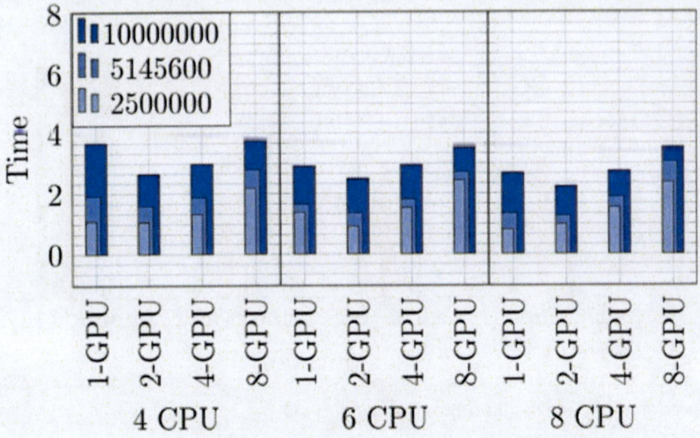

Fig. 4 Scalability of the block algorithms CGwO by CPU and GPU(continue)

In the standard PCG algorithm, three dot products need to be done per iteration, with each one requiring a global collective communication that may substantially degrade the scalability at scale. In order to reduce the global communication overhead, we employ a reformulated but mathematically equivalent variant of the basic PCG algorithm, the pipelined Block PCG.

As shown in Algorithm 2, the pipelined PCG method has two advantages. First, only one global reduction is required for each iteration. Second, the global reduction can be overlapped with the matrix-vector product and with the application of the preconditioner.

Figure 6 presents the results of accelerating the block algorithms of the conjugate gradient method, when divided into a larger number of blocks, accordingly 8, 12 and

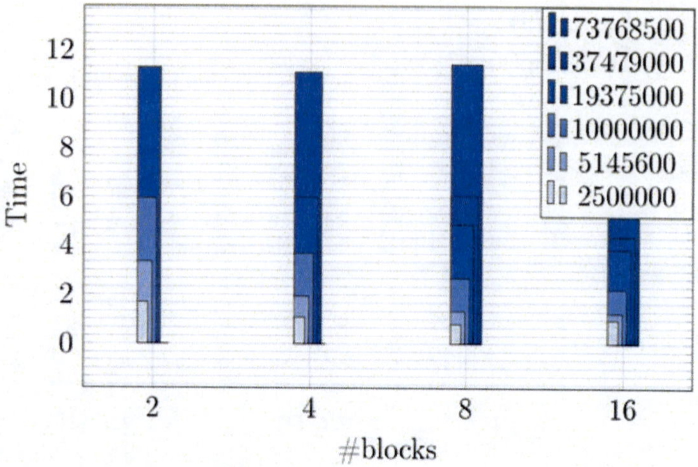

Fig. 5 Scalability of the block algorithms CGwO by number blocks

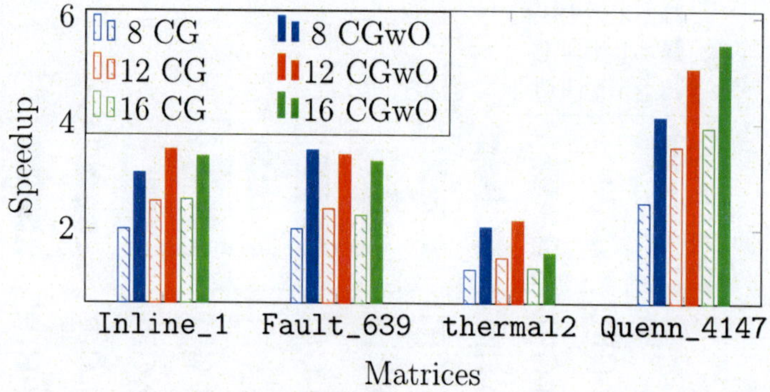

Fig. 6 Speedup of the block algorithms CG and CGwO

16. To compute the speedup, parallel application was run repeatedly with different mapping of subdomains to several CPUs and GPUs. For example, in the case of 12 subdomains, variants were considered: 2 CPUs with 6 GPUs, 3 CPUs with 4 GPUs, 6 CPUs with 2 GPUs. The best time is shown.

The results of comparing two algorithms of the conjugate gradient method on SLAEs containing test matrices are presented in Table 1. The results are given for several types of computing nodes using a single graphics accelerator.

The matrices are ordered by increasing the order of the system of equations (N) and the number of nonzero elements (nnz). Bold indicates the best time to solve the system in each case. The pipelined algorithm CGwO showed a reduction in execution time on small SLAEs which are characterized by a small computing load, due to which a reduction in communications provides less time. Note that the classic

Table 2 Time of solving by the block algorithms CG and CGwO on CPU/GPU, s

	CG/#blocks		CGwO/#blocks	
Matrix/DT	2	3	2	3
Plat362/M2090	1.55E+00		**1.22E+00**	
/K40m	1.92E+00	1.56E+00	1.28E+00	1.31E+00
1138_bus/M2090	1.84E+00		**9.28E-01**	
/K40m	1.90E+00	1.85E+00	1.03E+00	1.04E+00
/debug	1.25E+01		**5.36E+00**	
Muu/M2090	6.12E-01		2.29E-01	
/K40m	6.59E-01	5.64E-01	2.89E-01	**2.88E-01**
Kuu/M2090	1.30E+00		**6.43E-01**	
/K40m	1.29E+00	1.36E+00	6.81E-01	7.95E-01
Pres_Poisson/M2090	1.55E+00		**9.57E-01**	
/K40m	1.60E+00	1.66E+00	1.02E+00	1.19E+00
G3_circuit/M2090	4.27E+00		3.99E+00	
/K40m	4.04E+00	3.510E+00	3.27E+00	**2.77E+00**

CG algorithm was implemented based on CUBLAS, while the CGwO variant uses matrix and vector operations of its own GPU implementation.

For systems `Inline_1` and `Fault 639`, the execution time of the pipelined algorithm is 10 and 13.5% longer than the block version of CG, which is associated with additional vector operations that are not blocked by reduced communications. With a decrease in the number of iterations, for example, for solving a large system with `G3_circuit`) with an approximately equal number of equations with `thermal2`, the execution time of the CG and CGwO algorithms on one GPU increases slightly. For the system (`thermal2` and `G3_circuit`) the increase in costs becomes more significant.

Table 2 presents the results of the block variant of the algorithms for computing on several computing nodes for systems with small dimension matrices. Here are the results for 2 and 3 subdomains. Each subdomain was considered on a separate computing node. Communications were carried out using MPI technology. A significant influence of network characteristics on the performance of block methods can be seen in Table 2 for system of equations with matrix `1138_bus`. Computations for these SLAEs were performed at various computing nodes with different throughput and latency of the network. In numerical experiments on the CNs (partition “debug”) connected by a Gigabit network, communication costs significantly increase the execution time of the CG algorithm.

For example, in the variant `1138_bus` on the cluster partition “debug”, the execution time of the pipeline algorithm is 3.6 times less (the line “debug” in Table 2 and any row in Table 1). Using the Infiniband 20 Gb/s communication network reduces the execution time for all presented systems of equations (lines “M2090” and “K40m”).

When reducing the computational load, a decrease in the number of synchronization points and the consolidation of transfers per transaction is more pronounced. This shows a comparison of systems with matrices Kuu and Muu. Both systems have an equal number of equations and nonzero elements, but the conditionality of these matrices is significantly different and, as a consequence, the number of iterations in the conjugate gradient method is different. Table 1 shows that using the pipeline algorithm for the matrix Muu gives speedup by 70 times, compared with the matrix Kuu, where the speedup is only 2.8.

The speedup was considered relative to the option on one GPU from Table 1. An application that implements this algorithm was executed in the exclusive mode of the computing node but not of the network.

As can be seen from the presented results, the pipelined CG shows the speedup greater than the classic version of conjugate gradient method. Wherein, for the largest of the considered matrices Quenn_4147, the speedup achieves 5.49 times, while the classical version gives 3.92 as maximum. For the strongly sparse matrix thermal2, block algorithms don't give high speedup (maximum is 1.56), since the computational load depends mainly on the number of the nonzero elements.

An analysis of the results showed that reducing the data size due to the matrix partitioning and reducing the synchronization points slightly decrease the impact of communication costs on the total algorithm performance. Only the use of computing nodes connected by Infiniband allowed us to get speedup when computing on several computing nodes. The matrix partitioning into blocks allowed to decrease the execution time of the pipelined block algorithm in comparison with the conjugate gradients on one node on the matrices Inline_1, Fault_639 by reducing the computational load on one GPU.

Large systems thermal2, G3_circuit, solved by the block of the CGwO algorithm, as well as the reduction in communications costs and synchronization points, do not overlap the increasing costs of additional vector operations.

5 Conclusion

The heterogeneous computing platforms containing and sharing CPU + GPUs provide an effective solution to a wider range of problems with high energy efficiency when CPU and GPUs are uniformly loaded.

The parallel implementation of the solution of systems of linear algebraic equations on a heterogeneous platform was considered. The performance of parallel algorithms for classical conjugate gradient method is significantly limited by synchronization points when using the CPU and GPU together. A pipelined algorithm of the conjugate gradient method with one synchronization point was proposed. Also, it is provided the possibility of asynchronous computations, load balancing between several GPUs located both on the same computing node and for a GPU cluster when solving systems of large-dimensional equations. To further increase the efficiency of calculations, it is supposed to study not only the communication load of the algorithms but also the distributing of the computational load between the CPU and GPU. To obtain more reliable evaluation of communications costs, it is necessary to conduct a series of computational experiments on supercomputer with a completely exclusive mode of operation and a large number of heterogeneous nodes.

The following conclusions can be drawn from the analysis of data obtained during numerical experiments: the use of a pipeline algorithm reduces communication costs, but increases computational ones. For systems of small sizes or with a small number of iterations, this reduces the execution time of the algorithm when using a single GPU. For systems of large dimensions, a reduction in execution time, in comparison with CG, is possible only with a sufficiently small partition of the matrix into blocks, in which the increased computing costs overlap the communication decrease.

The proposed block algorithms, in addition to reducing the execution time, allow solving large linear systems that requires memory resources not provided by one GPU or computing node. At the same time, the pipelined block algorithm reduces the overall execution time by reducing synchronization points and combining communications into one message.

Acknowledgments The work was carried out with the financial support of Udmurt State University in the contest of the grants "Scientific Potential", project No. 2020-04-03.

References

1. Bosner, N., Bujanovi, Z., Drma, Z.: Parallel solver for shifted systems in a hybrid CPU–GPU framework. SIAM Journal on Scientific Computing **40**(4), C605–C633 (2018)
2. Chronopoulos, A., Gear, C.: s-step iterative methods for symmetric linear systems. Journal of Computational and Applied Mathematics **25**(2), 153–168 (1989)
3. Collignon, T., Gijzen, M.V.: Two implementations of the preconditioned conjugate gradient method on heterogeneous computing grids. International Journal of Applied Mathematics and Computer Science **20**(1), 109–121 (01 Mar 2010)
4. Davis, T.A., Hu, Y.: The university of florida sparse matrix collection **38**(1) (2011)
5. Dongarra, J., Heroux, M.A., Luszczek, P.: A new metric for ranking high-performance computing systems. National Science Review **3**(1), 30–35 (01 2016)
6. Hestenes, M.R., Stiefel, E.: Methods of conjugate gradients for solving linear systems. Journal of research of the National Bureau of Standards **49**, 409–436 (1952)
7. Jamal, A., Baboulin, M., Khabou, A., Sosonkina, M.: A hybrid CPU/GPU approach for the Parallel Algebraic Recursive Multilevel Solver pARMS. In: 2016 18th International Symposium on Symbolic and Numeric Algorithms for Scientific Computing (SYNASC). pp. 411–416 (2016)
8. Kasmi, N., Zbakh, M., Haouari, A.: Performance analysis of preconditioned conjugate gradient solver on heterogeneous (multi-CPUs/multi-GPUs) architecture. Lecture Notes in Networks and Systems **49**, 318–336 (2019).
9. Kopysov, S., Kuzmin, I., Nedozhogin, N., Novikov, A., Sagdeeva, Y.: Scalable hybrid implementation of the schur complement method for multi-gpu systems. Journal of Supercomputing **69**(1), 81–88 (2014)

10. Mittal, S., Vetter, J.S.: A survey of cpu-gpu heterogeneous computing techniques. ACM Comput. Surv. **47**(4) (Jul 2015).
11. Kadyrov, I.R., Kopysov, S.P., Novikov, A.K.: Partitioning of triangulated multiply connected domain into subdomains without branching of inner boundaries. Uchenye Zapiski Kazanskogo Universiteta. Seriya Fiziko-Matematicheskie Nauki **160**(3), 544–560 (2018)
12. Zhang, X., Yang, C., Liu, F., Liu, Y., Lu, Y.: Optimizing and scaling HPCG on Tianhe-2: early experience, vol. 8631, part I, p. 28–41. Springer, Dalian, China (aug 2014),

Accumulation of Microdamages During Cyclic Loading of CFRP Structure Elements

Vitaly N. Paimushin, Rashit A. Kayumov, and Sergey A. Kholmogorov

Abstract A review devoted to the problem of describing the microdamages accumulation is given. The results of the author's experimental studies are presented. They are demonstrating the features of the deformation processes of specimens with cross-ply ±45° lay-up made of ELUR-P unidirectional carbon fiber and cold-curing binder XT-118 under cyclic loading. When modeling these processes, it is assumed that strain consists of elastic, viscoelastic, viscoplastic parts, as well as a part formed as a result of microdamages accumulation. A phenomenological approach is used to describe the process of microdamages accumulation. When identifying the parameters of the constitutive relations, based on the test results, which relate the above-mentioned parts of the strain to stresses, some hypotheses are used that make it possible to simplify the solution of this problem. The results of solving the problems of determining the mechanical characteristics included in the proposed variants of the constitutive relations are presented.

1 Introduction

In the process of loading, structures made of fiber reinforced plastics (FRP), in addition to elastic ones, inelastic strains sometimes occur. They can be caused by the creep of the material, plastic strains (especially in composites with a metal matrix), the motion of dislocations, the stability loss of the phases of the composite, their failure at the microlevel (development of microcracks, micropores). Under a single static loading, some of strains can be interpreted as nonlinear elastic and viscoplastic using well-known relations for describing such deformation models. The processes

V. N. Paimushin (✉) · R. A. Kayumov
Kazan National Research Technical University named after A.N. Tupolev-KAI, Kazan, Russia

Kazan Federal University, Kazan, Russia

S. A. Kholmogorov
Kazan National Research Technical University named after A.N. Tupolev-KAI, Kazan, Russia

I. B. Badriev et al. (eds.), *Mesh Methods for Boundary-Value Problems and Applications*, Lecture Notes in Computational Science and Engineering 141,
https://doi.org/10.1007/978-3-030-87809-2_28

caused by microdamage are usually called the microdamages accumulation, and a number of well-known theories have been proposed for them [1–4]. Starting from work [1], the measure (parameter) of damage under axial tension is understood as the value

$$\omega(t) = 1 - A_{eff}(t)/A_0, \tag{1}$$

where A_0, $A_{eff}(t)$ is initial and effective (actually bearing load) of cross-section area, t—time. Effective stress was determined as

$$\sigma_{eff}(t) = P/A_{eff}(t). \tag{2}$$

To describe the damage evolution, the following relation was proposed [1]:

$$d\omega/dt = 1 - C(\sigma/(1-\omega)^n),\ \omega(0) = 0. \tag{3}$$

The equality was taken as a condition for material failure:

$$\sigma_{eff}(t^*) = \sigma^*, \tag{4}$$

where σ^* is the ultimate strength of the material. The condition was often used in a simpler form:

$$\omega(t^*) = 1. \tag{5}$$

The development of these ideas for the case of a complex stress state was the main subject of the continuum theory of damage, while various scalar and tensor measures of damage were proposed for isotropic and anisotropic media [5–10]. For example, the criterion of long-term static strength within the framework of this approach was proposed in [9], in which, to determine the effective stresses, an approach was used, according to which, in the criterion of short-term strength, the nominal stresses are replaced by the effective ones

$$f(\pi, \sigma_{eff}^{ij}(t)) = 1, \tag{6}$$

where π is set of strength parameters; σ_{eff}^{ij} is effective stress. This is a natural generalization of the failure condition (4) for the case of a complex stress state.

In most cases, when describing the processes of microdamages accumulation, the hypothesis proposed by Kachanov in [2] is used, namely, it is assumed that the formation of microdamages does not affect the creep process and vice versa. With complex types of impact of external factors, including cyclic loading, it is necessary to take into account the peculiarities of the microdamages accumulation in the FRP, depending on the structure and type of FRP components. Therefore, the models used to describe them differ from those that are usually used to analyze the deformation of bodies made of metals and other traditional structural materials. The simplest models of this kind of deformation processes under tension in transversal direction to the fibers of the lamina and its shear of laminates were proposed in [3, 4]. In this case, the relations that connect the stresses with the total strains of the lamina sometimes depend both on the states of "loading" or "unloading" and the sign of the stresses. In the axes of orthotropy of the lamina, it is assumed that strains caused by microdamages appear only when tension transverse the reinforcement and in shear. With a decrease in tensile or tangential stresses, unloading proceeds according to a linear law. When the stress changes from tensile to compressive, the hypothesis of the absence of linear strains caused by microdamages (due to the closure of microcracks) is accepted. During shear, the accumulation of microdamages is considered to be independent of the sign of shear stresses. Since it is assumed that only elastic strains and strains caused by microdamages are present, the latter no longer change after unloading the laminate and reloading to the same values of stresses. However, under cyclic loading, as will be seen from the results of experiments on specimens made of ELUR-P unidirectional carbon fiber, this assumption is no longer valid.

In the general case, damage measures can be included both in the strength criterion and in the constitutive relations for creep, plasticity, viscoelasticity, and vice versa. For example, in [1], the constitutive relation for the damage parameter contains creep strain. The author refers to this case to the processes of loading metals at high temperatures. An example of solving the problem of the theory of plastic flow related to the continuum theory of damage for the volumetric stress state of an isotropic body is given, for example, in [5].

There are theories based on other concepts as well. In [11], the intensity of accumulated creep strains was used as a measure of material damage. In [12], when considering a nonlinear-viscoelastic isotropic material, the constitutive strain relations were proposed to be written in the form (in what follows, the indices of tensor quantities will be omitted for ease of notation):

$$\varepsilon = \varepsilon^{el} + \varepsilon^{v} + \varepsilon^{\omega}. \tag{7}$$

Here ε^{el} is an elastic strain, ε^{v} is viscoelastic strains, ε^{ω} is strains due to the damage accumulation. They were determined by the following relationships:

$$\varepsilon^{\mathrm{v}} = \int_0^t H(t-\theta)\,\varphi(\sigma)d\theta,\ \varepsilon^{\omega} = \int_0^t M(t-\theta)\,\psi(\sigma)d\theta. \tag{8}$$

When formulating the material strength criterion, it was proposed to take the sums of only elastic strains and strains arising from the damage accumulation as independent variables

$$F(\varepsilon^{el} + \varepsilon^{\omega}) = 1. \tag{9}$$

In papers [13] and [14], a form of the kinetic relationship different from (3) was proposed in the form:

$$\omega(t) = 1 - \sigma / \left(\sigma + \int_0^t M(t - \theta)\, \sigma \, d\theta \right). \tag{10}$$

The Abel kernel was used as a difference kernel.

In [15], when considering an isotropic viscoelastic material, the kinetic relation of the hereditary type with the Abelian kernel was also used

$$\omega(t) = (1 + m) \int_0^t \frac{(t - \theta)^m}{\zeta^{1+m}} \, d\theta.$$

Here $\zeta = \zeta(\sigma)$ is the some function of equivalent stress.

In some works, the relative change in the modulus of elasticity [16], as well as the ratio of densities at the initial and current times [17], were taken as the damage parameter during cyclic loading. Such approaches are also used in which the level of damage is assessed by both residual stiffness and residual strength [18].

The work [19] is devoted to the method of predicting the behavior of mechanical systems under conditions of damage accumulation based on probabilistic models, which also contains significant experimental material.

In the works of Novozhilov V.V. an approach was proposed to the analysis of irreversible processes of deformation and accumulation of damage based on the introduction of the concept of microstress (see, for example, in [20]).

To describe the deformation process taking into account the damage accumulation, variants of the theory of the endochronic type with several internal times "triggered" for each component of the composite, associated with various mechanisms of damage accumulation, are also proposed [21].

In the experimental determination of the damage level in composite materials, many different approaches have also been proposed. A review of some methods for assessing material damage and its evolution by non-destructive testing methods can be found, in particular, in [22]. We also note the work [23], which concludes that the most practical are simple methods, namely, optical microscopy and acoustic emission. The latter is used quite often (see, for example, [24, 25]).

The results of a fairly large series of experiments carried out to study the damage patterns accumulation and fracture in highly filled polymer materials under loading of various types are given in [26].

All experiments are described within the framework of the nonlinear viscoelasticity model, and the criterion fracture parameter is the sum of the partial increments of the strain intensity in the active parts of the loading process.

In some works, when formulating strength criteria, the relationship between the moment of failure of specimens and the change in the nature of heat generation caused by an irreversible strain of the material is used [27].

Today, there are a fairly large number of publications devoted to other approaches to the analysis of damage accumulation and failure processes, methods of solving problems on determining the degree of material damage (including their numerical modeling), and taking it into account when assessing the load capacity of structural elements. Some of them can be found, for example, in [28–39].

Below we investigate the problem of assessing the damage level of composite specimens made by cross-ply ±45° lay-up under cyclic loading. A phenomenological approach is used to describe the process of microdamages accumulation. The problem of constructing constitutive relations is considered in a one-dimensional formulation in the specimen axes.

2 Experimental Results

In order to analyze the process of deformation of the composite material at sufficiently high levels of load, a number of experiments on cyclic tension and tension-compression were carried out. For testing, we used test specimens made of cross-ply reinforced fiber composites with ±45° lay-up based on ELUR-P unidirectional carbon fiber and XT-118 binder with average thickness $h = 0.56$ mm (four laminas with thickness 0.14 mm), width $b = 24.60$ mm and gage length $l = 110$ mm. Specimens from unidirectional carbon fiber HSE 180 REM prepreg was also used. The tests were carried out at room temperature on an Instron ElectroPuls E10000 servo-electric universal testing machine. To measure axial strains, an Instron mounted extensometer with a measurement base of 50 mm and accuracy class B-1 according to ASTM E83 was used. Such tests are regulated by standard ASTM D3518. During the tests, stresses and strains in the specimen axes are measured at each time point. The tests were carried out eight months and 3 years after the manufacture of test specimens, when the polymerization of the binder could be considered complete. In Fig. 1 we can see, that strain increments are decreased on each cycle.

If the level of maximum stresses is high enough, then the strain increments first decrease, and, starting from a certain cycle number, their increase is observed Fig. 2. This can be explained by the microdamages accumulation causing additional strain. As noted above, this process cannot be described by the models outlined in [3, 4], since in them these additional strains depend only on a certain level of stresses, similar to what is accepted in the theory of plasticity. Those, upon reaching this level, a strain can grow indefinitely without increasing stresses, but unlike the theory of plasticity, this approach assumes that there is no residual strain during unloading,

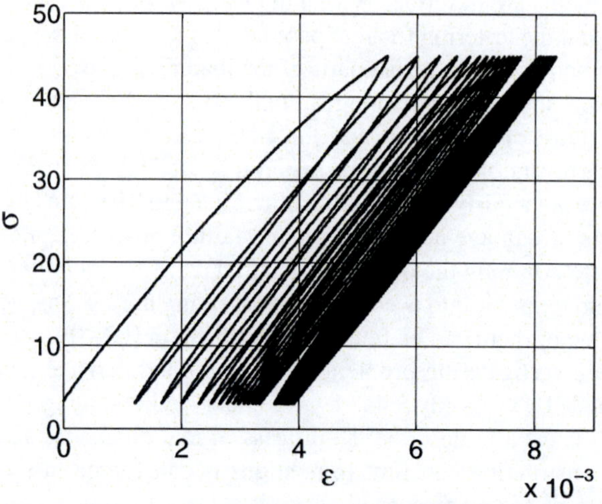

Fig. 1 Cycling tension of ELUR-P specimen under 45 MPa

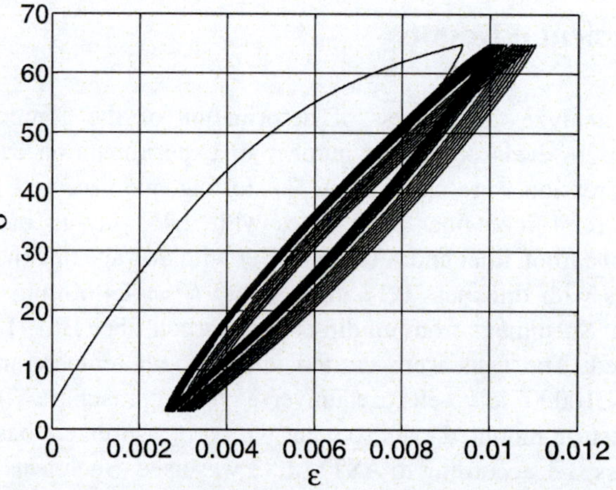

Fig. 2 Cycling tension of ELUR-P specimen under 65 MPa

which means that the secant modulus is less than the initial one. Models using the concept of effective stress and damage parameters [1, 2] are also not applicable here because they neglect additional strain caused by damage.

It should also be noted that the microdamages accumulation for composite materials can be much more intense during compression, which also contradicts the approach proposed in [3, 4]. For example, Fig. 3 shows a stress-strain curve of a symmetrical cyclic load test of a cross-ply HSE 180 REM specimen with lay-up ±45°. It can be seen that the hysteresis loops increase with each cycle, and this occurs much more intensively during compression, which cannot be described

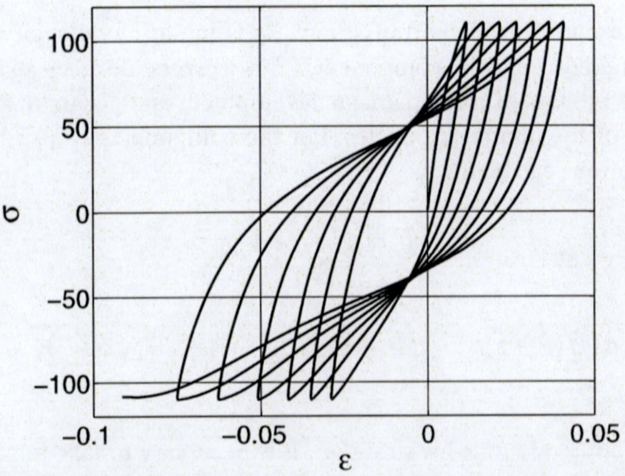

Fig. 3 Symmetric cyclic loading of HSE 180 REM ±45° specimen

by the usual relations of the theories of creep, viscoelasticity and relations [3, 4] for the strain caused by the appearance of micro-damages such as microcracks. It can be assumed that the reason for such a difference in the response to tension and compression may be the stability loss of the phases of the composite at the microlevel [40].

3 Constitutive Relations and Identification of Their Parameters

To describe the process of microdamages accumulation for the material under consideration, as in [3, 4, 12], we introduce the assumption that additional strain appears during the microdamages accumulation. We will also assume that these additional strain will also accumulate at low stresses, but at a lower rate. This does not contradict the hypothesis of a decrease in the effective area [1, 2] due to the appearance of microdamages such as micropores, microcracks, since it can be assumed that microdamages reduce not only the effective area, but also the geometric stiffness of the representative body elements containing them. This is also true under the assumption that additional strain during the microdamages accumulation is caused by stability loss of the FRP phases and the buckling of the fibers. In what follows, for simplicity, we consider only the one-dimensional case of tension. For total strain, we assume that it consists of the following terms:

$$\varepsilon = \varepsilon^{el} + \varepsilon^{cr} + \varepsilon^{v} + \varepsilon^{\omega},\ \varepsilon^{el} = \sigma/E_0. \tag{11}$$

Here ε^{el} is the linear elastic part of strain, E_0 is initial modulus of elasticity, ε^v is viscoelastic (heredity elastic) component, ε^{cr} is irreversible creep strain, ε^ω is strain caused by the process of microdamage development and accumulated by a certain point in time or the number of cycles. For the components ε^{cr}, ε^v, as in [41], the following relations can be taken:

$$d\varepsilon^r/dt = \chi_0\sigma/(1+\chi_1\varepsilon^{cr})^m, \tag{12}$$

$$\varepsilon^v = \int_0^t f(\sigma)H(t-\tau)d\tau,\ H(t-\tau) = \frac{B}{(t-\tau)^\alpha},\ 0 < \alpha < 1,\ B > 0. \tag{13}$$

When formulating relations for strain ε^ω, it is necessary to take into account (as in works [1, 2]) that it increases at a certain rate, which depends on both stresses and the level of these strains. In addition, in the case of cyclic loading, the accumulation rate ε^ω may depend on both the number of loading cycles and their type. Third, ε^ω must have different accumulation rates in tension and compression. Taking these assumptions into account, the constitutive relations in a general form can be represented as:

$$d\varepsilon^\omega = F_1\left(\sigma, sign(\sigma), \varepsilon^\omega, t\right)dt + F_2\left(\sigma, sign(\sigma), \varepsilon^\omega, \theta\right)d\theta,\ \theta = t/T. \tag{14}$$

Here T is period of cyclic loading. In fact, θ is an endochronous parameter with which a discrete variable (number of cycles) is replaced by a continuous one. As can be seen from (14), and as noted above, in contrast to works [3, 4], here the accumulation of strain caused by microdamages occurs at low stresses.

Next, we will consider a particular version of loading, namely, cyclic tension, therefore, for this case, we can take, for example, the following (one of the simplest) form of relation (14):

$$d\varepsilon^\omega = F_2 = \left(1 + a\varepsilon^\omega\right)^k\left(\sigma/\sigma_0\right)^n d\theta,\ k > 0,\ n > 0. \tag{15}$$

Here, the value σ_0 is the value of the stress, the excess of which at $\sigma > \sigma_0$ and large n leads to a sharp increase in the rate of strain accumulation caused by microdamages. The parameters $\sigma_0,\ a,\ k,\ n$ are determined by identification methods (see, for example, [42–44]) based on the results of experiments. One of the difficulties in solving these problems in the case under consideration is the problem of isolating from the experimental data various parts of the strain included in (11). For this, the approach used in [41] can be applied. Namely, we will assume that at small times, the strain ε^ω caused by the microdamages accumulation will be much less of strain $\varepsilon^{cr} + \varepsilon^v$. Therefore, they can be neglected for the first few cycles. Next, we need to separate the strain ε^{cr} and ε^v. This can be done in the same way as was proposed in [39], namely, it can be assumed that the rate of viscoplastic strain decays much faster than the rate of viscoelastic deformations. This makes it possible, at considerable time values (in our case, at considerable numbers of cycles) to assume, that the increments of the total strain $\Delta\varepsilon$ mainly consist of increments $\Delta\varepsilon^{v}$. Then, for large values of the numbers of cycles at some points in time that differ by values that are multiples of the period T, the increments of viscoelastic strain can be expressed through the experimental values of strain $\Delta\varepsilon^{v}$. For example, it is convenient to use those times at which the stress reach their maximum values. Then at these moments we can write the following relations

$$\Delta\varepsilon^{cr} + \Delta\varepsilon^{\mathrm{v}} \approx \Delta\varepsilon^{\mathrm{v}} \approx \Delta\varepsilon_{\exp}, \quad (16)$$

$$\begin{aligned} \Delta\varepsilon^{\mathrm{v}} &= \varepsilon^{\mathrm{v}}(j_1 T + T/2) - \varepsilon^{\mathrm{v}}(j_2 T + T/2), \\ \Delta\varepsilon_{\exp} &= \varepsilon_{\exp}(j_1 T + T/2) - \varepsilon_{\exp}(j_2 T + T/2). \end{aligned} \quad (17)$$

Here $j_1, j_2 \gg 1$—numbers of cycles.

When used in (13) for the $H(t-\tau)$ Abel kernel, the problem of finding the parameters α, B from (16), (17) can be divided into two specially obtained systems of equations [43], which make it possible to sequentially find the first α and then B. This approach is based on the peculiarities of relations (13) with the Abel kernel under cyclic loading [43]. In the general case, the mechanical characteristics included in $H(t-\tau)$ can be found from system (16), (17), for example, by the method of minimizing its quadratic residual. However, it should be noted that this may introduce difficulties that are inherent in identification problems, since they belong to the class of incorrect ones (see reviews, for example, in [42, 44–46]. After that, for small values j_1, j_2, you can find the increments $\Delta\varepsilon^{cr}$:

$$\Delta\varepsilon^{cr} \approx \Delta\varepsilon - \Delta\varepsilon^{v} \approx \Delta\varepsilon_{\exp} - \Delta\varepsilon^{v}. \quad (18)$$

Since $\Delta\varepsilon^{v}$ can already be found by relation (13), then (18) will contain only the required characteristics χ_0, χ_1, m. They can also be found by the method of minimizing the quadratic residual of the system obtained from (18) for different values $j_1, j_2 \sim 1$.

The modulus of elasticity E_0 can then be determined. To do this, it is necessary to use the "stress-strain" curve for any cycle at small values of the cycle numbers j, but taking into account the already found laws of viscoelastic and viscoplastic strain. Then an expression of the following form can be used to determine E_0:

$$\Delta(\varepsilon^{\exp} - \varepsilon^{v} - \varepsilon^{cr}) = \Delta\sigma/E_0. \quad (19)$$

According to the described technique, the rheological characteristics of the material were determined based on the results of the experiment shown in Fig. 2. The "stress-time" dependence on a half-cycle was considered linear. In this case, the calculation of the values ε^{v}, ε^{cr} was carried out numerically. As a result of processing test data at $\sigma_{\max} = 65$ MPa, $\sigma_{\min} = 3$ MPa, $T = 266$ s the following values of the required quantities were obtained:

$$E_0 = 9210 \text{ MPa}, B = 4.5 \cdot 10^{-6} \text{s}^{\alpha-1}/\text{MPa}, \alpha = 0.94,$$
$$\chi_0 = 5.5 \cdot 10^{-2}/(\text{MPa} \cdot \text{s}), \chi_1 = 993, m = 26.3. \tag{20}$$

At the last stage, one can find the parameters of relation (15). To do this, we can compose the following system of equations for cycles with large numbers:

$$\Delta\varepsilon^{cr} + \Delta\varepsilon^{v} + \Delta\varepsilon^{\omega} = \Delta\varepsilon_{\exp}, \ \Delta\varepsilon^{\omega} = \varepsilon^{\omega}(j_1 T + T/2) - \varepsilon^{\omega}(j_2 T + T/2). \tag{21}$$

The identification procedure gave the following results

$$\sigma_0 = 4.42 \cdot 10^8 \text{ MPa}, \ n = 1, \ k = 14.5, \ a = 586. \tag{22}$$

Figure 4 shows the results obtained in the experiment (indicated by round markers) and calculated using relations (11)–(13), (15). The strain values ε^{ω} calculated in accordance with (21) are shown in Fig. 5. It is seen that at $j_1, j_2 < 10$ values ε^{ω} are less than 5% of $\varepsilon^{cr} + \varepsilon^{v} + \varepsilon^{\omega}$. Therefore, the assumption that at small times the strains caused by microdamages accumulation will be much less than the sum of viscoelastic and viscoplastic strain can be considered confirmed. Below the results of the analysis of another specimen (also made of ELUR-P unidirectional carbon fiber, but after holding for three years), under tensile cyclic load at $\sigma_{\min} = 0$,

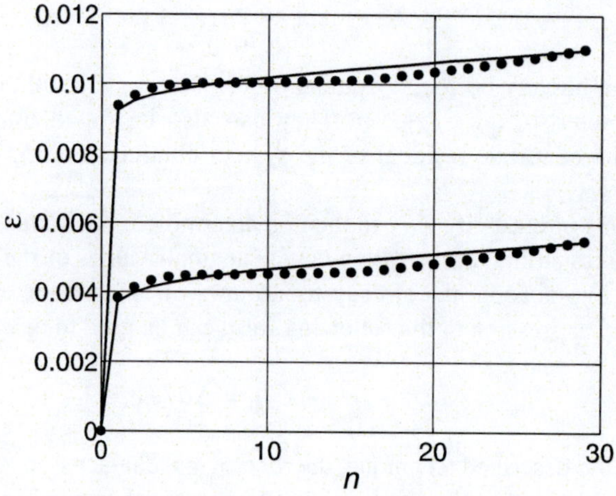

Fig. 4 Top curve—total strain value, bottom curve—sum of strain $\varepsilon^{cr} + \varepsilon^{v} + \varepsilon^{\omega}$ value, determined from the test for $\sigma = \sigma_{\max} = 65$ MPa (round markers), and calculated by relations (11)–(13), (15) of cycle number

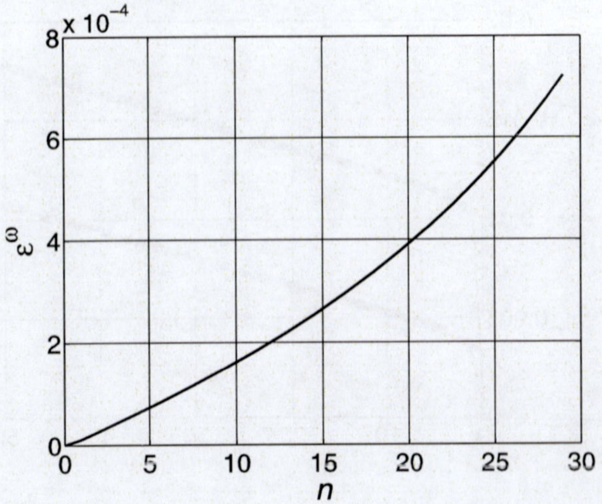

Fig. 5 Values of strain ε^{ω} of cycle number for $\sigma = \sigma_{max} = 65$ MPa

$\sigma_{max} = 75$ MPa are given. The procedure outlined above gave the following mechanical characteristics:

$$E_0 = 10100\,\text{MPa},\ B = 2.01 \cdot 10^{-6} \text{s}^{\alpha-1}/\text{MPa},\ \alpha = 0.68,$$
$$\chi_0 = 0.0116/(\text{MPa} \cdot \text{s}),\ \chi_1 = 693,\ m = 21.3,\ \sigma_0 = 2.475 \cdot 10^8,\ n = 1, \qquad (23)$$
$$k = 4.1,\ a = 101.$$

The values of the total ε and the sum of strain $\varepsilon^{cr} + \varepsilon^{v} + \varepsilon^{\omega}$ are shown in Fig. 6. Figure 7 shows the values of microdamage strain ε^{ω}. The difference obtained in the mechanical characteristics of these specimens, apparently, can be explained by the large difference in the times elapsed after their manufacture and the moment of the experiments.

Evaluation of the performance of structures must be carried out on the basis of the criteria of strength, rigidity, stability of their elements. Due to the presence of rheological properties of the material and the accumulation of microdamages over time, their stress-strain state usually changes, as well as a drop in the strength characteristics of the material. Therefore, in addition to assessing the level of stress-strain state, a large number of works are devoted to the development of criteria for the failure of materials, including taking into account the microdamages accumulation (see, for example, in [1, 2, 5–10, 13, 14]. When using relations of the type (15), as the critical one, one can take the time at which the rate of microdamages accumulation in some elements becomes dangerous (formally, when the curve $\varepsilon^{\omega} = \varepsilon^{\omega}(t)$ reaches the vertical asymptote).

If we use the approach proposed in [1, 2], then the results obtained above can also be used, assuming that the effective stress can be calculated through the initial elastic

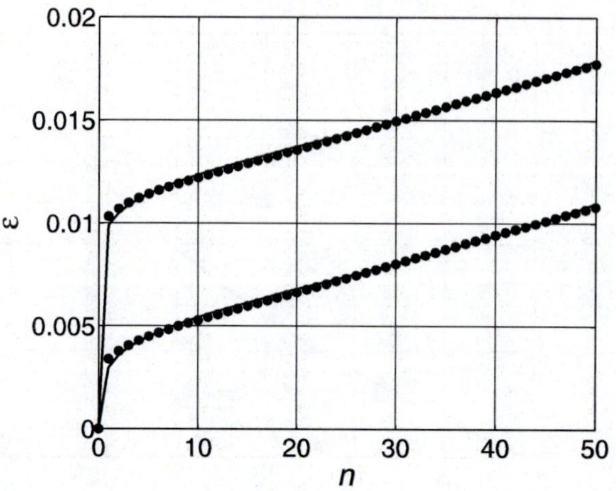

Fig. 6 Top curve—total strain value, bottom curve—sum of strain $\varepsilon^{cr} + \varepsilon^{v} + \varepsilon^{\omega}$ value, determined from the test for $\sigma = \sigma_{\max} = 75$ MPa (round markers), and calculated by relations (11)–(13), (15) of cycle number

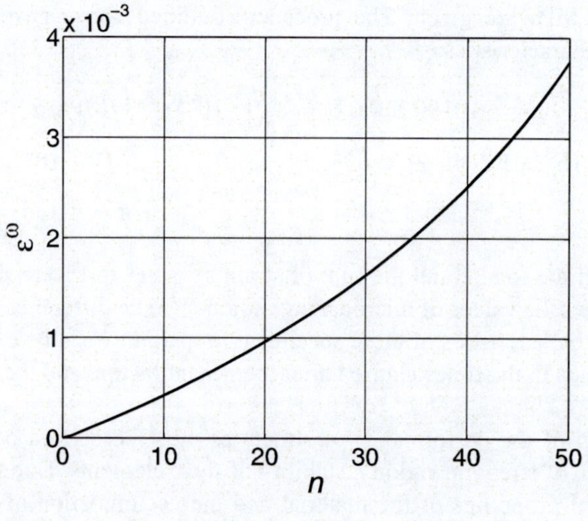

Fig. 7 Values of strain ε^{ω} of cycle number for $\sigma = \sigma_{\max} = 75$ MPa

characteristics and the sum of strain $\varepsilon^{el} + \varepsilon^{\omega}$. For example, in the one-dimensional case, similarly to [12], this condition can be written as follows:

$$\varepsilon^{el} + \varepsilon^{\omega} = \sigma^{*}/E_0. \tag{24}$$

Here σ^{*} is the ultimate strength of the material. However, verification of relation (24), on the one hand, was not included in the task of this study. On the other hand, when carrying out cyclic loading experiments, situations often arise in which the contact extensometers reach their limit values, after which they may fail. Therefore, it is not possible to bring the specimens to failure.

4 Conclusion

A review of works devoted to the problem of describing the processes of microdamages accumulation shows that for the experimental assessment of the level of this damage, as well as for the mathematical modeling of this phenomenon, a large number of methods and constitutive relationships have been proposed, which are selected depending on both the material and the type of load. Based on the analysis of the results of the author's experiments, some features of the deformation processes of specimens made of unidirectional carbon fiber by cross-ply $\pm 45^\circ$ layup are shown under cyclic loading. It is assumed that these features are caused by the microdamage accumulation. When modeling it, it is assumed that the strain consists of elastic, viscoelastic, viscoplastic parts, as well as a part formed as a result of the microdamages accumulation. Thus, a phenomenological approach is used to describe the process of microdamage accumulation. The well-known hypothesis is accepted that the constitutive relations describing the development of each part of the strain, except for this part itself, include only stress and time.

Some hypotheses are accepted to isolate the various parts of the strain from the experimental data. Namely, it is believed that at short times, the strain caused by the microdamages accumulation will be much less than other parts of the strain. Therefore, they are neglected in the initial cycles. Further, it is assumed that the rate of viscoplastic strain decays much faster than the rate of viscoelastic strain. This makes it possible, at considerable time values, to assume that the increments of total strain consist of increments of only viscoelastic strain. The accepted hypotheses make it possible to consistently determine the mechanical characteristics included in the constitutive relations.

Variants of physical relationships for the indicated parts of the strain are proposed. The results of solving the problems of identifying the constitutive relations of the parameters included in the proposed forms based on the results of cyclic tension tests are presented. It is concluded that the proposed hypotheses allow a fairly good description of the behavior of the specimens from the considered material.

Acknowledgments The work was carried out within the framework of the Russian Science Foundation (project No. 19-19-00059, Sect. 1, No. 19-79-10018, Sect. 2) and has been supported by the Kazan Federal University Strategic Academic Leadership Program ("PRIORITY-2030") (Sect. 3).

References

1. Rabotnov, Yu.N.: Polzuchest' elementov konstruktsiy [Creep of structural elements]. Nauka, Moscow (1966) [In Russian]
2. Kachanov, L.M.: O vremeni razrusheniya v usloviyakh polzuchesti [On the time of failure under creep conditions]. Izv. AN SSSR. OTN. **8**. 26–31 (1958) [In Russian]
3. Obrazcov, I.F., Vasilyev, V.V., Bunakov, V.A.: Optimalnoye armirovaniye obolochek vrashcheniya iz kompozitsionnykh materialov [Optimal reinforcement of shells of revolution made of composite materials]. Mashinostroenie, Moscow (1977) [In Russian]
4. Alfutov, N.A., Zinovev, P.A., Popov, B.G.: Raschet mnogosloynykh plastin i obolochek iz kompozitsionnykh materialov [Calculation of laminate plates and shells made of composite materials]. Mashinostroenie, Moscow (1984) [In Russian]
5. Astafev, V.I., Radaev, Yu.N., Stepanova, L.V.: Nelineynaya mekhanika razrusheniya [Nonlinear fracture mechanics]. Izd-vo Samarskii universitet, Samara (2001) [In Russian]
6. Bolotin, V.V.: K teorii zamedlennogo razrusheniya [To the theory of delayed failure]. Mechanics of solids.**1**. 137–146 (1981) [In Russian]
7. Murakami, S., Radaev Yu.N.: Matematicheskaya model' trekhmernogo anizotropnogo sostoyaniya povrezhdennosti [Mathematical model of a three-dimensional anisotropic state of damage]. Mechanics of Solids. **4**. 93–110 (1996) [In Russian]
8. Nazarov, S.A.: Tenzor i mery povrezhdennosti. 1. Asimptoticheskiy analiz anizotropnoy sredy s defektami [Damage tensor and measures. 1. Asymptotic analysis of an anisotropic media with defects]. Mechanics of Solids. **3**. 113–124 (2000) [In Russian]
9. Teregulov, I.G. Kriteriy prochnosti ortotropnogo tela i yego svyazs protsessom nakopleniya mikropovrezhdeniy [Strength criterion of an orthotropic body and its relationship with the process of microdamages accumulation]. Applied problems of strength and plasticity / Mezvuzovskii sbornik. Analysis and optimization of structures. **51**. 32–39 (1994)
10. Ilushin, A.A., Pobedrya B.E.: Osnovy matematicheskoy teorii termovyazkouprugosti [Foundations of the mathematical theory of thermoviscoelasticity]. Nauka, Moscow (1970) [In Russian]
11. Cocks, A.C.F., Ashby, M.F.: The growth of dominant crack in a creeping material. Scripta Mettallurgica. **16**. 109–114 (1982)
12. Ackhundov, M.B.: Povrezhdayemost' i deformirovaniye nelineynykh nasledstvennykh sred pri slozhnonapryazhennom sostoyanii [Damage and deformation of nonlinear hereditary media in a complexly stressed state]. Mechanics of composite materials. **2**. 235–239 (1991) [In Russian]
13. Suvorova, Yu.V.: O kriterii prochnosti, osnovannom na nakoplenii povrezhdeniy, i yego prilozheniyakh k kompozitam [On strength criterion based on damage accumulation and its applications to composites]. Mechanics of Solids. **4**. 107–111 (1979) [In Russian]
14. Dumanskii, A.M., Finogenov G.N.: Metodika otsenki povrezhdennosti polimernykh voloknistykh kompozitov pri dlitel'nom staticheskom nagruzhenii [A technique for assessing the damage of polymer fiber composites under prolonged static loading]. Factory laboratory. **4**. 60–62 (1993) [In Russian]
15. Moskvitin, V.V.: Nekotoryye voprosy dlitel'noy prochnosti vyazko-uprugikh tel [Some issues of long-term strength of viscoelastic bodies]. Strength problems. **2**. 55–58 (1972) [In Russian]
16. Arutunyan, A.R., Arutunyan, R.A.: Kriteriy ustalostnoy prochnosti kompozitsionnykh materialov, osnovannyy na kontseptsii povrezhdennosti [Fatigue strength criterion of composite materials based on the damage concept. XII Russian Congress on Fundamental Problems of Theoretical and Applied Mechanics: a collection of works in 4 volumes]. **3**. 556–558 (2019) [In Russian]
17. Arutyunyan, R.A.: High-Temperature Embrittlement and Long-Term Strength of Metallic Materials. Mechanics of Solids. **50(2)**. 191–197 (2015)
18. Van Paepegem, W., Degrieck, J.: A new coupled approach of residual stiffness and strength for fatigue of fibre-reinforced composites. International Journal of Fatigue. **24**. 747–762 (2002)
19. Bogdanoff, J., Kozin, F.: Veroyatnostnyye modeli nakopleniya povrezhdeniy [Probabilistic damage accumulation models]. Mir, Moscow (1989) [In Russian]

20. Rybakina, O.G.: O rabotakh V.V. Novozhilova v oblasti fenomenologicheskogo opisaniya pervoy stadii razrusheniya (nakopleniya povrezhdeniy) [About the works of V.V. Novozhilov in the field of phenomenological description of the first stage of failure (accumulation of damage)] Proceedings of the Central Research Institute. acad. A.N. Krylova. **53-1(337-1)**. 127–137 (2010) [In Russian]
21. Golovin, N.N., Kyvirkin, G.N.: Matematicheskiye modeli deformirovaniya uglerod-uglerodnykh kompozitov [Mathematical models of deformation of carbon-carbon composites]. Mechanics of Solids. **5**. 127–123 (2016) [In Russian]
22. Kluev, V.V.: Nerazrushayushchiy kontrol': spravochnik in 8 vol. [Non-Destructive Testing: A Handbook]. (2006) [In Russian]
23. Varna, J., Asp, L.: Microdamage in composite laminates: experiments and observation. Applied Mechanics and Material. **518**. 84–89 (2014).
24. Matvienko, Yu.G., Vasilev, I.E., Pankov, M.A.: Rannyaya diagnostika zon povrezhdeniya i razrusheniya kompozitsionnykh materialov s ispolzovaniyem khrupkikh tenzoindikatorov i akusticheskoy emissii [Early diagnosis of damage and failure zones of composite materials using brittle strain gauges and acoustic emission]. Factory laboratory. **1**. 45–56 (2016) [In Russian]
25. Wildemann, V.E., Spaskova, E.V., Shilova, A.I.: Research of the damage and failure processes of composite materials based on acoustic emission imonitoring and method of digital image correlation problems of deformation and fracture in materials and structures. Solid State Phenomena. **243**. 163–170 (2016)
26. Bukov D.L., Kazakov A.V., Konovalov D.N., Melnikov V.P., Peleshko V.A., Sadovnichii D.N. Law of damage accumulation and fracture criteria in highly filled polymer materials. Mechanics of Solids. **5**. 76–97 (2014)
27. Iziumova A. Yu, Vshivkov A.N., Prokhorov A.E., Plehov O.A., Venkatraman B.: Study of heat source evolution during elastic-plastic deformation of titanium alloy Ti-0.8Al-0.8Mn based on contact and non-contact measurements. PNRPU mechanics bulletin. **1**. 68–81 (2016) doi: https://doi.org/10.15593/perm.mech/2016.1.05
28. Kapustin, S.A., Churilov, Yu.A., Gorohov, V.A.: Modelirovaniye nelineynogo deformirovaniya i razrusheniya konstruktsiy v usloviyakh mnogofaktornykh vozdeystviy na osnove MKE [FEM-based modeling of nonlinear deformation and failure of structures under multifactorial effects]. Izd-vo Nizegorodskogo gos. un-ta im. N.I.Lobachevskogo, Nizhniy Novgorod (2015) [In Russian]
29. Volkov, I.A., Korotkich, Yu.G: Uravneniya sostoyaniya vyazkouprugoplasticheskikh sred s povrezhdeniyami [Equations of state for damaged viscoelastoplastic media]. Fizmatliz, Moscow (2008) [In Russian]
30. Volkov, I.A., Igumnov, L.A., Kazakov, D.A., Shishulin, D.N., Tarasov, I.S., Smetanin, I.V.: Constitutive relations of the mechanics of a damaged medium for evaluating the long-term strength structural alloys. Journal of Applied Mechanics and Technical Physics. **1**. 181–194 (2019). doi: 10.15372/PMTF20190119
31. Mohamed, S.L., Varna, J.: Effective shear modulus of a damaged ply in laminate stiffness analysis: Determination and validation. Journal of Composite Materials. **54(9)**. 1161–1176 (2019). doi: 10.1177/0021998319874369
32. Talreja, R., Singh, C.V.: Damage and failure of composite materials. Cambridge University Press, New York (2012)
33. Loukil, M. S., Varna, J.: Effective shear modulus of a damaged ply in laminate stiffness analysis: Determination and validation. Journal of Composite Materials. **54(9)**. 1–16 (2019). doi: 10.1177/0021998319874369.
34. Kayumov, R.A., Nezdanov, R.O.: Kriteriy dlitelnoy prochnosti dlya voloknistogo kompozita [Criterion for long-term strength for fiber composite]. Izv. TulGU. Seria Matematica. Mechanica. Informatica. **10(2)**. 111–123 (2004) [In Russian]
35. Van Paepegem, W.: Fatigue damage modelling of composite materials with the phenomenological residual stiffness approach [Fatigue Life Prediction of Composites and Composite Structures]. **1**. 102–138 (2010) [In Russian]

36. Berezin, A.V., Kozinkina, A.I.: Osobennosti diagnostiki povrezhdeniy i otsenki prochnosti kompozitov [Features of damage diagnosis and assessment of the strength of composites]. Mechanics of composite materials and structures. **5(1)**. 99–122 (1999) [In Russian]
37. Belov, P.A., Dudchenko, A.A., Lure, S.A., Semerin, A.M., Khardman, H.: Ob odnom algoritme ucheta povrezhdennosti v mekhanike materialov [On one algorithm for accounting for damage in material mechanics]. Mechanics of composite materials and structures. **12(4)**. 566–578 (2006) [In Russian]
38. Polilov, A.N.: Etyudy po mekhanike kompozitov [Composite Mechanics Studies]. Fizmatlit, Moscow (2015) [In Russian]
39. Fedulov, B.N., Fedorenko, A.N., Kantor, A.M., Lomakin, E.V.: Failure analysis of laminated composites based on degradation parameters. Meccanica. **53(1–2)**. 359–372 (2018)
40. Paimushin, V.N., Gazizullin, R.K., Shishov, M.A.: Spatial buckling modes of a fiber (fiber bundle) of composites with a $[\pm 45^\circ]_{2s}$ stacking sequence under the tension and compression of test specimens. Mechanics of Composite Materials. **55(6)**. 743–760 (2020). doi: 10.1007/s11029-020-09855-9
41. Paimushin, V.N., Kayumov, R.A., Kholmogorov, S.A.: Deformation Features and Models of $[\pm 45]_{2s}$ Cross-Ply Fiber-Reinforced Plastics in Tension. Mechanics of Composite Materials. **55(2)**. 141–154 (2019). doi: 10.1007/s11029-019-09800-5.
42. Kayumov, R.A.: Rasshirennaya zadacha identifikatsii mekhanicheskikh kharakteristik materialov po rezultatam ispytaniy konstruktsiy [Extended problem of identifying the mechanical characteristics of materials based on the results of structural tests]. Mechanics of Solids. **2**. 94–105 (2004) [In Russian]
43. Paimushin, V.N., Kayumov, R.A., Kholmogorov, S.A.: Features of Inelastic Behavior of a Composite Under Cyclic Loading. Experimental and Theoretical Investigations. Mechanics of Composite Materials. **56(4)**. 411–422 (2020). doi: 10.1007/s11029-020-09893-3.
44. Grop, D.: Metod identifikatsii sistem [System identification method]. Mir, Moscow (1979) [In Russian]
45. Nordin, L-O., Varna, J.: Methodology for parameter identification in nonlinear viscoelastic material model. Mechanics of Time-Dependent Materials. **9(4)**. 259–280 (2005). doi: 10.1007/s11043-005-9000-z
46. Tiknonov, A.N., Arsinin, V.Ya.: Metody resheniya nekorrektnykh zadach [Methods for solving ill-posed problems]. Nauka, Moscow (1979) [In Russian]

Two-Dimensional Integrating Matrices for Solving Elasticity Problems in a Rectangular Domain by the Finite Sum Method

Vitaly N. Paimushin and Maksim V. Makarov

Abstract Using the Poisson equation and equations describing the plane stress state of plates as an example, the method of finite sums (two-dimensional integrating matrices) for the numerical solution of two-dimensional boundary value problems of the theory of elasticity is presented. According to this method, the original differential problem is preliminarily reduced to integral equations of the Volterra type, and then their approximation is carried out based on the replacement of the integrand by the Lagrange interpolation polynomial over Gaussian nodes. Two-dimensional integrating matrices are constructed. Numerical estimates of the accuracy of various test problems are carried out. It is shown that the convergence is exponential.

1 Introduction

Most of the numerical methods used in practice for solving boundary value problems, as is known, are based on the reduction of the original problems to the systems of algebraic equations of one structure or another. To date, a fairly wide arsenal of methods for constructing systems of algebraic equations (difference schemes) that approximate the original differential problems has been developed. Of these (in particular, in the mechanics of a deformable solid), the most common are the methods of finite differences and finite elements, discrete orthogonalization, and some other methods that require the formulation of the initial problems in the form of differential or variational equations.

The natural desire to expand the arsenal of research methods led to the formulation of boundary value problems in the form of integral or integro-differential equations [1–4], since the reduction of differential equations to integral equations

V. N. Paimushin · M. V. Makarov (✉)
Kazan Federal University, Kazan, Russia

Kazan National Research Technical University named after A.N. Tupolev–KAI, Kazan, Russia

I. B. Badriev et al. (eds.), *Mesh Methods for Boundary-Value Problems and Applications*, Lecture Notes in Computational Science and Engineering 141,
https://doi.org/10.1007/978-3-030-87809-2_29

allows in some cases to formulate boundary value problems more compactly, leads to more robust computational procedures that allow achieving the required accuracy of results with less computation.

The advantage of using the integral representations of the original problems also lies in the fact that they do not contain derivatives of functions and, therefore, do not impose large restrictions on the smoothness of these functions. In addition, when problems are formulated in the form of integral equations, it is often easy to construct solutions in the class of discontinuous functions.

A reflection of the significant expansion of applications of integral equations is an increasing interest in the theory and methods for their solution in many applied areas of scientific research. To date, there are many works on the study of the properties of various types of integral equations, as well as methods for their solution.

Boundary integral equations formulated by various methods, which are equivalent in the formulation of boundary value problems for ordinary differential equations, have found the greatest application in problems of solid mechanics.

The use of approximate numerical methods for solving integral equations was the impetus for the development of the already classical finite sum method, Runge-Kutta, iterative methods, etc.

When solving integral equations numerically by any methods, one inevitably has to replace the integrals included in them by finite sums. In this case, the resulting final ratios can be auxiliary or have an independent character as the final calculated expressions of the finite sum method. The finite sum method (the method of quadratures, mechanical quadratures, quadrature formulas) consists in the compilation and direct use of calculated expressions obtained by replacing integral operators with finite sums based on the use of various quadrature formulas.

The choice of a quadrature formula for solving integral equations is not easy to carry out, since there are no ready-made recommendations, complete for practice, in the scientific literature, which is explained by the insufficient knowledge of the calculations of integrals with variable boundaries. This gave rise to many approaches and ways of applying the finite-sum method using various approximations of the integrand: formulas for rectangles, trapeziums, Simpson, Gauss, etc.

Significant advances in the application of finite sums for solving one-dimensional problems of mechanics have so far been achieved based on the use of the sliding interpolation of the integrand in a version of the method proposed by M. B. Vakhitov and called the apparatus of integrating matrices [2, 3]. The sliding interpolation by a polynomial of the third degree used by him, as shown by numerous studies, provides high accuracy of the numerical integration operation and the efficiency of the method he proposed.

In the general case, the error of the finite sum method is due to the error of replacing integrals by finite sums, i.e. is determined by the accuracy of the approximation of the integrands (the accuracy of the quadrature formulas). In this regard, for the interpolation of integrands, it seems appropriate to use splines (spline functions), the apparatus of which is an effective method for solving many problems of computational mathematics [5–8]. The advantages of splines, first of all, should be attributed to the ability to ensure high quality of approximation and efficiency of implementation on a computer of algorithms associated with their application. These properties have recently contributed to the expansion of the application of spline functions to the solution of integral equations [4], including the finite sum method in the version [2] using integrating matrices using spline approximations [9].

Further development of the method of integrating matrices was given in [10, 11], in which the use of unsaturated algorithms for approximating functions was proposed. In [10], integrating matrices were constructed in which the nodes associated with the zeros of the corresponding Legendre polynomials are used as collocation nodes. This turns out to be sufficient to obtain an unsaturated algorithm with exponential accuracy and to ensure the symmetry of the matrix generated by the self-adjoint problem, which is especially important when solving spectral problems. In [10], using the example of a self-adjoint fourth-order equation, a detailed study of the stability and accuracy of the proposed version of the integrating matrix method was carried out. In particular, it is shown that the conditionality of the system matrix does not deteriorate with the growth of collocation nodes. The undoubted advantages of this method include the fact that it allows a strong local refinement of the grid nodes and very accurately describes solutions with large gradients in very short sections.

This article discusses the application of the finite sum method to the solution of two-dimensional boundary value problems. Some aspects of its application were previously considered in articles [12, 13].

2 Two-Dimensional Integrating Matrices

Consider a boundary value problem in the form of Poisson's equation in a rectangular domain $[\Omega = \{X = (x, y), x \in (0, a), y \in (0, b)\}]$, bounded by a contour $\Gamma = \Gamma_1 \bigcup \Gamma_2 \bigcup \Gamma_3 \bigcup \Gamma_4$, where $\Gamma_1 = \{X = (x, 0), x \in (0, a)\}$, $\Gamma_2 = \{X = (0, y), y \in (0, b)\}$, $\Gamma_3 = \{X = (x, b), x \in (0, a)\}$, $\Gamma_4 = \{X = (a, y), y \in (0, b)\}$

$$-\Delta u = f(X), \ \ X \in \Omega, \tag{1}$$

under boundary conditions

$$u|_{\Gamma_1} = 0, \ u|_{\Gamma_2} = 0, \ \left.\frac{\partial u}{\partial y}\right|_{\Gamma_3} = \varphi_3, \ \left.\frac{\partial u}{\partial x}\right|_{\Gamma_4} = \varphi_4, \tag{2}$$

Integration of equation (1) over x from ξ to a, taking into account the boundary condition (2) for Γ_4, leads to the equality

$$-\varphi_4 + \frac{\partial u}{\partial x}(\xi, y) - \int_{\xi}^{a} \frac{\partial^2 u}{\partial y^2}(x, y)dx = \int_{\xi}^{a} f(x, y)dx. \tag{3}$$

Further, integrating equality (3) over y from η to b and taking into account the boundary condition (2) for Γ_3, we obtain for $\xi \in (0, a)$, $\eta \in (0, b)$

$$\int_{\eta}^{b} \frac{\partial u}{\partial x}(\xi, y)dy + \int_{\xi}^{a} \frac{\partial u}{\partial y}(x, \eta)dx = \int_{\eta}^{b}\int_{\xi}^{a} f(x, y)dxdy + \varphi_4(b-\eta) + \varphi_3(a-\xi). \tag{4}$$

Thus, the differential equation (1), subject to the boundary conditions (2) on the contour Γ_3, Γ_4, is reduced to equation (4) concerning the vector function $U = (\partial u/\partial x, \partial u/\partial y)^T$ with integral operators of the Volterra type. To close this equation with respect U we use the remaining boundary conditions (2) $u|_{\Gamma_1} = 0$, $u|_{\Gamma_2} = 0$. For this purpose, integrating $\frac{\partial u}{\partial x}(x, \eta)$ over x from 0 to ξ, $\frac{\partial u}{\partial y}(\xi, y)$ over y from 0 to η $\forall \xi \in (0, a)$, $\forall \eta \in (0, b)$, we obtain the dependencies

$$\int_{0}^{\xi} \frac{\partial u}{\partial x}(x, \eta)dx = u(\xi, \eta) - u(0, \eta), \int_{0}^{\eta} \frac{\partial u}{\partial y}(\xi, y)dy = u(\xi, \eta) - u(\xi, 0), \tag{5}$$

which, taking into account the boundary conditions $u(0, \eta) = u(\xi, 0) = 0$ reduce to the equation

$$\int_{0}^{\xi} \frac{\partial u}{\partial x}(x, \eta)dx - \int_{0}^{\eta} \frac{\partial u}{\partial y}(\xi, y)dy = 0. \tag{6}$$

Note that the resulting system of integral equations (4), (5) for functions $\frac{\partial u}{\partial x}(X)$, $\frac{\partial u}{\partial y}(X)$, is equivalent to the original problem (5), (2), after solving which the solution to the original boundary value problem (1), (2) is restored using equations (5).

We introduce into consideration the integral operators $\mathcal{L}_x$, $\mathcal{L}_y$,$\mathcal{R}_x$,$\mathcal{R}_y$:

$$\mathcal{L}_x v = \int_{0}^{\xi} v(x, \eta)dx, \mathcal{L}_y v = \int_{0}^{\eta} v(\xi, y)dy, \mathcal{R}_x v = \int_{\xi}^{a} v(x, \eta)dx, \mathcal{R}_y v = \int_{\eta}^{b} v(\xi, y)dy$$

and operator $\mathcal{A} = \begin{pmatrix} \mathcal{R}_y & \mathcal{R}_x \\ \mathcal{L}_x & -\mathcal{L}_y \end{pmatrix}$. Then problem (4), (6) can be written in the operator form

$$\mathcal{A}U = F, \tag{7}$$

where $F = (\mathcal{R}_y\mathcal{R}_x f + \varphi_4(b-\eta) + \varphi_3(a-\xi), 0)^T$, $\mathcal{A} : H \times H \to H \times H$, $H = L_2(\Omega)$—is the space of square-integrable functions on Ω.

To approximate problem (7), it is necessary to construct the corresponding two-dimensional integrating matrices L_x^h, L_y^h,R_x^h,R_y^h, which are grid analogs of the operators $\mathcal{L}_x$, $\mathcal{L}_y$, $\mathcal{R}_x$, $\mathcal{R}_y$ in problem (7). The first results on the construction of one-dimensional integrating matrices in the framework of the finite-sum method were obtained by Vakhitov [2] and were further developed, in particular, in the works of R.Z. Dautov and V.N. Paimushin [10]. For the approximation of one-dimensional integral equations of the Volterra type, a method of collocations along with Gaussian nodes and a method for constructing the corresponding integrating matrices were proposed in [10]. To construct two-dimensional integrating matrices, we first introduce into consideration integral operators $\mathcal{L}^c g = \int\limits_0^{\xi} g(x)dx$, $\mathcal{R}^c g = \int\limits_{\xi}^{c} g(x)dx$, $\mathcal{D}^c g = \int\limits_0^{c} g(x)dx$, $\xi \in (0, c)$, which we replace by their finite-dimensional analogs in the form of integrating matrices L_h^c, R_h^c, D_h^c, respectively (here and below, italics denote operators acting in the space of measurable functions, in roman type operators acting in the space of grid functions, the index c of operators was introduced in order to select the integration area and further construct two-dimensional integrating matrices for an arbitrary rectangular area Ω). For this purpose, on the segment $[0, c]$ we introduce a grid $\omega = \{x_i,\ i = 1, 2, \ldots, N\}$ according to the Gauss quadrature formula: $D^c g = \sum\limits_{i=1}^{N} d_i g(x_i)$, where $\{d_i\}$, $\{x_i\}$ are, respectively, the weights and collocation nodes associated with the roots of the Legendre polynomial of degree N. Denote by $g_i = g(x_i)$ and approximate g on the segment $[0, c]$ using interpolation function $g(x) \approx \sum\limits_{i=1}^{N} g_i l_i(x)$. As an interpolating function, we choose the basis Lagrange functions $\{l_i\}$ by nodes $\{x_i\}$. By expanding the functions l_i in terms of Legendre polynomials in [10], the integrating matrices L_h^c, R_h^c were constructed.

First, we write down the one-dimensional integrating matrix L_h^c in component-wise form $L_h^c = \left\{ l_{ij} = \int\limits_0^{x_i} l_j(x)dx \right\}_{i,j=1}^{N}$. Then

$$[\mathcal{L}^c g(x)]_h \approx L_h^c g_h = \left(\sum_{i=1}^{N} l_{1i} g_i, \sum_{i=1}^{N} l_{2i} g_i, \ldots, \sum_{i=1}^{N} l_{Ni} g_i \right)^T,$$

where $g_h = (g_1, g_2, \ldots g_N)^T$, $[g]_h$—is mesh projection of the function g.

In order to construct two-dimensional integrating matrices L_x^h, L_y^h, R_x^h, R_y^h we introduce a grid ω_h on the region Ω. Let the partition $\{x_i\}_{i=1}^{N_x}$, $\{y_j\}_{j=1}^{N_y}$ define an orthogonal mesh into Ω, moreover $\{x_i\}$, $\{y_j\}$ nodes associated with the roots of the Legendre polynomial on the segments $[0, a]$ and $[0, b]$, respectively. Let us denote $v_{ij}^h = v(x_i, x_j)$, $\mathbf{v}_{\bullet,j}^h = \left(v_{1j}, v_{2j}, \ldots, v_{N_x j}\right)^T$, $\mathbf{v}_{i,\bullet}^h = \left(v_{i1}, v_{i2}, \ldots, v_{iN_y}\right)^T$ and introduce the end-to-end numbering of grid nodes as follows: each pair (i, j) is assigned a unique number $k = i+(j-1)N_x$, which corresponds to the lexicographic order "from left to right", "from bottom to top". Introducing further notation $\mathbf{v}^h = (v_1, v_2, \dots, v_k, \dots v_{N_x \cdot N_y})^T$, we construct a two-dimensional integrating matrix L_x^h

$$[\mathcal{L}_x v\,(x, y)]_h = \begin{pmatrix} [\mathcal{L}^a v(x, y_1)]_h \\ [\mathcal{L}^a v(x, y_2)]_h \\ \vdots \\ [\mathcal{L}^a v(x, y_{N_y})]_h \end{pmatrix} = \begin{pmatrix} \int_0^{x_1} v(x, y_1)dx \\ \int_0^{x_2} v(x, y_1)dx \\ \vdots \\ \int_0^{x_{N_x}} v(x, y_1)dx \\ \int_0^{x_1} v(x, y_2)dx \\ \vdots \\ \int_0^{x_{N_x}} v(x, y_{N_y})dx \end{pmatrix} \approx \begin{pmatrix} \sum_{i=1}^{N_x} l_{1i} v_{i1} \\ \sum_{i=1}^{N_x} l_{2i} v_{i1} \\ \vdots \\ \sum_{i=1}^{N_x} l_{N_x i} v_{i1} \\ \sum_{i=1}^{N_x} l_{1i} v_{i2} \\ \vdots \\ \sum_{i=1}^{N_x} l_{N_x i} v_{i N_y} \end{pmatrix} =$$

$$= \begin{pmatrix} L_h^a \mathbf{v}_{\cdot,1} \\ L_h^a \mathbf{v}_{\cdot,2} \\ \vdots \\ L_h^a \mathbf{v}_{\cdot,N_y} \end{pmatrix} = \begin{bmatrix} L_h^a & 0 & \dots & 0 \\ 0 & L_h^a & \dots & 0 \\ \vdots & \vdots & \ddots & \vdots \\ 0 & 0 & \dots & L_h^a \end{bmatrix}_{N_x N_y \times N_x N_y} \begin{pmatrix} \mathbf{v}_{\cdot,1} \\ \mathbf{v}_{\cdot,2} \\ \vdots \\ \mathbf{v}_{\cdot,N_y} \end{pmatrix}_{N_x N_y} = \left(E_{N_y} \otimes L_h^a\right) \mathbf{v}^h.$$

Then $L_x^h = E_{N_y} \otimes L_h^a$ and by analogy, matrices are constructed $R_x^h = E_{N_y} \otimes R_h^a$, $L_y^h = L_h^b \otimes E_{N_x}$, $R_y^h = R_h^b \otimes E_{N_x}$, where $\otimes$ is the Kronecker product over matrices, E_N is an identity matrix of size N.

An approximate solution to problem (7) is called $U_h = (u_{h,x}, u_{h,y})$, that satisfies the equation

$$A^h U_h = F_h, \tag{8}$$

where $A^h = \begin{pmatrix} R_y^h & R_x^h \\ L_x^h & -L_y^h \end{pmatrix}$, $F_h = (R_y^h R_y^h [f]_h + \varphi_4(b-x) + \varphi_3(a-y), 0)^T$, $(x, y) \in \omega_h$, $u_{h,x} = (u_{1,x}, u_{2,x}, \dots, u_{k,x}, \dots u_{N_x \cdot N_y,x})^T$, $u_{h,y} = (u_{1,y}, u_{2,y}, \dots, u_{k,y}, \dots u_{N_x \cdot N_y,y})^T$ according to the introduced lexicographic order. An approximate solution to the original boundary value problem (1), (2) will be determined using equalities (5), as $u_h = L_x^h u_{h,x}$. Let us present the results of the numerical solution of some test problems (1), (2) that have exact solutions for a square domain ($a = 1, b = 1$)with homogeneous boundary conditions ($\varphi_3 = 0$, $\varphi_4 = 0$) and calculate for these problems the error of the method ($r_h(N) = \|[u]_h - u_h\| / \|[u]_h\|$, where $[u]_h$—is the grid projection of the exact solution, $N = N_x = N_y$, as a infinity grid norm was chosen):

1. $f(x, y) = -\pi^2 \sin(\pi/2x)\sin(\pi/2y)/2$ with exact solution $u(x, y) = \sin(\pi/2x)\sin(\pi/2y)$;
2. $f(x, y) = xe^{-x}(xe^{-y} - 2e^{-y}) + ye^{-y}(xe^{-x} - 2e^{-x})$ with exact solution $u(x, y) = xe^{-x}ye^{-y}$, $r_h(3) = 0.0054$, $r_h(4) = 3.4763 \cdot 10^{-4}$, $r_h(8) = 7.9110 \cdot 10^{-10}$, $r_h(16) = 1.8305 \cdot 10^{-14}$;
3. $f(x, y) = x(1-x)^2(6y-4) + y(1-y)^2(6x-4)$ with exact solution $u(x, y) = x(1-x)^2 y(1-y)^2$, $r_h(3) = 1.471 \cdot 10^{-15}$;
4. $f(x, y) = 4x(1-x)^4(1-y)^2(5y-2) + 4y(1-y)^4(1-x)^2(5x-2)$ with exact solution $u(x, y) = x(1-x)^4 y(1-y)^4$, $r_h(3) = 0.3360$, $r_h(4) = 0.0742$, $r_h(5) = 1.5943 \cdot 10^{-15}$.

It is seen that in the first Fig. 1 and second problems, where the exact solutions are given in the form of transcendental functions, the error is exponential. The third and fourth test problems, in which the exact solutions in the form of polynomials are chosen, emphasize the highest order of the algebraic accuracy of the method: starting with $N = 3$ and $N = 5$, respectively, the errors $r_h(N)$ have order values 10^{-15}, which corresponds to the exact solution.

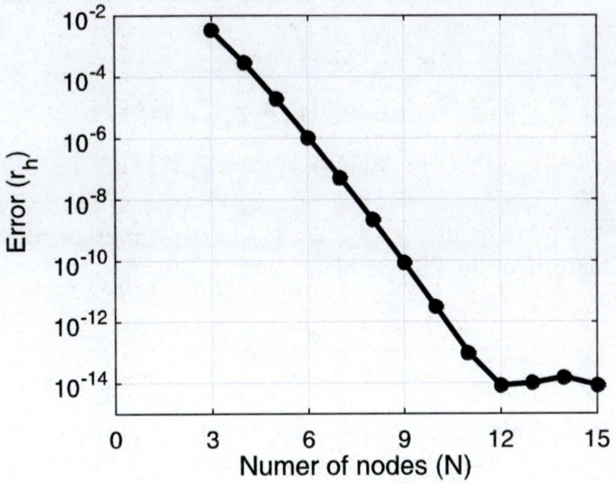

Fig. 1 Dependence of the relative error on the number of nodes

3 Application of Two-Dimensional Integrating Matrices for Solving a Plane Stress Problem of the Theory of Elasticity in a Rectangular Domain

Let us consider, as an example, the classical two-dimensional problem of deformation of a rectangular plate under plane stress state (PSS) conditions. The area occupied by the plate is denoted in the same way as in the first section, behind $\Omega \subset \mathbb{R}^2$, $\Omega = \{X = (x_1, x_2), x_1 \in (0, a), x_2 \in (0, b)\}$, which is bounded by the contour. $\Gamma = \Gamma_1 \bigcup \Gamma_2 \bigcup \Gamma_3 \bigcup \Gamma_4$. Here $\Gamma_1 = \{X = (x_1, 0), x_1 \in (0, a)\}$, $\Gamma_2 = \{X = (0, x_2), x_2 \in (0, b)\}$, $\Gamma_3 = \{X = (x_1, b), x_1 \in (0, a)\}$, $\Gamma_4 = \{X = (a, x_2), x_2 \in (0, b)\}$. Let us denote by u_1, u_2, the components of the displacement vector $\mathbf{u} : \Omega \to \mathbb{R}^2$. We take the kinematic relations in the form known in the linear theory of elasticity, which are described by the Cauchy strain tensor $\boldsymbol{\varepsilon}(\mathbf{u}) := \frac{1}{2}(\nabla \mathbf{u}^{\mathbf{T}} + \nabla \mathbf{u})$. Let us assume that the deformation of the plate is carried out only due to the surface forces applied to its boundaries. We will assume that the plate material is elastic and isotropic. Then the equilibrium equations of the plate can be represented in the following form

$$div \boldsymbol{\sigma} = 0, [\boldsymbol{\sigma}] = \begin{bmatrix} \sigma_{11} \\ \sigma_{22} \\ \sigma_{12} \end{bmatrix} = \frac{E}{1-\nu^2} \begin{bmatrix} 1 & \nu & 0 \\ \nu & 1 & 0 \\ 0 & 0 & \frac{1-\nu}{2} \end{bmatrix} \begin{bmatrix} \varepsilon_{11} \\ \varepsilon_{22} \\ 2\varepsilon_{12} \end{bmatrix}. \tag{9}$$

For equations (9), we formulate the boundary conditions of the form

$$\begin{aligned} &\sigma_{11}(x) = p_1, \ \sigma_{12}(x) = p_{12}, x \in \Gamma_2, \\ &\sigma_{22}(x) = p_2, \ \sigma_{12}(x) = p_{21}, x \in \Gamma_3, \\ &\mathbf{u}(x) = \mathbf{v}, \ x \in \Gamma_1 \bigcup \Gamma_4, \end{aligned} \tag{10}$$

we write out the first equation of system (9) in the Cartesian coordinate system within the framework of the PSS problem

$$\frac{\partial \sigma_{11}}{\partial x_1} + \frac{\partial \sigma_{12}}{\partial x_2} = 0 \tag{11}$$

and use the finite sum method outlined above. For this purpose, we will integrate Eq. (11) from x_1 to a, using the introduced integral operator $\mathcal{R}_1$, and use the first boundary condition from (10). Then

$$p_{11} - \sigma_{11} + \mathcal{R}_1 \frac{\partial \sigma_{12}}{\partial y} = 0.$$

Next, we integrate the obtained equality from x_2 to b, using the introduced integral operator $\mathcal{R}_2$ for this, and use the second boundary condition from (10)

$$\mathcal{R}_2\left(p_{11}-\sigma_{11}\right)+\mathcal{R}_1\left(p_{21}-\sigma_{12}\right)=0.$$

Let us write out the second equation of system (9) in the Cartesian coordinate system

$$\frac{\partial \sigma_{12}}{\partial x}+\frac{\partial \sigma_{22}}{\partial y}=0. \tag{12}$$

By analogy with the above, we will integrate Eq. (12) and take into account the third and fourth boundary conditions of system (10), transferring the applied loads to the right side. As a result, we obtain a system of two equations

$$\begin{aligned} \mathcal{R}_2\sigma_{11}+\mathcal{R}_1\sigma_{12}&=\mathcal{R}_2 p_{11}+\mathcal{R}_1 p_{21},\\ \mathcal{R}_2\sigma_{12}+\mathcal{R}_1\sigma_{22}&=\mathcal{R}_2 p_{12}+\mathcal{R}_1 p_{22}. \end{aligned} \tag{13}$$

Next, we write out the system (13) in displacements using the kinematic relations and relations of the generalized Hooke's law

$$\begin{aligned} \mathcal{R}_2\left[\frac{E}{1-\nu^2}\left(\frac{\partial u_1}{\partial x_1}+\nu\frac{\partial u_2}{\partial x_2}\right)\right]+\mathcal{R}_1\left[G\left(\frac{\partial u_1}{\partial x_2}+\frac{\partial u_2}{\partial x_1}\right)\right]&=\mathcal{R}_2 p_{11}+\mathcal{R}_1 p_{21},\\ \mathcal{R}_2\left[G\left(\frac{\partial u_1}{\partial x_2}+\frac{\partial u_2}{\partial x_1}\right)\right]+\mathcal{R}_1\left[\frac{E}{1-\nu^2}\left(\nu\frac{\partial u_1}{\partial x_1}+\frac{\partial u_2}{\partial x_2}\right)\right]&=\mathcal{R}_2 p_{12}+\mathcal{R}_1 p_{22}. \end{aligned} \tag{14}$$

Let us introduce the vector function of unknowns $U=\left(\frac{\partial u_1}{\partial x_1},\frac{\partial u_2}{\partial x_2},\frac{\partial u_1}{\partial x_2},\frac{\partial u_2}{\partial x_1}\right)$. To close equations (14) with respect U, we use the remaining boundary conditions (10) $u_1|_{\Gamma_1}=v_1$, $u_2|_{\Gamma_1}=v_{21}$, $u_2|_{\Gamma_4}=v_2$, $u_1|_{\Gamma_4}=v_{12}$. For this purpose, integrating $\frac{\partial u_1}{\partial x_1},\frac{\partial u_2}{\partial x_1}$, over x_1 and $\frac{\partial u_1}{\partial x_2},\frac{\partial u_2}{\partial x_2}$, over x_2, we obtain the dependencies

$$\begin{aligned} \mathcal{L}_1\frac{\partial u_1}{\partial x_1}&=u_1-v_1,\ \mathcal{L}_1\frac{\partial u_2}{\partial x_1}=u_2-v_{21},\\ \mathcal{L}_2\frac{\partial u_1}{\partial x_2}&=u_1-v_{12},\ \mathcal{L}_2\frac{\partial u_2}{\partial x_2}=u_2-v_2, \end{aligned} \tag{15}$$

which reduce to two equations

$$\mathcal{L}_1\frac{\partial u_1}{\partial x_1}-\mathcal{L}_2\frac{\partial u_1}{\partial x_2}=v_{12}-v_1,\ \mathcal{L}_1\frac{\partial u_2}{\partial x_1}-\mathcal{L}_2\frac{\partial u_2}{\partial x_2}=v_2-v_{12}. \tag{16}$$

Note that the resulting system of integral equations (14), (16) concerning the vector function $U = \left(\frac{\partial u_1}{\partial x_1}, \frac{\partial u_2}{\partial x_2}, \frac{\partial u_1}{\partial x_2}, \frac{\partial u_2}{\partial x_1}\right)$ is equivalent to the original problem (9), (10), after solving which the solution to the original boundary value problem (9), (10) is restored using equations (15). An approximate solution to problem (14), (15) is a vector function $U_h = \left(u_{1,1}, u_{2,2}, u_{1,2}, u_{2,2}\right)$, that, after applying two-dimensional integrating matrices, satisfies the equation

$$A^h U_h = F_h, A^h = \begin{pmatrix} BR_2^h & \nu BR_2^h & GR_1^h & GR_1^h \\ \nu BR_1^h & BR_1^h & GR_2^h & GR_2^h \\ L_1^h & 0 & -L_2^h & 0 \\ 0 & -L_2^h & 0 & L_1^h \end{pmatrix},$$

$F_h = (p_{11}(b-y) + p_{21}(a-x)\,, p_{12}(b-y) + p_{22}(a-x)\,, v_{12} - v_1, v_2 - v_{12})^T$, $(x, y) \in \omega_h$.

An approximate solution to the original boundary value problem (9), (10) will be determined using equalities (15), as, for example, $[u_1]_h = L_1^h u_{1,1} + v_1$.

Results of Numerical Experiments on Test Problems

1. Consider a plate rigidly fixed $\mathbf{u}(x) = 0$ at points $x \in \Gamma_1 \bigcup \Gamma_4$ under the action of a uniformly distributed load $\sigma_{11}(x) = p_1,\ \sigma_{12}(x) = 0$ on the right boundary $x \in \Gamma_2$ with a free upper boundary $\sigma_{22}(x) = 0,\ \sigma_{12}(x) = 0, x \in \Gamma_3$. When carrying out the calculations, the geometric, grid and elastic parameters were taken as follows: $a = 2$, $b = 1$, $N_{x_1} = 15$, $N_{x_2} = 30$, $E = 200 \cdot 10^4$, $\nu = 0$. Figure 2 shows the form of the deformed state of the plate under the action of the load $\sigma_{11}(x) = p_1$, the color scale corresponds to the values of the formed normal stresses $\sigma_{11}(x)$, $x \in \Omega$.

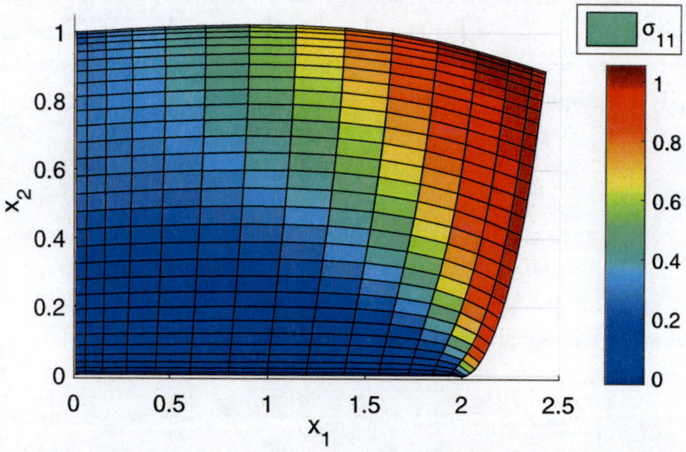

Fig. 2 Deformed state of the plate under the action of a uniformly distributed load $\sigma_{11}(x) = p_1,\ \sigma_{12}(x) = 0$

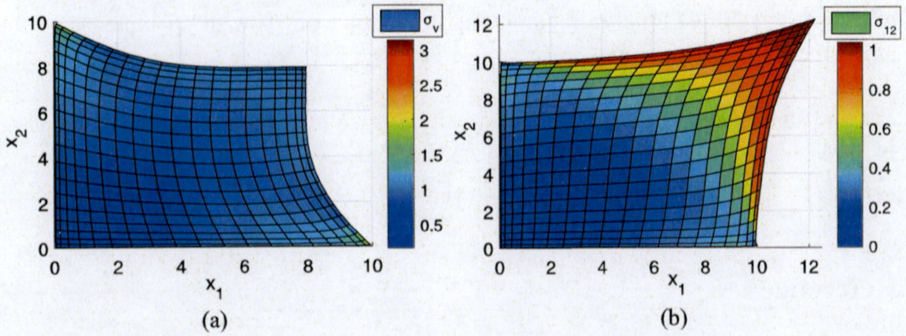

Fig. 3 Deformed states of the plate: **(a)** $\sigma_{11}(x) = \sigma_{22}(y) = -1$; **(b)** $\sigma_{12}(x) = \sigma_{12}(y) = 1$

2. The plate, which is also in conditions of rigid fixation $\mathbf{u}(x) = 0$ at the points $x \in \Gamma_1 \bigcup \Gamma_4$, is acted upon by a uniformly distributed load $\sigma_{11}(x) = \sigma_{22}(y) = -1$, applied at the boundaries $x \in \Gamma_2$, $y \in \Gamma_3$. When carrying out the calculations, the geometric, grid and elastic parameters were taken as follows: $a = 0.1$ m, $b = 0.1$ m, $N_{x_1} = N_{x_2} = 20$, $E = 200$ GPa, $\nu = 0.3$. In Fig. 3a shows a view of the deformed state of the plate, the color scale corresponds to the values of the formed stresses according to von Mises $\sigma_v = \sqrt{\sigma_{11}^2 - \sigma_{11}\sigma_{22} + \sigma_{22}^2 + 3\sigma_{12}^2}$.
3. The plate, which is also in the conditions of rigid fixation $\mathbf{u}(x) = 0$ at the points $x \in \Gamma_1 \bigcup \Gamma_4$, is acted upon by uniformly distributed tangential forces $\sigma_{12}(x) = \sigma_{12}(y) = 1$, applied at the boundaries $x \in \Gamma_2$, $y \in \Gamma_3$. When carrying out the calculations, the geometric, grid and elastic parameters are taken the same as in case 2. In Fig. 3b shows a view of the deformed state of the plate, the color scale corresponds to the values of the generated shear stresses σ_{12}.

4 Conclusion

On several model problems, formulated based on the Poisson equation and the plane problem of the theory of elasticity in a rectangular domain for given types of boundary conditions, an algorithm is presented for their reduction to two-dimensional integral equations containing Volterra-type operators. To find their numerical solutions, two-dimensional integrating matrices are constructed, based on which numerical solutions of the formulated problems are found. Numerical estimates of the accuracy of the found numerical solutions of the considered test problems are carried out. It is shown that the convergence of the method is exponential. Due to the use of the Gauss quadrature formula for approximation on test problems, the highest order of the algebraic accuracy of the method was demonstrated. It should also be noted that the construction of the proposed two-dimensional integrating matrices is based on the idea of constructing unsaturated algorithms, in connection with which the conditionality of the matrix of the resolving system of algebraic equations does not worsen with an increase in the amount of collocation nodes.

Acknowledgments This paper has been supported by the Kazan Federal University Strategic Academic Leadership Program ("PRIORITY-2030").

References

1. Birger, I.A.: Nekotoryye matematicheskiye metody resheniya inzhenernykh zadach [Some mathematical methods for solving engineering problems]. Oborongiz, Moscow (1956) [in Russian]
2. Vakhitov, M.B.: Integriruyushchiye matritsy – apparat chislennogo resheniya differentsial'nykh uravneniy stroitel'noy mekhaniki [Integrating matrices – apparatus for the numerical solution of differential equations of structural mechanics]. Russian Aeronautics. **3**, 50–61 (1966) [in Russian]
3. Vakhitov, M.B., Safariyev, M.S., Snigirev, V.F.: Raschet kryl'yevykh ustroystv sudov na prochnost' [Strength calculation of wing devices of ships]. Tat. kn. izd-vo, Kazan (1975) [in Russian]
4. Verlan', A.F., Sizikov, V.S: Integral'nyye uravneniya: metody, algoritmy, programmy [Integral equations: methods, algorithms, programs]. Nauk. dumka, Kiyev (1986) [in Russian]
5. Ahlberg, J.H., Nilson, E.N., Walsh, J. L.: The theory of splines and their applications. Academic Press, New York (1967)
6. Vasilenko, V.A.: Splayn-funktsii: teoriya, algoritmy, programma [Spline functions: theory, algorithms, program]. Nauka. Sib. Otd-niye, Novosibirsk (1983) [in Russian]
7. Krylov, V.I., Bobkov, V.V., Monastyrskiy, P.I.: Vychislitel'nyye metody: V 2 t. [Computational methods: In 2 volumes]. Nauka, Moscow (1977) [in Russian]
8. Stechkin, S.B., Subbotin, YU.N.: Splayny v vychislitel'noy matematike [Splines in computational mathematics]. Nauka, Moscow (1976) [in Russian]
9. Paimushin, V.N., Firsov, V.A.: Obolochki iz Stekla. Raschet napryazhenno-deformirovannogo sostoyaniya [Glass shells. Calculation of the stress-strain state]. Mashinostroyeniye, Moscow (1993) [in Russian]
10. Dautov, R. Z., Paimushin, V. N.: On the method of integrating matrices for solving boundary value problems for ordinary equations of the fourth order. Russian Math. **40**(10), 11–23 (1966)
11. Dautov, R. Z., Karchevsii, M.M., Paimushin, V. N.: On the method of integrating matrices for systems of ordinary differential equations. Russian Math. **47**(7), 16–24 (2003)
12. Stepanova, Ye.M., Petrushenko Yu.Ya.: Algebraicheskiy analog zadachi Puassona na osnove integriruyushchikh matrits, baziruyushchikhsya na polinomakh [Algebraic analogue of the Poisson problem based on integrating matrices based on Lagrange polynomials]. Matematicheskoye modelirovaniye i krayevyye zadachi. Trudy Vserossiyskoy nauchnoy konferentsii. Samara. Samarskiy gosudarstvennyy tekhnicheskiy universitet. 212–214 (2004) [in Russian]
13. Stepanova, Ye.M., Dautov, R.Z., Petrushenko Yu.Ya.: Resheniye krayevykh zadach, opisyvayemykh dvumernymi ellipticheskimi uravneniyami vtorogo poryadka, metodom integriruyushchikh matrits [Solution of boundary value problems described by two-dimensional second-order elliptic equations by the method of integrating matrices]. Matematicheskoye modelirovaniye i krayevyye zadachi. Trudy Vtoroy Vserossiyskoy nauchnoy konferentsii. Samara. Samarskiy gosudarstvennyy tekhnicheskiy universitet. 218–221 (2005) [in Russian]

On Resonant Effects in the Semi-Infinite Waveguides with Barriers

Nikolai Pleshchinskii, Garnik Abgaryan, and Bulat Vildanov

Abstract The problems of electromagnetic wave diffraction by thin conductive barriers in a semi-infinite parallel-plate waveguide are reduced to infinite sets of linear algebraic equations concerning the expansion coefficients of the field by its eigen waves. Values of resonant frequencies are obtained for which there is a sharp increase in the characteristics of the electromagnetic field in the area between the barrier and the metal wall.

1 Introduction

In the design of radiotechnical devices with optimal characteristics the situations when there is a resonant growth of certain parameters of the electromagnetic field are of particular interest. Barriers in waveguide structures are widely used in the production of converters, filters, splitters and other elements.

In this paper, we explore the resonant effects that occur when the diffraction of eigen electromagnetic wave, which attacks a thin conductive barriers in a semi-infinite parallel-plate waveguide with metal walls.

As it is known [1], any electromagnetic field in the parallel-plate waveguide can be presented as a sum of its eigen waves propagating or damping in different directions. The theory of equivalent chains was used in [2] as a simple model of the process of electromagnetic wave diffraction. In recent years, during the investigation of the resonant properties of waveguides with heterogeneities, the method of equivalent circuits [3], the method of moments [4] and more rigorous methods as well as method of the Riemann-Hilbert problem [5] and method of integral equations [6] are used. Some numerical results can be found in the works [7, 8].

N. Pleshchinskii · G. Abgaryan (✉) · B. Vildanov
Kazan Federal University, Kazan, Russia
e-mail: pnb@kpfu.ru

I. B. Badriev et al. (eds.), *Mesh Methods for Boundary-Value Problems and Applications*, Lecture Notes in Computational Science and Engineering 141,
https://doi.org/10.1007/978-3-030-87809-2_30

In this paper, the method of integral-series identities is used to reduce the paired series functional equations of diffraction problems by the screens to regular infinite sets of linear algebraic equations (ISLAE) [9]. Early we investigated the resonant properties of the diaphragms in the semi-infinite waveguides [10, 11].

2 Lateral Barrier in a Waveguide

Let us consider the two-dimensional problem of TE-wave diffraction by a lateral barrier in a half-infinite parallel-plate waveguide $0 < x < a,\ z < d$. The barrier is located in the plane $z = 0$. The part $\mathcal{M} = (\alpha, \beta)$ of the cross-section $[0, a]$ of the waveguide corresponds to it (Fig. 1). Let's denote by $\mathcal{N}$ supplement of $\mathcal{M}$ up to $[0, a]$.

Let free currents and charges be absent, the medium be homogeneous and isotropic, electromagnetic field harmoniously depend on time ($\exp(-i\omega t)$). Denote

$$\varphi_n(x) = \sqrt{2/a}\,\sin\frac{\pi n x}{a}, \quad \gamma_n = \sqrt{\kappa^2 - (\pi n/a)^2},$$

where κ is wave number, $\operatorname{Re}\gamma > 0$ or $\operatorname{Im}\gamma_n > 0$ and $n = 1, 2, \dots$

From the region $z < 0$ on the barrier runs its eigen wave

$$u^0(x, z) = \varphi_l(x)\, e^{i\gamma_l z}.$$

We will look for the wave reflected to the left in the form of

$$u^1(x, z) = \sum_{n=1}^{+\infty} a_n \varphi_n(x)\, e^{-i\gamma_n z},$$

Fig. 1 Lateral barrier in a plane waveguide

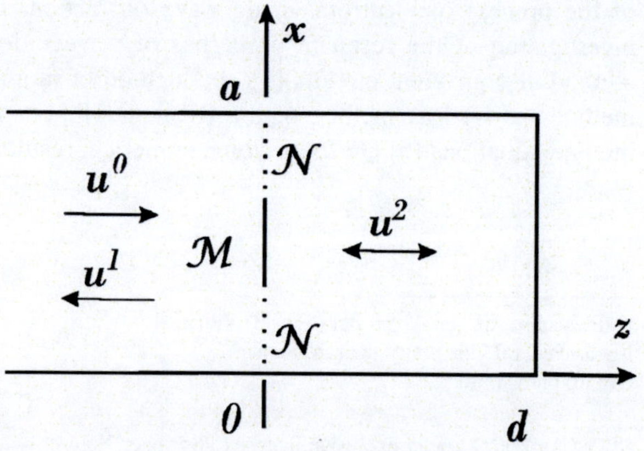

and we will look for the wave passed to the right in the form of

$$u^2(x,z) = \sum_{n=1}^{+\infty} b_n \varphi_n(x)\left(e^{i\gamma_n z} - e^{2i\gamma_n d}\, e^{-i\gamma_n z}\right).$$

For the wave $u^2(x,z)$ a boundary condition is fulfilled on the metal wall $z = d$, and this wave is bounded for $n \to +\infty$.

Let's write down the boundary conditions on the $\mathcal{M}$:

$$\varphi_l(x) + \sum_{n=1}^{+\infty} a_n\, \varphi_n(x) = 0, \quad \sum_{n=1}^{+\infty} b_n\left(1 - e^{2i\gamma_n d}\right)\varphi_n(x) = 0$$

and the conjugation conditions on the $\mathcal{N}$:

$$\varphi_l(x) + \sum_{n=1}^{+\infty} a_n\, \varphi_n(x) = \sum_{n=1}^{+\infty} b_n\left(1 - e^{2i\gamma_n d}\right)\varphi_n(x),$$

$$\gamma_l\, \varphi_l(x) - \sum_{n=1}^{+\infty} a_n\, \gamma_n\, \varphi_n(x) = \sum_{n=1}^{+\infty} b_n\, \gamma_n\left(1 + e^{2i\gamma_n d}\right)\varphi_n(x).$$

It follows that $1 + a_l = b_l\left(1 - e^{2i\gamma_l d}\right)$ and $a_n = b_n\left(1 - e^{2i\gamma_n d}\right),\ n \neq l$. We exclude the unknowns a_n.

To regularize the pair series functional equations, we use an integral-series identity

$$\int_0^a \left(\sum_{n=1}^{+\infty} b_n\left(1 - e^{2i\gamma_n d}\right)\varphi_n(t)\right) K(t,x)\, dt = \sum_{n=1}^{+\infty} b_n \gamma_n\, \varphi_n(x),$$

here

$$K(t,x) = \sum_{m=1}^{+\infty} \frac{\gamma_m}{1 - e^{2i\gamma_m d}}\, \varphi_m(t)\, \varphi_m(x).$$

It is assumed that $\gamma_m d \neq \pi j$.

Finally, after projecting on the function $\varphi_k(x)$ we get ISLAE ($k = 1, 2, \ldots$)

$$b_k \gamma_k - \sum_{n=1}^{+\infty} b_n\left(1 - e^{2i\gamma_n d}\right) \sum_{m=1}^{+\infty} \frac{\gamma_m}{1 - e^{2i\gamma_m d}}\, J_{nm}\, I_{mk} = \gamma_l\, J_{lk},$$

where

$$I_{nm} = \int\limits_{\mathcal{M}} \varphi_n(t)\,\varphi_m(t)\,dt, \quad J_{nm} = \int\limits_{\mathcal{N}} \varphi_n(t)\,\varphi_m(t)\,dt.$$

3 Computing Experiments, I

The computing experiments are based on the multiple solving the truncated ISLAE in the case when the frequency of the excitatory wave changes with a small step. We will look for an approximate solution of ISLAE by truncation method. It is enough to take the parameter of truncated method $N = 30$. As a wave incoming on the barrier, we will consider the first mode of the waveguide.

Let's choose the following parameters: $a = 1.1$, $d = 1.3$; $\alpha = 0.1$; $\beta = 1.0$ in dimensionless quantities. As the computing experiment has shown, the modules of coefficients $b_1, b_2, \ldots$ have sharp local maximums, with correspond to wave number ≈ 3.7410, ≈ 5.6135, $\approx 7.7910, \ldots$. These values are close to the eigen wave numbers ≈ 3.7412, ≈ 5.6140, ≈ 7.7921 of a two-dimensional rectangular region of the size $a \times d$.

The dependence of coefficient b_1 module on wave number κ in the neighborhoods of resonant values is shown in Figs. 2 and 3.

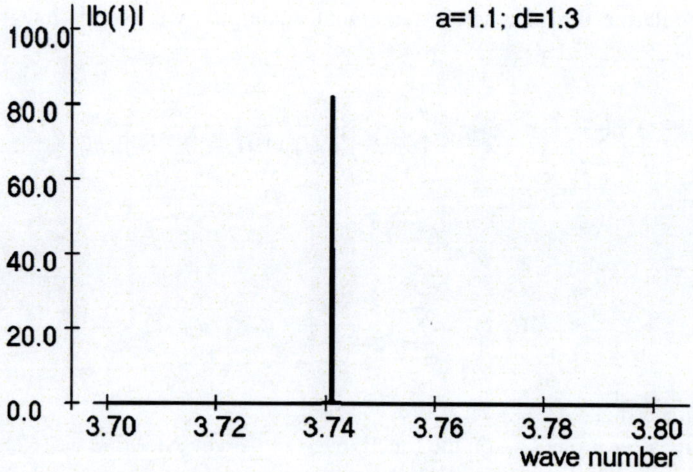

Fig. 2 Dependence of the modulus b_1 on the wave number κ near the eigen frequency k_{11}. Waveguide parameters: a=1.1, d=1.3

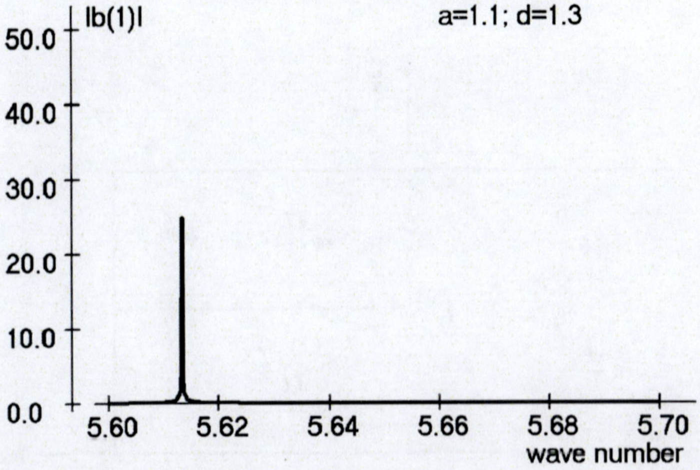

Fig. 3 Dependence of the modulus b_1 on the wave number κ near the eigen frequency k_{12}. Waveguide parameters: a=1.1, d=1.3

If the size of barrier decreases, then the resonant values of k decrease slightly also.

The dependence of the conditional number cond $A = ||A|| \cdot ||A^{-1}||$ on the parameter κ is also resonant.

If we balance the equations in SLAE (divide each equation by the largest in-module coefficient for the unknowns), then the conditioned number will be significantly reduced, but the solution of the SLAE will not change. But after balancing, it becomes possible to study the dependence on the parameter κ of the values of the determinant of the matrix of the SLAE coefficients. Now the modules of these values in the neighborhood of the resonant point do not exceed one. Before balancing, they had an order of 10^{45} or more.

As in the case of diaphragm in the semi-infinite waveguide [11], the resonant values of the parameter κ can be found: (1) when solving the SLAE of diffraction problem; (2) when calculating the conditioned number of the matrix of its coefficients; (3) when analyzing the values of the determinant of this matrix.

4 Longitudinal Barrier in a Waveguide

Now let the thin conductive barrier with a length of d be placed at the height of b from the lower wall of the waveguide (Fig. 4).

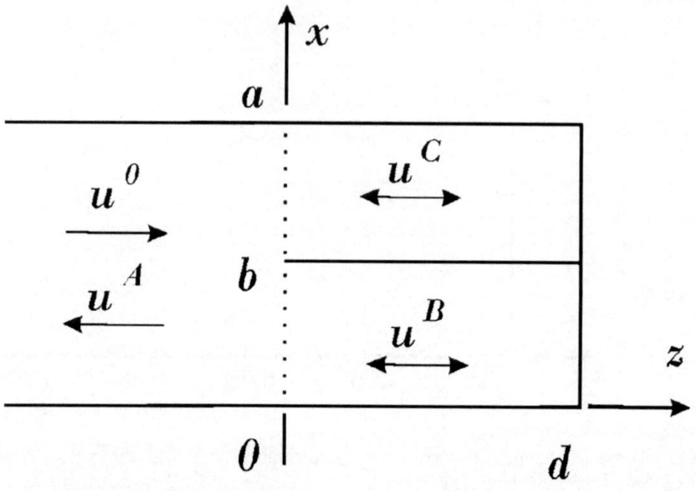

Fig. 4 Longitudinal barrier in a waveguide

We will use the following notations:

$$\varphi_n^a(x) = \sqrt{2/a}\,\sin\frac{\pi n x}{a}, \quad \gamma_n^a = \sqrt{\kappa^2 - (\pi n/a)^2},$$

$$\varphi_n^b(x) = \sqrt{2/b}\,\sin\frac{\pi n x}{b}, \quad \gamma_n^b = \sqrt{\kappa^2 - (\pi n/b)^2},$$

$$\varphi_n^c(x) = \sqrt{2/(a-b)}\,\sin\frac{\pi n (x-b)}{a-b},$$

$$\gamma_n^c = \sqrt{\kappa^2 - (\pi n/(a-b))^2}, \quad n = 1, 2, \ldots$$

Let the eigen wave

$$u^0(x, z) = \varphi_l^a(x) e^{i\gamma_l^a z}.$$

run on the barrier. We will look for the wave reflected to the left in the form of

$$u^A(x, z) = \sum_{n=1}^{+\infty} a_n \varphi_n^a(x) e^{-i\gamma_n^a z}$$

and we will look for the wave in the regions B: $0 < x < b,\ 0 < z < d$ and C: $b < x < a,\ 0 < z < d$ in the form of

$$u^B(x,z) = \sum_{n=1}^{+\infty} b_n \varphi_n^b(x)\left(e^{i\gamma_n^b z} - e^{2i\gamma_n^b d} e^{-i\gamma_n^b z}\right),$$

$$u^C(x,z) = \sum_{n=1}^{+\infty} c_n \varphi_n^c(x)\left(e^{i\gamma_n^c z} - e^{2i\gamma_n^c d} e^{-i\gamma_n^c z}\right).$$

The equalities on the $(0, b)$

$$2\varphi_l^a(x) + \sum_{n=1}^{+\infty} d_n \varphi_n^a(x) = \sum_{n=1}^{+\infty} b_n \varphi_n^b(x)\left(1 - e^{2i\gamma_n^b d}\right),$$

$$-\sum_{n=1}^{+\infty} d_n \gamma_n^a \varphi_n^a(x) = \sum_{n=1}^{+\infty} b_n \gamma_n^b \varphi_n^b(x)\left(1 + e^{2i\gamma_n^b d}\right),$$

and on the (b, a)

$$2\varphi_l^a(x) + \sum_{n=1}^{+\infty} d_n \varphi_n^a(x) = \sum_{n=1}^{+\infty} b_n \varphi_n^c(x)\left(1 - e^{2i\gamma_n^c d}\right),$$

$$-\sum_{n=1}^{+\infty} d_n \gamma_n^a \varphi_n^a(x) = \sum_{n=1}^{+\infty} c_n \gamma_n^c \varphi_n^c(x)\left(1 + e^{2i\gamma_n^c d}\right)$$

should be fulfilled if $z = 0$. Here $d_l = a_l - 1$, $d_n = a_n,\ n \neq l$.

Let's exclude unknowns d_n using the integral-series identity

$$\sum_{n=1}^{+\infty} d_n \varphi_n^a(x) = \int_0^a \left(\sum_{n=1}^{+\infty} d_n \gamma_n^a \varphi_n^a(t)\right) K(t,x)\, dt, \quad x \in (0, a),$$

$$K(t,x) = \sum_{m=1}^{+\infty} \frac{1}{\gamma_m^a} \varphi_m^a(t) \varphi_m^a(x).$$

Replace x by t in the equalities of the second pair, multiply both parts by $K(t, x)$ and integrate from 0 to a. Then we get

$$-\sum_{n=1}^{+\infty} d_n \varphi_n^a(x) = \sum_{n=1}^{+\infty} b_n \gamma_n^b (1 + e^{2i\gamma_n^b d}) \sum_{m=1}^{+\infty} \frac{1}{\gamma_m^a} \varphi_m^a(x)\, I_{mn}^b$$

$$+ \sum_{n=1}^{+\infty} c_n \gamma_n^c (1 + e^{2i\gamma_n^c d}) \sum_{m=1}^{+\infty} \frac{1}{\gamma_m^a} \varphi_m^a(x)\, I_{mn}^c, \quad x \in (0, a).$$

Let's add the equations of the first pair and new equality. The equation

$$2\varphi_l^a(x) = \sum_{n=1}^{+\infty} b_n \varphi_n^b(x) \big(1 - e^{2i\gamma_n^b d}\big)$$

$$+ \sum_{n=1}^{+\infty} b_n \gamma_n^b \big(1 + e^{2i\gamma_n^b d}\big) \sum_{m=1}^{+\infty} \frac{1}{\gamma_m^a} \varphi_m^a(x)\, I_{mn}^b$$

$$+ \sum_{n=1}^{+\infty} c_n \gamma_n^c \big(1 + e^{2i\gamma_n^c d}\big) \sum_{m=1}^{+\infty} \frac{1}{\gamma_m^a} \varphi_m^a(x)\, I_{mn}^c, \quad x \in (0, b),$$

we multiply by $\varphi_k^b(x)$ and integrate from 0 to b. A similar equation on (b, a) is multiplied by $\varphi_k^c(x)$ and integrated from b to a. Then

$$2I_{lk}^b = b_k (1 - e^{2i\gamma_k^b d}) + \sum_{n=1}^{+\infty} b_n \gamma_n^b (1 + e^{2i\gamma_n^b d}) \sum_{m=1}^{+\infty} \frac{1}{\gamma_m^a} I_{mn}^b\, I_{mk}^b$$

$$+ \sum_{n=1}^{+\infty} c_n \gamma_n^c (1 + e^{2i\gamma_n^c d}) \sum_{m=1}^{+\infty} \frac{1}{\gamma_m^a} I_{mn}^c\, I_{mk}^b, \quad k = 1, 2, \ldots$$

$$2I_{lk}^c = c_k (1 - e^{2i\gamma_k^c d}) + \sum_{n=1}^{+\infty} b_n \gamma_n^b (1 + e^{2i\gamma_n^b d}) \sum_{m=1}^{+\infty} \frac{1}{\gamma_m^a} I_{mn}^b\, I_{mk}^c$$

$$+ \sum_{n=1}^{+\infty} c_n \gamma_n^c (1 + e^{2i\gamma_n^c d}) \sum_{m=1}^{+\infty} \frac{1}{\gamma_m^a} I_{mn}^c\, I_{mk}^c, \quad k = 1, 2, \ldots$$

where

$$I_{mn}^{b} = \int_{0}^{b} \varphi_m^a(t)\varphi_n^b(t)\,dt, \quad I_{mn}^{c} = \int_{b}^{a} \varphi_m^a(t)\varphi_n^c(t)\,dt.$$

So, the ISLAE to determine the coefficients b_n and c_n consists of two groups of equations. When truncated, we leave N unknown in each equation and N equations in each group.

5 Computing Experiments, II

A computational experiment has shown that at some values of electromagnetic oscillation frequencies there is a resonant increase in a field expansion coefficients in regions B and C. The dependencies of the coefficient b_1 module on the frequency (more precisely, when the wave number k changes in the truncated ISLAU) are shown on Figs. 5, 6, and 7.

Resonant frequencies depend significantly on the value of the b. It's easy to see that the highest peak of the lines on the charts are observed when the frequencies are close to the eigen values $\pi\sqrt{1/b^2 + 1/d^2}$ of the frequencies of rectangular domain of the size $b \times d$. At low values of d resonances are not observed.

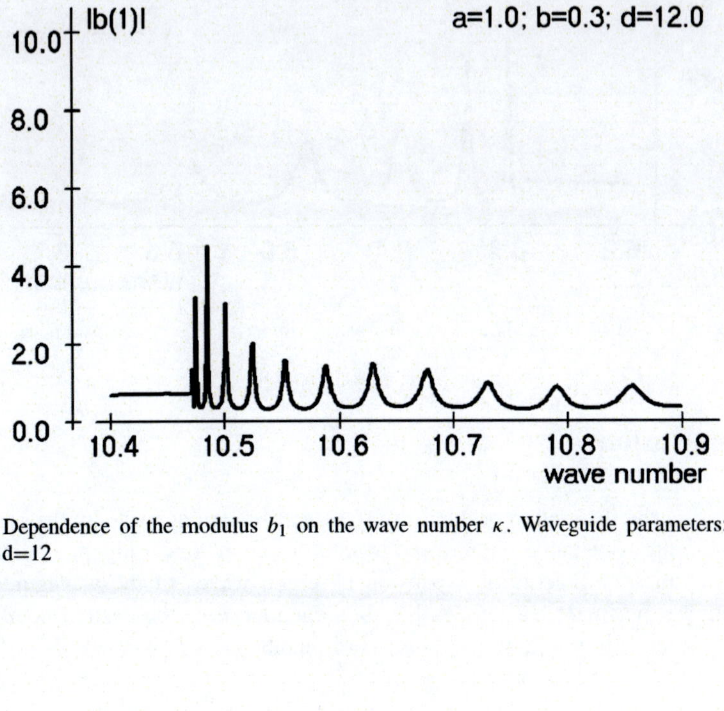

Fig. 5 Dependence of the modulus b_1 on the wave number κ. Waveguide parameters: a=1.0, b=0.3, d=12

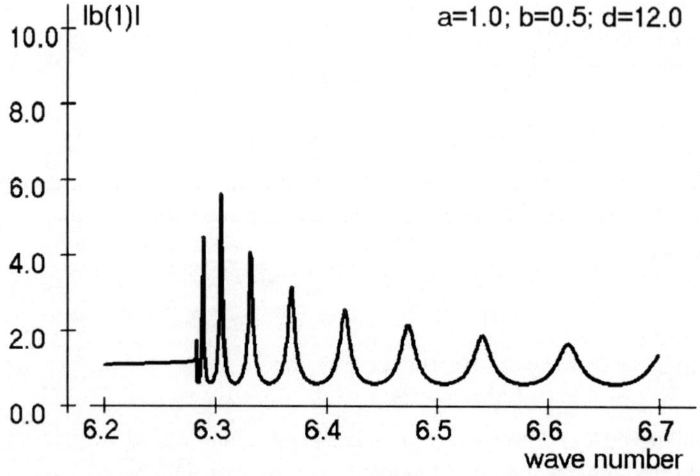

Fig. 6 Dependence of the modulus b_1 on the wave number κ. Waveguide parameters: a=1.0, b=0.5, d=12

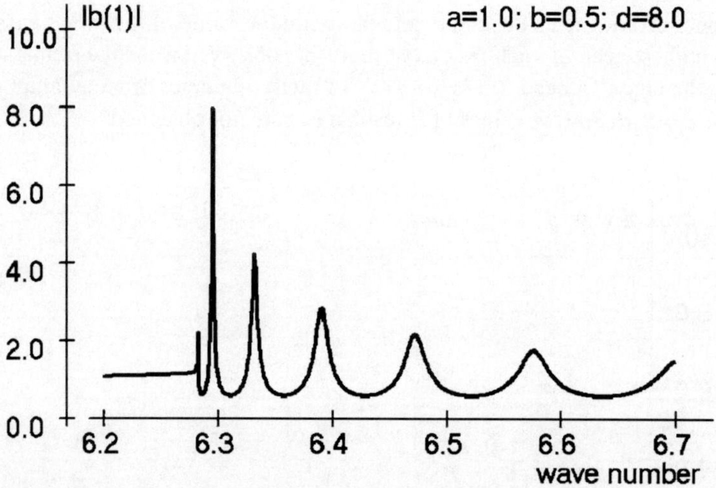

Fig. 7 Dependence of the modulus b_1 on the wave number κ. Waveguide parameters: a=1.0, b=0.5, d=8

6 Conclusion

In this paper the diffraction problems of the electromagnetic wave by the barrier in a semi-infinite waveguide are reduced to infinite sets of linear algebraic equations relative to the coefficients of expansion by eigen waves of the waveguide. The computing experiment has shown that the dependence of the desired coefficients on the frequency of the excitatory wave is resonant.

Acknowledgments This paper has been supported by the Kazan Federal University Strategic Academic Leadership Program ("PRIORITY-2030").

References

1. Lewin, L.: Theory of Waveguides. Newnes-Butterworths, London (1975).
2. Schwinger, Yu.: Inhomogeneities in waveguides (lecture notes). Zarubezhnaya Electronika **3**, 3–106 (1970) [in Russian]
3. Usanov, D.A., Gorbatov, S.S., Orlov, V.E., Venig S.B.: Resonances in semi-infinite waveguide with diaphragm, concerned with exitation of the high type waves. Pis'ma v Zhurnal Tekhnicheskoi Fiziki **26** (18), 47–49 (2000)
4. Datta, A., Chakraborty, A., Das, B.N.: Analysis of a strip loaded resonant longitudinal slot in the broad wall of a rectangular waveguide. IEE Proceedings H (Microwaves, Antennas and Propagation) **140** (2), 135–140 (1993)
5. Shestopalov, V.P.: Riemann-Hilbert Metod in the Theory of Diffraction and of Propagation of Electromagnetic Waves. Izdatel'stvo Kharkovskogo Universiteta, Kharkov (1971) [in Russian]
6. Lewin, L.: On the resolution of a class of waveguide discontinuity problems by the use of singular integral equations. IRE Transactions on Microwave Theory and Techniques **9** (4), 321–332 (1961)
7. Nesterenko, M.V.: Electromagnetic wave scattering by a resonant iris with the slot arbitrary oriented in a rectangular waveguide. Radiofizika i Radioastronomiya **9** (3), 274–285 (2004) [in Russian]
8. Chernousov, Yu.D., Levichev, A.E., Pavlov, V.M., Shamuilov, G.K.: Thin diaphragm in the rectangular waveguide. Vestnik NGU. Seriya: fizika **6** (1), 44–49 (2011) [in Russian]
9. Pleshchinskii, N.B.: Models and Methods of Waveguide Electrodynamics. Kazanskii Universitet, Kazan (2008) [in Russian]
10. Abgaryan, G.V., Pleshchinskii, N.B.: On the eigen frequencies of rectangular resonator with a hole in the wall. Lobachevskii Journal of Mathematics **40** (10), 1631–1639 (2019) https://doi.org/10.1134/S1995080219100020
11. Abgaryan, G.V., Pleshchinskii, N.B.: On resonant frequencies in a semi-infinite waveguide. Lobachevskii Journal of Mathematics **41** (7), 1325–1336 (2020) https://doi.org/10.1134/S1995080220070033

Numerical Simulation of Composite Structures Polymerization and Determination of Residual Deformations

Evgeniy A. Puzyretskiy, Leonid P. Shabalin, Igor N. Sidorov, and Azat M. Girfanov

Abstract This paper study numerical modeling of the technological process of composite structures polymerization. A full-scale and numerical experiments of curing process were conducted to determine the geometry distortion of the obtained structures from the nominal shape. A proven approach is proposed that allows to make a product with high accuracy. A comparative analysis of computational experiments using personal workstations and a high-performance cluster is carried out.

1 Introduction

Numerical modeling of real technological processes has become an essential part of prototyping in aviation, engineering, and medical industries. Without it, the manufacture of any particularly important structural element of an aircraft, ground transport or implant is not complete.

Products made of polymer composite materials, which play an important role in the operation of a particular design, require high manufacturing accuracy. The technological process often involves curing and post-curing at high temperatures, which leads to distortion of the shape and geometric parameters, which negatively affects their further exploitation. Numerical modeling of the composite product molding process is necessary to ensure the required manufacturing accuracy.

The problem of accumulation of residual stresses and the accuracy of composite products manufacturing is studied in detail in [1–13]. One group of researches is devoted to the development and verification of the material model and analysis of the modeling speed [3–5]. Several approaches for creating a material model are described. They differ mainly in the number of parameters required to describe the material state. Another group of studies is devoted to the problem of the influence

E. A. Puzyretskiy (✉) · L. P. Shabalin · I. N. Sidorov · A. M. Girfanov
Kazan National Research Technical University named after A. N. Tupolev – KAI (KNRTU-KAI), Kazan, Russia

I. B. Badriev et al. (eds.), *Mesh Methods for Boundary-Value Problems and Applications*, Lecture Notes in Computational Science and Engineering 141,
https://doi.org/10.1007/978-3-030-87809-2_31

of boundary conditions on the results of numerical modeling of technological stresses [6–12]. Calculation models are presented that differ in the number of factors used in the stress-strain state analysis.

The paper [1] describes the main causes and consequences of the residual stress-strain state in composite curing process. Major types of residual strain were described.

The radius of corner fillet of a composite product does not affect the value of the "Spring-In" [2]. It is shown that the size of the angle itself has a significant influence.

Work [3] is devoted to the study of the influence of technological stresses on the mechanical characteristics of composite products by the finite element method. This article provides examples of the destruction of products at the stage of their manufacture, due to the occurrence of internal stresses in them that exceed the limit.

Modified viscoelastic model is proposed in [4] for predicting residual stresses caused by curing in polymer matrix composites. Modifications are based on using the inverse of the Deborah number. A multi-layer composite plate was modeled and the evolution of residual stresses was further predicted. The analysis showed that in order to accurately model the residual stresses of a composite, the software and material model must take into account thermal strain and chemical shrinkage.

Work [5] is devoted to comparing the classical elastic model of a material with the CHILE (cure hardening instantaneous linear elastic) model. There is a widespread opinion that the CHILE model is more accurate, because it takes into account the dependence of the characteristics of the composite material on the temperature and degree of polymerization. However, the results of this research, using two samples of different geometric parameters and shapes, showed that the elastic model has a better convergence with the experiment.

In [6] presented experimental results which show that the residual strain can be highly influenced by a number of factors such as cure cycle, contact with technological mold, geometry and layup scheme. The results of the study can be obtained using the method of numerical modeling. This requires models of the materials used (including mold)

The article [7] describes the results of experiments on the production of prototypes and determination of residual deformations. Numerical simulations were performed. The results of the calculations were 20% higher than experimental ones, but the method predicted all the phenomena observed during the experiment.

In the "Spring-In" study [8], the effect of a composite bracket with three different thicknesses was calculated by sequentially solving the problem of non-stationary heat transfer and determining the stress-strain state using the finite element method. The analytical method is used to obtain the material properties necessary for numerical modeling. The values of the "Spring-In" effect obtained by the finite element method have a good convergence with the results of analytical analysis.

The paper [9] describes the process of creating and verifying a simplified material model, which reduces the amount of information that needs to be processed by a computer by 54 times. The material behavior is described by a linear viscoelastic model with the following assumptions: coefficients of linear thermal expansion in the cured and gel states do not depend on the degree of cure; in none of the states (liquid, gel, glass), the stiffness of the material does not depend on the temperature or degree of cure. The results of the analysis of technological stresses using a simplified model did not show any differences from the results obtained on the model that takes into account the dependence of the resin parameters on the temperature and degree of curing.

In [10], a method for modeling a composite product in the form of discrete layers of homogeneous plates at the macro level is proposed. At each integration point of the homogeneous model, the defining relationships of the plate are determined using micromechanics using the Extended Concentric Cylinder Assembly (ECCA) model. The proposed model can be used to predict the magnitude of residual deformations and stresses of a composite product.

A similar study was conducted in [11]. The simulation was performed in the same way, except that contact with the tooling was taken into account. Compared to other models, the proposed one provides better convergence with the experimental results.

The article [12] describes the process of developing a model for predicting the "Spring-In" effect of the angle of an L-shaped composite sample made by RTM technology. The composite product consists of 16 layers of AS-4 carbon fiber preform and EPON 862 epoxy resin. In this study, the composite is modeled as separate layers of orthotropic material using the finite element method. The proposed model is verified by comparing the predicted value of the "Spring-In" effect of angles with experimental values.

In [13] attention is paid to the study and verification of material properties using the CHILE model and the viscoelastic model. It was shown that the CHILE model is described by fewer variables and has less influence on the speed of numerical simulation of the process of accumulation of residual stresses.

The analysis of publications has shown that numerous of factors must be taken into account to determine the technological stresses. It is necessary to develop and verify the model of the material to achieve convergence of the simulation results with the experiment.

2 Numerical Simulation

Modeling of the technological process is demonstrated by the example of a thin-walled V-shaped composite body (hereinafter referred to as a «sample»). The analysis was performed using the Ansys software package.

To perform the analysis, a special model of the sample was created using hexahedral elements. The material model uses calibrated physical and mechanical characteristics. The boundary conditions for modeling the polymerization process are: surface temperature change during heat treatment, vacuum bag pressure, and various constraints on model displacement at different stages of the molding cycle.

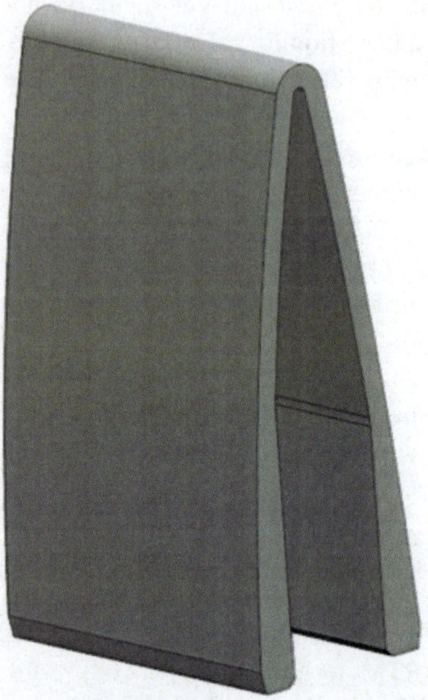

Fig. 1 CAD model of the sample

2.1 Geometry and Model Parameters of the Sample

The sample is a V-shaped structure whose geometry does not depend on the longitudinal coordinate (Fig. 1). However, the sample model is set as a three-dimensional object, since the effect of twisting on the geometry is not excluded. The main initial geometric parameters are: product angle α=23°, the distance between two parallel internal faces b = 16 mm, and the height h = 81 mm.

The computational model of the sample consists of hexahedral finite elements. In areas containing fillets, the mesh was refined to describe the sample geometry. This affects the convergence of the problem and the accuracy of the analysis results.

2.2 Material Model Description

The material model used in the calculations describes its behavior in three phases—liquid, gelled, and glassy. In each phase, the material has different elastic, physical, thermal, and chemical parameters. In this case, an orthotropic material HexPly M56/40%/193PW/AS4-3K was used. Also the following parameters were added: linear temperature expansion coefficient (CTE), heat capacity, and the Ansys Composite Cure Simulation polymerization model.

The characteristics of the material describing its properties were obtained by the laboratory tests. The necessary characteristics for the analysis of the polymerization process were obtained using the method of differential scanning calorimetry [14,

15]. Test sample is placed in the calorimeter chamber and maintained at a constant temperature during the exothermic reaction.

As a result, the total heat of the reaction and the heat flow are obtained. Then, based on the major parameters of the experiment, the parameters of the autocatalytic reaction equation (1) are selected until the data on the heat flow and total heat converge with the results of machine calculation. The pre-exponential factor, activation energy, and coefficients of the autocatalytic reaction equation were obtained. The accuracy of this method is determined by the accuracy of measuring the sample temperature with thermocouples and the accuracy of the autocatalytic reaction model used (1).

$$f(T, a) = A_1 \cdot \exp\left(-\frac{E_1}{T}\right) a^m (1-a)^n \tag{1}$$

here $A_1[1/s]$—is the pre—exponential factor. Indicate the number of molecules collisions of the interacting substances in a second;

$E_1[J]$—activation energy. The energy required by the system of interacting molecules for the reaction to occur;

$T[K]$—absolute temperature;

a—degree of polymerization of the resin;

m, n—coefficients determined experimentally

Chemical shrinkage was determined by measuring the density of the material in the liquid state and polymerized using helium pycnometer. It is able to make measurements with high accuracy in various directions. Thus, the numerical value of chemical shrinkage will be the ratio of densities.

2.3 Boundary Conditions

In this type of analysis, external factors are: time-varying temperature of the sample surface, external pressure from the vacuum bag, and displacement constraints of the product on the molding surface [6, 9, 16].

Sample layup with the reinforcement scheme $[0;90]_s$ was used. Total material layers—14. The thickness of one layer—0.214 mm.

The temperature conditions of prepreg polymerization was set according to the material specification.

3 Analysis of the Curing Process and Stress-Strain State

The polymerization process of the sample is described by the autocatalytic reaction equation (1). The result of the curing analysis is a degree of polymerization and the resin phase in each finite element of the computational domain. Later, this data is imported into the solver to analyze the residual stress-strain state of the sample.

The total finite element strain is the combination of chemical shrinkage (2), thermal expansion/compression (3) and boundary conditions.

$$\begin{aligned} \varepsilon_X &= \alpha \varepsilon_X^{'sh} \\ \varepsilon_Y &= \alpha \varepsilon_Y^{'sh} \\ \varepsilon_Z &= \alpha \varepsilon_Z^{'sh} \end{aligned} \tag{2}$$

$$\begin{aligned} \varepsilon_X^T &= \alpha_{LX} \Delta T \\ \varepsilon_Y^T &= \alpha_{LY} \Delta T \\ \varepsilon_Z^T &= \alpha_{LZ} \Delta T \end{aligned} \tag{3}$$

Here α- is the degree of curing of the resin

$\alpha_{LX,Y,Z}$- coefficient of linear temperature expansion of the material

$\varepsilon_{X,Y,Z}^{'sh}$- chemical shrinkage strain in the corresponding direction

$\varepsilon_{X,Y,Z}^{T}$- thermal strain

The result of the analysis is node displacement plot (Fig. 2).

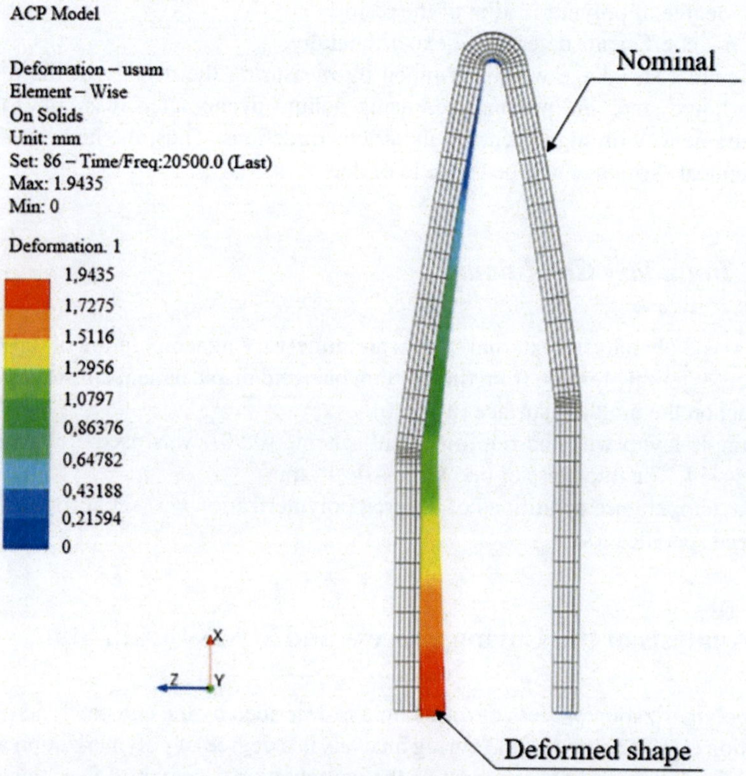

Fig. 2 Deformed and nominal shapes of the sample

It may be noticed that only one half of the model is deformed. This is because the sample was fixed to one of the lower faces to prevent rigid body movement.

3.1 Mesh Convergence

The major parameter of the mesh model is the number of elements through the thickness and the number of elements describing the fillet at the corners (at the top and on the sides of the model). Iteratively, it was found that the number of elements through the thickness and at the fillets of the sides has minor effect on the simulation results. The number of elements dividing the corner fillet of the model vertex has a significant influence on the results.

A series of calculations were performed with a different number of elements in the corner fillet. If the number of elements is set to 27, a further decrease element size changes the maximum distortion of the lower edge less than 0.001 mm. The diagram of lower edge maximum displacement by the number of elements is presented at Fig. 3.

Thus, the lower edges distortion is equivalent to 2.028 mm. The distortion angle is 0.69°.

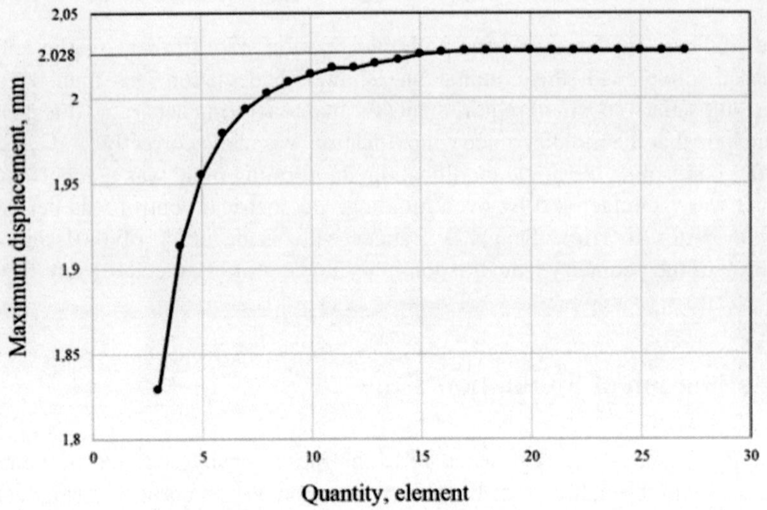

Fig. 3 Dependence of maximum displacement by the number of elements in a corner

4 Production and Geometry Control of Full-Scale Samples

To verify the calculation model, three full-scale samples were made and measured. Production was carried out by the method of vacuum bag molding in oven. The samples were measured using an electronic caliper. As a result, the values of the lower edge displacement were obtained. Average value of the distortion angle of the samples is $\Delta\alpha = 0,71°$, the error of the FE simulation results is:

$$\delta = \frac{0,71 - 0,69}{0,71} \cdot 100\% = 2,8\%$$

This result indicates a good convergence of the FE simulation with experimental data.

4.1 Accounting for Predicted Distortion of the Sample in a Mold Geometry

Based on technological simulation results it is possible to design the forming surface of the mold taking into account the distortion. To do this, the angle increased by 0.69°.

Re-analysis of the geometry with anticipation and further overlay of the deformed model with the nominal one showed a deviation less than 0.1 mm. This result satisfy specified requirements for manufacturing accuracy. It is possible to conclude that the mold geometry modification was made correctly.

After confirming the angle modification, the forming mold was made. Next, the product was manufactured by oven molding. Geometrical control was performed using the Atos II Triple Scan 3D scanner with an accuracy of 0.01 mm. The deviation of the geometry from the nominal value does not exceed 0.1 mm.

5 Estimation of Simulation Time

This type of analysis is very sensitive to the mesh density and size of elements. It requires a high-quality discretization of the geometry in corners. Also, analysis often needs to be performed for different configurations of the same element. It should also be taken into account that the design may contain many different details that require separate analysis. If the product has a complex shape and its dimensions is more than 500 mm, the analysis can take from several hours to several days. In this regard, a computing cluster or various hardware configurations of a workstation can be used to speed up the analysis.

A computing cluster is a combination of computing nodes connected by high-speed communication channels. In the case of solving the problem using the finite element method, the analysis is performed as follows: the Ansys solver splits the FE model into several domains and distributes them between processors. After that, each processor solves a local problem (simultaneously exchanging data with other processors) and sends solution data to the solver. Further, the solver, from the combination of data received from processors, makes a general "picture" of the solution. At the same time, communication between individual computing nodes occurs via a communication network (interconnect), the bandwidth of which has a serious impact on the cluster performance [17].

For this type of calculation, the KNRTU-KAI computing cluster can be used. Cluster based on six-core Intel Xeon "Westmere" X5650 processors including 192 cores, 768 GB of RAM and a performance equal to 510.72 Gflops/s. The cluster is running the Novell SuSe Linux operating system. Also, a high-speed QDR Infiniband network is connected. Analysis speed data is shown in Fig. 4.

The process of numerical simulation on a workstation containing 4 cores (8 threads), 12 GB RAM and an Intel Core i7–8750H processor lasts 135 min. Using a workstation with 4 cores (8 threads), 48 GB RAM, and an Intel Core i7–9700K processor reduces this time to 40 min and 5 s. The estimated analysis time using a supercomputer will take approximately 5 min.

It can be noted that the speed of numerical simulation varies non-linearly depending on the number of cores. This is due to the fact that workstations are used that differ in the type of processors and hardware configuration. Also, it is worth

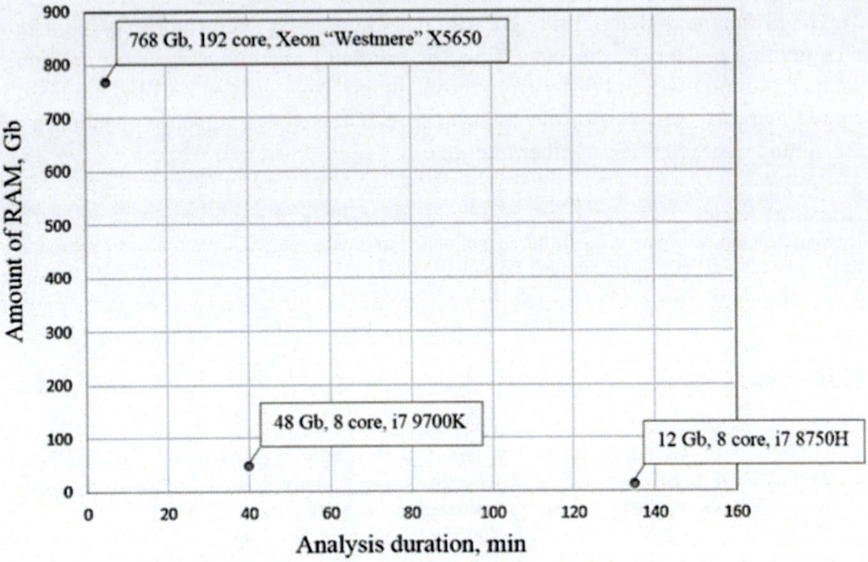

Fig. 4 Analysis speed data

considering that in the case of a computing cluster, there are serious power losses associated with data transfer between computing nodes. In the case of analyzing this numerical experiment, the amount of RAM does not play a role in solution performance, since the model described by a small number of finite elements.

6 Conclusion

A proven approach for finite element modeling of technological stresses in composite products and changes in the geometry of a mold to minimize residual deflection is described.

A comparative analysis of the computing capabilities of various computing stations for this task is carried out.

The described approach for minimizing residual deformations in a composite product has a huge potential in the field of numerical experiments. This approach allows conducting virtual tests to assess the quality of product manufacturing. This leads to a significant increase in manufacturing accuracy, increases production efficiency, reduces the number of defective products and helps to save labor and material resources.

The approach error varies from 0.5% to 6.5% depending on the coordinate measuring machine used, the production technology, and the quality of the mesh model. The error can be obtained from the specific geometric parameter of the product. The conducted research and production of samples showed that numerical modeling can reduce the amount of residual deformations by 69 %.

The problems of determining the residual stress-strain state can be solved using a computing cluster. At the same time, the results of the analysis performed using a supercomputer will not differ from those obtained on a personal workstation with shared memory. Analysis of the solution speed showed a nonlinear dependence of the solution speed on the number of cores.

Acknowledgments The work was supported by the Ministry of education and science of the Russian Federation as part of the fulfillment of obligations under Agreement number 075-03-2020-051/3 dated 09.06.2020 (topic number fzsu-2020-0021).

References

1. Puzyretskiy E.A.: Analiz problemy korobleniya izdeló iz kompozicionnyh materialov. Tekhnika i tekhnologii: puti innovacionnogo razvitiya: sbornik nauchnyh trudov 9-j Mezhdunarodnoj nauchno-tekhnicheskoj konferencii. Kursk. Vol. 2. 107–110 (2020)
2. Huang C.K., Yang S.Y.: Short communication—warping in advanced composite tools with varying angles and radii. Composites Part A. Vol. 28. Issue 9–10. 891–893 (1997)
3. D'Mello R.J., Maiarú M., Waas A. M.: Virtual manufacturing of composite aerostructures. The Aeronautical Journal. Vol. 120. 61–81 (2016)

4. Zhang J.T., Zhang M., Li S.X., Pavier M.J., Smith D.J.: Residual stresses created during curing of a polymer matrix composite using a viscoelastic model. Composites Science and Technology. Vol. 130. 20–27 (2016)
5. Galińska A.: Material Models Used to Predict Spring-in of Composite Elements: a Comparative Study Applied Composite Materials. Vol. 24. 159–170 (2016)
6. Fernlund G., Rahman N., Courdji R., Bresslauer M., Poursartip A., Willden K., Nelson K.: Experimental and numerical study of the effect of cure cycle, tool surface, geometry, and the lay-up on the dimensional stability of autoclave-processed composite parts. Composites Part A: Manufacturing. Vol. 13(3). 341–351 (2002)
7. Fernlund G, Poursartip A.: The effect of tooling material, cure cycle, and tool surface finish on spring-in of autoclave procesed curved composite parts. Proceedings of the 12th International Conference on Composite Materials (ICCM12). paper 690 (1999)
8. Rennick T., Radford D.W.: Components of manufacturing distortion in carbon fiber/epoxy angle brackets. Proceedings of the 28th International SAMPE Technical Conference. 189–197 (1996)
9. Svanberg J.M., Holmberg J.A.: Prediction of shape distortions. Part I. FE implementation of a path dependent constitutive model. Composites: Part A. 35(6):7. 11–21 (2004)
10. Chen W., Zhang D.: Improved prediction of residual stress induced warpage in thermoset composites using a multiscale thermo-viscoelastic processing model. Composites Part A: Applied Science and Manufacturing. Vol. 126 (2019)
11. Chen W., Zhang D.: A Micromechanics-Based Processing Model for Predicting Residual Stress in Fiber-Reinforced Polymer Matrix Compositesû Composite Structures. Vol. 204. 153–166 (2018)
12. Chen W., Li C., Zhang D.: A Multi-Physics Processing Model for Predicting Sping-In Angle of a Resin Transfer Molded Composite Flange. AIAA SciTech Forum. (2018)
13. Ding A., Wang J., Li S.: Computationally efficient pseudo-viscoelastic models for evaluation of residual stresses in thermoset polymer composites during cure. Composites:Part A 41. 247–256 (2010)
14. GOST R 57996-2017. Kompozity polimernye, Differencial'naya skaniruyushchaya kalorimetriya. Opredelenie energii aktivacii, predeksponencial'nogo mnozhitelya i poryadka reakcii. (2017)
15. Kulawik J., Szeglowski Z., Czaplal T., Kulawik J.P.: Determination of glass transition temperature, thermal expansion and, shrinkage of epoxy resins. Colloid and Polymer Science. Vol. 267. 970–975 (1989)
16. Bondarchuk D., Fedulov B.: Process modeling of carbon-epoxy composites: residual stress development during cure and analysis of free edge effects. Aviation. Vol. 23. 15–22 (2019)
17. ZHirkov A.: Superkomp'yutery: razvitie, tendencii, primenenie. Obzor HPC – reshenij Eurotech. ZHurnal CTA (Sovremennye tekhnologii avtomatizacii) 16–20 (2014)

Numerical Analysis of One Two-layer Completely Conservative Difference Scheme of Gas Dynamics in Eulerian Variables with Adaptive Viscosity

Orkhan Rahimly, Yury Poveshchenko, Viktoriia Podryga, and Parvin Rahimly

Abstract For the equations of gas dynamics in Euler variables, using the operator approach, a family of two-layer in time completely conservative difference schemes with space-profiled weighted factors used to approximate the momentum and energy fluxes over time has been constructed and numerically studied. Schemes have a second order of accuracy and are implemented using simple iterative processes. The regularization of the flux terms of the gas dynamics equations using adaptive artificial viscosity is proposed and numerically investigated by the example of the well-known Einfeldt problem. This regularization effectively eliminates unphysical oscillations of the solution, entropy peaks, and preserves the property of complete conservatism of schemes of this class.

1 Introduction

The present study is devoted to the numerical analysis of a family of two-layer in time completely conservative difference schemes (CCDS) with space-profiled time weights for the system of gas dynamics equations in Euler variables using adaptive artificial viscosity [1–3]. The goal of this work was to improve numerical approaches that correctly model the entropy evolution of the system and determine the quantitative characteristics of unstable fluxes in the form of spatially distributed viscous accumulations in a discrete medium.

The regularization of divergent fluxes of momentum mass and internal energy of gas dynamics equations using adaptive artificial viscosity that does not violate the properties of complete conservatism of schemes of this class is proposed. The analysis of the amplitude of this regularization and the possibility of its use on non-uniform grids are considered.

O. Rahimly · Yu. Poveshchenko · V. Podryga (✉) · P. Rahimly
Keldysh Institute of Applied Mathematics of RAS, Moscow, Russia
e-mail: orxan@rehimli.info; pervin@rehimli.info

I. B. Badriev et al. (eds.), *Mesh Methods for Boundary-Value Problems and Applications*, Lecture Notes in Computational Science and Engineering 141,
https://doi.org/10.1007/978-3-030-87809-2_32

Regularized fluxes make the scheme quasimonotonic and ensure coordination of momentum, kinetic and internal energy balances (with correct entropy evolution) while maintaining the property of complete conservatism. Moreover, in the constructed class of divergent difference schemes, following the laws of thermodynamics, there are no approximating permanently acting sources (or sinks) of internal energy. Schemes have a second order of accuracy. The paper describes an approximation to the introduction of artificial viscosity of CCDS. The mechanism of iterative CCDS algorithms with dynamically generated viscous accumulations in a discrete medium is described. Testing of the developed algorithm was carried out on the basis of the well-known Einfeldt problem [4–7] on the propagation of two symmetric rarefaction waves in opposite directions. Numerical calculations with the class of divergent adaptive viscosities developed for CCDS showed a significant improvement in the quality of the obtained approximate solutions in terms of their "high-frequency" monotonization and preservation of entropy properties. Entropic peaks in temperature profiles disappeared with a refinement of the spatial grid.

After some natural generalizations, the developed schemes can be used for gasdynamic calculations with more complex models, for example, in the case of fast processes in plasma with a separation of the temperatures of the electronic and ionic components, when one equation of the total energy balance of the medium is not enough to calculate, as the case, particularly, in plasma dynamics models [8].

2 Completely Conservative Differential Difference Scheme

Omitting the initial system of Euler equations for the flow of a medium (see [2, 3, 9]), we immediately write out the corresponding two-layer in time completely conservative difference scheme in Euler variables. Figure 1 shows the difference grid where ω are the nodes of the difference grid, Ω are its cells. Thermodynamic quantities as density ρ, pressure P, internal energy $E = \rho\varepsilon$, as well as cell volume V and its mass $M = \rho V$ will be attributed to the cells Ω. Velocity $\mathbf{u}$, nodal mass m and volume v will be assigned to nodes ω.

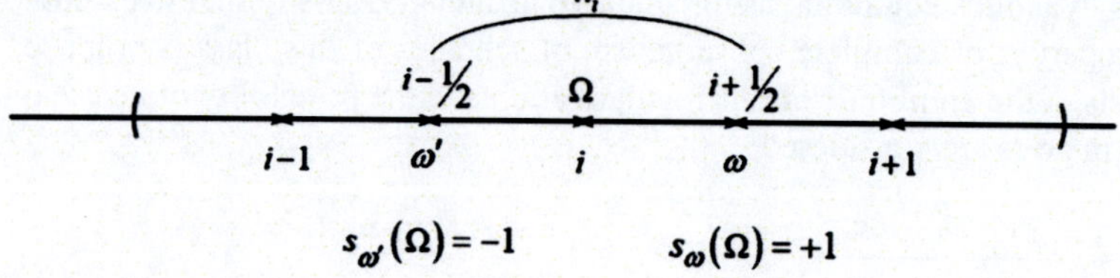

Fig. 1 Difference grid

Obviously: $m_\omega = 0.5 \sum\limits_{\Omega(\omega)} M_\Omega$, $V_\Omega = h_i$, $v_\omega = 0.5 \sum\limits_{\Omega(\omega)} V_\Omega = h_{i+0.5}$ and $\mu_D^{\sim} = 0.5 \sum\limits_{\omega(\Omega)} \mu_\omega^{\sim}$, $\rho_{v\omega} = \frac{m_\omega}{v_\omega} = \rho_{vi+0.5}$, where $\mu_\omega^{\sim}$ and $\mu_D^{\sim}$ respectively introduced nodal and cell mass fluxes ($\boldsymbol{\mu} = \rho\mathbf{u}$). Also, by the momentum assigned to the node, we mean the quantity $I_\omega = \rho_{v\omega} u_\omega$.

Next for the continuous operations of vector analysis $div\,\mathbf{u}$, $grad P$, $div(\boldsymbol{\mu} \cdot \mathbf{u})$ we introduce their difference analogues $DIV : (\omega) \rightarrow (\Omega)$, $GRAD : (\Omega) \rightarrow (\omega)$, to approximate transfer processes $DIV_D : (\Omega) \rightarrow (\omega)$ and divergence of the dyad $DIT_D : (\Omega) \rightarrow (\omega)$. Accordingly, we get:

$DIV\mathbf{u} = \frac{1}{V} \sum\limits_{\omega(\Omega)} s_\omega(\Omega) u(\omega)$, $GRADP = \frac{\Delta P}{v}$,

where $\Delta P = - \sum\limits_{\Omega(\omega)} s_\omega(\Omega) P_\Omega + S_{\partial\omega} P_{\partial\omega}$,

$DIV_D\, \boldsymbol{\mu}_D = -\frac{1}{v} \sum\limits_{\Omega(\omega)} s_\omega(\Omega) \mu_D(\Omega)$,

$DIT_D(\boldsymbol{\mu}_D \cdot \mathbf{u}_D) = -\frac{1}{v} \sum\limits_{\Omega(\omega)} s_\omega(\Omega) \mu_D(\Omega) \mathbf{u}_D(\Omega)$.

In the expression for ΔP, if the node $\omega = \partial\omega$ is a boundary one, the term with a value $P_{\partial\omega}$ at the boundary and a sign function $S_{\partial\omega} = \pm 1$ depending on the direction of the boundary normal is added.

We write out a completely conservative [9] difference scheme in Euler variables:

$$m_t = -vDIV_D \boldsymbol{\mu}_D^{\sim} \tag{1}$$

$$(mu)_t = -vGRADP^{\sim} - vDIT_D(\boldsymbol{\mu}_D^{\sim} \cdot \mathbf{u}_D^{\sim}) \tag{2}$$

$$(M\varepsilon)_t = -P^{\sim} VDIV\mathbf{u}^{\sim}) - VDIV[(\rho\varepsilon\mathbf{u})_\omega^{\sim}] \tag{3}$$

$$(m\frac{\mathbf{u}^2}{2})_t = -v(u^{\sim}, GRADP^{\sim}) - vDIV_D(\boldsymbol{\mu}_D^{\sim} \frac{\mathbf{u}_D^{2\sim}}{2}) \tag{4}$$

By $\boldsymbol{\mu}_{E\omega} = (\rho\varepsilon\mathbf{u})_\omega$ we mean some approximation of the internal energy flux in a node ω. Also in the cell formed by the nodes ω and ω', the quantities: $\mathbf{u}_D^{\sim} = 0.5(\mathbf{u}_\omega^{(\delta_\omega)} + \mathbf{u}_{\omega'}^{(\delta_{\omega'})})$, $\mathbf{u}_D^{2\sim} = 0.5(\mathbf{u}_\omega^{(\delta_\omega)}, \mathbf{u}_{\omega'}^{(\delta_{\omega'})})$ are introduced.

On time layers t and $\hat{t} = t + \tau$ ($\tau > 0$ is time step), time differential derivatives and spatially point (i.e., in grid nodes ω) time interpolations are introduced as $a_t = (\hat{a} - a)/\tau$, $a^{(\delta)} = \delta\hat{a} + (1 - \delta)a$.

Here, the interpolation weight δ may depend on the spatial mesh node ω, for example, according to the law: $\delta = \sqrt{\hat{m}}/(\sqrt{\hat{m}} + \sqrt{m})$.

Also, by arbitrary time interpolation of the grid functions a, $\hat{a}$ between the layers t and $\hat{t}$, we mean a certain grid value $a^{\sim}$.

3 Approximation and Introduction of Artificial Viscosity

We introduce a one-dimensional non-uniform grid over cells $\Omega_i \cup \partial\Omega$ and nodes $\omega_{i+0.5} \cup \partial\omega$ along a spatial variable $x_{i+0.5}$ (see Fig. 2).

The symbol ∂ in Fig. 2 identifies boundary cells and nodes. $\omega^0 = \omega/\partial\omega$, $\Omega^0 = \Omega/\partial\Omega$.

The equations for density ρ and internal energy E are reduced to the form: $A_i y_{i-1} - C_i y_i + B_i y_{i+1} = -F_i$, $i = 1, \ldots, N-1$, $y_0 = a_1 y_1 + b_1$, $y_N = a_2 y_{N-1} + b_2$ and are solved by the modified Newton method in combination with the sweep algorithm. In Eqs. (1)–(4) on the right-hand side, the implicitness is embedded in nodal and cell mass fluxes. After applying the Newton method and introducing the notation for the increments $\delta y_i = y_i^{s+1} - y_i^s$, omitting the calculations, we immediately write out the running coefficients (A, B, C, F) for the $\rho -$, $u -$ and $E -$ iterative groups. In this case, the upper symbol $\approx$ will indicate that the time dependence of the value on the implicit layer is taken at a known s-th iteration.

$$A_{\rho i} = \beta_{i-0.5} h_{i-0.5}, \quad B_{\rho i} = \beta_{i+0.5} h_{i+0.5}, \quad C_{\rho i} = h_i + A_{\rho i} + B_{\rho i}, \quad F_{\rho i} = -f^{\approx}_{\rho i}.$$

Similarly for velocity u we have:

$$A_{u_{i+0.5}} \delta_\Delta u_{i-0.5} - C_{u_{i+0.5}} \delta_\Delta u_{i+0.5} + B_{u_{i+0.5}} \delta_\Delta u_{i+1.5} = -F_{u_{i+0.5}},$$

$$A_{u_{i+0.5}} = \Bigg\{ -\tau \Bigg\{ -\Bigg[-\beta_i \frac{h_i}{\tau}(-\rho^{s+1}_{vi-0.5}) \Bigg]\Bigg\} - \tau[-\{\frac{1}{2}\{ \Big[-K_{i+0.5}(\rho^{s+1}_{i+1} - \rho^{s+1}_i)\Big] + \Big[-K_{i-0.5}(\rho^{s+1}_i - \rho^{s+1}_{i-1})\Big] \Big\} \frac{1}{2} \delta^{s+1}_{i-0.5}\}] \Bigg\} / \rho^{s+1}_{vi-0.5},$$

$$C_{u_{i+0.5}} = \Bigg\{ m^{s+1}_{i+0.5} + \tau \Bigg\{ [-\beta_{i+1}\frac{h_{i+1}}{\tau}(-\rho^{s+1}_{vi+0.5})] - \Bigg[-\beta_i \frac{h_i}{\tau}(\rho^{s+1}_{vi+0.5})\Bigg]\Bigg\} + \tau\Bigg[\Bigg\{\frac{1}{2}\{\Big[-K_{i+1.5}(\rho^{s+1}_{i+2} - \rho^{s+1}_{i+1})\Big] + \Big[-K_{i+0.5}(\rho^{s+1}_{i+1}-\rho^{s+1}_i)\Big]\Big\}\frac{1}{2}\delta^{s+1}_{i+0.5}\Bigg\} -$$

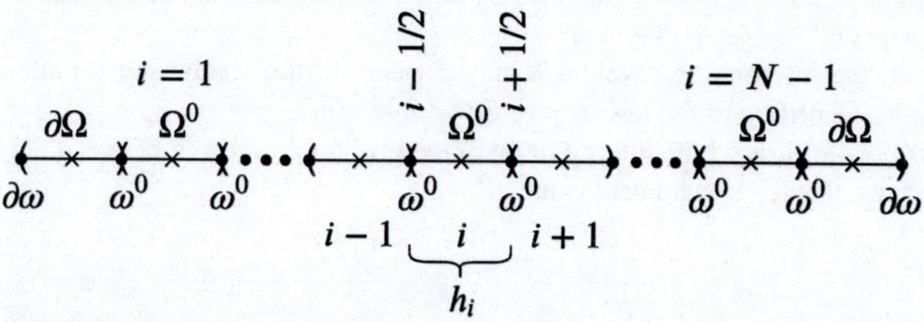

Fig. 2 One-dimensional non-uniform grid

$$-\left\{\frac{1}{2}\{\left[-K_{i+0.5}(\rho_{i+1}^{s+1}-\rho_i^{s+1})\right]+\left[-K_{i-0.5}(\rho_i^{s+1}-\rho_{i-1}^{s+1})\right]\}\frac{1}{2}\delta_{i+0.5}^{s+1}\right\}\Bigg]\Bigg\}/\rho_{vi+0.5}^{s+1},$$

$$B_{u_{i+0.5}}=\left\{-\tau\left\{[-\beta_{i+1}\frac{h_{i+1}}{\tau}(\rho_{vi+1.5}^{s+1})]\right\}-\tau[\{\frac{1}{2}\{\left[-K_{i+1.5}(\rho_{i+2}^{s+1}-\rho_{i+1}^{s+1})\right]+\right.$$

$$\left.+\left[-K_{i+0.5}(\rho_{i+1}^{s+1}-\rho_i^{s+1})\right]\}\frac{1}{2}\delta_{i+1.5}^{s+1}\}]\right\}/\rho_{vi+1.5}^{s+1},$$

$$F_{u_{i+0.5}}=-f_{ui+0.5}^{\approx},\quad K_{i+0.5}=\beta_{i+0.5}\frac{h_{i+0.5}}{\tau}.$$

For internal energy E we have:

$$A_{Ei}=\beta_{Ei-0.5}h_{i-0.5},\quad B_{Ei}=\beta_{Ei+0.5}h_{i+0.5},\, C_{Ei}=h_i+A_{Ei}+B_{Ei},\quad F_{Ei}=-f_{Ei}^{\approx},$$

where the notation is used

$$f_{\rho_i}=h_i(\hat{\rho}_i-\rho_i)+\tau(\mu_{i+0.5}^{\sim}-\mu_{i-0.5}^{\sim}),$$

$$f_{u_{i+0.5}}=(\hat{I}_{i+0.5}-I_{i+0.5})+\tau\{(P_{i+1}^{\sim}-P_i^{\sim})+[\mu_{D_{i+1}}^{\sim}u_{D_{i+1}}^{\sim}-\mu_{D_i}^{\sim}u_{D_i}^{\sim}]\},$$

$$f_{E_i}=h_i(\hat{E}_i-E_i)+\tau\{P_i^{\sim}[u_{i+0.5}^{\sim}-u_{i-0.5}^{\sim}]+[\mu_{E_{i+0.5}}^{\sim}-\mu_{E_{i-0.5}}^{\sim}]\},$$

$$\delta_\Delta u_{i+0.5}=\rho_{vi+0.5}^{s+1}\delta u_{i+0.5},$$

$$\delta_{i+0.5}^{s+1}=\sqrt{m_{i+0.5}^{s+1}}/(\sqrt{m_{i+0.5}^{s+1}}+\sqrt{m_{i+0.5}}).$$

Here, β is an adaptive viscous accumulation coefficient. In the expression $f_{ui+0.5}^{\approx}$ and for both the terms $p(\rho^{s+1},E)$ and $\delta(\rho^{(s+1)})$ in the velocity group on the implicit layer by time ρ is taken at the iteration $s+1$, but u, E at the s-th iteration. Also in the energy block for $f_{Ei}^{\approx}$, $p(\rho^{s+1},E)$ and $\delta(\rho^{(s+1)})$ on the implicit layer in time, ρ and u are taken at the iteration $s+1$.

Adaptive viscosity with coefficients $\{\nu,\nu_E,\nu_I\}$ is presented in Eqs. (1)–(4) in nodal mass fluxes $\mu_\omega=u_\omega\rho_\omega-(\nu\cdot GRAD\ \rho)_\omega$ (here ρ_w is the nodal approximation of density), as well as an additive to the transferred internal energy in the nodes $\boldsymbol{\mu}_{E\omega}=(\rho\varepsilon\mathbf{u})_\omega-(\nu_E\cdot GRAD\ E)_\omega$, and as an additive to pressure $P_\Omega-(\nu_I DIV(\rho_v^{s+1}\mathbf{u}))$. These viscosities $\{\nu,\nu_E,\nu_I\}$ are proportional, taken on an implicit layer in time $\hat{t}$, while other terms are taken with a time-symmetric approximation with weight 0.5.

Viscous diffusion coefficients are selected as follows:

$$\left(\frac{\nu}{h}\right)_\omega=\beta_\omega\frac{h_\omega}{\tau},\quad\left(\frac{\nu_E}{h}\right)_\omega=\beta_{E\omega}\frac{h_\omega}{\tau},\quad\left(\frac{\nu_I}{h}\right)_\Omega=\beta_\Omega\frac{h_\Omega}{\tau},$$

where

$$\beta_\omega = l^{\approx}_{\nu\omega} q_\nu^{n_{\nu\omega}} k_{r\omega}, \ \ \beta_{E\omega} = l^{\approx}_{\nu E\omega} q_{\nu E}^{n_{\nu E\omega}} k_{r\omega}, \ \ \beta_\Omega = l^{\approx}_{u\Omega} q_{\nu u}^{n_{\nu\Omega}} k_{r\Omega}.$$

Here, the Courant numbers at the nodes and the cell are introduced as:

$$k_{r\omega} = \frac{|u_\omega| + \varepsilon}{2} \frac{\tau}{h_\omega}, \ \ k_{r\Omega} = \frac{|u^s_\Omega| + \varepsilon}{2} \frac{\tau}{h_\Omega},$$

where ε is a small addition to the Courant number, which is significant at a velocity close to zero. u^s_Ω is a cellular interpolation of velocity at a known s-th iteration. Understanding by the viscous accumulation β in the scheme the quantity $\{\beta_\omega, \beta_{E\omega}, \beta_\Omega\}$, and by the viscosity coefficient ν one of the quantities $\{\nu, \nu_E, \nu_I\}$ and $l^{\approx} = \{l^{\approx}_\nu, l^{\approx}_{\nu E}, l^{\approx}_u\}$, we can briefly write:

$$\frac{\nu}{h} = \beta \frac{h}{\tau}, \ \ \beta = l^{\approx} q^n kr, \ \ kr = \frac{|u| + \varepsilon}{2} \frac{\tau}{h}, \ \ q > 1, \ n = 0, 1, 2, \ldots .$$

The adaptive viscous accumulation coefficients β include a template grid functional $l^{\approx}$ equal to 1 in the presence of artificial viscosity or 0 in its absence, depending on the criterion used to monotonize density ρ, internal energy E, and momentum $I = \rho_\nu u$ in the corresponding nodes ω and cells Ω.

At the initial local switching-on the artificial viscosity (due to the arising nonmonotonicity), it is assumed that $l^{\approx} = 1$, $n = 0$, which means "viscous" upwind approximation of the corresponding transfer process. Further, locally viscous accumulation β can increase with increasing n if the nonmonotonicity in the cell or node does not disappear. Values $\nu, h, \beta, l^{\approx}, n, kr$ have a spatially local meaning, while the magnitudes q and values of the limiters $\beta_0 = \{\beta_{\rho 0}, \beta_{E0}, \beta_{u0}\}$ (i.e. $\beta < \beta_0$) for density, internal energy, and momentum are global attributes of the problem throughout the grid.

4 About the Iterative CCDS Algorithm Mechanism

In the iterative algorithm used, to calculate the increments of physical quantities on the implicit layer in time $\hat{t}$, in addition to the non-stationary terms in Eqs. (1)–(3), only terms proportional to viscosities $\{\nu, \nu_E, \nu_I\}$ are taken in the implicit $s + 1$-th iteration. Therefore, the convergence of the iterative algorithm is determined by Courant numbers kr associated with its explicit iterative part. The correctness of the coefficients A, B, C in separate runs for increments $\delta\rho$, $\delta_\Delta u$ and δE is determined by viscous accumulations $\{\beta_\omega, \beta_{E\omega}, \beta_\Omega\}$. Thus, in general, at each s iteration, the algorithm consists of three groups dependent on each other. Besides, special blocks for monotonizing (smoothing) the solution work for every group. In each of the groups, the necessary conditions for non-negativity ($A \geq 0, B \geq 0, C > 0$) and diagonal predominance ($C - A - B > 0$) of running coefficients must be satisfied.

Before the initial iteration ($s = 0$), $l^{\approx} \equiv 0, k \equiv 1$ are set locally over the grid. Here, control parameter $k = \{k_{\nu\omega}, k_{\nu E\omega}, k_{\nu u\Omega}\}$ can take three values locally in corresponding node ω or cell Ω: $k = 1$ means that change of the corresponding β is allowed; $k = 0$ means that β can only be reduced; $k = -1$ means that we can only reduce β and until the iterations converge on the time layer $\hat{t}$ the parameter $k = -1$ is read-only. Before non-initial iterations ($s > 0$), if locally $k \geq 0$, then the corresponding $l^{\approx} = 0$ and $k = 1$ are set. Otherwise (at $k = -1$) corresponding values $l^{\approx}$ and β are not changed.

The block diagram of s-th iteration consists of the following groups:

$(< \rho^{s+1} - calculation >, < \beta_\rho - accumulator >), < \beta_{\rho u} - corrector >,$

$(< u^{s+1} - calculation >, < \beta_u - accumulator >),$

$(< E^{s+1} - calculation >, < \beta_E - accumulator >).$

$\{< \rho^{s+1} - calculation >:<$ Density ρ calculation on the $s+1$-th layer $>\}$.

$\{< \beta_\rho - accumulator >:<$ Adaptive viscous accumulation β_ω is formed, not exceeding the limiter $\beta_{\rho 0}$, with the control parameter $k_{\nu\omega} \neq 1$ if $\beta_\omega = \beta_{\rho 0}$ (with repeating $< \rho^{s+1} - calculations$), so that as a result there is no nonmonotonicity in the profile $\rho^{s+1} >\}$.

$\{< \beta_{\rho u} - corrector >:$

$<$ 1. It ensures the fulfillment of diagonal dominance $D_{ui+0.5} = C_{ui+0.5} - A_{ui+0.5} - B_{ui+0.5} > 0$ in nodes ω with a possible proportional decrease of the time step τ and all viscous accumulations β. When the time step is reduced at the nodes ω, where the condition $D_{ui+0.5} > 0$ is violated, $k_{\nu\omega} = -1$ is set, the values ρ, u, E at the s-th iteration are assigned the values from the explicit t - layer and the return to $< \rho^{s+1} - calculation >$ is performed$>$.

$<$ 2. If there were $l^{\approx}_{u\Omega} = 0$ in the cells $(\Omega(\omega^0))$ around the nodes ω^0, where $l^{\approx}_{\nu\omega} = 1$, then they rely on $l^{\approx}_{u\Omega} = 1, n_{\nu\Omega} = 0 >$.

$<$ 3. Due to the selection of viscous accumulations β_Ω and β_ω, the non-negativity of the running coefficients $A_{ui+0.5} \geq 0$, $B_{ui+0.5} \geq 0$ is ensured with a possible local achievement of the limiters ($\beta_\Omega = \beta_{u0}, k_{\nu u\Omega} \neq 1$) and a proportional decrease of the time step τ with viscous accumulations $\{\beta_\omega, \beta_{E\omega}\}$. At nodes ω, in which the values β_ω led to negative $A_u < 0$ and $B_u < 0$, $k_{\nu\omega} = -1$ is set. When the time step τ decreases, the values ρ, u, E at the s-th iteration are assigned the values from the explicit t - layer and the return to $< \rho^{s+1} - calculation >$ is performed$>\}$.

Iterative blocks $(< u^{s+1} - calculation >, < \beta_u - accumulator >)$ and $(< E^{s+1} - calculation >, < \beta_E - accumulator >)$ with monotonization in momentum $I = \rho_\nu u$ and internal energy E with the formation of viscous accumulations β_Ω and $\beta_{E\Omega}$ are similar to the block described above for density $(< \rho^{s+1} - calculation >, < \beta_\rho - accumulator >)$. If upon reaching the maximum possible number $nsmax$ the iterations did not converge, then a proportional decrease of the time step τ and all viscous accumulations β occurs. Assumed $k = -1$, the values ρ, u, E at the known s-th iteration are assigned values from the explicit t -

layer and the return to $< \rho^{s+1} - calculation >$ is performed. If the convergence criterion $\{|\delta\rho| < \varepsilon_1|\rho^s| + \varepsilon_2, |\delta I| < \varepsilon_1|I^s| + \varepsilon_2, |\delta E| < \varepsilon_1|E^s| + \varepsilon_2\}$ is satisfied, then iterations are stopped and relied upon $\{\hat{\rho} = \rho^{s+1}, \hat{I} = I^{s+1}, \hat{E} = E^{s+1}\}$.

5 Algorithm Testing

To test the algorithm, the Einfeldt problem [4–7] was chosen on the propagation of two symmetric rarefaction waves in opposite directions. Here, the feature is related to the behavior of internal energy on a numerical solution. The problem is solved as a special case of the disintegration problem. The segment $[-1, 1]$ is used as the computational domain. The gap is located in the center of this segment at the point $x = 0$. The initial conditions are presented in Table 1. SI is taken as the system of units of measurement.

Over time, an expanding fixed section (plateau) is formed in the center of the region with constant values of gas density and pressure, which are quite small. Since the state equation of an ideal gas is satisfied, the specific internal energy remains constant in this region, and the entropy also remains constant in the entire computational domain at $t > 0$ (isoentropic process). Numerical solutions of this problem based on many well-known methods unsatisfactorily convey the behavior of specific internal energy. We show that the algorithm proposed above significantly improves the approximation of these thermodynamic quantities in comparison with the most well-known methods.

The analitical and numerical solutions of the problem under consideration in the Euler variables are shown in Figs. 3, 4, 5, and 6 for $N = 2000$ calculated cells and Fig. 7 for $N = 500$. Blue and orange curves correspond to analytical and numerical solutions of the corresponding quantities. Calculations were performed for $N = 500, 1000$ and $N = 2000$ points. In the first three figures, there is more than a good agreement between the analytical and numerical solutions, not only for 2000 but also for $N = 500$ and 1000. The temperature graphs (see Figs. 6 and 7) are of particular interest. In Fig. 7, a noticeable deviation of the numerical solution in the vicinity of the plateau is observed, but it resolves fairly well when compacting the mesh (see Fig. 6). In most of the numerical algorithms currently used, conservative variables are traditionally used as such, while the amplitude of the entropy peak is about 70% of the solution. One of the most well-known works on optimizing the entropy wake is the work [7] using the discrete Galerkin method, where a noticeable improvement in the entropy peak is observed, but the effect of the final smoothing of the entropy peak is not observed when the mesh is refined to cells $N = 5000$.

Table 1 Initial conditions

Left side ($x < 0$)			Right side ($x > 0$)		
ρ	u	p	ρ	u	p
1	−2	0.4	1	2	0.4

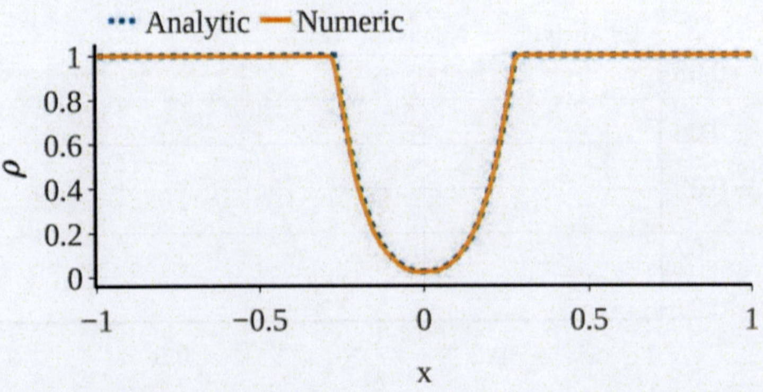

Fig. 3 Density distribution, $N = 2000$

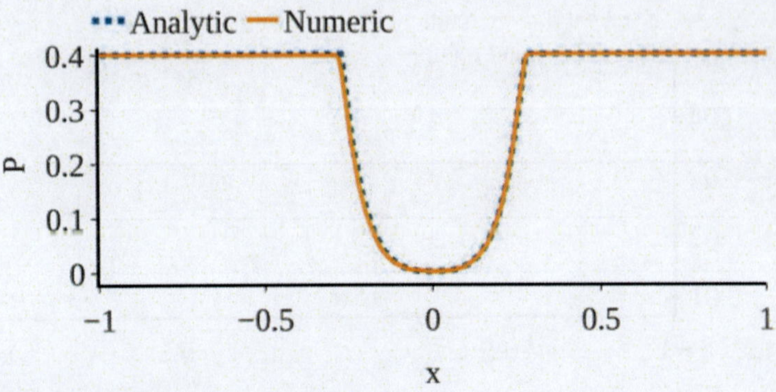

Fig. 4 Pressure distribution, $N = 2000$

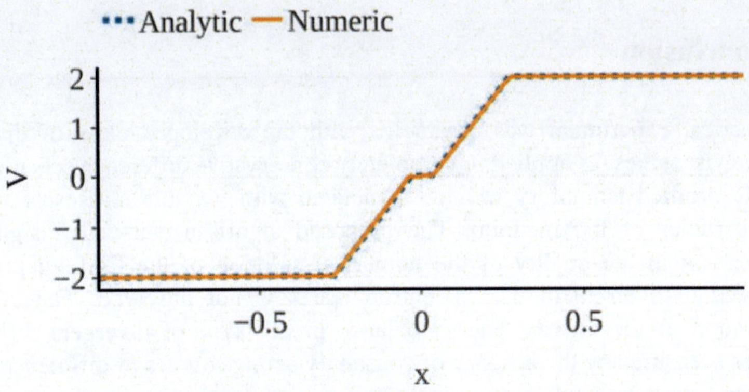

Fig. 5 Velocity distribution, $N = 2000$

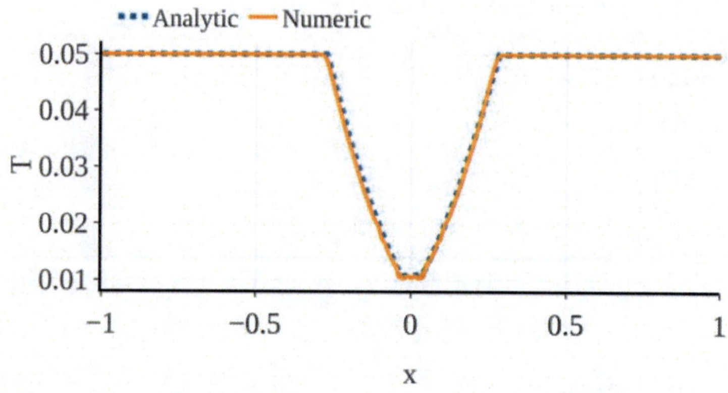

Fig. 6 Temperature distribution,$N = 2000$

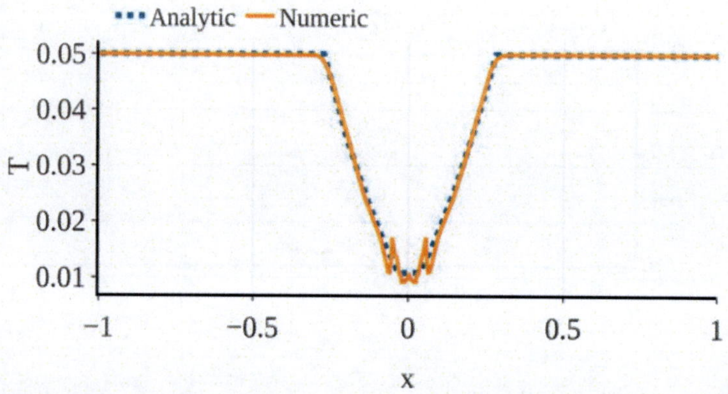

Fig. 7 Temperature distribution, $N = 500$

6 Conclusion

A numerical experiment was conducted with the developed class of divergent adaptive viscosities, as applied to completely conservative difference schemes with spatially profiled temporary weights associated with variable masses of moving nodal particles of the medium. The proposed algorithm showed a significant improvement in the quality of the numerical solution of the Einfeldt problem. Compared to other methods, the entropy peak is not observed. The effective preservation of the internal energy balance in this type of divergent difference schemes is ensured by the absence of constantly acting sources of difference origin producing computational entropy (including on singular features of the solution).

Acknowledgments The work was supported by the Russian Science Foundation (project No 17-71-20118-P).

References

[1] Popov, I.V., Fryazinov, I.V.: Method of Adaptive Artificial Viscosity for the Numerical Solution of Gas Dynamics Equations. Krasand, Moscow (2015) [in Russian]

[2] Poveschenko, Yu.A., Ladonkina, M.Y., Podryga, V.O., Rahimly, O.R., Sharova, Yu.S.: On a two-layer completely conservative difference scheme of gas dynamics in Euler variables with adaptive regularization. Preprints of the Keldysh Institute of Applied Mathematics **14** (2019) [In Russian]

[3] Rahimly, O., Podryga, V., Poveshchenko, Yu., Rahimly, P., Sharova, Yu.: Two-layer completely conservative difference scheme of gas dynamics in Eulerian variables with adaptive regularization of solution. In: Lirkov, I., Margenov, S. (eds) Large-Scale Scientific Computing. LSSC 2019. Lecture Notes in Computer Science, **11958**, pp. 618–625. Springer, Cham (2020)

[4] Bragin, M.D., Kriksin, Yu.A., Tishkin, V.F.: Ensuring the entropy stability of the discontinuous Galerkin method in gas-dynamic problems. Preprints of the Keldysh Institute of Applied Mathematics **51** (2019) [in Russian]

[5] Kriksin, Yu.A., Tishkin, V.F.: Variational entropy regularization of the discontinuous Galerkin method for equations of gas dynamics. Mathematical modeling **31**(5), 69–84 (2019) [in Russian]

[6] Cockburn, B.: An introduction to the discontinuous Galerkin method for convection-dominated problems. In: Lecture Notes in Mathematics. **1697**, pp. 150–268 (1997)

[7] Kriksin, Yu.A., Tishkin, V.F.: Numerical solution of the Einfeldt problem based on the discontinuous Galerkin method. Preprints of the Keldysh Institute of Applied Mathematics **90** (2019) [in Russian]

[8] Morozov, A.I.: Introduction to Plasma Dynamics. Fizmatlit, Moscow (2006) [in Russian]

[9] Popov, Yu.P., Samarskii, A.A.: Completely conservative difference schemes. USSR Comput. Math. Math. Phys. **9**(7), 296–305 (1969)

Accurate Simulation of On-Threshold Modes of Microcavity Lasers with Active Regions Using Galerkin Method

Anna I. Repina, Alina O. Oktyabrskaya, Ilya V. Ketov, and Evgenii M. Karchevskii

Abstract The current paper investigates a parametric eigenvalue problem for the Helmholtz equation on the plane specially tailored for accurate mathematical modeling of lasing modes of 2-D microcavity lasers with active regions. We reduce the original problem to a nonlinear eigenvalue problem for a system of boundary integral equations (BIEs) with weakly singular kernels known as Muller BIEs. For a less complicated problem for fully active lasers, it is known that there is no full spectral equivalence between the original problem and the eigenvalue problem for the system of Muller BIEs. In the present work, we clarify the connection between the spectra of the more complicated problem for microcavity lasers with active regions and the eigenvalue problem for the system of Muller BIEs. After that, for the numerical solution of the obtained nonlinear problem, we propose a Galerkin method, prove its convergence, and derive error estimates in the eigenvalue approximation. Previous numerical experiments show that holes of eccentric microring lasers located in well-defined places contribute to a significant increase in the directivity of lasing emission. In this paper, we strive to obtain the highest directivity possible while maintaining low lasing thresholds, for this purpose, we vary the radius of the hole.

A. I. Repina (✉)
Department of System Analysis and Information Technologies, Institute of Computational Mathematics and Information Technologies, Kazan Federal University, Kazan, Russia

A. O. Oktyabrskaya · I. V. Ketov · E. M. Karchevskii
Department of Applied Mathematics, Institute of Computational Mathematics and Information Technologies, Kazan Federal University, Kazan, Russia

I. B. Badriev et al. (eds.), *Mesh Methods for Boundary-Value Problems and Applications*, Lecture Notes in Computational Science and Engineering 141,
https://doi.org/10.1007/978-3-030-87809-2_33

1 Introduction

Various two-dimensional (2-D) microdisk and microring lasers (see, e.g., [1, 2]) can be investigated with the aid of an electromagnetic eigenvalue problem assigned to calculate the threshold values of gain in addition to the emission frequencies and called the Lasing Eigenvalue Problem (LEP) [3–6]. For 2-D microcavity lasers with uniform gain, LEP was reduced to a nonlinear spectral problem for the system of Muller boundary integral equations (BIEs) (see [7] and references therein). For 2-D homogeneous dielectric objects with smooth boundaries, the system obtained by Muller in [8] is widely used for the analysis of electromagnetic fields. Similarly, the eigenmodes of fully active [7] and passive [9] microcavities can be calculated using Muller BIEs. Many authors have described and used a physical model called the Complex-Frequency Eigenvalue Problem (CFEP) [7]. It is based on the search of complex-valued natural frequencies of open passive resonators. To be able to build a general theory for both LEP and CFEP models, a generalized model was proposed in [7]. It has received the following name: Generalized Complex-Frequency Eigenvalue Problem (GCFEP) [7]. The reason for reducing GCFEP to the Muller BIEs was to get a system of weakly singular integral equations [10] on the boundary of the microcavity laser. But there was no full equivalence between GCFEP and the eigenvalue problem for the system of Muller BIEs [11]. Namely, it was proved in [12] that for each eigenfunction of GCFEP there is a corresponding eigenvector of the system of Muller BIEs. But the assertion in the opposite direction is not true: there is one more problem that is reduced to the Muller BIEs. It was called GCFEP "turned inside out" [12]. GCFEP is equivalent to the eigenvalue problem for the system of Muller BIEs on a domain in the complex plane if GCFEP turned inside out does not have eigenvalues in this domain. If GCFEP and GCFEP turned inside out together have only the trivial solutions, then the system of Muller BIEs has only the trivial solution [12], and the resolvent set of the corresponding operator-valued function is not empty. This result is important for the theoretical investigation of the spectrum of the eigenvalue problem. Using it and fundamental results of the theory of projection methods for holomorphic Fredholm operator-valued functions [13, 14], the convergence of a Nystrom method was proved in [7].

Recently, a modified version of the Muller BIEs, together with a trigonometric Galerkin discretization technique, has been proposed for numerical simulation of more complicated microcavity lasers with active regions [15–18]. Mathematically, this means that there is an additional region inside the computational domain, and hence, an additional boundary in the integral formulation. It makes the theoretical analysis more difficult compared to [7, 11], and [12], where problems with one boundary were investigated, as it was proposed originally by Muller [8].

The main idea of the present work is to conduct a thorough mathematical study of GCFEP for lasers with active regions as well as to provide a rigorous proof of convergence of the Galerkin method. First of all, we clarify the connection between GCFEP and the eigenvalue problem for the system of Muller BIEs in this complicated situation. Our consideration is based on the fundamental results of the theory of holomorphic operator-valued functions in a pair of Banach spaces. After that, using the Galerkin method, we build a sequence of finite-dimensional holomorphic operator-valued functions that regularly approximates the original holomorphic Fredholm operator-valued function. This enables us to apply the results of [13, 14] to the numerical analysis of the proposed method.

Based on the presented numerical approach, we investigate directivities, spectra, and thresholds of laser modes of eccentric microring lasers. For circular microcavity lasers with non-concentric circular active regions, we obtain explicit expressions for the matrix elements [18]. This makes calculations much faster. Analysis of numerical experiments demonstrates the exponential convergence of the Galerkin method [18]. Our numerical results coincide well with exact solutions previously obtained by the method of separation of variables for circular microcavities with concentric circular holes [18]. The computational experiments show that holes located at certain places and having suitable radiuses can lead to a notable growth of the directivity of the lasing emission with the conservation of the low thresholds [18]. In this work, numerical results were obtained for holes of a relatively small radius. In the current paper, we demonstrate similar effects for a hole with a relatively large radius, which corresponds to the physical experiments described in [19].

2 GCFEP and Turned Inside Out GCFEP

The formulation of GCFEP for 2-D microcavity lasers with active regions is given in [15]. The geometry of enquired microcavities is shown in Fig. 1. The hole is designated by the domain Ω_1, the active region of the resonator is designated by Ω_2, and the environment of the resonator is Ω_3. The boundaries Γ_1 and Γ_2 separate

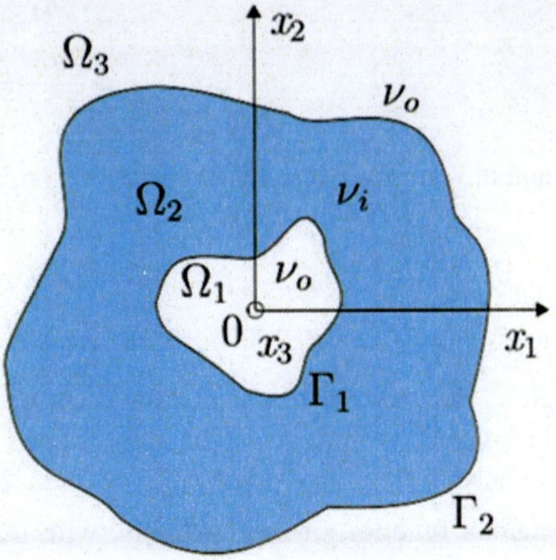

Fig. 1 Geometry of a 2-D microcavity laser with an active region

these regions. We suppose that the boundaries Γ_1 and Γ_2 are twice continuously differentiable, and n_1 and n_2 are the outer normal unit vectors to them, respectively.

We assume that a positive refractive index ν_o of the hole Ω_1 and the environment around the resonator Ω_3 is given. The complex-valued refractive index of the domain Ω_2 is $\nu_i = \alpha_i - i\gamma$. We denote the given real part of ν_i by $\alpha_i > 0$ and the parameter of GCFEP, which is real-valued, by $\gamma \in \mathbb{R}$. We take $\gamma = 0$, when the cavity is passive and without losses, another case is lossy cavities with $\gamma < 0$. The last case is when the region Ω_2 is filled in with a gain material, then $\gamma > 0$.

We assume that the electromagnetic field does not depend on the variable x_3 and depends on time as $\sim exp(-ikct)$. Herein, the speed of light in a vacuum is designated by c. As far as the wavenumber k is the eigenvalue of GCFEP, we guess that it is complex-valued. We are looking for values of k on the Riemann surface $\mathbb{L}$ of the function $\ln k$ (following [7]). Due to the independence of the electromagnetic field on the x_3 variable, we are dealing with the scalar eigenfunctions of GCFEP $u \in U \setminus \{0\}$, each of which is the third element of the density vector E or H for E- and H-polarization, respectively. We use the designation U of the space of functions which are complex-valued and continuous on $\overline{\Omega}_1$, $\overline{\Omega}_2$, and $\overline{\Omega}_3$ and twice continuously differentiable on Ω_1, Ω_2, and Ω_3.

For each $\gamma \in \mathbb{R}$, the eigenvalues $k \in \mathbb{L}$ and the eigenfuctions $u \in U \setminus \{0\}$ of GCFEP have to satisfy the Helmholtz equations,

$$\Delta u + k_o^2 u = 0, \quad x \in \Omega_1, \tag{1}$$

$$\Delta u + k_i^2 u = 0, \quad x \in \Omega_2, \tag{2}$$

$$\Delta u + k_o^2 u = 0, \quad x \in \Omega_3, \tag{3}$$

the transmission conditions,

$$u^- = u^+, \quad \eta_o \frac{\partial u^-}{\partial n_1} = \eta_i \frac{\partial u^+}{\partial n_1}, \quad x \in \Gamma_1, \tag{4}$$

$$u^- = u^+, \quad \eta_i \frac{\partial u^-}{\partial n_2} = \eta_o \frac{\partial u^+}{\partial n_2}, \quad x \in \Gamma_2, \tag{5}$$

and the outgoing Reichardt radiation condition [7, 19],

$$u(\rho, \varphi) = \sum_{l=-\infty}^{\infty} a_l H_l^{(1)}(k_o \rho) \exp^{il\varphi}, \quad \rho \geq R_0. \tag{6}$$

Here, the polar coordinates of the point x are denoted by (ρ, φ), $k_o = k\nu_o$, $k_i = k\nu_i$. In (4) and (5) we have the dependence of the coefficients on the polarization; namely, $\eta_{o,i} = \nu_{o,i}^{-2}$ and $\eta_{o,i} = 1$ for H- and E-polarization, respectively. The Hankel function of the first kind, with the index l is designated by $H_l^{(1)}(z)$. The functions $u \in U$ in (4) and (5), which are related to the boundary conditions, have the following limit values (see, e.g., [20, p. 68]):

$$\frac{\partial u^{\pm}}{\partial n_i}(x) = \lim_{h \to +0} (n_i(x), \operatorname{grad} u(x \pm hn(x))), \quad x \in \Gamma_i, \quad i = 1, 2, \tag{7}$$

which are expected to exist uniformly on $\Gamma_{1,2}$. The series in (6) converges uniformly and absolutely for any eigenfunction of GCFEP, also it is important to note that it is infinitely termwise differentiable [7].

We designate the major sheet of $\mathbb{L}$ by $\mathbb{L}_0$ and suppose that it is cut along the negative imaginary semi-axes by branch. At this point, we note that three types of GCFEP eigenfunctions dependent on the location of the eigenvalue $k \in \mathbb{L}_0$ exist. Equation (6) is interchangeable to the common Sommerfeld radiation condition in the case of $\operatorname{Im} k = 0$,

$$\left(\frac{\partial}{\partial \rho} - ik_o\right)u = o\left(\frac{1}{\sqrt{\rho}}\right), \quad \rho \to \infty. \tag{8}$$

The case of $\operatorname{Im} k > 0$ corresponds to the situation when u exponentially decays as $\rho \to \infty$. In the alternative case, $\operatorname{Im} k < 0$ is when the eigenfunction u grows exponentially at infinity. An important note for our consideration is that the incoming equality is true [15, 19] for any $k \in \mathbb{L}$, $\gamma \in \mathbb{R}$, and u, which satisfies (6):

$$\int_{\Gamma_R} u^-(y) \frac{\partial G_o(x, y)}{\partial n(y)} dl(y) - \int_{\Gamma_R} G_o(x, y) \frac{\partial u^-(y)}{\partial n(y)} dl(y) = 0. \tag{9}$$

Here, $x \in \Omega_3$, $G_o = (i/4)H_0^{(1)}(k_o\,|x - y|)$. We denote by Γ_R the circle with a big enough radius R which center is located at x. This fact helps us to explore all the eigenfunction types in the one scope.

We need to remember about the dependence of the imaginary part of $k \in \mathbb{L}_0$ on $\gamma \in \mathbb{R}$ [7]. In the case of a passive cavity, where $\gamma \le 0$, without losses or with them, the GCFEP statement conforms with the usual statement of CFEP [9]. At this point, $\operatorname{Im} k < 0$ for all the eigenvalues $k \in \mathbb{L}_0$. The alternative case is an active cavity, when $\gamma > 0$, and the imaginary part of $k \in \mathbb{L}_0$ can be equal to or greater than zero. The pair (k, γ), where γ and k are positive, and the corresponding eigenfunction u satisfy all the conditions of LEP [6].

Theorem 1 *[15] For each $\gamma \in \mathbb{R}$ and $k \in \mathbb{I}_+$ problem* (1)–(6) *has only the trivial solution $u = 0$, $x \in \mathbb{R}^2$.*

By $\mathbb{I}_+$ we denote the strictly positive imaginary semi-axis of $\mathbb{L}_0$. Theorem 1 was proved in [15] using the second Green's theorem (see, e.g., [20, p. 68]).

Arguing as in [11, 12], now we introduce GCFEP turned inside out that will be used later for investigation of connections between solutions of problem (1)–(6) and a spectral problem for Muler BIEs.

Now we assume that the refractive index in the domain Ω_2 is ν_o and the refractive index in the domain Ω_3 and in the hole Ω_1 is $\nu_i = \alpha_i - i\gamma$. As before, we suppose that ν_0 and α_i are positive and given. For any value of the parameter $\gamma \in \mathbb{R}$, a nonzero function $u \in U$ is referred to as an eigenfunction of the E-polarized GCFEP turned inside out corresponding to an eigenvalue $k \in \mathbb{L}$ if the following relations are satisfied:

$$\Delta u + k_i^2 u = 0, \quad x \in \Omega_1, \tag{10}$$

$$\Delta u + k_o^2 u = 0, \quad x \in \Omega_2, \tag{11}$$

$$\Delta u + k_i^2 u = 0, \quad x \in \Omega_3, \tag{12}$$

$$u^+(x) = u^-(x), \quad \frac{\partial u^-(x)}{\partial n_j(x)} = \frac{\partial u^+(x)}{\partial n_j(x)}, \quad x \in \Gamma_j, \quad j = 1, 2, \tag{13}$$

$$u = \sum_{l=-\infty}^{\infty} a_l H_l^{(1)}(k_i \rho) \exp(il\varphi), \quad \rho \geq R_0. \tag{14}$$

Theorem 2 *For each* $\gamma \in \mathbb{R}$ *and* $k \in \mathbb{I}_+$, *problem* (10)–(14) *has only the trivial solution* $u = 0$, $x \in \mathbb{R}^2$.

If $k = i\sigma$, where $\sigma > 0$, then the imaginary part of k_i is positive, the eigenfunction u decays exponentially at infinity, and the proof of the theorem is analogous to the proof of Theorem 2 [12].

3 Eigenvalue Problem for Muller BIEs

Following [15], we use the integral representations of the eigenfunctions of the problem (1)–(6) in the domains Ω_1, Ω_2, and Ω_3, respectively:

$$u(x) = -\int_{\Gamma_1} \frac{\partial G_o(x, y)}{\partial n_1(y)} u^-(y) dl(y) + \int_{\Gamma_1} G_o(x, y) \frac{\partial u^-(y)}{\partial n_1(y)} dl(y), \tag{15}$$

$$\begin{aligned} u(y) = &\int_{\Gamma_1} \frac{\partial G_o(x, y)}{\partial n_1(y)} u^+(y) dl(y) - \int_{\Gamma_1} G_o(x, y) \frac{\partial u^+(y)}{\partial n_1(y)} dl(y) \\ &- \int_{\Gamma_2} \frac{\partial G_i(x, y)}{\partial n_2(y)} u^-(y) dl(y) + \int_{\Gamma_2} G_i(x, y) \frac{\partial u^-(y)}{\partial n_2(y)} dl(y), \end{aligned} \tag{16}$$

$$u(y) = \int_{\Gamma_2} \frac{\partial G_o(x, y)}{\partial n_2(y)} u^+(y) dl(y) - \int_{\Gamma_2} G_o(x, y) \frac{\partial u^+(y)}{\partial n_2(y)} dl(y) = 0, \tag{17}$$

where $G_i = (i/4)H_0^{(1)}(k_i|x-y|)$. We know the equalities (15) and (16) well (see, e.g., [20], p. 68). Equality (17) also holds since we have (9) for each $k \in \mathbb{L}$ and $\gamma \in \mathbb{R}$. Let

$$u_j(x) = u^+(x) = u^-(x), \quad x \in \Gamma_j, \quad j = 1, 2, \tag{18}$$

$$v_1 = \frac{\eta_i + \eta_o}{2\eta_o}\frac{\partial u^+}{\partial n_1} = \frac{\eta_i + \eta_o}{2\eta_i}\frac{\partial u^-}{\partial n_1}, \quad x \in \Gamma_1, \tag{19}$$

$$v_2 = \frac{\eta_i + \eta_o}{2\eta_i}\frac{\partial u^+}{\partial n_2} = \frac{\eta_i + \eta_o}{2\eta_o}\frac{\partial u^-}{\partial n_2}, \quad x \in \Gamma_2, \tag{20}$$

and let us denote the space of continuous on Γ_j, $j = 1, 2$, functions with the maximum norm by $C_j = C(\Gamma_j)$, $j = 1, 2$, $C = C_1 \times C_2$, and $W = C \times C$. Let us indicate the identical operator in the space W by I. Then any solution of GCFEP (1)–(6) in terms (18)–(20) satisfy the following nonlinear eigenvalue problem for the set of Muller BIEs [15]:

$$A(k,\gamma)w = (I + B(k,\gamma))w = 0, \tag{21}$$

$$B = \begin{pmatrix} B_1^{(1,1)} & B_1^{(1,2)} & B_1^{(1,3)} & B_1^{(1,4)} \\ D_1^{(2,1)} & D_1^{(2,2)} & D_1^{(2,3)} & D_1^{(2,4)} \\ B_2^{(3,1)} & B_2^{(3,2)} & B_2^{(3,3)} & B_2^{(3,4)} \\ B_2^{(4,1)} & B_2^{(4,2)} & B_2^{(4,3)} & B_2^{(4,4)} \end{pmatrix}, \quad w = \begin{pmatrix} u_1 \\ v_1 \\ u_2 \\ v_2 \end{pmatrix},$$

$$\left(B_j^{(l,m)}(k,\gamma)g\right)(x) = \int_{\Gamma_j} K_j^{(l,m)}(k,\gamma;x,y)g(y)dl(y).$$

Here, we designate u_j or v_j, $j = 1, 2$ by the function g. The kernels have the following form [15]:

$$K_j^{1,1} = -K_j^{3,3} = \frac{\partial(G_o(x,y) - G_i(x,y))}{\partial n_j(y)}, \quad j = 1, 2,$$

$$K_1^{1,2} = -K_2^{3,4} = \frac{2(\eta_o G_i(x,y) - \eta_i G_o(x,y))}{\eta_i + \eta_o},$$

$$K_1^{1,3} = \frac{\partial G_i(x,y)}{\partial n_2(y)}, \quad K_2^{3,1} = -\frac{\partial G_i(x,y)}{\partial n_1(y)}, \quad K_1^{1,4} = -\frac{2\eta_o G_i(x,y)}{\eta_o + \eta_i},$$

$$K_j^{2,1} = -K_j^{4,3} = \frac{\partial^2(G_o(x,y) - G_i(x,y))}{\partial n_j(x)\partial n_j(y)}, \quad j = 1, 2,$$

$$K_j^{2,2} = -K_j^{4,4} = \frac{2\eta_o}{\eta_o + \eta_i} \frac{\partial G_i(x, y)}{\partial n_j(y)} - \frac{2\eta_i}{\eta_o + \eta_i} \frac{\partial G_o(x, y)}{\partial n_j(y)}, \quad j = 1, 2,$$

$$K_1^{2,3} = -K_2^{4,1} = \frac{\partial^2 G_i(x, y)}{\partial n_1(x) \partial n_2(y)}, \quad K_2^{3,2} = \frac{2\eta_o G_i(x, y)}{\eta_o + \eta_i},$$

$$K_j^{2,4} = -K_j^{4,2} = -\frac{2\eta_o}{\eta_o + \eta_i} \frac{\partial G_i(x, y)}{\partial n_j(y)}, \quad j = 1, 2.$$

Several of the Kernels $K_j^{q,s}$ have logarithmic singularities and the others are continuous [10]. Consequently, the operator $B(k, \gamma) : W \to W$ is compact for every $k \in \mathbb{L}$ and $\gamma \in \mathbb{R}$. The following theorem is proved analogously to Theorem 3 [12].

Theorem 3 *If $u \in U$ is an eigenfunction of the problem* (1)–(6) *corresponding to an eigenvalue $k \in \mathbb{L}$ for a value of the parameter $\gamma \in \mathbb{R}$, then defined in* (18)–(20) *functions u_j and v_j belong to the Banach spaces C_j, $j = 1, 2$, respectively, and form a nontrivial solution $w \in W$ of* (21) *with the same values of k and γ.*

The assertion in the opposite direction relative to the statement of Theorem 3 is not true (see, e.g., [18]) since we do not substitute representations (15)–(17) into (4) and (5), but add term by term the limit values of them and their normal derivatives from both sides of the boundaries Γ_1 and Γ_2 [7, 12]. However, the following result holds true.

Theorem 4 *For each $\gamma \in \mathbb{R}$ and $k \in \mathbb{L}_+$ problem* (21) *has only the trivial solution $w = 0$, $w \in W$.*

Theorem 4 is proved analogously to Theorem 4 [12] using Theorems 1 and 2.

Theorem 5 *The following statements are true.*

1. *For each $\gamma \in \mathbb{R}$, the resolvent set of the operator-valued function $A(k)$ is not empty, namely, $\mathbb{L}_+ \subset \rho(A)$.*
2. *For each $\gamma \in \mathbb{R}$, the spectrum $\sigma(A)$ of the operator-valued function $A(k)$ can be only a set of isolated points on $\mathbb{L}$, which are the eigenvalues of $A(k)$ of finite algebraic multiplicities.*
3. *Each eigenvalue k of the operator-valued function $A(k)$ depends continuously on $\gamma \in \mathbb{R}$ and can appear and disappear only on the boundary of its analyticity domain, i.e., at zero and infinity on $\mathbb{L}$.*

The first assertion of Theorem 5 follows from Theorem 4, the compactness of the operator $B(k)$, and the Fredholm alternative. For each $\gamma \in \mathbb{R}$, the operator-valued function $B(k)$ is holomorphic in $k \in \mathbb{L}$ [7]. Therefore, all the other statements of the theorem follow from the results of the theory of holomorphic operator-valued functions (see, e.g., Appendix in [21], and paper [22]).

Statements of Theorem 5 conform to CFEP, if γ is equal to or less than zero. The next corollary from Theorem 5 characterizes the eigenvalues of LEP.

Corollary 1 *If for some $\gamma > 0$ the intersection of the spectrum $\sigma(A)$ and the positive real semi-axis of $\mathbb{L}_0$ is not empty, then it can be only a set of isolated points $k > 0$, which are the eigenvalues of $A(k)$ of finite algebraic multiplicities.*

4 Galerkin Method

In the current section, we present the Galerkin method for the numerical solution of the problem (21). Assume that each contour Γ_j has a parameterization $\rho_j(t) = (\rho_j^1(t), \rho_j^2(t))$, where $\rho_j^1(t) = f_j(t)\cos t$, $\rho_j^2(t) = f_j(t)\sin t$, $t \in [0, 2\pi]$, $j = 1, 2$. Then, for any given $\gamma \in \mathbb{R}$, we have

$$\left(B_j^{(l,m)}(k,\gamma)w^{(m)}\right)(t) = \frac{1}{2\pi}\int_0^{2\pi} K_j^{(l,m)}(k;t,\tau)w^{(m)}d\tau.$$

Here, $l, m = 1, 2, 3, 4$, $y = y(\tau) \in \Gamma_j$, $j = 1, 2$,

$$K_j^{(l,m)}(t,\tau) = 2\pi K_j^{(l,m)}(x,y)\left|\rho_j'(\tau)\right|^{-1}.$$

For construction and investigation of the Galerkin method, it is convenient to consider problem (21) in the Hilbert space $H = (L_2)^4$ where L_2 denotes the space of square integrable functions with the inner product

$$(u,v) = \frac{1}{2\pi}\int_0^{2\pi} u(\tau)\overline{v(\tau)}d\tau, \quad u, v \in L_2.$$

By $T_n \subset L_2$ we denote the subspace of all trigonometric polynomials of order no greater than n with complex coefficients. Then $H_n = (T_n)^4 \subset H$ is the subspace with elements of the form

$$\mathbf{w}_n = \begin{pmatrix} w_n^{(1)} \\ w_n^{(2)} \\ w_n^{(3)} \\ w_n^{(4)} \end{pmatrix}, \quad w_n^{(1)}, w_n^{(2)}, w_n^{(3)}, w_n^{(4)} \in T_n.$$

By $p_n : H \to H_n$ we define the following projection operator:

$$p_n w = \begin{pmatrix} \Phi_n w^{(1)} \\ \Phi_n w^{(2)} \\ \Phi_n w^{(3)} \\ \Phi_n w^{(4)} \end{pmatrix}, \quad w^{(1)}, w^{(2)}, w^{(3)}, w^{(4)} \in L_2.$$

Here, $\Phi_n : L_2 \to T_n$ is the Fourier operator,

$$(\Phi_n w^{(m)})(t) = \sum_{q=-n}^{n} c_q(w^{(m)})\varphi_q(t), \quad m = 1, 2, 3, 4.$$

For $q = -n, \ldots, n$, the vectors $\varphi_q(t) = \exp(iqt)$ form the orthonormal basis in the space T_n, and

$$c_q(w^{(m)}) = (w^{(m)}, \varphi_q) = \frac{1}{2\pi}\int_0^{2\pi} w^{(m)}(t)\exp(-iqt)dt,$$

are the Fourier coefficients of the function $w^{(m)}$. We rewrite Eq. (21) as follows

$$w^{(l)} + \sum_{m=1}^{4} B^{(l,m)}(k)w^{(m)} = 0, \quad l = 1, 2, 3, 4. \tag{22}$$

We find approximate solutions $w_n^{(1)}, w_n^{(2)}, w_n^{(3)}, w_n^{(4)} \in T_n$ of the system of equations (22) in the form

$$w_n^{(m)}(t) = \sum_{q=-n}^{n} \alpha_q^{(m)}\varphi_q(t), \quad n \in N, \quad m = 1, 2, 3, 4.$$

Therefore, we have

$$w_n^{(l)} + \sum_{m=1}^{4} B^{(l,m)}(k)w_n^{(m)} = 0, \quad l = 1, 2, 3, 4, \quad n \in N.$$

We calculate unknowns $\alpha_q^{(m)}$ using the Galerkin method,

$$\left(w_n^{(l)}, \varphi_p\right) + \sum_{m=1}^{4} \left(B^{(l,m)}(k)w_n^{(m)}, \varphi_p\right) = 0, \quad p = -n, \ldots, n, \tag{23}$$

where $l = 1, 2, 3, 4$. Since the trigonometric functions are orthonormal, we can rewrite equations (23) in the form of the following system of linear algebraic equations:

$$\alpha_p^{(l)} + \sum_{m=1}^{4}\sum_{q=-n}^{n} h_{pq}^{(l,m)}(k)\alpha_q^{(m)} = 0, \quad p = -n, \ldots, n, \tag{24}$$

where $l = 1, 2, 3, 4$,

$$h_{pq}^{(l,m)}(k) = \frac{1}{4\pi^2} \int_0^{2\pi} \int_0^{2\pi} K^{(l,m)}(k; t, \tau) \exp(-ipt) \exp(iq\tau) dt d\tau.$$

Denote by w_n the solution of system (24), by $A_n(k)$ its matrix, and by $\sigma(A_n)$ the spectra of $A_n(k)$. We investigate the convergence of the Galerkin method using ideas of [7] and fundamental results of [13, 14].

Theorem 6 *For any given $\gamma \in \mathbb{R}$, the following statements are true.*

1. *For every eigenvalue k_0 of $A(k)$, there exists a sequence $\{k_n\}_{n \in \mathbb{N}}$ converging to k_0 with the eigenvalues k_n of $A_n(k)$.*
2. *If $\{k_n\}_{n \in \mathbb{N}}$ and $\{w_n\}_{n \in \mathbb{N}}$ are some sequences of eigenvalues k_n of $A_n(k)$ and normalized eigenfunctions w_n of $A_n(k)$, so that $k_n \to k_0 \in \mathbb{L}$ $(n \in \mathbb{N})$, then*
   - *i) k_0 is an eigenvalue of $A(k)$,*
   - *ii) $\{w_n\}_{n \in \mathbb{N}}$ is a discretely compact sequence and its cluster points are normalized eigenfunctions of $A(k_0)$.*
3. *For every compact $L_0 \subset \rho(A)$, the sequence $\{A_n(k)\}_{n \in \mathbb{N}}$ is stable on L_0, i.e., there exist $n(L_0)$ and $c(L_0)$ such that $L_0 \subset \rho(A_n)$, $\|A_n(k)^{-1}\| \leq c(L_0)$ for all $k \in L_0$ and $n \geq n(L_0)$.*

The proof of this theorem is based on the general results of the discrete convergence theory [23] applied for investigation of approximate methods in the eigenvalue problem where the parameter appears non-linearly [13].

The next theorem follows from [14].

Theorem 7 *Assume that $\gamma \in \mathbb{R}$ is given, k_0 is an eigenvalue of $A(k)$, and $L_0 \subset \mathbb{L}$ is a compact set with the boundary $\Gamma_0 \subset \rho(A)$ so that $L_0 \cap \sigma(A) = \{k_0\}$. Let us denote by ε_n the maximum of the approximation error over $k \in \Gamma_0$ and $w \in G(A, k_0)$,*

$$\varepsilon_n = \sup\{\|A_n(k) p_n w - p_n A(k) w\|_{W_n} : k \in \Gamma_0,\ w \in G(A, k_0),\ \|w\|_W = 1\}. \tag{25}$$

Here, $G(A, k_0)$ is the generalized eigenspace, i.e., the closed linear hull of all the generalized eigenfunctions of $A(k)$ corresponding to k_0. Then $\varepsilon_n \to 0$ $(n \in \mathbb{N})$ and the following estimations hold for almost all $n \in \mathbb{N}$:

- *i) $|k_n - k_0| \leq c\varepsilon_n^{1/\kappa}$ for all $k_n \in \sigma(A_n) \cap L_0$, where $\kappa = \kappa(k_0, A)$ is the order of the pole k_0 of the operator-valued function $A^{-1}(k)$;*
- *ii) $|\bar{k}_n - k_0| \leq c\varepsilon_n$, where $\bar{k}_n$ is the weighted (proportionally to their algebraic multiplicities) mean of all the eigenvalues of $A_n(k)$ in L_0, $\bar{k}_n = \sum_{k \in \sigma(A_n) \cap L_0} \mu_k \cdot k$, $\mu_k = \nu(k, A_n)/\nu(k, A)$, where $\nu(\cdot, \cdot)$ is the algebraic multiplicity of the corresponding eigenvalue k;*

iii) $\max\{|k_n - k_0| : k_n \in \sigma(A_n) \cap L_0\} \le c\varepsilon_n^{1/l_n}$, *where* l_n *is the number of the different eigenvalues of* $A_n(k)$ *in* L_0.

We solve the nonlinear eigenvalue problem (24) using the residual inverse iteration algorithm [24]. If the boundaries of the active cavity and the round piercing hole are nonconcentric circles, then the entries of the Galerkin's matrix have the explicit expressions calculated carefully in [18]. We use them in the next section.

5 Numerical Results

In the systematic computations by the Galerkin method, we assume that the environment in the original problem (1)–(6) is air with $\nu_3 = \nu_1 = 1$. We look for the H-polarized modes of the active microcavity with the real part of the refractive index $\alpha_i = 2.63$. This is known as the effective refractive index for a 200-nm GaAs layer in the infrared spectrum [4]. In the following, we will use the normalized notations, $\kappa = ka_2, d = |O_1 - O_2|/a_2$, and $r = a_1/a_2$. Boundaries Γ_1 and Γ_2 are circles with centers at points O_1 and O_2 with radiuses a_1 and a_2, respectively.

We use a common notation for the modes of a circular cavity (see, for instance [18]). The values of (κ, γ) of the mode (11,1,e) for $(r, d) = (0.505, 0.170)$ are shown in Fig. 2 together with ones for all other modes with sufficiently small

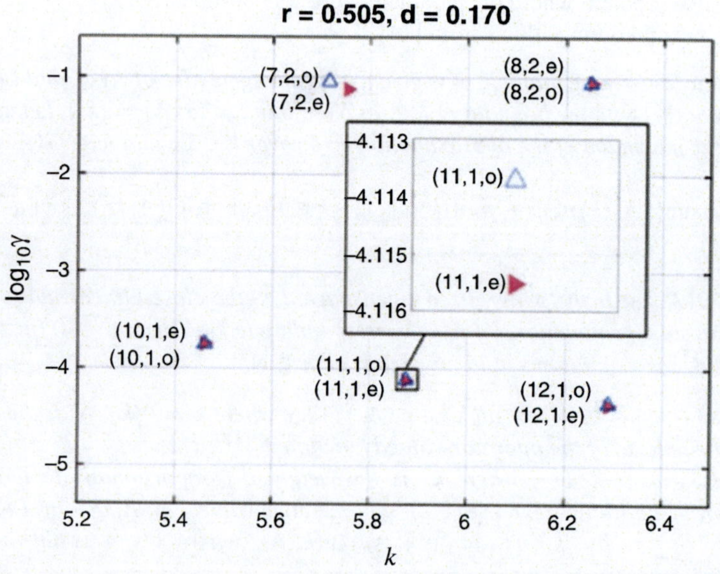

Fig. 2 Values of the normalized frequency of lasing κ and the threshold gain γ of all modes with sufficiently small thresholds for $\kappa \in [5.2, 6.4]$ and $r = 0.505$, $d = 0.170$

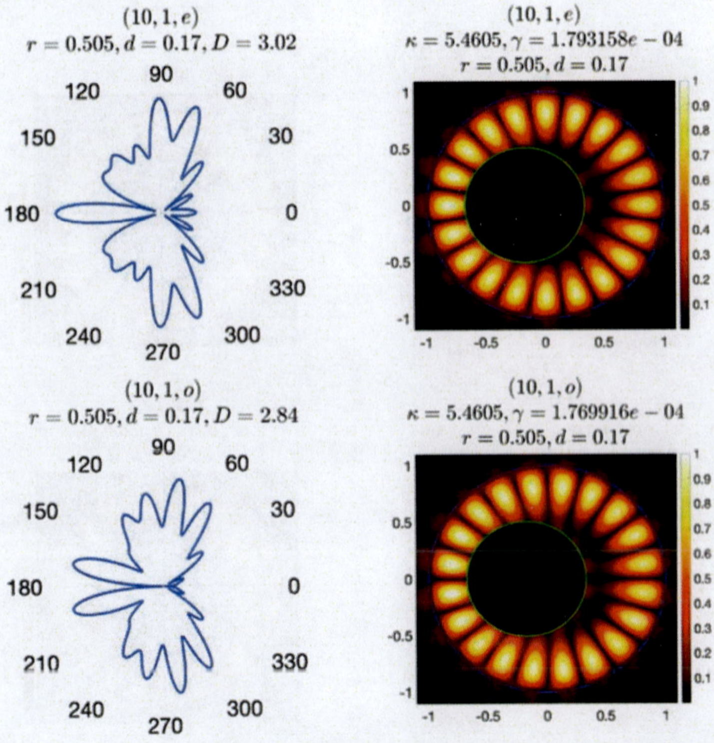

Fig. 3 The near-field (right panels) and the far-field (left panels) patterns for x_1-even and x_1-odd $H_{10,1}$ modes for $r = 0.505$, $d = 0.170$

thresholds and $\kappa \in [5.2, 6.4]$. We see that this mode is working since it has the lowest threshold. All the field patterns of the modes with $n = 1$ shown in Fig. 2 are presented in Figs. 3, 4, and 5. What is interesting that if the hole is relatively big, then the main beam directs to the same side where the hole is shifted. For relatively small holes, we obtain the unidirectional emission, and the direction of the main beam is opposite to the hole shift direction [18]. These effects for small and large holes were experimentally demonstrated in [25]. It is important to note that quasi-unidirectional emission can also be explained with the help of Photonic Jet effect described in works [26–28].

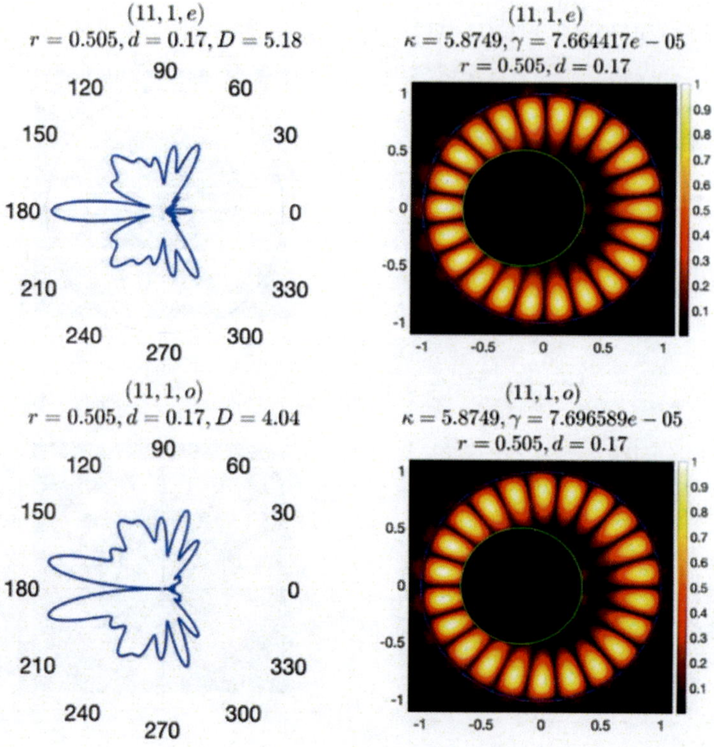

Fig. 4 The near-field (right panels) and the far-field (left panels) patterns for x_1-even and x_1-odd $H_{11,1}$ modes for $r = 0.505$, $d = 0.170$

6 Conclusion

We have used the Muller BIEs for the analysis of the spectrum of GCFEP with the help of the general results of the theory of operator-values functions depending on the parameter. We have also shown the main steps in the reduction of GCFEP for a 2-D laser with a partial active region to a set of four coupled boundary integral equations of the Muller type. We have further explained the discretization of these equations with the Galerkin method and proved its convergence.

Finally, we have calculated the on-threshold characteristics of a lasing mode of an eccentric microcavity with the shifted hole. In the numerical experiments, we have varied the position of the piercing hole on the x_1 axis in the cavity and the radius of the hole and measured the changes in the lasing frequencies, directionalities, and thresholds, and have presented in the current paper only the results for the highest directivity for a relatively big piercing hole.

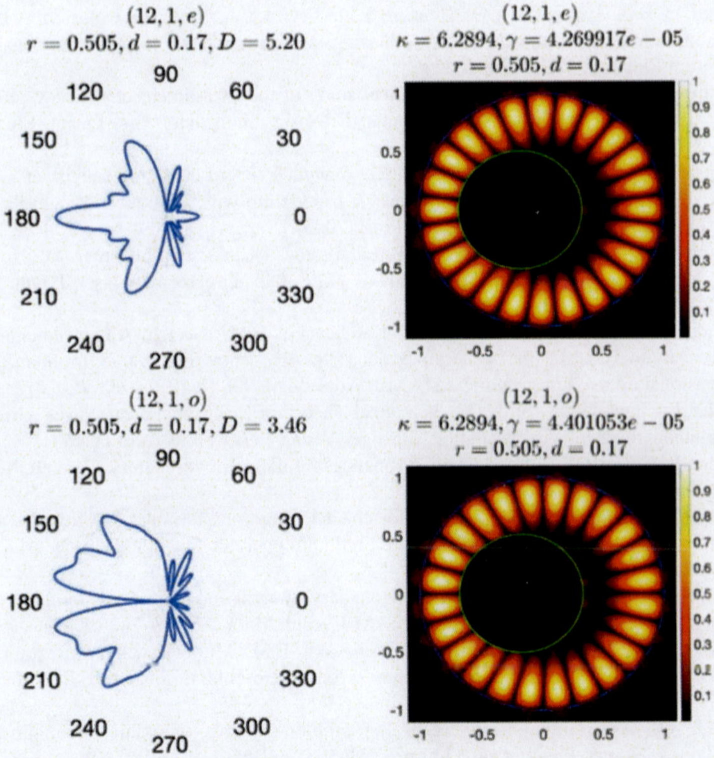

Fig. 5 The near-field (right panels) and the far-field (left panels) patterns for x_1-even and x_1-odd $H_{12,1}$ modes for $r = 0.505, d = 0.170$

Our numerical investigation has shown that a hole of a suitable radius and located at a certain place can lead to a notable growth of the directivity of lasing mode with the conservation of its low threshold. Hence, a big piercing hole's radius and position in the 2-D eccentric microcavity laser can be used as an engineering tool to control efficiently the directivity of emission.

Acknowledgments This paper has been supported by the Kazan Federal University Strategic Academic Leadership Program ("PRIORITY-2030").

References

1. Du, W., Li, C., Sun, J., Xu, H., Yu, P., Ren, A., Wu, J., Wang, Z.: Nanolasers based on 2D materials. Laser Photonics Rev., 2000271 (2020)
2. Wiersig, J., Hentschel, M.: Unidirectional light emission from high-Q modes in optical microcavities. Phys. Rev. A **73**, 031802(R) (2006)

3. Smotrova, E.I., Byelobrov, V.O., Benson, T.M., Ctyroky, J., Sauleau,R., Nosich, A.I.: Optical theorem helps understand thresholds of lasing in microcavities with active regions. IEEE J. Quantum Electron. **47**, 20–30 (2011)
4. Smotrova, E.I., Nosich, A.I.: Mathematical study of the two-dimensional lasing problem for the whispering-gallery modes in a circular dielectric microcavity. Opt. Quant. Electron. **36**, 213–221 (2004)
5. Smotrova, E.I., Nosich, A.I., Benson, T.M., Sewell, P.: Cold-cavity thresholds of microdisks with uniform and nonuniform gain: quasi-3-d modeling with accurate 2-d analysis. IEEE J. Sel. Top. Quantum Electron. **11**, 1135–1142 (2005)
6. Smotrova, E.I., Tsvirkun, V., Gozhyk, I., Lafargue, C., Ulysse, C., Lebental, M., Nosich, and A.I.: Spectra, thresholds, and modal fields of a kite-shaped microcavity laser. J. Opt. Soc. Am. B **30**, 1732–1742 (2013)
7. Spiridonov, A.O., Oktyabrskaya, A.O., Karchevskii, E.M., Nosich, A.I.: Mathematical and numerical analysis of the generalized complex-frequency eigenvalue problem for two-dimensional optical microcavities. SIAM J. Appl. Math. **80**(4), 1977–1998 (2020)
8. Muller, C.: Foundations of the Mathematical Theory of Electromagnetic Waves. Grundlehren der mathematischen Wissenschaften, Springer-Verlag Berlin Heidelberg (1969)
9. Heider, P.: Computation of scattering resonances for dielectric resonators. Comput. Math. with Appl. **60**(6), 1620–1632 (2010)
10. Spiridonov, A.O., Karchevskii, E.M., Nosich, A.I.: Rigorous formulation of the lasing eigenvalue problem as a spectral problem for a Fredholm operator function. Lobachevskii J. Math. **39**(8), 1148–1157 (2018)
11. Misawa, R., Niino, K., Nishimura, N.: Boundary integral equations for calculating complex eigenvalues of transmission problems. SIAM J. Appl. Math. **77**, 770–788 (2017)
12. Oktyabrskaya, A.O., Spiridonov A.O., Karchevskii, E.M.: Muller boundary integral equations for solving generalized complex-frequency eigenvalue problem. Lobachevskii J. Math. **41**(7), 1377–1384 (2020)
13. Karma, O.: Approximation in eigenvalue problems for holomorphic Fredholm operator functions I. Numer. Funct. Anal. Optim. **17**, 365–387 (1996)
14. Karma, O.: Approximation in eigenvalue problems for holomorphic Fredholm operator functions II (convergence rate). Numer. Funct. Anal. Optim. **17**, 389–408 (1996)
15. Spiridonov, A.O., Karchevskii, E.M., Nosich, A.I.: Mathematical and numerical modeling of on-threshold modes of 2-D microcavity lasers with piercing holes. Axioms **8**(3), 1–16 (2019)
16. Spiridonov, A.O., Karchevskii, E.M., Benson, T.M., Nosich, A.I.: Why elliptic microcavity lasers emit light on bow-tie-like modes instead of whispering-gallery-like modes. Opt. Comm. **439**, 112–117 (2019)
17. Spiridonov, A.O., Karchevskii, E.M.: Mathematical and numerical analysis of the spectral characteristics of dielectric microcavities with active regions. Proc. Int. Conf. Days on Diffraction (DD-2016), Saint-Petersburg, art. no. 7756880, 390–395 (2016)
18. Oktyabrskaya, A.O., Repina, A.I., Spiridonov, A.O., Karchevskii, E.M., Nosich, A.I.: Numerical modeling of on-threshold modes of eccentric-ring microcavity lasers using the Muller integral equations and the trigonometric Galerkin method. Opt. Commun. **476**, 126311 (2020)
19. Reichardt, H.: Ausstrahlungsbedingungen fur die wel-lengleihung. Abh. Math. Sem. Hamburg **24**, 41–53 (1960)
20. Colton, D., Kress, R.: Integral Equation Methods in Scattering Theory. SIAM: Philadelphia (2013)
21. Kozlov, V., Maz'ya, V.: Differential Equations with Operator Coefficients with Applications to Boundary Value Problems for Partial Differential Equations. Springer, Heidelberg (1999)
22. Steinberg, S.: Meromorphic families of compact operators. Arch. Rational Mech. Anal. **31**, 372–379 (1968)
23. Vainikko, G.: Multidimensional Weakly Singular Integral Equations. Springer, Heidelberg (1993)
24. Neumaier, A.: Residual inverse iteration for the nonlinear eigenvalue problem. SIAM J.Numer. Anal., **22**, 914–923 (1985)

25. Zhang, S., Li, Y., Hu, P., Li, A., Zhang, Y., Du, W., Du, M., Li, Q., Yun, F.: Unidirectional emission of GaN-based eccentric microring laser with low threshold. Opt. Express **28**(5), 6443–6451 (2020)
26. Heifetz, A., Kong, S.-C. , Sahakian, A.V., Taflove, A., Backman,V.: Photonic Nanojets. J. Comput. Theor. Nanosci. **6**, 1979–1992 (2009)
27. Luk'yanchuk, B.S., Paniagua-Domi'nguez, R., Minin, I., Minin, O., Wang, Z.: Refractive index less than two: photonic nanojetsyesterday, today and tomorrow. Opt. Mater. Express **7**, 1820–1847 (2017)
28. Dukhopelnykov, S.V., Sauleau, R., Garcia-Vigueras, M., Nosich, A. I.: Combined plasmon-resonance and photonic-jet effect in the THz wave scattering by dielectric rod decorated with graphene strip. J. Appl. Phys. **126**, 023104 (2019)

A Solution of Inverse Problem in the Theory of Supercritical Fluid Extraction of Oil from Ground Plant Material

Artur A. Salamatin and Andrey G. Egorov

Abstract In supercritical fluid extraction, the problem of experimental determination of apparent transport properties of the ground plant material is coupled with analysis of the histogram of the particle size distribution function. Thus, both, the diffusion coefficient, and the distribution function, have to be inferred simultaneously. An inverse procedure which relates experimentally available overall extraction curve and particle size distribution function is introduced in the present study. In a particular case of flat particles, the problem is solved in a closed form. The forward problem for the flat particles and the explicit analytical solution of the inverse problem have been tested as a preconditioner in the iterative solution of the inverse problem in case of spherical particles. Convergence of the computational procedure is demonstrated in a series of test runs for pre-defined distribution functions. Limitations of the overall algorithm applicability to the solution of inverse problems are discussed.

1 Introduction

The supercritical fluid extraction (SFE) is a topical technology since it is in line with the "green chemistry" policy and principals of sustainable development [1]. As the major application, SFE is used for extraction of valuable natural compounds, such as vegetable oil (lipids) and essential oil from ground raw plant material. Using green solvents such as CO_2 at supercritical conditions, SFE is less harmful for

A. A. Salamatin (✉)
Institute of Mechanics and Engineering, FRC Kazan Scientific Center, Russian Academy of Sciences, Kazan, Russia

Kazan Federal University, Kazan, Russia
e-mail: salamatin@imm.knc.ru

A. G. Egorov
Institute of Mechanics and Engineering, FRC Kazan Scientific Center, Russian Academy of Sciences, Kazan, Russia

I. B. Badriev et al. (eds.), *Mesh Methods for Boundary-Value Problems and Applications*, Lecture Notes in Computational Science and Engineering 141,
https://doi.org/10.1007/978-3-030-87809-2_34

the environment and consumers of the final product, which is free of toxic organic solvent residues.

Mathematical model of supercritical fluid extraction (SFE) of lipids from plant material with high initial oil content is considered in the present paper. This application is of general interest for the biofuel industry, in particular, for production of natural nutritional supplements, and many others. The extraction takes place in a cylindrical column (the extractor) with a stationary packed bed of ground plant material particles. Sometimes sieving is applied to single out particle fraction with a narrow particle size distribution. However, generally, the ensemble of particles used for extraction is essentially polydisperse [2, 3]. The particle size distribution is characterized by a volumetric cumulative distribution function $F(a)$. It is the volume (mass) fraction of particles in the bed with the size smaller than a. The packed bed is assumed to be homogeneous, and the polydisperse particles are randomly mixed. Typically, a spherical symmetry of mass transfer processes is assumed on the particle scale. Sometimes planar approximation, or flat particles, is considered. Thus, a stands for the particle radius or half-thickness, respectively [4].

The solvent, which dissolves the oil (solute), is pumped through the bed at a given volumetric flow rate. The concentration gradient of solute is the driving force of extraction on the particle scale. Once the oil has diffused along the particle transport channels to the particle surface and is exposed to the interparticle pore volume, the solvent carries it out of the extractor column. Two stages can be distinguished in the SFE process based on the dependence of accumulated oil mass on time t, which is called the overall extraction curve (OEC) $Y(t)$ [5]. At the first stage the OEC is linear with respect to time, and the solution leaving the extractor column is saturated by the solute up to a maximum value θ_*. This limiting concentration depends on temperature and pressure in the system. After the initial linear part, the rate of oil accumulation slows down, the outlet solute concentration steadily decreases, and the OEC varies non-linearly with time. Eventually, the packed bed is depleted, and the OEC takes a constant value, which represents the total mass of oil initially (at $t = 0$) available for extraction.

Further development of SFE process directly depends on the ability to predict (simulate) its dynamics under various conditions. Thus, distinct mass transfer models have been suggested recently to explain and describe the observed multistage process dynamics [6–10]. One of the widely applied approaches (Sect. 2) consists of the shrinking core (SC) model for the particle-scale mass transfer [2, 11, 12], and the quasi-stationary model of convection in the packed bed. The latter submodel takes into account the polydisperse nature of particles, while it does not take into account the diffusion boundary layer and macroscopic backmixing effects. They are shown to be not important for typical SFE conditions [13]. As a result, the model has a single adjustable parameter—apparent diffusion coefficient D_{eff} of solute mass transfer on the particle-scale. This parameter is used to scale-up the process, to evaluate the economic efficiency and performance of suggested SFE implementation and conditions [14]. In principle, the value of D_{eff} can be inferred from the experimental OEC.

However, the extraction dynamics are essentially affected by the particle size distribution as well [15, 16]. As a rule, the distribution function is unavailable, and D_{eff} should be inferred together with the $F(a)$-histogram. As suggested in the present paper, both characteristics of the packed bed, D_{eff} and $F(a)$, can be inferred one after another. Simply, the problem is analyzed in dimensionless variables, and the $F(a)$-histogram is obtained in dimensionless variables as well. Then, typical particle size of the packed bed is used to imposed a correspondence between the dimensional and dimensionless quantities. A corresponding inverse procedure for the determination of $F(a)$ is introduced in Sect. 3 in dimensionless form and discussed further in Sect. 4. The problem is shown to be ill-posed. Its analytical solution is obtained in a closed form for flat particles. This solution is used as a preconditioner in an iterative computational algorithm to solve the inverse problem for spherical particles.

2 Forward Problem

2.1 Problem Formulation

Hereinafter we use the following dimensionless quantities: time t, spatial coordinate z varying along the vessel from its inlet ($z = 0$) to outlet ($z = 1$), solute concentration $0 < c(t, z) < 1$ in the pores of the packed bed, fraction $0 < s(t, z, a) < 1$ of oil extracted from an individual particle of size a to the moment t at cross-section z. The scales relating the corresponding dimensional and dimensionless characteristics are

$$t \sim \frac{\theta_0}{\theta_*}\frac{H(1-\varepsilon)}{v}, \quad z \sim H, \quad a^2 \sim a_{sc}^2 \equiv 2mD_{eff}\frac{H(1-\varepsilon)}{v}, \quad c \sim \theta_*, \tag{1}$$

where θ_0 is the density of oil originally stored in the plant material per its unit volume, H—vessel height, ε—packed bed porosity, v—superficial velocity of the fluid passing through the packed bed, and m is the shape factor, which is 1 for flat particles and 3 for spherical ones. Note that only the particle size scale a_{sc} depends on D_{eff}. Thus, the analyzed data can be presented in dimensionless quantities to infer $F(a)$. Then, a typical (dimensional) particle size must be used to infer the value of D_{eff}.

SFE is essentially a multiscale process. It is governed by different mass-transfer mechanisms on the particle-scale and on the scale of packed bed pore volume. Thus, the overall SFE model consists of two, macro- and micro-scale, sub-models [10]. The first one describes the convective solvent flow through a porous medium, composed of randomly mixed polydisperse ground plant material particles. After [5, 12], the mass balance is described in one-dimensional quasi-stationary approximation. The master-equation and the inlet boundary condition take the following forms

$$\frac{\partial c}{\partial z} = \frac{\partial}{\partial t}\int_0^\infty s(t, z, a) f(a) da, \quad c|_{z=0} = 0, \tag{2}$$

where $f(a)$ is the density of the overall particle size distribution function $F(a)$, and $dF(a) = f(a)da$.

The mass balance on the particle scale is described within the framework of SC [4, 12] approach. Here, two zones can be distinguished in the particle during extraction: the inner oil-containing core, and the outer transport zone. The zones are separated by a sharp boundary that shrinks towards the center ($m = 3$) or the center plane ($m = 1$) of the particle with the extraction progress. The oil is depleted in the outer zone, which is the diffusive transport path for the oil dissolved in the solvent near the moving boundary. The conductivity of the transport zone to the solute diffusion is the major mechanism that hinders and controls the extraction process. The conductivity $\lambda_m(s)$ depends on the current level of particle depletion s and reflects the particle shape and symmetry of the mass transfer process on the particle-scale

$$\lambda_m : \quad \lambda_1 = \frac{1}{2s}, \quad \lambda_3 = \frac{0.5(1-s)^{1/3}}{1-(1-s)^{1/3}}.$$

Here, $m = 1$ and 3 for flat and spherical particles, respectively.

Finally, the particle mass-balance equation and initial condition become

$$\frac{\partial s}{\partial t} = \frac{\lambda_m(s)}{a^2}(1-c), \quad s|_{t=0} = 0. \tag{3}$$

Importantly, mass balance Eq. (3) is valid until the complete depletion of the particle, i.e., $s < 1$. Once $s = 1$, it remains unity afterwards. Detailed derivation and discussions of the problem (2)–(3) are given in Ref. [13]. In polydisperse packed bed, Eq. (3) is solved for every particle fraction of size a at every cross-section z.

2.2 Problem Solution

The OEC $Y(t)$ is of primary interest for the theoretical analysis since it is typically the only one experimentally observed characteristic of the process. It represents the extracted fraction of oil in the packed bed as a function of time t

$$Y(t) = \int_0^t c(\tau, 1) d\tau,$$

where $c(t, 1)$ is the solute concentration of solution leaving the extraction column, or the rate of oil accumulation. Along with OEC it is convenient to consider the zonal oil fraction $0 < y(t, z) < z$ extracted from the packed bed interval $[0; z]$ to the time t

$$y(t, z) = \int_0^t c(\tau, z)d\tau, \tag{4}$$

where $0 < Y(t) = y(t, 1) < 1$.

Integration of Eqs. (2) and (3) with respect to time t with account of the initial condition for s and the definition (4) yields the following system of equations

$$\frac{\partial y}{\partial z} = \int_0^\infty sf(a)da, \quad y|_{z=0} = 0, \tag{5}$$

$$\varphi_m(s) = \min\left\{1, \frac{t-y}{a^2}\right\}. \tag{6}$$

Here, the min-operator assumes that $s < 1$, and $s = 1$ is its maximum value indicating the complete depletion of the particle. Due to scaling, the function

$$\varphi_m(s) = \int_0^s \frac{d\omega}{\lambda_m(\omega)} = \begin{cases} s^2, & m = 1; \\ 3\left(1-(1-s)^{2/3}\right) - 2s, & m = 3; \end{cases} \tag{7}$$

monotonously varies with s from zero to unity, as shown in Fig. 1. Hence, Eqs. (6) and (7) define s as a function, inverse to $\varphi_m(s)$

$$s\left(\frac{t-y}{a^2}\right) = \varphi_m^{-1}\left(\min\left\{1, \frac{t-y}{a^2}\right\}\right), \tag{8}$$

and Eq. (5) renders the first-order Ordinary Differential Equation (ODE) with respect to y as a function of z with another argument t as a parameter. Numerical integration of Eq. (5) with respect to z from 0 to 1 yields $Y(t)$ at any given moment t.

Substituting Eq. (8) in Eq. (5), the ODE can be solved for an implicit dependence of $Y(t)$ on time for any $f(a)$. With the definition

$$k(\tau) = \int_0^\infty s\left(\frac{\tau}{a^2}\right) f(a)da, \tag{9}$$

Eq. (5) reads as

$$\frac{\partial y}{\partial z} = k(t-y),$$

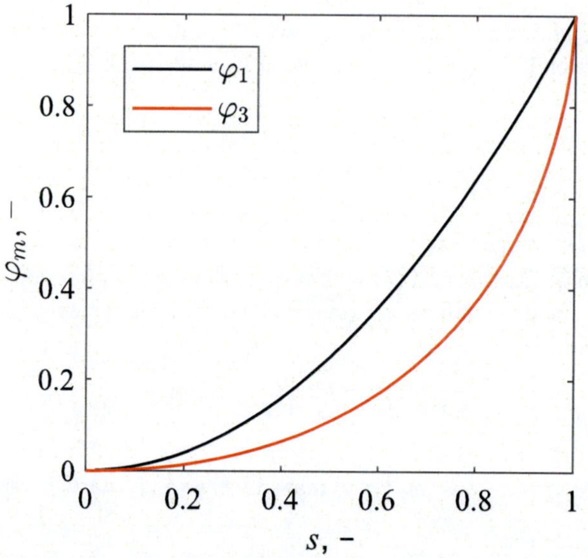

Fig. 1 The monotonous functions $\varphi_1(s)$ and $\varphi_3(s)$ for flat and spherical particles respectively. Due to scaling both functions vary between zero and unity for $0 < s < 1$

which is an ODE with separable variables. Integrating with respect to $0 < z < 1$ yields an integral equation with respect to $Y(t)$

$$\int_{t-Y}^{t} \frac{d\tau}{k(\tau)} = 1. \qquad (10)$$

A set of OECs for different discrete particle size distributions is demonstrated in Fig. 2. Figure 2a shows a set of OECs corresponding to packed beds of monodisperse particles. Note that the extraction rates and the duration of initial linear stage decrease with the particle size, and ultimately more time is required to attain the complete extraction of the packed bed. Figure 2b shows the extraction dynamics for packed beds made of two discrete fractions of spherical particles, with sizes $a_1 \ll a_2$ and corresponding volume fractions α and $1 - \alpha$. The size of small particle fraction only affects the short-term intermediate stage between the initial linear one and the subsequent non-linear one, while the volume fraction α determines the duration of the linear stage. In Fig. 2b, compare the curves corresponding to the same volume fraction α. They differ on a very small segment, where the OEC abruptly bends. This time interval determines the resolution limit of the distribution function by the experimentally obtained, discrete OEC. There always exists the lower bound of the particle-size resolution. If the mesh of discrete time moments does not resolve the transition stage in the OEC then any particle fraction of sufficiently small size approximates the transition region equally well.

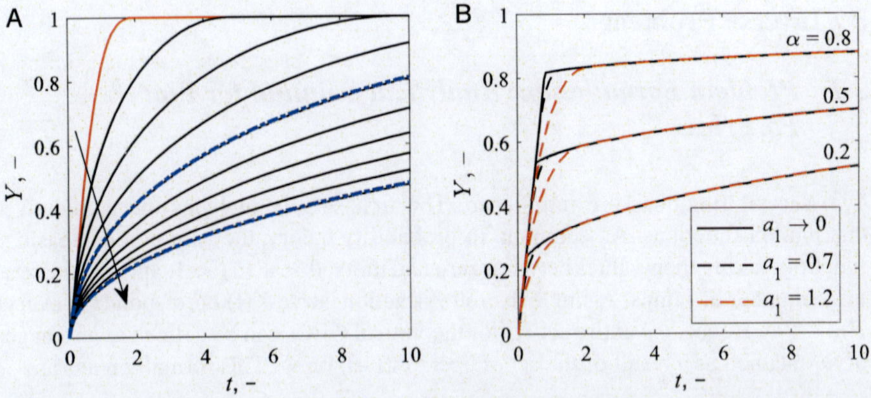

Fig. 2 The overall extraction curves $Y(t)$ for various discrete particle size distributions of spherical particles. (**a**) The monodisperse ensembles with particle sizes $1 \leq a \leq 10$ varying at fixed step 1. The arrow shows the increase of particle size. The red curve shows the OEC corresponding to $a = 1$. The blue dashed curves demonstrate convergence of the iterative algorithm described in Sect. 3.2; (**b**) the ensembles are a mixture of two monodisperse fractions of particles. The particle size distribution is described by the volume fraction α of small-size fraction a_1, (α, a_1), the volume fraction $1-\alpha$ of large-size fraction, $(1-\alpha, a_2)$, $a_1 \ll a_2$. The values of $\alpha = \{0.8, 0.5, 0.2\}$ are shown in the plot, and $a_2 = 12$ is fixed. The black solid curves are at $a_1 \to 0$, the black dashed curves are at larger $a_1 = 0.7$, and the red dashed curves are at $a_1 = 1.2$

There are two reference points in the OEC plot that indicate the ends of the respective extraction stages. During the first stage, the OEC is linear with time, and the solute leaves the extraction column at the maximum, saturation, concentration $c = 1$. The duration of this phase is designated as t_-. At $t > t_-$, the solute concentration in the solvent at the outlet cross-section drops down, $c < 1$, and the OEC varies nonlinearly with time. The second stage ends at $t = t_+$, when the packed bed is depleted, and Y reaches its maximum value, $Y = 1$. Finally, the following properties are essential for each OEC

$$Y(t) = t, \quad t < t_-; \quad Y(t) = 1, \quad t > t_+. \tag{11}$$

After [17], both time moments, t_- and t_+, can be found from Eq. (10)

$$t_+[F] = 1 + a_{max}^2, \quad \int_0^{t_-[F]} \frac{d\tau}{k(\tau)} = 1. \tag{12}$$

Here, $a_{\max}$ is the maximum size of particles in the packed bed. Thus, $F(a) = 1$ at $a > a_{\max}$.

3 Inverse Problem

3.1 Problem Formulation: Analytical Solution for Flat Particles

Two sets of functions are introduced. The first set, Φ, includes the cumulative distribution functions. As common, in probability theory, they are non-decreasing functions taking their values between zero and unity, $0 < F(a) < 1$, and may have a finite number of jumps. A single overall extraction curve $Y(t)$ corresponds to every $F(a) \in \Phi$. Hence, the entire set Λ_m of theoretical OECs can be defined as an image of Φ obtained as a result of the act of operator (10) on Φ. This formally reads as

$$A_m(F) = Y, \quad A_m : \Phi \to \Lambda_m, \quad m = 1, 3,$$

where $A_{m=1,3}$ are operators of the two forward problems described by Eqs. (7)–(10). Here, the subscript m indicates that a unique set Λ_m is obtained for a prescribed mass transfer symmetry on the particle scale. Whether $\Lambda_1 = \Lambda_3$ or not is an open question.

Λ_m is a subset of continuous functions with bounded first derivative, $0 \leq dY/dt \equiv c(t, 1) \leq 1$. Since, the outlet solute concentration is a non-increasing function, the curvature of Y does not change sign, i.e., $d^2Y/dt^2 = dc(t, 1)/dt \leq 0$. Other constraints on the admissible theoretical OECs are given in Eq. (11). Thus, the proper OEC must demonstrate an initial linear segment corresponding to $c(t, 1) \equiv 1$ at $t \leq t_-$ followed by a nonlinear segment with $dc(t, 1)/dt < 0$ at $t_- < t \leq t_+$. Apparently, when a forward problem is solved, the two reference time moments, as well as the entire OEC, are functionals of a distribution function, $t_\pm = t_\pm[F]_m$, where the subscript m indicates the implied symmetry of the particle-scale mass transfer processes.

Evidently, with these constraints, the set Λ_m is "narrow", and may not include every experimentally obtained OEC Y_{exp}. The experimental curves are presented as discrete sets of points given with some uncertainty. Hence, an interpolation should be introduced to define Y_{exp} as a continuous function of time. Thus, any Y_{exp} could hardly have a corresponding admissible generating function F from Φ even after a proper interpolation between discrete moments of time. One reason for this is the experimental uncertainty which breaks the curvature constraint on the admissible OEC, and the other—is that the forward model (2) and (3) is only an approximation of the real process. Moreover, a continuous interpolated analog of Y_{exp} has an infinite set of corresponding generating distribution functions $F(a)$. This is due to a decrease of the time frame of the process transition stage with the decrease of the smallest particle fraction size. Ultimately, it falls between adjacent time moments when the OEC was recorded and is smeared off by interpolation. Thus, the inverse problem of inferring the overall distribution function F based on experimentally available, uncertain and discrete OEC Y_{exp} should be considered as an ill-posed problem. Therefore, as the preliminary step of inverse problem analysis, in simulations and test runs instead of Y_{exp} we used proxy OECs calculated according to Eqs. (5) and (6) for a pre-defined distribution function $F(a)$. In this case, the "experimental" input data in the inverse problem analysis was self-consistent, without experimental uncertainty.

The formal notation for the inverse operator A_m^{-1} reads as

$$F = A_m^{-1}(Y_{exp}), \quad A_m^{-1} : \Lambda_m \to \Phi, \quad m = 1, 3. \tag{13}$$

It is assumed in the solution of inverse problem that the linear and non-linear stages of the process can be clearly distinguished based on OEC, thus the reference time moments are functionals of Y_{exp}, i.e., $t_\pm = t_\pm[Y_{exp}]$. This assumption is essential for the suggested algorithm. Finally, the calculation of the right-hand side of Eq. (13) can be reduced to the inversion of an integral equation which is derived below.

Consider the function

$$G(t) = t_+ - \int_t^{t_+} \frac{d\tau}{k(\tau)}, \quad G(t \geq t_+) = t. \tag{14}$$

With the definition (14) the forward operator (10) takes the following equivalent form

$$G\left(t - Y_{exp}(t)\right) = G(t) - 1, \quad t > t_-[Y_{exp}]. \tag{15}$$

The observed Y_{exp}-OEC defines the function $G(t - Y_{exp})$. It can be shown that $G(t - Y_{exp}) = t - 1$ at $t > t_+$, since $Y_{exp}(t > t_+) = 1$. Equation (15) is used to calculate $G(t - Y_{exp})$ for $t_- < t < t_+$, or, equivalently, $G(x)$ at $0 < x < t_+ - 1$. Particularly, $G(0) = G(t_-) - 1$. The details of the recursive algorithm of calculation of $G(x)$ are given in Ref. [17].

With the use of this algorithm, the continuous function $G(x)$ can be calculated for any $x > 0$ at a given Y_{exp}-function. Then, $k(t)$ is obtained as

$$\left(\frac{dG(t)}{dt}\right)^{-1} = k(t) \equiv \int_0^\infty s\left(\frac{t}{a^2}\right) f(a)da. \tag{16}$$

Integrating the right-hand side of Eq. (16) by parts, one arrives at the integral equation with respect to F with the right-hand side depending solely on Y_{exp}:

$$\int_0^1 F\left(\sqrt{\frac{t}{\varphi_m(s)}}\right) ds = \left(\frac{dG(t)}{dt}\right)^{-1}. \tag{17}$$

Equation (17) can be solved in a closed form for flat particles at $m = 1$ and $\varphi_1(s) = s^2$. The change of variables $\xi = t^{1/2}s^{-1}$ results in

$$\int_{\sqrt{t}}^\infty F(\xi)\frac{d\xi}{\xi^2} = \left(\sqrt{t}\frac{dG(t)}{dt}\right)^{-1}.$$

Differentiating the above equation with respect to t, after substitution a for $t^{1/2}$ one obtains

$$F(a) = -2a^2 \frac{d}{da}\left(\frac{dG(a^2)}{da}\right)^{-1}. \tag{18}$$

Equation (18) is a direct confirmation that the formulated inverse procedure is an ill-posed problem. The searched function F for the packed bed of flat particles is a result of double differentiation of function $G(x)$ which in turn depends on the experimentally obtained OEC Y_{exp}. Apparently, a function $F(a)$ straightforwardly deduced from using Eq. (18) for an arbitrary experimental (noisy) curve Y_{exp}, which does not belong to Λ_m, is not, in general, a cumulative distribution function at all. It may take negative values, be greater than unity, or be non-monotonous. Thus, a regularization procedure is required to explicitly impose these constraints and to arrive at a physically reasonable approximation of F based on experimental OEC.

3.2 *Numerical Algorithm for the Case of Spherical Particles*

We do not have an explicit solution of Eq. (17) for spherical particles, $m = 3$, since there is no analytical expression for φ_3^{-1} in Eq. (8). The non-linear equation is solved numerically, using an iterative algorithm with solution (18) for flat particles suggested and tested as a pre-conditioner.

For the numerical scheme, it is assumed that the searched distribution function F belongs to the set $C_N(\xi; \Xi)$ of continuous piece-wise linear functions with the domain $[\xi; \Xi]$, where $\Xi > (t_+ - 1)^{0.5}$ is an arbitrary fixed constant, $t_+ = t_+[Y_{exp}]$, and ξ is the smallest particle size in the packed bed that can be deduced from the available (discrete) Y_{exp}-data. The lower bound for Ξ assumes, see Eq. (12), that the complete OEC is known, a complete extraction of the packed bed was reached at the experiment. The subscript N is the number of subsegments of the uniform mesh introduced on the segment $[\xi; \Xi]$. Corresponding nodes are $\xi = a_0 < a_1 < \cdots < a_N = \Xi$, where $a_i = \xi + i(\Xi - \xi)/N$. Distribution functions are defined in a_i-nodes with linear interpolation in between. It is assumed that $F(a_0) = 0$ and $F(a_N) = 1$. Finally, the set $C_N(\xi; \Xi)$ is comprised of functions with bounded derivatives and satisfies the following conditions

$$C_N(\xi; \Xi) = \left\{ F(a) | F(\xi) = 0 \le F \le 1 = F(\Xi), \quad 0 \le \frac{dF}{da} \le \frac{N}{\Xi - \xi} \right\}.$$

While every function F from $C_N(\xi; \Xi)$ is analyzed solely in the domain $[\xi; \Xi]$, to make it a cumulative distribution function, the following extrapolation outside the $[\xi; \Xi]$-domain is assumed

$$F(a) = 0, \quad 0 \le a \le \xi, \quad F(a) = 1, \quad \Xi \le a < +\infty.$$

The defined set of functions $C_N(\xi;\,\Xi)$ is included in the normed space $L_1(\xi;\,\Xi)$ of integrable functions, and is a compact according to the Arzela-Ascoli theorem [18]. The norm $\|\bullet\|_{L_1}$ of $L_1(\xi;\,\Xi)$-space can be introduced to measure the "distance" between two arbitrary elements from $C_N(\xi;\,\Xi)$.

The experimental OECs Y_{exp} are represented as continuous, piece-wise linear functions known in the nodes $t_i = i(1+\Xi^2)/M$ of the uniform mesh $0 = t_0 < t_1 < \cdots < t_M = 1+\Xi^2$ over the segment $[0;\,1+\Xi^2]$. The lower bound constraint on Ξ suggests that $t_+ < t_M$.

The discretized quasi-solution of the integral equation (17) at $m = 3$ is searched iteratively after [19], as a solution of the following non-linear functional equation

$$\frac{A_1\left(\bar{F}^{(k+1)}\right) - A_1\left(F^{(k)}\right)}{\sigma} = Y_{exp} - A_3\left(F^{(k)}\right), \quad k = 1, 2 \ldots.$$

Here, A_1 is used as a preconditioner in the algorithm, σ is the relaxation parameter, and k is the number of current iteration. The auxiliary function $\bar{F}^{(k+1)}$ is the solution of Eq. (18) with the OEC given as $Y_1^{(k+1)}$

$$Y_1^{(k+1)} = A_1\left(F^{(k)}\right) + \sigma\left(Y_{exp} - A_3\left(F^{(k)}\right)\right), \quad Y_1^{(1)} = Y_{exp}. \tag{19}$$

As mentioned above, $\bar{F}^{(k+1)}$ does not necessarily belong to the set $C_N(\xi;\,\Xi)$, and

$$\bar{F}^{(k+1)} = A_1^{-1}\left(Y_1^{(k+1)}\right), \quad F^{(k+1)} = \Pr\left(\bar{F}^{(k+1)}\right), \tag{20}$$

where the operator Pr projects $\bar{F}^{(k+1)}$ onto the set $C_N(\xi;\,\Xi)$.

The cumulative distributions are non-decreasing and bounded functions, $0 \le F^{(k+1)} \le 1$ inherited by the Pr-operator, which regularizes the ill-posed inverse problem (13). At the first stage, an intermediate function $\hat{F}^{(k+1)}$ is defined as a solution of the minimization problem

$$\hat{F}^{(k+1)} = \arg\min_F \left\|\bar{F}^{(k+1)} - F\right\|_{L_1}. \tag{21}$$

Construction of $\hat{F}^{(k+1)}$ based on $\bar{F}^{(k+1)}$ is demonstrated in Fig. 3. First, an integrated function

$$\mathrm{I}\left[\bar{F}^{(k+1)}\right](a) = \int_\xi^a \bar{F}^{(k+1)}(a)da$$

is introduced and can be evaluated explicitly, since the integrand $\bar{F}^{(k+1)}$ is a piece-wise linear function. Thus, the integrand values $\mathrm{I}\left[\bar{F}^{(k+1)}\right]$ are known at the same set of a-nodes as $\bar{F}^{(k+1)}$. The integrand is convex in the regions where $\bar{F}^{(k+1)}$ is monotonous, and non-convex otherwise. This explicitly indicates the regions where

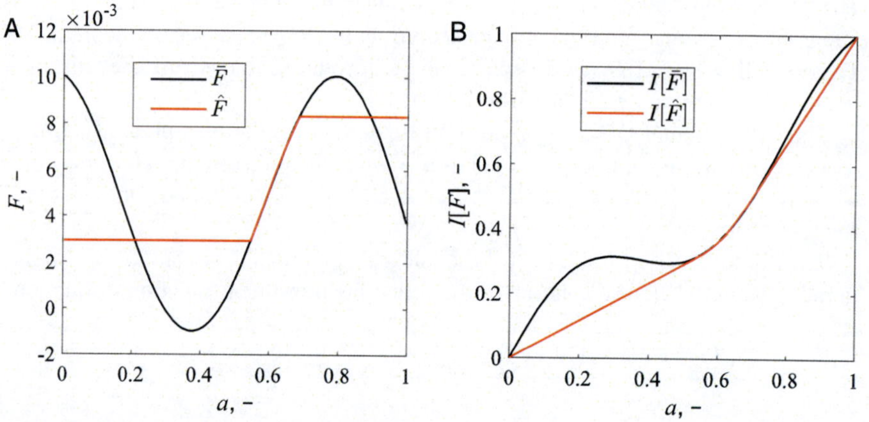

Fig. 3 The act of operator Pr on an arbitrary non-monotonous function. For demonstration, $\mathrm{I}\left[\bar{F}\right] = a + a(1-a)\cos(5a)$ is assumed. **(a)** The original function $\bar{F}$ and its non-decreasing L_1-analog $\hat{F}$. By construction, $\hat{F} = \bar{F}$ where $\bar{F}$ is monotonous, and $\hat{F}$ takes a constant value where $\bar{F}$ is not monotonous. (**b**) The integrals $\mathrm{I}\left[\bar{F}\right]$ and $\mathrm{I}\left[\hat{F}\right]$ of $\bar{F}$ and $\hat{F}$, respectively. Note that $\mathrm{I}\left[\hat{F}\right]$ is the lower convex-hull of $\mathrm{I}\left[\bar{F}\right]$

the original function $\bar{F}^{(k+1)}$ has to be "repaired" to restore its global monotonicity. Consider an integral function

$$\mathrm{I}\left[\hat{F}^{(k+1)}\right](a) = \int_{\xi}^{a} \hat{F}^{(k+1)}(a)da$$

of the searched monotonous function $\hat{F}^{(k+1)}$. It can be shown that $\mathrm{I}\left[\hat{F}^{(k+1)}\right]$ is a lower convex hull of $\mathrm{I}\left[\bar{F}^{(k+1)}\right]$. Since, each function is defined on a discrete, finite set of points, the lower convex hull can be found using standard algorithms, like Graham's scan and Jarvis' march [20]. Once the lower convex hull is deduced, it's derivative is calculated numerically to find the searched solution $\hat{F}^{(k+1)}$ of Eq. (21). For instance, in MatLab this procedure is implemented using a built-in operator "diff". Finally, the projection $F^{(k+1)}$ is delivered by

$$F^{(k+1)} = \min(1, \max(0, \hat{F}^{(k+1)})).$$

4 Test Runs

To test the algorithm convergence, an "experimental" OEC Y_{exp} was generated based on a predefined "experimental" cumulative distribution function F_{exp} with a known maximum particle size $a_{\max}$. By definition, $Y_{exp} = A_3(F_{exp})$, and, thus, the F_{exp}-function was the proxy of the inverse-problem solution to be inferred from the values $Y_{exp}(t_i)$ calculated in advance in a discrete set of time moments $t_i, i = 0..M$.

The number of mesh nodes for Y_{exp} and F_{exp} was chosen as $N = 50$ and $M = 30$, respectively. The minimum particle size ξ was fixed as 0.3, while $\Xi = 2(a_{\max}^2 + 1)$ if not specified otherwise. The iterations of numerical algorithm (19) and (20) were carried out until both conditions

$$\left\| F^{(k+1)} - F^{(k)} \right\|_{L_1} < \varepsilon_1, \qquad \max_{t_i, i=1..M} \left| Y_{exp} - A_3\left(F^{(k+1)} \right) \right| < \varepsilon_2$$

were satisfied. It was assumed that $\varepsilon_1 = \varepsilon_2 = 10^{-3}$, and typically 10–15 iterations were sufficient to attain the desired accuracy of the solution.

Convergence at different ξ was, first, investigated for a monodisperse packed bed of particle size $a = 1$. Corresponding original particle size distribution function F_{exp} was zero at $a < 1$ and unity otherwise, i.e. the density function was the Dirac's delta-function $f(a) = \delta(a - 1)$. The function F_{exp} generates the OEC Y_{exp}, which is shown by the red curve in Fig. 2a. Robustness of the algorithm is demonstrated by Fig. 4. For relatively large $\xi = 0.5$ (Fig. 4a), the iterations promptly tend to the "experimental" distribution function. The domain of non-zero values of $f(a)$ is approx. [0.9; 1.1]. The recursive algorithm of calculation of $G(x)$

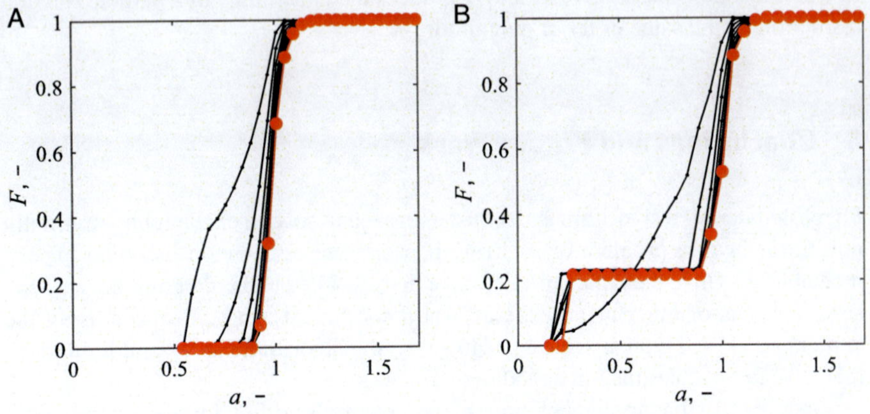

Fig. 4 The convergence of the iterative algorithm for monodisperse packed bed of particle size $a = 1$ at $N = 50$, $M = 30$, $\Xi = 4$, and **(a)** $\xi = 0.5$, **(b)** $\xi = 0.1$. The jump in **(b)** at $a \sim 0.15$ indicates that the transition part of OEC is insufficiently resolved by the given uniform mesh of $N = 50$ nodes to correctly infer the particle size distribution of smaller particles, $a < 0.3$

requires the values of OEC at intermediate time-moments, between the nodes t_i. The interpolation introduces an error, though relatively small. Thus, the algorithm does not converge to an exact F_{exp}.

At smaller $\xi \sim 0.1$, an artifact is observed near $a \sim 0.15$ (Fig. 4b). It shows that $\sim 25\%$ of packed bed volume is occupied by a false fraction of particles of size $a \sim 0.15$, while the other fraction representing the real particles is detected near $a = 1$. While the two final approximations of distribution functions, $F^{(5)}$ and $F^{(6)}$, are close in L_1-space, the final approximation $F^{(6)}$ is far from the original distribution F_{exp}. Thus, the algorithm did not converge at $\xi = 0.1$. Computational experiments showed that the artefact remains at $\xi < 0.25$ and disappears otherwise at fixed values of M and N.

We attribute the nature of this artifact with the resolution limitations of the approach to the inverse problem solution. As we explained in Sect. 2.2 and Fig. 2b, the small particle fraction of size $a \sim \xi$ determines the rate of the extraction process transition from an initial linear stage to the final non-linear one. The smaller the particle size the more rapid the transition is. Thus, for a coarse temporal mesh, the linearly interpolated OECs render indistinguishable once the smallest particle fraction size is reduced below a certain limit. Corresponding transition stage is so short that it could not be resolved by the current temporal mesh. The number M of time-nodes must be increased in this case.

At $\xi = 0.3$, the iterative algorithm reveals reliable convergence for a variety of experimental particle size distributions and fixed $N = 50$ and $M = 30$. This is confirmed by a series of tests for two monodisperse and one bimodal distribution functions $F_{exp}(a)$ illustrated by Fig. 5. Comparison of Fig. 5b and d demonstrates the effect of M and N on the algorithm output. A monodisperse packed bed with $f(a) = \delta(a - 10)$ was studied at different discretization numbers M and N. A 10-fold simultaneous increase of M and N reduces the uncertainty of a particle fraction resolution by the same order of magnitude.

5 Conclusions and Future Work

Suggested iterative algorithm is a robust approach to solve a challenging essentially non-linear inverse problem of inferring particle size distribution based on experimental OEC. Once the typical particle size in the packed bed is determined by either sieve analysis or other means, the diffusion coefficient can be deduced using the expression (1) for typical scale a_{sc}. However, the algorithm has several limitations that have to be underlined in concluding remarks.

First of all, the implementation of recursive algorithm for calculating $G(x)$ crucially depends on the measurement accuracy of the t_+-moment of complete extraction of the packed bed. Therefore, it was assumed that the discrete OEC is known at least for the entire time frame $[0; t_+]$ with a uniform time step, and a complete extraction of the packed bed was reached during the experiment. This provides the resolution of the largest particle fraction, while the temporal

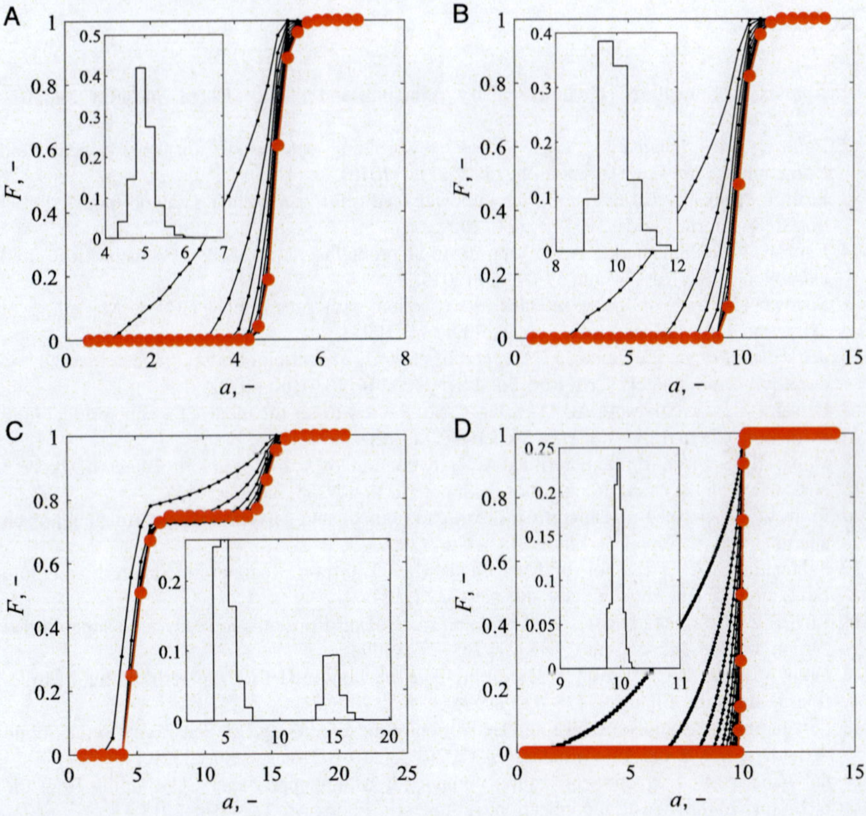

Fig. 5 The convergence of the iterative algorithm at $N = 50$, $M = 30$, for (**a**) $a = 5$, (**b**, **d**) $a = 10$, (**c**) bimodal distribution with $f(a) = 0.75\delta(a-5)+0.25\delta(a-15)$. (**d**) is obtained at $N = 500$, and $M = 300$. A better resolution of a single particle fraction is observed if compared with (**b**). Inserts are histograms of final particle size distributions

discretization sets up the resolution limit (given by ξ) for the smallest particle fraction. Such long-term experiments are expensive and rare in practice. Therefore, the approach should be further developed to take into account that the largest particle fraction is not resolved by the data.

An accurate specification of t_+ as well as t_- is challenging even for complete OEC since the data is typically noisy. To deduce a way to project experimental, noisy OECs onto the set Λ_m is the main challenge in this problem since it drastically affects the inferred values of t_- and t_+.

Acknowledgments The research was supported by the Russian Foundation for Basic Research through the grant No. 19-31-60013. This paper has been partially supported by the Kazan Federal University Strategic Academic Leadership Program ("PRIORITY-2030"). Egorov A.G. is thankful for the partial financial support from the government assignment for FRC Kazan Scientific Center of RAS for the analytical study of the forward problem.

References

1. Anastas, P., Eghbali, N.: Green chemistry: principles and practice. Chem. Soc. Rev. **39**, 301–312 (2010).
2. Egorov, A.G., Salamatin, A.A.: Bidisperse shrinking core model for supercritical fluid extraction. Chem. Eng. Technol. **38**, 1203–1211 (2015).
3. Fiori, L.: Supercritical extraction of sunflower seed oil: Experimental data and model validation. J. Supercrit. Fluids. **50**, 218–224 (2009).
4. Egorov, A.G., Salamatin, A.A.: Optimization problems in a theory of supercritical fluid extraction of oil. Russ. Math. **59**, 48–56 (2015).
5. Sovova, H.: Rate of the vegetable oil extraction with supercritical CO_2–I. Modelling of extraction curves. Chem. Eng. Sci. **49**, 409–414 (1994).
6. del Valle, J.M., de La Fuente, J.C.: Supercritical CO_2 extraction of oilseeds: review of kinetic and equilibrium models. Crit. Rev. Food Sci. Nutr. **46**, 131–160 (2006).
7. Oliveira, E.L.G., Silvestre, A.J.D., Silva, C.M.: Review of kinetic models for supercritical fluid extraction. Chem. Eng. Res. Des. **89**, 1104–1117 (2011).
8. Duba, K.S., Fiori, L.: Supercritical CO_2 extraction of grape seed oil: Effect of process parameters on the extraction kinetics. J. Supercrit. Fluids. **98**, 33–43 (2015).
9. Fiori, L.: Grape seed oil supercritical extraction kinetic and solubility data: Critical approach and modeling. J. Supercrit. Fluids. **43**, 43–54 (2007).
10. Salamatin, A.A.: Detection of Microscale Mass-Transport Regimes in Supercritical Fluid Extraction. Chem. Eng. Technol. **40**, 829–837 (2017).
11. Roy, B.C., Goto, M., Hirose, T.: Extraction rates of oil from tomato seeds with supercritical carbon dioxide. Jpn. J. Chem. Eng. **27**, 768–772 (1994).
12. Goto, M., Roy, B.C., Hirose, T.: Shrinking-core leaching model for supercritical-fluid extraction. J. Supercrit. Fluids. **9**, 128–133 (1996).
13. Salamatin, A.A.: Supercritical Fluid Extraction of the Seed Fatty Oil: Sensitivity to the Solute Axial Dispersion. Ind. Eng. Chem. Res. (2020) doi: 10.1021/acs.iecr.0c03329.11.
14. del Valle, J.M.: Extraction of natural compounds using supercritical CO_2: Going from the laboratory to the industrial application. J. Supercrit. Fluids. **96**, 180–199 (2015).
15. Salgin, U., Korkmaz, H.: A green separation process for recovery of healthy oil from pumpkin seed. J. Supercrit. Fluids. **58**, 239–248 (2011).
16. del Valle, J.M., Carrasco, C. V., Toledo, F.R., Nunez, G.A.: Particle size distribution and stratification of pelletized oilseeds affects cumulative supercritical CO_2 extraction plots. J. Supercrit. Fluids. **146**, 189–198 (2019).
17. Egorov, A.G., Salamatin, A.A., Maksudov, R.N.: Forward and inverse problems of supercritical extraction of oil from polydisperse packed bed of ground plant material. Theor. Found. Chem. Eng. **48**, 39–47 (2014).
18. Dieudonnee, J. ed: Foundations of Modern Analysis. (1969).
19. Il'inskii, N.B., Mardanov, R.F., Solov'ev, S.A.: Combined method for solving an inverse boundary value problem of aerohydrodynamics for an axisymmetric body. Comput. Math. Math. Phys. **48**, 1234–1242 (2008).
20. Preparata, F.P., Shamos, M.I.: Computational Geometry. Springer New York, New York, NY (1985).

Mathematical Model of a Dynamically Loaded Thrust Bearing of a Compressor and Some Results of Its Calculation

Nikolay V. Sokolov, Mullagali B. Khadiev, Pavel E. Fedotov, and Eugeny M. Fedotov

Abstract The basic equations of the full three-dimensional periodic thermoelastohydrodynamic (PTEHD) model of stationary and dynamic modes of operation of a thrust sliding bearing with fixed pads of a centrifugal or screw compressor are presented. The dynamic loading of the bearing is created by directly using the equation of axial displacement of the thrust collar of the compressor rotor within the bearing's operating clearance. Some results of calculations of a loaded thrust bearing are presented, showing the fundamental capabilities of the numerically implemented Sm2Px3Txτ program.

1 Introduction

Thrust bearings (TB) are most widely used in designs of centrifugal and screw compressors used in various industries. They are designed to absorb the axial load from gas forces and/or helical gearing and transfer it to the compressor stator. The working conditions of the TB can change significantly throughout the entire period of their operation. This is primarily due to transient modes of compressor operation [1]. These modes can occur when the compressor is connected to the discharge network or when it is disconnected from it by transferring to the bypass line, as well as when the characteristic of a centrifugal compressor changes, for example, when changing from one rotor speed to another. They can also occur when the characteristics of the suction and discharge networks change, when additional sections with their hydraulic resistance are connected or disconnected.

N. V. Sokolov · M. B. Khadiev
Kazan National Research Technological University, Kazan, Russia

P. E. Fedotov
Kazan Federal University, Kazan, Russia

E. M. Fedotov (✉)
AST Volga Region LTD, Kazan, Russia

I. B. Badriev et al. (eds.), *Mesh Methods for Boundary-Value Problems and Applications*, Lecture Notes in Computational Science and Engineering 141,
https://doi.org/10.1007/978-3-030-87809-2_35

The most significant influence on the operation of a thrust bearing is also exerted by non-stationary gas-dynamic processes occurring in the flow path of a centrifugal compressor (CC), which include stall, rotating stall and surge [2, 3]. So, when the CC surges, significant fluctuations occur with a frequency of (1... 10) Hz in the volume of compressed gas filling the flow path of the compressor and the network. As a result, the axial load, which acts on the rotor, changes significantly until the sign changes and can reach the maximum value for the bearing. Under the action of the load, the thrust collar of the rotor moves within the axial clearance, which leads to a dynamic loading of the thrust bearing and a change in its local and integral characteristics over time. An important local characteristic is the maximum lubricant temperature in the hydrodynamic film of the bearing.

Of the integral characteristics, the most important are the bearing capacity (load capacity), friction power losses, lubricant consumption through the inlet and outlet sections of the lubricating film, heat fluxes through the sections of structural elements and the lubricant film, etc. The results of experimental studies carried out by the authors also confirm the dynamic nature of processes in lubricating bearing films [4]. At large amplitudes of collar displacement, an increase in the maximum lubricant temperature to the limit value, axial displacement of the rotor and subsequent contact and, consequently, failure of the UPS can occur. It follows from this that the study of the dynamic loading of the thrust bearing is directly related to the increase in the reliability of centrifugal and screw compressors during transient operating conditions.

In the present studies, a new three-dimensional periodic thermoelastohydrodynamic (PTEHD) model of stationary and dynamic modes of operation of a thrust bearing with fixed pads is used [5]. Based on the numerical experiments carried out using the developed program Sm2Px3Txτ [6], some results of the dependence of the local, distributed and integral characteristics of the TB on time during axial displacement of the collar are presented, showing the fundamental capabilities of the calculation program.

Studies of a dynamically loaded thrust bearing with a different profile of the working surfaces of fixed pads were also carried out by the authors of [7, 8]. However, the researchers either do not take into account the mutual influence of the pads on each other during the flow of the lubricant in the direction of rotation of the collar, taking into account the supply of fresh fluid from the inter-pad channels, or they neglect the joint heat exchange processes in the lubricating films and elements of the bearing structure.

2 Formulation of the Problem: Non-stationary PTEHD Model

The design of the thrust bearing of the compressor in accordance with the design diagram shown in Fig. 1a–c, implies the presence of bearing fixed pads 1 with

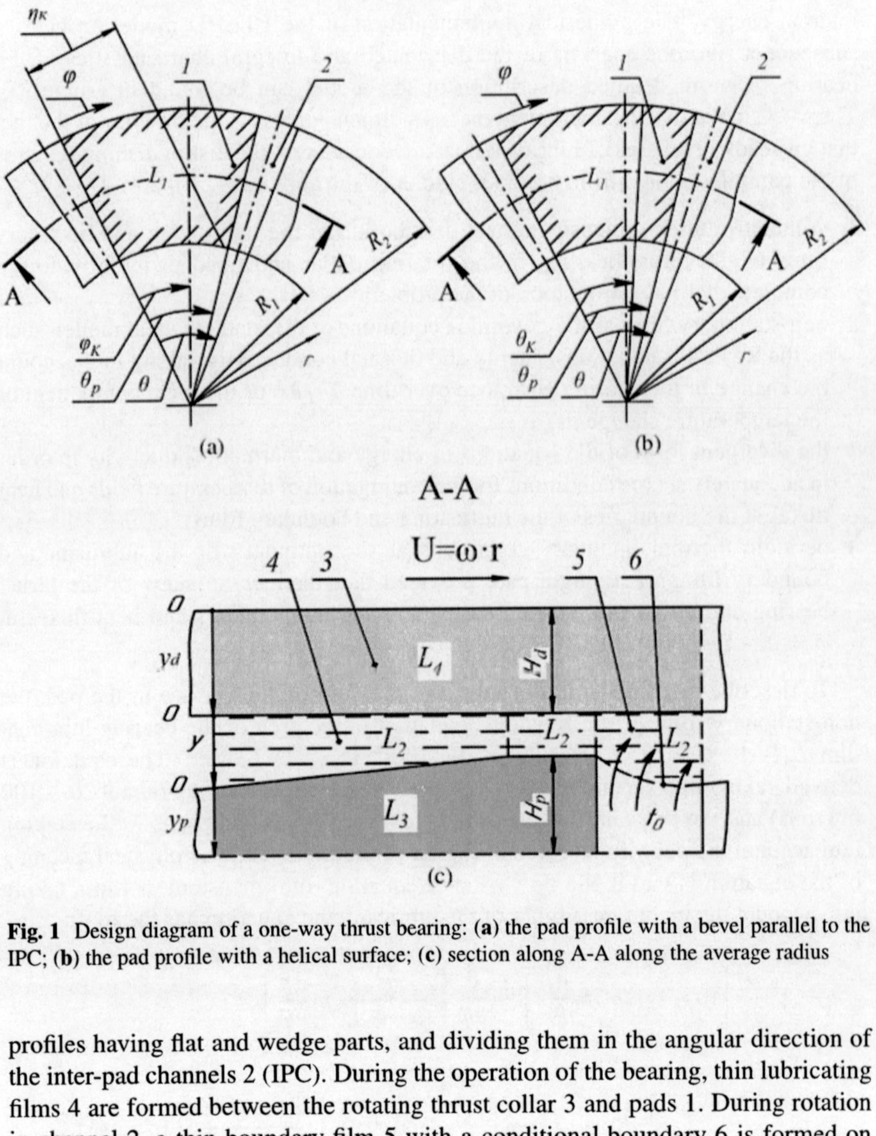

Fig. 1 Design diagram of a one-way thrust bearing: **(a)** the pad profile with a bevel parallel to the IPC; **(b)** the pad profile with a helical surface; **(c)** section along A-A along the average radius

profiles having flat and wedge parts, and dividing them in the angular direction of the inter-pad channels 2 (IPC). During the operation of the bearing, thin lubricating films 4 are formed between the rotating thrust collar 3 and pads 1. During rotation in channel 2, a thin boundary film 5 with a conditional boundary 6 is formed on the surface of the collar, interacting with the lubricating films of adjacent pads and fresh fluid of channel 2. Fresh liquid enters through the conditional boundary 6 of the boundary film from the inter-pad channel 2 and further, mixing with the hot waste stream of lubricant of the previous pad, it enters the inlet section at $\varphi = 0$ of the lubricating film 4 of the next pad.

The mathematical description of the operation of a thrust bearing with fixed pads was constructed using a non-stationary periodic TEHD model, which showed the greatest convergence with the results of a physical experiment [9]. The mathematical model is based on the fundamental laws of conservation of mass, momentum and internal energy. The numerical implementation of the PTEHD model provides a subsequent rigorous analysis of the distributed and integral characteristics of the bearing. A more detailed description of the model can be found in article [5]. The «—» sign above the coefficient means a dimensionless value. In view of the fact that unsteady processes of lubricant flow are considered, the distinguishing features of the compiled model from previous studies of the authors [9, 10] are:

1. volumetric three-dimensional heat distribution in the lubricating and boundary films of the thrust bearing, in the rotating collar and fixed pads, providing a complete thermal formulation of the problem;
2. non-stationary form of the governing equations of the mathematical model, such as the Reynolds equations, energy and thermal conductivity, taking into account the change in the bearing clearance over time $\partial h/\partial\tau$ or the local component of the temperature change $\partial t/\partial\tau$;
3. the divergent form of the equations of energy and thermal conductivity in order to adequately set the conditions for the conjugation of temperature fields and heat fluxes at the boundaries of the lubricating and boundary films;
4. periodic thermal boundary conditions at the boundary of the lubricant and boundary films of a single pad, provided that there is no skew of the thrust bearing and collar runout, i.e., the equality of temperatures and heat fluxes at $\overline{\varphi} = 0, -1 \le \overline{r} \le 1, 0 \le \overline{y} \le 1$.

To describe the distribution of lubricant pressure on the surface of the pad, the non-stationary form of the Reynolds equation in the area of the bearing lubricant film $\overline{L}_1(-1 \le \overline{r} \le 1, 0 \le \overline{\varphi} \le \overline{\theta}_n, 0 \le \overline{y} \le 1)$ is used. The equation is derived, taking into account the low clearance height in the bearing (about 20…100 microns) and the constancy of pressure across the film thickness, as well as taking into account the centrifugal force of inertia of the lubricant. The physical meaning of the equation is that it is a flow balance equation. In dimensionless form, taking into account the incompressibility of the lubricant, the equation has the form:

$$-\lambda^2\frac{\partial}{\partial\bar{r}}\left[(\sigma\bar{r}+1)\bar{h}^3\bar{f}_0\frac{\partial\bar{p}}{\partial\bar{r}}\right]-\frac{\partial}{\partial\bar{\varphi}}\left[\frac{\bar{h}^3}{(\sigma\bar{r}+1)}\bar{f}_0\frac{\partial\bar{p}}{\partial\bar{\varphi}}\right]=$$
$$=-Re\psi\sigma\lambda^2\frac{\partial(\bar{h}^3\bar{f}_1)}{\partial\bar{r}}+\omega(\sigma\bar{r}+1)\frac{\partial(\bar{h}\bar{f}_2)}{\partial\bar{\varphi}}+Sh(\sigma\bar{r}+1)\bar{A}, \tag{1}$$

where $\bar{A} = \frac{\partial}{\partial\bar{\tau}}\left(\bar{h}\int_0^1\bar{\rho}d\bar{y}\right) - \bar{\rho}_{\bar{y}=1}\frac{\partial\bar{h}}{\partial\bar{\tau}}$ —the non-stationary multiplier; $\bar{r} = (r - R_{av})$, $\bar{\varphi} = \varphi/\theta$ —dimensionless coordinates; $\bar{p} = ph_{20}^2/\left(\mu_0\omega_* R_{av}^2\theta\right)$ —local dimensionless pressure; $\lambda = (2R_{av}\theta)/(R_2 - R_1)$, $\sigma = (R_2 - R_1)/(R_2 + R_1)$ —the relative length and width of the pad, respectively; $\psi = h_{20}/(R_{av}\theta)$ – relative thickness of the lubricating film; $Re = \rho_0\omega_* R_{av}h_{20}/\mu_0$—Reynolds criterion;

$Sh = (R_{av}\theta)/(\omega R_{av}\tau^*)$—Strouhal criterion; h_{20} – characteristic thickness, for example, half of the total clearance of a double-sided bearing.

The dimensionless functions $\bar{f}_0$, $\bar{f}_1$, $\bar{f}_2$ included in Eq. (1), which take into account the variability of the lubricant viscosity and the constancy of pressure over the film thickness, have the form:

$$\bar{f}_0 = \int_0^1 \bar{\rho}\left(\bar{i}_1 - \frac{\bar{m}_1}{\bar{m}_0}\bar{i}_0\right)\mathrm{d}\bar{y},\ \bar{f}_1 = \int_0^1 \bar{\rho}\left(\bar{j} - \frac{\bar{n}}{\bar{m}_0}\bar{i}_0\right)\mathrm{d}\bar{y},\ \bar{f}_2 = \int_0^1 \bar{\rho}\left(1 - \frac{\bar{i}_0}{\bar{m}_0}\right)\mathrm{d}\bar{y} \tag{2}$$

Equation (1) is supplemented with boundary conditions along the entire contour of the pad:

1. at $\bar{r} = -1$ and $\bar{r} = 1$, $(0 \le \bar{\varphi} \le \bar{\theta}_p, 0 \le \bar{y} \le 1)$, pressures $\bar{p}_{R_1}$ and $\bar{p}_{R_2}$ are set;
2. at $\bar{\varphi} = 0$ and $\bar{\varphi} = \bar{\theta}_p$, $(-1 \le \bar{r} \le 1,)$, the pressure gradient $\partial\bar{p}/\partial\bar{\varphi} = 0$ is specified (Neumann condition) or calculated by interpolation between $\bar{p}_{R_1}$ and $\bar{p}_{R_2}$ pressure $\bar{p}_{\varphi=0}$ and $\bar{p}_{\varphi=\theta_p}$ (Dirichlet condition).

To describe the temperature distribution of the lubricant in the area of the lubricant and boundary films $\overline{L}_2(-1 \le \bar{r} \le 1, 0 \le \bar{\varphi} \le 1, 0 \le \bar{y} \le 1)$, a three-dimensional non-stationary internal energy equation is used, which in dimensional divergent form has the form:

$$c_p\left(\rho\frac{\partial t}{\partial\tau} + t\frac{\partial\rho}{\partial\tau}\right) + \frac{1}{r}\frac{\partial}{\partial r}(c_p\rho r V_r t) + \frac{\partial}{\partial\varphi}\left(\frac{c_p\rho}{r}V_\varphi t - \frac{\lambda_{oil}}{r^2}\frac{\partial t}{\partial\varphi}\right) + \\ +\frac{\partial}{\partial y}\left(c_p\rho V_t - \lambda_{oil}\frac{\partial t}{\partial}\right) = \mu\left[\left(\frac{\partial V_\varphi}{\partial y}\right)^2 + \left(\frac{\partial V_r}{\partial y}\right)^2\right], \tag{3}$$

where t—the local temperature, c_p, λ_{oil}—isobaric heat capacity and thermal conductivity of the lubricant.

Equation (3) was reduced to a dimensionless form at the stage of numerical implementation using a dimensionless temperature $\bar{t} = c_{po}\rho_0 h_{20}^2(t - t_0)/(\mu_0\omega_* R_{av}^2\theta)$ and a Jacobi matrix. It should be noted that in the transition to a dimensionless variable $\bar{y} = y/(h_{20}\bar{h})$, the dimensional area of L_2 the thrust bearing (Fig. 1a–c) is transformed from a curved view into a dimensionless area $\overline{L}_2$ having a rectangular view. This simplifies the specification of the nodes of the approximating mesh in the numerical implementation and analysis of the temperature level of the areas $\overline{L}_2$, $\overline{L}_3$, $\overline{L}_4$. The lubricant flow rates are derived from the truncated Navier-Stokes equations and have the form:

$$\bar{V}_r = \frac{V_r}{\omega_* R_{av}} = \lambda\bar{h}^2\frac{\partial\bar{p}}{\partial\bar{r}}\left(\bar{i}_1 - \frac{\bar{m}_1}{\bar{m}_0}\bar{i}_0\right)\delta_1 - \frac{\mathrm{Re}\,\psi\sigma\lambda\bar{h}^2}{(\sigma\bar{r}+1)}\left(\bar{j} - \frac{\bar{n}}{\bar{m}_0}\bar{i}_0\right) \tag{4}$$

$$\bar{V}_\varphi = \frac{V_\varphi}{\omega_* R_{av}} = \frac{\bar{h}^2}{(\sigma\bar{r}+1)}\frac{\partial \bar{p}}{\partial \bar{\varphi}}\left(\bar{i}_1 - \frac{\bar{m}_1}{\bar{m}_0}\bar{i}_0\right)\delta_1 + \bar{\omega}(\sigma\bar{r}+1)\left(1 - \frac{\bar{i}_0}{\bar{m}_0}\right) \quad (5)$$

In order to adequately set the thermal boundary conditions at the boundaries between the lubricating film and the surfaces of the pad and collar, the velocity $\bar{V}_y$ is calculated using two equations, where the transition boundary $\bar{y}_0$ is approximately equal to half the film thickness h:

$$\bar{V}_y = \frac{V}{\omega_* R_{av}} = -\frac{\psi}{\sigma\bar{\rho}}\int_0^{\bar{y}}\left[Sh\frac{\partial\bar{\rho}}{\partial\bar{\tau}} + \frac{1}{(\sigma\bar{r}+1)}\left(\bar{h}\left\{\theta\frac{\partial}{\partial\bar{r}}((\sigma\bar{r}+1)\bar{\rho}\bar{V}_r) + \right.\right.\right.$$

$$\left.\left.\left. +\sigma\frac{\partial}{\partial\bar{\varphi}}(\bar{\rho}\bar{V}_\varphi)\right\} - \bar{y}\left\{\bar{h}'_{\bar{r}}\theta\frac{\partial}{\partial\bar{y}}((\sigma\bar{r}+1)\bar{\rho}\bar{V}_r) + \bar{h}'_{\bar{\varphi}}\sigma\frac{\partial}{\partial\bar{y}}(\bar{\rho}\bar{V}_\varphi)\right\}\right)\right]d\bar{y}, \quad (6)$$

$$0 \le \bar{y} \le \bar{y}_0;$$

$$\bar{V}_y = \frac{V}{\omega_* R_{av}} = \frac{\psi}{\sigma\bar{\rho}}\int_{\bar{y}}^{1}\left[Sh\frac{\partial\bar{\rho}}{\partial\bar{\tau}} + \frac{1}{(\sigma\bar{r}+1)}\left(\bar{h}\left\{\theta\frac{\partial}{\partial\bar{r}}((\sigma\bar{r}+1)\bar{\rho}\bar{V}_r) + \right.\right.\right.$$

$$\left.\left.\left. +\sigma\frac{\partial}{\partial\bar{\varphi}}(\bar{\rho}\bar{V}_\varphi)\right\} - \bar{y}\left\{\bar{h}'_{\bar{r}}\theta\frac{\partial}{\partial\bar{y}}((\sigma\bar{r}+1)\bar{\rho}\bar{V}_r) + \bar{h}'_{\bar{\varphi}}\sigma\frac{\partial}{\partial\bar{y}}(\bar{\rho}\bar{V}_\varphi)\right\}\right)\right]d\bar{y}, \quad (7)$$

$$\bar{y}_0 \le \bar{y} \le 1,$$

where $\delta_1 = \{1,\ 0 \le \bar{\varphi} \le \bar{\theta}_p \text{ and } 0,\ \bar{\theta}_p \le \bar{\varphi} \le 1\}$—the unit function. In order to take into account, the inflowing fresh lubricant from the inter-pad channel over the thickness of the boundary film, the velocity $\bar{V}_y$ is calculated using Eq. (6), which is completely ignored over the film thickness from 0 to h.

The initial condition for Eq. (3) is the calculated temperature of the stationary process. At the boundaries $\bar{\varphi} = 0$ and $\bar{\varphi} = 1$, a periodic condition of equality of temperatures and heat fluxes is set, which largely determines the operating mode of the bearing [11]. Along the coordinate $\bar{r}$, when the velocity $\bar{V}_r$ is directed inside the area $\bar{L}_2$, the temperature $\bar{t} = 0$ of the inflowing lubricant is set; otherwise, the condition is not set. At the boundaries of contact of the lubricant with the surfaces of the pad and the thrust collar, the condition of equality of temperatures and heat fluxes is also set. At the boundary of the boundary film and the IPC at $(-1 \le \bar{r} \le 1,\ \bar{\theta}_p \le \bar{\varphi} \le 1,\ \bar{y} = 1)$, the temperature $\bar{t} = 0$ of the inflowing lubricant or the condition $\partial\bar{t}/\partial\bar{y} = 0$ when heat transfer occurs only by convection can be specified.

To describe the temperature distribution in the thickness of the pad in the area $\overline{L}_3(-1 \le \bar{r} \le 1,\ 0 \le \bar{\varphi} \le \bar{\theta}_p,\ 0 \le \bar{y}_p \le \Psi_p)$ and the thickness of the thrust collar in the region (Fig. 1a–c), three-dimensional unsteady equations of thermal conductivity are used, where $\Psi_{p,d} = H_{p,d}/(R_{av}\theta)$ is the relative thickness of the pad or collar. At the outer boundaries of the areas $\overline{L}_3$ and $\overline{L}_4$ to take into account the heat transfer, the Newton-Richman boundary conditions are set. Thus, Eqs. (3)–(6)

with the corresponding initial and boundary conditions simulate a three-dimensional temperature field. It allows one to consider the picture of heat distribution in the lubricant and boundary films and solid elements of the thrust bearing under various operating conditions.

The thrust collar is the only element through which mechanical energy is supplied from the outside to the bearing. In addition, the collar plays the role of a heat accumulator, which gives off or absorbs heat when it comes into contact with different parts of the lubricating and boundary films [11, p. 56]. Consequently, the use of a three-dimensional equation of thermal conductivity of the collar with appropriate boundary conditions is an improvement of the previously compiled and studied mathematical model [9, 10]. This model improves the accuracy of the calculations compared to previous studies. It also allows you to analyze the influence of the external environment and collar material on bearing characteristics, check the constancy of the collar temperature in the direction of rotation, and clarify the temperature $\bar{T}_{dp}$ of the working surface at $\bar{y}_d = \Psi_d$.

The governing differential equations of the mathematical model and their boundary conditions are interconnected with the help of such physical properties of the working lubricant as viscosity, density, heat capacity and thermal conductivity, as well as with the shape of the bearing clearance, which includes the geometric profile of the pad working surface and some operating parameters. At the same time, different profiling of the working surface of the pads is considered (Fig. 1a–c).

To describe the axial displacement of the thrust collar along the rotor axis and formulate the direct problem, the equation in dimensional form is used [12]:

$$y_{d.disp.} = y_{st} + y_d \tag{8}$$

where y_{st}, y_d—constant and dynamic components, respectively. In the case of soft surge [2, p. 15], the component y_d can be specified in the form of a harmonic law:

$$y_d = A \sin \Omega\tau \tag{9}$$

where A, Ω—the amplitude and frequency of the disturbing force during the surge, τ—the time.

In the case of hard surging, the displacement of the collar can be modeled as a curve with a sawtooth shape [2, p. 15]. The specified displacement $y_{d.disp.}$ as a term is substituted into the equation of the thrust bearing clearance shape, which also takes into account thermal deformations of the pad, and is reduced to a dimensionless form.

As a result of the numerical implementation, the local, distributed and integral characteristics of the thrust bearing are determined, the most important of which are the maximum lubricant temperature t_{max} in the film and the bearing capacity:

$$P = zP_i = z\int_0^{\theta_p}\int_{R_1}^{R_2} p_i r \mathrm{d}\varphi \mathrm{d}r, \tag{10}$$

where z—the number of pads, P_i—the bearing capacity of a single pad. The dimensionless bearing capacity coefficient is

$$\bar{P} = \frac{Ph_{20}^2}{\mu_0 \omega R_{av}^3 \theta^2 (R_2 - R_1)} \tag{11}$$

3 Calculation Results

One of the sides and some geometric dimensions of a thrust bearing with fixed pads of the research stand located in the laboratory of the Compressor Machines and Installations Department of the KNITU [4] was taken as the object of research. The stand is based on a compressor unit with a single-stage centrifugal compressor of the multiplier type, designed for air compression.

The stationary removable disc with fixed thrust bearing pads has pad bevels, made parallel to the inter-pad channel [5, 10], with the following dimensions: inner and outer diameters $D_1 = 70$ mm and $D_2 = 115$ mm; number of pads $z = 8$; angular length $\theta_p = 38.8°$; bevel width and depth—$h_k = 20$ mm and $\Delta h = 0.05$ mm; thrust collar thickness $H_d = 25$ mm; pad thickness $H_p = 5$ mm. The coordinate of an arbitrary center for a one-way bearing is taken $h_{20} = 0.5$ mm. The lubricant supply temperature is $t_0 = 40°$C; oil Tp-22S (ISO VG 32) was used as a lubricant. For the Reynolds equation, the Dirichlet boundary conditions are accepted at all edges of the pad. On the conditional boundary of the boundary film, the temperature of the lubricant in the inter-pad channel is set, i.e. $t = t_0$. The rotational speed of the thrust collar is taken to be $n = 5000$ rpm, i.e., the angular velocity of rotation is $\omega = 523.6$ rad/s. The minimum thickness of the lubricating film is defined as $h_{min} = (h_{20} - y_{st}) = 50\,\mu m$. The frequency of the perturbing displacement of the thrust collar during surge is taken from the interval $\Omega = 2\pi f = (6.28\ldots62.8)$ rad/s $= 31.4$ rad/s, where $f = (1\ldots10)$ Hz is the surge frequency. The amplitude of the disturbing displacement is taken equal to $A = 7.5\,\mu m$ or 15% of the minimum thickness of the lubricating film. In the stationary mode, the amplitude is equal to $A = 0$.

The diagram of the distribution of isobars over the working surface of the pad in the stationary mode is shown in Fig. 2. When the radial $\bar{r}$ and angular $\bar{\varphi}$ coordinates change, the pressure increases monotonically, reaching the maximum value $p_{max} = 1.11$ MPa inside the pad, which has a wedge and flat parts, and then monotonically decreases. The center of pressure pmax is displaced closer to the trailing edge and the outer radius due to the curvature in the plan of the pad at $\sigma = 0.24$ and the influence of centrifugal inertia forces. In this case, the profile of the working surface of the pad with a bevel parallel to the IPC turns the pressure extrema (dashed line) as the coordinate $\bar{\varphi}$ increases to the outer radius of the pad due to the variability of the gap height along the coordinate $\bar{r}$ [10].

The distribution of isotherms in the section of the lubricant and boundary films at the average radius of the pad in the stationary mode is shown in Fig. 3.

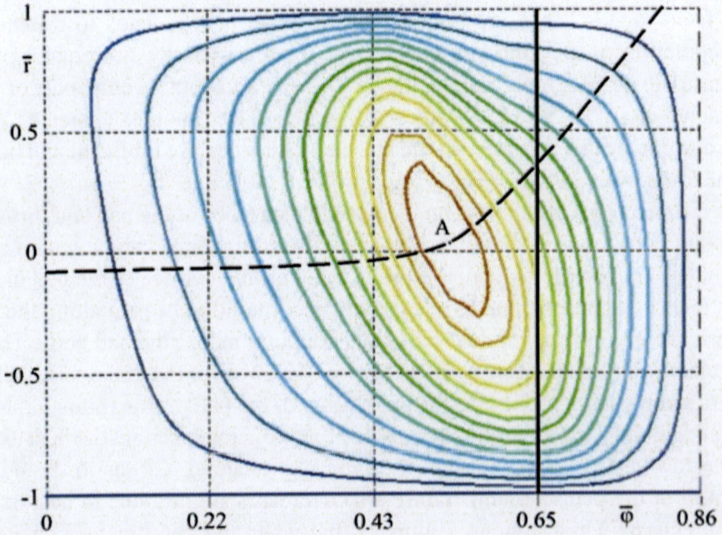

Fig. 2 Diagram of isobar distribution over the working surface of the pad (solid line is the boundary of the transition from the wedge to the flat part at the variable $\bar{\varphi}_k$)

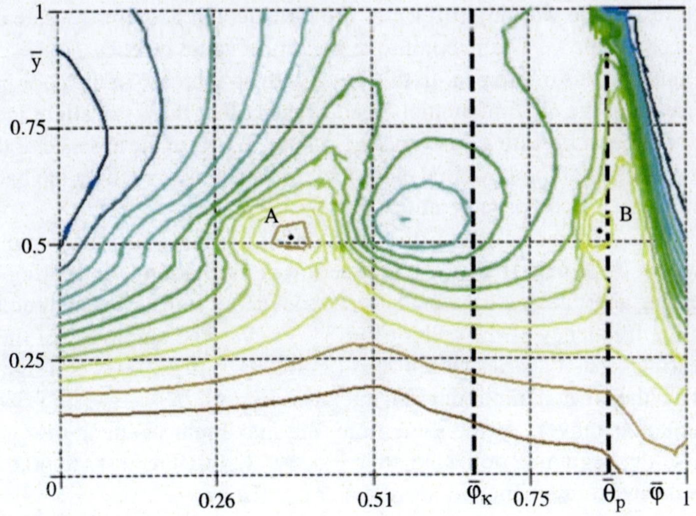

Fig. 3 Distribution of isotherms in the section of the lubricant and boundary films at the average radius of the pad at $\bar{r} = 0$

The temperature levels (isotherms) of the lubricant increase as the coordinate $\bar{\varphi}$ increases, which is associated with the release and accumulation of heat due to the dissipation of mechanical energy. Through the conditional boundary of the boundary film ($-1 \leq \bar{r} \leq 1$, $\bar{\theta}_p \leq \bar{\varphi} \leq 1$, $\bar{y} = 1$) fresh lubricant with a low temperature enters from the IPC into the boundary film. In turn, this leads to a decrease in the temperature of the lubricant in the inlet section with the subsequent pad. The maximum lubricant temperature is inside the area above the wedge part of the pad and reaches $t_{max} = 59.85\,°C$ (point A). In this case, another center of elevated temperature (point B) is formed at the boundary between the lubricant and boundary films near the boundary $\bar{\varphi} = \bar{\theta}_p$.

The distribution of isotherms on the working surfaces of the pad and thrust collar in the stationary mode is shown in Fig. 4a, b. From Fig. 4a, it can be seen that as the coordinate $\bar{\varphi}$ increases, the temperature of the working surface of the pad increases. This is due to an increase in the release and accumulation of heat along the flow of the lubricant and its transfer due to thermal conduction into the pad body. The value of the maximum temperature reaches $t_{max} = 49.02\,°C$ in the area located closer to the outer radius of the pad due to the peculiarity of the pad profile (point C). Near the trailing edge, the pad temperature drops due to taking into account the heat exchange with the lubricant flowing through the inter-pad channel. Closer to the outer and inner radii of the pad, the temperature also decreases slightly due to heat exchange with the external environment. Figure 4b that when the coordinate $\bar{\varphi}$ changes, the thrust collar temperature remains practically constant due to significant convective heat transfer along the collar rotation (coordinate $\bar{\varphi}$) compared to conductive heat transfer with lubricant and boundary films in the transverse direction (coordinate $\bar{y}_d$). As a result, the working surface of the thrust collar acquires a value averaged over the coordinate $\bar{\varphi}$. The maximum temperature value reaches $T_{cw} = 56.73\,°C$ near the outer radius of the pad. In this case, with an increase in the coordinate due to an increase in the circumferential speed of the collar, there is a slight increase in the collar temperature with a difference of $\Delta T_c = 3.4\,°C$ in the absence of the effect of heat exchange with the external environment. When approaching the back of the collar, this effect becomes more noticeable.

Under the harmonic action of the thrust collar, that is, when the gap h changes depending on the time τ (Fig. 5, the first period of oscillation), the bearing capacity of the bearing also changes harmonically without delay with a frequency equal to the perturbation frequency of the collar $\Omega = 31.4$ rad/s. The amplitude of the change in the bearing capacity varies within $A = (2596.99\ldots2974.04)$ N and significantly depends o n the created minimum film thickness $h_{min} = (h_{20} - y_{d.disp.}) = 42.5\,\mu m$ when the collar moves. At the same time, the maximum temperature t_{max} of the lubricant at the beginning of the dynamic process slightly increases to 60.02 °C and then sets during subsequent periods of collar oscillation.

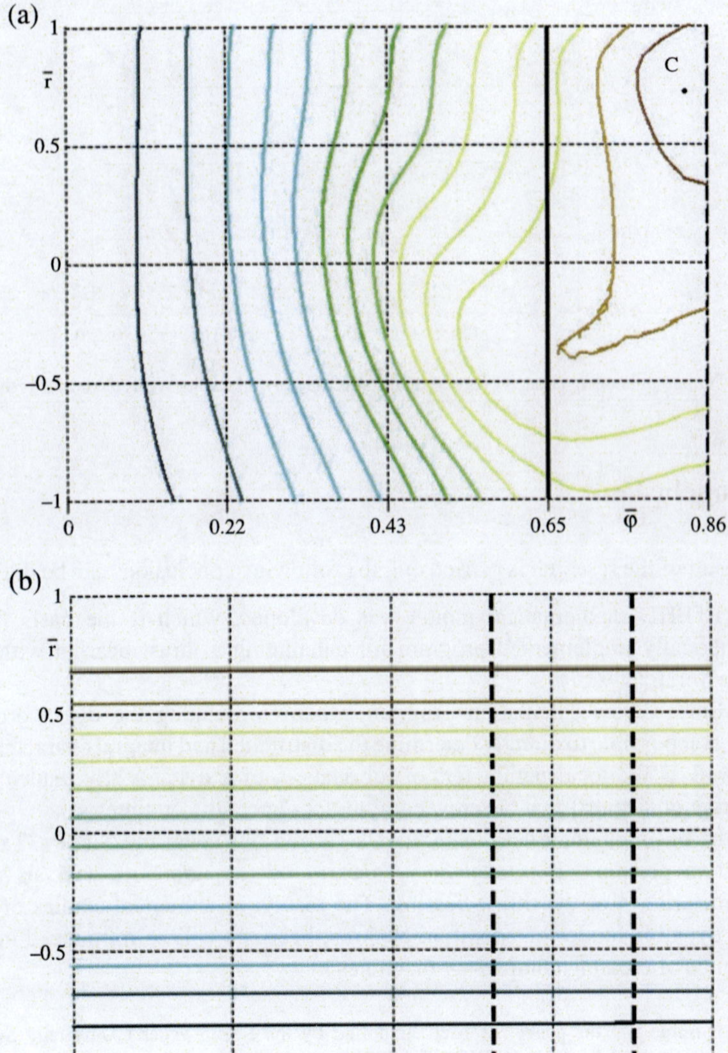

Fig. 4 Distribution of isotherms on the working surfaces: (**a**) of the pad at $\bar{y}_p = 0$; (**b**) of the thrust collar at $\bar{y}_d = \Psi_d$

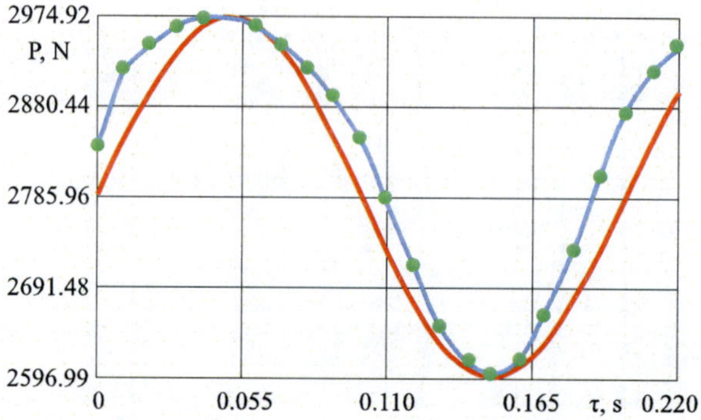

Fig. 5 Change in bearing capacity (line with dots) of the thrust bearing when the collar is displaced

4 Conclusion

As a result of the research is carried out, the following conclusions can be drawn:

1. the PTEHD mathematical model was developed, which is the basis for the numerically implemented program for calculating a thrust bearing with fixed pads Sm2Px3Txτ [6];
2. the Sm2Px3Txτ calculation program, when formulating the direct problem, makes it possible to directly determine the distributed and integral characteristics, as well as the local parameters of lubrication, of a dynamically loaded thrust bearing of a centrifugal or screw compressor depending on time;
3. on the basis of numerical experiments carried out using the Sm2Px3Txτ calculation program, it is necessary to carry out a parametric analysis of the characteristics of the thrust bearing. The results of numerical studies of static and dynamic modes of operation of thrust bearings will be introduced into the practice of creating compressor machines.

Acknowledgments This paper has been supported by the Kazan Federal University Strategic Academic Leadership Program ("PRIORITY-2030").

References

1. Seleznev, K.P., Galerkin. Yu. B.: Centrifugal compressors. L. Mashinostroenie, Leningrad department. **271** (1982)
2. Kazakevich, V.V.: Autooscillations (pumping) in the compressors. M. Mashinostroenie. **192** (1959)
3. Gravdahl, J.T., Egeland, O.: Compressor surge and rotating stall : modeling and control. Springer-Verlag London Limited, London. **232** (1999)

4. Sokolov, N.V., Maksimov, T.V., Khadiev, M.B.: Conditions of dynamic loading of the thrust bearing of the centrifugal compressor of the multiplier type. Compressors and pneumatics **1**, 16–21 (2020)
5. Sokolov, N.V., Khadiev, M.B., Maksimov, T.V., Fedotov, E.M., Fedotov, P.E.: Mathematical modeling of dynamic processes of lubricating layers thrust bearing turbochargers. J. Phys.: Conf. **1158** (2019)
6. Fedotov, P.E., Fedotov, E.M., Sokolov, N.V., Khadiev, M.B.: Sm2Px3Txτ - Dynamically loaded thrust plain bearing when setting a direct problem. Certificate of the state registration of a computer program No. 2020615227. 2020
7. Alyokhin, A.V.: Load capacity and dynamic characteristics of thrust bearings with fluid friction. Abstract. Diss. Cand. tehn. Sciences **23**
8. Zhu, Q., Zhang, W.J.: A preliminary nonlinear analysis of the axial transient response of the sector-shaped hydrodynamic thrust bearing-rotor system. ASME Journal of Tribology. **125**(4), 854–858 (2003)
9. Maksimov, V.A., Khadiev, M. B., Fedotov, E.M.: Determination of hydrodynamic and thermal characteristics of thrust bearings by mathematical modeling. Vestnik mashinostroeniya **6**, 39–45 (2004)
10. Khadiev, M.B., Sokolov, N.V., Fedotov, E.M.: Hydrodynamic, thermal and deformation characteristics of the lubricating films of thrust bearings with a bevel parallel to the radial inter-pad channel. Vestnik Mashinostroeniya **6**, 54–59 (2014)
11. Podolsky, M.E. 1981 Thrust plain bearings: Theory and calculation L. Mashinostroenie, Leningrad department. 261
12. Sokolov, N.V., Khadiev, M.B., Khavkin, A.L., Khusnutdinov, I.F.: The nature of axial vibrations of the rotor at variable operating modes of a centrifugal compressor unit. Compressors and pneumatics **4**, 29–32 (2018)

Approximation of Positive Semidefinite Nonlinear Eigenvalue Problems

Pavel S. Solov'ev, Diana M. Korosteleva, and Sergey I. Solov'ev

Abstract A positive semidefinite symmetric eigenvalue problem in an infinite-dimensional Hilbert space with nonlinear dependence on the spectral parameter is investigated. The existence of eigenvalues and eigenelements is established. The initial infinite-dimensional nonlinear eigenvalue problem is approximated by a nonlinear eigenvalue problem in a finite-dimensional subspace of the Hilbert space. The convergence and accuracy of approximate eigenvalues, eigenelements, and eigensubspaces are investigated.

1 Introduction

Linear and nonlinear differential eigenvalue problems apply in mathematical modeling of complex technical processes and systems. A weak statement of the linear differential eigenvalue problem is formulated as a variational eigenvalue problem $a(u, v) = \lambda b(u, v)$ in an infinite-dimensional Hilbert space V. Suppose that the bilinear form $a(.,.)$ is symmetric, positive definite, and bounded and the bilinear form $b(.,.)$ is symmetric, nonnegative, and compact. Denote $K = \{v : v \in V, b(v, v) = 0\}$ and suppose that $\operatorname{codim} K = \infty$. Then the formulated eigenvalue problem has a sequence of positive eigenvalues λ_k, $k = 1, 2, \ldots$, of finite multiplicity with the limit point at infinity. To the sequence of eigenvalues, there corresponds a complete in the space $K^{\perp}$ orthonormal system of eigenelements u_k, $k = 1, 2, \ldots$ Define finite-dimensional subspaces V_h of the space V satisfying the limit density condition. Approximate the original eigenvalue problem by the following finite-dimensional problem $a(u^h, v^h) = \lambda^h b(u^h, v^h)$ in the space V_h. This problem has positive eigenvalues λ_k^h, $k = 1, 2, \ldots, N_h$, $N_h = \dim V_h -$

P. S. Solov'ev · S. I. Solov'ev (✉)
Kazan (Volga Region) Federal University, Kazan, Russian Federation

D. M. Korosteleva
Kazan State Power Engineering University, Kazan, Russian Federation

I. B. Badriev et al. (eds.), *Mesh Methods for Boundary-Value Problems and Applications*, Lecture Notes in Computational Science and Engineering 141,
https://doi.org/10.1007/978-3-030-87809-2_36

dim K_h, $K_h = \{v^h : v^h \in V_h, b(v^h, v^h) = 0\}$, and corresponding eigenfunctions u_k^h, $k = 1, 2, \ldots, N_h$, forming a complete orthonormal system in the space $K_h^\perp$. For sufficiently small h, the following error estimates are valid

$$0 \leq \lambda_k^h - \lambda_k \leq c\,(\varepsilon^h)^2, \quad \|u_k^h - u_k\| \leq c\,\varepsilon^h, \quad \varepsilon^h = \sup_{u \in U_k, \|u\|=1} \inf_{v^h \in V_h} \|u - v^h\|,$$

where c is a constant independent of h, U_k is the eigensubspace corresponding to the eigenvalue λ_k, $\|\cdot\|$ is the norm on the space V.

In the present paper, the formulated results are generalized for nonlinear eigenvalue problems. Eigenvalue problems with the nonlinear dependence on the spectral parameter arise in various fields of science and engineering, for example, in plasma physics [1–5], construction mechanics [6–8], numerical algorithms for mesh equations [9–11], and eigenvibration modeling [12, 13]. Spectral approximations for compact operators are investigated in the papers [14–17]. Generalizations of spectral approximations for holomorphic Fredholm operator functions are derived in the papers [18, 19]. Preconditioned iterative methods for solving linear spectral problems are proposed and investigated in the papers [20–27]. Numerical methods for solving nonlinear matrix eigenvalue problems were constructed and investigated in the papers [28–38]. Error estimates for the finite difference methods for differential eigenvalue problems with the nonlinear entrance of the spectral parameter were derived in [39, 40]. The finite element method for solving nonlinear eigenvalue problems was investigated in [5, 41], and estimations of the effect of numerical integration in finite element eigenvalue and eigenfunction approximations were established in [42–44]. The investigations of approximate methods for solving variational eigenvalue problems with the nonlinear entrance of the spectral parameter in a Hilbert space were carried out in the paper [41] with the help of general results for linear variational and operator eigenvalue problems [42–44]. Numerical algorithms without saturation for solving problems of mathematical physics and mechanics were constructed and investigated in [45–52]. This paper develops and generalizes the theoretical results of the papers [41, 42].

2 Statement of the Problem

Let V be a real infinite-dimensional Hilbert space with the norm $\|\cdot\|$, let $\mathbb{R}$ be the real line, and let $\Lambda = (\nu_1, \nu_2)$, $0 \leq \nu_1 < \nu_2 \leq \infty$. Let us introduce mappings $a : \Lambda \times V \times V \to \mathbb{R}$ and $b : \Lambda \times V \times V \to \mathbb{R}$, that, for fixed $\mu \in \Lambda$, are symmetric bilinear forms $a(\mu) = a(\mu, ., .) : V \times V \to \mathbb{R}$ and $b(\mu) = b(\mu, ., .) : V \times V \to \mathbb{R}$. Suppose that, for fixed $\mu \in \Lambda$, the bilinear form $a(\mu, ., .)$ is positive definite and bounded; i.e., there exist positive continuous functions $\alpha_1(\mu)$ and $\alpha_2(\mu)$ such that

$$\alpha_1(\mu)\|v\|^2 \leq a(\mu, v, v) \leq \alpha_2(\mu)\|v\|^2 \quad \forall v \in V.$$

Suppose that, for fixed $\mu \in \Lambda$, the bilinear form $b(\mu, ., .)$ is nonnegative and compact; i.e., $b(\mu, v, v) \geq 0$ for any $v \in V$ and $b(\mu, v_i, v_i) \to b(\mu, v, v)$ as $i \to \infty$ for $v_i \rightharpoonup v$ in V as $i \to \infty$. The symbol $\rightharpoonup$ denotes the weak convergence of a sequence in a Hilbert space V. Denote $K(\mu) = \ker b(\mu)$, $\ker b(\mu) = \{v : v \in V,\ b(\mu, v, v) = 0\}$, $K(\mu, \eta) = K(\mu) \cup K(\eta)$, and assume that $\operatorname{codim} K(\mu) = \infty$. Note that $b(\mu, v, v) > 0$ for $v \in V \setminus K(\mu)$ and there exists a positive constant $\beta_2(\mu)$ such that

$$b(\mu, v, v) \leq \beta_2(\mu)\|v\|^2 \quad \forall v \in V.$$

Let us formulate the nonlinear eigenvalue problem: find $\lambda \in \Lambda$, $u \in V \setminus K(\lambda)$ such that

$$a(\lambda, u, v) = \lambda b(\lambda, u, v) \quad \forall v \in V. \tag{1}$$

A number λ and an element u satisfying (1) are called an eigenvalue and eigenelement of the problem (1). The set $U(\lambda)$ of eigenelements corresponding to an eigenvalue λ and the zero element form a closed subspace in V, which is called an eigensubspace corresponding to the eigenvalue λ. The dimension of this subspace is called a multiplicity of the eigenvalue λ. If the dimension of an eigensubspace is equal to unity, then the corresponding eigenvalue is said to be simple.

Define the norm of a symmetric bilinear form $c : V \times V \to \mathbb{R}$ by the formula

$$\|c\| = \sup_{v \in V,\ \|v\|=1} |c(v, v)|.$$

Assume that the bilinear forms $a(\mu) = a(\mu, ., .)$ and $b(\mu) = b(\mu, ., .)$ satisfy the conditions $\alpha(\mu, \eta) \to 0$, $\beta(\mu, \eta) \to 0$ as $\mu \to \eta$, $\mu, \eta \in \Lambda$, where

$$\alpha(\mu, \eta) = \|a(\mu) - a(\eta)\|, \quad \beta(\mu, \eta) = \|b(\mu) - b(\eta)\|.$$

Introduce the Rayleigh functional by the following relation

$$R(\mu, v) = \frac{a(\mu, v, v)}{b(\mu, v, v)} \quad \forall v \in V \setminus K(\mu), \mu \in \Lambda,$$

and assume that

$$R(\mu, v) \geq R(\eta, v), \quad \mu < \eta, \mu, \eta \in \Lambda, \quad v \in V \setminus K(\mu, \eta).$$

3 Parameter Problem

If W is a subspace of V, then we denote

$$W^{\perp}_{a(\mu)} = \{v : v \in V,\ a(\mu, v, w) = 0\ \forall w \in W\}.$$

For fixed $\mu \in \Lambda$, introduce the parameter-dependent linear eigenvalue problem: find $\gamma = \gamma(\mu) \in \mathbb{R}$, $y = y(\mu) \in V \setminus K(\mu)$ such that

$$a(\mu, y, v) = \gamma b(\mu, y, v) \quad \forall v \in V. \tag{2}$$

There exists a sequence of positive eigenvalues of finite multiplicity numbered concerning to multiplicity $\gamma_k = \gamma_k(\mu)$, $k = 1, 2, \ldots$, $0 < \gamma_1 \le \gamma_2 \le \ldots \le \gamma_k \le \ldots$, $\gamma_k \to \infty$ as $k \to \infty$, and corresponding eigenelements $y_k = y_k(\mu)$, $k = 1, 2, \ldots$, $a(\mu, y_i, y_j) = \gamma_i \delta_{ij}$, $b(\mu, y_i, y_j) = \delta_{ij}$, $i, j = 1, 2, \ldots$ The eigenelements y_k, $k = 1, 2, \ldots$ form a complete system in the space $(K(\mu))^{\perp}_{a(\mu)}$.

Put

$$E_k(\mu) = \operatorname{span}\{y_1(\mu), y_2(\mu), \ldots, y_k(\mu)\},$$

$k = 1, 2, \ldots$, $E_0(\mu) = \{0\}$, $(E_0(\mu))^{\perp}_{a(\mu)} = V$. By $\mathcal{E}_k(W)$ we denote the set of all k-dimensional subspaces of the space W for $k \ge 1$. The set $\mathcal{E}_0(W)$ consists only of $E_0(\mu)$. Put $\mathcal{E}_k = \mathcal{E}_k(V)$ for $k \ge 0$.

The eigenvalues of the problem (2) are characterized by the following variational properties

$$\gamma_k = \min_{v \in (E_{k-1}(\mu))^{\perp}_{a(\mu)} \setminus K(\mu)} R(\mu, v) = \max_{v \in E_k(\mu) \setminus K(\mu)} R(\mu, v),$$

$$\gamma_k = \max_{W \in \mathcal{E}_{k-1}} \min_{v \in W^{\perp}_{a(\mu)} \setminus K(\mu)} R(\mu, v) = \min_{W \in \mathcal{E}_k} \max_{v \in W \setminus K(\mu)} R(\mu, v),$$

for $k = 1, 2, \ldots$ These variational relations imply the inequalities $\gamma_k(\mu) \ge \gamma_k(\eta)$ for $\mu < \eta$, $\mu, \eta \in \Lambda$, $k = 1, 2, \ldots$, since

$$\begin{aligned}
\gamma_k(\eta) &= \min_{W \in \mathcal{E}_k} \max_{v \in W \setminus K(\eta)} R(\eta, v) \\
&\le \max_{v \in E_k(\mu) \setminus K(\eta)} R(\eta, v) = \max_{v \in E_k(\mu) \setminus K(\mu, \eta)} R(\eta, v) \\
&\le \max_{v \in E_k(\mu) \setminus K(\mu, \eta)} R(\mu, v) \le \max_{v \in E_k(\mu) \setminus K(\mu)} R(\mu, v) = \gamma_k(\mu).
\end{aligned}$$

Denote by $c = c(\mu, \eta)$ various constants continuously depending on $\mu, \eta \in \Lambda$.

Lemma 1 *If* $\mu, \eta \in \Lambda$ *and* $|\mu - \eta|$ *is sufficiently small, then there exists a constant* c *such that*

$$|\gamma_k(\mu) - \gamma_k(\eta)| \le c\,(\alpha(\mu, \eta) + \beta(\mu, \eta)).$$

Proof By taking into account the variational properties of eigenvalues and by applying the following equality

$$R(\mu, v) - R(\eta, v) = \frac{\dfrac{a(\mu, v, v) - a(\eta, v, v)}{a(\mu, v, v)} R(\eta, v) + \dfrac{b(\eta, v, v) - b(\mu, v, v)}{a(\mu, v, v)} R^2(\eta, v)}{1 - \dfrac{a(\mu, v, v) - a(\eta, v, v)}{a(\mu, v, v)} - \dfrac{b(\eta, v, v) - b(\mu, v, v)}{a(\mu, v, v)} R(\eta, v)}$$

for sufficiently small $|\mu - \eta|$ and for $v \in V \setminus K(\mu, \eta)$, $\mu, \eta \in \Lambda$, we derive

$$\begin{aligned}
\gamma_k(\mu) &= \min_{W \in \mathcal{E}_k} \max_{v \in W \setminus K(\mu)} R(\mu, v) \\
&\le \max_{v \in E_k(\eta) \setminus K(\mu)} R(\mu, v) = \max_{v \in E_k(\eta) \setminus K(\mu,\eta)} R(\mu, v) \\
&\le \max_{v \in E_k(\eta) \setminus K(\mu,\eta)} R(\eta, v) + \max_{v \in E_k(\eta) \setminus K(\mu,\eta)} |R(\mu, v) - R(\eta, v)| \\
&\le \gamma_k(\eta) + \varkappa(\mu, \eta),
\end{aligned}$$

where

$$\varkappa(\mu, \eta) = \frac{\dfrac{\alpha(\mu, \eta)}{\alpha_1(\mu)} \gamma_k(\eta) + \dfrac{\beta(\mu, \eta)}{\alpha_1(\mu)} \gamma_k^2(\eta)}{1 - \dfrac{\alpha(\mu, \eta)}{\alpha_1(\mu)} - \dfrac{\beta(\mu, \eta)}{\alpha_1(\mu)} \gamma_k(\eta)} \le c\,(\alpha(\mu, \eta) + \beta(\mu, \eta)).$$

As a result, we obtain the following inequalities

$$\begin{aligned}
\gamma_k(\mu) - \gamma_k(\eta) &\le c\,(\alpha(\mu, \eta) + \beta(\mu, \eta)), \\
\gamma_k(\eta) - \gamma_k(\mu) &\le c\,(\alpha(\mu, \eta) + \beta(\mu, \eta)),
\end{aligned}$$

which complete the proof of the lemma. □

Let $\gamma_i(\mu)$, $i = 1, 2, \ldots$ be the sequence of eigenvalues of the problem (2), and let $y_i(\mu)$, $i = 1, 2, \ldots$ be the orthonormal system of corresponding eigenelements.

For fixed $\mu \in \Lambda$, assume that $\gamma_k = \gamma_k(\mu)$ is an eigenvalue of the problem (2) of multiplicity s such that

$$\gamma_{k-1} < \gamma_k = \gamma_{k+1} = \ldots = \gamma_{k+s-1} < \gamma_{k+s},$$

where $k \geq 1$, $\gamma_0 = 0$. For fixed $\eta \in \Lambda$, put

$$Y_k(\eta) = \text{span}\{y_k(\eta), y_{k+1}(\eta), \ldots, y_{k+s-1}(\eta)\}.$$

Let W_1 and W_2 be two closed subspaces of the Hilbert space V, $\dim W_1 = \dim W_2 < \infty$, and let P_1 and P_2 be the operators of orthogonal projections onto W_1 and W_2, respectively. The gap between subspaces W_1 and W_2 of the Hilbert space V is defined by the following relations

$$\vartheta(W_1, W_2) = \max_{u \in W_2, \|u\|=1} \|u - P_1 u\| = \max_{u \in W_1, \|u\|=1} \|u - P_2 u\|.$$

Lemma 2 *If $\mu, \eta \in \Lambda$ and $|\mu - \eta|$ is sufficiently small, then there exists a constant c such that*

$$\vartheta(Y_k(\mu), Y_k(\eta)) \leq c\,(\alpha(\mu, \eta) + \beta(\mu, \eta)).$$

Proof For some element $y_0(\mu) \in K(\mu)$, the element $y \in Y_k(\mu)$ can be represented in the form

$$y = Q_k(\eta)y + v_k(\eta) + w_k(\eta)$$

for

$$Q_k(\eta)y = \sum_{i=k}^{k+s-1} \beta_i(\eta) y_i(\eta), \quad v_k(\eta) = \sum_{i=1}^{k-1} \beta_i(\eta) y_i(\eta),$$

$$w_k(\eta) = y_0(\mu) + \sum_{i=k+s}^{\infty} \beta_i(\eta) y_i(\eta), \quad \beta_i(\eta) = b(\eta, y, y_i(\eta)), \quad i = 1, 2, \ldots$$

Here any sum $\sum_{i=m}^{n} \beta_i(\eta) y_i(\eta)$ for $n < m$ is zero by convention.

By Lemma 1, we choose $\eta \in \Lambda$ such that

$$\gamma_k(\mu) - \gamma_{k-1}(\eta) > 0, \quad \gamma_{k+s}(\eta) - \gamma_k(\mu) > 0.$$

Denote

$$\delta(\eta, v) = \sup_{w \in V\setminus\{0\}} \frac{|a(\eta, v, w) - \gamma_k b(\eta, v, w)|}{\|w\|_{a(\eta)}}.$$

For $k \geq 1$, the following estimate holds

$$\|v_k(\eta)\|_{a(\eta)} \leq \frac{\gamma_{k-1}(\eta)}{\gamma_k(\mu) - \gamma_{k-1}(\eta)} \delta(\eta, y).$$

This estimate is obviously true for $k = 1$. For $k \geq 2$, we get the relations

$$\begin{aligned} a(\eta, y, v_k(\eta)) &= a(\eta, v_k(\eta), v_k(\eta)), \\ b(\eta, y, v_k(\eta)) &= b(\eta, v_k(\eta), v_k(\eta)), \\ a(\eta, v_k(\eta), v_k(\eta)) &\leq \gamma_{k-1}(\eta) b(\eta, v_k(\eta), v_k(\eta)), \end{aligned}$$

and hence

$$\begin{aligned} &- a(\eta, y, v_k(\eta)) + \gamma_k b(\eta, y, v_k(\eta)) \\ &= -a(\eta, v_k(\eta), v_k(\eta)) + \gamma_k b(\eta, v_k(\eta), v_k(\eta)) \\ &\geq (\gamma_k(\mu) - \gamma_{k-1}(\eta)) b(\eta, v_k(\eta), v_k(\eta)) \\ &\geq \frac{\gamma_k(\mu) - \gamma_{k-1}(\eta)}{\gamma_{k-1}(\eta)} a(\eta, v_k(\eta), v_k(\eta)), \quad k \geq 2. \end{aligned}$$

These relations imply the desired estimate.

Now let us prove the following estimate

$$\|w_k(\eta)\|_{a(\eta)} \leq \frac{\gamma_{k+s}(\eta)}{\gamma_{k+s}(\eta) - \gamma_k(\mu)} \delta(\eta, y).$$

for $k \geq 1$. One can readily see that

$$\begin{aligned} a(\eta, y, w_k(\eta)) &= a(\eta, w_k(\eta), w_k(\eta)), \\ b(\eta, y, w_k(\eta)) &= b(\eta, w_k(\eta), w_k(\eta)), \\ a(\eta, w_k(\eta), w_k(\eta)) &\geq \gamma_{k+s}(\eta) b(\eta, w_k(\eta), w_k(\eta)). \end{aligned}$$

Then we obtain the relations

$$\begin{aligned} &a(\eta, y, w_k(\eta)) - \gamma_k b(\eta, y, w_k(\eta)) \\ &= a(\eta, w_k(\eta), w_k(\eta)) - \gamma_k b(\eta, w_k(\eta), w_k(\eta)) \\ &\geq \frac{\gamma_{k+s}(\eta) - \gamma_k(\mu)}{\gamma_{k+s}(\eta)} a(\eta, w_k(\eta), w_k(\eta)) \\ &\geq (\gamma_{k+s}(\eta) - \gamma_k(\mu)) b(\eta, w_k(\eta), w_k(\eta)), \quad k \geq 1. \end{aligned}$$

which imply the desired estimate.

As a result, we derive

$$\begin{aligned}\|y - Q_k(\eta)y\|_{a(\eta)} &\le \|v_k(\eta)\|_{a(\eta)} + \|w_k(\eta)\|_{a(\eta)} \\ &\le \left(\frac{\gamma_{k-1}(\eta)}{\gamma_k(\mu) - \gamma_{k-1}(\eta)} + \frac{\gamma_{k+s}(\eta)}{\gamma_{k+s}(\eta) - \gamma_k(\mu)}\right)\delta(\eta, y) \\ &\le c\,\delta(\eta, y) \le c\,(\alpha(\mu, \eta) + \beta(\mu, \eta))\|y\|_{a(\eta)}.\end{aligned}$$

Consequently, we conclude

$$\begin{aligned}\vartheta(Y_k(\mu), Y_k(\eta)) &= \sup_{y \in Y_k(\mu)\setminus\{0\}} \frac{\|y - P_k(\eta)y\|}{\|y\|} \\ &\le c \sup_{y \in Y_k(\mu)\setminus\{0\}} \frac{\|y - Q_k(\eta)y\|_{a(\eta)}}{\|y\|_{a(\eta)}} \\ &\le c\,(\alpha(\mu, \eta) + \beta(\mu, \eta)),\end{aligned}$$

where $P_k(\eta)$ is the operator of orthogonal projection onto $Y_k(\eta)$. The proof of the lemma is completed. □

4 Existence and Properties of Solutions

Let us formulate the existence results for the nonlinear eigenvalue problem (1). Put $\min\{i : i \in I\} = 0$ for $I = \varnothing$,

$$\gamma_i(\nu_j) = \lim_{\mu \to \nu_j} \gamma_i(\mu), \quad j = 1, 2, \quad i = 1, 2, \ldots$$

Theorem 1 *Suppose* $0 \le \nu_1 < \nu_2 < \infty$, $1 \le m \le n$ *and denote*

$$\begin{aligned}m &= \min\{i : \nu_1 - \gamma_i(\nu_1) < 0,\ i \ge 1\}, \\ n &= \max\{i : \nu_2 - \gamma_i(\nu_2) > 0,\ i \ge 0\}.\end{aligned}$$

Then the problem (1) *has the positive eigenvalues* λ_k, $k = m, m+1, \ldots, n$, *numbered concerning to multiplicities,*

$$\nu_1 < \lambda_m \le \lambda_{m+1} \le \ldots \le \lambda_n < \nu_2.$$

Each eigenvalue λ_i, $m \le i \le n$ *is the unique root of the equation*

$$\mu - \gamma_i(\mu) = 0, \quad \mu \in \Lambda, \quad m \le i \le n.$$

The eigensubspace $U(\lambda_i)$ of the problem (1) is the eigensubspace $Y(\mu)$ corresponding to the eigenvalue $\gamma_i(\mu)$ of the linear eigenvalue problem (2) for $\mu = \lambda_i$.

Theorem 2 *Suppose $\nu_1 \geq 0$, $\nu_2 = \infty$, $m \geq 1$, and denote*

$$m = \min\{i : \nu_1 - \gamma_i(\nu_1) < 0, i \geq 1\}.$$

Then the problem (1) has the positive eigenvalues λ_k, $k = m, m+1, \ldots$, numbered concerning to multiplicities,

$$0 < \lambda_m \leq \lambda_{m+1} \leq \ldots \leq \lambda_k \leq \ldots, \qquad \lim_{k\to\infty} \lambda_k = \infty.$$

Each eigenvalue λ_i, $i \geq m$ is the unique root of the equation

$$\mu - \gamma_i(\mu) = 0, \quad \mu \in \Lambda, \quad i \geq m.$$

The eigensubspace $U(\lambda_i)$ of the problem (1) is the eigensubspace $Y(\mu)$ corresponding to the eigenvalue $\gamma_i(\mu)$ of the linear eigenvalue problem (2) for $\mu = \lambda_i$.

Theorem 3 *Let λ_i be the eigenvalue of the problem (1) of multiplicity s such as*

$$\lambda_{i-1} < \lambda_i = \lambda_{i+1} = \ldots = \lambda_{i+s-1} < \lambda_{i+s}.$$

If $\mu \in \Lambda$ and $|\lambda_i - \mu|$ is sufficiently small, then there exists a constant c such that

$$|\lambda_i - \gamma_i(\mu)| \leq c\,(\alpha(\lambda_i, \mu) + \beta(\lambda_i, \mu)),$$
$$\vartheta(U(\lambda_i), Y_i(\mu)) \leq c\,(\alpha(\lambda_i, \mu) + \beta(\lambda_i, \mu)).$$

The proofs of Theorems 1–3 generalize the corresponding results from the paper [41].

5 Approximation of the Problem

Introduce the finite-dimensional subspaces V_h of dimension M_h in the Hilbert space V satisfying the limit density condition:

$$\varepsilon_h(v) = \inf_{v^h \in V_h} \|v - v^h\| \to 0$$

as $h \to 0$ for any element v from V. The limit density condition implies that $M_h \to \infty$ as $h \to 0$.

Denote $K_h(\mu) = \{v^h : v^h \in V_h,\ b(\mu, v^h, v^h) = 0\}$, $N_h = \operatorname{codim} K_h(\mu)$. Note that $b(\mu, v^h, v^h) > 0$ for $v^h \in V_h \setminus K_h(\mu)$, $N_h = \operatorname{codim} K_h(\mu) = \dim V_h \setminus K_h(\mu) = \dim V_h - \dim K_h(\mu)$, $N_h \to \infty$ as $h \to 0$.

The nonlinear eigenvalue problem (1) is approximated by the following finite-dimensional problem: find $\lambda^h \in \Lambda$, $u^h \in V_h \setminus K_h(\lambda^h)$ such that

$$a(\lambda^h, u^h, v^h) = \lambda^h b(\lambda^h, u^h, v^h) \quad \forall v^h \in V_h. \tag{3}$$

A number λ^h satisfying (3) is referred to as an approximate eigenvalue, and an element u^h is referred to as an approximate eigenelement corresponding to the eigenvalue λ^h. The set $U_h(\lambda^h)$ of eigenelements corresponding to the eigenvalue λ^h and the zero element form a closed subspace in the space V_h, which is referred to as the eigensubspace corresponding to the eigenvalue λ^h.

6 Existence of Approximate Solutions

If W_h is a subspace of V_h, then we denote

$$(W_h)^{\perp}_{a(\mu)} = \{v^h : v^h \in V_h,\ a(\mu, v^h, w^h) = 0\ \forall w^h \in W_h\}.$$

For fixed $\mu \in \Lambda$, introduce parameter eigenvalue problem: find $\gamma^h = \gamma^h(\mu) \in \mathbb{R}$, $y^h = y^h(\mu) \in V_h \setminus K_h(\mu)$ such as

$$a(\mu, y^h, v^h) = \gamma^h b(\mu, y^h, v^h) \quad \forall v^h \in V_h. \tag{4}$$

This problem has positive eigenvalues $\gamma_k^h = \gamma_k^h(\mu)$, $k = 1, 2, \ldots, N_h$, of finite multiplicity numbered concerning to multiplicities, $0 < \gamma_1^h \le \gamma_2^h \le \ldots \le \gamma_{N_h}^h$, and the corresponding orthonormal system of eigenelements $y_k^h = y_k^h(\mu)$, $k = 1, 2, \ldots, N_h$ such that $a(\mu, y_i^h, y_j^h) = \gamma_i^h \delta_{ij}$, $b(\mu, y_i^h, y_j^h) = \delta_{ij}$, $i, j = 1, 2, \ldots, N_h$. The eigenelements y_k^h, $k = 1, 2, \ldots, N_h$ form a complete system in the space $(K_h(\mu))^{\perp}_{a(\mu)}$.

Put

$$E_k^h(\mu) = \operatorname{span}\{y_1^h(\mu), y_2^h(\mu), \ldots, y_k^h(\mu)\},$$

$k = 1, 2, \ldots, N_h$, $E_0^h(\mu) = \{0\}$, $(E_0^h(\mu))^{\perp}_{a(\mu)} = V_h$. By $\mathcal{E}_k^h(W_h)$ we denote the set of all k-dimensional subspaces of the space W_h for $1 \le k \le N_h$. The set $\mathcal{E}_0^h(W_h)$ consists only of $E_0^h(\mu)$. Put $\mathcal{E}_k^h = \mathcal{E}_k^h(V_h)$ for $0 \le k \le N_h$.

The eigenvalues of the problem (4) are characterized by the following variational properties

$$\gamma_k^h = \min_{v^h \in (E_{k-1}^h(\mu))^{\perp}_{a(\mu)} \setminus K_h(\mu)} R(\mu, v^h) = \max_{v^h \in E_k^h(\mu) \setminus K_h(\mu)} R(\mu, v^h),$$

$$\gamma_k^h = \max_{W_h \in \mathcal{E}_{k-1}^h} \min_{v^h \in (W_h)^{\perp}_{a(\mu)} \setminus K_h(\mu)} R(\mu, v^h) = \min_{W_h \in \mathcal{E}_k^h} \max_{v^h \in W_h \setminus K_h(\mu)} R(\mu, v^h),$$

$$\gamma_k^h = \max_{W_h \in \mathcal{E}_{N_h-k+1}^h} \min_{v^h \in W_h \setminus K_h(\mu)} R(\mu, v^h),$$

where $k = 1, 2, \ldots, N_h$. These variational relations imply the inequalities $\gamma_k^h(\mu) \geq \gamma_k^h(\eta)$ for $\mu < \eta$, $\mu, \eta \in \Lambda$, $k = 1, 2, \ldots, N_h$.

If $\mu, \eta \in \Lambda$ and $|\mu - \eta|$ is sufficiently small, then there exists a constant c such that

$$|\gamma_k^h(\mu) - \gamma_k^h(\eta)| \leq c\,(\alpha(\mu, \eta) + \beta(\mu, \eta)).$$

Denote $\gamma_i^h(\nu_j) = \lim_{\mu \to \nu_j} \gamma_i^h(\mu)$, $j = 1, 2$, $i = 1, 2, \ldots, N_h$.

Theorem 4 *Suppose* $0 \leq \nu_1 < \nu_2 < \infty$, $1 \leq m \leq n$, *and denote*

$$m = \min\{i : \nu_1 - \gamma_i^h(\nu_1) < 0,\ i \geq 1\},$$

$$n = \max\{i : \nu_2 - \gamma_i^h(\nu_2) > 0,\ i \geq 0\}.$$

Then the problem (3) *has the positive eigenvalues* λ_k^h, $k = m, m+1, \ldots, n$, *numbered concerning to multiplicities,* $\nu_1 < \lambda_m^h \leq \lambda_{m+1}^h \leq \ldots \leq \lambda_n^h < \nu_2$. *Each eigenvalue* λ_i^h, $m \leq i \leq n$ *is the unique root of the equation*

$$\mu - \gamma_i^h(\mu) = 0, \quad \mu \in \Lambda, \quad m \leq i \leq n.$$

The eigensubspace $U_h(\lambda_i^h)$ *of the problem* (3) *is the eigensubspace* $Y_h(\mu)$ *corresponding to the eigenvalue* $\gamma_i^h(\mu)$ *of the linear eigenvalue problem* (4) *for* $\mu = \lambda_i^h$.

Theorem 5 *Suppose* $\nu_1 \geq 0$, $\nu_2 = \infty$, $m \geq 1$, *and denote*

$$m = \min\{i : \nu_1 - \gamma_i^h(\nu_1) < 0,\ i \geq 1\}.$$

Then the problem (3) *has the positive eigenvalues* λ_k^h, $k = m, m+1, \ldots, N_h$, *numbered concerning to multiplicities,* $0 < \lambda_m^h \leq \lambda_{m+1}^h \leq \ldots \leq \lambda_{N_h}^h$. *Each eigenvalue* λ_i^h, $m \leq i \leq N_h$ *is the unique root of the equation*

$$\mu - \gamma_i^h(\mu) = 0, \quad \mu \in \Lambda, \quad m \leq i \leq N_h.$$

The eigensubspace $U_h(\lambda_i^h)$ of the problem (3) is the eigensubspace $Y_h(\mu)$ corresponding to the eigenvalue $\gamma_i^h(\mu)$ of the linear eigenvalue problem (4) for $\mu = \lambda_i^h$.

The proofs of the Theorems 4 and 5 can be carried out by analogy with the proofs of Theorems 1 and 2, respectively, with regard to the finite dimension of the problem (3).

7 Convergence Analysis

Let λ_k be an eigenvalue of the problem (1) of multiplicity s such that

$$\lambda_{k-1} < \lambda_k = \lambda_{k+1} = \ldots = \lambda_{k+s-1} < \lambda_{k+s},$$

where λ_{k-1} for $k > m$ and λ_{k+s} for $k \geq m$ are eigenvalues of the problem (1) and m is the number defined in Theorems 1 and 2; if $k = m$, then we set $\lambda_{k-1} = \lambda_{m-1} = 0$, $U_k = U(\lambda_k)$ is the eigensubspace corresponding to the eigenvalue λ_k, $\dim U_k = s$, $U_k^h = \operatorname{span}\{y_k^h, y_{k+1}^h, \ldots, y_{k+s-1}^h\}$, and y_i^h, $i = k, k+1, \ldots, k+s-1$ are the eigenelements of the approximate scheme (4) for $\mu = \lambda_k^h$.

For $\mu \in \Lambda$, introduce an operator $P_h(\mu) : V \to V_h$ by the rule

$$a(\mu, u - P_h(\mu)u, v^h) = 0 \quad \forall v^h \in V_h,$$

where $u \in V$. Note that $P_h(\mu)u \to u$ as $h \to 0$, $u \in V$. Denote $P_h = P_h(\lambda_k)$ and set

$$\varepsilon^h = \sup_{u \in U_k, \|u\|=1} \varepsilon_h(u).$$

Note that $\varepsilon^h \to 0$ as $h \to 0$.

Theorem 6 *For sufficiently small h, the following error estimate*

$$0 \leq \lambda_k^h - \lambda_k \leq c\,(\varepsilon^h)^2$$

is valid, where c is a constant independent of h.

Proof Since $\mathcal{E}_k^h \subset \mathcal{E}_k$, $K_h(\mu) \subset K(\mu)$, we derive

$$\gamma_k(\mu) = \min_{W \in \mathcal{E}_k} \max_{v \in W \setminus K(\mu)} R(\mu, v) \leq \min_{W_h \in \mathcal{E}_k^h} \max_{v^h \in W_h \setminus K_h(\mu)} R(\mu, v^h) = \gamma_k^h(\mu).$$

To prove the inequality $\lambda_k^h \geq \lambda_k$, we suppose the contrary: $\lambda_k^h < \lambda_k$. Then we arrive at a contradiction

$$\lambda_k = \gamma_k(\lambda_k) \leq \gamma_k^h(\lambda_k) \leq \gamma_k^h(\lambda_k^h) = \lambda_k^h.$$

Now, for sufficiently small h, we obtain

$$0 \leq \lambda_k^h - \lambda_k = \gamma_k^h(\lambda_k^h) - \gamma_k(\lambda_k) \leq \gamma_k^h(\lambda_k) - \gamma_k(\lambda_k) \leq c\,(\varepsilon^h)^2.$$

Here we have applied the error estimate for the linear eigenvalue problem from the paper [42]. □

Theorem 7 *Suppose that λ_k^h is the eigenvalue of the approximate scheme (3), u_k^h is the corresponding eigenelement such that $b(\lambda_k^h, u_k^h, u_k^h) = 1$. Then the convergence $\lambda_k^h \to \lambda_k$ as $h \to 0$ is valid, and each sequence $h' \to 0$ contains a subsequence $h'' \to 0$ such that $u_k^h \to u_k$ in V as $h = h'' \to 0$, where λ_k and u_k are the eigenvalue and corresponding eigenelement of the problem (1). If λ_k is the simple eigenvalue and the sign of the eigenelements u_k and u_k^h are chosen so as ensure that $b(\lambda_k^h, u_k^h, P_h u_k) > 0$, then $u_k^h \to u_k$ in V as $h \to 0$.*

Theorem 8 *The convergence $\vartheta(U_k, U_k^h) \to 0$ as $h \to 0$ holds.*

The proofs of Theorems 7 and 8 generalize the corresponding results from the paper [41].

8 Error Investigation

Let λ_k be an eigenvalue of the problem (1) of multiplicity s such that

$$\lambda_{k-1} < \lambda_k = \lambda_{k+1} = \ldots = \lambda_{k+s-1} < \lambda_{k+s},$$

where λ_{k-1} for $k > m$ and λ_{k+s} for $k \geq m$ are eigenvalues of the problem (1) and m is the number defined in Theorems 1 and 2; if $k = m$, then we set $\lambda_{k-1} = \lambda_{m-1} = 0$, $U_k = U(\lambda_k)$ is the eigensubspace corresponding to the eigenvalue λ_k, $\dim U_k = s$, $U_k^h = \text{span}\{y_k^h, y_{k+1}^h, \ldots, y_{k+s-1}^h\}$, y_i^h, $i = k, k+1, \ldots, k+s-1$ are eigenelements of the approximate scheme (4) for $\mu = \lambda_k^h$. Denote $\delta^h = \alpha(\lambda_k, \lambda_k^h) + \beta(\lambda_k, \lambda_k^h)$.

Theorem 9 *For sufficiently small h, the following error estimate*

$$\vartheta(U_k, U_k^h) \leq c\,(\varepsilon^h + \delta^h)$$

is valid, where c is a constant independent of h.

Proof Denote $\beta_i^h = b(\lambda_k^h, P_h u, y_i^h)$, $i = 1, 2, \ldots, N_h$, where y_i^h, $i = 1, 2, \ldots, N_h$ eigenelements corresponding to an eigenvalue of the problem (4) for $\mu = \lambda_k^h$, $u \in U_k$, $\|u\| = 1$. Since eigenelements y_i^h, $i = 1, 2, \ldots, N_h$ of the scheme (4) for $\mu = \lambda_k^h$ form an orthonormal basis in the space $(K_h(\mu))^{\perp}_{a(\mu)}$, for some element y_0^h from $K_h(\lambda_k)$ the element $P_h u \in V_h$ can be represented in the form $P_h u = Q_k^h u + v_k^h + w_k^h$, where

$$Q_k^h u = \sum_{i=k}^{k+s-1} \beta_i^h y_i^h, \quad v_k^h = \sum_{i=1}^{k-1} \beta_i^h y_i^h, \quad w_k^h = y_0^h + \sum_{i=k+s}^{N_h} \beta_i^h y_i^h.$$

For $k \geq m$, by analogy with [41] the following estimates are valid:

$$\|v_k^h\| \leq c\,(\varepsilon^h + \delta^h), \quad \|w_k^h\| \leq c\,(\varepsilon^h + \delta^h),$$

for sufficiently small h. Hence

$$\begin{aligned} \vartheta(P_h U_k, U_k^h) = & \sup_{u \in U_k \setminus \{0\}} \frac{\|P_h u - P_k^h P_h u\|}{\|P_h u\|} \\ & \leq c \sup_{u \in U_k, \|u\|=1} \|P_h u - Q_k^h u\| \\ & \leq c \sup_{u \in U_k, \|u\|=1} (\|v_k^h\| + \|w_k^h\|) \\ & \leq c\,(\varepsilon^h + \delta^h), \end{aligned}$$

where P_k^h is the operator of orthogonal projection onto U_k^h. Consequently, we conclude $\vartheta(U_k, U_k^h) \leq \vartheta(U_k, P_h U_k) + \vartheta(P_h U_k, U_k^h) \leq c\,(\varepsilon^h + \delta^h)$. The proof of the theorem is complete. □

Theorems 6 and 9 imply the following results.

Theorem 10 *Suppose that u_k^h is the eigenelement of the scheme (3), λ_k^h is the corresponding eigenvalue, $b(\lambda_k^h, u_k^h, u_k^h) = 1$. Then there exists an eigenelement $u = u(u_k^h) \in U_k$ of the problem (1) such that for sufficiently small h the following error estimate $\|u_k^h - u\| \leq c\,(\varepsilon^h + \delta^h)$ is valid, where c is a constant independent of h.*

Theorem 11 *Suppose that $\delta^h = O(\varepsilon^h)$ as $h \to 0$, u_k^h is the eigenelement of the scheme (3), λ_k^h is the corresponding eigenvalue, $b(\lambda_k^h, u_k^h, u_k^h) = 1$, λ_k is the eigenvalue of the problem (1). Then there exists an eigenelement $u = u(u_k^h) \in U_k$ of the problem (1) such that for sufficiently small h the following error estimates $0 \leq \lambda_k^h - \lambda_k \leq c\,(\varepsilon^h)^2$, $\|u_k^h - u\| \leq c\,\varepsilon^h$, are valid, where c is a constant independent of h.*

Acknowledgments The reported study was supported by Russian Foundation for Basic Research according to the research projects nos. 20-31-90087, 20-31-90086 (Sect. 1–4) and Russian Science Foundation according to the research project no. 16-11-10299 (Sect. 5–8).

References

1. Abdullin, I.Sh., Zheltukhin, V.S., Kashapov, N.F.: Radio-Frequency Plasma-Jet Processing of Materials at Reduced Pressures: Theory and Practice of Applications. Izd. Kazan. Univ., Kazan (2000) [in Russian]
2. Zheltukhin, V.S., Solov'ev, S.I., Solov'ev, P.S., Chebakova, V.Yu.: Existence of solutions for electron balance problem in the stationary high-frequency induction discharges. IOP Conf. Series: Materials Science Engin. **158**(1), Art. 012103, 1–6 (2016)
3. Zheltukhin, V.S., Solov'ev, S.I., Solov'ev, P.S., Chebakova, V.Yu., Sidorov, A.M.: Third type boundary conditions for steady state ambipolar diffusion equation. IOP Conf. Series: Materials Science Engin. **158**(1), Art. 012102, 1–4 (2016)
4. Solov'ev, S.I., Solov'ev, P.S., Chebakova, V.Yu.: Finite difference approximation of electron balance problem in the stationary high-frequency induction discharges. MATEC Web Conf. **129**, Art. 06014, 1–4 (2017)
5. Solov'ev, S.I., Solov'ev, P.S.: Finite element approximation of the minimal eigenvalue of a nonlinear eigenvalue problem. Lobachevskii J. Math. **39**(7), 949–956 (2018)
6. Goolin, A.V., Kartyshov, S.V.: Numerical study of stability and nonlinear eigenvalue problems. Surv. Math. Ind. **3**, 29–48 (1993)
7. Betcke, T., Higham, N.J., Mehrmann, V., Schröder, C., Tisseur, F.: NLEVP: A collection of nonlinear eigenvalue problems. ACM Trans. Math. Software **39**(2), Art. 7 (2013)
8. Kozlov, V.A., Maz'ya, V.G., Rossmann, J.: Spectral Problems Associated with Corner Singularities of Solutions to Elliptic Equations. AMS, Providence (2001)
9. Lyashko, A.D., Solov'ev, S.I.: Fourier method of solution of FE systems with Hermite elements for Poisson equation. Russ. J. Numer. Anal. Math. Modelling **6**(2), 121–130 (1991)
10. Solov'ev, S.I.: Fast direct methods of solving finite-element grid schemes with bicubic elements for the Poisson equation. J. Math. Sciences **71**(6), 2799–2804 (1994)
11. Solov'ev, S.I.: A fast direct method of solving Hermitian fourth-order finite-element schemes for the Poisson equation. J. Math. Sciences **74**(6), 1371–1376 (1995)
12. Solov'ev, S.I.: Eigenvibrations of a bar with elastically attached load. Differ. Equations **53**(3), 409–423 (2017)
13. Samsonov, A.A., Solov'ev, S.I.: Eigenvibrations of a beam with load. Lobachevskii J. Math. **38**(5), 849–855 (2017)
14. Osborn, J.E.: Spectral approximation for compact operators. Math. Comp. **29**(131), 712–725 (1975)
15. Bramble, J.H., Osborn, J.E.: Rate of convergence estimates for nonselfadjoint eigenvalue approximations. Math. Comp. **27**(123), 525–549 (1973)
16. Knyazev, A.V., Osborn, J.E.: New a priori FEM error estimates for eigenvalues. SIAM J. Numer. Anal. **43**(6), 2647–2667 (2006)
17. Sakurai, T., Sugiura, H.: A projection method for generalized eigenvalue problems using numerical integration. J. Comput. Appl. Math. **159**, 119–128 (2003)
18. Karma, O.: Approximation in eigenvalue problems for holomorphic Fredholm operator functions I. Numer. Funct. Anal. Optim. **17**, 365–387 (1996)
19. Karma, O.: Approximation in eigenvalue problems for holomorphic Fredholm operator functions II: Convergence rate. Numer. Funct. Anal. Optim. **17**, 389–408 (1996)
20. Knyazev, A.V., Neymeyr, K.: A geometric theory for preconditioned inverse iteration III: A short and sharp convergence estimate for generalized eigenvalue problems. Linear Algebra Appl. **358**(1–3), 95–114 (2003)

21. Knyazev, A.V., Neymeyr, K.: Efficient solution of symmetric eigenvalue problems using multigrid preconditioners in the locally optimal block conjugate gradient method. Electron. Trans. Numer. Anal. **15**, 38–55 (2003)
22. Neymeyr, K.: A geometric theory for preconditioned inverse iteration I: Extrema of the Rayleigh quotient. Linear Algebra Appl. **322**(1–3), 61–85 (2001)
23. Neymeyr, K.: A geometric theory for preconditioned inverse iteration II: Convergence estimates. Linear Algebra Appl. **322**(1–3), 87–104 (2001)
24. Ovtchinnikov, E.E.: Computing several eigenpairs of Hermitian problems by conjugate gradient iterations. J. Comput. Phys. **227**(22), 9477–9497 (2008)
25. Ovtchinnikov, E.E.: Jacobi correction equation, line search, and conjugate gradients in Hermitian eigenvalue computation I: Computing an extreme eigenvalue. SIAM J. Numer. Anal. **46**(5), 2567–2592 (2008)
26. Ovtchinnikov, E.E.: Jacobi correction equation, line search, and conjugate gradients in Hermitian eigenvalue computation II: Computing several extreme eigenvalues. SIAM J. Numer. Anal. **46**(5), 2593–2619 (2008)
27. Ovtchinnikov, E.E.: Sharp convergence estimates for the preconditioned steepest descent method for Hermitian eigenvalue problems. SIAM J. Numer. Anal. **43**(6), 2668–2689 (2006)
28. Gulin, A.V., Kregzhde, A.V.: On the Applicability of the Bisection Method to Solve Nonlinear Difference Eigenvalue Problems. Preprint no. 8. Inst. Appl. Math., USSR Science Academy, Moscow (1982) [in Russian]
29. Gulin, A.V., Yakovleva, S.A.: On a numerical solution of a nonlinear eigenvalue problem. In: Marchuk, G.I. (ed.), Computational Processes and Systems, vol. 6, pp. 90–97. Nauka, Moscow (1988) [in Russian]
30. Ruhe, A.: Algorithms for the nonlinear eigenvalue problem. SIAM J. Numer. Anal. **10**, 674–689 (1973)
31. Tisseur, F., Meerbergen, K.: The quadratic eigenvalue problem. SIAM Rev. **43**(2), 235–286 (2001)
32. Mehrmann, V., Voss, H.: Nonlinear eigenvalue problems: a challenge for modern eigenvalue methods. GAMM–Mit. **27**, 1029–1051 (2004)
33. Kressner, D.: A block Newton method for nonlinear eigenvalue problems. Numer. Math. **114**(2), 355–372 (2009)
34. Huang, X., Bai, Z., Su, Y.: Nonlinear rank-one modification of the symetric eigenvalue problem. J. Comput. Math. **28**(2), 218–234 (2010)
35. Schwetlick, H., Schreiber, K.: Nonlinear Rayleigh functionals. Linear Algebra Appl. **436**(10), 3991–4016 (2012)
36. Beyn, W.-J.: An integral method for solving nonlinear eigenvalue problems. Linear Algebra Appl. **436**(10), 3839–3863 (2012)
37. Leblanc, A., Lavie, A.: Solving acoustic nonlinear eigenvalue problems with a contour integral method. Eng. Anal. Bound. Elem. **37**(1), 162–166 (2013)
38. Qian, X., Wang, L., Song, Y.: A successive quadratic approximations method for nonlinear eigenvalue problems. J. Comput. Appl. Math. **290**, 268–277 (2015)
39. Gulin, A.V., Kregzhde, A.V.: Difference Schemes for Some Nonlinear Spectral Problems. Preprint no. 153. Inst. Appl. Math., USSR Science Academy, Moscow (1981) [in Russian]
40. Kregzhde, A.V.: On difference schemes for the nonlinear Sturm–Liouville problem. Differ. Uravn. **17**(7), 1280–1284 (1981) [in Russian]
41. Solov'ev, S.I.: Approximation of nonlinear spectral problems in a Hilbert space. Differ. Equations **51**(7), 934–947 (2015)
42. Solov'ev, S.I.: Approximation of positive semidefinite spectral problems. Differ. Equations **47**(8), 1188–1196 (2011)
43. Solov'ev, S.I.: Approximation of operator eigenvalue problems in a Hilbert space. IOP Conf. Series: Materials Science Engin. **158**(1), Art. 012087, 1–6 (2016)
44. Solov'ev, S.I.: Quadrature finite element method for elliptic eigenvalue problems. Lobachevskii J. Math. **38**(5), 856–863 (2017)

45. Algazin, S.D.: Localization of eigenvalues of closed linear operators. Siber. Math. J. **24**(2), 155–159 (1983)
46. Algazin, S.D.: Discretization of linear equations of mathematical physics with separable variables. Comp. Math. Math. Phys. **35**(3), 321–330 (1995)
47. Algazin, S.D.: Discretization of linear equations of mathematical physics with separable variables. Comp. Math. Math. Phys. **35**(3), 321–330 (1995)
48. Algazin, S.D.: Calculating the eigenvalues of ordinary differential equations. Comp. Math. Math. Phys. **35**(4), 477–482 (1995)
49. Algazin, S.D.: High-precision calculation of the eigenvalues of the Laplace operator. Dokl. Math. **78**(2), 675–678 (2008)
50. Algazin, S.D.: Computational experiments in the problem on eigenvalues for the Laplace operator in the polygonal domain. Math. Models Comp. Simulat. **5**(6), 520–526 (2013)
51. Algazin, S.D.: Cosserat spectrum of the first boundary-value problem of elasticity. J. Appl. Mech. Tech. Phys. **54**(2), 287–294 (2013)
52. Algazin, S.D.: High-accuracy calculation of eigenvalues of the laplacian in an ellipse (with Neumann boundary condition). Dokl. Math. **99**(3), 260–262 (2019)

Numerical Modelling of the Hydraulic Fracturing Through Microseismic Monitoring

Polina Stognii, Nikolay Khokhlov, and Igor Petrov

Abstract Hydraulic fracturing is the method, used during the seismic works on hydrocarbon's extraction. It is necessary to control the properties of the fracture to provide the maximum safety and productivity of the works. The main problem is to obtain the data from the direct modelling, based on which the inverse problem is lately solved. In this paper we present the approach to the direct numerical solution for the problem of hydraulic fracturing with the use of microseismic monitoring. The grid-characteristic method of the third order of accuracy is applied for the numerical modelling of the microseisms spread through the homogeneous medium with the hydraulic fracture. The obtained wave fields and seismogramms demonstrate the possibility of solving the problem of hydraulic fracturing using the microseismic monitoring through the direct computer modelling. Later on, based on the data obtained, the inverse problem of the fracture properties determining can be solved.

1 Introduction

Hydrocarbon deposits are often situated in such locations of the geological areas, where they are hard to extract. The traditional methods of collecting oil deposits, when the well is drilled, are rather costly. In addition, the well flow rate decreases at the final stages of the deposits extraction. Then, the method of hydraulic fracturing can help to increase the oil-bearing bed delivery [1].

The hydraulic fracturing [2–4] is described by creating the artificial fault. For this, the special fracturing fluid is injected into the well, which stimulates oil, gas or any other collected deposit flow more freely. The problems, appearing while the method of hydraulic fracturing, include the possible leakage of methane, triggering

P. Stognii (✉) · N. Khokhlov · I. Petrov
Moscow Institute of Physics and Technology (National Research University), Moscow, Russian Federation
e-mail: stognii@phystech.edu; k_h@inbox.ru; petrov@mipt.ru

I. B. Badriev et al. (eds.), *Mesh Methods for Boundary-Value Problems and Applications*, Lecture Notes in Computational Science and Engineering 141,
https://doi.org/10.1007/978-3-030-87809-2_37

earthquakes and fracture size control [5]. In this paper we examine the problem of controlling the hydraulic fracture size by fixing the required fracture size and investigating further results using the methods of direct computer modelling.

The common method of seismic prospecting of the investigated territory includes the use of the seismic source and the signal receivers [6]. In this case the seismic waves, spreading from the impulse source, need to go straight to the investigated object and backwards, as a consequence, extra reflections from different heterogeneties add the sort of seismic noise to the seismogramms. One of the ways to avoid it is the use of the microseisms as a natural seismic source. Microseisms are the kind of small earthquakes, occurring in the geological area as a natural source of seismic impulses [7]. The method of exploring the geological area using the natural microsesms is known as microseismic monitoring of the researched area.

The method of seismic monitoring is often used for observing the size and form of the hydraulic fracture [8]. In this case, the microseisms receivers are established along the hydraulic fracture in order to detect the reflections from the fracture directly. One of the main advantages of this method is the absence of the head wave on the seismograms, which makes it easier to interpret them.

In this paper we present the results of modelling the microsesms spread through the homogeneous medium with the hydraulic fracture. The grid-characteristic method, well suitable for solving the direct seismology problems, was used in the computations. The fracture was modelled using the model of a two-shore extremely thin fracture. As a result, we obtained the wave fields and the seismogramms, demonstrating the possibility of solving the problems of the hydraulic fracture seismic monitoring .

The paper is organized as follows. The Introduction part presents the overall view of the hydraulic fracturing, the possible ways of controlling the fracture form. In the second part, we describe the numerical method used in the computations. In the third part, we demonstrate the results of solving the problem of seismic monitoring of the hydraulic fracturing using the methods of the direct computer modelling. The problems of modelling the hydraulic fracture and the possible ways of their solution are discussed in the fourth part of the paper. The last part concludes the presented work.

2 Numerical Method

For describing the dynamic behaviour of the microseisms in the homogeneous medium, we used the system for the linear-elastic medium [9]:

$$\rho \frac{\partial}{\partial t} \boldsymbol{v} = (\nabla \cdot \sigma)^T, \tag{1}$$

$$\frac{\partial}{\partial t} \sigma = \lambda(\nabla \cdot \boldsymbol{v}) I + \mu(\nabla \otimes \boldsymbol{v} + (\nabla \otimes \boldsymbol{v})^T). \tag{2}$$

In (1), (2) $\boldsymbol{v}$ is the seismic waves velocity, ρ is the medium density, $\boldsymbol{\sigma}$ is the Cauchy stress tensor, t is the time, λ and μ are the Lame parameters.

We solved the Eqs. (1), (2) with the use of the grid-characteristic method of the third order of accuracy [10]. For this, we present the system (1), (2) in a view:

$$\frac{\partial \mathbf{q}}{\partial t} + \mathbf{A_x}\frac{\partial \mathbf{q}}{\partial x} + \boldsymbol{A}_y\frac{\partial \boldsymbol{q}}{\partial y} = 0. \tag{3}$$

In (3) $\boldsymbol{q} = \{\upsilon_x, \upsilon_y, \sigma_{xx}, \sigma_{yy}, \sigma_{xy}\}^T$ The matrixes A_x , A_y are constructed out of the coefficients of the system (1), (2):

$$A_x = \begin{pmatrix} 0 & 0 & -\frac{1}{\rho} & 0 & 0 \\ 0 & 0 & 0 & 0 & -\frac{1}{\rho} \\ -\lambda - 2\mu & 0 & 0 & 0 & 0 \\ 0 & -\lambda - 2\mu & 0 & 0 & 0 \\ 0 & -\mu & 0 & 0 & 0 \end{pmatrix}, \tag{4}$$

$$A_y = \begin{pmatrix} 0 & 0 & 0 & 0 & -\frac{1}{\rho} \\ 0 & 0 & 0 & -\frac{1}{\rho} & 0 \\ 0 & -\lambda & 0 & 0 & 0 \\ -\lambda & 0 & 0 & 0 & 0 \\ -\mu & 0 & 0 & 0 & 0 \end{pmatrix}. \tag{5}$$

Then, the splitting method in the space coordinates is applied to (3), and two 1D systems of equations are obtained:

$$\frac{\partial \boldsymbol{q}}{\partial t} + \boldsymbol{A}_i\frac{\partial \boldsymbol{q}}{\partial i} = 0, i = x, y. \tag{6}$$

Now, we examine the system (6) for the x-axis:

$$\frac{\partial \boldsymbol{q}}{\partial t} + \boldsymbol{A}_x\frac{\partial \boldsymbol{q}}{\partial x} = 0. \tag{7}$$

The system (7) is hyperbolic, then it can be represented it as:

$$\frac{\partial \boldsymbol{q}}{\partial t} + \boldsymbol{\Omega}_x{}^{-1}\boldsymbol{\Lambda}_x\boldsymbol{\Omega}_x\frac{\partial \boldsymbol{q}}{\partial x} = 0. \tag{8}$$

In (8) $\boldsymbol{\Omega}_x$ is constructed out of the eigen vectors of the matrix A_x, $\boldsymbol{\Lambda_x}$ is the diagonal matrix with the eigen values $\{c_s, 0, -c_p, c_p, -c_s\}$ on the diagonal. In (8) c_p is the longitudinal sound velocity, c_s is the transverse sound velocity:

$$c_p = \sqrt{(\lambda + 2\mu)/\rho}, \tag{9}$$

$$c_s = \sqrt{\mu/\rho}. \tag{10}$$

The analogical equations can be derived for the y-axis.

Then, after the variable change $\boldsymbol{v} = \boldsymbol{\Omega q}$ the system (8) will transfer to:

$$\frac{\partial \boldsymbol{v}}{\partial t} + \boldsymbol{\Lambda}\frac{\partial \boldsymbol{v}}{\partial x} = 0. \tag{11}$$

The system (11) consists of five equations, each of which can be solved using any differential scheme. We used the Rusanov scheme of the third order of accuracy [11].

When all the components ν are transported, we can find the final solution:

$$\boldsymbol{q}^{n+1} = \boldsymbol{\Omega}^{-1}\boldsymbol{v}^{n+1}. \tag{12}$$

For calculating the points on the model boundaries, we applied the following equation:

$$\boldsymbol{B}\boldsymbol{q}^{n+1} = \boldsymbol{b}. \tag{13}$$

In (13) $\boldsymbol{B}$ is the matrix of the 3x9 size, $\boldsymbol{b}$ is the three-dimensional vector, $\boldsymbol{q}^{n+1}$ is the vector of the velocity and the stress tensor meanings in the examined point on the model boundary on the next time step t^{n+1}.

The solution (12) on the next time step t^{n+1} will be:

$$\boldsymbol{q}^{n+1} = \boldsymbol{\Omega}^{(in)}\boldsymbol{v}^{n+1(in)} + \boldsymbol{\Omega}^{(out)}\boldsymbol{v}^{n+1(out)} = \boldsymbol{q}^{n+1(in)} + \boldsymbol{\Omega}^{(out)}\boldsymbol{v}^{n+1(out)}. \tag{14}$$

In (14) $\boldsymbol{\Omega}^{(in)}$ and $\boldsymbol{\Omega}^{(out)}$ are the matrices, constructed out of the columns, corresponding to the ingoing or outgoing characteristics of the matrix $\boldsymbol{\Omega}^{-1}$. The vector $\boldsymbol{v}^{n+1(in)}$ can be calculated the same way as the vector $\boldsymbol{v}^{n+1}$ for the inner points The vector $\boldsymbol{v}^{n+1(out)}$ can be obtained from the boundary conditions (13):

$$\boldsymbol{v}^{n+1(out)} = (\boldsymbol{B}\boldsymbol{\Omega}^{(out)})^{-1}(\boldsymbol{b} - \boldsymbol{B}\boldsymbol{q}^{n+1(in)}). \tag{15}$$

If we unite the Eqs. (14) and (15), we will obtain the overall formula for calculating the points on the model boundaries:

$$\boldsymbol{q}^{n+1} = \boldsymbol{q}^{n+1(in)} + \boldsymbol{\Omega}^{(out)}(\boldsymbol{B}\boldsymbol{\Omega}^{(out)})^{-1}(\boldsymbol{b} - \boldsymbol{B}\boldsymbol{q}^{n+1(in)}). \tag{16}$$

For describing the non-reflecting boundary condition, we used the equation:

$$\boldsymbol{B} = \boldsymbol{\Omega}^{(*)}, \boldsymbol{b} = \boldsymbol{0}. \tag{17}$$

In (17) the matrix $\boldsymbol{\Omega}^{(*)}$ is constructed out of the columns of the matrix with the eigen values on the diagonal $\boldsymbol{\Lambda}$, corresponding to the outgoing characteristics.

For modelling the fracture, we used the model of a two-shore extremely thin fracture [12], filled with fluid. Then, in order to calculate the points on the contact boundary of the fracture, the free slip contact conditions were used:

$$\boldsymbol{v}_n^l = \boldsymbol{v}_n^r, \tag{18}$$

$$\boldsymbol{f}_n^l = -\boldsymbol{f}_n^r, \tag{19}$$

$$\boldsymbol{f}_\tau^l = \boldsymbol{f}_\tau^r = \boldsymbol{0}. \tag{20}$$

In (18)–(20) the indexes l and r indicate the points on the different sides of the fracture, $\boldsymbol{f}$—is the density of the outdoor forces.

3 The Results of the Numerical Modelling for the Hydraulic Fracturing Problem

In this section we present the results of the numerical solution for the problem of the hydraulic fracture seismic monitoring. For this, we carried out the numerical modelling of the seismic waves spread in a homogeneous medium in the presence of the microfracture for the two-dimensional case. The model consisted of the medium with the following parameters. The velocity of the longitudinal waves was equal to 5000 m/s, the velocity of the transverse waves was equal to 3000 m/s, the density of the medium was 2500 kg/m^3. The overall size of the model was 4500×4500 m^2.

The microfracture was modelled as a vertical fluid-filled extremely thin fracture. The length of the microfracture was 30 m.

The schematic image of the model is presented in Fig. 1. The source of microseismic impulses is situated at the distance of 300 m from the fracture center. The microseismic source is presented as a black cross in Fig. 1. The hydraulic fracture and the row of the seismic receivers are shown as black lines in Fig. 1.

The consistent explosions from the seismic source were modelled with the period of $5 * 10^{-3}$ sec. We modelled six consistent explosions from the seismic source. Twenty signal receivers were situated in the front of the fracture at the depth of 800 m. The receivers are shown as direct triangles, the fracture is shown as a vertical black line in the center of Fig. 1.

The parameters of the calculations were the following.The time step was equal to $7 * 10^{-5}$ s, the grid step was 0.5 m. All in all, we computed 14,000 steps in time.

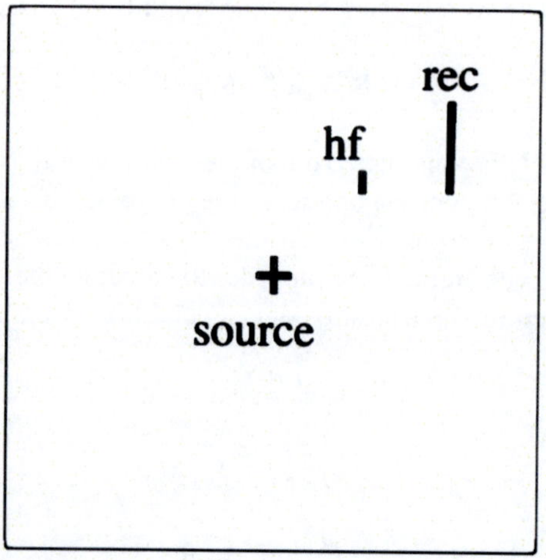

Fig. 1 The schematic representation of the computed model. The seismic source is depicted as a black cross, the seismic receivers and the fracture are shown as a short and long line, respectively

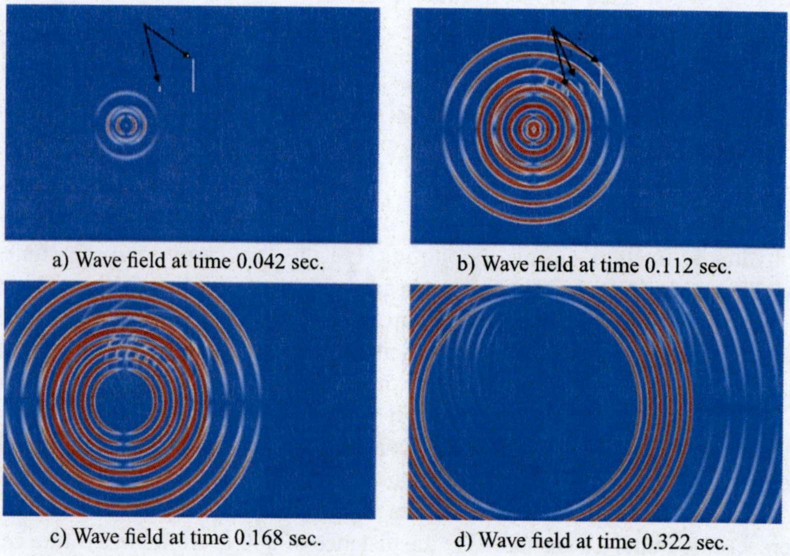

Fig. 2 Wave field for the described formulation of the problem for the model with the microfracture at different moments of time. **(a)** Wave field at time 0.042 s. **(b)** Wave field at time 0.112 s. **(c)** Wave field at time 0.168 s. **(d)** Wave field at time 0.322 s

The wave fields for the described problem are presented in Fig. 2a–d at 0.042, 0.112, 0.168, 0.322 s. moments of time, accordingly. The images in Fig. 2a–d demonstrate the overall view of the consequent spread of the seismic waves from the microseismic impulse source. The reflections from the fracture can be hardly distinguished due to the small size of the fracture (30 m) comparing with the distance of the waves spread (300 m). The fracture and the row of the receivers are depicted as the vertical white lines in Fig. 2a,b. The black pointers 1, 2 indicate the fracture and the receivers, accordingly. The pointer 3 in Fig. 2b indicates the seismic reflections just from the fracture, where they can be clearly seen.

Figure 3 presents the seismograms for the given formulation of the problem. Figure 3a demonstrates the seismogram for the x-component of the velocity, Fig. 3b presents the seismogram for the y-component of the velocity for the twenty seismic receivers. The consistent signals are clearly shown. The use of the microseisms makes the seismograms distinct as they are free of the head wave.

The meanings of the Vx-components of the velocity for the first receiver, located in the lowest point of the receivers row, are depicted in Fig. 4a, the meanings of the Vy-components of the velocity for the same receiver are presented in Fig. 4b. The consistent reflections with the same amplitude can be well seen, according to the described formulation of the problem of consistent seismic explosions.

In reality, it is rather hard and costly to use a row of the receivers, the real experience includes 2–3 receivers. Then, the graphs in Fig. 5a and b demonstrate the meanings of the Vx-components of the velocity and the meanings of the Vy-components of the velocity for the three receivers, which are situated at the distance of 10 m from each other along the vertical row of the receivers. The amplitudes of the signals demonstrate the same periodic behaviour with the time displacement due to their location. The consequent peaks demonstrate the consequent reflections from the fracture.

4 Discussion

The main problem, connected with the numerical solving of the hydraulic fracturing, is the large difference between the fracture size and the distance between the microseisms source and the fracture—30 m against 300 m, accordingly. In order to speed up the computations, we need to increase the grid step [13]. However, the accurate solution demands approximately 10–30 nodes for the fracture size. Therefore, in our computations the grid step was equal to 0.5 m.

In addition, in reality the fracture contains special fluid inside with the characteristic parameters, which makes the fracture increase in width and length [14]. Then, the width of the fracture is not equal to zero. However, the characteristic width of the fracture is 20–30 sm, which is small in comparison to the height of the fracture (30 m) and tiny if compare with the distance between the fracture and the microseismic source. Therefore, we applied the model of an extremely thin fracture in the computations.

One of the possible ways of solving the indicated problem of considering the fracture width is the use of the hierarchical grids [15], where the grid step can be decreased while approaching the fracture. However, we are not interested in the fracture width consideration, but in the characteristic parameters of the fluid inside it. Then, the Shoenberg model of fracture [16] is the best solution in this case as

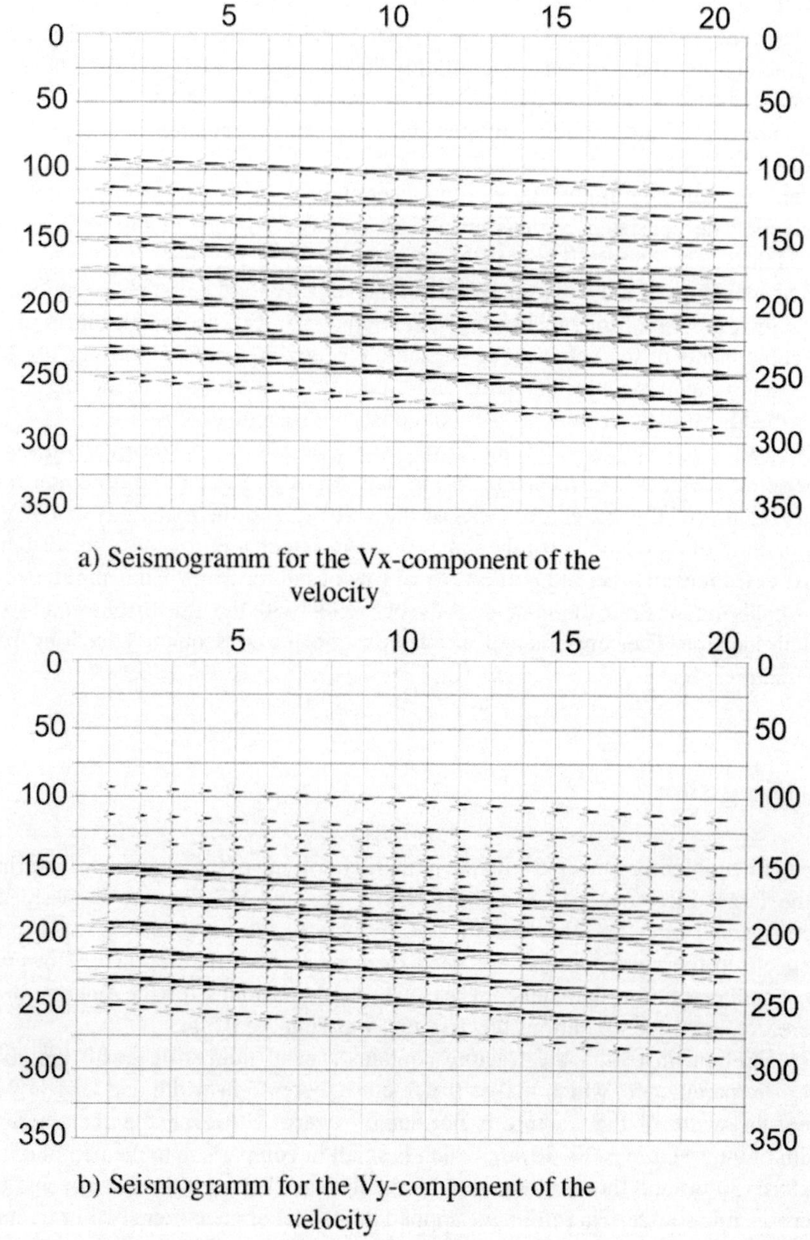

Fig. 3 Seismogramm for the Vx- and Vy-components of the velocity for the model with the microfracture. (**a**) Seismogramm for the Vx-component of the velocity. (**b**) Seismogramm for the Vy-component of the velocity

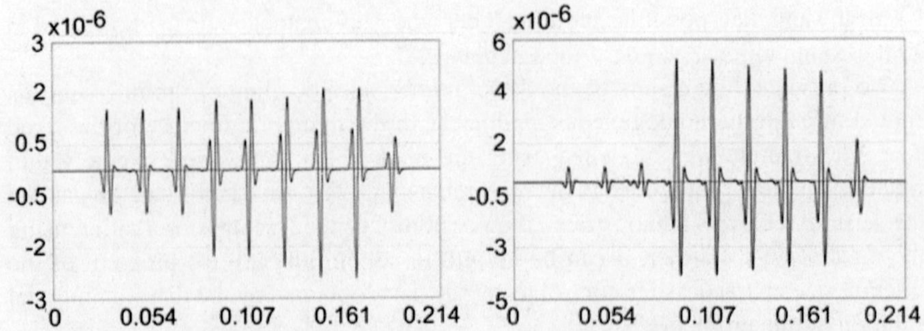

a) The graph of the Vx meanings on the first receiver.

b) The graph of the Vy meanings on the first receiver.

Fig. 4 The graph of the Vx and Vy meanings on the first receiver. (**a**) The graph of the Vx meanings on the first receiver. (**b**) The graph of the Vy meanings on the first receiver

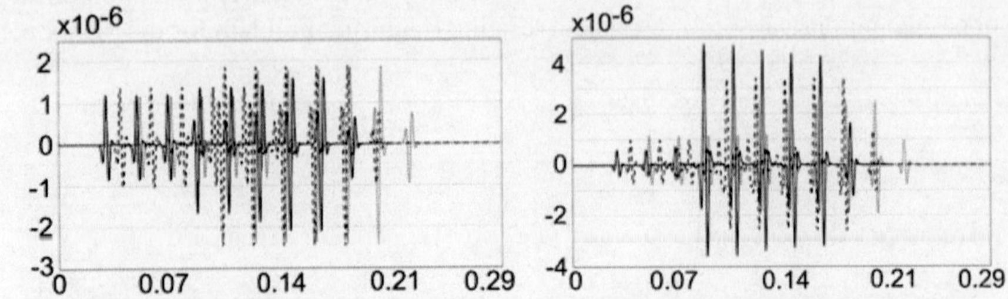

a) The graph of the Vx meanings on the receivers

b) The graph of the Vy meanings on the receivers

Fig. 5 The graphs of the obtained data on the receivers at distance of 10 m between each other. (**a**) The graph of the Vx meanings on the receivers. (**b**) The graph of the Vy meanings on the receivers

the fracture in this model is characterized by the special parameter of disclosure, depending on the characteristic parameters of the fluid inside the fracture and on the width of the fracture [17]. And the fracture width should not be considered while building the mesh grid, which is the key point for this very problem.

The demonstrated results aimed to present the possible way of solving the hydraulic fracturing problem using the grid-characteristic method on structured grids of the third order of accuracy. The further research on the theme of the direct numerical solving of the pointed problem is considering the fracture width as well as the fluid filling it and the expansion of the numerical solution to the 3D case.

5 Conclusions

In this paper we presented the solution to the problem of the hydraulic fracturing seismic monitoring. We discussed the main problems connected with the hydraulic fracturing and their possible solutions. Then, the grid-characteristic method, used in all the computations, was described in detail.

We presented the results of the direct numerical modelling of the microseisms spread through the homogeneous medium with the hydraulic fracture of the fixed size. The fracture was modelled using the model of a two-shore extremely thin fracture, previously approved in the research papers. The wave fields and the data on the seismic receivers demonstrated the possibility of the fracture observation using the microseisms. Further, it can be helpful in solving the inverse problem of the fracture size characteristics control using the data, obtained by the direct numerical solution of this problem.

We discussed the problem of the computational grid step choice and connected with it the chosen model of an extremely thin fracture, applied in the computations. Later on, the complications of the presented model can be made by using either the hierarchical grids or the Shoenberg fracture model in order to consider the fracture width and the fluid parameters inside it. In addition, the logical continuation of the work is the numerical solution of the hydraulic fracturing problem for the 3D case.

Acknowledgments The reported study was funded by RFBR, project number 20-01-00261.

References

1. Esipov, D.V., Kuranakov, D.S., Lapin, V.N., Cherny, S.G.: Mathematical models of hydraulic fracturing. Comput. Technologies. **19**, No. 2, 33–61 (2014).
2. Bakulin, A., Grechka, V., Tsvankin, I.: Estimation of fracture parameters from reflection seismic data–Part I: HTI model due to a single fracture set. Geophysics. **65**, No. 6, 1788–1802 (2000).
3. Bakulin, A., Grechka, V., Tsvankin, I.: Estimation of fracture parameters from reflection seismic data–Part II: Fractured models with orthorhombic symmetry. Geophysics. **65**, No. 6, 1803–1817 (2000).
4. Bakulin, A., Grechka, V., Tsvankin, I.: Estimation of fracture parameters from reflection seismic data–Part III: Fractured models with monoclinic symmetry. Geophysics. **65**, No. 6, 1818–1830 (2000).
5. Howarth, R.: Methane emissions and climatic warming risk from hydraulic fracturing and shale gas development: implications for policy. Energy and Emission Control Technologies. **3**, 45–54 (2015).
6. Yilmaz, Oz (2001). Seismic data analysis. Publisher: Society of Exploration Geophysicists.
7. Carl W.Ebeling: Advances in Geophysics. **53**, 1–33 (2012).
8. Mirko van der, B., Eaton, D., Dusseault, M. (2013). Microseismic Monitoring Developments in Hydraulic Fracture Stimulation. In *ISRM International Conference for Effective and Sustainable Hydraulic Fracturing 2013* (pp.439–66). International Society for Rock Mechanics
9. Savin, G.N., Rushchitskii, Y.Y. Novatskii, V.: The theory of elasticity. Soviet Applied Mechanics. **7**, 808–811 (1971).
10. Favorskaya, A.V., Breus, A.V., Galitskii, B.V. (2019). Application of the Grid-Characteristic Method to the Seismic Isolation Model. Proceedings of the Conference 50 Years of the Development of Grid-Characteristic Method. In: *Smart Modeling for Engineering Systems.* **133** (pp. 167–181).

11. Ivanov, A., Khokhlov, N. (2019). Efficient Inter-process Communication in Parallel Implementation of Grid-Characteristic Method. Proceedings of the Conference 50 Years of the Development of Grid-Characteristic Method. In *Smart Modeling for Engineering Systems* **133** (pp. 91–102) .
12. Khokhlov, N., Stognii, P.: Novel Approach to Modeling the Seismic Waves in the Areas with Complex Fractured Geological Structures. Minerals. **10**, 122 (2020).
13. Nikitin, I.S., Burago, N.G., Golubev, V.I., Nikitin, A.D.: Mathematical modeling of the dynamics of layered and block media with nonlinear contact conditions on supercomputers. Journal of Physics: Conference Series. **1392**, 012057 (2019).
14. Montgomery, C.T. Fracturing Fluid Components (2013).
15. Breus, A., Favorskaya, A., Golubev, V., Kozhemyachenko, A., Petrov, I.: Investigation of seismic stability of high-rising buildings using grid-characteristic method. Procedia Computer Science. **154**, 305–310 (2019).
16. Schoenberg, M.: Elastic wave behavior across linear slip interfaces. The Journal of the Acoustical Society of America. **68**, No.5, 1516–1521 (1980).
17. Stognii, P.V., Khokhlov, N.I., Petrov, I.B.: Modelling of wave processes in fractured geological media using Shoenberg model. Prikladnaya Matematika i Mehanika. **84**, No.3, 375–386 (2020).

Modeling of Deformation of Solids with Material Damage

Lenar U. Sultanov and Almaz M. Kadirov

Abstract The work is devoted to solving problems of large elastoplastic deformations taking into account the material damage. The resolving equation is obtained from the equation of the principle of virtual work in velocity terms. The stress state is described using the Cauchy stress tensor. Constitutive equations are obtained using the potential energy of deformation. Modelling of plastic deformations is based on the method of projecting stresses onto the yield surface with iterative refinement of the current stress-strain state. For solving a nonlinear equation an incremental method is used. The numerical implementation is based on the finite element method. The necking problem of plate with plastic deformation and material damage is solved.

1 Introduction

Fracture processes are described based on the concept of fracture as a loss of the ability of a material to resist deformation due to a violation of internal bonds with an increase in the concentration of microcracks [1–5]. Fracture mechanics studies the mechanisms that are involved in the destruction of a material. At the microlevel, destruction is the accumulation of microstresses between adjacent microdefects or interfaces and the breaking of bonds, which damages the material. At the mesoscale, this is the growth and fusion of microcavities and microcracks, which grows into a crack. At the macro level, this is the growth and propagation of a crack. Processes at the micro- and meso-levels can be studied using the damage variables of continuum mechanics, while the macro-level processes are studied using fracture mechanics with the variables defined at the micro level.

At the mesoscale, fracture can manifest itself in various ways, depending on the nature of the material, the type of load and temperature, therefore, there are:

L. U. Sultanov (✉) · A. M. Kadirov
Kazan Federal University, Kazan, Russia
e-mail: Lenar.Sultanov@kpfu.ru

I. B. Badriev et al. (eds.), *Mesh Methods for Boundary-Value Problems and Applications*, Lecture Notes in Computational Science and Engineering 141,
https://doi.org/10.1007/978-3-030-87809-2_38

brittle fracture, plastic fracture, creep fracture, low-cycle fatigue fracture, high-cycle fatigue fracture.

To obtain a mathematically correct description of the destruction processes, various kinds of regularizers, for example, Scalar parameter of damage are introduced into the constitutive equations. Scalar parameter of damage is responsible for the ability of the medium to resist deformation. With an increase in damage, which occurs when the criterion for the onset of fracture is met, the resistance of the medium decreases: the effective elastic moduli decrease with an increase in damage.

2 Solving Algorithm

The deformation gradient tensor $\mathbf{F}$ is introduced, which shows the change in an elementary oriented segment during deformation. To describe the finite strains and strain rate, the Finger tensor $\mathbf{B} = \mathbf{F} \cdot \mathbf{F}^T$, the velocity gradient tensor $\mathbf{h} = \dot{\mathbf{F}} \cdot \mathbf{F}^{-1}$ and the rate of deformation tensor $\mathbf{d} = 0.5\left[\mathbf{h} + \mathbf{h}^T\right]$ are used [6].

Solving a problem of plastic deformation involves reducing the nonlinear problem to a sequence of linearized problems. When using the incremental method, the resolving equations are constructed by time differentiation of the equation of the principle of virtual work in the actual configuration [6–10]:

$$\int_{\Omega} \boldsymbol{\sigma} \cdot \cdot \delta \mathbf{d} d\Omega = \int_{\Omega} \mathbf{f} \cdot \delta \mathbf{v} d\Omega + \int_{S^{\sigma}} \mathbf{t}_n \cdot \delta \mathbf{v} dS, \tag{1}$$

where $\boldsymbol{\sigma}$—Cauchy stress tensor, Ω—current volume, $\mathbf{f}$—body forces per unit volume, $\mathbf{t}_n = p\mathbf{n}$—traction forces per unit area acting on the S^{σ}, $\mathbf{n}$—surface normal, $\mathbf{v}$—velocity. A linearized equation (1) has the form:

$$\begin{aligned} &\int_{\Omega} \left[\dot{\boldsymbol{\sigma}} \cdot \cdot \delta \mathbf{d} + \boldsymbol{\sigma} \cdot \cdot \delta \dot{\mathbf{d}} + \tfrac{\dot{J}}{J} \boldsymbol{\sigma} \cdot \cdot \delta \mathbf{d}\right] d\Omega + \int_{S^{\sigma}} \left\{\mathbf{t}_n \cdot \mathbf{h} - \tfrac{\dot{J}}{J} \mathbf{t}_n\right\} \cdot \delta \mathbf{v} dS = \\ &= \int_{\Omega} \left[\dot{\mathbf{f}} + \mathbf{f} \tfrac{\dot{J}}{J}\right] \delta \mathbf{v} d\Omega + \int_{S^{\sigma}} \dot{\mathbf{t}}_n \cdot \delta \mathbf{v} dS, \end{aligned} \tag{2}$$

where $J = \det(\mathbf{F})$—volume change, $\dot{\mathbf{t}}_n = \dot{p}\mathbf{n}$ [6, 7, 10, 11].

The constitutive relations are obtained using the potential energy of elastic deformation function W. The components of the Finger tensor $\mathbf{B}$ are accepted as arguments of the function W, i.e.

$$W = W(\mathbf{B}), \tag{3}$$

then the Cauch stress tensor will be expressed in the following form [6]:

$$\boldsymbol{\sigma} = \frac{2}{J}\mathbf{B} \cdot \frac{\partial W}{\partial \mathbf{B}}. \tag{4}$$

For an isotropic material the potential energy function is depend on the invariants of the Finger tensor [6]:

$$W = W(I_{1\mathbf{B}}, I_{2\mathbf{B}}, I_{3\mathbf{B}}), \tag{5}$$

where $I_{1\mathbf{B}}, I_{2\mathbf{B}}, I_{3\mathbf{B}}$—the corresponding invariants of the Finger tensor **B**. The stress tensor can be expressed as

$$\boldsymbol{\sigma} = \frac{2}{J}\mathbf{B} \cdot \left\{ \frac{\partial W}{\partial I_{1\mathbf{B}}}\mathbf{I} + \frac{\partial W}{\partial I_{2\mathbf{B}}}[I_{1\mathbf{B}}\mathbf{I} - \mathbf{B}] + \frac{\partial W}{\partial I_{3\mathbf{B}}} I_{3\mathbf{B}}\mathbf{B}^{-1} \right\}. \tag{6}$$

Linearizing relation (4), the rate of the Cauchy stress is obtained:

$$\dot{\boldsymbol{\sigma}} = \frac{\partial \boldsymbol{\sigma}}{\partial \mathbf{B}} \cdot \cdot \dot{\mathbf{B}} \tag{7}$$

and

$$\begin{aligned} \dot{\boldsymbol{\sigma}} &= 2\left\{ \tfrac{1}{J}\dot{\mathbf{B}} \cdot \tfrac{\partial W}{\partial \mathbf{B}} + \tfrac{1}{J}\left[\mathbf{B} \cdot \tfrac{\partial^2 W}{\partial \mathbf{B}^2}\right] \cdot \cdot \dot{\mathbf{B}} - \tfrac{1}{J}\mathbf{B}\tfrac{\partial W}{\partial \mathbf{B}} I_{1\mathbf{d}} \right\} = \\ &= \boldsymbol{\Lambda}_{\sigma} \cdot \cdot \mathbf{d} + \mathbf{h} \cdot \boldsymbol{\sigma} + \boldsymbol{\sigma} \cdot \mathbf{h}^T - \boldsymbol{\sigma} I_{1\mathbf{d}}, \end{aligned} \tag{8}$$

where the notation is introduced

$$\boldsymbol{\Lambda}_{\sigma} = \frac{4}{J}\mathbf{B} \cdot \frac{\partial^2 W}{\partial \mathbf{B} \partial \mathbf{B}} \cdot \mathbf{B}. \tag{9}$$

As a result, the constitutive relationship of elastic deformation was obtained in the form of a linear equation:

$$\boldsymbol{\sigma}^{Tr} = \boldsymbol{\Lambda}_{\sigma} \cdot \cdot \mathbf{d}, \tag{10}$$

where $\boldsymbol{\sigma}^{Tr} = \dot{\boldsymbol{\sigma}} + \mathbf{h} \cdot \boldsymbol{\sigma} + \boldsymbol{\sigma} \cdot \mathbf{h}^T - I_{1d}\boldsymbol{\sigma}$—the Truesdell derivative of the stress tensor $\boldsymbol{\sigma}$.

Total deformations are represented as the sum of elastic and plastic components [7, 8].

$$\mathbf{d} = \mathbf{d}^e + \mathbf{d}^p. \tag{11}$$

The associative flow law is used:

$$\mathbf{d}^p = \dot{\lambda}\frac{\partial \Phi}{\partial \boldsymbol{\sigma}}, \tag{12}$$

where $\dot{\lambda}$—plastic strain rate, Φ—flow function. The von Mises condition as a criterion for elastic deformation is used:

$$\Phi = \sigma_i - \sigma_y(\chi) \leq 0 \tag{13}$$

where $\sigma_i = \sqrt{\frac{3}{2}\boldsymbol{\sigma}' \cdot \cdot \boldsymbol{\sigma}'}$—stress intensity, $\boldsymbol{\sigma}'$—deviatric part of $\boldsymbol{\sigma}$, $\sigma_y(\chi)$—hardening function, χ—hardening parameter. Using (13), the plastic deformation rate can be written as follows

$$\mathbf{d}^p = \dot{\lambda}\frac{\partial \Phi}{\partial \boldsymbol{\sigma}} = \dot{\lambda}\frac{\partial \sigma_i}{\partial \boldsymbol{\sigma}'} = \frac{3}{2}\dot{\lambda}\frac{\boldsymbol{\sigma}'}{\sigma_i}. \tag{14}$$

For modelling plastic deformations the method of projecting stresses onto the yield surface is used [7, 8, 10, 11, 16, 17]. Let all process parameters, including configuration, stress state, values of elastic and inelastic deformations, etc. for the k-th state are known. The parameters of the k+1-th state are determined by the formula [7, 11]

$$\begin{aligned} {}^{k+1}\boldsymbol{\sigma} &= {}^{k}\boldsymbol{\sigma} + {}^{k}\dot{\boldsymbol{\sigma}}\Delta t = \\ &= {}^{k}\boldsymbol{\sigma} + \left[{}^{k}\boldsymbol{\sigma}^{Tr} + {}^{k}\mathbf{h} \cdot {}^{k}\boldsymbol{\sigma} + {}^{k}\boldsymbol{\sigma} \cdot {}^{k}\mathbf{h}^T - I_{1d}\,{}^{k}\boldsymbol{\sigma}\right]\Delta t = \\ &= {}^{k}\boldsymbol{\sigma} + \left\{\boldsymbol{\Lambda} \cdot \cdot \left[{}^{k}\mathbf{d} - \tfrac{3}{2}\dot{\lambda}\tfrac{{}^{k+1}\boldsymbol{\sigma}'}{{}^{k+1}\sigma_i}\right] + {}^{k}\mathbf{h} \cdot {}^{k}\boldsymbol{\sigma} + {}^{k}\boldsymbol{\sigma} \cdot {}^{k}\mathbf{h}^T - I_{1d}\,{}^{k}\boldsymbol{\sigma}\right\}\Delta t, \end{aligned} \tag{15}$$

where Δt is the parameter (time) increment, which determines the transition from the previous state to the next.

The last equation can be written in the following form

$${}^{k+1}\boldsymbol{\sigma} + \frac{3\dot{\lambda}}{2\sigma_i}\boldsymbol{\Lambda} \cdot \cdot\, {}^{k+1}\boldsymbol{\sigma}' = {}^{k+1}\tilde{\boldsymbol{\sigma}}, \tag{16}$$

where ${}^{k+1}\tilde{\boldsymbol{\sigma}} = {}^{k}\boldsymbol{\sigma} + \left\{\boldsymbol{\Lambda} \cdot \cdot\, {}^{k}\mathbf{d} + {}^{k}\mathbf{h} \cdot {}^{k}\boldsymbol{\sigma} + {}^{k}\boldsymbol{\sigma} \cdot {}^{k}\mathbf{h}^T - I_{1d}\,{}^{k}\boldsymbol{\sigma}\right\}\Delta t$—trial stresses tensor. Equation (16) defines the projection of stresses to the yield surface.

The deformation process is represented as a sequence of equilibrium states. The transition from the previous equilibrium k-th state to the next equilibrium k+1-th state occurs by a loading increment. When plastic deformations occur, the method of projecting the stress onto the yield surface is applied. The resolving equation at the k-th step has the form [6, 7]:

$$\int\limits_{\Omega_k} \left\{ {}^k\mathbf{d} \cdot\cdot {}^k\mathbf{\Lambda} \cdot\cdot \delta\mathbf{d} + \frac{1}{2}\, {}^k\boldsymbol{\sigma} \cdot\cdot \left[\delta\mathbf{h}^T \cdot {}^k\mathbf{h} + {}^k\mathbf{h}^T \cdot \delta\mathbf{h} \right] - \frac{{}^k_{}j}{{}^kJ}\mathbf{f} \cdot \delta\mathbf{v} \right\} d\Omega +$$
$$+ \int\limits_{S^\sigma} \left\{ {}^k\mathbf{t}_n^* \cdot {}^k\mathbf{h} - \frac{{}^k_{}j}{{}^kJ}\, {}^k\mathbf{t}_n^* \right\} \cdot \delta\mathbf{v} dS = \int\limits_{\Omega_k} {}^k\dot{\mathbf{f}}^* \cdot \delta\mathbf{v} dV + \int\limits_{S^\sigma} {}^k\dot{\mathbf{t}}_n^* \cdot \delta\mathbf{v} dS -$$
$$- \frac{1}{\Delta t} \left\{ \int\limits_{\Omega_k} {}^k\boldsymbol{\sigma} \cdot\cdot \delta\mathbf{d} dV - \int\limits_{\Omega_k} {}^k\mathbf{f}^* \cdot \delta\mathbf{v} dV - \int\limits_{S^\sigma} {}^k\mathbf{t}_n^* \cdot \delta\mathbf{v} dS \right\}. \tag{17}$$

Due to a quasi-static problem is considered, it is possible to pass from rates to increments, for example $\mathbf{v} = \Delta\mathbf{u}/\Delta t$, where $\mathbf{u}$—displacement. Solving Eq. (17), we obtain the displacement vector $\mathbf{u}$, which can be used to determine the configuration and stress state at k+1 loading step:

$${}^{k+1}\mathbf{R} = {}^k\mathbf{R} + \Delta^k\mathbf{u}, \tag{18}$$

$${}^{k+1}\tilde{\boldsymbol{\sigma}} = {}^k\boldsymbol{\sigma} + \Delta^k\boldsymbol{\sigma}. \tag{19}$$

As a result of using the projection of stresses to the yield surface method (16), the defined stress state does not satisfy the resolving system of equations (17). Therefore, the iterative refinement of the stress-strain state is used. This iterative procedure is based on introducing the power of an additional stresses on virtual deformation rates into the resolving Eq. (17), where an additional stresses are defined as the difference between true stresses ${}^{k+1}\boldsymbol{\sigma}$ (16) and trial stresses ${}^{k+1}\tilde{\boldsymbol{\sigma}}$ (19) [7, 11].

In according of the continuum damage mechanics, to characterize the damage to the elementary volume of the material at the microlevel, the scalar parameter ω is introduced [1]:

$$\omega = \frac{S - S_\omega}{S}, \tag{20}$$

where S is the cross-sectional area of the plane of the elementary volume, and S_ω is the area of intersection of microdefects with this plane. ω takes values from 0 (material is not damaged) to 1 (material is completely destroyed).

Material damage at a given time can be determined by various mechanisms of resource depletion (deformation, transitional creep, fatigue, brittle fracture, etc.), therefore, the total material damage is the sum of the damage state functions ω_n for each class of defects [12]:

$$\omega = \sum_n \omega_n. \tag{21}$$

Determination of damage ω_n of the corresponding type presupposes a description of the mechanics of the behavior of the environment within the framework of the phenomenon under study, the construction of an evolutionary equation for the accumulation of damage. Integrating it and conducting basic experiments to determine the appropriate material characteristics.

At the final stage of the accumulation of damage scattered throughout the volume, the effect of damage on the viscoplastic behavior of the material is observed. This influence can be taken into account on the basis of the concept of a degrading continuum (introduction of effective stresses) [12–15]:

$$\tilde{\sigma} = \frac{\sigma}{1 - \omega}. \tag{22}$$

The value of the increment of the current damage $\Delta\omega^k$ is determined as the sum of the damage of those destruction mechanisms that are taken into account in the calculation. For example, under plastic deformation, the damage can be defined as [2, 3, 5]:

$$\Delta^k \omega = \frac{-\Delta^k \lambda}{1 - \omega}\left(-\frac{{}^k Y}{a}\right)^b, \tag{23}$$

where

$$-Y = \frac{\sigma_i^2}{2E(1-\omega)^2}\left[\frac{2}{3}(1+\nu) + 3(1-2\nu)\left(\frac{\sigma_0}{\sigma_i}\right)^2\right],$$

$$\sigma_0 = \frac{1}{3}(\sigma_{11} + \sigma_{22} + \sigma_{33}),$$

Y—damage energy release rate, E—elasticity modulus, ν—Poisson's ratio, a, b—material parameters.

3 Numerical Examples

The problem of stretching a rectangular strip with dimensions: $h = 266.67$ mm, $w = 64.13$, $d = 5$ mm is considered. The von Mises condition is used as the plasticity criterion. A material with nonlinear isotropic hardening is used [16, 17]. Table 1 lists the properties of the material [3]. An example of the construction of physical relations for the following potential of elastic deformations is considered [18]:

$$W = \frac{\lambda + 2\mu}{8}(I_{1B} - 3)^2 + \mu(I_{1B} - 3) - \frac{\mu}{2}(I_{2B} - 3), \tag{24}$$

where λ, μ—Lame parameters, $\lambda = \frac{E\nu}{(1+\nu)(1-2\nu)}$, $\mu = \frac{E}{2(1+\nu)}$.

Table 1 Material parameters

Parameter	Value
Hardening law	$\sigma_y(\chi) = \sigma_y + h\chi + (\sigma_\infty - \sigma_0)(1 - e^{-\delta\chi})$
E, MPa	180,000
ν	0.32
σ_∞, MPa	715
σ_y, MPa	450
h	0.129
δ	16.93
ε_{pcr}	0.44
a	2.4
b	2.4
ω_{cr}	0.2

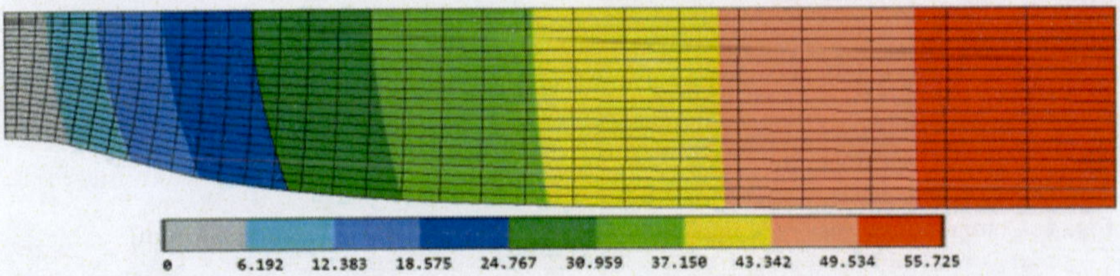

Fig. 1 Distribution of displacements along the OX axis at u = 55.725 mm

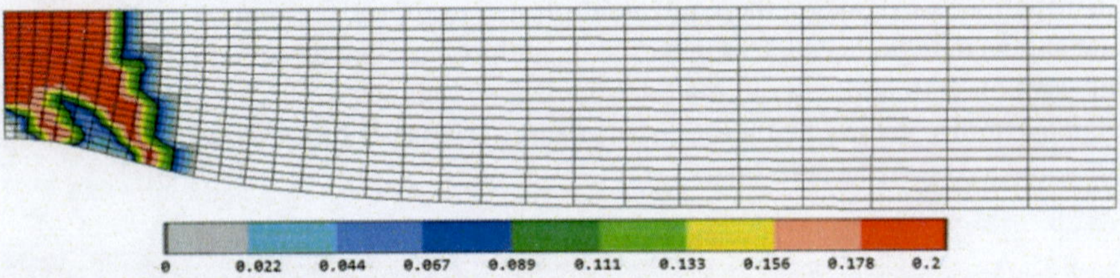

Fig. 2 The distributions of damage at u = 55.725 mm

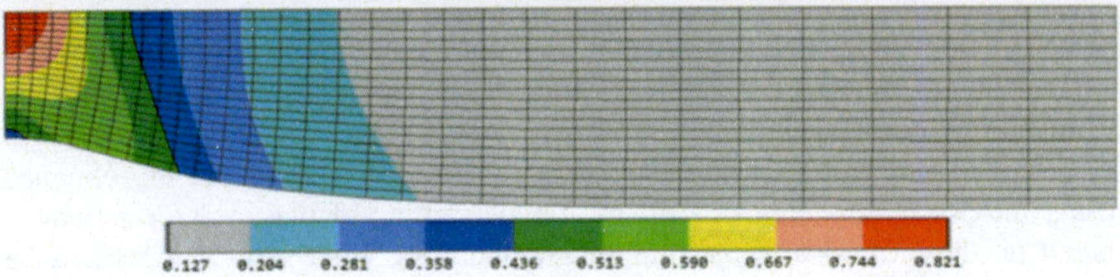

Fig. 3 Equivalent plastic strain at u = 55.725 mm

An eight-node 3D finite element with linear approximation is used to calculate numerical examples [19]. The problem is symmetric, therefore, the calculation is performed for 1/8 of the strip, with the imposition of symmetry conditions and

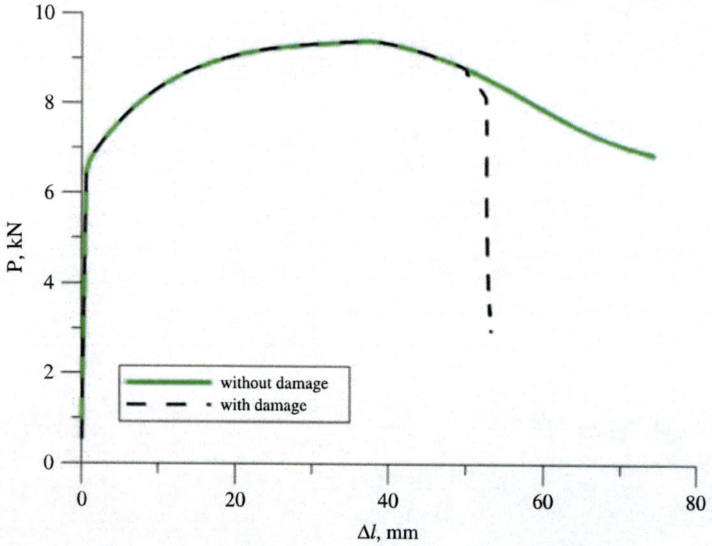

Fig. 4 Computational results of applied force P [kN] versus axial elongation Δl [mm]

a kinematic boundary condition. The displacements along the tensile axis, the distributions of damage, the equivalent plastic strain for the current configuration are shown in Figs. 1, 2, 3. Computational results of applied force P [kN] versus axial elongation with and without material damage are shown in the Fig. 4. Inclusion of damageability in the calculation entails a fast decrease of force, which means a drop of a material resistance to load.

4 Conclusion

A method for numerical investigation of the deformations of solids with large plastic strains and the material damage is developed. The resolving equation is obtained from the equation of the principle of virtual work in velocity terms. The stress state is described using the Cauchy stress tensor. Constitutive equations are obtained using the potential energy of deformation. Modelling of plastic deformations is based on the method of projecting stresses onto the yield surface with iterative refinement of the current stress-strain state. Numerical calculation is based on a finite element method. The necking problem of plate with plastic strains and material damage is solved.

Acknowledgments This paper has been supported by the Kazan Federal University Strategic Academic Leadership Program ("PRIORITY-2030").

References

1. Kachanov, L.M.: Introduction to continuum damage mechanics. Martinus Nijhoff (1986)
2. Lemaitre, J .: A course of damage mechanics. Berlin, Springer (1996)
3. Lemaitre, J., Desmorat, R.: Engineering damage mechanics. Berlin, Springer (2005)
4. Chaboche, J.L.: Continuous damage mechanics – a tool to describe phenomena before crack initiation. Nuclear Engeneering Design. **64**, 233–247 (1981)
5. Broumand, P., Khoei, A.R.: The extended finite element method for large deformation ductile fracture problems with a non–local damage–plasticity model. Engineering fracture mechanics. **112–113**, 97–125 (2013)
6. Bonet, J., Wood, R.D.: Nonlinear continuum mechanics for finite element analysis. USA, Cambridge University Press (1997)
7. Davydov, R.L., Sultanov, L.U.: Numerical algorithm for investigating large elasto-plastic deformations. Journal of engineering physics and thermophysics. **88**, 1280–1288 (2015)
8. Golovanov, A.I.: Numerical modeling of large deformations of elastoplastic solids in terms of logarithms of principal stretches. Computational continuum mechanics. **4**, 25–35 (2011)
9. Golovanov, A.I.: Finite element analysis of large deformations of hyperelastic solids in principal axes. Computational continuum mechanics. **2**, 19–37 (2009)
10. Sultanov, L.U.: Analysis of large elastic-plastic deformations: Integration algorithm and numerical examples. Uchenye zapiski kazanskogo universiteta. Seriya fiziko-matematicheskie nauki. **159**, 509–517 (2017)
11. Sultanov, L.U.: Analysis of finite elastic-plastic deformations. Kinematics and constitutive equations. Uchenye zapiski kazanskogo universiteta. Seriya fiziko-matematicheskie nauki. **157**, 158–165 (2015)
12. Kapustin, S.A., Likhacheva, S.Y.: Modeling the processes of deformation and destruction of materials with a periodically repeating structure. N. Novgorod, NNGASU Publishing (2012)
13. Bouchard, P-O, Bourgeon, L., Fayolle, S., Mocellin, K.: An enhanced Lemaitre model formulation for materials processing damage computation. International journal of material forming. **4**, 299–315 (2011)
14. Celentano, D.J., Tapia, P.E., Chaboche, J–L.: Experimental and numerical characterization of damage evolution in steels. Mecanica computacional. **23**, 1–14 (2004)
15. Kintzel, O., Mosler, J.: A coupled isotropic elasto–plastic damage model based on incremental minimization principles. Technische mechanik. **30**, 177–184 (2010)
16. Eidel, B., Gruttmann, F.: Elastoplastic orthotropy at finite strains: multiplicative formulation and numerical implementation. Computational materials science. **28**, 732–742 (2003)
17. Schröder, J., Gruttmann, F., Löblein, J.: A simple orthotropic finite elasto-plasticity model based on generalized stress-strain measures. Computational mechanics. **30**, 48–64 (2002)
18. Chernykh, K.F.: Nonlinear theory of elasticity in machine-building calculations. Leningrad, Mashinostroenie (1986)
19. Bathe K.J.: Finite element procedures in engineering analysis. Prentice Hall (1996)

Prediction of Temperature-Dependent Processes in Multicomponent Fluid Flow Through Porous Media

Marina A. Trapeznikova, Natalia G. Churbanova, and Antonina A. Chechina

Abstract The research deals with the development of efficient tools for the simulation of thermal processes in porous media when flows of multiphase multicomponent slightly compressible fluids are considered. Such flows occur in the subsurface during the hydrocarbon recovery or during remediation of contaminated soils, fluid filtration also takes place in various industrial installations. For an adequate description of non-isothermal processes the transfer of mass and energy between phases should be reproduced, therefore the multicomponent composition of fluids cannot be neglected. The classic model is modified to be implemented by explicit difference schemes with sufficient accuracy and mild stability conditions. The experience of constructing the hyperbolic quasi-gas dynamic system of equations was transferred to flows in porous media. Conservation laws are formulated for the components in terms of the mass concentrations of components in phases. The mass balance equation for each component contains the second time derivative and a dissipative term with small parameters having the sense of minimum reference sizes in time and in space. Constants of phase equilibrium are used to close the system of equations. To verify the developed approach test calculations of two- and three-phase flows were performed, physically correct results were obtained.

1 Introduction

The solution of many industrial and environmental problems involves the calculation of fluid flows in porous media. Among such applications there are problems of flows in the subsurface, in particular, a wide range of problems associated with modeling the hydrocarbon recovery. The simulation of contaminant infiltration into the underground space when solving problems of the soil remediation and prevention of groundwater contamination is also worth mentioning. In addition,

M. A. Trapeznikova (✉) · N. G. Churbanova · A. A. Chechina
Keldysh Institute of Applied Mathematics RAS, Moscow, Russia
e-mail: mtrapez@yandex.ru; nataimamod@mail.ru; chechina.antonina@yandex.ru

I. B. Badriev et al. (eds.), *Mesh Methods for Boundary-Value Problems and Applications*, Lecture Notes in Computational Science and Engineering 141,
https://doi.org/10.1007/978-3-030-87809-2_39

filtration of fluids serves as the basis for technological processes in various industrial installations, for example, in the equipment for processing of organic fuels while cleaning oil and gas from impurities. In all these problems the system under consideration is essentially multiphase: liquid (aqueous and non-aqueous) and gaseous phases are mobile, they are assumed to be immiscible; an immobile solid phase, namely, the porous skeleton, is also taken into account. The studied processes depend significantly on temperature: promising technologies for the recovery of high-viscosity oil suppose the use of thermal methods (heat carrier pumping into the stratum, in-situ combustion), and technologies for the soil restoration employ the exposure by hot steam. For adequate description of non-isothermal processes, it is necessary to reproduce the transfer of mass and energy between phases, therefore, the multicomponent composition of the phases cannot be neglected.

The present paper develops an approach to modeling flows of multiphase multicomponent slightly compressible fluids in porous media in view of possible heat sources. The original idea was to modify the classic model of fluid flow in a porous medium [1–4] in such a way that it would be possible to implement it by logically simple computational algorithms, namely, by explicit finite difference schemes with sufficient accuracy and mild stability conditions. For this purpose the experience of constructing a hyperbolic version of the known quasi-gas dynamic (QGD) system of equations [5–7] was borrowed to describe flows in porous media. For the first time the mathematical model created by analogy to the QGD system was presented in [8], the hyperbolic form of the model was proposed and substantiated in [9, 10], then it was generalized to the multicomponent case in [11, 12]. Since the numerical solution of the applied problems of interest is extremely time consuming and practically unrealizable without high performance computing, the authors pay attention mainly to explicit-type algorithms that allow efficient parallel implementation, including realization on hybrid architectures.

To verify the developed approach numerous test calculations have been carried out. The most interesting problem solutions are presented in this paper—we illustrate convection of a three-phase fluid in a porous medium in a reservoir with differently heated walls as well as phase transition between the water and gas phases in a heat pipe. The numerical results obtained are in good agreement with results of other authors [13].

2 State of the Art in the Research Field

The present work corresponds to modern scientific trends and is comparable with the achievements of other research teams as evidenced by publications in the literature.

Nowadays techniques called Enhanced Oil Recovery (EOR) methods [14] are applied to improve the oil recovery in a hydrocarbon reservoir. EOR covers secondary and tertiary recovery and is preceded by the waterflooding technique that almost always has to be considered. Currently, all over the world, the concept of intelligent oil and gas fields is being applied; it implies the use of both innovative mining technologies and advanced information technologies. Despite the great experience in hydrodynamic modeling and many corresponding software packages, there remains a need for their improvement, and mainly concerning the reservoir models and algorithms for an adequate description of multiscale thermo-hydro-mechanical-chemical processes [15].

Among publications on compositional modeling of multiphase compressible flow in porous media paper [16] should be noted where the traditional strategy of switching the primary variables (see [13], e.g.) is criticized. Instead, a general approach based on the molar masses of components is developed. It is close to the development of the present paper promoting also some general approach but the mass concentrations instead of the molar concentrations are used here to simplify the equations and computational algorithm when chemical reactions are not taken into account. Article [16] considers the black oil model only, while the present paper proposes a universal model that allows an arbitrary number of components and any reasonable composition of phases.

Implicit, semi-implicit and explicit-implicit algorithms are mainly used in modeling filtration processes, in particular, the traditional IMPES method [1] still remains popular [17]. However the present paper substantiates advantages of explicit schemes — not only high efficiency of parallelization but a gain when performing calculations with critical accuracy [11, 18].

Quasi-gas and quasi-hydrodynamic models are also used by other teams to develop algorithms for modeling flows in porous media, especially on the core scale [19].

Note that the hyperbolization technique implied in the present paper to increase the stability of difference schemes is a modern trend in CFD [6, 7, 20].

We must pay tribute that interesting multiprocessor implementations including those for GPUs are presented in periodicals [21–23]. Nevertheless, there are not so many works on modeling three-phase compositional flows. Basically, the literature reflects the simulation of two-phase flows.

Simulation of flows in the subsurface as any large-scale computational problem is connected with processing, storage and interpretation of huge amounts of data. Appropriate hardware and software are invested in the promising Big Data technology [24]: massively parallel processing (MPP) systems, uniting hundreds and thousands of computational nodes, have to be used for the data treatment.

The actual performance of supercomputers and relevant architectures are introduced by the list of the most powerful general purpose systems of the world—the TOP 500 rating [25], the most powerful systems of CIS are introduced by TOP 50 [26]. Now the world top system is Fugaku manufactured by Fujitsu and installed at RIKEN Center for Computational Science in Kobe, Japan. It turned in a High Performance Linpack result of 415.5 petaflops. This is a rare example of a supercomputer based on traditional CPU-only architecture: it involves 158,976 CPUs, each has 48 cores. Many modern supercomputers (144 systems on TOP 500) have hybrid architectures and include various computing accelerators, mainly GPUs. Therefore, software developers have to ensure not only the scalability of supercomputer applications, but also their portability in a wide class of hybrid architectures. In this sense, algorithms based on explicit difference schemes are perfect.

Let us summarize with a quote from [27]:

> The first exascale systems are expected to be available in about one year. For sure, there is still a lot of work to be done to let cutting-edge science applications fully exploit their potential.

3 Background of the Research

The QGD system of equations is the basis of the present research. The QGD system is a differential approximation of the kinetically consistent finite difference (KCFD) schemes [5] that belong to promising kinetic algorithms of gas- and hydrodynamics. The QGD equations are based on the generalization of the Navier-Stokes equations and differ from them by additional terms with a small parameter acting as the solution stability regularizers. For the first time, the QGD system was proposed in the 1980s by a group of researchers of Keldysh Institute of Applied Mathematics (KIAM) under the leadership of Prof. Boris Chetverushkin. The convergence of the QGD algorithm was illustrated numerically on a set of test problems [28].

The derivation of KCFD schemes and QGD system is guided by the so-called principle of minimum sizes [29]. This principle states that for the numerical solution of continuum mechanics problems it makes no sense to consider scales smaller than some minimal reference values. In gas dynamics, for example, the minimal reference length is the mean free path of a molecule. Applying this principle to porous media flows one can derive that the scale of averaging at which the filtering rock microstructure is negligible is such a length. The order of this value is a hundred rock grain sizes. The concept of the minimum reference time scale is also introduced: in gas dynamics this is the time interval between collisions of molecules, and in the porous medium flow problems, the parameter can be interpreted as the time for establishing internal equilibrium in a volume of the above mentioned reference size.

In [8], for the first time, a mathematical model of single compressible fluid filtration was constructed by analogy to the QGD system taking into account the principle of minimum sizes. The continuity equation acquired an additional dissipative term (a regularizer), which provided stability of the explicit scheme with central differences for the convective term approximation. This was followed by a series of works proposing a hyperbolized version of the model [9], its generalization to the case of multiphase fluids [30, 31], and also implementation on MPP systems and hybrid clusters [31, 32]. Paper [10] summarizes these developments for isothermal processes and contains an alternative to the manner of obtaining the modified continuity equation from [8]. A number of works are devoted to the simulation of non-isothermal multiphase flows in porous media via the new model including the total energy conservation equation modified after QGD system and approximated by an explicit scheme [11, 31]. A detailed description of the complete model and the computational algorithm as well as attempts to generalize them to the case of a multicomponent composition of fluids are reflected in [11]. Paper [12] introduces the compositional model and the algorithm formulated for the simulation of non-isothermal two-phase flow of water and gas where the gas phase consists of two components—water (as steam) and air.

4 Governing Model

Flow of multiphase multicomponent fluid through a non-deformable homogeneous isotropic porous medium is under consideration. By a component a single chemical compound or a mixture of compounds is assumed. Let n_α be the number of phases (usually no more than three mobile phases) and n_κ be the number of components (an arbitrary number). Further speaking of phases we mean mobile phases, subscripts w, n and g denote water, NAPL (Non-Aqueous Phase Liquid) and gas phases respectively. Liquids are considered as slightly compressible, gas is ideal, the rock skeleton is incompressible. The temperature of all phases and the rock is considered identical.

Obviously the same component may be present in different phases. In the current version of the model, the relative amount of the component in the phase is expressed as the mass concentration:

$$C_\alpha^\kappa = \frac{m_\alpha^\kappa}{m_\alpha}, \quad \kappa = 1, \ldots, n_\kappa, \quad \alpha = w, n, g \tag{1}$$

where m_α^κ is the mass of component κ in phase α, m_α is the mass of phase α.

Based on conservation laws for components, one can formulate the next compositional model:

$$\phi \frac{\partial}{\partial t} \sum_\alpha \rho_\alpha S_\alpha C_\alpha^\kappa + \tau \frac{\partial^2}{\partial t^2} \sum_\alpha \rho_\alpha S_\alpha C_\alpha^\kappa + \sum_\alpha \operatorname{div}\left(\rho_\alpha C_\alpha^\kappa \mathbf{u}_\alpha\right) = \\ = Q^\kappa + \sum_\alpha \operatorname{div} \frac{l c_\alpha}{2} \operatorname{grad}\left(\rho_\alpha S_\alpha C_\alpha^\kappa\right), \quad \kappa = 1, \ldots, n_\kappa, \tag{2}$$

$$\mathbf{u}_\alpha = -K \frac{k_\alpha}{\mu_\alpha} \left(\operatorname{grad} P_\alpha - \rho_\alpha \mathbf{g}\right), \quad \alpha = w, n, g, \tag{3}$$

$$\frac{\partial}{\partial t}\left[\phi \sum_\alpha \rho_\alpha S_\alpha E_\alpha + (1-\phi)\, \rho_r E_r\right] + \sum_\alpha \operatorname{div}\left(\rho_\alpha H_\alpha \mathbf{u}_\alpha\right) + \\ + \operatorname{div}\left(-\lambda_{\text{eff}} \operatorname{grad} T\right) = \sum_\alpha \operatorname{div} \frac{l c_\alpha}{2} \rho_\alpha \operatorname{grad} T, \tag{4}$$

$$\rho_\alpha = \rho_\alpha\left(P_\alpha, T, C_\alpha^\kappa\right), \quad \kappa = 1, \dots, n_\kappa, \quad \alpha = w, n, g, \tag{5}$$

$$\begin{aligned} P_n - P_w &= P_{c\,nw}\left(S_w\right), \\ P_g - P_n &= P_{c\,gn}\left(S_g\right), \end{aligned} \tag{6}$$

$$\sum_\alpha S_\alpha = 1, \tag{7}$$

$$\sum_\kappa C_\alpha^\kappa = 1, \quad \alpha = w, n, g, \tag{8}$$

$$\frac{C_\alpha^\kappa}{C_\beta^\kappa} = K_{\alpha\beta}^\kappa\left(P, T\right), \quad \kappa = 1, \dots, n_\kappa, \quad \alpha = w, n, g. \tag{9}$$

The following notations are used: S_α is the saturation (of α-phase), P_α is the pressure, ρ_α is the density, $\mathbf{u}_\alpha$ is the Darcy velocity, T is the temperature, E_α is the internal energy, H_α is the enthalpy, λ_{eff} is the effective coefficient of heat conductivity, ϕ is the porosity, K is the absolute permeability, k_α is the relative phase permeability, μ_α is the dynamic viscosity, Q^κ is the source of component κ, $\mathbf{g}$ is the gravity vector, ρ_r = const—the rock density (subscript r denotes the rock), c_α is the sound speed in α-phase, small parameters l and τ are the minimal reference length and time respectively, $P_{c\,nw}\left(S_w\right)$ and $P_{c\,gn}\left(S_g\right)$ are capillary pressures, $K_{\alpha\beta}^\kappa$ is the constant of phase equilibrium (so-called "K-value").

The above model consists of:

- the mass balance equation for each component (2),
- the extended Darcy's law (3),
- the total energy conservation law (4),
- the state equations (5),
- differences between the pressures (the capillary pressures) (6),
- constitutive relations (7) and (8),
- constants of phase equilibrium (9).

As it was mentioned in previous sections of the article this model naturally follows from the hyperbolized QGD-based model of multiphase fluid flow in a porous medium. The reasoning behind the model can be found in [11]. Attention should be paid to the second time derivative with the small parameter τ in (2) as well as to the right-hand sides of (2) and (4) including regularizers with the small parameter l.

Note that in the multicomponent case:

$$H_\alpha = \sum_\kappa H_\alpha^\kappa \tag{10}$$

where components′ enthalpies can be found via heat capacities of the components. And the next connection between the internal energy and the enthalpy is used in computations:

$$E_\alpha = H_\alpha - \frac{P_\alpha}{\rho_\alpha}, \qquad E_r = H_r. \tag{11}$$

Since local thermal equilibrium is assumed, the heat conductivity of the fluid-filled porous medium is averaged from the known heat conductivities of the phases and the solid matrix. In the current research the linear mixing model is used and the effective coefficient of heat conductivity is expressed as follows:

$$\lambda_{\text{eff}} = \phi \sum_\alpha S_\alpha \lambda_\alpha + (1 - \phi)\, \lambda_r. \tag{12}$$

There are more complicated expressions for the averaging (see [13], e.g.) using nonlinear mixing models and taking into account the change in heat conductivity with increasing temperature.

Formally, the universal way to describe the thermodynamic state of a fluid is to set equations of state for each component. It is possible to calculate the required phase densities on their basis if the components′ concentrations are known. The concentrations themselves are determined by the equations of phase equilibrium but their solution is a computationally expensive task.

This work employs the technique that is often used in practice:

- the calculation of densities is performed via equations of state;
- phase equilibrium is calculated using K-values.

Currently we consider linear state equations for liquids and assume the validity of the ideal gas law for all components in the gas phase. The phase densities are determined as the weighted harmonic mean of the components densities. The phase viscosity $\mu_\alpha = \mu_\alpha \left(\mu_\alpha^\kappa(T), C_\alpha^\kappa\right)$ also depends on the component composition.

Generally speaking K-values are complicated functions of the pressure, the temperature, the composition, the porous material, they should be specified for all components to be condense. Correct setting of these functions can significantly improve the accuracy of calculations. The most famous formula has been presented by Wilson for weak solutions and low pressures [33], some K-values can be found in [34], but all of them depend only on properties of the given component and do not depend on properties of other components of the solution, what can lead to errors in calculations for real mixtures.

The porous medium flow equations contain strongly nonlinear coefficients of the relative phase permeability and the capillary pressures (6) depending on the phase saturations. One can find different relations fitted to experimental data to describe these functions. In the present research for the simulation of three-phase flow we choose Parker's model [35] to describe capillary pressures, the relative phase permeability is presented by Stone's Model I [1], these models are also cited in [10, 11]. For the two-phase flow van Genuchten constitutive relationships [4] are used.

5 Computational Algorithm

A computational algorithm of the explicit type was developed for numerical implementation of the QGD-based model of multiphase fluid flow in a porous medium [10, 11, 32]. When constructing the algorithm, rectangular computational domains covered by Cartesian grids were considered. High efficiency of parallel implementation on CPU cores as well as on GPUs of a hybrid supercomputer was demonstrated while solving infiltration problems [32].

The algorithm is naturally generalized to predict multicomponent fluid flow. The set of primary variables usually includes the temperature, one phase saturation and the other phase pressure in the two-phase case or two phase saturations and the third phase pressure in the three-phase case. Now this set is enlarged due to the concentrations of components in the phases (1). Their number depends on the problem statement. In [12] the algorithm was formulated in the special case when the two-phase two-component flow was considered, the phases were water and gas, the components were water and air. Then the next primary variables were chosen: T, S_w, P_g and C_g^w (the steam concentration).

At each time step j the next main stages of the algorithm are fulfilled:

- Calculation of the term
$$\left(\sum_{\alpha} \rho_\alpha S_\alpha C_\alpha^\kappa\right)^{j+1} \tag{13}$$
from (2) for all components via the three-level explicit scheme with central differences for convective term approximation (the scheme has the second order in time and in space).
- Calculation of the internal energy
$$\left[\phi \sum_{\alpha} \rho_\alpha S_\alpha E_\alpha + (1-\phi)\,\rho_r E_r\right]^{j+1} \tag{14}$$
from (4) via the explicit scheme with central differences.
- Calculation of primary variables by solving a system of nonlinear algebraic equations locally at each point of the computational grid. This system is formed using state equations (5), expressions for internal energy (11) and K-values (9). Values (13) and (14) obtained at the previous stages are used in the right-hand sides of the given algebraic equations. The system is solved by Newton′s method that takes only a few iterations.

Parallel implementation of the algorithm is based on the principle of geometric parallelism (i.e. data partitioning is used): the computational domain is divided into subdomains in different directions depending on the geometry of the considered problem. On inner boundaries of subdomains data exchange takes place. The developed algorithm does not require inversion of the full matrix of unknowns and is parallelized as explicit schemes. The system solution by Newton's method does not cause additional exchanges — only one two-way data exchange operation occurs on each inner boundary at each time step.

6 Test Predictions

6.1 Heat Pipe Effect Simulation

The problem of the heat pipe effect [13] has been used to verify the compositional model (2)–(9) and the algorithm discussed in Sect. 5. This is a two-phase two-component flow problem: a thin tube is filled by a porous medium saturated with water and air. The liquid phase consists of the water component only, the gas phase can consists of two components—steam and air. Figure 1 illustrates the statement.

The initial conditions are as follows:

$$P_g^0 = 1\,\text{atm}, \quad S_w^0 = 0.5, \quad T^0 = 343.15\,\text{K}, \quad \left(C_g^w\right)^0 = 0.05\,.$$

A constant heat flux inside the domain as well as zero fluxes for all mass components are given at the right-hand boundary (Neumann conditions). At the left-hand boundary Dirichlet conditions are set:

$$P_{g\,1} = 1\,\text{atm}, \quad S_{w\,1} = 0.98, \quad T_1 = 343.15\,\text{K}, \quad \left(C_g^w\right)_1 = 0.29\,.$$

The following physical process must be observed as a result of simulation: due to the heat flux the water-air system is heated until water turns into steam, thus the phase transition happens—some amount of the water component passes from the water phase to the gas phase.

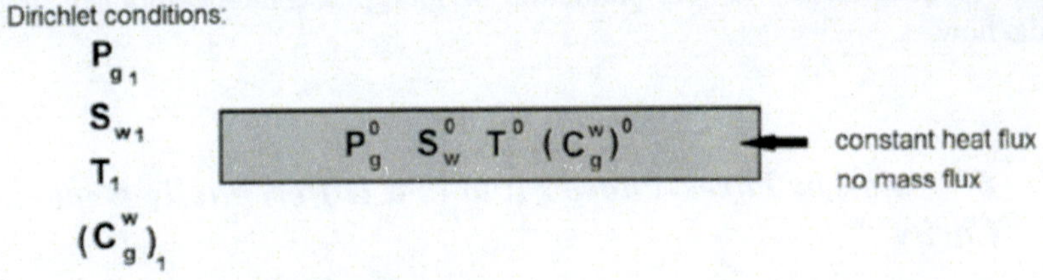

Fig. 1 The heat pipe problem statement

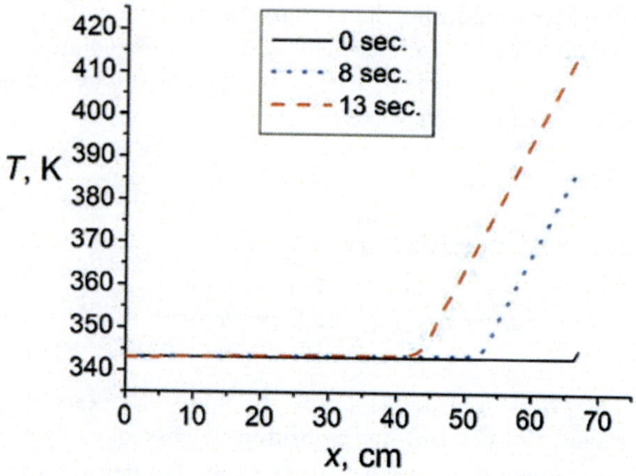

Fig. 2 Temperature in the heat pipe at different time moments

In computations the following K-value is used [36]:

$$K_{gw}^{w} = \frac{C_g^w}{C_w^w} = \exp\left[5.373\,(1+\omega_w)\left(1-\frac{T_c}{T}\right)\right]\frac{P_c}{P_g} \tag{15}$$

where ω_w is the water molecule acentric factor, T_c and P_c are known critical values of the temperature and the pressure at which water turns into steam and they have identical properties,

$$\omega_w = 0.76949, \quad T_c = 647.1\,\text{K}, \quad P_c = 221.15\,\text{bar}.$$

Figures 2, 3, 4, 5 illustrate the beginning of the process until the expected phase transition occurs. The temperature and the pressure are increased gradually at the right-hand boundary, where starting from some time moment the water phase saturation is decreased while the concentration of water steam in the gas phase is increased. Water also penetrates inside the domain from the left-hand boundary due to capillary effects. One can conclude that the proposed computational technique ensures a qualitatively correct prediction of multiphase multicomponent porous media flow.

6.2 Filtration of Three-Phase Fluid in a Differentially Heated Cavity

One of the popular test problems in CFD is the problem of natural convection of air in a differentially heated square cavity [37]. The similar problem is solved at the assumption that the cavity is filled by a porous medium, see [38] e.g. In the present paper we suppose that the medium is saturated with three-phase (water-oil-air) fluid.

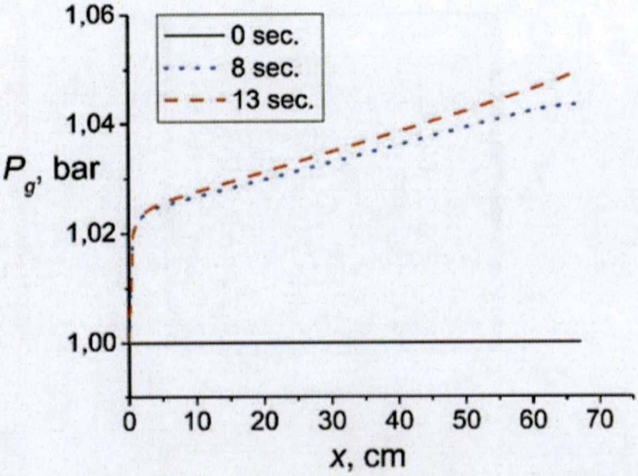

Fig. 3 Gas phase pressure in the heat pipe at different time moments

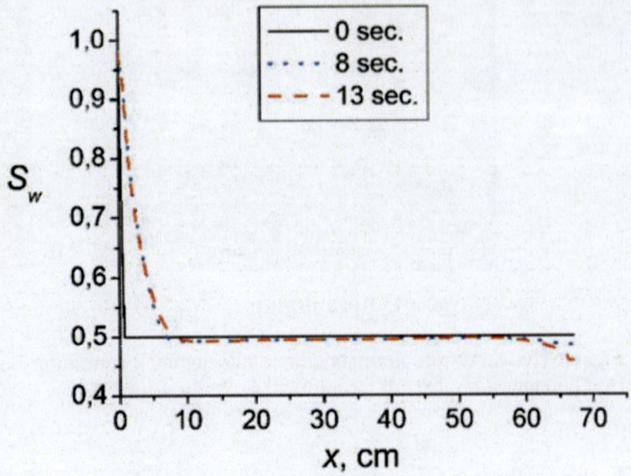

Fig. 4 Water saturation in the heat pipe at different time moments

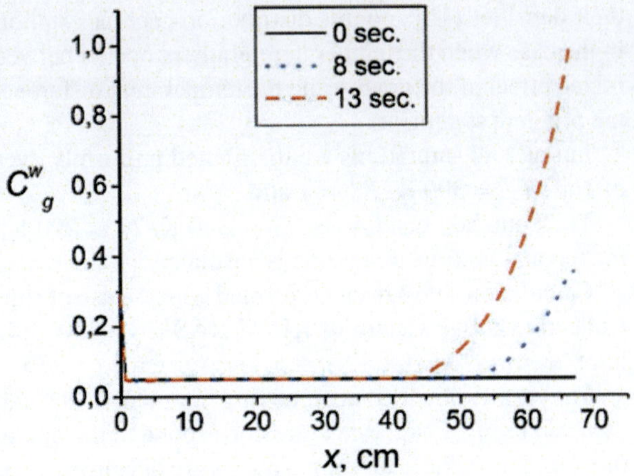

Fig. 5 Mass concentration of the water component in the gas phase inside the heat pipe at different time moments

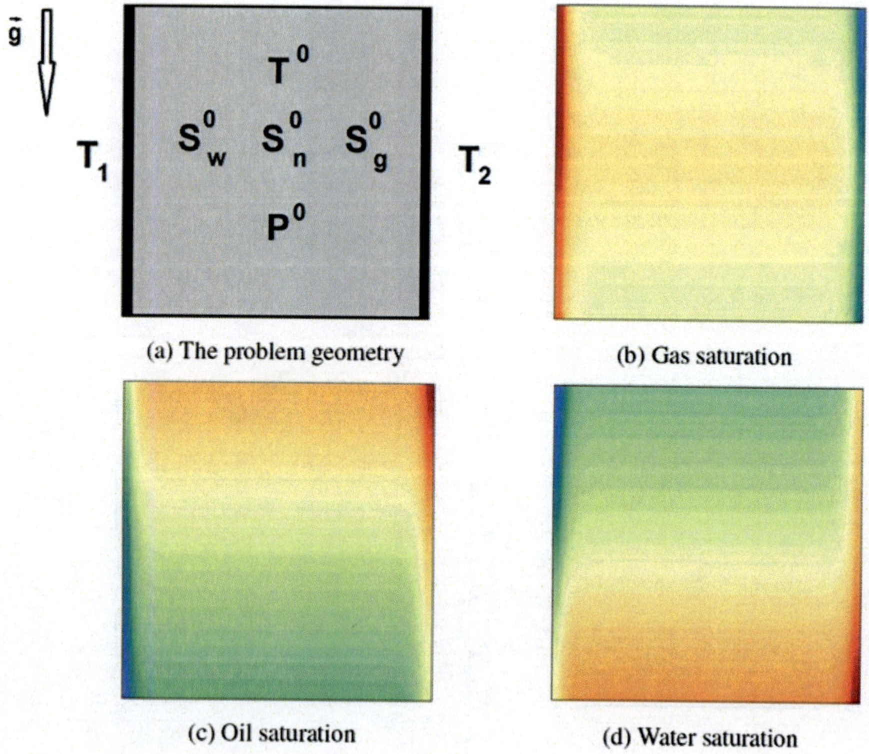

Fig. 6 The cavity problem statement and obtained saturation fields. (**a**) The problem geometry. (**b**) Gas saturation. (**c**) Oil saturation. (**d**) Water saturation

In the isothermal case fluids are eventually distributed over the domain according to their densities [10] and this distribution becomes stationary. Now we are interested in the case when the temperature gradient occurs between the vertical walls and the mixed effect of the gravity and the temperature difference appears, Fig. 6a illustrates the problem statement.

Initially all saturations are distributed uniformly over the reservoir with the size of 1m^2, $T^0 = 300$ K, $P^0 = 1$ atm.

The boundary conditions: $T_1 = 310$ K, $T_2 = 290$ K, all walls are impermeable, the top and bottom are thermally insulated.

Calculations have been performed with the use of supercomputers installed in the Collective Usage Centre of KIAM RAS [39]: 200 CPU cores of the K100 cluster have been exploited.

The results obtained at some time moment of the early stage of calculations are depicted in Fig. 6b–d. Red areas correspond to the maximum values, blue areas—to the minimum. The saturation patterns are asymmetric, interesting behavior of fluids is observed. Gradually gas accumulates at the top of the warm wall, water moves down to the bottom of the cold wall and displaces oil that clings to the cold wall top.

7 Conclusion

Robust computational tools including the QGD-based hyperbolic model of non-isothermal flow of multiphase multicomponent fluid in a porous medium as well as parallel algorithm of its implementation have been formulated and verified by test predictions. Physically correct behavior of the fluid obtained by the simulation indicates the adequacy of the developed approach.

At present, the approach is approved via the fourth SPE comparative solution project involving three steam injection problems [34]. The first problem is the cyclic steam injection in a heavy-oil reservoir. The problem is solved in the $(r-z)$-geometry, therefore the equations are expressed now in the cylindrical coordinate system. The given formulation is also useful for modeling processes in different technological installations because cylindrical tanks are often included in their constructions. In the future, the authors plan to simulate the processing of organic fuels during the purification of crude oil or natural gas from various contaminants.

References

1. Aziz, K., Settari, A.: Petroleum Reservoir Simulation. Applied Science Publ., London (1979)
2. Helmig, R.: Multiphase Flow and Transport Processes in the Subsurface - A Contribution to the Modelling of Hydrosystems. Springer, Berlin (1997)
3. Chen, Z.: Reservoir Simulation: Mathematical Techniques in Oil Recovery. SIAM, Philadelphia (2007)
4. Pinder, G.F., Gray, W.G.: Essentials of Multiphase Flow and Transport in Porous Media. John Wiley & Sons, Hoboken, NJ (2008)
5. Chetverushkin, B.N.: Kinetic Schemes and Quasi-Gas Dynamic System of Equations. CIMNE, Barcelona (2008)
6. Davydov, A.A., Chetverushkin, B.N., Shil′nikov, E.V.: Simulating flows of incompressible and weakly compressible fluids on multicore hybrid computer systems. Comp. Math. Math. Phys. **50**(12), 2157–2165 (2010) doi: 10.1134/S096554251012016X
7. Chetverushkin, B., D'Ascenzo, N, Ishanov, S, Saveliev, V.: Hyperbolic type explicit kinetic scheme of magneto gas dynamics for high performance computing systems. Rus. J. Num. Anal. Math. Model. **30**(1), 27–36 (2015) doi: 10.1515/rnam-2015-0003
8. Trapeznikova, M.A., Belocerkovskaja, M.S., Chetverushkin, B.N.: Analog kineticheski-soglasovannyh shem dlja modelirovanija zadachi fil'tracii (Analog of kinetically consistent schemes for simulation of a filtration problem). Matematicheskoe modelirovanie (Math modeling) **14**(10), 69–76 (2002). (In Russian)
9. Chetverushkin, B.N., Morozov, D.N., Trapeznikova, M.A., Churbanova, N.G., Shil′nikov, E.V.: An explicit scheme for the solution of the filtration problems. Math. Mod. and Comp. Sim. **2**(6), 669–677 (2010) doi: 10.1134/S2070048210060013
10. Chetverushkin, B., Churbanova, N., Kuleshov, A., Lyupa, A., Trapeznikova, M.: Application of kinetic approach to porous medium flow simulation in environmental hydrology problems on high-performance computing systems. Rus. J. Numer. Anal. Math. Modelling **31**(4), 187–196 (2016) doi: 10.1515/rnam-2016-0019
11. Trapeznikova, M., Churbanova, N., Lyupa, A.: CMMSE 2019: An explicit algorithm for the simulation of non-isothermal multiphase multicomponent flow in a porous medium. J. Math. Chem. **58**, 595–611 (2020) doi: 10.1007/s10910-019-01088-z

12. Trapeznikova, M., Churbanova, N., Lyupa, A.: Non-isothermal compositional model for simulation of multiphase porous media flows. In: A. Nadykto et al. (eds.) EPJ Web of Conferences **224**, IV International Conference Modeling of Nonlinear Processes and Systems, Article No. 02010 (2019) https://doi.org/10.1051/epjconf/201922402010
13. Class, H., Helmig, R., Bastian, P.: Numerical simulation of non-isothermal multiphase multicomponent processes in porous media. 1. An efficient solution technique. Advances in Water Resources **25**, 533–550 (2002)
14. Lake, L.W.: Enhanced Oil Recovery. Prentice Hall, New Jersey (1989)
15. Thermo-Hydro-Mechanical-Chemical Processes in Fractured Porous Media: Modelling and Benchmarking, Kolditz O. et al. (Eds.), Springer (2016)
16. Amooie, M.A., Moortgat, J.: Higher-order black-oil and compositional modeling of multiphase compressible flow in porous media. Int. J. of Multiphase Flow **105**, 45–59 (2018)
17. Chen H., Kou J., Sun S., Zhang, T.: Fully mass-conservative IMPES schemes for incompressible two-phase flow in porous media. Computer Methods in Applied Mechanics and Engineering **350**, 641–663 (2019)
18. Fedorenko, R.P.: Introduction to Computational Physics: Study Guide for Universities. Moscow Inst. Phys. Tech. Publ., Moscow (1994). (In Russian)
19. Balashov, V., Savenkov, E.B.: Direct Numerical Simulation of Single and Two-Phase Flows at Pore-Scale. In: Karev V., Klimov D., Pokazeev K. (eds.) Physical and Mathematical Modeling of Earth and Environment Processes. Springer Proceedings in Earth and Environmental Sciences, pp. 374–379. Springer, Cham (2018) https://doi.org/10.1007/978-3-030-11533-3_37
20. Myshetskaya, E.E., Tishkin, V.F.: On the Solution of Evolution Equations Based on Multigrid and Explicit Iterative Methods. Comp. Math. Math. Phys. **55**(8), 1270–1275 (2015) doi: 10.1134/S0965542515080138
21. Wang, K., Liu, H., Chen, Z.: A scalable parallel black oil simulator on distributed memory parallel computers. J. of Computational Physics **301**, 19–34 (2015)
22. Buesing, H.: Efficient Solution Techniques for Multi-phase Flow in Porous Media. In: Lirkov, I., Margenov, S. (eds.) Large-Scale Scientific Computing. LSSC 2017. LNCS **10665**, pp. 572–579. Springer, Cham (2018)
23. Teja-Juarez, V.L., de la Cruz, L.M.: A GPU based implementation of an incompressible two-phase flow model in porous media. Geofisica Internacional **57**(3), 205–222 (2018)
24. Chen, M., Mao, S., Zhang, Y, Leung, V.C.M.: Big Data. Related Technologies, Challenges and Future Prospects. In: Springer Briefs in Computer Science. Spinger (2014) doi: 10.1007/978-3-319-06245-7
25. TOP 500. The List. https://www.top500.org/
26. TOP 50. http://top50.supercomputers.ru/list
27. Software for Exascale Computing - SPPEXA 2016–2019, In: Bungartz, H.-J. et al. (eds.) LNCSE **136**. Springer (2020) https://doi.org/10.1007/978-3-030-47956-5
28. Elizarova, T.G., Shilnikov, E.V.: Capabilities of a quasi-gasdynamic algorithm as applied to inviscid gas flow simulation. Comp. Math. Math. Phys. **49**(3), 549–566 (2009) doi: 10.1134/S0965542509030142
29. Chetverushkin, B.N.: Resolution limits of continuous media mode and their mathematical formulations. Math. Mod. and Comp. Sim. **5**(3), 266–279 (2013)
30. Morozov, D.N., Trapeznikova, M.A., Chetverushkin, B.N., Churbanova, N.G.: Application of explicit schemes for the simulation of the two-phase filtration process. Math. Mod. and Comp. Sim. **4**(1), 62–67 (2012) doi: 10.1134/S2070048212010085
31. Lyupa, A.A., Morozov, D.N., Trapeznikova, M.A., Chetverushkin, B.N., Churbanova, N.G.: Three-phase filtration modeling by explicit methods on hybrid computer systems. Math. Mod. and Comp. Sim. **6**(6), 551–559 (2014) doi: 10.1134/S2070048214060088

32. Trapeznikova, M.A., Churbanova, N.G., Lyupa, A.A., Morozov, D.N.: Simulation of Multiphase Flows in the Subsurface on GPU-based Supercomputers. In: M. Bader et al. (eds.) Parallel Computing: Accelerating Computational Science and Engineering (CSE), Advances in Parallel Computing **25**, pp. 324–333. IOS Press, Amsterdam (2014) doi: 10.3233/978-1-61499-381-0-324
33. Wilson, G.M.,: A modified Redlich-Kwong EOS. Application to General Physical Data Calculations. Paper No. 15C presented at the 1969 AlChE Natl.Meeting, Cleveland, Ohio.
34. Aziz, K., Ramesh, A.B., Woo, P.T.: Fourth SPE comparative solution project: comparison of steam injection simulators. J. Pet. Technol. **39**(12), 1576–1584 (1987)
35. Parker, J.C., Lenhard, R.J., Kuppusami, T.: A parametric model for constitutive properties governing multiphase flow in porous media. Water Resources Research **23**(4), 618–624 (1987)
36. Brusilovsky, A.I.: Fazovye prevrashhenija pri razrabotke mestorozhdenij nefti i gaza (Phase Transitions in the Development of Oil and Gas Fields). Graal Publ., Moscow (2002). (In Russian)
37. De Vahl Davis, G.: Natural convection of air in a square cavity: A bench mark numerical solution. Int. J. Numer. Methods Fluids **3**, 249–264 (1983)
38. Misra, D., Sarkar, A.: A comparative study of porous media models in a differentially heated square cavity using a finite element method, Int. J. of Numerical Methods for Heat & Fluid Flow **5**(8), 735–752 (1995) https://doi.org/10.1108/EUM0000000004124
39. KIAM – The official site of Keldysh Institute of Applied Mathematics. https://www.kiam.ru/MVS/resourses/

A Difference Scheme Based on the Schwarz Method for a Time-Dependent Singular Perturbation Problem in a Doubly Connected Domain

Irina V. Tselishcheva and Grigorii I. Shishkin

Abstract The object of our study is an initial–boundary value problem for a singularly perturbed parabolic reaction–diffusion equation. The highest derivatives are multiplied by a small perturbation parameter ε taking any values in the half-open interval (0,1]. The Dirichlet problem is considered in the space-time domain $\overline{G} = \overline{D} \times [0, T]$, where $\overline{D}$ is a doubly connected domain in space, i.e., a rectangle $\overline{D}_1$ with a removed circle D_2. As $\varepsilon \to 0$, boundary layers of different types arise in neighborhoods of smooth parts of the lateral boundary and lateral edges. The boundary layers decrease exponentially with the distance from the outer and inner lateral boundaries. We discuss an approach to develop a reliable numerical method based on the techniques for simply connected domains. Our aim is to construct an iterative Schwarz method on overlapping subdomains that cover either the boundary of the parallelepiped or the boundary of the cylinder. It is shown that the method converges ε-uniformly in the maximum norm with increasing the number of iterations (and the number of mesh points in the case of a difference scheme). We use the Shishkin meshes that condense in the boundary layers and are piecewise uniform along the normal to the smooth parts of the boundaries. To construct meshes in regions near the outer and inner lateral boundaries, the Cartesian and cylindrical coordinate systems are used, respectively.

1 Introduction

Recently, interest among numerical analysts grows in the development of reliable numerical methods for solving singular perturbation problems in domains with complex geometry and with quite complicated boundary layers. Reliable methods are intended for the accurate resolution of boundary layers in the maximum norm

I. V. Tselishcheva (✉) · G. I. Shishkin
Krasovskii Institute of Mathematics and Mechanics, Ural Branch of the Russian Academy of Sciences, Yekaterinburg, Russia
e-mail: tsi@imm.uran.ru; shishkin@imm.uran.ru

I. B. Badriev et al. (eds.), *Mesh Methods for Boundary-Value Problems and Applications*, Lecture Notes in Computational Science and Engineering 141,
https://doi.org/10.1007/978-3-030-87809-2_40

for all possible values of the perturbation parameter. For this, it is attractive to apply overlapping domain decomposition methods [1] that allow us to reduce the solution of the problem to a sequence of problems on simpler subdomains containing singularities of the same type and further to parallelize the solution process. We discuss an approach to constructing a reliable numerical method for the case of a doubly connected domain based on the techniques developed earlier for simply connected domains. The aim of our study is to develop iterative schemes based on the alternating Schwartz method. To construct difference schemes, we use classical monotone approximations and the simplest piecewise uniform meshes condensing in a neighborhood of the boundary layers in each coordinate, i.e., meshes piecewise uniform along the normal to the smooth parts of the boundaries of the subdomains.

2 Problem Formulation

In the biconnected space-time domain (see Fig. 1)

$$\overline{G} = \overline{D} \times [0, T], \tag{1}$$

where $\overline{D}$ is the rectangle $\overline{D}_1$ with the removed circle D_2 (Fig. 2), $D = D_1 \setminus D_2$, $D_1 = (-d_1 < x_1 < d_1) \times (-d_2 < x_2 < d_2)$; $D_2 = \{x : (x_1^2 + x_2^2)^{1/2} < d\}$; $d_1, d_2 > d > 0$, we consider the Dirichlet problem for the singularly perturbed parabolic reaction–diffusion equation[1]

$$\begin{aligned} L_{(2)}u(x,t) \equiv \varepsilon^2 \Delta u - a(x,t)u - p(x,t)\frac{\partial u}{\partial t} = f(x,t), \quad (x,t) \in G, \\ u(x,t) = \varphi(x,t), \quad (x,t) \in S. \end{aligned} \tag{2}$$

Here, $S = \overline{G} \setminus G$, the coefficients and the right-hand side $a(x,t)$, $p(x,t)$, $f(x,t)$, and also the boundary function $\varphi(x,t)$ are assumed to be bounded and sufficiently smooth on $\overline{G}$ and on S (on smooth parts of the boundary); $a(x,t) \geq 0$, $p(x,t) \geq p_0 > 0$. The parameter ε takes arbitrary values in the interval (0,1].

A similar singularly perturbed problem is known as Hemker's model problem [2] but considered in an unbounded domain exterior of the unit disc.

As $\varepsilon \to 0$, boundary layers of different types appear in neighborhoods of the smooth parts of the lateral boundary S^L and of the lateral edges. In a neighborhood of the outer boundary, i.e., the lateral faces of the parallelepiped but outside regions near its edges, there appear parabolic boundary layers; in a neighborhood of the edges, the layer is angular. In a neighborhood of the inner lateral boundary (the

[1] The notation $L_{(k)}$ ($m_{(k)}$, $M_{(k)}$, $D_{h(k)}$) means that this operator (constant, grid) was introduced in formula (k). By M (m), we denote sufficiently large (small) positive constants independent of ε and the stencils of difference schemes.

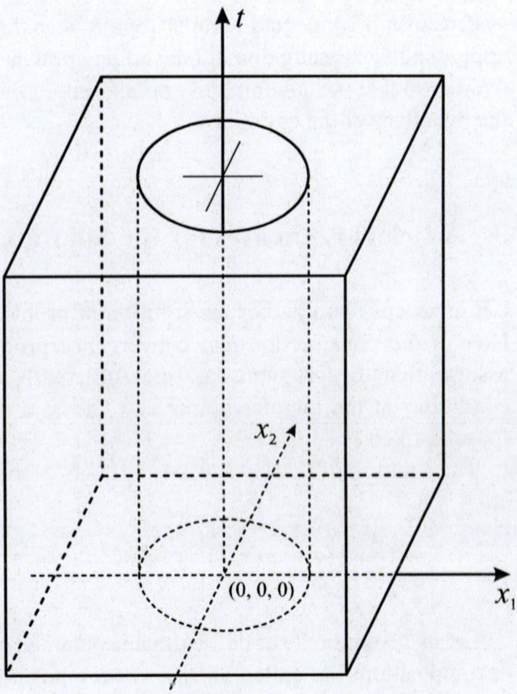

Fig. 1 Doubly connected domain $\overline{G}$

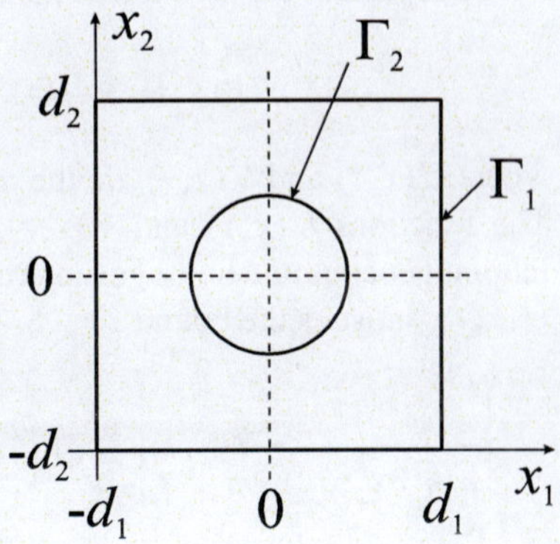

Fig. 2 Doubly connected domain $\overline{D}$; Γ_1 is the boundary of the rectangle, Γ_2 is the boundary of the circle

cylindrical surface), there appears a circular boundary layer that is regular. The boundary layers decrease exponentially with the distance from the outer and inner parts of S^L. The boundary layers of such structure give rise to difficulties in constructing ε-uniform grid approximations to the differential problem (2), (1) and motivate the necessity for using special "connected" grids matched with the boundaries and an overlapping Schwarz-type method for interfacing between the grid subdomains. To construct grids in the neighborhoods of the outer lateral boundary, where the boundary layer is sufficiently smooth, it is reasonable to use the Cartesian coordinate system, while it is proposed to apply a cylindrical coordinate system in a biconnected annular region near the inner lateral boundary. Next, we apply the overlapping domain decomposition method with a sufficient width of the overlap so that the subdomains contains either the boundary of the parallelepiped or the boundary of the cylinder.

3 A Priori Estimates for the Solution and Derivatives

Let us discuss bounds for the solution of problem (2), (1) and its derivatives used later in the construction and convergence proof of schemes. For simplicity, we assume that the problem data are sufficiently smooth and satisfy compatibility conditions at the angular points that ensure a sufficiently smooth solution on $\overline{G}$ for each fixed ε.

The solution of (2), (1) satisfies a rough "standard" estimate

$$\left|\frac{\partial^{k+k_0}}{\partial x_1^{k_1}\partial x_2^{k_2}\partial t^{k_0}}u(x,t)\right| \le M\,\varepsilon^{-k}, \quad (x,t)\in\overline{G}, \quad k+2k_0\le K. \tag{3}$$

Let us give more "subtle" estimates obtained on the basis of asymptotic solution decompositions (so-called Shishkin decompositions).

We represent the solution as the sum of functions

$$u(x,t) = U(x,t) + V(x,t), \quad (x,t)\in\overline{G}, \tag{4}$$

where $U(x,t)$ and $V(x,t)$ are the regular and singular components of the solution. The functions $U(x,t)$ and $V(x,t)$, $(x,t)\in\overline{G}$, are solutions of the appropriate inhomogeneous and homogeneous equations, respectively. The smooth component, $U(x,t)$, satisfies the bound

$$\left|\frac{\partial^{k+k_0}}{\partial x_1^{k_1}\partial x_2^{k_2}\partial t^{k_0}}U(x,t)\right| \le M\left[1+\varepsilon^{2-k}\right], \quad (x,t)\in\overline{G}, \quad k+2k_0\le K. \tag{5}$$

The function $V(x,t)$ can be represented in its turn as the sum of boundary layers of dimensions 1 and 2:

$$V(x,t) = V_0(x,t) + \sum_{j=1}^{4} V_j(x,t) + \sum_{i,j=1}^{4} V_{ij}(x,t), \quad (x,t)\in\overline{G}. \tag{6}$$

Here $V_0(x,t)$ is the regular circular layer in a neighborhood of the inner boundary S_2^L; $V_j(x,t)$ and $V_{ij}(x,t)$ are the regular and angular parabolic boundary layers in a neighborhood of the outer boundary S_1^L.

The singular components from (6) satisfy the estimates

$$\left|\frac{\partial^{k+k_0}}{\partial n \partial t^{k_0}} V_0(x,t)\right| \le M\varepsilon^{-k}\exp(-m\varepsilon^{-k}r(x,\Gamma_2)),$$

$$\left|\frac{\partial^{k+k_0}}{\partial x_1^{k_1}\partial x_2^{k_2}\partial t^{k_0}} V_j(x,t)\right| \le M\left[\varepsilon^{-k_j}+\varepsilon^{1-k}\right]\exp(-m\varepsilon^{-1}r(x,\Gamma_1^j)),\quad 1\le j\le 4,$$

$$\left|\frac{\partial^{k+k_0}}{\partial x_1^{k_1}\partial x_2^{k_2}\partial t^{k_0}} V_{ij}(x,t)\right| \le M\varepsilon^{-k}\min\left[\exp(-m\varepsilon^{-1}r(x,\Gamma_1^i)),\exp(-m\varepsilon^{-1}r(x,\Gamma_1^j))\right],$$

$$(x,t)\in\overline{G},\quad k\le K, \tag{7}$$

where $r(x,\Gamma)$ is the distance from x to the boundary Γ; $\partial^k/\partial n$ is the derivative along the normal to the boundary Γ_2; $k_j=1$ for $j=1,3$ and $k_j=2$ for $j=2,4$;

$$0<m<m^0,\qquad m^0=\min_{\overline{G}}\left[a^{1/2}(x,t)\right].$$

By Γ_1^j, $j=1,2,3,4$, we denote the sides of the rectangle D_1. Assume that Γ_1^1 and Γ_1^3 are orthogonal to the x_1-axis, while Γ_1^2 and Γ_1^4 are orthogonal to the x_2-axis; the sides Γ_1^1 and Γ_1^2 contain the vertex $(-d_1,-d_2)$.

It is easy to see from (7) that the boundary-layer functions decrease exponentially with the distance from the corresponding boundaries. In (5) and (7), $K\ge 4$.

4 Difference Schemes

4.1 Simply Connected Case

Let $D_2=\emptyset$.

On $\overline{G}$, we introduce the rectangular grid

$$\overline{G}_h=\overline{D}_h\times\overline{\omega}_0,\quad \overline{D}_h=\overline{\omega}_1\times\overline{\omega}_2, \tag{8}$$

where $\overline{\omega}_0$ is a uniform mesh in $[0,T]$; $\overline{\omega}_s$ is a mesh, generally nonuniform, in $[-d_s,d_s]$ on the x_s-axis, $s=1,2$. Define $h_s^i=x_s^{i+1}-x_s^i$, $x_s^i, x_s^{i+1}\in\overline{\omega}_s$, $h_s=\max_i h_s^i$, $h=\max_s h_s$, $s=1,2$. We denote by N_0 the number of mesh intervals in $\overline{\omega}_0$ ($h_0=TN_0^{-1}$ is the step of the grid $\overline{\omega}_0$) and by N_s the number of mesh interval in $\overline{\omega}_s$, $N=\min_s N_s$, $s=1,2$. We assume that $h\le MN^{-1}$.

On the grid $\overline{G}_h$, we approximate problem (2) by the difference scheme [3]

$$\Lambda_{(9)} z(x,t) = f(x,t), \quad (x,t) \in G_h, \quad z(x,t) = \varphi(x,t), \quad (x,t) \in S_h. \tag{9}$$

Here $G_h = G \cap \overline{G}_h$, $S_h = S \cap \overline{G}_h$,

$$\Lambda_{(9)} = \varepsilon^2 \sum_{s=1,2} \delta_{\overline{x}_s \widehat{x}_s} - a(x,t) - p(x,t)\delta_{\overline{t}},$$

$\delta_{\overline{x}_s \widehat{x}_s} z(x,t)$ are the second difference derivatives in x_s on a nonuniform mesh:

$$\delta_{\overline{x}_s \widehat{x}_s} z(x,t) = 2\left(h_s^i + h_s^{i-1}\right)^{-1} \left[\delta_{x_s} z(x,t) - \delta_{\overline{x}_s} z(x,t)\right],$$

and $\delta_{\overline{t}} z(x,t)$ is the first backward difference derivative in t.

Scheme (9), (8) is monotone [3] ε-uniformly on meshes with arbitrarily distributed nodes. The scheme converges only for fixed values of the parameter ε, namely, under the condition $h = o(\varepsilon)$, with an error bound given by

$$| u(x,t) - z(x,t) | \le M\left[\left(\varepsilon + N^{-1}\right)^{-1} N^{-1} + N_0^{-1}\right], \quad (x,t) \in \overline{G}_h. \tag{10}$$

Let us introduce the special grid [4]

$$\overline{G}_h^c = \overline{D}_h^c \times \overline{\omega}_0, \quad \overline{D}_h^c = \overline{\omega}_1^c \times \overline{\omega}_2^c, \tag{11a}$$

where $\overline{\omega}_s^c = \overline{\omega}_s^c(\sigma_s)$ is a piecewise uniform mesh condensing near the endpoints of $[-d_s, d_s]$, σ_s is a mesh parameter depending on ε and N, $\sigma_s \le 4^{-1} d_s$. To construct the mesh $\overline{\omega}_s^c(\sigma_s)$, we divide the interval $[-d_s, d_s]$ in three subintervals $[0, \sigma_s]$, $[\sigma_s, d_s - \sigma_s]$, and $[d_s - \sigma_s, d_s]$. We place a uniform grid in each part with $N_s/4$ intervals on $[0, \sigma_s]$ and $[d_s - \sigma_s, d_s]$ and $N_s/2$ intervals on $[\sigma_s, d_s - \sigma_s]$. The mesh parameter σ_s is defined by

$$\sigma_s = \sigma_s(\varepsilon, N_s) = \min\left[\, d_s/4, M\varepsilon \ln N_s \,\right], \quad \text{where } M \ge 2(m_{(7)})^{-1}. \tag{11b}$$

The difference scheme (9), (11) converges ε-uniformly:

$$| u(x,t) - z(x,t) | \le M\left[N^{-1} \ln N + N_0^{-1}\right], \quad (x,t) \in \overline{G}_h^c. \tag{12}$$

Theorem 1 *Suppose that the data of problem* (2) *are sufficiently smooth and satisfy the compatibility conditions that yield the required smoothness of the solution. Let the solution and its components satisfy estimates* (3), (5), *and* (7) *with* $K = 4$. *Then the solution of scheme* (9) *on grid* (11) *converges* ε*-uniformly with bound* (12).

4.2 Doubly Connected Case

In a similar way, we construct an ε-uniformly convergent scheme if $D_2 \neq \emptyset$.

In the case of the doubly connected ring in space

$$\overline{D}_3^* = \overline{D}_3 \setminus D_2, \tag{13}$$

where $\overline{D}_3$ is some r_0-neighborhood of D_2, passing to polar coordinates r and ψ such that $x_1 = r\cos\psi$, $x_2 = r\sin\psi$ (Fig. 3), we consider the boundary value problem in cylindrical coordinates (r, ψ, t)

$$\begin{aligned} L_{(2)}^{r,\psi}\, u(r,\psi;t) &= f(r,\psi;t), \quad (r,\psi) \in D_3^*, \quad t \in [0,T], \\ u(r,\psi;t) &= \varphi(r,\psi;t), \quad (r,\psi) \in \Gamma_3^*, \quad t \in [0,T]. \end{aligned} \tag{14}$$

The problem data in the cylindrical coordinates are assumed to be sufficiently smooth; the differential operator $L_{(2)}^{r,\psi}$ is given by the expression

$$L_{(2)}^{r,\psi} = \varepsilon^2 \left(\frac{\partial^2}{\partial r^2} + \frac{1}{r}\frac{\partial}{\partial r} + \frac{1}{r^2}\frac{\partial^2}{\partial \psi^2} \right) - a(r,\psi;t) - p(r,\psi;t)\frac{\partial}{\partial t}$$

or in the divergence form

$$L_{(2)}^{r,\psi} = \varepsilon^2 \left(\frac{1}{r}\frac{\partial}{\partial r}\left(r\frac{\partial u}{\partial r} \right) + \frac{1}{r^2}\frac{\partial^2 u}{\partial \psi^2} \right) - a(r,\psi;t) - p(r,\psi;t)\frac{\partial}{\partial t}.$$

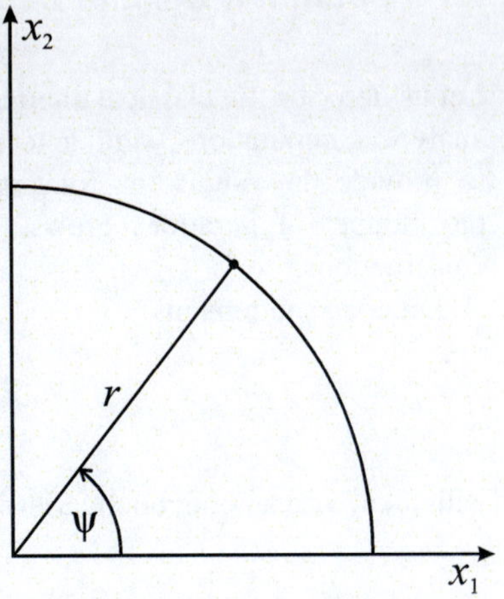

Fig. 3 Plane polar coordinates

We approximate problem (12) by the implicit difference scheme

$$\Lambda_{(9)}^{r,\,\psi}\, z(r,\psi;t) = f(r,\psi;t), \quad (r,\psi) \in D^*_{3h}, \quad t \in \overline{\omega}_0,$$
$$z(r,\psi;t) = \varphi(r,\psi;t), \quad (r,\psi) \in \Gamma^*_{3h}, \quad t \in \overline{\omega}_0. \tag{15}$$

For the solution of this scheme, we obtain an error bound similar to (10)

$$|\, u(r,\psi;t) - z(r,\psi;t)\,| \le M\big[\,\big(\varepsilon + N_*^{-1}\big)^{-1} N_*^{-1} + N_0^{-1}\big],$$
$$(r,\psi) \in \overline{D}^*_{3h}, \quad t \in \overline{\omega}_0, \tag{16}$$

where $N_* + 1$ is the minimal number of mesh points in r and ψ.

On the grid piecewise uniform in r and uniform in ψ and t, we have the error bound

$$|\, u(r,\psi;t) - z(r,\psi;t)\,| \le M\left[N_*^{-1}\ln N_* + N_0^{-1}\right], \quad (r,\psi) \in \overline{D}^*_{3h}, \quad t \in \overline{\omega}_0. \tag{17}$$

5 Domain Decomposition Schemes

To implement the numerical solution of problem (2), we use the overlapping Schwarz method successfully applied to singular perturbation problems in [5–8].

5.1 Continual Domain Decomposition Scheme

Let us describe the classical alternating Schwarz method that allows us to perform analytic computations, similar to that considered in [9, 10]. We give conditions to provide the ε-uniform convergence of a sequence of iterative solutions as the number of iterations grows. The techniques from [4, 11] are used in the constructions.

Let open subdomains

$$D^1, D^2, \ldots, D^K, \tag{18a}$$

with piecewise smooth boundaries $\Gamma^k = \overline{D}^k \setminus D^k$, cover D: $D = \bigcup_{k=1}^{K} D^k$, and let

$$G^k = D^k \times (0,T], \quad k = 1, \ldots, K. \tag{18b}$$

The subdomains D^k are assumed convex. By $D^{[k]}$ we denote the union of the subdomains $D^1, \dots, D^K$ that do not contain the set D^k:

$$D^{[k]} = \bigcup_{i=1,\, i \neq k}^{K} D^i. \tag{18c}$$

The minimal width of overlapping the sets D^k and $D^{[k]}$ is denoted by δ^k. The minimal overlap δ of the subdomains from (18) is the smallest of δ^k; i.e.

$$\delta = \min_{k, x^1, x^2} \rho(x^1, x^2), \quad x^1 \in \overline{D}^k, \ x^2 \in \overline{D}^{[k]},$$

$$x^1, \ x^2 \notin \{ D^k \cap D^{[k]} \}, \ k = 1, 2,$$

where $\rho(x^1, x^2)$ is the distance between points x^1 and $x^2 \in \overline{D}$. The value δ, generally speaking, may depend on the parameter ε: $\delta = \delta(\varepsilon)$.

Let

$$u^0(x, t), \quad (x, t) \in \overline{G} \tag{19a}$$

be an arbitrary initial function in the iteration process that satisfies the boundary condition from (2). The auxiliary functions $u^{n+\frac{k}{K}}(x, t)$, $(x, t) \in \overline{G}$, $k = 1, \dots, K$, $n = 1, 2, \dots,$ can be found by a sequential solving of the problems

$$\begin{gathered} L_{(19)}(u^{n+\frac{k}{K}}(x, t)) \equiv L_{(2)}u(x, t) - f(x, t) = 0, \quad (x, t) \in G^k, \\ u^{n+\frac{k}{K}}(x, t) = u^{n+\frac{k-1}{K}}(x, t), \quad (x, t) \in \overline{G} \setminus G^k, \ k = 1, \dots, K, \\ u^{n+1}(x, t) = u^{n+\frac{K}{K}}(x, t), \ (x, t) \in \overline{G}, \quad n = 0, 1, 2, \dots. \end{gathered} \tag{19b}$$

Each function $u^{n+\frac{k}{K}}(x, t)$, $(x, t) \in \overline{G}$, is the solution of the Dirichlet problem on $\overline{G}^k$ and coincides with $u^{n+\frac{k-1}{K}}(x, t)$ on $\overline{G} \setminus G^k$. To compute the function $u^n(x, t)$, $(x, t) \in \overline{G}$, we solve problems (19b) on $\overline{G}^k$ successively.

The sequence of the functions $u^n(x, t)$, $(x, t) \in \overline{G}$, $n = 1, 2, \dots$, is called the solution of iteration process (19), (18), i.e., the Schwarz alternating method.

The condition on the minimal overlap width of the subdomains

$$\delta = \delta(\varepsilon) > 0, \quad \varepsilon \in (0, 1], \quad \inf_{\varepsilon \in (0,1]} \left[\varepsilon^{-1} \delta(\varepsilon) \right] > 0, \tag{20}$$

which is equivalent to the condition $\delta = \delta(\varepsilon) \geq m_{(20)}\varepsilon$, is sufficient for the solutions $u^n(x,t)$ of the Schwarz method to converge ε-uniformly to the solution $u(x,t)$ of the initial-boundary value problem as $n \to \infty$:

$$\left|u^n(x,t) - u(x,t)\right| \leq Mq^n, \quad (x,t) \in \overline{G}, \quad \text{where } q \leq 1 - m. \tag{21}$$

Note that the quantity q is, generally speaking, $q(\varepsilon, \delta)$ and tends to unity as $\delta \to 0$ for fixed ε. For $\delta = 0$, the functions $u^n(x,t)$ do not converge as $n \to \infty$ even for fixed ε.

Condition (20) is also necessary. If (20) is violated and the quantity δ satisfies the condition

$$\delta = \delta(\varepsilon) > 0, \quad \varepsilon \in (0,1], \quad \inf_{\varepsilon\in(0,1]} \left[\varepsilon^{-1}\delta(\varepsilon)\right] > 0, \tag{22}$$

then the functions $u^n(x,t)$ as $n \to \infty$ do not converge ε-uniformly. Under condition (22), for any arbitrarily large number n, there is a value of the parameter ε, $\varepsilon = \varepsilon(n)$, such that the functions $u^n(x,t) = u^n(x,t;\ \varepsilon(n), \delta(\varepsilon))$ and $u(x,t) = u(x,t;\ \varepsilon(n))$ satisfy the inequality

$$\max_{\overline{G}} \left|u(x,t) - u^n(x,t)\right| \geq m,$$

where m is independent of n.

Theorem 2 *Condition (20) is necessary and sufficient for the solution $u^n(x,t)$ of the iterative Schwarz method (19), (18) to converge ε-uniformly as $n \to \infty$ to the solution $u(x,t)$ of problem (2). Let the hypotheses of Theorem 1 hold. Then, under condition (20), the solution of Schwarz method (19), (18) satisfies error bound (21).*

5.2 *Difference Domain Decomposition Scheme*

In a similar way, we construct iterative difference schemes of the Schwarz method on piecewise uniform grids condensing in the boundary layers.

On each of the sets $\overline{G}^k$, we introduce the special grids

$$\overline{G}_h^k \equiv \overline{G}_h^{k\,c} = \overline{G}^k \bigcap \overline{G}_{h(11)}^c. \tag{23}$$

We assume that the boundaries of $\overline{G}^k$ pass through nodes of the grid $\overline{G}_h$.

Let the function $z^0(x,t)$, $(x,t) \in \overline{G}_h$, be an arbitrary function satisfying the condition

$$z^0(x,t) = \varphi(x,t), \quad (x,t) \in S_h. \tag{24a}$$

We find the sequence of auxiliary functions $z^{n+\frac{k}{K}}(x,t)$, $k = 1,\ldots,K$, $n = 1,2,\ldots$, by solving the grid problems

$$\Lambda_{(24)}(z^{n+\frac{k}{K}}(x,t)) \equiv \Lambda_{(9)} z^{n+\frac{k}{K}}(x,t) - f(x,t) = 0, \quad (x,t) \in G_h^k,$$

$$z^{n+\frac{k}{K}}(x,t) = z^{n+\frac{k-1}{K}}(x,t), \quad (x,t) \in \overline{G}_h \setminus G^k, \quad k = 1,\ldots,K, \tag{24b}$$

$$z^{n+1}(x,t) = z^{n+\frac{K}{K}}(x,t), \quad (x,t) \in \overline{G}_h, \quad n = 0,1,2,\ldots.$$

Each function $z^{n+\frac{k}{K}}(x,t)$ is defined on the set $\overline{G}_h$, being a solution of the grid Dirichlet problem, and it coincides with the function $z^{n+\frac{k-1}{K}}(x,t)$ on the set $\overline{G}_h \setminus G^k$. The function $z^n(x,t)$, $(x,t) \in \overline{G}_h$, $n = 1,2,\ldots$, is called the solution of the iterative grid Schwarz method (24), (23).

Taking bounds (12), (17), and (21) into account, we find that the solution $z^n(x,t)$ of the grid iterative Schwarz method on overlapping subdomains containing either the boundary of the parallelepiped or the boundary of the cylinder converges ε-uniformly:

$$\left|u(x,t) - z^n(x,t)\right| \le M\left(N^{-1}\ln N + N_*^{-1}\ln N_* + N_0^{-1} + q^n\right), \quad (x,t) \in \overline{G}_h, \tag{25}$$

as the numbers of mesh points N, N_*, N_0 and the number n of iterations in the Schwarz method grow. In (25), $q \le 1 - m$.

Theorem 3 *Let the hypotheses of Theorem 1 hold. Then, under condition (20), the solution of the iterative grid Schwarz method (24) on the grid (23) converges ε-uniformly as N, N_*, N_0, $n \to \infty$ to the solution of the initial-boundary value problem (2) with error bound (25).*

6 Conclusion

In the case of Dirichlet's initial-boundary value problem for a singularly perturbed parabolic reaction-diffusion equation in a doubly connected domain, we have constructed and investigated a continual and difference (on piecewise-uniform grids condensing in the boundary layers) schemes of the overlapping domain decomposition method with subdomains containing either the boundary of the parallelepiped or the boundary of the cylinder. A priori estimates for the solution of the problem and its derivatives are obtained, showing that the derivatives of the singular components of the solution in neighborhoods of the boundary layers grow unboundedly as the perturbation parameter ε tends to zero; the boundary layers decrease exponentially with distance from the outer and inner lateral boundaries. Necessary and sufficient conditions are given that provide the ε-uniform convergence of solutions of the decomposition schemes as $n \to \infty$, where n is the number of iterations. It is shown that the iterative Schwarz method converges ε-uniformly in the maximum norm as the number of iterations (and the number of mesh points in the case of difference schemes) grows. We used the Shishkin meshes, i.e., piecewise uniform meshes along the normal to the smooth parts of the boundaries of the subdomains. In the case of meshes with an arbitrary distribution of nodes, in particular, uniform meshes, the grid method converges only for fixed values of the parameter ε, namely, under the condition $h = o(\varepsilon)$, where h is the maximum effective step of the space grid.

The present paper is a continuation of our study in [12, 13]. Note that the classical Schwarz method was constructed in a such way that we have to repeatedly solve the subproblems at each point of the domain G. An open question is to develop a modified Schwarz method in which the problem is repeatedly solved only at the intersection of the subdomains when the subdomains alternate.

Acknowledgments The work is supported by the Russian Foundation for Basic Research, grant no. 20-01-00650.

References

1. A. Quarteroni, A. Valli, *Domain Decomposition Methods for Partial Differential Equations* (Oxford Univ. Press, Oxford, 1999)
2. P.W. Hemker, A singularly perturbed model problem for numerical computation. J. Comp. Appl. Math. **76** (1–2), 277–285 (1996)
3. A.A. Samarskii, *The Theory of Difference Schemes* (Marcel Dekker, New York, 2001)
4. G.I. Shishkin, *Discrete Approximations of Singularly Perturbed Elliptic and Parabolic Equations* (UrO RAN, Ekaterinburg, 1992) [in Russian]
5. M. Garbey, A Schwarz alternating procedure for singular perturbation problems. SIAM J. Sci. Comput. **17** (5), 1175–1201 (1996)
6. I. Boglaev, On a domain decomposition algorithm for a singularly perturbed reaction–diffusion problem. J. Comput. Appl. Math. **98** (2), 213–232 (1998)
7. I. Boglaev, Domain decomposition in boundary layers for singularly perturbed problems. Appl. Numer. Math. **34** (2), 145–166 (2000)
8. S. Kumar, S.C.S. Rao, A robust overlapping Schwarz domain decomposition algorithm for time-dependent singularly perturbed reaction–diffusion problems. J. Comput. Appl. Math. **261**, 127–138 (2014)
9. G.I. Shishkin, I.V. Tselishcheva, Parallel methods of solving singularly perturbed boundary value problems for elliptic equations. Mat. Model. **8** (3), 111–127 (1996)
10. G.I. Shishkin, Acceleration of the process of the numerical solution to singularly perturbed boundary value problems for parabolic equations on the basis of parallel computations. Russ. J. Numer. Anal. Math. Model. **12** (3), 271–291 (1997)
11. G.I. Shishkin, L.P. Shishkina, *Difference Methods for Singular Perturbation Problems* (CRC Press, Boca Raton, 2009)

12. I.V. Tselishcheva, G.I. Shishkin, Development of a reliable numerical method for solving the Dirichlet boundary value problem for a singularly perturbed elliptic reaction–diffusion equation in a doubly connected domain, in *Proc. Int. Conf. "Marchuk Scientific Readings – 2017"*, Novosibirsk, June 25 – July 14, 2017, pp. 961–967 [in Russian]
13. I.V. Tselishcheva, G.I. Shishkin, A difference scheme for a singularly perturbed elliptic reaction–diffusion equation with the third-kind boundary conditions in a doubly connected domain, in *Mesh Methods for Boundary-Value Problems and Applications, Proc. 12th Int. Conf.* (Kazan Univ., Kazan, 2018), pp. 185–191 [in Russian]

Two Finite Volume Schemes for Advection Equation

Alexander V. Vyatkin, Vladimir V. Shaydurov, and Elena V. Kuchunova

Abstract Two finite volume schemes for two-dimensional advection equation are compared. First one is based on Gauss-Ostrogradsky theorem for volume bounded by a small rectangle at upper time level, four sides formed by characteristic trajectories issued out backward in time from boundary of this rectangle, and curvilinear quadrangle carved by these trajectories at the previous time level. The curvilinear quadrangle at the previous time level is approximated by straight-sided quadrangle. The solution is sought in the class of piecewise constant functions on a rectangular grid. The substantiation of the first order of approximation and the convergence for the obtained grid problem is carried out. In the second scheme, two-dimensional advection operator is decomposed in two one-dimensional operators. The justifying the approximation and the convergence for this scheme is obtained by a simple generalization of these properties for one-dimensional discrete operators. Comparison of algorithmic realization of these schemes demonstrates the different properties. The first one is more complicated for assembling but is more appropriate for implementation to the problems with high velocities.

Keywords Advection equation · Finite volume method · Gauss-Ostrogradsky theorem · Approximation · Convergence

A. V. Vyatkin (✉)
Institute of Computational Modeling SB RAS, Akademgorodok, Russia
e-mail: vyatkin@icm.krasn.ru

V. V. Shaydurov · E. V. Kuchunova
Institute of Computational Modeling SB RAS, Akademgorodok, Russia

Siberian Federal University, Krasnoyarsk, Russia
e-mail: shaidurov04@mail.ru; kuchunova@sfu-kras.ru

I. B. Badriev et al. (eds.), *Mesh Methods for Boundary-Value Problems and Applications*, Lecture Notes in Computational Science and Engineering 141,
https://doi.org/10.1007/978-3-030-87809-2_41

1 Introduction

Nowadays a semi-Lagrangian approximation of the advection operator is intensively developed in fluid dynamics [1–4]. The main feature of the initial semi-Lagrangian approaches consists in the approximation of advection operator as the directional (Lagrangian) derivative in the motion direction (see [3, 5] and the references therein).

In this paper, an initial-boundary value problem is considered for the two-dimensional advection equation. We start with the integral balance equality between two neighboring time levels [6]. To construct the first discrete problem, the finite volume method is used with the approximation of each integral in this balance equality. In the second discrete scheme, two-dimensional advection operator is decomposed in two one-dimensional operators and the separate approximation is used for them.

Algorithmic realizations of these schemes have the different properties. The first scheme is more complicated for assembling but is more appropriate for implementation to the problems with high velocities than the second one.

2 The Differential Problem

Let $D = (0, 1) \times (0, 1)$ be the unit square with the boundary Γ. Denote $\bar{D} = D \cup \Gamma$. In the closed domain $[0, T] \times \bar{D}$, $T > 0$, consider the two-dimensional advection equation

$$\frac{\partial \rho}{\partial t} + \frac{\partial (\rho u)}{\partial x} + \frac{\partial (\rho v)}{\partial y} = 0. \tag{1}$$

Here $\rho(t, x, y)$ is an unknown function (such as a density or a concentration); $u(t, x, y)$ and $v(t, x, y)$ are the known components of a velocity vector $U = (u, v)$. Let the boundary Γ consist of three parts: $\Gamma = \Gamma_{\text{in}} \cup \Gamma_{\text{out}} \cup \Gamma_{\text{rigid}}$. At the inlet boundary $\Gamma_{\text{in}} = \{(0, y) : 0 \le y \le 1\}$ we suppose that

$$\mathbf{U} \cdot \mathbf{n} \le 0 \quad \forall\ (t, x, y) \in [0, T] \times \Gamma_{\text{in}} \tag{2}$$

where $\mathbf{n} = \big(n_x(x, y), n_y(x, y)\big)$ is the outward normal to Γ; $\mathbf{U} \cdot \mathbf{n}$ is scalar product of two vectors. At the outlet boundary $\Gamma_{\text{out}} = \{(1, y) : 0 \le y \le 1\}$ we suppose

$$\mathbf{U} \cdot \mathbf{n} \ge 0 \quad \forall\ (t, x, y) \in [0, T] \times \Gamma_{\text{out}}. \tag{3}$$

Finally, at the rigid boundary $\Gamma_{\text{rigid}} = \{(x, y) : x \in [0, 1], y = 0, 1\}$ we impose no-slip condition

$$\mathbf{U} = (0, 0) \quad \forall\ (t, x, y) \in [0, T] \times \Gamma_{\text{rigid}}. \tag{4}$$

For the function ρ, the initial and the boundary conditions are specified

$$\rho(0, x, y) = \rho_{\text{init}}(x, y) \quad \forall\ (x, y) \in \bar{D}, \tag{5}$$

$$\rho(t, 0, y) = \rho_{\text{in}}(t, y) \quad \forall\ (t, y) \in [0, T] \times \Gamma_{\text{in}}. \tag{6}$$

Let functions u, v and ρ be bounded on $[0, T] \times \bar{D}$:

$$|u| \leq u^{\max}, \quad |v| \leq v^{\max}, \quad |\rho| \leq \rho^{\max}. \tag{7}$$

Suppose also boundedness of its all first and second partial derivatives on $[0, T] \times \bar{D}$.

3 The Local Conservation Law

To construct the semi-Lagrangian method for problem (1)–(7), we put integers $N_x, N_y \geq 1$, and define a uniform grid $\bar{D}_h$ for mesh-sizes $h_x = 1/N_x$ and $h_y = 1/N_y$ in the x- and y-directions:

$$\bar{D}_h = \left\{(x_i, y_j) : x_i = ih_x, y_j = jh_y; \quad i = 0, \ldots, N_x, j = 0, \ldots, N_y\right\}$$

$$\text{and } D_h = \bar{D}_h \cap D.$$

Introduce also the cells $\omega_{i,j} = \left[x_{i-1/2}, x_{i+1/2}\right) \times \left[y_{j-1/2}, y_{j+1/2}\right) \cap \bar{D}$ with the auxiliary points $x_{i\pm1/2} = x_i \pm h_x/2, \ \ y_{j\pm1/2} = y_j \pm h_y/2$. On the time segment $[0, T]$ introduce $K+1$ time levels $t_k = \tau k$ for $k = 0, \ldots, K$ with the step $\tau = T/K$.

To simplify theoretical justification, first we suppose

$$\tau \leq \min\left\{h_x/2u^{\max}, \ h_y/2v^{\max}\right\}, \ \max\left\{h_x, h_y\right\} = h, \ \text{and } h \leq c\tau. \tag{8}$$

We will discuss disturbance of inequalities (8) in the conclusion.

Hereinafter we use the notation $f_{i,j}^k = f\left(t_k, x_i, y_j\right)$. For any node $\left(x_i, y_j\right) \in \bar{D}_h$ introduce the basis function $\varphi_{i,j}(x, y)$ which equals 1 in $\omega_{i,j}$ and 0 at any other point of $\bar{D}_h$. At each time level t_k, we find the approximate solution $\rho_h^k(x, y)$ in the form of piecewise constant function

$$\rho_h^k(x, y) = \sum_{i=0}^{N_x} \sum_{j=0}^{N_y} \rho_{h,i,j}^k \varphi_{i,j}(x, y). \tag{9}$$

For this function, put the initial condition

$$\rho_h^0\left(x_i, y_j\right) = \rho_{\text{init}}\left(x_i, y_j\right) \quad \forall\ \left(x_i, y_j\right) \in \bar{D}_h \tag{10}$$

and the boundary one

$$\rho_h^k\left(0, y_j\right) = \rho_{\text{in}}\left(t_k, y_j\right) \quad \forall k = 0, \ldots, K \quad \forall \left(0, y_j\right) \in \Gamma_{\text{in}} \cap \bar{D}_h. \tag{11}$$

Let us fix the cell $\omega_{i,j}$ at level t_k and construct the trajectories from its boundary back in time to t_{k-1}. The trajectory $\left(\hat{x}(t), \hat{y}(t)\right)$ from point $(t_k, \bar{x}, \bar{y})$ is constructed as a solution of the Cauchy problem for the system of ordinary differential equation

$$\hat{x}'(t) = u\left(t, \hat{x}(t), \hat{y}(t)\right), \quad \hat{y}'(t) = v\left(t, \hat{x}(t), \hat{y}(t)\right) \quad \forall\, t \in [t_{k-1}, t_k] \tag{12}$$

with the initial condition

$$\hat{x}(t_k) = \bar{x}, \quad \hat{y}(t_k) = \bar{y}. \tag{13}$$

Denote solution of problem (12)–(13) by $\tilde{x}\,(t; \bar{x}, \bar{y})$, $\tilde{y}\,(t; \bar{x}, \bar{y})$. Thus, at the plane $t = t_{k-1}$ we get a curvilinear quadrangle $Q_{i,j}^{k-1}$ (Fig. 1). Let $V_{i,j}^{k-1}$ be a volume bounded by the following surfaces: $\omega_{i,j}$ at the plane $t = t_k$, $Q_{i,j}^{k-1}$ at the plane $t = t_{k-1}$, and the trajectories those issue out of all boundary points of $\omega_{i,j}$ between time levels t_{k-1} and t_k. We integrate equality (1) over $V_{i,j}^{k-1}$ and with help of divergence theorem we arrive the following statement.

Statement 1 *For a smooth solution of problem* (1)–(7) *the equality is valid*

$$\int_{\omega_{i,j}} \rho\,(t_k, x, y)\,\mathrm{d}x\mathrm{d}y = \int_{Q_{i,j}^{k-1}} \rho\,(t_{k-1}, x, y)\;\mathrm{d}x\mathrm{d}y \tag{14}$$

$$\forall\, i = 1, \ldots, N_x; \;\; j = 0, \ldots, N_y.$$

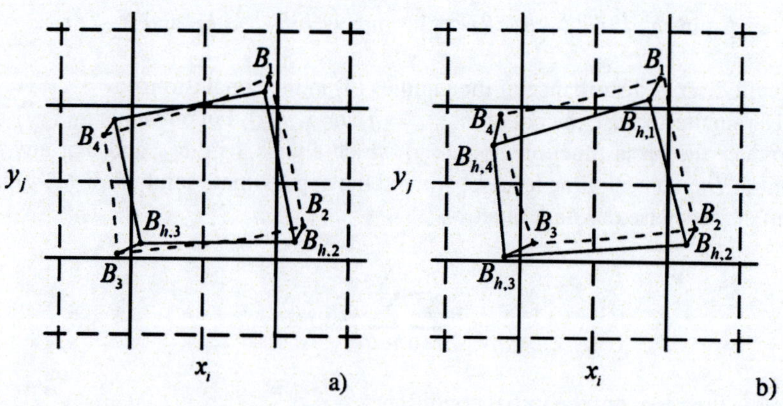

Fig. 1 The curvilinear quadrangles $Q_{i,j}^{k-1}$ and the straight-sided quadrangles $P_{i,j}^{k-1}$ (**a**) x- and y-irregular cases; (**b**) x- and y-regular cases

4 The First Scheme: Approximation of the Curvilinear Domain

To construct the first scheme, we approximate each term of equality (14).

1. In the left-hand side of (14) we replace the exact solution $\rho(t_k, x, y)$ by an approximate one $\rho_h^k(x, y)$ and get

$$\int_{\omega_{i,j}} \rho(t_k, x, y)\, \mathrm{d}x\mathrm{d}y \approx \int_{\omega_{i,j}} \rho_h^k(x, y)\, \mathrm{d}x\mathrm{d}y = \rho_{h,i,j}^k \,\mathrm{meas}\left(\omega_{i,j}\right).$$

Here $\mathrm{meas}\left(\omega_{i,j}\right)$ is a square of $\omega_{i,j}$. If $i \neq 0$, $i \neq N_x$, $j \neq 0$, or $j \neq N_y$, then $\mathrm{meas}\left(\omega_{i,j}\right) = h_x h_y$. In other cases, it equals half or quarter of this value.

2. To calculate integral in the right-hand side of (12), we substitute numerical solution $\rho_h^{k-1}(x, y)$ instead of exact solution $\rho(t_{k-1}, x, y)$. Let $A_n = \left(A_n^x, A_n^y\right)$, $n = 1, \ldots, 4$, be vertices of rectangle $\omega_{i,j}$ at the level t_k.We put $B_n = \left(B_n^x, B_n^y\right)$ are vertices of the curvilinear quadrangle $Q_{i,j}^{k-1}$ at the level t_{k-1}. To compute the coordinates of B_n approximately, we solve system (12) backward in time on $[t_{k-1}, t_k]$ with corresponding initial conditions $\hat{x}(t_k) = A_n^x$, $\hat{y}(t_k) = A_n^y$ by Euler method:

$$B_{h,n}^x = A_n^x - \tau u\left(t_k, A_n^x, A_n^y\right), \quad B_{h,n}^y = A_n^y - \tau v\left(t_k, A_n^x, A_n^y\right). \tag{15}$$

Lemma 1 *The following inequalities are valid:*

$$\left|B_{h,n}^x - B_n^x\right| \leq O(\tau^2), \quad \left|B_{h,n}^y - B_n^y\right| \leq O(\tau^2). \tag{16}$$

Proof Firstly, we prove the left inequality. We issue out trajectory from point $A_n = \left(A_n^x, A_n^y\right)$ backward in time from time level t_k to level t_{k-1}. We expand the function $\hat{x}(t)$ in the neighborhood of point $t = t_k$ into a Taylor series

$$\hat{x}(t_k - \tau) = \hat{x}(t_k) - \tau \frac{d\hat{x}}{dt}(t_k) + \frac{\tau^2}{2}\frac{d^2\hat{x}}{dt^2}(t_k) + O(\tau^3).$$

□

Here $\hat{x}(t_k - \tau) = B_n^x$, $\hat{x}(t_k) = A_n^x$, and $d\hat{x}/dt(t_k) = u\left(t_k, A_n^x, A_n^y\right)$. Therefore

$$B_n^x = A_n^x - \tau u\left(t_k, A_n^x, A_n^y\right) + \frac{\tau^2}{2}\frac{d^2\hat{x}}{dt^2}(t_k) + O(\tau^3). \tag{17}$$

We subtract from this equality the left relation from (15) and get

$$B_n^x - B_{h,n}^x = \frac{\tau^2}{2}\frac{d^2\hat{x}}{dt^2}(t_k) + O(\tau^3).$$

The following equalities are valid for $d^2\hat{x}/dt^2\left(t, \hat{x}(t), \hat{y}(t)\right)$:

$$\frac{d^2\hat{x}}{dt^2}(t) = \frac{du}{dt}\left(t, \hat{x}(t), \hat{y}(t)\right) = \left(\frac{\partial u}{\partial t} + u\frac{\partial u}{\partial x} + v\frac{\partial u}{\partial y}\right)\left(t, \hat{x}(t), \hat{y}(t)\right). \tag{18}$$

Therefore, with the help of $\left|\hat{x}(t) - x_i\right| \le h_x$ and $\left|\hat{y}(t) - y_j\right| \le h_y$ we get

$$B_n^x - B_{h,n}^x = \frac{\tau^2}{2}\left(\frac{\partial u}{\partial t} + u\frac{\partial u}{\partial x} + v\frac{\partial u}{\partial y}\right)\Bigg|_{(t_k, x_i, y_j)} + O(\tau^3).$$

Thus, taking into account smoothness of functions u and v we prove the first inequality in (16). Second one is verified in the same way.

Recall that $P_{i,j}^{k-1}$is the straight-sided quadrangle with 4 vertices $B_{h,n}$ (Fig. 2). Then we put

$$\rho_{h,i,j}^k \text{meas}\left(\omega_{i,j}\right) = \int_{P_{i,j}^{k-1}} \rho_h^{k-1}(x, y)\, dxdy. \tag{19}$$

Combine these equalities for $i = 1, \ldots, N_x,\ j = 0, \ldots, N_y$ with discrete boundary conditions

$$\rho_{h,0,j}^k = \rho_{\text{in}}\left(t_k, y_j\right),\ j = 0, \ldots, N_y. \tag{20}$$

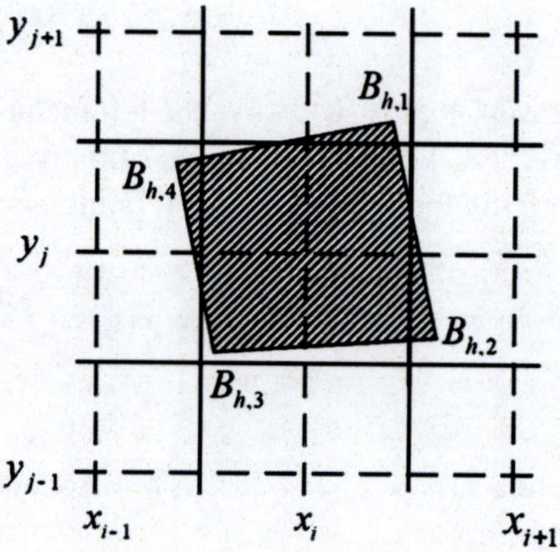

Fig. 2 The straight-sided quadrangle $P_{i,j}^{k-1}$

As the result, we get the system of linear algebraic equation to compute $\rho^k_{h,i,j}$ for all $i = 0, \ldots, N_x;\ j = 0, \ldots, N_y$. Taking into consideration the initial conditions

$$\rho^0_{h,i,j} = \rho_{\text{init}}\left(x_i, y_j\right) \forall\ i = 0, \ldots, N_x;\ \ j = 0, \ldots, N_y, \tag{21}$$

we get the explicit monotone difference scheme [7] for computing $\rho_h(t, x, y)$.

Give a sketch of proof for the convergence of this scheme. Firstly, evaluate the square of the lune between the segments of $P^{k-1}_{i,j}$ and the corresponding arcs of the curvilinear polygon $Q^{k-1}_{i,j}$ (Fig. 1). For example, consider the segment $\overline{B_{h,3}B_{h,2}}$ and the arc $\widehat{B_3B_2}$ at Fig. 1a. They form a curved quadrangle $B_2B_{h,2}B_{h,3}B_3$ with two self-intersecting sides. We place this quadrangle into a rectangle for which we prove its square equal $O(\tau^2 h_x)$. Each point of arc $\widehat{B_3B_2}$ is the mark $\left(\hat{x}(t_{k-1}; \bar{x}),\ \hat{y}(t_{k-1}; \bar{x})\right)$ of the solution of system (10) with the initial condition $\hat{x}(t_k; \bar{x}) = \bar{x},\ \hat{y}(t_k; \bar{x}) = y_{j-1/2}$ with a parameter $\bar{x} \in \left[x_{i-1/2}, x_{i+1/2}\right]$. Each point of the segment $\overline{B_{h,2}B_{h,3}}$ is also the mark $\left(\tilde{x}(t_{k-1}; \bar{\bar{x}}),\ \tilde{y}(t_{k-1}; \bar{\bar{x}})\right)$ at level t_{k-1} of the solution of the system $\forall\ t \in [t_{k-1}, t_k]$

$$\begin{aligned}
\tilde{x}'(t; \bar{\bar{x}}) &= u\left(t_k, x_{i-1/2}, y_{j-1/2}\right)\left(x_{i+1/2} - \bar{\bar{x}}\right)/h_x + \\
&\quad u\left(t_k, x_{i+1/2}, y_{j-1/2}\right)\left(\bar{\bar{x}} - x_{i-1/2}\right)/h_x, \\
\tilde{y}'(t; \bar{\bar{x}}) &= v\left(t_k, x_{i-1/2}, y_{j-1/2}\right)\left(x_{i+1/2} - \bar{\bar{x}}\right)/h_x + \\
&\quad v\left(t_k, x_{i+1/2}, y_{j-1/2}\right)\left(\bar{\bar{x}} - x_{i-1/2}\right)/h_x
\end{aligned} \tag{22}$$

with the initial condition

$$\tilde{x}(t_k; \bar{\bar{x}}) = \bar{\bar{x}},\ \ \tilde{y}(t_k; \bar{\bar{x}}) = y_{j-1/2} \tag{23}$$

with some parameter $\bar{\bar{x}} \in \left[x_{i-1/2}, x_{i+1/2}\right]$.

Let $P = \left(P^x, P^y\right)$ and $P_1 = \left(P_1^x, P_1^y\right)$ be two points in plane Oxy. We define distance between them in the following form:

$$|P - P_1| = \left(\left(P^x - P_1^x\right)^2 + \left(P^y - P_1^y\right)^2\right)^{1/2}.$$

Also we define distance between point P and a set M as dist $(P, M) = \min\limits_{P_1 \in M} |P - P_1|$

Let point $P_{\max} \in \widehat{B_2B_3}$ be located most far from segment $\overline{B_{h,2}B_{h,3}}$. Point $P_{\max}$ corresponds to a parameter $\bar{x}$ in such conception that $P_{\max}$ is trace of $\left(\hat{x}(t_{k-1}; \bar{x}),\ \hat{y}(t_{k-1}; \bar{x})\right)$. Consider the solution of problem (22)–(23) with initial data $\tilde{x}(t_k; \bar{x}) = \bar{x},\ \ \tilde{y}(t_k; \bar{x}) = y_{j-1/2}$ with parameter $\bar{\bar{x}} = \bar{x}$. We put P_1 is trace of solution $(\tilde{x}(t;\bar{x}),\ \tilde{y}(t;\bar{x}))$ at time level $t = t_{k-1}$. Due to Lemma 1 we get $|P_{\max} - P_1| = O(\tau^2)$. Since $P_1 \in \overline{B_{h,2}B_{h,3}}$, then

$$\forall\, P \in \widehat{B_2 B_3}\ \ \mathrm{dist}\,(P, \overline{B_{h,2}B_{h,3}}) \le \mathrm{dist}\,(P_{\max}, \overline{B_{h,2}B_{h,3}}) \le |P_{\max} - P_1| = O(\tau^2). \tag{24}$$

In general case points of arc $\widehat{B_2 B_3}$ can be located on both sides of segment $\overline{B_{h,2}B_{h,3}}$. Therefore, the width of the rectangle containing $\overline{B_{h,2}B_{h,3}}$ is twice greater than numerical estimation of $|P_{\max} - P_1|$ in (24). Despite this reasoning, the width of this rectangle is $O(\tau^2)$. Due to restriction (8), the length of this rectangle is limited by $2h$. Thus, the square of this rectangle containing curvilinear quadrangle $B_2 B_{h,2} B_{h,3} B_3$ is $O(\tau^2 h)$.

There are four such quadrangles. Finally, we summarize errors together and get the following estimation:

$$\mathrm{meas}\left(Q_{i,j}^{k-1} \backslash P_{i,j}^{k-1}\right) = O(\tau^2 h) \text{ and } \mathrm{meas}\left(P_{i,j}^{k-1} \backslash Q_{i,j}^{k-1}\right) = O(\tau^2 h). \tag{25}$$

Unfortunately, estimation (25) is not enough to prove first order of convergence. Although, convergence is confirmed by numerical experiments [8]. Actually, we can improve evaluation (25) in same cases. For instance, consider curvilinear quadrangle $B_2 B_{h,2} B_{h,3} B_3$ with intersecting sides, see Fig. 1a. We prove that square of $B_2 B_{h,2} B_{h,3} B_3$ is $O(\tau^2 h^2)$. For this purpose, put $P_{int} = (P_{int}^x, P_{int}^y)$ be the point of across of quadrangle sides. This point is trace of a trajectory issued out from point $(t_k, \bar{x}, y_{j-1/2})$. Therefore, similar to (17) the following Taylor expansions are valid:

$$P_{int}^x = \bar{x} - \tau u\left(t_k, \bar{x}, y_{j-1/2}\right) + \frac{\tau^2}{2}\frac{d^2\hat{x}}{dt^2}(t_k) + O(\tau^3), \tag{26}$$

$$P_{int}^y = y_{j-1/2} - \tau v\left(t_k, \bar{x}, y_{j-1/2}\right) + \frac{\tau^2}{2}\frac{d^2\hat{y}}{dt^2}(t_k) + O(\tau^3). \tag{27}$$

Another point $\left(P_{\mathrm{Eul}}^x, P_{\mathrm{Eul}}^y\right)$ issued out from $(t_k, \bar{x}, y_{j-1/2})$ and computed by Euler method on segment $B_{h,2}B_{h,3}$ has the following coordinates:

$$P_{\mathrm{Eul}}^x = \bar{x} - \tau u\left(t_k, \bar{x}, y_{j-1/2}\right) \text{ and } P_{\mathrm{Eul}}^y = y_{j-1/2} - \tau v\left(t_k, \bar{x}, y_{j-1/2}\right). \tag{28}$$

From Lemma 1 we get

$$P_{int}^x - P_{\mathrm{Eul}}^x = O(\tau^2). \tag{29}$$

Now we show that incline of segment $B_{h,2}B_{h,3}$ to axis Ox is $O(\tau)$. To prove it, we consider subtraction $B_{h,3}^y - B_{h,2}^y = \tau v\left(t_k, x_{i+1/2}, y_{j-1/2}\right) -$

$\tau v\left(t_k, x_{i-1/2}, y_{j-1/2}\right) = O(\tau h_x)$. Due to (8) we see that $B^x_{h,3} - B^x_{h,2} \geq h_x/2$. Therefore, incline of segment $B_{h,2}B_{h,3}$ to axis Ox is evaluated by relation

$$\left(B^y_{h,3} - B^y_{h,2}\right) / \left(B^x_{h,3} - B^x_{h,2}\right) = O(\tau). \tag{30}$$

Thus, from (27) we see that

$$P^y_{int} - P^y_{\text{Eul}} = O(\tau^3). \tag{31}$$

From definition of P_{int} and P_{Eul} we get

$$P^y_{int} - P^y_{\text{Eul}} = \frac{\tau^2}{2}\frac{d^2\hat{y}}{dt^2}(t_k) = O(\tau^3).$$

Therefore,

$$\frac{d^2\hat{y}}{dt^2}(t) = \frac{dv}{dt}\left(t, \hat{x}(t), \hat{y}(t)\right) = \left(\frac{\partial u}{\partial t} + u\frac{\partial u}{\partial x} + v\frac{\partial u}{\partial y}\right)\left(t, \hat{x}(t), \hat{y}(t)\right) = O(\tau).$$

Due to boundedness of the second derivatives, the last equalities mean that

$$\left(\frac{\partial u}{\partial t} + u\frac{\partial u}{\partial x} + v\frac{\partial u}{\partial y}\right)(t, x, y) = O(\tau + h) \text{ on } [t_{k-1}, t_k] \times \omega_{i,j}. \tag{32}$$

Now we repeat our reasoning from (22) to (25) and with the help of (30), (32) we get more accurate estimation

$$\text{meas}\left(Q^{k-1}_{i,j} \backslash P^{k-1}_{i,j}\right) \leq c_1\tau^2h^2 \text{ and meas}\left(P^{k-1}_{i,j} \backslash Q^{k-1}_{i,j}\right) \leq c_1\tau^2h^2 \tag{33}$$

for curvilinear quadrangle with intersecting sides.

Also we construct the similar evaluation for quadrangle without intersecting sides. In the Fig. 3 at the plane $t = t_{k-1}$ are presented curvilinear quadrangles $B_2B_{h,2}B_{h,3}B_3$ and $B_1B_{h,1}B_{h,4}B_4$, obtained from segments with vertices $(x_{i\pm1/2}, y_{j-1/2})$ and $(x_{i\pm1/2}, y_{j+1/2})$. Quadrangle $B_{12}B_{h,12}B_{h,34}B_{34}$ is designed in the same way from segment with vertices $(x_{i\pm1/2}, y_j)$. We make parallel transfer of $B_2B_{h,2}B_{h,3}B_3$ to $\overrightarrow{B_2B_{h,2}B_{h,3}B_3}$ by changing of variables in the following way:

$$\mathbf{x}' = x + \frac{\tau h_y}{2}\frac{\partial u}{\partial y}\left(t_k, x_i, y_j\right), \quad \mathbf{y}' = y + \frac{h_x}{2} + \frac{\tau h_y}{2}\frac{\partial v}{\partial y}\left(t_k, x_i, y_j\right). \tag{34}$$

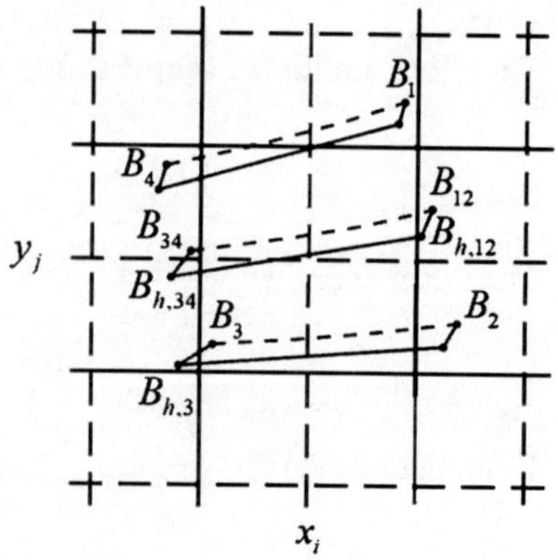

Fig. 3 Three curvilinear quadrangles

Hereinafter the arrow in $\overrightarrow{B_2 B_{h,2} B_{h,3} B_3}$ means the parallel transfer and do not indicate vector. Now we show that the curvilinear quadrangle $\overrightarrow{B_2 B_{h,2} B_{h,3} B_3}$ differs slightly from $B_{12} B_{h,12} B_{h,34} B_{34}$ in the following sense:

$$\text{meas}\left(B_{12} B_{h,12} B_{h,34} B_{34}\right) - \text{meas}\left(\overrightarrow{B_2 B_{h,2} B_{h,3} B_3}\right) = O(\tau^2 h^2). \tag{35}$$

Firstly, show that distance between corresponding quadrangles vertices is $O(\tau h^2)$. We consider subtraction $\overrightarrow{B_{h,2}} - B_{h,12} = \left(\overrightarrow{B^x_{h,2}} - B^x_{h,12}, \overrightarrow{B^y_{h,2}} - B^y_{h,12}\right)$. With the help of Taylor expansion along Oy and Ox axes we obtain

$$\left|\overrightarrow{B^x_{h,2}} - B^x_{h,12}\right| = \Big| - \tau u(t_k, x_{i+1/2}, y_{j-1/2}) +$$
$$\frac{\tau h_y}{2}\frac{\partial u}{\partial y}(t_k, x_i, y_j) + \tau u(t_k, x_{i+1/2}, y_j)\Big| = \tag{36}$$
$$\left|\frac{\tau h_y}{2}\frac{\partial u}{\partial y}(t_k, x_i, y_j) - \frac{\tau h_y}{2}\frac{\partial u}{\partial y}(t_k, x_{i+1/2}, y_j) + O(\tau h_y^2)\right| = O(\tau h^2).$$

By the same way we get

$$\left|\overrightarrow{B^x_{h,3}} - B^x_{h,34}\right| = O(\tau h^2). \tag{37}$$

Therefore, all points of $\overrightarrow{B_{h,34} B_{h,12}}$ and $\overrightarrow{B_{h,3} B_{h,2}}$ are located at distance $O(\tau h^2)$.

Now we consider two points $B = (B^x, B^y) \in \widehat{B_{34}B_{12}}$ and $B_1 = \left(B_1^x, B_1^y\right) \in \widehat{B_3B_2}$. They are traces at plane $t = t_{k-1}$ of trajectories issued out from points $A = \left(x, y_j\right)$ and $A_1 = \left(x, y_{j-1/2}\right)$ at the plane $t = t_k$ for $x \in [x_{i-1/2}, x_{i+1/2}]$. We use Taylor expansion along Ot axis and receive

$$\left|\overrightarrow{B_1^x} - B^x\right| = \left|-\tau u(t_k, x, y_{j-1/2}) + \frac{\tau h_y}{2}\frac{\partial u}{\partial y}(t_k, x_i, y_j) + \tau u(t_k, x, y_j) + \right.$$

$$\left.\frac{\tau^2}{2}\frac{d^2\hat{x}}{dt^2}(t_k; A_1) - \frac{\tau^2}{2}\frac{d^2\hat{x}}{dt^2}(t_k; A) + O(\tau^3)\right|.$$

First three terms in the right-hand side are evaluated by $O(\tau h^2)$ similar to (36). Additional arguments A and A_1 in two penultimate members mean corresponding initial condition for trajectories (12)–(13). To evaluate them, we use formula (18) and get

$$\frac{d^2\hat{x}}{dt^2}(t_k; A_1) - \frac{d^2\hat{x}}{dt^2}(t_k; A) = \left(\frac{\partial u}{\partial t} + u\frac{\partial u}{\partial x} + v\frac{\partial u}{\partial y}\right)\Bigg|_{(t_k, x, y_j)}^{(t_k, x, y_{j-1/2})}.$$

Taking into account the difference of arguments in the right-hand side we obtain

$$\frac{\tau^2}{2}\left|\frac{d^2\hat{x}}{dt^2}(t_k; A_1) - \frac{d^2\hat{x}}{dt^2}(t_k; A)\right| \le O(\tau^2 h_y).$$

Combining it together with previous equality we get $\left|\overrightarrow{B_1^x} - B^x\right| \le O(\tau^2 h)$. In the same way we prove inequality $\left|\overrightarrow{B_1^y} - B^y\right| \le O(\tau^2 h)$. Therefore,

$$\left|\overrightarrow{B_1} - B\right| \le O(\tau^2 h). \tag{38}$$

So, any point of boundary $B_{12}B_{h,12}B_{h,34}B_{34}$ is located no further than $O(\tau^2 h)$ from boundary $\overrightarrow{B_2B_{h,2}B_{h,3}B_3}$. Due to (8), length of boundary do not exceed $O(h)$. It leads to evaluation (35) which is equivalent to approximate equality meas $\left(B_2B_{h,2}B_{h,3}B_3\right)$ = meas $\left(B_{12}B_{h,12}B_{h,34}B_{34}\right) + O(\tau^2h^2)$. Similar reasoning leads to approximate equality of squares meas $\left(B_1B_{h,1}B_{h,4}B_4\right)$ = meas $\left(B_{12}B_{h,12}B_{h,34}B_{34}\right) + O(\tau^2h^2)$. Finally, from the transitivity we get

$$\text{meas}\left(B_1B_{h,1}B_{h,4}B_4\right) = \text{meas}\left(B_2B_{h,2}B_{h,3}B_3\right) + O(\tau^2h^2). \tag{39}$$

We consider "regular" case when both curvilinear quadrangles are located on the same side of corresponding straight segments (in previous example $\overline{B_{h,1}B_{h,4}}$ and $\overline{B_{h,2}B_{h,3}}$). What happened when they are located on different sides of corresponding straight segments? In this case proof of estimation (29) is the same. Therefore, any point of arc $\overleftarrow{B_1 B_4}$ is located no further than $O(\tau^2 h)$ from a point on arc $\overrightarrow{B_2 B_3}$. It means that images of both curvilinear quadrangles $\overleftarrow{B_1 B_{h,1} B_{h,4} B_4}$ and $\overrightarrow{B_2 B_{h,2} B_{h,3} B_3}$ are located in rectangle with width $O(\tau^2 h)$. Since the length of this rectangle is no further than $2h_x$, we get equalities

$$\begin{aligned} &\text{meas}\left(B_1 B_{h,1} B_{h,4} B_4\right) = O(\tau^2 h^2) \text{ and} \\ &\text{meas}\left(B_2 B_{h,2} B_{h,3} B_3\right) = O(\tau^2 h^2). \end{aligned} \tag{40}$$

Summarizing our reasoning about estimation of squares of regular and irregular curvilinear quadrangles we receive the following statement.

Lemma 2 *Square of an irregular curvilinear quadrangle is evaluated by* $O(\tau^2 h^2)$. *For every adjacent pair of a regular curvilinear quadrangle difference between their squares is* $O(\tau^2 h^2)$.

These estimations together with property of function smoothness lead to following evaluation of approximation error.

Lemma 3 *For smooth functions* u, v, ρ *the estimate holds*

$$\left| \int_{Q_{i,j}^{k-1}} \rho\,(t_{k-1}, x, y)\ \text{d}x\text{d}y - \int_{P_{i,j}^{k-1}} \rho\,(t_{k-1}, x, y)\ \text{d}x\text{d}y \right| = O(\tau^2 h^2). \tag{41}$$

Proof Considered integral subtraction for case of irregular curvilinear quadrangles, for instance shown in the Fig. 1a, can be changed and evaluated in the following way:

$$\left| \int_{Q_{i,j}^{k-1} \backslash P_{i,j}^{k-1}} \rho\,(t_{k-1}, x, y)\ \text{d}x\text{d}y - \int_{P_{i,j}^{k-1} \backslash Q_{i,j}^{k-1}} \rho\,(t_{k-1}, x, y)\ \text{d}x\text{d}y \right| \le$$

$$\rho^{\max}\text{meas}\left(Q_{i,j}^{k-1} \backslash P_{i,j}^{k-1}\right) + \rho^{\max}\text{meas}\left(P_{i,j}^{k-1} \backslash Q_{i,j}^{k-1}\right) = O(\tau^2 h^2).$$

In case of regular curvilinear quadrangles, we additionally use the relation $\rho\,(t_{k-1}, x, y) = \rho\,(t_{k-1}, x_i, y_j) + O(h)$. Therefore,

$$\left| \int_{Q_{i,j}^{k-1} \backslash P_{i,j}^{k-1}} \rho\,(t_{k-1}, x, y)\ \text{d}x\text{d}y - \int_{P_{i,j}^{k-1} \backslash Q_{i,j}^{k-1}} \rho\,(t_{k-1}, x, y)\ \text{d}x\text{d}y \right| \le$$

$$\left|\rho_{i,j}^{k-1}\text{meas}\left(Q_{i,j}^{k-1} \backslash P_{i,j}^{k-1}\right) - \rho_{i,j}^{k-1}\text{meas}\left(P_{i,j}^{k-1} \backslash Q_{i,j}^{k-1}\right)\right| +$$

$$\left|O(h)\ \text{meas}\left(Q_{i,j}^{k-1} \backslash P_{i,j}^{k-1}\right) + O(h)\ \text{meas}\left(P_{i,j}^{k-1} \backslash Q_{i,j}^{k-1}\right)\right| = O(\tau^2 h^2).$$

Combination in a curvilinear quadrangle regular and unregular pairs does not change the main meaning of proof. □

From geometrical reasoning we see that union of quadrangles satisfies the following properties

$$\bigcup_{\substack{i=0,\dots,N_x\\ j=0,\dots,N_y}} P_{i,j}^{k-1} = \bar{\Omega}, \quad P_{i,j}^{k-1} \cap P_{i',j'}^{k-1} = \emptyset \ \forall i \neq i',\ j \neq j'.$$

Introduce the norm of grid function s defined on grid $\bar{D}_h$

$$\|s\| = \sum_{\substack{i=0,\dots,N_x\\ j=0,\dots,N_y}} \text{meas}\left(\omega_{i,j}\right) \left|s_{i,j}\right|.$$

Due to initial condition (21), we have $\left\|\rho_{i,j}^0 - \rho_{h,i,j}^0\right\| = 0$. Now, suppose that at the time level t_{k-1} we have already proved evaluation

$$\left\|\rho^{k-1} - \rho_h^{k-1}\right\| < C_2 \tau t_{k-1}. \tag{42}$$

Confirm this inequality at time level t_k. Firstly, we use the simple estimation carried out from limitation of second derivatives

$$\int_{\omega_{i,j}} \left|\rho\left(t_k, x, y\right) - \rho\left(t_k, x_i, y_j\right)\right| \, dx dy = \tilde{O}(h^4). \tag{43}$$

Here $\tilde{O}(h^4)$ means $O(h^4)$ when $\omega_{i,j}$ lies inside D and $O(h^3)$ when it lies along the boundary Γ. Therefore, due to (8) from (43) we get

$$\int_{\Omega} \left|\rho\left(t_{k-1}, x, y\right) - \rho^h\left(t_{k-1}, x, y\right)\right| \, dx dy \leq c_2 \tau t_{k-1}. \tag{44}$$

We decompose the difference of these functions at time level t_k into several parts

$$\text{meas}\left(\omega_{i,j}\right) \left|\rho_{i,j}^k - \rho_{h,i,j}^k\right| \leq \left|\int_{\omega_{i,j}} \rho_{i,j}^k - \rho\left(t_k, x, y\right) \, dx dy\right| +$$

$$\left|\int_{\omega_{i,j}} \rho\left(t_k, x, y\right) \, dx dy - \int_{Q_{i,j}^{k-1}} \rho\left(t_{k-1}, x, y\right) \, dx dy\right| +$$

$$\left|\int_{Q_{i,j}^{k-1}} \rho\left(t_{k-1}, x, y\right)\, dxdy - \int_{P_{i,j}^{k-1}} \rho\left(t_{k-1}, x, y\right)\, dxdy\right| +$$
$$\left|\int_{P_{i,j}^{k-1}} \rho\left(t_{k-1}, x, y\right)\, dxdy - \int_{P_{i,j}^{k-1}} \rho_h^{k-1}\left(x, y\right)\, dxdy\right| +$$
$$\left|\int_{P_{i,j}^{k-1}} \rho_h^{k-1}\left(x, y\right)\, dxdy - \rho_{h,i,j}^{k} \text{meas}\left(\omega_{i,j}\right)\right|.$$

Due to (14) and (19) two members in the right hand sides are equal to zero. Two other terms are evaluated in (41), (44). Finally,

$$\text{meas}\left(\omega_{i,j}\right)\left|\rho_{i,j}^{k} - \rho_{h,i,j}^{k}\right| \le \left|\int_{P_{i,j}^{k-1}} \rho\left(t_{k-1}, x, y\right)\, dxdy - \right.$$
$$\left.\int_{P_{i,j}^{k-1}} \rho_h^{k-1}\left(x, y\right)\, dxdy\right| + \tilde{O}(h^4).$$

Summing up these inequalities over all elements on Ω and due to (42) we get

$$\left\|\rho^k - \rho_h^k\right\| \le \int_{\Omega} \left|\rho\left(t_{k-1}, x, y\right) - \rho_h^{k-1}\left(x, y\right)\right| dxdy + \sum_{\substack{i = 0, \dots, N_x \\ j = 0, \dots, N_y}} \tilde{O}(h^4) \le c_2 \tau t_k.$$

So, we prove convergence of numerical scheme (19)–(21) with the first order of accuracy.

5 The Second Scheme: Decomposition of Operator

Decompose the initial operator in two one-dimensional ones

$$\frac{\partial \rho}{\partial t} + \frac{\partial\left(\rho u\right)}{\partial x} + \frac{\partial\left(\rho v\right)}{\partial y} = \left(\frac{1}{2}\frac{\partial \rho}{\partial t} + \frac{\partial\left(\rho u\right)}{\partial x}\right) + \left(\frac{1}{2}\frac{\partial \rho}{\partial t} + \frac{\partial\left(\rho v\right)}{\partial y}\right). \tag{45}$$

After that we use have already considered above method with a simplifications for one-dimensional operators [9]. For the first operator in right hand side of (45), the equation holds

$$\frac{1}{2}\frac{\partial \rho}{\partial t} + \frac{\partial\left(\rho u\right)}{\partial x} = f_1 \text{ with right hand side } f_1 = -\frac{1}{2}\frac{\partial \rho}{\partial t} - \frac{\partial\left(\rho v\right)}{\partial y}. \tag{46}$$

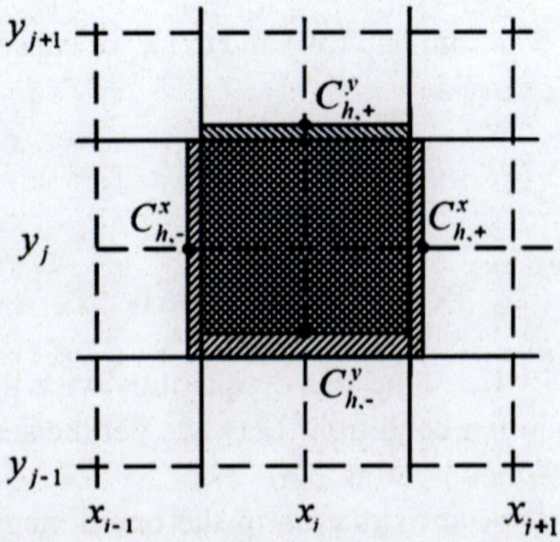

Fig. 4 Location of nodes $C^x_\pm$, $C^y_\pm$ and corresponding rectangles

For cell $\omega_{i,j}$ located at time level t_k we construct two trajectories issued out from points $(t_k, x_{i\pm1/2}, y_j)$ backward in time to level t_{k-1}. Thus, we get two nodes C^x_- and C^x_+ in the direction Ox at the level t_{k-1}, see Fig. 4. Note that equations described these trajectories are transformed from (12) to

$$\hat{x}'(t) = 2u\left(t, \hat{x}(t), y_j\right) \quad \forall\, t \in [t_{k-1}, t_k]. \tag{47}$$

After applying one-dimensional divergence theorem we have

$$\int_{x_{i-1/2}}^{x_{i+1/2}} \rho\left(t_k, x, y_j\right) \mathrm{d}x = \int_{C^x_-}^{C^x_+} \rho\left(t_{k-1}, x, y_j\right) \mathrm{d}x + \int_{\overline{\sqcup}} f_1 \, \mathrm{d}x\, \mathrm{d}t, \tag{48}$$

where $\overline{\sqcup}$—is curvilinear quadrangle in plane Otx bounded by $\left[x_{i-1/2}, x_{i+1/2}\right]$ at time level t_k, segment $\left[C^x_-, C^x_+\right]$ at level t_{k-1} and two trajectories connected ends of these segments. To compute C^x_- and C^x_+, we usually use one step of Euler method. Therefore, instead of C^x_- and C^x_+ we appeal to

$$C^x_{h,\pm} = C^x_\pm - 2\tau u\left(t_k, C^x_\pm, y_j\right). \tag{49}$$

Therefore, we change $C^x_\pm$ to $C^x_{h,\pm}$ in (48) and employ quadrature formula in point $\left(t_k, x_i, y_j\right)$ for last integral. After dividing by h_x we get the difference equation

$$\rho^k_{h,i,j} = \int_{C^x_-}^{C^x_+} \rho_h\left(t_{k-1}, x, y_j\right) \mathrm{d}x + \tau f_1\left(t_k, x_i, y_j\right). \tag{50}$$

In the same way we get the difference equation for the second operator with the right hand side $f_2 = -1/2 \cdot \partial\rho/\partial t - \partial\left(\rho u\right)/\partial x$:

$$\rho^k_{h,i,j} = \int_{C^y_-}^{C^y_+} \rho_h\left(t_{k-1}, x_i, y\right) \mathrm{d}y + \tau f_2\left(t_k, x_i, y_j\right). \tag{51}$$

Summing up (50) and (51), we cancel sum $f_1 + f_2$ and get

$$\rho^k_{h,i,j} = \frac{1}{2}\int_{C^x_-}^{C^x_+} \rho_h\left(t_{k-1}, x, y_j\right) \mathrm{d}x + \frac{1}{2}\int_{C^y_-}^{C^y_+} \rho_h\left(t_{k-1}, x_i, y\right) \mathrm{d}y, \tag{52}$$

$$i = 1, \ldots, N_x, \quad j = 0, \ldots, N_y.$$

Combine these equalities with the discrete boundary conditions (20) and the initial conditions (21), we get the second explicit monotone difference scheme [9] for computing $\rho_h(t, x, y)$.

Approximation of the one-dimensional schemes with the first order is justified in [9]. The further proof of convergence is similar to that carried out for the first scheme and therefore we will not dwell on their proof here.

6 Conclusion

Let us compare two considered algorithms. Obviously, the second scheme is substantially comfortable for forming grid equations than the first one. Especially, it is valid for explicit format of coefficients of difference equation [9]. However, the first scheme has advantage appeared during computations with high values of velocity functions. It has already mentioned in literature that semi-Lagrangian approximations allow to avoid Courant-Friedrichs-Lewy condition for time step. For this purpose, adaptive templates with time shift along flow are used. In described approaches cancelling restriction $\tau \leq \min\{h_x/u^{\max}, h_y/v^{\max}\}$ should be accompanied by more accurate computation of nodes $B_{h,i}$ in first scheme and $C^x_\pm, C^y_\pm$ in the second method. We can achieve it by using, for instance, one full step of Runge-Kutta method.

In the first scheme, the location of nodes $B_{h,i}$ is compact. Moreover, these nodes are moving along flow. More often the solution of problem is smoother along flow than in transverse to flow directions. For example, transfer of stair under high constant velocity makes zero derivative along flow and infinity item of time and space derivatives in point of gap. Therefore, approximation used along flow leads to sufficiency small errors than approximations along time or space axes.

In the second scheme in case of high velocities, nodes $C^x_\pm, C^y_\pm$ diverge in different paths. It brings to significant growth of scheme template. Finally, it reduces approximation possibilities of scheme in compared with the first scheme during cancelling of Courant-Friedrichs-Lewy condition.

The second scheme is more convenience for three-dimensional problem because the corresponding operator can be easy decomposed into three one-dimensional operators. In contradistinction to the first scheme, the second one requires some algebraic complications and sufficient efforts for theoretical justification. At the same time it keeps valid for three-dimensional problem the useful property of low dependence from Courant-Friedrichs-Lewy condition.

Acknowledgments This work is supported by Project 20-01-00090 of Russian Foundation for Basic Research.

References

1. Ewing, R.E., Wang, H.: A summary of numerical methods for timedepended advection-dominated partial differential equations. J. Comput. Appl. Math. 128 423–445 (2001).
2. Bonaventura, L.: An introduction to semi-Lagrangian methods for geographical scale flows. Lecture Notes, Zurich (2004).
3. Iske, A.: Conservative semi-Lagrangian advection on adaptive unstructured meshes. Numer. Meth. Part. Diff. Eq. 20, 388–411 (2004).
4. Arbogast, T., Wen-Hao Wang: Convergence of a fully conservative volume corrected characteristic method for transport problems. placecountry-regionSIAM. J. Numer. Anal. 48(3), 797–823 (2010).
5. Celledoni, E., Kometa, B., Verdier, O.: High order semi-Lagrangian methods for the incompressible Navier-Stokes equations. J. Sci. Comput. 66(1), 91–115 (2016).
6. Phillips, T.N., Williams, A.J.: Conservative semi-Lagrangain finite volume schemes. Numer. Meth. Part. Diff. Eq. 17, 403–425 (2001).
7. Quarteroni, A., Valii, A.: Numerical Approximation of partial differential equations. Springer, Heidelberg (1994).
8. Efremov, A., Karepova, E., Shaydurov, V., Vyatkin, A.: A computational realization of a Semi-Lagrangian for solving the advection equation. J. Appl. Math. 2014, 610398 (2014).
9. Shaidurov, V., Vyatkin, A., Kuchunova, E.: Semi-Lagrangian difference approximations with different stability requirements. Russ. J. Numer. Anal. Math. Modelling 33(2), 123–135 (2018).

Mathematical Modeling and Diagnostics Using Neural Networks and a Genetic Algorithm for Epilepsy Patients

Tatiana V. Yakovleva, Vitalii V. Dobriyan, Tatiana Yu. Yaroshenko, and Vadim A. Krysko-jr

Abstract Medical and biological electroencephalogram (EEG) signals are widely used for diagnosis, further medical support and treatment of such disease as epilepsy. A method for detecting of epileptiform activity presence in patients has been developed based on analysis of EEG signals taken during the 1st, 2nd and 3rd sleep phases. To implement this task, an approach was designed with use of genetic algorithm and artificial neural network (ANN), which can be classified according to the following criteria: the network is analog; self-organizing; a direct distribution network, static. The input neurons count in the neural network is equal to the channels count in the EEG recording, and in this study it is equal to 21, 15 neurons in the hidden layer, 1 output neuron. The input layer accepts data in the form of numbers representing the calculated characteristics for each channel. the largest Lyapunov exponent calculated by the Rosenstein, Wolf, Sano-Sawada methods, for men and women with epilepsy and from the control group. Test sample: 33 patients, of which 6 were healthy and 27 patients with epilepsy with different diagnoses. Some of the combinations made it possible to obtain 100% accuracy in determining the presence or absence of disease in patients.

1 Introduction

At present, dynamic processes studies of a very different nature (mechanical, natural, medico-biological, social, historical) based on neural networks are gaining in popularity. Human body can be viewed as a complex, nonlinear biological shell, consisting of nervous, bone, muscular, cardiovascular and other systems, and is a continuous medium. Therefore, when studying it, one should apply all the variety of

T. V. Yakovleva (✉) · V. V. Dobriyan · T. Yu. Yaroshenko
Yuri Gagarin State Technical University Of Saratov, Saratov, Russia
e-mail: yan-tan1987@mail.ru; dobriy88@yandex.ru

V. A. Krysko-jr.
Lodz University of Technology, Lodz, Poland

I. B. Badriev et al. (eds.), *Mesh Methods for Boundary-Value Problems and Applications*, Lecture Notes in Computational Science and Engineering 141,
https://doi.org/10.1007/978-3-030-87809-2_42

mathematical and probabilistic methods, including the human brain study. Medical and biological EEG signals are widely used in the diagnosis, further medical support and treatment for some forms of such disease as epilepsy. According to the WHO (World Health Organization), about fifty million people suffer from this neurological disease. Electroencephalography plays an important role in diagnosis of this disease and in monitoring the brain activity of patients with epilepsy. EEG recordings are time-varying signals of brain activity—time series, so they have a non-linear nature. Analysis of signals by methods of non-linear dynamics makes it possible to describe quantitatively EEG recordings, since signal characteristics can be measured. To diagnose a disease in the process of changing of the EEG signal, it is important to highlight the features (characteristic patterns) that accompany it.

For this kind of research, several neural networks types are used: artificial (ANN), probabilistic (PNN), convolutional (CNN). ANN can be considered as a directed graph with weighted connections and nodes—artificial neurons. According to the architecture of connections, artificial neural networks can be grouped into two classes: feedforward networks, recurrent networks, or feedback networks (RNN). In feedforward networks (multilayer perceptrons), neurons are arranged in layers and have unidirectional connections between layers. These networks are static, that is, for a given input they generate one set of output values that do not depend on the previous network state. Recurrent networks are dynamic, due to feedbacks in them the inputs of neurons are modified, and thereby the network state changes. They consist of a straight line neural network with circular connections.

In a number of papers [1–9], the authors propose an ANN-based automatic detection system for characteristic epileptic patterns that can work with the complexities associated with EEG signals to predict the most appropriate solution. In the research, the analysis is carried out using an artificial neural network of EEG obtained from an epileptic and healthy brain. To minimize the root mean square error (MSE) of the network, a genetic algorithm is used [1]. In [2], the epilepsy signs classification is carried out by the EEG signal based on the genetic algorithm (GA) and artificial neural network (ANN). Epileptic EEG signals are pre-processed using a discrete wavelet transform to be divided into frequency subbands (delta, theta, alpha, beta, gamma) using such feature as entropy. The results comparisons of studying the healthy person EEG signals and patients suffering from epilepsy in different periods—ictal and postictal, are carried out. ANN multilayer neural network with feedback and integrated GA improves classification accuracy when diagnosing and grouping EEG signals. In [3], the analysis is carried out using an artificial neural network (ANN), where FFT coefficients are used training of a values. The aim of the work was to select the most accurate method for training of a multilayer network for the qualitative classification of the EEG epilepsy features. For this, several learning methods comparisons are carried out: Levenberg-Marquardt (LM), Quickprop (QP), Delta-bar delta (DBD), Momentum and Conjugate Gradient (CG), genetic algorithm (GA). The best performance was achieved by optimizing the learning rate weights using GA. In [4], EEG segments are analyzed using a time-frequency distribution, and then, for each segment, several features are identified that represent the energy distribution in the time-frequency plane. Functions are used to train a neural network. The Fast Fourier Transform and multiple time-frequency distributions are compared. In [5], ANN acts as a features classifier that have been sorted from signals using a combination of discrete wavelet transform (DWT) and fast Fourier transform (FFT). Using this methods combination, a good accuracy of 98.889% can be achieved. In [6], the features analysis of epilepsy EEG signals is carried out using the genetic algorithm GAFDS. Several types of classifiers are compared. For this, the frequency domain elements are identified and combined with non-linear characteristics. Classifiers such as k-nearest neighbor, linear discriminant analysis, decision tree, AdaBoost, multilayer perceptron, naive Bayesian method are used. Combined with GAFDS, the accuracy is 99% and 97%. When conducting cross-sectional analyzes, the authors found that GAFDS performs well in identifying effective features for EEG classification. Therefore, the proposed model for characteristics selection and optimization can improve the classification accuracy. In [7], an artificial neural network (ANN) with feedback is used to identify the epileptic EEG signal. The classification criteria are the wavelet coefficients of the studied signals. GA is used for training. Harmonic weights are used to improve the classification accuracy, thus achieving an accuracy of 99.19%. In [8], the ictal state identification by calculating the maximum Lyapunov exponent (STLmax) is carried out according to the Kantz method. The proposed approach is based on dividing the EEG signal into periods corresponding to epileptic and non-epileptic activity. The STLmax values are used to classify the EEG signal. Segmentation and calculation of STLmax values are performed using a trained neural network. For the study, EEG signals of 5 healthy volunteers with open and closed eyes and 5 epilepsy patients were used. In [9], the EEG signal is first preprocessed using discrete wavelet transform (DWT) to remove noise and extract features. The processed data are the input values of the RNN for classification. Several experiments were carried out to obtain the optimal parameter for the model. The model then compared with Logistic Regression (LR), Support Vector Machine (SVM), K-Nearest Neighbor (KNN), Random Forest (RF), and Decision Tree (DT).

In a series of works [10, 11], a probabilistic neural network (PNN) is used to classify the EEG signals features. A probabilistic neural network PNN is a kind of neural networks for classification and pattern recognition problems, where the class membership probability density is estimated by means of a kernel approximation. It is one of the so-called Bayesian networks type. This neural network type was derived from Bayesian network and Fisher's statistical algorithm. With this method, the misclassification probability is minimized. In [10], comparisons are made between analysis results for probabilistic (PNN) and recurrent (RNN) neural networks. In general, according to the proposed methodology, the authors found that the recurrent network analysis results are more accurate than for feedforward network models. The Lyapunov exponents have become the attribute on which the classification is based. The probabilistic neural network has shown that it can be useful in the analysis of long-term EEG signals for the electroencephalographic changes early detection. In [11], the algorithm is constructed in such a way that decision-making is carried out in two stages: the Lyapunov exponents calculation in the feature vectors form and classification using classifiers trained on the extracted features. The combination of the Lyapunov exponent values and the probabilistic neural network was aimed to identify the optimal classification algorithm for the epileptic EEG in order to identify characteristic patterns and their possible regularities.

A convolutional neural network (CNN) is suitable for recognizing patterns characteristic of an epileptic EEG. Recognition is considered as the neural network ability to extract the necessary features and at the same time be invariant to various kinds of interference and image distortions. A number of works are devoted to this [12–15].

The Lyapunov exponents technique is well suited for the study of nonlinear biomedical signals [16–18]. The epilepsy diagnosis is based on the EEG chaotic behavior assessment. Lyapunov exponents are a quantitative measure for distinguishing of orbits in phase space according to their sensitivity to initial conditions and are used to determine the stability of any steady-state time series, as well as to determine the system dynamics complexity [19]. Previously, the use of the Lyapunov exponents analysis was very effective for studying the chaotic dynamics of distributed mechanical structures [20, 21].

In well-known works, studies were carried out for EEGs taken while patients were awake, and the values of wavelet transforms, Fourier transforms, k-nearest neighbor were most often used as a features classifier. There are separate works where only one Kantz [22] method was used in the study of the largest Lyapunov exponent. Considering that there is currently no universal method for calculating Lyapunov exponents, it is necessary to apply several methods for calculating Lyapunov exponents to obtain reliable results.

In the present work, it was possible to develop a method for detecting the epileptiform activity presence in patients based on the analysis of EEGs taken during various sleep stages, using methods for calculating the largest Lyapunov exponent and further training of the neural network using a genetic algorithm.

2 Neural Network

To implement this task, a neural network was designed, which can be classified according to the following criteria: belongs to the ANN class (artificial neural networks); according to the input information, the network is analog (information is presented in the form of real numbers); by the training form, the network is self-organizing (it forms the output solutions space only on the input actions basis); by the connections nature, the network is a direct propagation network (all connections are directed strictly from input neurons to output neurons) and static (each neuron output is connected to all inputs of the next layer neurons and there are no dynamic connections) (Fig. 1).

The designed neural network has three layers. The number of neurons in the hidden layer is configurable. The hidden and output layers neurons have different combinations of three different activation functions types (Fig. 2).

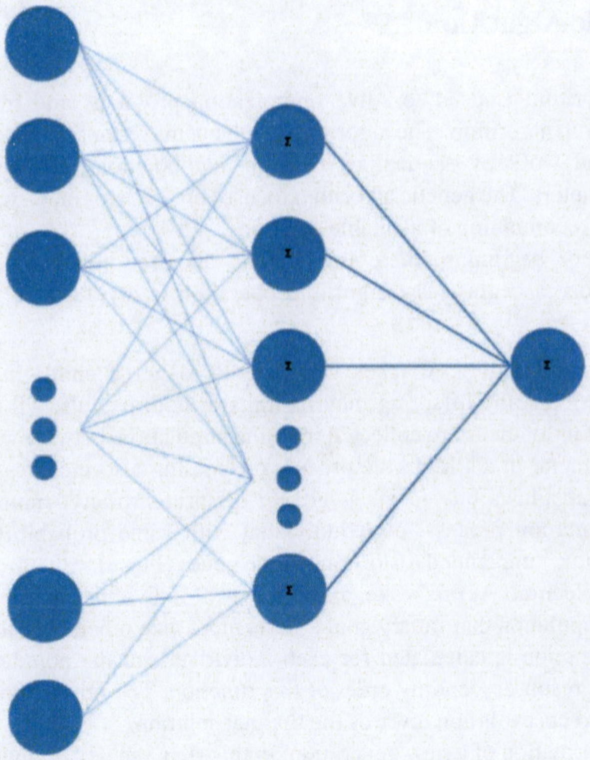

Fig. 1 Three-layer neural network

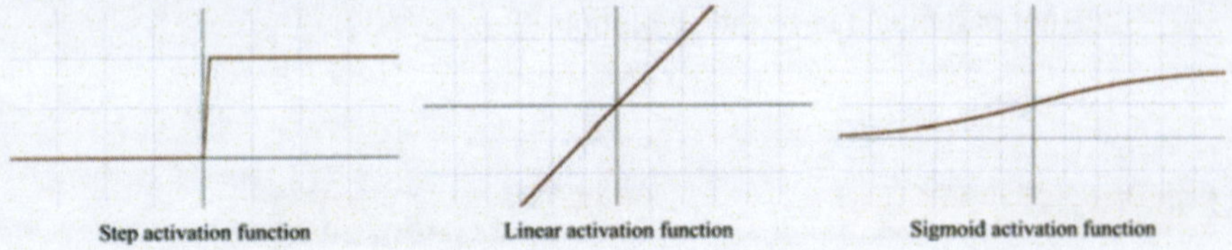

Fig. 2 Activation functions

1. Step activation function. It is represented by the function $f(x) = \left\{\begin{matrix} 0, & x < 0 \\ 1, & x \geq 0 \end{matrix}\right\}$ and has a derivative $f'(x) = \left\{\begin{matrix} 0, & x \neq 0 \\ ?, & x = 0 \end{matrix}\right\}$.
2. Linear activation function. It is represented by the function $f(x) = Cx$ and has a derivative $f'(x) = C$.
3. Sigmoid activation function. It is represented by the function $f(x) = \sigma(x) = \frac{1}{1+e^{-x}}$ and has a derivative $f'(x) = f(x)(1 - f(x))$.

3 Genetic Algorithm

Genetic algorithm is used to solve optimization problems and in essence is a heuristic search algorithm. The algorithm mechanisms resemble those of biological evolution and work by sequential selection, combination, and variation of the sought parameters. The genetic algorithm focuses on the "crossing", which performs operation of recombining of available solutions.

At the very beginning, there is a certain ancestor population, from which evolution process begins. The algorithm operation is divided into the following stages.

Stage 1. Crossbreeding. It takes two parents to get a child. In the crossing process, the offspring inherits traits of both parents. All possible pairs of individuals are collected from the population, between which crossing occurs. In addition, the process of crossing also includes a mechanism that allows one to get a greater offspring variety—mutations. In the mutation process, each individual with some probability can receive some "unplanned distortion" in the genes (Fig. 3).

Stage 2. Selection. At this stage, a limited set of individuals is selected from the population that satisfy goal criteria more than others. For this, the fitness function is calculated for each individual and the population is sorted in result descending order of this function. The fitness function, in fact, directs evolution towards the optimal solution.

Stage 3. Formation of a new generation. In this step, the next individuals population is created, which is based on the "best" individuals from the previous generation. Individuals not included in the new generation "die" and do not participate in further evolution.

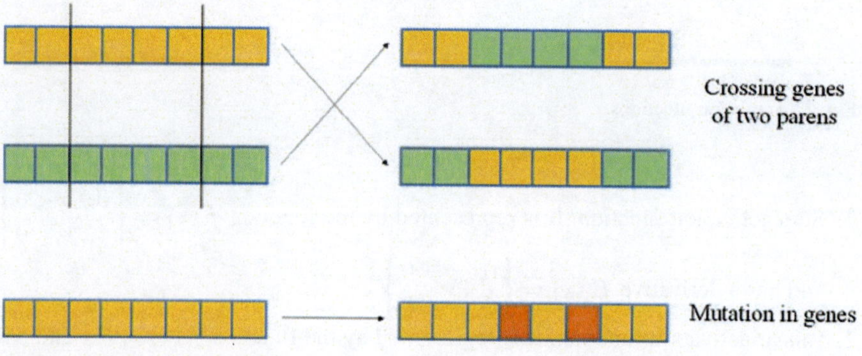

Fig. 3 Gene crossing and mutation

4 Approach Implementation

To increase efficiency of problem solving, a neural network, described in paragraph 1, was taken as the population individuals for the genetic algorithm. This approach will significantly speed up the neural networks training.

The algorithm was trained on a sample that included people with a known diagnosis (there is epilepsy, there is no epilepsy).

4.1 Object of Study

Patients EEG recording was performed at the Epineiro Medical Center for Neurology, Epilepsy Diagnosis and Treatment in Saratov city using 21 channels: O2, O1, P4, P3, C4, C3, F4, F3, Fp2, Fp1, T6, T5, T4, T3 , F8, F7, Pz, Cz, Fz, A2, A1 with the electrode arrangement shown in Fig. 4. On average, one signal duration is 10 seconds, sampling rate is 250 Hz. A neurophysiologist cleared the artifacts. EEGs taken during the first, second and third stages of sleep were analyzed for patients with epilepsy with different diagnoses (headaches, focal tonic spasms, generalized and focal seizures, absences) and in the control group.

Fig. 4 Arrangement of EEG electrodes

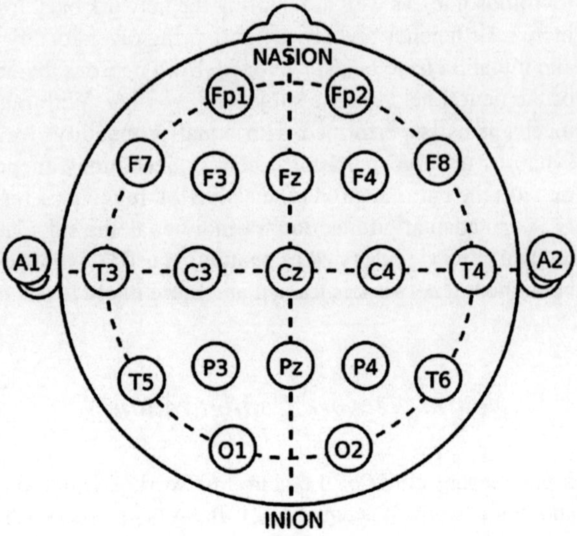

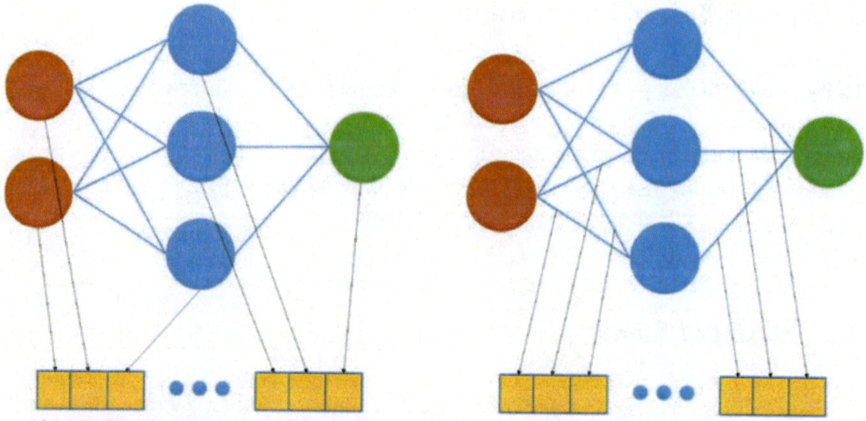

Fig. 5 Linearization of synapses and neurons into a gene sequence

4.2 Crossbreeding and Mutations

In order to solve the problem of crossing and mutating of neural networks, a method was developed that allows to linearize the neural network into a genes sequence (chromosome), as well as restoring the network back from this sequence (Fig. 5). To increase efficiency and broaden offspring diversity, several mechanisms of crossing and mutation have been developed. Both neurons themselves and synapses weights of the neural network are subject of crossing. With mutation, one of the following mechanisms is performed with equal probability: weights in the neural network synapses undergo mutations; neuron activation function parameters are mutated; mutates the entire neuron (one activation function is replaced by another).

As an adaptation function for individual, the calculation number of the correctly defined EEG signals (sick or healthy) is used. If more correctly defined EEGs, then better neural network is trained and more likely for its offspring to "survive".

4.3 Neural Network Configuration

Input neurons count of the neural network is equal to channels in the EEG signal, and in this study, it is equal to 21. Input layer receives data as numbers representing calculated characteristics for each channel. The following characteristics were used: the largest Lyapunov exponent calculated by Rosenstein method [23]; the largest Lyapunov exponent according to Wolf method [24]; the largest Lyapunov exponent according to Sano-Sawada method [25].

Test sample: 33 patients, of which 6 are healthy and 27 patients with epilepsy with different diagnoses (headaches, generalized, focal). For each EEG channel, the

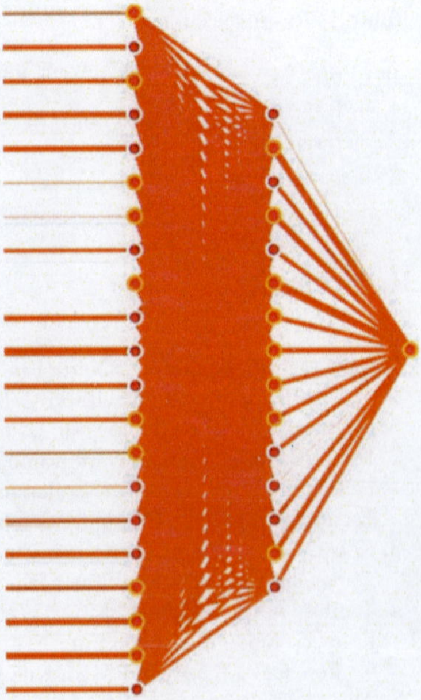

Fig. 6 Visualization of the trained neural network during the EEG analysis

first Lyapunov exponent was calculated by the methods of Sano-Sawada, Rosenstein and Wolf.

Training: For training, a neural network of the following configuration was used: 21 input neurons (the channels count in the EEG), 15 neurons in hidden layer, 1 output neuron. The largest Lyapunov exponents were fed to the neurons input of the input layer along all EEG channels (Fig. 6).

5 Results and Conclusions

In this work, a comprehensive approach has been developed for detecting of epilepsy presence for patients based on the EEG analysis taken during the first, second and third stages of sleep. EEG analysis is carried out using three different methods for calculating of the largest Lyapunov exponent, namely Rosenstein, Wolf, Sano-Sawada, and further training of neural network using a genetic algorithm.

In the studies course, it was possible to find out that the following combinations turned out to be the most accurate in determining of the epilepsy presence: EEG analysis of the first sleep stage using Sano-Sawada and Wolf methods, as well as the EEG of the third sleep stage by Sano-Sawada method. In general, Rosenstein method showed the worst results.

Some of the combinations made it possible to obtain 100% accuracy in determining the presence or absence of patients disease. Other combinations showed lower accuracy, which, however, was at least 85%. The data are presented in Table 1.

Table 1 Results accuracy

Sleep stage	Sano-Sawada method	Wolf method	Rosenstein method
1 sleep stage	100%	100%	85%
2 sleep stage	96%	92%	78%
3 sleep stage	100%	92%	97%

References

1. Saini, Jagriti, Maitreyee, Dutta: Epilepsy Disease Detection Using Artificial Neural Network and MSE Optimization with GA. International Journal of Innovative Research in Science, Engineering and Technology. **6**, Is. 7, (2017)
2. KumarBandil, Manoj, Wadhwanib, A.K.: Multi-Resolution EEG AND EEG Sub-Band Features Optimization for Epileptic Classification Using Hybrid Evolutionary Computing Technique. Procedia Computer Science. **152**, 243–251 (2019)
3. Kocer, Sabri, Rahmi Canal, M: Classifying Epilepsy Diseases Using Artificial Neural Networks and Genetic Algorithm. Med. Syst. **35(4)**, 489–498 (2011)
4. Tzallas, A.T., Tsipouras, M.G., Fotiadis D.I.:The Use of Time-Frequency Distributions for Epileptic Seizure Detection in EEG Recordings. Proceedings of the 29th Annual International Conference of the IEEE EMBS Cité Internationale, Lyon, France August 23–26 (2007)
5. Azian, Azamimi, Abdullah, Saufiah, Abdul, Rahim, Adira, Ibrahim: Development of EEG-based Epileptic Detection using Artificial Neural Network. International Conference on Biomedical Engineering (ICoBE), Penang (2012)
6. Wen, T, Zhang, Z.: Effective and extensible feature extraction method using genetic algorithm-based frequency-domain feature search for epileptic EEG multiclassification. Medicine (Baltimore) **96(19)**, (2017)
7. Patnaik, Lalit, Manyam, Ohil: Epileptic EEG detection using neural networks and post-classification. Computer methods and programs in biomedicine. **91**, (2008)
8. Golovko, V., Artsiomenka, S., Kisten, V., Evstigneev, V.: Towards automatic epileptic seizure detection in EEGs based on neural networks and largest Lyapunov exponent. International Journal of Computing. **14(1)**, 36–47 (2015)
9. Aliyu, Ibrahim, Lim, Yong, Lim, Chang: Epilepsy Detection in EEG Signal using Recurrent Neural Network. ISMSI 2019: Proceedings of the 2019 3rd International Conference on Intelligent Systems, Metaheuristics & Swarm Intelligence. 50–53 (2019)
10. Guler, N.F., Ubeyli, E.D., Guler, I.: Recurrent neural networks employing Lyapunov exponents for EEG signals classification. Expert Systems with Applications. **29**, 506–514 (2005)
11. Ubeyli, E.D.: Lyapunov exponents/probabilistic neural networks for analysis of EEG signals. Expert Systems with Applications. **37**, 985–992 (2011)
12. Omer, Turk, Mehmet, Sirac, Ozerdem: Epilepsy Detection by Using Scalogram Based Convolutional Neural Network from EEG Signals. Brain Sci. **9(15)**, Is. 115, (2019)
13. Mengni, Zhou, Cheng, Tian, Rui, Cao, Bin, Wang, Yan, Niu, Ting, Hu, Hao, Guo, Jie, Xiang: Epileptic Seizure Detection Based on EEG Signals and CNN. Front Neuroinform. **12(95)**, (2018)
14. Ali, Emami, Naoto, Kunii, Takeshi, Matsuo, Takashi, Shinozaki, Kensuke, Kawai, Hirokazu, Takahashi: Seizure detection by convolutional neural network-based analysis of scalp electroencephalography plot images. NeuroImage: Clinical. **22**, 101684 (2019)
15. Wenbin, Hu, Jiuwen, Cao, Xiaoping, Lai, Junbiao, Liu: Mean amplitude spectrum based epileptic state classification for seizure prediction using convolutional neural networks. Journal of Ambient Intelligence and Humanized Computing. doi.org/10.1007/s12652-019-01220-6 (2019)

16. Kutepov, I.E., Dobriyan, V.V., Zhigalov, M.V., Stepanov, M.F., Krysko, A.V., Krysko, V.A., Yakovleva, T.V.: EEG analysis in patients with schizophrenia based on Lyapunov exponents. Informatics in Medicine Unlocked. **18**, 100289 (2020)
17. Yakovleva, T.V., Kutepov, I.E., Karas, A.Yu., Yakovlev, N.M., Dobriyan, V.V., Papkova, I.V., Zhigalov, M.V., Saltykova, O.A., Krysko, A.V., Yaroshenko, T.Yu., Erofeev, N.P., Krysko, V.A.: EEG Analysis in Structural Focal Epilepsy Using the Methods of Nonlinear Dynamics (Lyapunov Exponents, Lempel-Ziv Complexity, and Multiscale Entropy). The Scientific World Journal. 8407872 (2020)
18. Krysko, V.A. et al : J. Phys.: Conf. Ser. 1260 072010 (2019)
19. Awrejcewicz, J., Krysko, A., Erofeev, N., Dobriyan, V., Barulina, M., Krysko, V.: Quantifying Chaos by Various Computational Methods. Part 1: Simple Systems. Entropy. **20(3)**, Is. 175 (2018)
20. Awrejcewicz, J., Krysko, V.A., Papkova, I.V., Krysko, A.V.: 'Routes to chaos in continuous mechanical systems. Part 3: The Lyapunov exponents, hyper, hyper-hyper and spatial-temporal chaos. Chaos, Solitons, Fractals. **45**, 721–736 (2012)
21. Krysko-Jr., V., Awrejcewicz, J., Yakovleva, T., Kirichenko, A., Szymanowska, O., Krysko, V.: Mathematical modeling of MEMS elements subjected to external forces, temperature and noise, taking account of coupling of temperature and deformation fields as well as a nonhomogenous material structure. Communications in Nonlinear Science and Numerical Simulation. **72**, 39–58 (2019)
22. Kantz, H.: A robust method to estimate the maximal Lyapunov exponent of a time series. Physics letters A. **185(1)**, 77–87 (1994)
23. Rosenstein, M.T., Collins, J.J., de Luca, C.J.: A practical method for calculating the largest Lyapunov exponent from small data sets. Physica D. **65**, Is. 117 (1993)
24. Wolf, A, Swift, JB, Swinney, HL, Vastano, JA.: Determining Lyapunov exponents from a time series. Physica. **16D**, 285–317 (1985)
25. Sano, M., Sawada, Y.: Measurement of Lyapunov spectrum from a chaotic time series. Phys. Rev. Lett. **55**, 1082 (1985)

A Two-Stage Cutting-Plane Method for Conditional Minimizing Function

Igor Zabotin, Oksana Shulgina, and Rashid Yarullin

Abstract A cutting-plane method is proposed for solving conditional minimization problem. In this method each main iteration point is constructed by two stages. At the first stage, a set is fixed which approximates the feasible set, and on the basis of some auxiliary points it is performed approximation of an epigraph of the objective function. The first stage is finished, when the approximation quality of the epigraph is quite good. At the second stage, the next main iteration point is constructed by cutting the previous ones from the set which approximates the feasible set and it is performed a process of updating the set which approximates the epigraph. Computational aspects of the proposed method are discussed.

1 Introduction

One of the classes of methods for solving mathematical programming problems is the so-called cutting-plane methods (for example, [1–19]). The method proposed here belongs to the mentioned class. When using approximations, it uses approximation by polyhedral sets of both feasible set and epigraph of the objective function.

The construction of each point of the main sequence occurs in two stages. At the first stage, the set is fixed which approximate the feasible set, and on the basis of some auxiliary points, the sets approximating the epigraph are sequentially constructed. When the approximation quality of the epigraph becomes in a certain sense acceptable, the first stage is completed. At the second stage, the main iteration point is found, and, by cutting this point, another set is constructed which approximates the feasible set. Note, that after finding the main iteration point, it is possible to update the approximating epigraph of the set by discarding any number of previously constructed cutting planes.

I. Zabotin · O. N. Shulgina · R. S. Yarullin (✉)
Kazan (Volga region) Federal University, Kazan, Russia

I. B. Badriev et al. (eds.), *Mesh Methods for Boundary-Value Problems and Applications*, Lecture Notes in Computational Science and Engineering 141,
https://doi.org/10.1007/978-3-030-87809-2_43

2 Problem Settings

Solve the problem

$$\min\{f(x) : x \in D\}, \tag{1}$$

where $D = \bigcap_{j \in J=\{1,\dots m\}} D_j$, the sets D_j, $j \in J$, are convex and closed from an n-dimensional Euclidian space R_n, it is assumed that $\text{int} D_j \neq \emptyset$ for each $j \in J$, and $f(x)$ is a convex function attained its minimum value f^* on the set D. Note that the equality $\text{int}\, D = \emptyset$ is admissible in (1).

Suppose $X^* = \{x \in D : f(x) = f^*\}$, $x^* \in X^*$, $\text{epi}(f, R_n) = \{(x, \gamma) \in R_{n+1} : x \in R_n, \gamma \geq f(x)\}$. Let $W(z, Q)$ be a set of normalized generalized support vectors at the point z for the set Q, $K = \{0, 1, \dots\}$.

3 Minimization Method

For solving problem (1) the proposed method constructs a sequence of points $\{x_k\}$, $k \in K$, as follows. Choose points $v^j \in \text{int}\, D_j$, $j \in J$, and $v \in \text{int}\,\text{epi}(f, R_n)$. Construct a convex bounded closed set $M_0 \subset R_n$ and a convex closed set $G_0 \subseteq R_{n+1}$ such that

$$x^* \in M_0, \quad \text{epi}(f, R_n) \subset G_0.$$

Define numbers $\bar{\gamma}$, ε_k, τ_k, $k \in K$ such that $\bar{\gamma} \leq f(x)$ for all $x \in M_0$, $\varepsilon_k > 0$, $\tau_k \geq 0$, $k \in K$,

$$\varepsilon_k \to 0, \quad k \to \infty, \quad \tau_k \to 0, \quad k \to \infty, \tag{2}$$

$1 \leq q < +\infty$. Assign $i = 0$, $k = 0$.

1. Find a solution $u_i = (y_i, \gamma_i)$, where $y_i \in R_n$, $\gamma_i \in R_1$, of the problem

$$\min\{\gamma : x \in M_k,\ (x, \gamma) \in G_i,\ \gamma \geq \bar{\gamma}\}. \tag{3}$$

If

$$y_i \in D, \quad f(y_i) = \gamma_i, \tag{4}$$

then $y_i \in X^*$, and the process of solving problem (1) is completed.

2. A point $\bar{u}_i \notin \operatorname{int}\operatorname{epi}(f, \mathrm{R_n})$ is found from the interval $(v, u_i]$ such that there exists a point

$$z_i \in \operatorname{epi}(f, \mathrm{R_n}) \tag{5}$$

which satisfies the inequality

$$\|u_i - z_i\| \leq q\|u_i - \bar{u}_i\|.$$

Choose a finite set $A_i \subset W(\bar{u}_i, \operatorname{epi}(f, \mathrm{R_n}))$.

3. If the inequality

$$\|u_i - \bar{u}_i\| > \varepsilon_k\|v - \bar{u}_i\| \tag{6}$$

is fulfilled, then

$$G_{i+1} = G_i \bigcap \{u \in \mathrm{R_{n+1}} : \langle a, u - \bar{u}_i\rangle \leq 0 \ \forall a \in A_i\}, \tag{7}$$

and go to Step 1 with incremented i. Otherwise, execute Step 4.

4. Choose a point $\tilde{y}_i \in M_k$ such that

$$f(\tilde{y}_i) \leq f(y_i) + \tau_k, \tag{8}$$

and assign

$$i_k = i, \quad x_k = \tilde{y}_i, \quad \sigma_k = \gamma_i,$$

$$G_{i+1} = G_{r_i} \bigcap \{u \in \mathrm{R_{n+1}} : \langle a, u - \bar{u}_i\rangle \leq 0 \ \forall a \in A_i\}, \tag{9}$$

where $0 \leq r_i \leq i$.

5. Form a set $J_k = \{j \in J : x_k \notin D_j\}$.
6. If $J_k = \emptyset$, then assign

$$M_{k+1} = M_k, \tag{10}$$

and go to Step 10. Otherwise, execute Step 7.

7. For each $j \in J_k$ the point $\bar{x}_k^j \notin \operatorname{int}D_j$ is chosen from the interval (v^j, x_k) such that there exists a point $z_k^j \in D_j$ satisfied the inequality

$$\|x_k - z_k^j\| \leq q\|x_k - \bar{x}_k^j\|. \tag{11}$$

8. Find an index $j_k \in J_k$ according to condition

$$\|x_k - \bar{x}_k^{j_k}\| = \max_{j \in J_k} \|x_k - \bar{x}_k^j\|. \tag{12}$$

9. Choose a finite set $B_k \subset W(\bar{x}_k^{j_k}, D_{j_k})$, and assign

$$M_{k+1} = M_k \bigcap \{x \in \mathrm{R_n} : \langle b, x - \bar{x}_k^{j_k} \rangle \le 0 \ \forall b \in B_k\}. \tag{13}$$

10. For all $j \in J \setminus J_k$ assign

$$z_k^j = \bar{x}_k^j = x_k. \tag{14}$$

Increase the values of i and k by one, and go to Step 1.

First, let's comment parameters settings which are installed at the initial step of the developed method.

Remark 1 It is natural to define M_0, G_0 as polyhedral sets. Then for finding auxiliary points u_i the linear programming problems will be solved. Moreover, if D is a polyhedron, then it is useful to put $M_0 = D$. In this case, for each $k \in K$ the equation $J_k = \emptyset$ is valid. Therefore, according to Step 6 we have $M_k = M_0$, $k \in K$, and cutting planes won't be constructed to approximate the constraint region D.

Remark 2 It is not difficult to obtain the point v. Namely, if we put $v = (b, f(b) + \tau)$, where $b \in \mathrm{R_n}$, $\tau > 0$, then according to definition of epi $(f, \mathrm{R_n})$ we get $v \in$ int epi $(f, \mathrm{R_n})$. Note that if int $D \neq \emptyset$ and there is some point $w \in$ int D, then it is conveniently to assign $v^j = w$ for all $j \in J$.

Remark 3 The number $\bar{\gamma}$ can be selected, for example, as a solution of the minimization problem

$$\gamma \to \min_{\gamma \in \mathrm{R_1}},$$

$$\langle c, x \rangle - \gamma \le \langle c, u \rangle - f(u),$$

$$x \in M_0,$$

where $u \in \mathrm{R_n}$, c is a subgradient of the function $f(x)$ at the point u. If problem (3) has a solution under $i = 0$, $k = 0$ without constraints, then the parameter $\bar{\gamma}$ can be considered as a large negative number.

The points x_k, $k \in K$, are constructed by the proposed method with approximating the epigraph epi $(f, \mathrm{R_n})$ by the sets G_i, $i \in K$, and with approximating the feasible set D by the sets M_k, $k \in K$, according to two stages. At the first stage the set G_{i_k} is constructed under the fixed set M_k with enough good approximation of the epigraph epi $(f, \mathrm{R_n})$. At the second stage, if it is necessary the set G_{i_k} is updated, the point x_k is found, and using this point the next approximating set M_{k+1} is constructed.

The above updates of the sets, which approximate the epigraph, can be performed in the iterations $i = i_k$ as follows. When inequality (6) is fulfilled, it is assumed that the quality of the epigraph approximation is considered unsatisfactory, and G_{i+1} is built on the basis of G_i. If we have

$$\|u_i - \bar{u}_i\| \leq \varepsilon_k \|v - \bar{u}_i\|, \tag{15}$$

then the main iteration point x_k is fixed, and the set G_{i_k+1} is constructed in accordance with (9) on the basis of any set $G_0, \ldots, G_{i_k}$. In case $r_{i_k} < i_k$ the cutting planes are dropped. Below it will be proved that for each $k \in K$ inequality (6) holds only for a finite number of numbers $i \in K$. This means that for each $k \in K$ the number i_k will be fixed, i.e. it will be possible to update the set G_{i_k+1}, and, in addition, the point x_k will be constructed.

Remark 4 We can put $r_{i_k} = i_k$ at Step 4 of the proposed method. In this case, the approximating set G_{i_k+1} is constructed on the basis of G_{i_k} without discarding cutting planes, and equality (7) is valid for each $i \in K$.

Remark 5 The selection of points $\bar{u}_i, \bar{x}_k^j$ is valid at Steps 2, 7. In particular, they can be chosen as boundary points of the sets epi$(f, \mathrm{R_n})$, D_j respectively. At Step 4 it is possible to put, for example, $x_k = \tilde{y}_{i_k} = y_{i_k}$.

According to technique [16], taking into account conditions of choosing the sets M_0, G_0 and the approach of constructing cutting planes, it is easy to prove on the basis of induction the following

Lemma 1 *The point* (x^*, f^*) *satisfies constraints of problem (3) for all* $i \in K$, $k \in K$.

Note that on the basis of Lemma 1 the estimation holds

$$\gamma_i \leq f^* \tag{16}$$

for the solution $u_i = (y_i, \gamma_i)$ of problem (3) under any $i \in K, k \in K$.

Now let's prove the stopping criterion which is represented at Step 1 of the developed method.

Theorem 1 *Suppose that expressions (4) are fulfilled. Then* y_i *is a solution of problem (1).*

Proof Since the point (x^*, f^*) is a admissible solution of problem (3) according to Lemma 1, and $u_i = (y_i, \gamma_i)$ is a minimum point of this auxiliary problem by construction, then from (4), (16) we obtain $f(y_i) \leq f^*$. But in accordance with conditions of the theorem $y_i \in D$, consequently, $f(y_i) \geq f^*$. Thus, $f(y_i) = f^*$, the statement of the theorem is proved. □

Further, let's research properties of the sequence $\{u_i\}$, $i \in K$. The following lemma is proved in [20].

Lemma 2 *Let $U \subset R_n$ be a convex set, L be its carrier subspace, Q be a bounded set defined in the affine shell of the set U, and $Q \cap \mathrm{ri}\, U = \emptyset$, where $\mathrm{ri}\, U$ is a relative interior of the set U. If the point $u \in R_n$ is determined according to $u \in \mathrm{ri}\, U$, then there exists a number $\delta > 0$ such that $z \in Q \setminus \mathrm{ri}\, U$ and the inequality $\langle a, u - z\rangle \leq -\delta$ is fulfilled for all $a \in L \bigcap W(z, U)$.*

Lemma 3 *The sequence $\{u_i\}$, $i \in K$, is bounded.*

Proof Since $u_i = (y_i, \gamma_i)$ is a solution of problem (3), and according to Steps 6, 9 of the proposed method the inclusion $M_{k+1} \subset M_k$ is defined, then we have $y_i \in M_0$ for all $i \in K$, $k \in K$. Consequently, taking into account boundness of the set M_0 there exists a number $\rho > 0$ such that

$$\|y_i\| \leq \rho \quad \forall i \in K. \tag{17}$$

Further, from (3), (16) it follows that $\bar{\gamma} \leq \gamma_i \leq f^*$ for all $i \in K$, $k \in K$. Therefore, the expression holds

$$0 \leq |\gamma_i - \bar{\gamma}| \leq |f^* - \bar{\gamma}| \quad \forall i \in K.$$

Hence and from (17) we get

$$\|u_i\| \leq \|y_i\| + |\gamma_i \pm \bar{\gamma}| \leq \rho + |f^* - \bar{\gamma}| + |\bar{\gamma}| \quad \forall i \in K.$$

The lemma is proved. □

Lemma 4 *Suppose that the sequence $\{u_i\}$, $i \in K$, is constructed by the way that the sets G_{i+1} have form (7) at Steps 3, 4 of the proposed method. Then*

$$\lim_{i \in K} \|\bar{u}_i - u_i\| = 0. \tag{18}$$

Proof Assume that the statement is not correct. Then there exists a subsequence $\{u_i\}$, $i \in K' \subset K$ such that

$$\|\bar{u}_i - u_i\| \geq \Delta > 0 \quad \forall i \in K'. \tag{19}$$

Select a convergence subsequence $\{u_i\}$, $i \in K'' \subset K'$, from the sequence $\{u_i\}$, $i \in K'$. Let i, $p_i \in K''$ be numbers such that $p_i > i$. In view of (7) we have $G_{p_i} \subset G_i$. But $u_{p_i} \in G_{p_i}$, and, more over, $A_i \subset W(\bar{u}_i, G_{p_i})$. Consequently,

$$\langle a, u_{p_i} - \bar{u}_i\rangle \leq 0 \quad \forall a \in A_i.$$

By construction for each $l \in K$ we have

$$\bar{u}_l = u_l + \alpha_l(v - u_l), \tag{20}$$

where $\alpha_l \in (0, 1)$. Hence and from the last inequality it follows that

$$\langle a, u_i - u_{p_i} \rangle \geq \alpha_i \langle a, u_i - v \rangle \quad \forall a \in A_i.$$

Since $v \in \text{int epi}\,(f, \mathrm{R_n})$, then according to Lemma 2 there exists a number $\sigma > 0$ such that for each $l \in K$ we get

$$\langle a, v - u_l \rangle \leq -\sigma \quad \forall a \in A_l.$$

In this regard,

$$\|u_i - u_{p_i}\| \geq \langle a, u_i - u_{p_i} \rangle \geq \alpha_i \sigma \quad \forall a \in A_i.$$

Hence and from convergence of the sequence $\{u_i\}$, $i \in K''$, we obtain $\alpha_i \to 0$, $i \in K''$. Consequently, from equality (20), where $l = i$, it follows that $\|\bar{u}_i - u_i\| \to 0$, $i \in K''$. This limit relation contradicts (19). The lemma is proved. □

Lemma 5 *Let $\{u_i\}$, $i \in K$, may be constructed by the proposed method. Then for each $k \in K$ there exists a number $i = i_k$ such that inequality (15) is fulfilled.*

Proof Note that the sequence $\{\bar{u}_i\}$, $i \in K$, corresponds to the sequence $\{u_i\}$, $i \in K$. Since $v \in \text{int epi}(f, \mathrm{R_n})$ by construction, and $\bar{u}_i \notin \text{int epi}(f, \mathrm{R_n})$ for all $i \in K$, then there exists a constant $\rho > 0$ such that

$$\|v - \bar{u}_i\| \geq \rho \quad \forall i \in K.$$

(1) Suppose $k = 0$. If inequality (15) is determined for $k = 0, i = 0$, then according to Step 4 of the method we get $i_0 = 0$, $x_0 = \tilde{y}_{i_0} = \tilde{y}_0$, where $\tilde{y}_0$ satisfies condition (8). Therefore, assume that $\|u_0 - \bar{u}_0\| > \varepsilon_0 \|v - \bar{u}_0\|$. Let's show that there exists a number $i = i_0 > 0$ which satisfies the inequality

$$\|u_{i_0} - \bar{u}_{i_0}\| \leq \varepsilon_0 \|v - \bar{u}_{i_0}\|.$$

Assume the inverse, i.e.

$$\|u_i - \bar{u}_i\| > \varepsilon_0 \|v - \bar{u}_i\| \geq \varepsilon_0 \rho, \quad \forall i \in K, \quad i > 0. \tag{21}$$

In this case, according to Step 3 of the method the set G_{i+1} has form (7) for each $i > 0$. Consequently, in view of 4 we have $\lim\limits_{i \in K} \|\bar{u}_i - u_i\| = 0$. Hence and from (21) we obtain the contradictory statement $0 < \varepsilon_0 \rho \leq 0$.

(2) Now suppose that the inequality (15) is fulfilled under some $k \geq 0$, i.e. $x_k = \tilde{y}_{i_k}$, where $\tilde{y}_{i_k}$ is chosen according to (8). Let's show that there exists a number $i_{k+1} > i_k$ such that

$$\|u_{i_{k+1}} - \bar{u}_{i_{k+1}}\| \leq \varepsilon_{k+1}\|v - \bar{u}_{i_{k+1}}\|.$$

Assume the inverse, i.e.

$$\|u_i - \bar{u}_i\| > \varepsilon_{k+1}\|v - \bar{u}_i\| \geq \varepsilon_{k+1}\rho \quad \forall i \in K, \quad i > i_k. \tag{22}$$

Consequently, according to the proposed method the sets G_{i+1} are given in form (7) for all $i > i_k$, and limit equality (18) holds by Lemma 4. Hence and from (22) we get the contradictory expression $0 < \varepsilon_{k+1}\rho \leq 0$. Thus, we have shown the existence of a number i_{k+1} which satisfied condition (15). The lemma is proved. □

Lemma 6 *Suppose that $\{x_k\}$, $k \in K' \subset K$, is a convergence subsequence of the sequence $\{x_k\}$, $k \in K$, and $\bar{x}$ is its limit point. Then we get the inclusion*

$$\bar{x} \in D. \tag{23}$$

Proof If we propose that the inclusion $x_k \in D$ is fulfilled for an infinite count of numbers $k \in K'$, then according to closeness of the set D statement (23) is obvious. Therefore, let's assume that $x_k \notin D$ for all numbers $k \in K'$ starting from the number $N \in K'$.

Let $l \in J$ be an index such that the number j_k satisfies condition (12) and equals to l for an infinite count of numbers $k \in K'$, $k \geq N$.

Put

$$K_l = \{k \in K' : j_k = l, \ k \geq N\},$$

and, firstly, let's prove the equality

$$\lim_{k \in K_l} \|\bar{x}_k^l - x_k\| = 0, \tag{24}$$

and taking into account the fact that the sequences $\{\bar{x}_k^j\}$, $\{z_k^j\}$, $k \in K_l$ are constructed for each $j \in J$ with the sequence $\{x_k\}$, $k \in K_l$.

Note that for all $k \in K$ and $j \in J$ we have

$$\bar{x}_k^j = x_k + \gamma_k^j(v^j - x_k), \tag{25}$$

by construction, where $\gamma_k^j \in [0, 1)$, moreover, $\gamma_k^l > 0$ for all $k \in K_l$.

Let's fix a number $p_k \in K_l$ for arbitrary $k \in K_l$ such that $p_k > k$. According to (10), (13) the inclusion $M_{p_k} \subset M_k$ is fulfilled. Moreover, in view of (3) $x_{p_k} \in$

M_{p_k}, and any element of the set B_k is generalized support ones for the set M_{p_k} at the point $\bar{x}_k^l$ too. Consequently,

$$\langle a, x_{p_k} - \bar{x}_k^l \rangle \le 0 \quad \forall a \in B_k.$$

Hence taking into account (25) under $j = l$ we have

$$\langle a, x_k - x_{p_k} \rangle \ge \gamma_k^l \langle a, x_k - v^l \rangle$$

for all $a \in B_k$.

According to Lemma 2 there exists a number $\delta_l > 0$ such that $\langle a, x_k - v^l \rangle \ge \delta_l$ for all $k \in K_l, \ a \in B_k$. Therefore, $\langle a, x_k - x_{p_k} \rangle \ge \gamma_k^l \delta_l$ for all $a \in B_k$, and since $\|a\| = 1$ for all $a \in B_k$, then

$$\|x_k - x_{p_k}\| \ge \gamma_k^l \delta_l \quad \forall k, p_k \in K_l, \quad p_k > k, \quad k \ge N. \tag{26}$$

Since the sequence $\{x_k\}$, $k \in K_l$, is convergence, then in accordance with (26) we get $\gamma_k^l \to 0, \ k \to \infty, \ k \in K_l$. Therefore, from (25) under $j = l$ taking into account boundness of the sequence $\{\|v^l - x_k\|\}$, $k \in K_l$, it follows equality (24).

Further, in view of condition (12) for all $k \in K_l$ the inequalities

$$\|\bar{x}_k^l - x_k\| \ge \|\bar{x}_k^j - x_k\|, \ j \in J_k$$

are valid. Moreover, according to Step 10 of the method we have $\|\bar{x}_k^j - x_k\| = 0$ for all $j \in J \setminus J_k, \ k \in K_l$. Consequently,

$$\|\bar{x}_k^l - x_k\| \ge \|\bar{x}_k^j - x_k\|$$

for any $k \in K_l, \ j \in J$. Then in view of (24) we get

$$\lim_{k \in K_l} \|\bar{x}_k^j - x_k\| = 0 \quad \forall j \in J. \tag{27}$$

In accordance with Steps 6, 10 of the algorithm for each $k \in K_l$ and $j \in J$ the point z_k^j either equals to x_k or satisfies condition (11). Since the sequence $\{x_k\}, \ k \in K_l$, is bounded, then taking into account (25) it follows that sequences $\{z_k^j\}, \ k \in K_l$, are bounded for all $j \in J$. Now for each $j \in J$ let's select the convergence subsequence $\{z_k^j\}$, $k \in K_l^j \subset K_l$, from the sequence $\{z_k^j\}$, $k \in K_l$, and let w_j be its limit point. Note that $w_j \in D_j$, $j \in J$, in view of closeness of the sets D_j. Hence for each $j \in J$ taking into account (11), (14) and (24) we obtain

$$\lim_{k \in K_l^j} \|x_k - z_k^j\| \le q \lim_{k \in K_l^j} \|x_k - \bar{x}_k^j\| = 0.$$

Consequently, $\bar{x} = w_j$ for all $j \in J$, i.e. $\bar{x} \in D$. The lemma is proved. □

Finally, let us formulate a convergence theorem for the proposed method.

Theorem 2 *For any limit point* $(\bar{x}, \bar{\sigma})$ *of the sequence* $\{(x_k, \sigma_k)\}$, $k \in K$, *constructed by the proposed method the expressions*

$$\bar{x} \in X^*, \quad \bar{\sigma} = f^*$$

are valid.

Proof Since according to Lemma 3 the sequence $\{u_i\}, i \in K$, is bounded, the points $z_i, \bar{u}_i, i \in K$, belong to the interval $(v, u_i]$, and the point $\tilde{y}_{i_k}$ satisfies condition (8) for each $i_k \in K$, then it is not difficult to show that the sequence $\{z_i\}, i \in K$, are bounded, and, moreover, taking into account (2) it is possible to prove that the sequences $\{(\tilde{y}_{i_k}, \gamma_{i_k})\}$, $\{(x_k, \sigma_k)\}$, $k \in K$, are bounded too.

Let $K' \subset K$ be a set of numbers such that the sequences $\{(x_k, \sigma_k)\}$, $\{(y_{i_k}, \sigma_k)\}$, $\{z_{i_k}\}, k \in K'$, are convergence, and $(\bar{x}, \bar{\sigma})$, $(\bar{y}, \bar{\sigma})$, $\bar{z}$ be their limit points respectively.

Since the set epi $(f, \mathrm{R_n})$ are closed, then

$$\bar{z} \in \text{epi}\,(f, \mathrm{R_n}). \tag{28}$$

Taking into account condition (2) of selection ε_k, boundedness of the sequence $\{\|v - \bar{u}_i\|\}$, $i \in K$, and the equalities $u_{i_k} = (y_{i_k}, \sigma_k)$ from the inequalities

$$\|u_{i_k} - z_{i_k}\| \le q\|\bar{u}_{i_k} - u_{i_k}\| \le q\varepsilon_k\|v - \bar{u}_{i_k}\|, \quad k \in K',$$

it follows that $(\bar{y}, \bar{\sigma}) = \bar{z} \in \text{epi}(f, \mathrm{R_n})$. Therefore, $f(\bar{y}) \le \bar{\sigma}$. But according to (8) we have $f(x_k) \le f(y_{i_k}) + \tau_k$, $k \in K$. Then taking into account (2) under $k \to \infty$, $k \in K'$, we obtain $f(\bar{x}) \le \bar{\sigma}$. And in view of $\sigma_k \le f^*$ we get $f(\bar{x}) \le f^*$.

On the other hand, $\bar{x} \in D$ by Lemma 6, i.e. $f(\bar{x}) \ge f^*$. Consequently, $f(\bar{x}) = f^*$ is valid. The theorem is proved. □

Acknowledgments This paper has been supported by the Kazan Federal University Strategic Academic Leadership Program ("PRIORITY-2030").

References

1. Balas, E.: An additive algorithm for solving linear programs with zero-one variables. J . ORSA. **13** (4), 517–546 (1965).
2. Bulatov, V.P.: Embedding Methods in Optimization Problems. Nauka, Novosibirsk (1977) [in Russian].
3. Bulatov, V.P., Khamisov O.V.: Cutting methods in En+1 for solving optimization problems in a class of functions. Comput. Math. Math. Phys. **47**, 1756–1767 (2007).
4. Demyanov, V.F., Vasilev, L.V.: Nondifferentiable Optimization. Mir, Moscow (1972) [in Russian].
5. Kelley, J.E.: The cutting-plane method for solving convex programs. J. Soc. Ind. Appl. Math. **8**, 703–712 (1960).

6. Kolokolov, A. A.: Regular partitions and cuts in integer programming. In: Discrete Analysis and Operations Research, Springer, Netherland, **355** 59–79 (1996).
7. Lemarechal, C., Nemirovskii, A., Nesterov Yu.: New variants of bundle methods. Mathematical Programming, **69**111–148 (1995).
8. Levitin, E.S., Polyak, B.T.: Constrained Minimization Methods. Zhurn. Vychisl. Matem. i Matem. Fiz. **6** (5), 787–823 (1966).
9. Nesterov, Yu. E.: Introductory Lectures on Convex Optimization. MTsMNO, Moscow (2010) [in Russian].
10. Nurminski, E.A.: The separating planes method with bounded memory for convex nonsmooth optimization. Vychisl. Metody Program. **7**, 133–137 (2006).
11. Polyak, B. T.: Introduction to optimization. Nauka, Moscow (1983) [in Russian].
12. Shulgina, O. N., Yarullin, R. S., Zabotin, I. Ya.: A Cutting Method with Approximation of a Constraint Region and an Epigraph for Solving Conditional Minimization Problems. Lobachevskii J Math. **39** (6), 847–854 (2018).
13. Topkis, D. M.: Cutting plane methods without nested constraint sets. Operat . Res., 3 (1970).
14. Veinot, A.F.: The supporting hyperplane method for unimodal programming. Operat . Res. **15** (1), 147–152 (1967).
15. Vorontsova, E.A.: A Projective Separating Plane Method with Additional Clipping for Non-Smooth Optimization. WSEAS Transactions on Mathematics. **13**, 115–121 (2014).
16. Zabotin, I.Ya., Yarullin, R.S.: A Cutting Plane Algorithm with an Approximation of an Epigraph. Uchen. Zap. Kazansk. Univ. Ser. Fiz.-Matem. Nauki. **144** (4), 48–54 (2013) [in Russian].
17. Zabotin, I. Ya., Kazaeva, K. E.: A variant of the penalty method with approximating an epigraph of auxiliary functions. Uchen. Zap. Kazansk. Univ. Ser. Fiz.-Matem. Nauki. **161** (2), 263 273 (2019).
18. Zabotin, I., Shulgina, O., Yarullin, R.: A minimization algorithm with approximation of an epigraph of the objective function and a constraint set. In: Proc DOOR 2016 CEUR-WS, September 19–23, Vladivostok, **1623** 321–324 (2016).
19. Zangwill, W. I.: Nonlinear Programming: A Unified Approach. Prentice-Hall, Englewood Cliffs (1969).
20. Zabotin, I. Ya.: Some Embedding-Cutting Algorithms for Mathematical Programming Problems. Izv. Irkutsk. Gos. Univ., Ser. Matem. **4**(2), 91–101 (2011) [in Russian].

Self-Consistent Model of Low Pressure Inductively Coupled RF Discharge

Viktor Zheltukhin, Aleksandr Shemakhin, Timur Terentev, and Ekaterina Samsonova

Abstract A new approach has been modeled for the description of low-pressure inductively coupled RF discharge. Within the framework of the investigated model, the system of differential equations is interpreted as an eigenvalue problem. The model takes into account the influence of electromagnetic fields and boundary conditions of various types. The developed approach makes it possible to obtain an internal self-consistent solution that requires a minimum number of input parameters.

1 Introduction

At the present time, it is difficult to imagine the study of complex physical phenomena without resorting to methods of mathematical modeling. Both experimental methods and mathematical modeling are equally effective tools for studying processes in RF plasmas [1–5], they complement each other.

Low-pressure radio-frequency discharges occur as a result of the action of an electromagnetic field with a frequency of $1.76 - 13.56$ MHz generated by either a solenoid or electrodes in a quartz tube of 1–5 cm in diameter at a plasma-forming gas pressure of $1.33 - 133$ Pa. The plasma generated by the discharge has the following properties: the degree of ionization $10^{-7} - 10^{-4}$, electron concentration $n_e = 10^{15} - 10^{19}$ m^{-3}, electron temperature $T_e = 1 - 4$ eV, temperature of atoms and ions in the center of the plasma $T_a = 0.2 - 0.3$ eV.

Taking into account the properties of a quasineutral plasma of an RF discharge at low pressure in inert gases, one can write a system of 15 nonlinear boundary-value

V. Zheltukhin (✉)
Kazan National Research Technological University, Kazan, Russia

A. Shemakhin · T. Terentev · E. Samsonova
Kazan Federal University, Kazan, Russia
e-mail: Terentevt@yandex.ru; samsonova.ek.s@yandex.ru

I. B. Badriev et al. (eds.), *Mesh Methods for Boundary-Value Problems and Applications*, Lecture Notes in Computational Science and Engineering 141,
https://doi.org/10.1007/978-3-030-87809-2_44

problems [5] in unknowns $H_L^2, E_L^2, H_C^2, E_C^2, \psi_H, \psi_E, \gamma_{H_L}, \gamma_{E_C}, n_e, T_e, n_a, T_a, \mathbf{v}$, which are the squared moduli of inductively part $\{\mathbf{H}_L, \mathbf{E}_L\}$, the squared moduli of capacitively part $\{\mathbf{H}_C, \mathbf{E}_C\}$, phases, and angular functions of the vectors of the magnetic and electric fields $\{\mathbf{H}, \mathbf{E}\}$, the concentration and temperature of electrons, the concentration and temperature of neutral atoms, and components of the plasma velocity vector, consequently. Here $\mathbf{H}_L = H_r\mathbf{i}_r + H_z\mathbf{i}_z, \mathbf{E}_L = E_\varphi\mathbf{i}_\varphi, \mathbf{H}_C = H_\varphi\mathbf{i}_\varphi, \mathbf{E}_C = E_r\mathbf{i}_r + E_z\mathbf{i}_z$, and $\mathbf{i}_r, \mathbf{i}_\varphi, \mathbf{i}_z$ are the unitary vectors of cylindrical coordinate system. Even taking into account the mathematical complexities of the system, the correctness of the solution depends on the adequate formulation of the boundary conditions for the set of equations. Unfortunately, there are currently no models that meet these requirements, the exception is work [6].

The purpose of this work is to find a self-consistent solution of the system of equations describing an inductively coupled RF (ICRF) discharge in a one-dimensional model at local approximation when diffusion and ionization coefficients depends from the reduced electric field (ratio E/N). In this case, the unknown parameters are the squared moduli of the strength of the magnetic and electric fields H_L^2, E_L^2, the electron density n_e.

The problem studied in this work is solved as an eigenproblem with boundary conditions of the third kind. Such approach reflects the most important property of the positive column of a gas-discharge plasma which is a self-adjusting, self-regulating system. It means that the electrons density and their energy (electron temperature) are automatically adjusted to changes in the energy introduced into the plasma, and in a nonlinear manner. Therefore, the existence of a solution to the eigenvalue problem is a condition for the existence of a steady-state of the discharge within the framework of the considered model.

2 Formulation of the Problem

Within the framework of this problem, the discharge chamber is interpreted as an infinite cylindrical tube with the infinite solenoid. Based on the shape of the discharge chamber, calculations will be performed in the polar coordinate system. It is assumed that the ionization frequency ν_i and the ambipolar diffusion coefficient D_a are functions of the ratio $E(r)/N$. The dependence of the coefficients of the equations is based on the results of calculations using the Bolsig+ program [7, 8].

For the first time the boundary value problem

$$-D_a \frac{1}{r}\frac{d}{dr}\left(r\frac{dn_e}{dr}\right) = \nu_i n_e, \tag{1}$$

$$\left.\frac{dn_e}{dr}\right|_{r=0} = 0, \quad \left.n_e\right|_{r=R} = 0, \tag{2}$$

at D_a = const, n_e = const was solved by Schottky in 1924 [9]. Schottky have considered the problem (1), (2) as an eigenproblem, and have interpreted the minimal eigenvalue $\lambda_0 = \sqrt{\nu_i/D_a}R = 2.405$ as a characteristic diffusion length. The solution made it possible to estimate the characteristic value of the electric field strength required to sustain the discharge, and the radial distribution of the relative electron density.

In the one-dimensional ICRF discharge the electromagnetic field is equal $\mathbf{H}_L = \dot{H}\,\mathbf{i}_z$, $\mathbf{E}_L = \dot{E}\,\mathbf{i}_\varphi$ and Maxwell's equations take the form:

$$\frac{d\dot{H}}{dr} = (\sigma + i\epsilon_0\epsilon\omega)\dot{E}, \tag{3}$$

$$\frac{d\dot{E}}{dr} = -i\mu_0\epsilon\omega\dot{H}, \tag{4}$$

where

$$\dot{H} = H \cdot \exp(i\omega t + \psi_H), \tag{5}$$

$$\dot{E} = E \cdot \exp(i\omega t + \psi_E). \tag{6}$$

Here μ_0 is the magnetic constant, ϵ_0 is the electric constant, $\omega = 2\pi f$ is the circular frequency of the electromagnetic field, f is generator frequency, i is the imaginary unit, $i^2 = -1$,, t is time,

$$\sigma = \frac{n_e e^2 \nu_c}{m_e\left(\nu_c^2 + \omega^2\right)}, \quad \epsilon = 1 - \frac{n_e e^2 \omega}{\epsilon_0 m_e\left(\nu_c^2 + \omega^2\right)}, \tag{7}$$

m_e is the electron mass, e is the elementary charge, ν_c is frequency of the electron elastic collision with atoms and ions.

Applying the generalized Sobolev rearrangement [10, 11] to Maxwell's equations, we obtain

$$\frac{1}{r}\frac{d}{dr}\left(\frac{r}{\sigma}\frac{dH^2}{dr}\right) = 2\sigma E^2, \tag{8}$$

$$\frac{1}{r}\frac{d}{dr}\left(\frac{1}{r}\frac{d\left(r^2E^2\right)}{dr}\right) = 2\left(\mu_0\omega\right)^2 H^2, \tag{9}$$

with boundary conditions:

$$\left.\frac{dH^2}{dr}\right|_{r=0} = 0, \quad H^2(R) = H_R^2, \tag{10}$$

$$E(0) = 0, \quad \left.\frac{d}{dr}\left(r^2E^2\right)\right|_{r=R} = 2\mu_0\omega R^2\,|E|\,|H|\,. \tag{11}$$

If the equation describing the distribution of the electrons density is added to the first kind boundary condition at $r = R$ like (2), then the electron velocity

$$v_e = \frac{D_a}{n_e}\frac{dn_e}{dr},$$

either grows infinitely or must have a finite limit at r lim R. The latter means that

$$-\left(D_a\frac{dn_e}{dr}\right)\Big|_{r=R} = \alpha n_e(R), \tag{12}$$

where α is the constant which is specified by equality of electron and ion fluxes on the tube wall [12]. Thus, the third kind boundary conditions (12) for n_e is more correct than (2).

3 System of Equations as a Self-Consistent Eigenvalue Problem

The mathematical model consists of the following system of equations with the corresponding boundary conditions:

$$\frac{1}{r}\frac{d}{dr}\left(\frac{r}{\sigma}\frac{dH^2}{dr}\right) = 2\sigma E^2, \tag{13}$$

$$\frac{dH^2}{dr}\Big|_{r=0} = 0, \quad H^2\Big|_{r=R} = H_R^2, \tag{14}$$

$$\frac{1}{r}\frac{d}{dr}\left[\frac{1}{r}\frac{d}{dr}\left(r^2E^2\right)\right] = 2\left(\mu_0\omega\right)^2 H^2, \tag{15}$$

$$E\Big|_{r=0} = 0, \quad \frac{d}{dr}\left(r^2E^2\right)\Big|_{r=R} = 2\mu_0\omega R^2\left|E\right|\left|H\right|, \tag{16}$$

$$-\frac{1}{r}\frac{d}{dr}\left(rD_a\frac{dn_e}{dr}\right) = \nu_i n_e, \tag{17}$$

$$\left(D_a\frac{dn_e}{dr}\right)\Big|_{r=0} = 0, \quad -D_a\frac{dn_e}{dr}\Big|_{r=R} = \alpha n_e\Big|_{r=R}. \tag{18}$$

The connection of equations in the system is specified through the dependence of the equation coefficients.

Based on the form of the equation, we note that one of the solutions of the Eq. (17) with boundary conditions (18) is the identity zero, $n_e \equiv 0$. However, in accordance with the physical sense of the problem, only a nontrivial solution is of interest, which, moreover, must be non-negative, $n_e \geq 0$. Note that the solution of the problem (17), (18) is determined up to an arbitrary factor: any function of the form $n_e(r) = n_e^* \cdot \overline{n}_e(r)$ where n_e^* is an arbitrary factor and $\overline{n}_e(r)$ is some solution, will also be another solution. This fact means that the problem (17), (18) is an eigenproblem, and the electron density $n_e(r)$ is an eigenfunction.

But there is no a parameter in the primary formulation of the problem (17), (18), it describes a balance between processes generation and loss of charged particles. Therefore, it is necessary to carry out the mutual agreement the Eq. (17) with Eqs. (13) and (15). It is known from the theory of boundary value problems [13] that the generalized eigenvalue problem has nontrivial solutions at the certain discrete set of the parameter $\lambda_n, n = 0, 1, 2, \ldots,$ which are complex in the general case [14]. Moreover, there is the unique positive eigenfunction, which corresponds to the real least eigenvalue λ_0.

The electron diffusion equation (17) can be rewrite in the dimensionless form

$$-\frac{1}{\rho}\left(\rho\overline{D}\frac{d\overline{n}}{d\rho}\right) = \overline{\lambda}\overline{\nu}\cdot\overline{n}, \tag{19}$$

$$\left(\overline{D}\frac{d\overline{n}}{d\rho}\right)\bigg|_{\rho=0} = 0, \quad \overline{D}\frac{d\overline{n}}{d\rho}\bigg|_{\rho=1} = \overline{\alpha n}\bigg|_{\rho=1}. \tag{20}$$

where $\rho = r/R, \overline{D} = D_a/\max(D_a), \overline{n} = n_e/\max(n_e), \overline{\nu} = \nu_i/\max(\nu_i), \overline{\alpha} = \alpha/\max(D_a), \overline{\lambda} = \max(\nu_i) R^2/\max(D_a)$.

As far as $\overline{n} \geq 0 \Rightarrow \overline{\lambda} \equiv \overline{\lambda}_0 = \min\{\overline{\lambda}_k\}, k = 0, 1, 2, \ldots$.

It is known that the minimum eigenvalue is the infimum of the Rayleigh quotient on the set of all possible functions that are not identically equal to 0, and is attained on the eigenfunction corresponding to the least eigenvalue [15, 16].

$$\overline{\lambda}_0[E^*] = \frac{\displaystyle\int_0^1 \overline{D}[E^*\cdot(\overline{E}/p)]\left(\frac{dn}{d\rho}\right)^2\rho d\rho}{\displaystyle\int_0^1 \overline{\nu}[E^*\cdot(\overline{E}/p)]\,n^2\rho d\rho}. \tag{21}$$

Go back to the primary equation (17) we obtain that the problem (17), (18) has the nonnegative nontrivial solution if and only if the condition

$$\lambda_0(E^*) = 1, \tag{22}$$

as has been fulfilled, where λ_0 is calculated in the same Rayleigh quotient as (21).

4 Results of Calculation

To calculate the task, a program was written in the Python programming language. Calculations were carried out with the following parameters: $p = 133$ Pa, $f = 1.76$ MHz, $R = 12$ mm, $n_e = 10^{18}$ m^{-3}, $\nu_c = 10^8$ Hz.

The system of boundary value problem (13)–(18) was solved by a finite-difference scheme using an iterative method.

The block diagram of the program is shown in the Fig. 1. The calculation results are presented in graphs Figs. 2, 3, 4.

The distribution of the electron density along the tube radius is shown in Fig. 2. At the center of the tube, the concentration has a maximum.

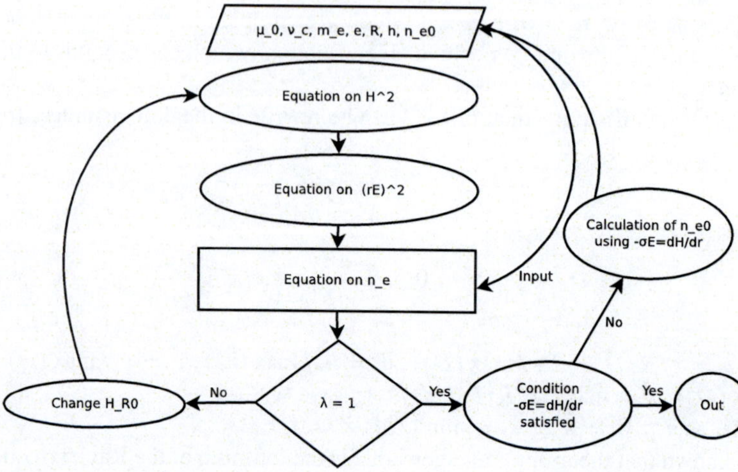

Fig. 1 Block diagram of the problem calculation

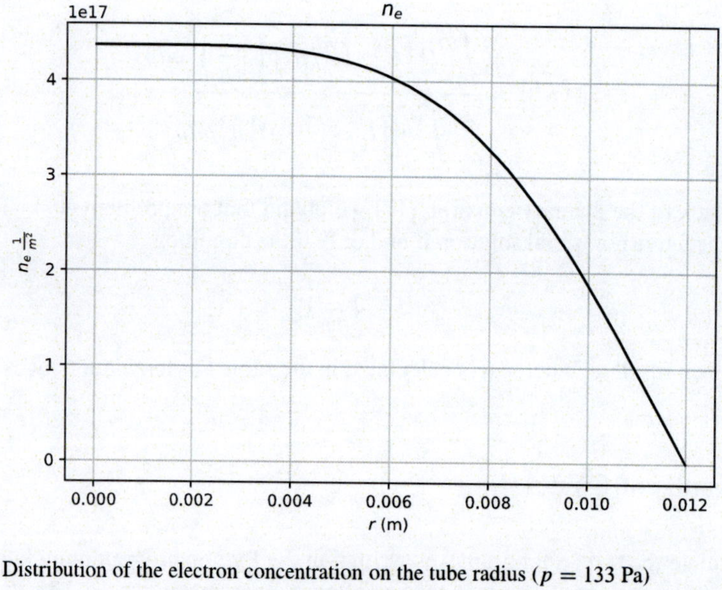

Fig. 2 Distribution of the electron concentration on the tube radius ($p = 133$ Pa)

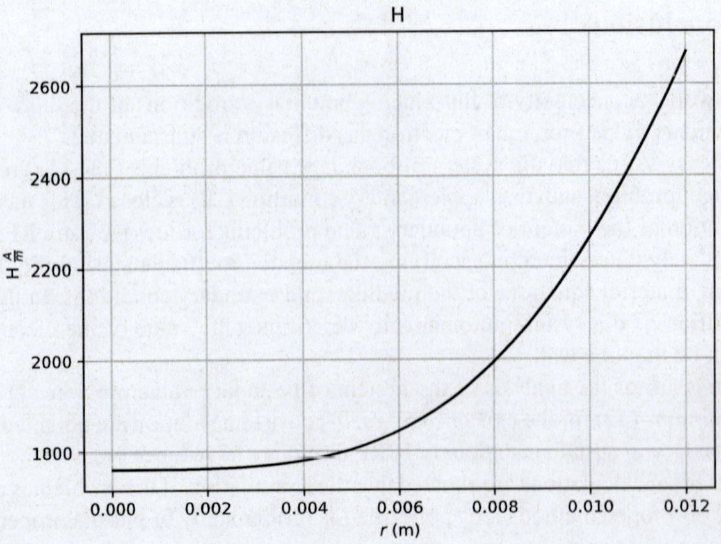

Fig. 3 Graph of the dependence of the magnetic field strength along the RF radius plasmatron. The dots indicate the experimental data. $p = 133$ Pa, $f = 1.76 \cdot 10^6$ Hz

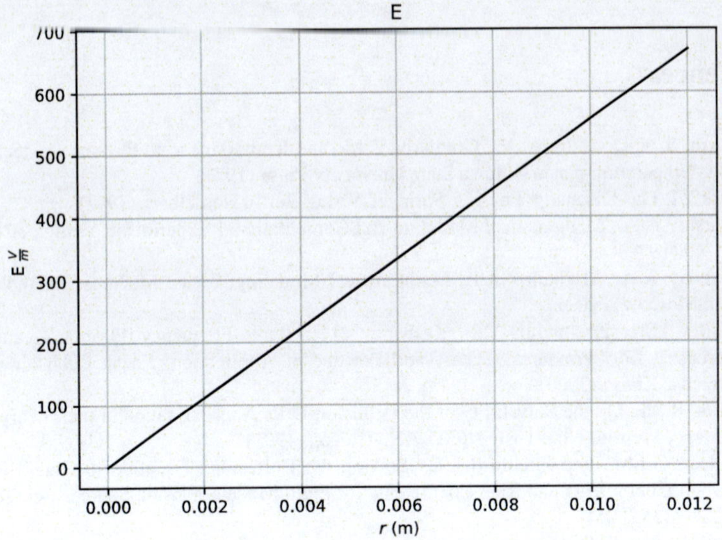

Fig. 4 The graph of the dependence of the modulus of the electric field strength on the tube radius ($p = 133$ Pa)

In Fig. 3 the r-dependence of the modulus of the magnetic field strength is shown. Calculations show that the modulus of the electric field strength increases linearly along the tube radius to values 668 V/m (Fig. 4).

5 Conclusions

In this work, the necessity of introducing boundary conditions of the third kind for the boundary value problem of electron gas diffusion is substantiated.

It is easy to see that the system of boundary value problems (13)–(18) reviewed as an eigenproblem with the supplementary condition (22) is closed. This means that the solution of the system of boundary value problems for low-pressure RF plasma is completely determined by specifying the transfer coefficients using the Bolsig+ program, material equations of the medium, and boundary conditions. In this case, the solution of the system automatically determines the value of the electric field strength on the tube wall.

As a result of the analysis of the system of boundary value problems (17), (18) the conditions (22) for the existence of a self-consistent solution are obtained, which are necessary to sustain a stationary ICRF discharge of reduced pressure.

Numerical calculations are showed that the formulation of the problem is correct, and the developed methods are applicable for further study of plasma properties.

Acknowledgments Work was supported by Russian Science Foundation, project No. 19-71-10055.

References

1. Dresvin, S. V. Donskoi, A. V. , Goldfarb, V. M., and Klubnikin, V. S. Physics and technology of low-temperature plasmas. Iowa State University Press (1977)
2. Raizer Y.P. Gas Discharge Physics. Springer-Verlag Berlin Heidelberg (1991)
3. Samarskiy A.A.: Mathematical Modeling and Computational Experiment, Vestnik AN SSSR, **5**, 38–49 (1979)
4. Samarskiy A.A., Mikhailov A.P.: Mathematical Modeling: Ideas. Methods. Examples, Fizmatlit, Moscow (2002)
5. Abdullin I.Sh., Zheltukhin V.S., Kashapov N.F.: Radio-Frequency Plasma Treatment of Materials at Low Pressures. Theory and Practice of Application. Kazan Publishing House University, Kazan (2000)
6. Islamov R.Sh.: On the Solvability of the Diffusion-Drift Approximation in the Theory of Gas Discharge. Preprint No. 91-81. NICTLAN, Troitsk, (1991)
7. Hagelaar, G.J.M., and Pitchford L.C.: Solving the Boltzmann Equation to Obtain Electron Transport Coefficients and Rate Coefficients for Fluid Models. Plasma Sources Sci. Technol., **14**, 722–733 (2005)
8. BOLSIG+, Ver. 03/2016, https://www.bolsig.laplace.univ-tlse.fr/
9. Schottky W.: Diffusionstheorie der Positiven Säule. Phys. Zheitschr., **XXV**, 635–640 (1924).
10. Gruzdev V.A., Rovinsky R.E., Sobolev A.P.: Approximate Solution of a Stationary Radio-Frequency Discharge in a Closed Volume. Zh. Prikl. mechanics and tech. fiz., **3**, 197–199 (1968)
11. Abdullin I.Sh., Zheltukhin V.S.: Mathematical Modeling of Plasma of an Inductive Diffuse Discharge. Bull. of Siberian Branch of the USSR Academy of Sci. Ser.Technical Sci., **16**(3), 106–109 (1985)

12. Zheltukhin, V.S., Solov'ev, S.I., Solov'ev, P.S., Chebakova, V.Yu., and Sidorov A.M. Third type boundary conditions for steady state ambipolar diffusion equation. IOP Conf. Series: Materials Science and Engineering, **158**, 012102 (2016)
13. Ladyzhenskaya O.A., Uraltseva N.N. : Linear and Quasilinear Equations of Elliptic Type. Nauka, Moscow, (1973)
14. Aslanyan A.G., Lidskiy V.B.: On the Spectrum of an Elliptic Equation. Matem. Notes. **7** (7), 495–502 (1970)
15. Dunford, N., and Schwartz J.T.: Linear Operators. Part II. Spectral Theory. Self Adjoint Operators in Hilbert Space. Intersci. Publ., N.Y., London (1963)
16. Parlett B.N.: The Symmetric Eigenvalue Problem. Prentice-Hall, Inc., Englewood Cliffs, N.J., 07632 (1980)

Robust One/Two-Grid Solver for Black-Box Software in the Computational Continuum Mechanics

Weixing Zhou and Sergey Martynenko

Abstract We present and analyse a linear one/two-grid algorithm for solving boundary value problems in black-box manner on globally/locally structured and unstructured grids. The key ingredient of the new algorithm is Robust Multigrid Technique used on the auxiliary structured grid to minimize the number of problem-dependent components. The theoretical analysis predicts h-independent convergence of the solver with close-to-optimal algorithmic complexity. In addition, comparison with a basic one-grid algorithm (Vanka-type smoother) gives all extra problem-dependent components of the proposed approach.

1 Introduction

Scientific and engineer software development can be approached in various ways. The most promising approach is to develop autonomous (black-box) codes, for which the user has to specify only the physical problem. We define software to be black-box if it does not require any additional input from the user apart from the physical problem specification consisting of the domain geometry, boundary and initial conditions, the enumeration of equations to be solved (heat conductivity equation, Navier–Stokes equations, Maxwell equations, etc.) and mediums. The

W. Zhou
Harbin Institute of Technology, Harbin, China
e-mail: zhouweixing@hit.edu.cn

S. Martynenko (✉)
Bauman Moscow State Technical University, Moscow, Russian Federation

Institute of Problems of Chemical Physics of Russian Academy of Sciences Academician Semenov, Chernogolovka, Russian Federation

Central Institute of Aviation Motors CIAM, Moscow, Russian Federation
e-mail: Martynenko@icp.ac.ru; Martynenko@ciam.ru

I. B. Badriev et al. (eds.), *Mesh Methods for Boundary-Value Problems and Applications*, Lecture Notes in Computational Science and Engineering 141,
https://doi.org/10.1007/978-3-030-87809-2_45

user does not need to know anything about numerical methods, or high-performance and parallel computing [1].

A promising and challenging trend in numerical simulation and scientific computing is development of new computational techniques for black-box software. In [2] it has been formulated the basic requirements for the numerical methods for black-box software:

- the least number of the problem-dependent components;
- close-to-optimal algorithmic complexity in a wide range of the problem parameters;
- the highest possible parallel efficiency (speed-up over the fastest sequential algorithm);
- high adaptability (opportunity to flexibly change the method and order of the government equation approximation);
- minimal memory requirement.

Robust multigrid technique (RMT) has been proposed and developed for solving (initial-)boundary value problems of the continuum mechanics in black-box software [3]. In order to minimize the number of the problem-dependent components, RMT is based on the application of the essential multigrid principle[1] in one-grid algorithm [1–3].

Previously, RMT has been used only to solve the boundary value problems of the continuum mechanics on globally structured grids [4–6]. Goal of this article is to analysis convergence and algorithmic complexity of the technique on locally structured and unstructured grids [7–9].

2 Computational Grids and Basic One-Grid Algorithm

Let N_{G_0} is the number of grid points of grid G_0. Assume that the original grid G_0 generates subgrids $G_i \in G_0$ so that

$$G_0 = \bigcup_{i=1}^{I} G_i \quad \text{and} \quad G_n \cap G_m = \varnothing, \quad n \neq m.$$

N_{G_i} are the number of grid points of subgrids G_i.

[1] The essential multigrid principle is to approximate the smooth (long wavelength) part of the error on coarser grids. The non-smooth or rough part is reduced with a small number (independent of h) of iterations with a basic iterative method on the fine grid [10].

All subgrids G_i, $i = 1, 2, \ldots, I$ form a grid level

$$\sum_{i=1}^{I} N_{G_i} = N_{G_0},$$

but the original grid G_0 forms zero level. These subgrids G_i may be generated in different ways, but we will use those subgrids that allow to minimize discretization error of the boundary value problem.

Definition 1 Computational grid G_1^0 is called globally structured, if its subgrids G_k^l, $l = 1, 2, \ldots, L^+$, $k = 1, 2, \ldots, K$ have the following properties:

Property 1. Each grid G_k^l ($l \neq L^+$, where L^+ is the coarsest level) can be represented as a union of K coarser subgrids of level $l + 1$. As a consequence, the original grid G_1^0 can be represented as a union of all subgrids of the same level l:

$$G_1^0 = \bigcup_{k=1}^{K} G_k^l, \quad l = 0, \ldots, L^+.$$

Property 2. Subgrids of the level l have no common points, i.e.

$$G_n^l \cap G_m^l = \varnothing, \quad n \neq m, \quad l = 1, \ldots, L^+.$$

Property 3. Each finite volume on the subgrids G_k^l can be represented as union of K finite volumes on the original grid G_1^0.

The set of all computational grids will be called a multigrid structure.

Definition 2 Assume that a domain Ω can be represented as union of subdomains Ω_m. Structured grids $G_0^{(m)}$ can be generated in each subdomain Ω_m, but the composite grid $G_0 = \bigcup_{m=1}^{M} G_0^{(m)}$ is not a globally structured grid. Such grids will be referred as a locally structured or a multi-block grids.

Let $\Omega \in \mathbb{R}^3$ is an arbitrary bounded domain with a boundary $\partial\Omega$. Our problem is to found a solution $\boldsymbol{u}(\boldsymbol{x}) = \left(u^{(1)}(\boldsymbol{x}), u^{(2)}(\boldsymbol{x}), \ldots, u^{(N_M)}(\boldsymbol{x})\right)^T$ of 3D boundary value problem for a system of linear partial differential equations

$$\sum_{j=1}^{N_M} \mathcal{L}_{\Omega}^{(ij)} u^{(j)}(\boldsymbol{x}) = f_{\Omega}^{(i)}(\boldsymbol{x}), \quad \boldsymbol{x} \in \Omega, \qquad i = 1, 2, \ldots, N_M, \tag{1a}$$

$$\sum_{j=1}^{N_M} \mathcal{L}_{\partial\Omega}^{(ij)} u^{(j)}(\boldsymbol{x}) = f_{\partial\Omega}^{(i)}(\boldsymbol{x}), \quad \boldsymbol{x} \in \partial\Omega, \qquad i = 1, 2, \ldots, N_B, \tag{1b}$$

Here $\boldsymbol{x} = (x_1, x_2, x_3)^T$, $\mathcal{L}_{\Omega}^{(ij)}$ and $\mathcal{L}_{\partial\Omega}^{(ij)}$ are linear differential operators and $f_{\Omega}^{(i)}$ and $f_{\partial\Omega}^{(i)}$ are known functions in the domain Ω and its boundary $\partial\Omega$ $(N_M \leq N_B)$. For brevity we will denote this boundary value problem as $\mathcal{L}\boldsymbol{u} = \boldsymbol{f}$ and assume that it has a unique solution $\boldsymbol{u} = \mathcal{L}^{-1}\boldsymbol{f}$.

First, we formulate a basic one-grid algorithm (Vanka-type smoother) for coupled solution of (1). Let a globally structured grid G_0 with N_{G_0} vertices has been generated for approximation of (1). Discrete analogue of (1) can be written in the matrix form

$$A_0\boldsymbol{\varphi}_0 = \boldsymbol{b}_0,$$

where the vector of unknowns $\boldsymbol{\varphi}_0$ approximates the vector $\boldsymbol{u}(\boldsymbol{x})$. Basic one-grid algorithm (Vanka-type smoother [11]) becomes

$$W_V\left(\boldsymbol{\varphi}_0^{(\nu+1)} - \boldsymbol{\varphi}_0^{(\nu)}\right) = \boldsymbol{b}_0 - A_0\boldsymbol{\varphi}_0^{(\nu)}, \quad \nu = 0, 1, 2, \ldots, \tag{2}$$

where W_V is a splitting matrix of the Vanka iterations. The basic one-grid algorithm (Vanka-type smoother [11, 12]) can be rewritten as

$$\boldsymbol{\varphi}_0^{(\nu+1)} = \left(I - W_V^{-1}A_0\right)\boldsymbol{\varphi}_0^{(\nu)} + W_V^{-1}\boldsymbol{b}_0. \tag{3}$$

An iterative method (3) whose related system $(I - S_V)\boldsymbol{\varphi}_0 = W_V^{-1}\boldsymbol{b}_0$ has a unique solution $\boldsymbol{\varphi}_0$ which is the same as the solution of $A_0^{-1}\boldsymbol{b}_0$ is said to be completely consistent [13]. It leads to

$$\boldsymbol{\varphi}_0 - \boldsymbol{\varphi}_0^{(\nu)} = S_V^{\nu}\left(\boldsymbol{\varphi}_0 - \boldsymbol{\varphi}_0^{(0)}\right),$$

where $S_V = I - W_V^{-1}A_0$ is real Vanka iteration matrix for the basic one-grid method.

Algorithmic complexity of iterations (2) is $\mathcal{W} = O\left(n_b^{-2}(N_{G_0}N_M)^{3+k/d}\right)$ arithmetic operations, where $1 \ll n_b \leqslant N_{G_0}N_M$, $d = 2, 3$, n_b is the number of unknown blocks, k is a constant, N_{G_0} is the number of points of the grid G_0, N_M is the number of equations in the system (1). The optimal (minimum) complexity of the fastest algorithm is $\mathcal{W}_{\text{opt}} = O\left(N_{G_0}N_M\right)$ arithmetic operations. Development of the robust computational technique for black-box software means reduction of the basic one-grid algorithm complexity down to close-to-optimal one using the least number of the problem-dependent components.

3 Linear One/Two-Grid Solver

From the multigrid point of view, unstructured grids are a complication. For a given unstructured grid, it is usually not difficult to define a sequence of finer grids, but it may be difficult to define a sequence of reasonable coarser grids [14]. To develop robust solver, one/two-grid preconditioning technique is used for computation of correction on an auxiliary structured grid and smoothing on the original (un)structured grid [15, 16]. Linear one/two-grid algorithm can be represented as

1. Computational of the residual on the original grid G_0:

$$\boldsymbol{b}_0 - A_0\boldsymbol{\varphi}_0^{(q)}.$$

2. Restriction of the residual $\boldsymbol{b}_0 - A_0\boldsymbol{\varphi}_0^{(q)}$ from the original grid G_0 onto the auxiliary grid G_A:

$$\mathcal{R}_{0\to A}\big(\boldsymbol{b}_0 - A_0\boldsymbol{\varphi}_0^{(q)}\big).$$

3. Computation of the correction $\boldsymbol{c}_A$ on the auxiliary grid G_A:

$$\boldsymbol{c}_A = A_A^{-1}\mathcal{R}_{0\to A}\big(\boldsymbol{b}_0 - A_0\boldsymbol{\varphi}_0^{(q)}\big).$$

4. Prolongation of the correction $\boldsymbol{c}_A$ from the auxiliary grid G_A onto the original grid G_0:

$$\boldsymbol{c}_0 = \mathcal{P}_{A\to 0}\boldsymbol{c}_A.$$

5. Vanka smoothing of the original grid G_0:

$$\boldsymbol{c}_0^{(\nu+1)} = S_V\boldsymbol{c}_0^{(\nu)} + (I - S_V)A_0^{-1}\big(\boldsymbol{b}_0 - A_0\boldsymbol{\varphi}_0^{(q)}\big), \qquad \nu = 0, 1, 2 \ldots, \mathcal{V}.$$

where starting guess is $\boldsymbol{c}_0^{(0)} = \mathcal{P}_{A\to 0}\boldsymbol{c}_A$.

6. Computation of new approximation to the solution $A_0^{-1}\boldsymbol{b}_0$:

$$\boldsymbol{\varphi}_0^{(q+1)} = \boldsymbol{\varphi}_0^{(q)} + \boldsymbol{c}_0^{(\mathcal{V}+1)}.$$

7. Check convergence, continue if necessary.

Here q is intergrid iteration counter, $\mathcal{R}_{0\to A}$ and $\mathcal{P}_{A\to 0}$ are the restriction and prolongation operators, S_V is Vanka smoothing iteration matrix, $\mathcal{V}$ is the number of the Vanka smoothing iterations, and subscripts 0 and A refer to the original grid G_0 and the auxiliary grid G_A, respectively.

Linear one/two-grid algorithm can be written in the matrix form

$$\boldsymbol{b}_0 - A_0\boldsymbol{\varphi}_0^{(q+1)} = M\big(\boldsymbol{b}_0 - A_0\boldsymbol{\varphi}_0^{(q)}\big), \tag{4}$$

where

$$M = A_0 S_V^{\nu}\big(A_0^{-1} - \mathcal{P}_{A\to 0} A_A^{-1} \mathcal{R}_{0\to A}\big) \tag{5}$$

is the iteration matrix. Note that main computational efforts are needed to solve the system of linear equation $A_A \boldsymbol{c}_A = \mathcal{R}_{0\to A}\big(\boldsymbol{b}_0 - A_0 \boldsymbol{\varphi}_0^{(q)}\big)$ on auxiliary grid G_A.

Convergence proof of the multigrid methods is a very difficult problem due to the complicated matrix of multigrid iterations [17]. In classical convergence analysis we need the following:

(a) Smoothing property: existence of a monotonically decreasing function $\eta(\nu)$: $\mathbb{R}_+ \to \mathbb{R}_+$ such that $\eta(\nu) \to 0$ for $\nu \to \infty$ and

$$\|A_0 S_V^{\nu}\| \leqslant \eta(\nu) \|A_0\|. \tag{6}$$

(b) Approximation property: existence of a constant $C_A > 0$ such that

$$\|A_0^{-1} - \mathcal{P}_{A\to 0} A_0^{-1} \mathcal{R}_{0\to A}\| \leqslant C_A \|A_0\|^{-1}. \tag{7}$$

The smoothing and approximation properties should be proved for each case.

Assume that the smoothing property (6) and approximation property (7) hold, then

$$\|M\| \leqslant \|A_0 S_V^{\nu}\| \cdot \|A_0^{-1} - \mathcal{P}_{A\to 0} A_A^{-1} \mathcal{R}_{0\to A}\| \leqslant C_A \eta(\nu) < 1 \tag{8}$$

with enough smoothing iterations ν. It results in h-independent convergence of the linear one/two-grid algorithm

$$\|\boldsymbol{b}_0 - A_0 \boldsymbol{\varphi}_0^{(q)}\| \leqslant \big(C_A \eta(\nu)\big)^q \|\boldsymbol{b}_0 - A_0 \boldsymbol{\varphi}_0^{(0)}\|. \tag{9}$$

The average reduction factor of the residual is defined as

$$\rho_q = \left(\frac{\|\boldsymbol{b}_0 - A_0 \boldsymbol{\varphi}_0^{(q)}\|}{\|\boldsymbol{b}_0 - A_0 \boldsymbol{\varphi}_0^{(0)}\|} \right)^{1/q}$$

with q the number of intergrid iterations [10]. Then (9) can be rewritten as

$$\rho_q \leqslant C_A \eta(\nu) < 1.$$

The linear one/two-grid algorithm can be considered as two level preconditioning technique with the preconditioning matrix $P^{-1} = A_0^{-1}(I - M)$

$$P^{-1} A_0 \boldsymbol{\varphi}_0 = P^{-1} \boldsymbol{b}_0.$$

Remark 1 Numerical solution of the auxiliary system $A_A \boldsymbol{c}_A = \mathcal{R}_{0\to A}(\boldsymbol{b}_0 - A_0\boldsymbol{\varphi}_0^{(q)})$ is more expensive than the Vanka smoothing on the original grid G_0. Computational cost of the Vanka smoothing iteration is $\mathcal{W}_{G_0} = O(n_b^{-2} N_{G_0}^3 N_M^3)$ arithmetic operations (ao), $1 \ll n_b \leqslant N_G N_M$. The complexity of solving the auxiliary system should not exceed $\mathcal{W}_{G_A} = O(n_b^{-2} N_{G_A}^3 N_M^3 \log N_{G_A}^{1/d})$ arithmetic operations, $1 \ll n_b \leqslant N_G N_M$. Since $N_{G_0} \approx N_{G_A}$, then computational cost of the intergrid iteration $\mathcal{W}^{(1)}$ is

$$\mathcal{W}^{(1)} = O(n_b^{-2} N_{G_0}^3 N_M^3 \log N_{G_0}^{1/d}) \quad \text{ao}, \quad 1 \ll n_b \leqslant N_G N_M.$$

Remark 2 The linear one/two-grid algorithm is a variant of well-known approach called defect correction [14]. The final solution $\boldsymbol{\varphi}_0 = A_0^{-1}\boldsymbol{b}_0$ is independent on the auxiliary system. Therefore, the linear one/two-grid algorithm allows to flexibly change the type and order of approximation, as well as the ordering of unknowns.

Remark 3 The linear one/two-grid algorithm has the following problem-dependent components:

(a) coefficient matrix A_0 and vector $\boldsymbol{b}_0$ of the resulting system of linear equations $A_0\boldsymbol{\varphi}_0 = \boldsymbol{b}_0$ obtained after approximation of the system (1) on the original grid G_0;
(b) auxiliary system $A_A \boldsymbol{c}_A = \mathcal{R}_{0\to A}(\boldsymbol{b}_0 - A_0\boldsymbol{\varphi}_0^{(q)})$;
(c) intergrid operators $\mathcal{R}_{A\to 0}$ and $\mathcal{P}_{A\to 0}$;
(d) iterative solution of the auxiliary system $A_1\boldsymbol{c}_1 = \bar{\mathcal{R}}_{0\to 1}(\boldsymbol{b}_0 - A_0\boldsymbol{\varphi}_0^{(q)})$;
(e) the number of smoothing iterations ν on the original grid G_0;
(f) the unknowns ordering;
(g) a stopping criterion iterations of the intergrid iterations.

Problem-dependent components (b), (c) and (d) of the linear one/two-grid algorithm have no analogues in the basic algorithm (2). Next, we will analyse the one/two-grid assuming that the Robust Multigrid Technique (RMT) is used to solve the auxiliary system.

3.1 Globally Structured Grids

Let G_0 a globally structured grid, i.e. G_0 generates the multigrid structure. Since RMT uses the essential multigrid principle in the single grid algorithm, this approach can be considered as a problem-independent technique of the Vanka iteration (2) convergence acceleration. In this case, the auxiliary grid G_1 coincides with the original one G_0. In addition, it is possible to avoid a smoothing on the original grid G_0

$$G_1 = G_0 \Rightarrow \mathcal{R}_{A\to 0} = \mathcal{P}_{A\to 0} = I \quad \text{and} \quad S_V = I.$$

As a result, the linear one/two-grid algorithm (4) transforms to a single grid solver with the iteration matrix

$$M = I - A_0 A_A^{-1},$$

where $A_0 \neq A_A$ defines the defect correction iterations. If $A_0 = A_A$ then we have one-grid pseudomultigrid solver (4) with the iteration matrix $M = A_0 \mathcal{Q}_0$ [1], where

$$\mathcal{Q}_l = \begin{cases} S_l^{\nu_l}\left(d_l \mathcal{R}_{0\to l} + \mathcal{P}_{l+1\to l}\mathcal{Q}_{l+1}\right), & l = 0, 1, 2, \ldots, L^+ - 2 \\ S_l^{\nu_l} d_l \mathcal{R}_{0\to l}, & l = L^+ - 1 \end{cases},$$

$$d_l = A_l^{-1} - \mathcal{P}_{l+1\to l} A_{l+1}^{-1} \mathcal{R}_{l\to l+1}.$$

Theorem 1 *Assume that the smoothing and approximation properties hold, and* $\|\mathcal{R}_{0\to l}\| \leq C_{\mathcal{R}}$. *Then RMT is a convergent iterative method and the multigrid iteration matrix norm is estimated by*

$$\|M\| \leqslant C_A \eta(\nu_0) + C_A C_{\mathcal{R}} \sum_{l=1}^{L^+-1} C^l \eta(\nu_l).$$

Remember that the coarse grid problems (A_l) and transfer operators ($\mathcal{P}_{l+1\to l}$ and $\mathcal{R}_{l\to l+1}$) are the problem-independent components of the RMT [1]. As compared with the basic one-grid algorithm (Vanka-type smoother) (3), the number of Vanka smoothing iteration is single problem-dependent component of the RMT. In fact, the amount of computational work is weakly dependent on the number of smoothing iterations [2]. Algorithmic complexity of RMT is $\mathcal{W} = O\left(n_b^{-2} N_G^3 N_M^3 \log N_G^{1/d}\right)$ arithmetic operations, $1 \ll n_b \leqslant N_G N_M$, in a wide range of the problem parameters.

3.2 Locally Structured Grids

Consider a linear BVP

$$L_{\Omega_1} g(\boldsymbol{x}) = f_{\Omega_1}(\boldsymbol{x}), \quad \boldsymbol{x} \in \Omega_1, \tag{10a}$$

$$L_{\Omega_2} g(\boldsymbol{x}) = f_{\Omega_2}(\boldsymbol{x}), \quad \boldsymbol{x} \in \Omega_2. \tag{10b}$$

Here $\boldsymbol{x} = (x_1, \ldots, x_d)^T$ and $\Omega \in \mathbb{R}^d$ is the given open domain with the boundary $\partial\Omega$, L_{Ω_1} and L_{Ω_2} are elliptic differential operators defined in the subdomains Ω_1 and Ω_2, f_{Ω_1} and f_{Ω_2} are known functions on Ω_1 and Ω_2. Boundary condition

$$L_{\partial\Omega} g(\boldsymbol{x}) = f_{\partial\Omega}(\boldsymbol{x}), \quad \boldsymbol{x} \in \partial\Omega, \tag{10c}$$

is given on the external boundary $\partial\Omega$ of the domain Ω, and

$$L_{\partial\Omega_1} g(\boldsymbol{x}) = L_{\partial\Omega_2} g(\boldsymbol{x}) \tag{10d}$$

on (internal) boundary $\partial\Omega_2$.

Let structured grids $G_0^{(1)}$ and $G_0^{(2)}$ be generated in domains Ω_1 and Ω_2 as shown on Fig. 1. Approximation of a BVP on the grid formed by $G_0^{(1)}$ and $G_0^{(2)}$ leads to the resulting system

$$\begin{pmatrix} B & C \\ D & F \end{pmatrix} \begin{pmatrix} \boldsymbol{u} \\ \boldsymbol{v} \end{pmatrix} = \begin{pmatrix} \boldsymbol{b}_u \\ \boldsymbol{b}_v \end{pmatrix}, \tag{11}$$

where $\boldsymbol{u}$ and $\boldsymbol{v}$ are discrete analogues of the functions g on $G_0^{(1)}$ and $G_0^{(2)}$, respectively. The matrices B and F are invertible, while C and D are generally rectangular and $C^T \neq D$. Pattern of the matrices C and D depends on the grids $G_0^{(1)}$ and $G_0^{(2)}$ (which may have a common boundaries or intersect) and the method of interpolation between blocks of grids.

Consider the simplest iterative method for solving (11)

$$W_B\left(\boldsymbol{u}^{(n+1)} - \boldsymbol{u}^{(n)}\right) = \boldsymbol{b}_u - B\boldsymbol{u}^{(n)} - C\boldsymbol{v}^{(n)}, \tag{12a}$$

$$W_F\left(\boldsymbol{v}^{(n+1)} - \boldsymbol{v}^{(n)}\right) = \boldsymbol{b}_v - D\boldsymbol{u}^{(n+1)} - F\boldsymbol{v}^{(n)}, \tag{12b}$$

where W_B and W_F are splitting matrices for B and F.

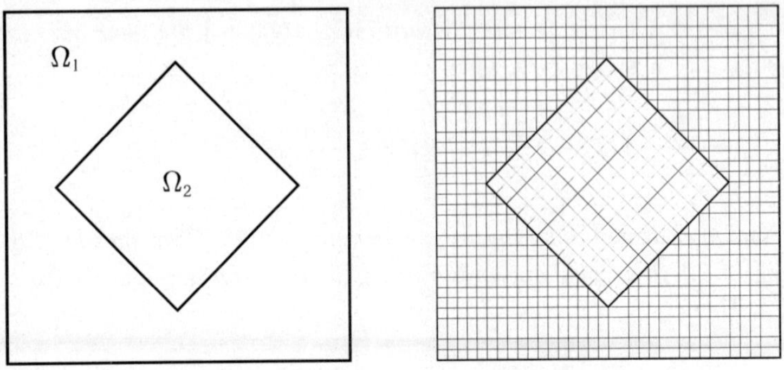

Fig. 1 Domain $\Omega = \Omega_1 \cup \Omega_2$ and locally structured (two block) grid

The iterations can be rewritten as

$$\boldsymbol{\psi}^{(n+1)} = \left(I - W^{-1}\tilde{A}\right)\boldsymbol{\psi}^{(n)} + W^{-1}\tilde{\boldsymbol{b}}, \tag{13}$$

where

$$\boldsymbol{\psi} = \begin{pmatrix} \boldsymbol{u} \\ \boldsymbol{v} \end{pmatrix}, \quad \tilde{A} = \begin{pmatrix} B & C \\ D & F \end{pmatrix} - \begin{pmatrix} 0 & 0 \\ 0 & DW_B^{-1} \end{pmatrix}\begin{pmatrix} 0 & 0 \\ B & C \end{pmatrix},$$

$$W = \begin{pmatrix} W_B & 0 \\ 0 & W_F \end{pmatrix}, \quad \tilde{\boldsymbol{b}} = \begin{pmatrix} I & 0 \\ -DW_B^{-1} & I \end{pmatrix}\begin{pmatrix} \boldsymbol{b}_u \\ \boldsymbol{b}_v \end{pmatrix}.$$

To simplify the analysis, we rewrite (13) as

$$\boldsymbol{\psi}^{(n+1)} = \left(I - \tilde{W}^{-1}A\right)\boldsymbol{\psi}^{(n)} + W^{-1}\tilde{A}A^{-1}\boldsymbol{b}.$$

where

$$\tilde{W} = A\tilde{A}^{-1}W.$$

In nonsymmetric case, the smoothing property of damped smoother had been analysed in [1]. Iteration (13) becomes

$$\boldsymbol{\psi}^{(\nu+1)} = S(\omega)\boldsymbol{\psi}^{(\nu)} + W^{-1}\tilde{A}A^{-1}\boldsymbol{b}, \tag{14}$$

where $S(\omega)$ is the iteration matrix

$$S(\omega) = I - \frac{1}{1+\omega}\tilde{W}^{-1}A,$$

and $\|S(\omega)\| < 1$, and $\omega \geqslant 0$ is some parameter.

Smoothing analysis is based on the following theorems:

Theorem 2 *Let a matrix $Q \in \mathbb{R}^{n\times n}$ satisfies to $\|Q\| < 1$ for some operator norm and $3 - 2\sqrt{2} \leqslant \omega \leqslant 3 + 2\sqrt{2}$, then*

$$\frac{1}{(1+\omega)^{\nu+1}}\|(I-Q)(\omega I + Q)^{\nu}\| \leqslant \frac{1}{\sqrt{e\omega\nu}}, \quad \nu = 1, 2, \ldots \tag{15}$$

Theorem 3 *Let a smoothing iteration matrix $S(0) \in \mathbb{R}^{n\times n}$ satisfies to $\|S(0)\| < 1$ and $\|\tilde{W}\| \leqslant C\|A\|$ in some operator norm, C is some constant and $3 - 2\sqrt{2} \leqslant \omega \leqslant 3 + 2\sqrt{2}$. Then*

$$\|AS^{\nu}(\omega)\| \leqslant C\frac{1+\omega}{\sqrt{e\omega\nu}}\|A\|. \tag{16}$$

Pseudomultigrid iterations of RMT (4) on this two block grid have the iteration matrix $M = A_0 Q_0$, where

$$Q_l = \begin{cases} S_l^{\nu_l}(\omega_l)\left(d_l \mathcal{R}^*_{0\to l} + \mathcal{P}_{l+1\to l} Q_{l+1}\right), & l = 0, 1, 2, \ldots, L_3^+ - 2 \\ S_l^{\nu_l}(\omega_l) d_l \mathcal{R}^*_{0\to l}, & l = L_3^+ - 1 \end{cases},$$

$$\mathcal{R}^*_{0\to l} = W_l^{-1} \tilde{A}_l A_l^{-1} \mathcal{R}_{0\to l}.$$

Convergence theorem on the two-block grid becomes

Theorem 4 *If the smoothing and approximation properties hold,* $\|I - \tilde{W}_l^{-1} A_l\| < 1$, $\|\mathcal{R}^*_{0\to l}\| \leqslant C_{\mathcal{R}}$ *and* $3 - 2\sqrt{2} \leqslant \omega_l \leqslant 3 + 2\sqrt{2}$. *Then iterations of RMT converges and*

$$\rho_q \leqslant \|\tilde{A}_0 Q_0\| \leqslant C_A C C_{\mathcal{R}} \sum_{l=0}^{L_3^+ - 1} C_*^l \frac{1 + \omega_l}{\sqrt{e\omega_l \nu_l}} < 1. \qquad (17)$$

This theorem predicts h-independent convergence of RMT.

4 Conclusion

In this article, a linear two-grid algorithm was analyzed. Three possible cases are considered:

(1) If the original grid G_0 is globally structured, then the auxiliary grid G_A coincides with the original one ($G_A = G_0$) resulting in the one-grid algorithm. Compared to the basic algorithm (3), proposed one-grid solver has one additional problem-dependent component (the number of smoothing iterations).
(2) If original grid G_0 is locally structured, then the auxiliary grid G_A coincides with the original one ($G_A = G_0$) resulting in the one-grid algorithm. Compared to the basic algorithm (3), proposed one-grid solver has two additional problem-dependent components (the number of smoothing iterations and interpolation between grid blocks).
(3) If the original grid G_0 is unstructured, then the auxiliary grid G_A may be a structured boundary unfitted grid ($G_A \neq G_0$). Compared to the basic algorithm (3), proposed two-grid solver has additional problem-dependent components: the number of smoothing iterations and the intergrid interpolation $G_A \neq G_0$.

Note that the total amount of computational work is weakly dependent on the number of smoothing iterations [2].

If the smoothing and approximation properties hold, then h-independent convergence of the linear one/two-grid solver is expected. Application of RMT for solving the auxiliary system of linear equations makes it possible to reduce algorithmic complexity of the basic algorithm (3) $\mathcal{W} = O\big(n_b^{-2}(N_{G_0}N_M)^{3+k/d}\big)$ arithmetic operations down to close-to-optimal value $\mathcal{W}_{G_A} = O\big(n_b^{-2}N_{G_A}^3 N_M^3 \log N_{G_A}^{1/d}\big)$ arithmetic operations.

References

1. Martynenko, S.I.: The robust multigrid technique: For black-box software. De Gruyter, Berlin (2017)
2. Martynenko, S.I.: Sequential software for robust multigrid technique. Triumph, Moscow (2020) `https://github.com/simartynenko/Robust_Multigrid_Technique_2020`
3. Martynenko, S.I.: Robust multigrid technique for solving partial differential equations on structured grids. Num. Meth. and Prog., 1, 83–100 (2000)
4. Frey, P., George, P.L., Mesh generation. Wiley, New York (2010)
5. George, P.L., Automatic mesh generation. Wiley, New York (1991)
6. Thompson, J.F., Soni, B.K., Weatherill, N.P., Handbook of grid generation. Boca Raton. CRC Press (1998)
7. Knupp, P., Steinberg, S., Fundamentals of grid generation. Boca Raton. CRC Press (1993)
8. Liseikin, V.D., Grid generation methods. Berlin. Springer (1999)
9. Liseikin, V.D., Layer resolving grids and transformations for singular perturbation problems. Utrecht. VSP (2001)
10. Wesseling, P.: An introduction to multigrid methods. Chichester, Wiley (1992)
11. Vanka, S.P.: Block-Implicit Multigrid Solution of Navier–Stokes Equations in Primitive Variables. J. Comput. Phys., 65(1), 138–158 (1986)
12. Benzi, M., Golub, G.H., Liesen., J.: Numerical solution of saddle point problems. Acta Numerica, 14, 1–137 (2006)
13. Hageman, L.A., Young., D.M.: Applied Iterative Methods. International Series of Numerical Mathematics. Academic Press, New York (1981).
14. Trottenberg, U., Oosterlee, C.W., Schüller, A.: Multigrid. Academic Press, London (2001)
15. Xu, J.: The auxiliary space method and optimal multigrid preconditioning techniques for unstructured grids. Computing., 56, 215–235 (1996)
16. Martynenko, S.I.: Potentialities of the Robust Multigrid Technique. Comp. Meth. in Appl. Math., 10(1), 87–94 (2010)
17. Hackbusch., W.: Multi-grid Methods and Applications. Springer, Berlin, Heidelberg (1985).

Large Deformations of Biaxial Tension-Compression of the Plate, Consisting Two Pre-deformed Layers Made of Incompressible Treloar Material

Konstantin M. Zingerman, Vladimir A. Levin, Leonid M. Zubov, Anton E. Belkin, and Danila R. Biryukov

Abstract For the case of large strains, the paper presents an exact analytical solution to the problem of the stress-strain state of a composite slab obtained by joining two pre-deformed layers. Each layer is obtained by straightening a curved panel originally shaped as a sector of a hollow circular cylinder. The cylinders are made of incompressible nonlinear elastic material—the Treloar (neo-Hookean) material. The axes of the cylinders are orthogonal before deformation. After joining, the slab is subjected to biaxial tension or compression in its plane. The problem is formulated on the basis of the theory of superimposed large deformations. An exact analytical solution to the problem is obtained. Nonlinear effects are investigated. The obtained solution can be used to verify the software designed for the numerical solution of problems on the stress-strain state of structural elements made by junction of pre-deformed parts.

1 Introduction

The production of structural elements may be accompanied by the connection of pre-deformed parts. The theory of superimposed large deformations is used to analyze the stress-strain state of such structural elements in the case of finite strains [1]. Some problems on the stress-strain state of bodies, made by connecting pre-

K. M. Zingerman (✉)
Tver State University, Tver, Russia

V. A. Levin
Lomonosov Moscow State University, Moscow, Russia

L. M. Zubov
Southern Federal University, Rostov-on-Don, Russia

A. E. Belkin · D. R. Biryukov
Tula State University, Tula, Russia
e-mail: belkin@saldlab.com; biryukov@saldlab.com

I. B. Badriev et al. (eds.), *Mesh Methods for Boundary-Value Problems and Applications*, Lecture Notes in Computational Science and Engineering 141,
https://doi.org/10.1007/978-3-030-87809-2_46

deformed parts, were solved earlier [2–4]. This article presents an exact analytical solution of the problem on the stress-strain state of the composite plate, obtained by joining two pre-deformed layers for the case of finite strains. Each layer is obtained by straightening a curved panel, initially shaped as a sector of a hollow circular cylinder. We assume that the cylinder axes before deformation are orthogonal. After connecting the layers, the composite plate undergoes biaxial tension or compression in its plane. The solution is obtained for the case when the cylinders are made of incompressible nonlinear elastic material—the Treloar (neo-Hookean) material [5–7].

2 The Generalized Statement of Problem

Let's consider two cylindrical panels, having mutually orthogonal axes (Fig. 1). For the upper panel let's define the cylindrical coordinates ρ, θ, ξ by formulas

$$x_2 = \rho\sin\theta\ ,\ y_2 = \xi,\ z_2 = \rho\cos\theta$$

For the lower panel the cylindrical coordinates r, φ, ζ are introduced by formulas

$$x_1 = r\cos\varphi\ ,\ y_1 = r\sin\varphi\ ,\ z_1 = \zeta$$

Here x, y, z are Cartesian coordinates of points in space.

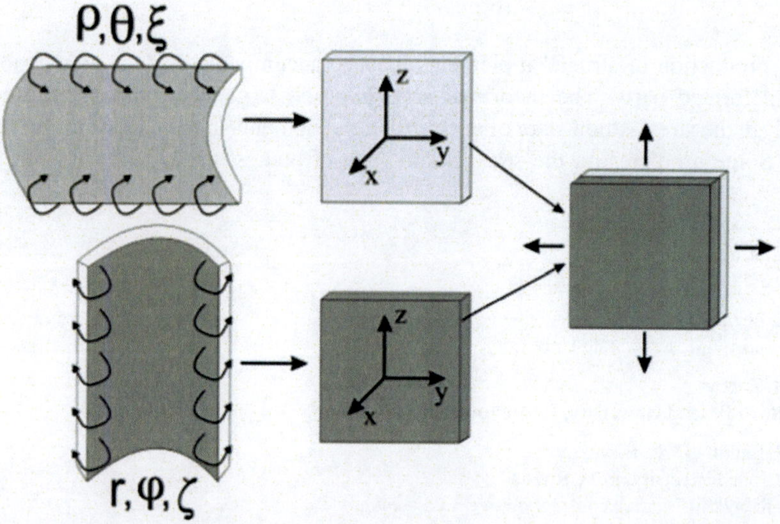

Fig. 1 Scheme of junction and loading of two prestrained plates

Coordinates ρ, θ, ξ and r, φ, ζ are the Lagrangian coordinates of particles in the initial configurations, which are assumed to be natural, unstressed. It is assumed that $\rho_0 \le \rho \le \rho_1$, and $r_1 \le r \le r_0$, where ρ_0, ρ_1, r_0 and r_1 are the radii of panels in the initial configuration.

Below we consider a composite prestressed rectangular plate consisting of two layers obtained by straightening (unbending) these cylindrical panels. This state of the two-layer plate is taken as an intermediate configuration. After that, the plate is subjected to biaxial tension-compression by forces parallel to the axes y, z and attached to the flat edges of the plate $y = \text{const}$ and $z = \text{const}$ and goes into the final state (configuration).

Treloar's constitutive relations for an incompressible elastic material have the form

$$\mathbf{T} = -p\mathbf{I} + \mu\mathbf{B}, \tag{1}$$

where $\mathbf{I}$ is the unit tensor, μ is the material constant, $\mathbf{B} = \mathbf{F} \cdot \mathbf{F}^T$ is the Finger strain measure, $\mathbf{F} = \mathbf{F}_{\text{init}} \cdot \mathbf{F}_{\text{add}}$ is the total deformation gradient; $\mathbf{F}_{\text{init}}$ is the initial deformation gradient, corresponding to the transition from the initial configuration to the intermediate one; $\mathbf{F}_{\text{add}}$ is the additional deformation gradient, which corresponds to the transition from the intermediate configuration to the final one; $\mathbf{T}$ is the true stress tensor in the final state; p is the Lagrange multiplier.

3 The Initial Deformation of Panels

Let us denote by x, y, z the Cartesian coordinates of body particles in the intermediate configuration and set the straightening deformation, that is, the transition of the lower and upper panels from the natural configuration to the intermediate one, respective to the formulas

$$\begin{aligned} x &= x(r), \quad y = \tau_1\varphi, \quad z = \alpha_1\zeta, \\ x &= x(\rho), \quad y = \alpha_2\xi, \quad z = \tau_2\theta, \end{aligned} \tag{2}$$

where $\tau_1,\ \alpha_1,\ \tau_2,\ \alpha_2$ are constants, $x(r), x(\rho)$ are unknown functions.

Hereinafter, functions with an argument r will be considered to belong to the lower panel, with an argument ρ—to the upper one. I.e., $x(r)$ and $x(\rho)$ are different functions.

Initial deformation gradients, corresponding to the transition from a natural reference configuration to an intermediate one, have the following form for the top and bottom panels [8–10]:

$$\mathbf{F}_{\text{init}}(r) = x'(r)\,\mathbf{e}_r \otimes \mathbf{i}_1 + \tfrac{\tau_1}{r}\mathbf{e}_\varphi \otimes \mathbf{i}_2 + \alpha_1 \mathbf{i}_3 \otimes \mathbf{i}_3$$
$$\mathbf{F}_{\text{init}}(\rho) = x'(\rho)\,\mathbf{e}_\rho \otimes \mathbf{i}_1 + \tfrac{\tau_2}{\rho}\mathbf{e}_\theta \otimes \mathbf{i}_3 + \alpha_2 \mathbf{i}_2 \otimes \mathbf{i}_2$$
$$x'(r) = \tfrac{df_1(r)}{dr} \quad x'(\rho) = \tfrac{df_2(\rho)}{d\rho} \tag{3}$$
$$\mathbf{e}_r = \mathbf{i}_1 \cos\varphi + \mathbf{i}_2 \sin\varphi \quad \mathbf{e}_\varphi = -\mathbf{i}_1 \sin\varphi + \mathbf{i}_2 \cos\varphi$$
$$\mathbf{e}_\rho = \mathbf{i}_3 \cos\theta + \mathbf{i}_1 \sin\theta \quad \mathbf{e}_\theta = -\mathbf{i}_3 \sin\theta + \mathbf{i}_1 \cos\theta$$

Here $\mathbf{i}_1, \mathbf{i}_2, \mathbf{i}_3$ are constant unit vectors of Cartesian coordinates.

In the case of an incompressible material the following ratio, allowing to determine functions $x(r)$ and $x(\rho)$ must be satisfied:

$$\det \mathbf{F}_{\text{init}} = 1 \tag{4}$$

The formula (4), taking into account the ratios (3), can be rewritten as

$$\alpha_1 \tau_1 x(r) = r, \quad \alpha_2 \tau_2 x(\rho) = \rho \tag{5}$$

The solutions of Eq. (5) must satisfy the conditions $x(\rho_1) = 0$, $x(r_0) = x(\rho_0)$. The last condition means that the straightened panels are connected without a gap and form a two-layer plate.

From (2) and (5) the dependence between x and r (for the lower panel, respectively) and ρ (for the upper panel, respectively) can be determined:

$$x(r) = \frac{r^2 - r_0^2}{2\alpha_1 \tau_1} + \frac{\rho_0^2 - \rho_1^2}{2\alpha_2 \tau_2}, \; x(\rho) = \frac{\rho^2 - \rho_1^2}{2\alpha_2 \tau_2} \tag{6}$$

4 The Additional Deformation of the Composite Plate

Next, let's consider the problem of tension-compression in two directions of a pre-stressed two-layer plate by forces distributed over its edges $y = \text{const}$, $z = \text{const}$. The deformation of the transition from the intermediate configuration to the final state will be considered in the form of

$$X = X(x), \; Y = \beta_2 y, \; Z = \beta_3 z$$
$$\beta_2 = \text{const}, \; \beta_3 = \text{const} \tag{7}$$

Here X, Y, Z are the Cartesian coordinates of body particles in the final state. The additional deformation gradient:

$$\mathbf{F}_{\text{add}} = \frac{dX(x)}{dx}\mathbf{i}_1 \otimes \mathbf{i}_1 + \beta_2 \mathbf{i}_2 \otimes \mathbf{i}_2 + \beta_3 \mathbf{i}_3 \otimes \mathbf{i}_3 \tag{8}$$

We can rewrite (8) for each layer using (2):

$$\mathbf{F}_{\text{add}}(r) = \frac{1}{x'(r)}\frac{dX(r)}{dr}\mathbf{i}_1 \otimes \mathbf{i}_1 + \beta_2\mathbf{i}_2 \otimes \mathbf{i}_2 + \beta_3\mathbf{i}_3 \otimes \mathbf{i}_3$$
$$\mathbf{F}_{\text{add}}(\rho) = \frac{1}{x'(\rho)}\frac{dX(\rho)}{d\rho}\mathbf{i}_1 \otimes \mathbf{i}_1 + \beta_2\mathbf{i}_2 \otimes \mathbf{i}_2 + \beta_3\mathbf{i}_3 \otimes \mathbf{i}_3 \quad (9)$$

In the case of an incompressible material, the following ratios allowing to determine the functions $X(r)$ and $X(\rho)$ must be satisfied:

$$\det \mathbf{F}_{\text{add}} = 1 \quad (10)$$

The formula (10), taking into account the ratios (6) and (9), can be rewritten as

$$\beta_2\beta_3\frac{dX(r)}{dr} = \frac{r^2 - r_0^2}{2\alpha_1\tau_1} + \frac{\rho_0^2 - \rho_1^2}{2\alpha_2\tau_2}, \quad \beta_2\beta_3\frac{dX(\rho)}{d\rho} = \frac{\rho^2 - \rho_1^2}{2\alpha_2\tau_2} \quad (11)$$

The solutions of Eq. (11) must satisfy the conditions $X(\rho_1) = 0$, $X(r_0) = X(\rho_0)$. The dependency between X and r (for the lower panel, respectively) and ρ (for the upper panel, respectively) we get in the form of:

$$X(r) = \frac{3\alpha_1\tau_1(\rho_0^2-\rho_1^2)(r-r_0)+\alpha_1\tau_1(\rho_0^3+2\rho_1^3-3\rho_0\rho_1^2)+\alpha_2\tau_2(r^3+2r_0^3-3rr_0^2)}{6\beta_2\beta_3\alpha_1\alpha_2\tau_1\tau_2}$$
$$X(\rho) = \frac{\rho^3 - 3\rho_1\rho + 2\rho_1^3}{6\beta_2\beta_3\alpha_2\tau_2} \quad (12)$$

From the above formulas, it can be seen that the components of the Finger strain measure tensor in this problem are independent of the Y and Z coordinates, but depend only on the X coordinate. Assuming that the Lagrange multiplier p also depends only on X, it is possible, based on formula (1), to conclude that the true stresses will also depend only on this coordinate. In this case, the equilibrium equations in the absence of mass forces are reduced to one equation $\frac{dT_{11}}{dx} = 0$, whence follows $T_{11} = \text{const}$. In the absence of stresses in the final state on the foundations of the plate $T_{11} = 0$.

Using (1) and the equality $T_{11} = 0$ in the final state, one can express the Lagrange multiplier p, substituting $X(r)$ and $X(\rho)$ with received expressions (12):

$$p(r) = \mu\left(\frac{dX(r)}{dr}\right)^2$$
$$p(\rho) = \mu\left(\frac{dX(\rho)}{d\rho}\right)^2 \quad (13)$$

Formulas (1) and (13) allow us to calculate the remaining components of the tensor **T**.

5 Numerical Results

Figures 2, 3, and 4 show the stress graphs for layers with the following boundary radii:

$$\rho_1/r_1 = 1.1, \rho_0/r_1 = 1, r_0/r_1 = 1.1.$$

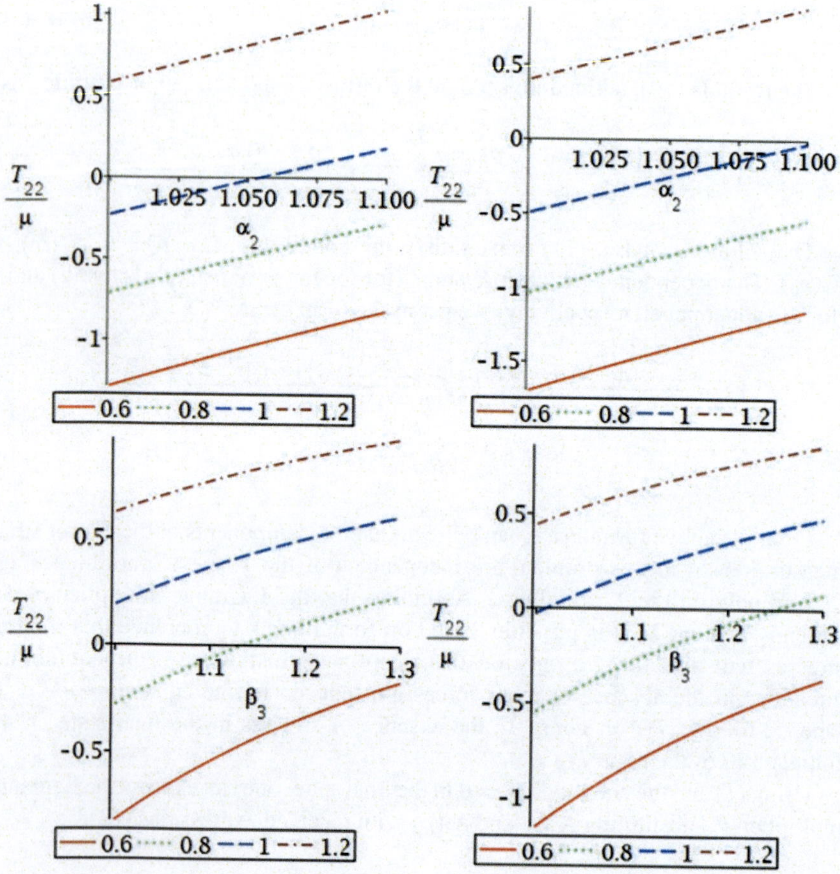

Fig. 2 Top left: the dependence of stress $\frac{T_{22}}{\mu}$ on the strain α_2 for $\rho = \rho_0$ and different β_2. Top right: the dependence of stress $\frac{T_{22}}{\mu}$ on the strain α_2 for $\rho = \rho_1$ and different β_2. Bottom left: the dependence of stress $\frac{T_{22}}{\mu}$ on the strain β_3 for $\rho = \rho_0$ and different β_2. Bottom right: the dependence of stress $\frac{T_{22}}{\mu}$ on the strain β_3 for $\rho = \rho_1$ and different β_2. The values of β_2 are shown below every graph

Fig. 3 Top: the dependence of stress $\frac{T_{22}}{\mu}$ on the strain τ_2 for $\rho = \rho_0$ and different β_2. Bottom: the dependence of stress $\frac{T_{22}}{\mu}$ on the strain τ_2 for $\rho = \rho_1$ and different β_2. The values of β_2 are shown below every graph

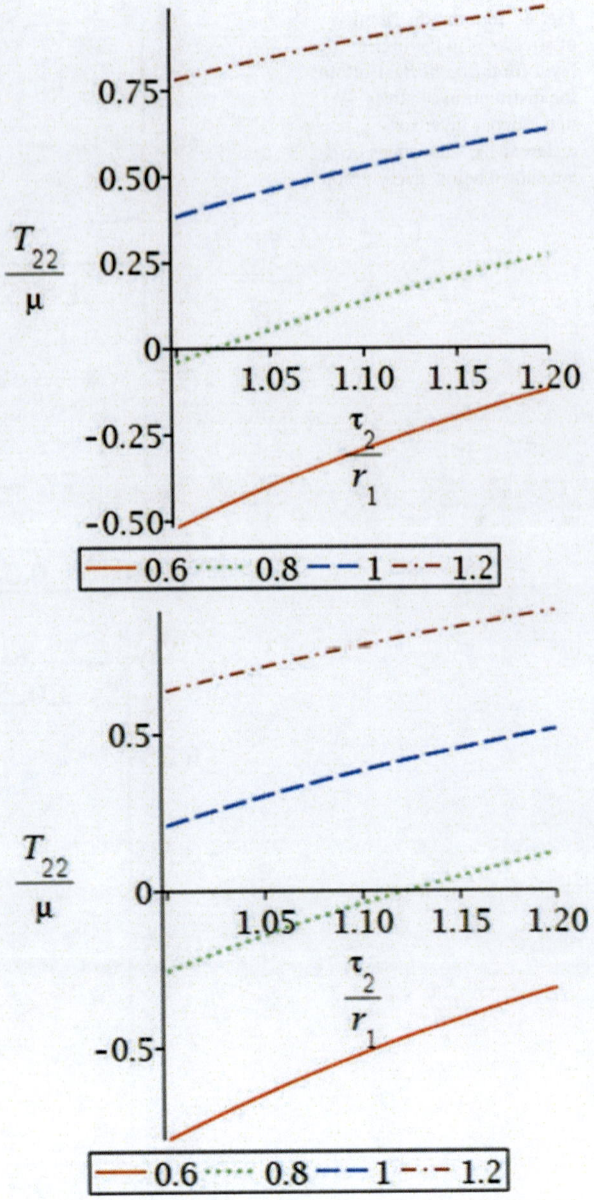

Fig. 4 Top: the distribution of stress $\frac{T_{22}}{\mu}$ in the upper layer for different β_2. Bottom: the distribution of stress $\frac{T_{33}}{\mu}$ in the upper layer for different β_2. The values of β_2 are shown below every graph

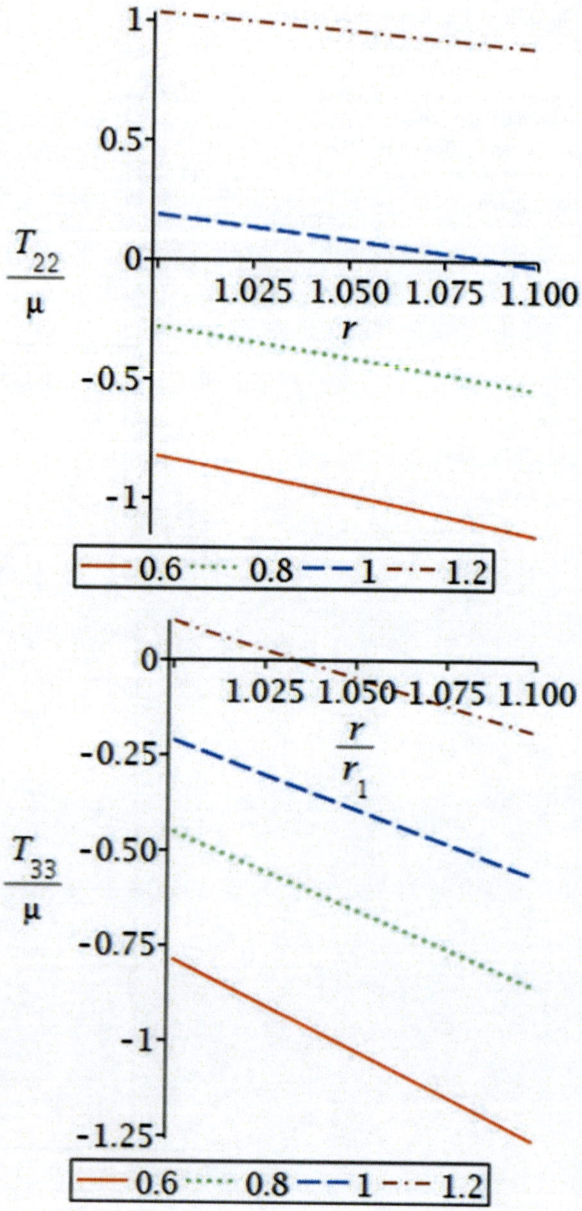

Figures 2 and 3 show the dependencies of the true stress $\frac{T_{22}}{\mu}$ at the boundaries of the first layer on the characteristics of deformations for different values of the strain parameter β_2. The following values of strain parameters are used:

Graphs on Fig. 2 (top): $\alpha_1 = \beta_3 = 1$, $\tau_1 = \tau_2 = 0.9r_1$.

Graphs on Fig. 2 (bottom): $\alpha_1 = 1$, $\alpha_2 = 1.1$, $\tau_1 = \tau_2 = 0.9r_1$.

Figure 3: $\alpha_1 = \beta_3 = 1$, $\alpha_2 = 1.1$, $\tau_1 = 0.9r_1$.

The material constant μ is the same for both layers.

Graphs on Fig. 4 show the stress distribution in the upper layer for different values of β_2 for the following values of strain parameters: $\alpha_1 = \beta_3 = 1$, $\alpha_2 = 1.1$, $\tau_1 = \tau_2 = 0.9r_1$. The values of β_2 for each line type of the graph are shown below the graphs.

6 Conclusion

The model of junction of two straightened panels and further biaxial stretch of the obtained composite plate is developed within the framework of the theory of superimposed large strains. The analysis is performed for the incompressible nonlinear-elastic material (Treloar's material). The exact analytical solution of the problem is obtained, and some numerical results are given. These results can be used to verify the software designed for the finite-element analysis of structural elements made by junction of pre-deformed parts [11–13]. In addition, the results may be applied for the modeling of composite plates and shells [14–16], in particular, sandwich plates [17, 18].

The solution can be further generalized for micropolar incompressible materials [19].

Acknowledgment This work is supported by the Russian Science Foundation under the grant 19-71-10008. (Sections 1–4) and by the Ministry of Education and Science of the Russian Federation as part of the program of the Mathematical Center for Fundamental and Applied Mathematics under the agreement 075-15-2019-1621 (Sects. 5, 6).

References

1. Levin, V.A., Tarasiev, G.S.: Superposition of large elastic deformations in the space of final-states. Doklady Akademii Nauk SSSR. **251**, 63–66 (1980)
2. Levin, V.A., Zubov, L.M., Zingerman, K.M.: The torsion of a composite, nonlinear-elastic cylinder with an inclusion having initial large strains. International Journal of Solids and Structures, **51**, 1403–1409 (2014)
3. Levin, V.A., Zubov, L.M., Zingerman, K.M.: An exact solution for the problem of flexure of a composite beam with preliminarily strained layers under large strains. International Journal of Solids and Structures, **67–68**, 244–249 (2015)
4. Levin, V.A., Zubov, L.M., Zingerman, K.M.: Multiple joined prestressed orthotropic layers under large strains. International Journal of Engineering Science, **133**, 47–59 (2018)

5. Lurie, A.I.: Non-linear theory of elasticity. North Holland, Amsterdam (1990).
6. Treloar, L.R.G.: The Physics of Rubber Elasticity. Oxford University Press (1975)
7. Mooney, M.: A theory of large elastic deformation. Journal of Applied Physics. **11**, 582 (1940) https://doi.org/10.1063/1.1712836
8. Ericksen, J.L.: Deformations possible in every isotropic, incompressible, perfectly elastic body. Journal of Applied Mathematics and Physics (ZAMP) **5**, 466–489 (1954)
9. Truesdell, C. A First Course in Rational Continuum Mechanics. The Johns Hopkins University. Baltimore, Maryland (1972)
10. Saccomandi, G.: Universal results in finite elasticity. In: Fu, Y. B., Ogden, R. W. (eds.) Nonlinear Elasticity. Theory and Applications. Cambridge University Press, Cambridge (2001)
11. Abramov, S.M., Amel'kin S.A., Kljuev, L.V., Krapivin, K.J., Nozhnickij, J.A., Servetnik, A.N., Chichkovskij, A.A.: Modeling the development of large plastic deformations in a rotating disk in the Fidesys program. Chebyshevskii Sbornik. **18**(3),15–27. (In Russ.)
12. Konovalov, D. A., Yakovlev, M. Ya.: Numerical estimation of effective elastic properties of elastomer composites under finite strains using spectral element method with CAE Fidesys. Chebyshevskii Sbornik. **18**(3), 316–329. (In Russ.)
13. Vershinin, A. V., Sabitov, D. I., Ishbulatov, S. Y., Myasnikov, A. V.: Hydrogeomechanical modeling of reservoir by external coupling of specialized computational software and universal CAE Fidesys. Chebyshevskii Sbornik. **18**(3), 154–186. (In Russ.)
14. Noor, A.K., Burton, W.S.: Assessment of shear deformation theories for multilayered composite plates. Applied Mechanics Review. **41**, 1–18 (1989)
15. Carrera, E.: Theories and finite elements for multilayered, anisotropic, composite plates and shells. ARCO **9**, 87–140 (2002).
16. Kulikov, G.M.: Computational models for multilayered composite shells with application to tires. Tire Science and Technology. **24** (1), 11–38 (1996)
17. Badriev, I. B., Makarov, M. V., Paimuhin, V. N. Longitudinal and transverse bending by a cylindrical shape of the sandwich plate stiffened in the end sections by rigid bodies. IOP Conference Series: Materials Science and Engineering. **158**, 012011 (2016)
18. Makarov, M. V., Badriev, I. B., Buyanov, V. Yu., Smirnova, E. V. On solving the geometrically nonlinear and linear problems of transverse bending of a hinged fixing sandwich plate with transversally soft core. Journal of Physics: Conference Series, **1158** (3), 032026 (2019)
19. Zubov, L.M.: Universal deformations of micropolar isotropic elastic solids. Mathematics and Mechanics of Solids. **21**, 152–167 (2016)

Index

I. B. Badriev et al. (eds.), *Mesh Methods for Boundary-Value Problems and Applications*, Lecture Notes in Computational Science and Engineering 141,
https://doi.org/10.1007/978-3-030-87809-2

Z